Academic Press Rapid Manuscript Reproduction

Proceedings of the International Conference
on Colloids and Surfaces—50th Colloid and Surface Science Symposium,
held in San Juan, Puerto Rico
on June 21-25, 1976

Colloid and Interface Science

Science

VOL. I

Plenary and Invited Lectures

EDITED BY

Milton Kerker
Clarkson College of Technology
Potsdam, New York

Albert C. Zettlemoyer
Lehigh University
Bethlehem, Pennsylvania

Robert L. Rowell
University of Massachusetts
Amherst, Massachusetts

Academic Press Inc.
New York   San Francisco   London   1977
A Subsidiary of Harcourt Brace Jovanovich, Publishers

ACADEMIC PRESS, INC.
111 Fifth Avenue, New York, New York 10003

United Kingdom Edition published by
ACADEMIC PRESS, INC. (LONDON) LTD.
24/28 Oval Road, London NW1

Library of Congress Cataloging in Publication Data

Main entry under title:

International Conference on Colloids and Surfaces, 50th,
    San Juan, P.R., 1976.
    Colloid and interface science.

    Includes bibliographical references.
    CONTENTS: v. 1. Overview, plenary lectures, invited
lectures.—v. 2. Aerosols, emulsions, and surfactants.—
v. 3. Adsorption, catalysis, solid surfaces, wetting,
surface tension, and water. [etc.]
    1.   Colloids—Congresses. 2. Surface chemistry—
Congresses. I. Kerker, Milton, II. Title.
QD549.I6 1976   541'.345   76-47668
ISBN 0-12-404501-4 (v.1)

Contents

FORCES AT INTERFACES

Plenary Lecture

Invited Lectures

CATALYSIS

Plenary Lecture

CONTENTS

AEROSOLS

SOLID SURFACES

CONTENTS

WATER AT INTERFACES

Plenary Lecture

Invited Lectures

RHEOLOGY OF DISPERSE SYSTEMS

Plenary Lecture

Invited Lectures

CONTENTS

STABILITY AND INSTABILITY

Plenary Lecture

Invited Lectures

MEMBRANES

Invited Lectures

CONTENTS

SURFACE THERMODYNAMICS

Plenary Lecture

Invited Lectures

LIQUID CRYSTALS

Plenary Lecture

Invited Lectures

List of Contributors

I. N. Aleinikova, The Institute of Physical Chemistry of the USSR Academy of Sciences, Moscow, USSR

Anthony P. Allen, School of Pharmaceutical Chemistry, University of Bradford, Bradford BD 7 1 DP, West Yorkshire, England

A. Bellemans, Université Libre de Bruxelles, Brussels, Belgium

Howard Brenner, Department of Chemical Engineering, Carnegie-Mellon University, Pittsburgh, Pennsylvania 15213

J. Bricard, University of Paris VI and Center of Nuclear Studies of Fontenay-aux-Roses, Paris, France

D. E. Brooks, Department of Pathology, University of British Columbia, Vancouver 8, Canada

Glenn H. Brown, Liquid Crystal Institute, Kent State University, Kent, Ohio 44242

Donald H. Buckley, Lewis Research Center, Cleveland, Ohio 44135

Robert L. Burwell, Jr., Ipatieff Laboratory, Department of Chemistry, Northwestern University, Evanston, Illinois 60201

M. Cabane, University of Paris VI and Center of Nuclear Studies of Fontenay-aux-Roses, Paris, France

R. Defay, Chimie-Physique II, Free University of Brussels, Campus Plaine, C. P. 231, Bd. Triomphe, Brussels 1050, Belgium

J. E. Demuth, IBM Thomas J. Watson Research Center, Yorktown Heights, New York 10598

B. V. Derjaguin, The Institute of Physical Chemistry of the USSR Academy of Sciences, Moscow, USSR

W. Drost-Hansen, Laboratory for Water Research, Department of Chemistry, University of Miami, Coral Gables, Florida 33124

Donald Eagland, School of Pharmaceutical Chemistry, University of Bradford, Bradford BD7 1 DP, West Yorkshire, England

LIST OF CONTRIBUTORS

Frederick R. Eirich, Department of Chemistry, Polytechnic Institute of New York, Brooklyn, New York 11201

J. J. Fripiat, Centre de Recherche sur les Solides à Organisation Cristalline Imparfaite, Rue de la Férollerie, 45045 Orleans Cedex, France

Gérard E. E. Gardes, Laboratoire de Thermodynamique et Cinétique Chimiques (L.A. N° 231 CNRS), Université Claude Bernard (Lyon I) 69621, Villeurbanne, France

Lawrence J. Gaydos, Department of Chemical Engineering, Carnegie-Mellon University, Pittsburgh, Pennsylvania 15213

W. Keith Hall, Department of Chemistry, University of Wisconsin, Milwaukee, Wisconsin 53201

D. A. Haydon, Physiological Laboratory, Downing Street, Cambridge, United Kingdom

Richard H. Heist, Department of Chemistry, University of California, Los Angeles, California 90024

J. J. Hermans, Chemistry Department, University of North Carolina, Chapel Hill, North Carolina 27514

Can Hoang-Van, Laboratoire de Thermodynamique et Cinétique Chimiques (L.A. N° 231 CNRS), Université Claude Bernard (Lyon I) 69621, Villeurbanne, France

Felix T. Hong, The Rockefeller University, New York, New York 10021

H. H. Hsing, Center for Surface and Coatings Research, Lehigh University, Bethlehem, Pennsylvania 18015

J. N. Israelachvili, Department of Applied Mathematics, Research School of Physical Sciences, Institute of Advanced Studies, Australian National University, Canberra, A. C. T. 2600, Australia

Kenneth G. Johnson, Division of Biological Sciences, National Research Council of Canada, Ottawa, Canada K1A OR6

Milton Kerker, Clarkson College of Technology, Potsdam, New York 13676

K. Klier, Center for Surface and Coatings Research, Lehigh University, Bethlehem, Pennsylvania 18015

L. G. Leal, Chemical Engineering, California Institute of Technology, Pasadena, California

Benjamin Y. H. Liu, Particle Technology Laboratory, Mechanical Engineering Department, University of Minnesota, Minneapolis, Minnesota 55455

Mariano Lo Jacono, Department of Chemistry, University of Wisconsin, Milwaukee, Wisconsin 53201

LIST OF CONTRIBUTORS

J. Lyklema, Laboratory for Physical and Colloid Chemistry of the Agricultural University, De Dreijen 6, Wageningen, The Netherlands

G. Madelaine, University of Paris VI and Center of Nuclear Studies of Fontenay-aux-Roses, Paris, France

J. David Margerum, Hughes Research Laboratories, Malibu, California 90265

Dean C. Marvin, Department of Chemistry, University of California, Los Angeles, California 90024

S. G. Mason, Pulp and Paper Research Institute of Canada and Department of Chemistry, McGill University, Montreal H3C 3G1, Canada

J. H. Mathews,* Department of Chemistry, University of Wisconsin, Madison, Wisconsin 53706

Egon Matijević, Institute of Colloid and Surface Science and Department of Chemistry, Clarkson College of Technology, Potsdam, New York 13676

Alain René Mazabrard, Laboratorie de Thermodynamique et Cinétique Chimiques (L.A. N° 231 CNRS), Université Claude Bernard (Lyon I) 69621, Villeurbanne, France

Leroy J. Miller, Hughes Research Laboratories, Malibu, California 90265

V. M. Muller, The Institute of Physical Chemistry of the USSR Academy of Sciences, Moscow, USSR

D. H. Napper, Department of Physical Chemistry, University of Sydney, N. S. W. 2006, Australia

B. E. Nieuwenhuys, Gorlaeus Laboratoria, Rijksuniversiteit Leiden, P. O. Box 75, Leiden, The Netherlands

B. W. Ninham, Department of Applied Mathematics, Research School of Physical Sciences, Institute of Advanced Studies, Australian National University, Canberra, A. C. T. 2600, Australia

Donald R. Olander, Materials and Molecular Research Division of the Lawrence Berkeley Laboratory and the Department of Nuclear Engineering of the University of California, Berkeley, California 94720

R. H. Ottewill, School of Chemistry, University of Bristol, Bristol BS8 1TS, England

J. Th. G. Overbeek, Van't Hoff Laboratory, University of Utrecht, Transitorium III, Padualaan 8, Utrecht, The Netherlands

Gérard Pajonk, Laboratorie de Thermodynamique et Cinétique Chimiques (L. A. N° 231 CNRS), Université Claude Bernard (Lyon I) 69621, Villeurbanne, France

*Deceased.

Demetrios Papahadjopoulos, Experimental Pathology, Roswell Park Memorial Institute, Buffalo, New York 14263

Carl F. Polnaszek, Division of Biological Sciences, National Research Council of Canada, Ottawa, Canada K1A OR6

I. Prigogine, Chimie-Physique II, Free University of Brussels, Campus Plaine, C. P. 231, Bd. Triomphe, Brussels 1050, Belgium

David Y. H. Pui, Particle Technology Laboratory, Mechanical Engineering Department, University of Minnesota, Minneapolis, Minnesota 55455

Howard Reiss, Department of Chemistry, University of California, Los Angeles, California 90024

J. Requena, Centro de Biofisica, Instituto Venezolano de Investigaciones Cientificas, Caracas, Venezuela

Lloyd H. Reyerson,* School of Chemistry, University of Minnesota, Minneapolis, Minnesota 55455

W. M. H. Sachtler, Gorlaeus Laboratoria, Rijksuniversiteit Leiden, P. O. Box 75, Leiden, The Netherlands

A. Sanfeld, Chimie-Physique II, Free University of Brussels, Campus Plaine, C. P. 231, Bd. Triomphe, Brussels 1050, Belgium

Alfred Saupe, Liquid Crystal Institute, Kent State University, Kent, Ohio 44242

Anthony M. Schwartz, Consultant in Chemistry, 2260 Glenmore Terrace, Rockville, Maryland 20850

S. J. Singer, Department of Biology, University of California at San Diego, La Jolla, California 92093

Donald M. Small, Boston University Medical School, Boston, Massachusetts 02118

Ian C. P. Smith, Division of Biological Sciences, National Research Council of Canada, Ottawa, Canada K1A OR6

G. A. Somorjai, Materials and Molecular Research Division, Lawrence Berkeley Laboratory and Department of Chemistry, University of California, Berkeley, California 94720

Kenneth S. Stec, Ipatieff Laboratory, Department of Chemistry, Northwestern University, Evanston, Illinois 60201

Gerald W. Stockton, Division of Biological Sciences, National Research Council of Canada, Ottawa, Canada K1A OR6

*Deceased.

F. J. Szalkowski, Science Center, Rockwell International, Thousand Oaks, California 91360

D. Tabor, Physics and Chemistry of Solids, Cavendish Laboratory, Cambridge, England

Stanislas Jean Teichner, Laboratoire de Thermodynamique et Cinétique Chimiques (L.A. N° 231 CNRS), Université Claude Bernard (Lyon I) 69621, Villeurbanne, France

Yu. P. Toporov, The Institute of Physical Chemistry of the USSR Academy of Sciences, Moscow, USSR

Alexander P. Tulloch, Division of Biological Sciences, National Research Council of Canada, Ottawa, Canada K1A OR6

A. C. Zettlemoyer, Center for Surface and Coatings Research, Lehigh University, Bethlehem, Pennsylvania 18015

Preface

This first volume of papers presented at the International Conference on Colloids and Surfaces, held in San Juan, Puerto Rico, June 21-25, 1976, consists of the plenary and invited papers, and a general overview of these papers by A. M. Schwartz. These papers were given during the morning sessions.

The afternoon sessions were devoted to 221 contributed papers, which appear as follows:

Volume II: Aerosols, Emulsions, and Surfactants
Volume III: Adsorption, Catalysis, Solid Surfaces, Wetting, Surface Tension, and Water
Volume IV: Hydrosols and Rheology
Volume V: Biocolloids, Polymers, Monolayers, Membranes, and General Papers

The conference was sponsored jointly by the Division of Colloid and Surface Chemistry of the American Chemical Society and the International Union of Pure and Applied Chemistry in celebration of the 50th Anniversary of the Division and the 50th Colloid and Surface Science Symposium.

The National Colloid Symposium originated at the University of Wisconsin in 1923 on the occasion of the presence there of The Svedberg as Visiting Professor. The interesting remarks of J. H. Mathews at the opening of the 40th National Colloid Symposium and also those of Lloyd H. Reyerson are also included in this volume. It was during his stay at Wisconsin that Svedberg developed the ultracentrifuge; he also made progress on moving boundary electrophoresis, which his student Tiselius brought to fruition.

The National Colloid Symposium is the oldest such divisional symposium within the American Chemical Society. There were no meetings in 1933 and during the war years 1943-1945, and this lapse accounts for the 50th National Colloid Symposium occurring on the 53rd anniversary. However, these circumstances brought the numerical rank of the Symposium into phase with the age of the Division of Colloid and Surface Chemistry. The Division was established in 1926, partly as an outcome of the Symposium. Professor Mathews gives an amusing account of this in the article presented here. The 50th anniversary meeting is also the first one bearing the new name Colloid and Surface Science Symposium, which reflects the true breadth of interest and participation.

PREFACE

An amusing historical review of colloid and surface chemistry from the dawn of history to the present time—FUN AND FOIBLES OF THE FRATERNITY—was given at the banquet by A. C. Zettlemoyer.

There were 476 participants, including many from abroad.

This program could not have been organized without the assistance of a large number of persons and I do hope that they will not be offended if all of their names are not acknowledged. Still, the Organizing Committee should be mentioned: Milton Kerker, Chairman, Paul Becher, Tomlinson Fort, Jr., Howard Klevens, Henry Leidheiser, Jr., Egon Matijević, Robert A. Pierotti, Robert L. Rowell, Anthony M. Schwartz, Gabor A. Somorjai, William A. Steele, Hendrick Van Olphen, and Albert C. Zettlemoyer.

Special appreciation is due to Robert L. Rowell and Albert C. Zettlemoyer, who served with me as an executive committee that made many of the difficult decisions. In addition, Dr. Rowell handled publicity and announcements, while Dr. Zettlemoyer worked zealously to raise funds among corporate donors to provide travel grants for some of the participants. Not the least of their efforts was the joint editorship of this volume, which first appeared in the *Journal of Colloid and Interface Science,* Volume 58, 1977.

Teresa Adelmann worked hard and most effectively both prior to the meeting and at the meeting as secretary, executive directress, editress, and general overseer. She made the meeting and these proceedings possible. We are indebted to her.

Milton Kerker

An Overview

The celebration of a traditionally important milestone can take many forms. Sentimental excursions to the past, comparisons of early simplicity to present complexity, prideful expositions of progress and promises of still greater achievements are all standard fare. The organizers of this fiftieth in the series of Colloid Symposia planned for the celebration to result in a memento of lasting value, a panorama of where colloid and surface science stands today and what its leading practitioners are doing. To this end a number of eminent researchers were invited to give lectures or present papers in their field of specialization, either at a plenary session or at one of the split morning sessions of the meeting. These plenary lectures and invited papers were superposed on the customary concentrated format of National Colloid Symposia and form the substance of this volume.

As might be expected, the forty-five individual contributions vary widely in style, scope, and approach. Some of the authors use the historical review form, in the style of Accounts of Chemical Research or Nobel acceptance lectures; others emphasize present problems and strategies in their areas of investigation; still others use the detail of a research paper to convey the broader picture by implication and reference. The total result is a uniquely authoritative compendium of special topics in present day colloid and surface science. The ten plenary lectures deal, respectively, with surface forces, heterogeneous catalysis, aerosols, solid surfaces, water at interfaces, rheology of suspensions, colloid stability, liquid crystals, biological membranes, and surface thermodynamics. These lectures encompass the domains of the invited lectures that follow them in the program, and the whole group can therefore be discussed conveniently under these headings.

It seems appropriate in an overview of this scope to describe the panoramic scene at least briefly before burdening the reader with the reviewer's personal reactions to it.

Surface Forces

In the view of many the modern era of colloid and surface science began in the 1940s, when DLVO focused attention on the fundamental importance of surface forces and the need for their precise evaluation. The first section of this symposium deals with the forces of attraction across the interface between two condensed phases. Tabor's plenary lecture briefly reviews the earlier work now recognized as classic. It then proceeds to summarize and analyse recent studies on solid–solid adhesion. The importance of contact mechanics and surface roughness is emphasized as well as that of molecular interaction energy.

The paper by Israelachvili and Ninham, presented with considerable aplomb at the meeting, traces and analyzes the progress of intermolecular force theory. Primary attention is focused on the period since Lifshitz' publications in the mid-1950s. The theory is viewed as having followed three distinct roads: short range forces, long range forces, and statistical mechanics. Using vesicle formation by phospholipids as an example, it is pointed out that either short range or long range forces may predominate, depending on the system being considered. Electrostatic force theory and the problems in describing double layer forces are reviewed briefly and optimistically.

Fripiat's paper is primarily experimental in content. It describes the use of proton NMR data to develop a physical model for the adsorption of hydroxyl-containing molecules on oxygen-containing surfaces. From these data and models the macroscopically measured thermodynamic properties can be recovered. Correlation times, the times between random translational or reorientation jumps, are measured for methanol adsorbed on a silica gel and a molecular sieve. These substrates differ in pore size and therefore in the freedom of motion accorded the adsorbate. Fourier transform pulsed NMR is used to measure these correlation times, which range from 10^{-4} to 10^{-10} secs and are assigned to definite motions and energy distributions among the adsorbed species. Similar measurements are described for water adsorbed on a layer lattice silicate.

Bellemans describes the procedure for deriving boundary effects in lattice models. He then illustrates its use in modeling and solving for the surface tension and adsorption of monomer–dimer mixtures. The technique is an adaptation of one previously developed for bulk properties of mixtures.

Buff's paper* is very largely a review of his own extensive and pioneering work on the statistical mechanics of phase interfaces, including surface and interfacial tension and interfacial thickness. Nucleation is accorded special attention. Mathematical derivations are presented for the common equilibrium interfacial shapes in and out of gravitational fields, as well as for capillary rise.

The paper by Requena, Brooks, and Haydon is the only one in this section to deal with the liquid–liquid interface. It is aimed at evaluating the van der Waals forces operative at the interface between a black lipid film and an adjacent aqueous phase. The film is formed by drainage of a solution of lipid in an alkane solvent. Agreement between the experimentally derived Hamaker coefficients and those calculated from Lifshitz theory is found to be excellent. It is also concluded that partial segregation of alkane from lipid in the thin film can be of crucial importance for calculation of the van der Waals forces.

The metal–metal interface and its behavior in adhesion, friction, and wear is a field where Buckley's work has long been notable. His paper reviews recent develop-

---

*Not submitted for publication in these proceedings.

ments from the experimental point of view. The techniques discussed include: high vacuum to produce and maintain clean surfaces; sophisticated manipulating systems to handle the specimens and measure frictional and adhesive forces; LEED, Auger spectroscopy, scanning electron microscopy, and field ion microscopy to examine surface composition and state. Among the specific topics considered are: interfaces between similar and dissimilar metals along different crystal planes; grain boundary and orientation effects on interfacial transfer; the noble metal–iron interface; the interfaces between metals of the platinum group, and their interfaces with the Group IV elements. Results are presented for several different metal and alloy pairs. The relationship between interfacial bonding and the character of the metallic bond is discussed.

The paper by Derjaguin* and co-workers was not presented at the meeting. It is a highly condensed and entirely theoretical discussion of the relationship between electrostatic and molecular components of the adhesive force between an elastic particle and a solid surface.

Catalysis

The four papers in the group on catalysis, including the plenary lecture, are all accounts of experimental research. The paper by Burwell and Stec deals with the behavior of hydrogen adsorbed on chromia at temperatures of -130 to $-196°$C. Of special interest is the weak chemisorption of H_2 at the lower temperature, presumably a molecular polarization effect at coordinatively unsaturated surface ("CUS") ion pairs of Cr^{3+} and O^{2-}. At $-196°$C D_2 preferentially displaces this adsorbed H_2 without formation of HD. At $-163°$C substantial quantities of HD are formed.

The paper by Nieuwenhuis and Sachtler reviews the conclusions of their extensive work on the physical adsorption of nitrogen and xenon by the transition metals. The field emission-probe hole technique (described and referenced in the paper) was used to study adsorption of N_2 on the various crystal faces of Pt and Ir. The adsorption is face specific. Xenon is adsorbed on specific sites of the atomically rough crystal faces of Ir, but forms a close packed monolayer on the smooth faces. From photoelectron emission measurements of adsorption heats and surface potentials conclusions can be drawn regarding the adsorption bond of xenon. On gold, copper, nickel, platinum, and rhodium this bond results partly from dispersion forces and partly from a charge transfer–no bond interaction.

LoJacono and Hall describe in detailed fashion the isomerization and disproportionation of cyclopropane over a series of thoroughly characterized molybdena-alumina catalysts. Isomerization apparently correlates with the concentration of Mo–OH groups on the surface. Olefinic metathesis to ethene and butene correlates with the concentration of coordinatively unsaturated Mo^{2+} centers. This type of catalyst is of industrial interest, being used for hydro-desulfurization of petroleum feed stocks.

*See p. 561.

Teichner and co-workers describe several hydrogenation reactions catalysed by alumina rendered catalytically active by hydrogen spillover. The pure alumina is originally in contact with an alumina-supported metal catalyst, platinum or nickel. The reactor is so designed that the unmetallized alumina can be separated and isolated after the system has been flooded with hydrogen. This metal-free alumina is now an effective catalyst for hydrogenation reactions, i.e., it has been activated by hydrogen "spilled over" from the active metal. It catalyses the isomerization of methylcyclopropane as well as the hydrogenation of olefines.

Aerosols

Kerker's plenary lecture on aerosols starts with a beautifully condensed summary of the necessary fundamentals of light scattering. It then proceeds to a discussion of three recently discovered unusual light scattering effects. The first is the remarkable decrease in opacity of a titania pigmented paint film when microvoids (air bubbles in the same size range as the pigment particles) are introduced. The microvoids encase the pigment particles. The ratio of void volume to pigment volume is critical, and the ratio of particle size to wavelength of light also affects the hiding power. A very complete and satisfactory theory is presented to explain this phenomenon. The second effect discussed is the internal heating of small particles by radiation, and its contravening effect on radiation pressure. The internal heating is uneven, and the particle therefore radiates thermal energy unsymmetrically. In a gaseous medium this produces a radiometer effect. The basis of a quantitative theory is presented. The third topic of discussion is the effect of particle size and refractive index on the intensity of Raman and fluorescent scattering when the particles containing the scattering molecules are in the same size range as a wavelength of light. This hitherto neglected effect seriously complicates the quantitative analysis of particulate air pollutants via Raman and fluorescence spectroscopy. A theoretical solution is presented from which the diminution of intensity can be calculated. This solution has been checked experimentally using monodisperse fluorescent latex particles.

The paper by Bricard, Cabane, and Madelaine deals with the nucleation of neutral ultrafine particles and small ions from gaseous impurities of the air. Reiss' paper, also on nucleation, contains a considerable proportion of previously unpublished research results on photo-induced nucleation and its use in studying gas phase reactions. The remaining paper in this section, by Liu and Pui, also presents previously unpublished work on the electrical charging of aerosol particles.

Solid Surfaces

The solid surfaces section of the symposium is largely concerned with the newer techniques for exploring surface structure and surface reactions. Somorjai's plenary lecture outlines our present knowledge of the structure of clean surfaces, and the structure and behavior of adsorbed layers, as revealed by low energy electron diffraction. Surface crystallography is discussed and exemplified. The new techniques make possible determination of the surface composition of metal alloys, which can differ considerably from the bulk composition. Molecular beam scattering is discussed with regard to both theory and technique. It is a valuable method for studying mechanism and kinetics of surface reactions.

Olander's paper on modulated molecular beam mass spectrometry is unusual in many ways. This technique for studying gas–solid reactions is the subject of considerable literature, but there are very few laboratories in which it is actually being practiced. This paper explains the numerous difficulties that are encountered, not only in the experimentation but also in deducing valid models of reaction mechanisms from the massive amount of data that the method affords. Details of the reduction of uranium dioxide by hydrogen are given as an illustration.

Demuth's paper is an excellently organized progress report on the use of low energy electron diffraction (LEED) to determine the locations and translational periodicities of surface atoms. Model calculations for surface crystallography are presented and discussed. Procedures used in determining the structure of sorbed oxygen and hydrogen on single nickel crystal faces are discussed, and results of the determinations are presented.

With the advent of ultrahigh vacuum technology, electron spectroscopic techniques have become practical and have proven to be sensitive probes for the first few layers of a solid surface. Szalkowski's lecture illustrates this statement abundantly, emphasizing X-ray photoelectron spectroscopy (ESCA or XPS) and Auger electron spectroscopy. Qualitative and quantitative chemical composition and chemical bonding states are among the properties that can be determined. Illustrative examples of each are presented.

Water at Interfaces

Klier's plenary lecture is an exposition of the behavior and structure of water at inorganic surfaces including metals, oxides, and silicates. Studies that have led to a picture of water on the (100) crystal face of iron are presented in detail. In the adsorption process it appears that a mobile water molecule becomes fixed by forming firmly bound hydroxyls. At the boundaries vacant spaces are left large enough to accommodate oxygen. The resulting lateral distribution of work function may play an important role in corrosion. The techniques used include LEED, Auger, and electron spectroscopy. Thermal decomposition of nickel hydroxide to high area nickel oxide leaves surface hydroxyls on the (111) plane, stabilizing the structure. Hot stage electron microscopy, electron diffraction, infrared absorption, adsorption, and calorimetry were all used in this study. The thermal removal of surface hydroxyls is difficult and leaves a strongly adsorbent surface. The binding of hydroxyls and water to silica and silicates is discussed in more general terms.

The last and longest part of this paper describes the author's work on the motions of adsorbed water molecules and clusters, especially their restricted rotations.

These are followed via the Fourier transforms of a combination near infrared band in a region where no interference occurs from surface hydroxyls. The theoretical background is developed for both bulk and surface water, and is applied to the heterogeneous nucleation of freezing on a variety of surfaces, especially the fluorine micas. The effect is evidently to induce in the water molecules vibrations with a frequency close to those of ice. Epitaxy helps, but is not of key importance.

Eagland and Allen present a research paper addressed to a specific problem: the effect of surface hydration on the zeta potential of poly (vinylacetate) in the presence of several different electrolytes. The extensive theoretical discussion as well as the lengthy bibliography make this an excellent source work as well as an account of the investigation.

Lyklema presents a characteristically thorough and convincing study of silver iodide sols, including electrokinetic, double layer, and stability measurements. He concludes that the zeta potential and the potential of the outer Helmholtz plane (OHP potential) coincide within experimental error. This result precludes the existence of any thick stagnant water layers at the interface.

Drost-Hansen's paper, "Effects of Viscinal Water on Colloidal Stability and Sedimentation Processes," starts with a review of evidences for long range (up to 0.1 micrometers) water structure at solid surfaces. Among those cited are increased viscosity, anomalies of ion conductance and ionic distribution near silica surfaces, compressional heats in clay–water systems, and the results of ultrasonic velocity studies. All of these effects change sharply at certain definite temperatures, indicating changes in molecular ordering. Energy differences between the different structures, however, are small. This suggests that studies of colloidal stability versus temperature be made in systems where high energy effects are minimal. Preliminary results are presented of the author's recent studies on kaolin sedimentation as a function of temperature. A very extensive review of literature in this area is presented.

The paper by Zettlemoyer and Hsing is in the same area as that of the plenary lecture. These investigators use near-infrared reflectance to study the reactions of surface hydroxyls with silanes, and then measure water vapor adsorption on the silane-coated surfaces. Two substrates are used: a partially hydroxylated and a fully hydroxylated silica. The silanes include a number of the so-called reactive or functional silanes used industrially as coupling agents in glass-reinforced plastics. One silazane is also included. The combined IR and adsorption data are interpreted in terms of total surface-adsorbate configuration. The exclusion of water is not prevented by the organic adduct on the originally fully hydroxylated silica.

Rheology

Mason's plenary lecture on the rheology of disperse systems summarizes some of his work on the motion of particles suspended in a liquid undergoing shear. Starting with two-body collisions of both neutral and interacting rigid spheres, he demonstrates the marked difference between particle behavior in pure shear flow and simple shear flow. The effects discussed in detail include the dispersion of fiber aggregates, the coiling and uncoiling of flexible threads, and various oscillatory effects noted in suspensions of disks. As we have come to expect of this investigator, the theoretical development is incisive and the experiments are works of art. An impressive feature of this paper is a table outlining and referencing the work of the author and his colleagues on orthokinetics. It includes a score or more of different particle types in systems of varying Peclet number. Gravitational, centrifugal, electrical, and interparticle fields are considered as well as rotational and irrotational shear fields in the liquid medium.

Hermans' paper is entirely theoretical, dealing with the effect of inertial forces on the motion of solids through liquids. The author considers both rotational and translational Brownian motion, and develops the contribution to viscosity of suspended linearly elastic dumbbells.

Leal presents a detailed discussion of recent theoretical studies on diffusive heat flux, or more generally on macroscopic transport properties, in a sheared suspension. A fundamental equation is derived relating microscale thermal properties of the suspension-to-bulk conductive heat flux. This latter quantity is shown to depend not only on the conductivity of the phases but also on velocity and type of flow and on particle shape, size, and rigidity. The equations appear applicable to all fluids of suspensionlike structure, including solutions of macromolecules. Specific evaluations of conductivity are worked out for dilute suspensions of rigid spheres.

Brenner and Gaydos analyze rigorously the combined diffusive and convective transport of an isolated, rigid, spherical, neutrally buoyant Brownian particle suspended in a Poiseuille flow. The radius of the particle is comparable in size to the tube radius. Both hydrodynamic and nonhydrodynamic interactions between wall and particle play dominant roles in the analysis. The results are also applicable to dilute suspensions of noninteracting particles. A more general theoretical treatment is proposed to handle nonspherical particles and noncylindrical walls.

Colloidal Stability

Ottewill's plenary lecture starts by briefly summarizing the historical development of the attractive–repulsive force balance concept as the key to colloid stability. The stability of lyophobic colloids is one of the oldest research areas in colloid science, and the theory was advanced in one great step thirty years ago by DLVO. Great advances, however, are noteworthy for begetting further advances. The major portion of the lecture reviews modern experimental methods for studying electrostatic repulsive forces and steric stabilization. Both macroscopic systems and fine particle dispersions are considered.

Matijevic's paper is a review of the extensive work, performed mainly in his laboratory, on the colloidal effects of chemical complexing. Complexes are effective in the preparation of monodisperse sols of the hydrous metal oxides. The interactions of hydrolysed metal ions and of chelates with latexes and with silver halide sols are used to illustrate the influence of complexing on colloidal stability.

Complexing in the bulk liquid phase, as well as at the liquid–solid interface, has an effect on stability. The last part of the paper describes quantitative experiments on the adhesion of chromium oxide particles to glass in aqueous media of varying pH and metallic cation content.

Steric stabilization, an effect familiar to earlier colloid scientists as "protective colloid action," is the subject of a paper by Napper, one of the pioneers who resuscitated interest in it some years after the triumphant advent of DLVO. The stabilizing agents considered in this paper are all nonionic macromolecules. Their stabilizing effect is interpreted in terms of the theory of polymer solutions. Theta conditions for the stabilizing chains correspond to the stability limit of the system.

Overbeek's lecture develops the point of view that coagulation is a rate process, the resultant of several simpler processes each occurring at its own rate. The inverse process repeptization is also the resultant of several rate processes. After reviewing the three types of interparticle force and the classical and modified collision theories of coagulation, the author enumerates and discusses the rate effects in a single collision. Repeptization is then considered from an experimental as well as theoretical point of view. Emulsions are discussed very briefly. The paper concludes with a list of current problems in the field of colloid stability.

Eirich's paper is concerned with a more specific aspect of steric stabilization, namely, the adsorption of macromolecules and their conformation at the liquid-solid interface. It is essentially a review of work performed in recent years, largely in the author's laboratory. Adsorption isotherms and hydrodynamic thickness measurements on a wide range of systems show the adsorbates to consist of monolayers of individual macromolecules. These are held by relatively few attachments, and have the form of solvent-pervaded coils of dimensions similar to those observed in free solution. This general picture applies to aqueous and organic solvents, and both polar and nonpolar adsorbents and adsorbates. Polyelectrolytes can be pulled onto the surface by salt formation. Oligomers with reactive end groups, peptides, and proteins may assume different configurations depending on their constitution and that of the substrate.

Biological Membranes

Three of the papers in the section on biological membranes can be classed as accounts of current research, and two as historical reviews. Danielli's plenary lecture* traces historically the development of the now firmly established concepts of cell membrane structure and function. Recent work comparing natural membranes with artificially prepared bilayer lipid structures receives major emphasis. It is shown by energy considerations that the bilayer should be the only type of lipid structure to occur naturally, as it indeed appears to be. Transfer across membranes and the structure of the protein portion of membranes are discussed extensively.

Smith and co-workers present an extended research paper and progress report on the use of deuterium NMR and spin label ESR in membrane research. These tech-

*Not submitted for publication in these proceedings.

niques are used in exploring the mobility and organization of lipids in model and natural membranes. They give results that agree qualitatively but not quantitatively, probably due to perturbations induced by the nitroxide label. Both methods report high order and low mobility for segments of the fatty acid chains near the carboxyl groups; less order and more rapid motion near the terminal methyls. Both methods indicate the gel–liquid crystal transition. The merits and demerits of each method are discussed.

Early in his paper on membrane proteins Singer states that although lipids provide the primary structural framework of biological membranes it is the proteins that carry out their specific chemical functions. An understanding of the structure and special characteristics of membrane proteins is therefore essential to an understanding of how membranes work. Singer's paper provides this understanding admirably, reviewing the work of over a decade that has led to it. The locations and attachments of peripheral and integral proteins are described, and the fluid mosaic model of membrane structure is presented. Current ideas on the mechanisms of transport are discussed.

Papahadjopoulos' research paper deals with the thermotropic transitions of phospholipid membranes from the solid to the liquid crystal state. The transition temperatures and enthalpies are profoundly affected by the presence of calcium and magnesium ions. Different classes of proteins also influence these transition parameters in very specific ways, depending on how they are bound to the lipid. The experimental results suggest that phase separations and density fluctuations, induced isothermally by divalent metal ions and proteins, are involved in membrane fusion and in the shape changes observed during endocytosis.

Hong's lecture, accepting the 1976 Victor K. LaMer award, is a tour de force in one of the most difficult areas of molecular biology, that of electric and magnetic effects. Photoelectric and magneto-orientation effects in pigmented biological membranes are analyzed in terms of levels of order in the membrane structure. An artificial bilayer lipid pigmented with a magnesium porphyrin is used to study photoelectric responses. The concept of a chemical capacitance is introduced and used to explain previously reported data on several different pigmented membranes. The theory is also applied to the photosynthesizing membranes of chloroplasts and to the membranes of visual structures. An extended bibliography ranges widely and thoroughly over this field of investigation.

Surface Thermodynamics

Defay, Prigogine, and Sanfeld's paper, introducing the section on surface thermodymanics, starts with a general discussion of irreversible thermodynamics in the treatment of phase interfaces. The concepts of local and global entropy production and local equilibrium are applied to adsorption, surface chemical reaction, and momentum and heat fluxes. The adsorption equations extrapolated to equilibrium become equal to the Gibbs equation. A second section of the paper is a well-illustrated discussion of the ways in which classical thermodynamics and statistical mechanics have been used to solve problems of surface stability and curvature effects, and to derive the surface phase rule. The major portion of the paper gives numerous examples of nonequilibrium interfacial situations that can be mathematically modeled. These include dynamic surface tension, surface diffusion, the Marangoni effect, and others, concluding with an extended discussion of interfacial instability. With well over 100 references this paper is at once the clearest discussion and the best literature source of its topic that this reviewer has seen.

Liquid Crystals

The section on liquid crystals is introduced by Brown's plenary lecture on the structures and properties of this state of matter. As the title indicates this is a general review of the type that might grace a first rate encyclopedia. In a gratifying number of passages it also refers to the author's own researches and those of his coworkers, especially in the fields of tilt angle determination and the structure of nematogenic compounds.

The lyotropic liquid crystals formed in water by soaps. surfactants, and similar amphiphilic compounds are of practical as well as academic interest. The detergent effects of these systems have long been known, and more recently their importance in biological systems has been recognized. Saupe's paper describes the various phases encountered among lyotropic liquid crystals, and points out their distinguishing features. Studies on myelin figures are presented. A theory is suggested to explain their formation and their tendency to coil. The stability of cylindrical vesicles is also discussed.

The paper by Margerum and Miller is one of the few originating from an industrial laboratory and devoted primarily to applications. The basic properties of liquid crystals are reviewed, stressing the important effects of dopants and surface alignment. The various categories of electro-optical effects and devices are outlined. Three of the more complicated applications of nematics are discussed in detail. This paper, like the others in this section, has abundant illustrative figures and an exceptional complement of literature references.

Small's paper is a review of liquid crystalline structures in biological systems. In healthy living systems there is a liquid crystalline order of molecules not only in membranes but also in many protein systems, notably those of muscle tissue. The role of liquid crystal build up in certain human diseases is discussed at considerable length.

Porter's paper* explores the relationship between mesophase temperature ranges and the entropies of transition in steroids.

In the course of preparing this overview the writer had occasion to reread the proceedings of the first National Colloid Symposium, held in 1923 on the occasion of a sabbatical visit by The Svedberg to The University of Wisconsin. In comparing the 1976 and 1923 symposia, it appears that the progress in colloid and surface science has been neither greater nor less than that in most other areas of science.

*Not submitted for publication in these proceedings.

AN OVERVIEW

Over the past fifty years developments in colloid and surface science (CSS for brevity; please pardon the acronym) have paralleled developments in the physical and biological sciences and in the technologies with which they are associated. Despite the impressive scope of the present compendium, the scope of CSS in 1923, relative to the science of its day, was just as broad. This should not be too surprising.

If from all the scientific disciplines we assemble those portions that involve phase interfaces we have the substance of *surface* science. By the time of the first Colloid Symposium several areas of physics, for example, capillarity and friction, had been adopted by CSS. *Colloid* science started with the observation that glue solutions diffuse more slowly than solutions of salt or sugar. By 1923 it included gels, surfactants, and a full range of lyophilic colloid systems both aqueous and nonaqueous. The physicochemical aspects of what eventually became macromolecular science were at that time considered part of colloid chemistry. Lyophobic dispersions, involving macroscopically recognizable phase interfaces, were definitely the province of colloid science. The intimate and extensive connection with technology was already well established. Equally intimate was the connection with biology, via proteins, lipids, plant and animal tissue structures, and single cell behavior. Individual sections of CSS thus formed an integral part of a host of other disciplines, to which they contributed and from which they borrowed. This situation has not changed.

Despite the breadth of this compendium several topics of perennial interest to colloid and surface scientists are not covered, even though they were reasonably well represented among contributed papers at the Meeting. Among others these topics include surfactants, emulsions, foams, spread monolayers, wetting, and capillarity. Rather than indicating that CSS is becoming narrower, this situation probably reflects a change in the fashionable areas of research within CSS, guided by similar changes within its related disciplines.

In the period before 1923, CSS had few specialized techniques. Svedberg in his address to the meeting that year stated "Almost every advance in colloid chemistry can be traced back to some new technical means especially designed for the study of colloids. Colloid chemistry requires its own technical facilities, different from the ordinary equipment of a physicochemical laboratory." It is evident that succeeding generations took this idea seriously. The specialized techniques developed by colloid and surface scientists match those of any other discipline in both number and effectiveness, and the advances afforded by them have not been disappointing.

Advances in theoretical development over the past fifty years have kept pace with experimental advances. This is amply evidenced by the sections on surface thermodynamics, interfacial forces, and solid surfaces within this compendium. In the period before 1923, CSS was for the most part an experimental and descriptive science.

Looking beyond the present scene, CSS will certainly grow along some lines suggested by the papers in this symposium; the growth will be an accretion on nuclei already formed. The general relationship between CSS and other areas of science will remain essentially unchanged, as it has since the first colloid symposium. CSS will continue to benefit from the results of others' studies on single phase phenomena, and will in turn furnish knowledge and stimulus from its own studies of the interface between phases. No one doubts, for example, that knowledge of the bulk liquid phase is advanced by studies on liquid crystals and micellar solutes. The relationship between CSS and technology has always been close and will inevitably remain so. As in the past much of the support for research in CSS will come from the advanced technology industries. Connections with biology and medicine will continue unabated.

Techniques and tools will continue to proliferate and make significant experimentation easier to perform. A good example is the approaching opportunity to carry out prolonged experiments in space vehicles at near-zero gravity and near-perfect vacuum. Some specialized areas of CSS will reach maturity, just as some have matured and become part of elementary knowledge over the past fifty years. Other areas will grow to powerful majority; still others will emerge from dormancy. It seems almost certain to this reviewer that the old dichotomies between lyophilic and lyophobic, surface and colloid, will disappear. This is already happening in the area of liquid crystals, membranes, and biological structures. Finally, advances made by the theorists will open up fields that have hitherto been either arcane or purely descriptive. Irreversible thermodynamics and the quantum chemistry of surfaces could lead to great advances in the understanding of heterogenous reactions, nucleation, solution processes, and other chemical processes where the solid phase is either consumed or generated. New light would be cast on corrosion, electrode reactions, crystal growth, and similar areas that have proved refractory in the past.

The reviewer firmly believes that most prophecy, like his own, is simply an early recognition of what is already here. The papers in this compendium tell much about the state of colloid and surface science.

Anthony M. Schwartz

Early History of the National Colloid Symposium

Forty-three years ago this month, Professor J. H. Mathews, Matty to his friends, called the first session of the First National Colloid Symposium to order. In fact, this occurred in the very room where the fortieth of these symposia begins. At that time Matty was chairman of the department, in Madison, and as such, he had prevailed upon Professor The Svedberg of Upsala University, Sweden, to spend the academic year 1922–23 in residence at the University of Wisconsin. No doubt Professor Svedberg's presence stimulated much of the thinking that resulted in organizing the first symposium on colloids ever to be held.

According to the best records available, one hundred and seventy-eight persons registered for the three-day sessions. Scarcely a dozen of these registrants are still living, but among them are Professors Mathews and Svedberg. Anyone present at the fortieth symposium who also attended the first symposium may from time to time glance to the rear of the room, hoping wishfully that they might again see Victor Lenher, of selenium oxychloride fame, and Louis Kahlenberg, or perhaps their ghosts coming through the door. Kahlenberg, one of the few remaining scientists at that time who did not believe in the ionization of acids, bases, and salts, would not begin to present his paper until students in the rear of the room gave him the usual loud huzzah. One must say that at this point, the symposium was starting with a *bang*.

Everyone present, when Matty gave his opening remarks, sensed that something new and big was happening to American science. Even though meeting on a university campus, here was gathered a happy blending of industrial and academic science. The industrial chemists came from such diverse laboratories as those of rubber, oil, glass, soap, adhesives, forest products, leather, paper, and photographic materials. Most of the university representatives were physical or chemical scientists, but there were a number from the biological, medical, and some related agricultural sciences. As Matty said, he and Harry Holmes, chairman of the Committee of Colloid Chemistry of the National Research Council, realized that colloid science had entered a quantitative stage and that it was time to try the experiment of organizing a

[1] Presented on the occasion of the Fortieth Annual Session at the University of Wisconsin at Madison. These remarks first appeared in the Journal of Colloid and Interface Science, 22, 412–418 (1966). The late Lloyd Hilton Reyerson was head of the School of Chemistry at the University of Minnesota from 1937–1954.

symposium that would attract those scientists from industry and education who had a real interest in this newer developing area.

As we can all attest, the experiment was an outstanding success. It showed American science that symposia such as this could be so planned that the speakers, as well as other registrants, would be attracted from a wide spectrum of chemical activity. The give and take in this and succeeding symposia has had a profound influence on chemistry in the New World. That Matty should have taken the lead in initiating a colloid symposium is not too surprising, for he had always shown himself to be a farsighted planner and organizer. Even while a student, here at Wisconsin, during the early years of the century, he became one of the prime movers in the formation of Alpha Chi Sigma, the National Chemical Fraternity. This organization, after early struggles to survive, made significant contributions to the growth of chemistry. Congratulations to Professor Mathews, one of the prime innovators in the American chemical scene during the first decades of this century and to all who joined him in the planning of the First National Colloid Symposium.

Professor Svedberg was the acknowledged foreign guest of honor at this meeting. Thus a precedent was established. At each succeeding symposium, with few exceptions, a distinguished scientist from abroad was invited to attend as the honored guest. Each symposium was, in part at least, organized around the colloid interests of this guest of honor. Professor Svedberg's paper at this first symposium really challenged the attending scientists. As its title, "Colloid Chemistry Technique," indicated, Svedberg pointed out that colloid science needed to originate and develop new experimental equipment which was especially designed for use in this new area of science.

Let us note Professor Svedberg's challenge, quoting from his paper, "We have to contemplate the properties of colloids, as far as we know them, and ask ourselves: what kind of instruments would one like to have at one's disposal to study such systems? I think that such a method would pay much better in the long run than just to take the instruments that one might happen to come across in the field of ordinary physics and chemistry and try to study colloids by those instruments." He went on to advise colloid scientists that it would save a lot of work, time, and money if they took a hard look at their equipment. Very often it would be found that the technical means used in studying colloids were not rational. Svedberg outlined and then explained many of the colloid chemistry techniques available to the scientists of that time. This paper showed, even then, the many contributions which he had already made to the development of new methods and original equipment, that could be used more effectively in the studies of colloidal systems. He also indicated possible new directions for such studies.

Many of the twenty-four additional papers given at this first symposium discussed the preparation, properties, and behavior of a number of colloidal systems. These included gels, jellies, emulsions, membranes, films as well as colloids obtained from naturally occurring materials. However, one paper had a marked effect on the direction of research in the young but expanding field of catalysis. In this work, entitled "The Problem of Adsorption from the Standpoint of Catalysis," Professor

Symposium number	Year held	Where held	Foreign guest of honor
1	1923	University of Wisconsin	The Svedberg
2	1924	Northwestern University	Leonor Michaelis
3	1925	University of Minnesota	Herbert Freundlich
4	1926	Massachusetts Institute of Technology	James W. McBain
5	1927	University of Michigan	H. R. Kruyt
6	1928	University of Toronto	William B. Hardy
7	1929	Johns Hopkins University	F. G. Donnan
8	1930	Cornell University	William L. Bragg
9	1931	Ohio State University	
10	1932	Ottawa, Ontario	Emil Hatschek
11	1934	University of Wisconsin	
12	1935	Cornell University	Arne Tiselius
13	1936	Washington University	P. Koets
14	1937	University of Minnesota	Herbert Freundlich
15	1938	Massachusetts Institute of Technology	Wolfgang Ostwald
16	1939	Stanford University	J. D. Bernal
17	1940	University of Michigan	
18	1941	Cornell University	
19	1942	University of Colorado	
20	1946	University of Wisconsin	The Svedberg
21	1947	Stanford University	
22	1948	Massachusetts Institute of Technology	
23	1949	University of Minnesota	Leonor Michaelis
24	1950	St. Louis	
25	1951	Cornell University	
26	1952	University of Southern California	R. M. Barrer and P. C. Carman
27	1953	Iowa State College	J. Th. G. Overbeek
28	1954	Rensselaer Polytechnic Institute	
29	1955	Rice Institute	W. Feitknecht
30	1956	University of Wisconsin	R. M. Barrer
31	1957	New York Schools, Statler Hotel	J. H. de Boer
32	1958	University of Illinois	
33	1959	University of Minnesota	
34	1960	Lehigh University	J. Th. G. Overbeek
35	1961	University of Rochester	H. L. Booij
36	1962	Stanford University	A. E. Alexander
37	1963	Carleton University, Ottawa	W. J. Dunning
38	1964	University of Texas	J. H. de Boer
39	1965	Clarkson College of Technology	Aharon Katchalski

Symposium number	Year held	Where held	Foreign guest of honor
40	1966	University of Wisconsin	M. Eigen
			B. V. Derjaguin
41[1]	1967	University of Buffalo	A. E. Alexander
42	1968	Illinois Institute of Technology	H. Fujita
43	1969	Case Western Reserve University	D. H. Everett
44	1970	Lehigh University	B. V. Derjaguin
			G. M. Schwab
45	1971	Georgia Institute of Technology	A. Bellemans
46	1972	University of Massachusetts	R. H. Ottewill
47	1973	Carleton University	K. S. W. Sing
48	1974	University of Texas	D. H. Everett
49	1975	Clarkson College of Technology	B. A. Pethica
			J. H. Block
50	1976	San Juan, Puerto Rico	See Plenary and Invited Lectures in this Volume.

[1] Data in this table for the 41st and following symposia have been added by the Editors.

H. S. Taylor stressed the need for new approaches to the problems that arise in dealing with catalytic reactions occurring at the solid–gas interface. Taylor clearly showed that specificity of adsorption of gases by active surfaces was one of the keys to the activity of such surfaces. Thus the nature and mechanism of the adsorption of gases by active catalysis became an ever-expanding study. Papers presented at succeeding symposia give ample proof that Taylor's ideas had a profound effect in directing investigation along fruitful lines in the search for the answers as to how catalysts act.

As the first symposium drew to a close, sentiment almost unanimously built up for a continuance of such meetings. When Professor Mathews read a letter from President Scott, inviting the symposium to meet on the Northwestern University campus in Evanston, during June 1924, there was prolonged applause from the registrants. It was agreed to turn the job of arranging future symposia over to the Colloid Committee of the National Research Council. Members of this committee at the symposium then met and voted unanimously to accept the invitation extended so cordially by Northwestern University. The registrants were also informed that the papers presented at this first symposium would appear in book form, including any discussion, and that Matty would edit this volume.

The Second National Colloid Symposium met on the campus of Northwestern University, June 18-20, 1924 under the chairmanship of Harry N. Holmes. He was chairman of the Colloid Committee of the National Research Council which organized and sponsored this symposium. Leonor Michaelis, the foreign guest of honor, opened the meetings by giving a paper entitled "General Principles of Ion Effects on Colloids." Here he stressed that there were two kinds of influences of ions on colloids. One must make clear the differences between the electrostatic effects and the lyotropic effect. The first is observed in the action of ions of an opposite charge to the colloid, and this raises all the problems involved in the electrical double layer. The second effect becomes more manifest the more the colloid is a hydrophilic one, and it depends more upon the power of the ions to attract water molecules. He again raised the long-standing problem of the absolute hydration of ions. Michaelis' discussion has been found more than useful in colloid chemistry. This he had sincerely hoped for when he concluded his presentation.

At this same meeting Dr. Ross Aitkin Gortner gave a monumental paper on "Physico-chemical Studies on Proteins. I. The Prolines—Their Composition in Relation to Acid and Alkali Binding." The great amount of laboratory work presented fills nearly the last half of the published Colloid Symposium Monograph Volume II. It is a masterly reply to certain implied criticisms of colloid chemistry found in Loeb's book on proteins and colloidal behavior. Many other important pieces of research were presented and discussed. This left little doubt about the success of such symposia. It convinced the colloid scientists of the day that such symposia should be held annually. At the close of the sessions the registrants were pleased to learn that the third symposium would be held during June of 1925 at the University of Minnesota.

The leaders of both of these two symposia felt strongly that colloid science had reached a point where it merited divisional status in the American Chemical Society. Many registrants argued that colloid science could better achieve its proper place among the physical sciences in the American scene if there was created a division of the rapidly growing American Chemical Society. Such a division would then hold sessions at the ACS national meetings. However, among its most important functions would be the sponsoring of national symposia, such as those already held. Members of the Committee of Colloid Chemistry of NRC agreed in general with this idea. The membership of this committee at that time consisted of the following:

Harry N. Holmes, *Chairman,* Oberlin College
Jerome Alexander, New York City
Wilder D. Bancroft, Cornell University
Floyd E. Bartell, University of Michigan
E. C. Bingham, Lafayette College
R. A. Gortner, University of Minnesota
J. H. Mathews, University of Wisconsin
S. E. Sheppard, Eastman Kodak Co.
E. B. Spear, Pittsburgh, Pennsylvania
Harry B. Weiser, Rice Institute (University)
J. A. Wilson, A. F. Gallun Sons Co.

Members of this committee, together with others deeply interested in this growing field of physical science, met and began to lay plans for the organization of such a division of the ACS. In those days they consulted with Charles L. Parsons, the able secretary of the ACS. With his advice, an informal committee began the preparation of the necessary documents for submission to the council of the ACS. By the time of the meeting of the Third National Colloid Symposium at the University of Minnesota, June 17–19, 1925, the proposal had jelled sufficiently to call for the naming of a small committee to draw up the necessary constitution, articles, and bylaws for submission to the council of ACS. Dr. Harry B. Weiser was appointed to succeed Harry N. Holmes as chairman of the Colloid Committee of NRC. For years thereafter, until his untimely death, Harry Weiser took charge of the successive symposia that were held.

Under the auspices of the School of Chemistry, University of Minnesota and with the cooperation of the Minnesota section of the ACS, the Third National Colloid Symposium was held June 17–19, 1925, in Minneapolis. At this symposium Professor Herbert Freundlich of the Kaiser Wilhelm Institute for Physical and Electrochemistry, Berlin, Germany, was the foreign guest of honor. His paper on "The Electrokinetic Potential" opened the sessions. He raised basic questions as to the source of this potential and showed rather forcefully that this potential at interfaces differed fundamentally from the Nernst thermodynamic potential at electrode surfaces. In his best example he compared the potentials of the glass electrode with the electrokinetic potentials of glass surfaces.

There were other outstanding presentations at this symposium. One notes especially the paper given by Irving Langmuir entitled "The Distribution and Orientation of Molecules." Here Langmuir developed the theory for the distribution and orientation of molecules both within liquids and at their interfaces. He then clinched his arguments by presenting experimental results, many of them his own, proving the theory. At the symposium a year earlier, W. D. Harkins had presented more qualitative evidence for similar orientation at interfaces. The importance of Langmuir's extensive contribution to the field of surface chemistry was evidenced by the awarding to him of the Nobel Prize. Students of the present can do no better, if they wish to grasp the fundamentals of molecular orientation, than to study Langmuir's paper in Volume III of Colloid Symposium Monographs.

Before leaving the third symposium one must also mention such important contributions as "Photographic Sensitivity" by S. E. Sheppard; "Colloidal Water and Ice" by Howard T. Barnes; "The Colloid Chemistry of Protoplasm" by L. V. Heilbrunn; "Elasticity and Some Structural Features of Soap Solutions" by William Seifriz. Other discussions concerned the colloidal character of soils, the mechanism of blood clotting, and the colloid chemistry of protoplasm. As the symposium neared a close, the registrants were informed that the fourth symposium would be held at the Massachusetts Institute of Technology during June of 1926. The proposal for divisional status had reached such a state that the group approved in principle and asked that the committee send the material on to Charles Parsons for submission to the council of the ACS.

The fourth symposium was held in Cambridge, Massachusetts, June 23-25, 1926 under the auspices of the Massachusetts Institute of Technology in cooperation with the Northeastern Section of the ACS. The foreign guest of honor was James W. McBain of the University of Bristol, Bristol, England. He gave the opening address of the sessions on the subject "A Survey of the Main Principles of Colloid Science." He pointed out that the two main problems of colloid science were structure and stability. What is the structure that places matter in this category? Whence do such structures derive such a measure of stability as to constitute nearly all the common materials met with in daily life? The answer to the first question is definitely that in most cases the unit of which colloids are built up is not the mere molecule, but a higher organized unit, the particle called by Nageli "the micelle." This was probably the first time that American colloid scientists had heard this term used and its value explained. Professor McBain went on to state that there are classes of substances the most stable state of which is the colloidal condition, i.e., they are thermodynamically most stable in this state. As the great leader in the study of soaps and detergents he illustrated the above problems in terms of the micelles which soaps form, pointing out that there are neutral and ionic micelles. The stability of these systems is determined by the nature of the surface of the micelles. He concluded by pleading that the phenomena of colloidal behavior should be described in terms of directly observed facts rather than of theories or inferences.

Again we find a number of important subjects being discussed by investigators in their respective areas. Hugh Taylor gave new evidence and stressed the importance of reaching an understanding of the character of the surfaces of solid catalysts. Bancroft raised the problems concerning the complexity of water by discussing water equilibrium. Professor E. F. Burton, of Toronto, kept alive, for colloid scientists, the questions involved in relating the Helmholtz double layer to ions and charged particles. H. R. Kruyt, of Utrecht, later in the symposium gave experimental evidence for the importance of cataphoretic studies in determining ionic effects. For the first time at these symposia, George L. Clark proved the usefulness of X-rays in the establishment of the structures of colloids and Harry B. Weiser pointed out the many ionic antagonisms found in colloid systems. The elucidation of the structures and colloidal properties of a number of natural products stressed the importance of quantitative measurements in these difficult studies.

As the symposium drew to a close it was announced that the fifth symposium would be held at the University of Michigan in June of 1927. It was hoped that by that time the principal sponsoring group would be the Division of Colloid Chemistry of the American Chemical Society. Up to this time the sponsoring organization continued to be the Colloid Committee of the NRC. At the Golden Jubilee Meeting of the ACS in Philadelphia in September 1926, the council voted unanimously to establish a Division of Colloid Chemistry in the ACS organization. Thus the colloid division was born just fifty years after the organization of the parent society in 1876; and the division will celebrate its golden anniversary when the ACS holds its centennial celebration. From 1926 on, the newly established division of the ACS sponsored the colloid symposia annually at different institutions of higher learning.

The Fifth National Colloid Symposium was held at the University of Michigan, Ann Arbor, Michigan, June 22–24, 1927 under the chairmanship of Harry B. Weiser, with Professor Floyd E. Bartell serving as local chairman. Professor H. R. Kruyt of the University of Utrecht was the foreign guest of honor. He made many illuminating contributions to the discussions in addition to opening the sessions with a remarkable discussion of "Unity in the Theory of Colloids." In this work he gave evidence that colloidal solutions, both of lyophobic and lyophilic sols, are built up of polymolecular particles, and that both groups have an electrokinetic nature. He went on to point out that any theory which seeks unity by identifying lyophilic colloids with ionic solutions is on the wrong track. The unity in colloids comes from the fact that all colloidal solutions are principally of the same type, and one of the first aims of colloid science should be a better understanding of the electrical double layer which is met so frequently in all sorts of colloid systems.

Additional investigations pointed directions that colloid science should take in order to develop a basic set of concepts for this area of science, once called the "field of neglected dimensions." W. D. Harkins discussed the stability of emulsions by considering monomolecular and multimolecular films. He went on to consider the cases for spreading and nonspreading of films on liquids and showed that spreading occurs if the interfacial work of adhesion between the liquids is greater than the surface work of cohesion in the liquid. Elmer O. Kraemer brought to the fore "Some Unsolved Problems in the Electrokinetic Behavior of Colloidal Suspensions" and L. Michaelis described "Investigations on Molecular Sieve Membranes." It should also be mentioned that William Blum of the Bureau of Standards discussed "Colloids in the Electrodeposition of Metals." This might have better been considered as unsolved problems dealing with the effects of colloids in such electrochemical work.

All in all, this symposium like the four that preceded it covered the main areas of research in surface and colloid chemistry. In every symposium unsolved problems were raised and the directions which future studies ought to take were indicated. This short history of the colloid symposium has made special reference to the first five meetings because these meetings largely determined the character of those held later. Certainly the first five meetings stimulated and gave direction to the whole of colloid science in this country.

This year the fortieth symposium is once again being held on the campus of the University of Wisconsin, where these symposia started forty-three years ago. One may ask why this is not the forty-third meeting? The answer lies in the fact that the first year in which a meeting was skipped was in 1933 at the worst part of the Great Depression. Then two meetings[1] had to be postponed during World War II. The tradition is now so well established that it will take some major event or change to force further postponement of any future meetings. In addition to the five meetings at Madison, Wisconsin, Cornell University, Ithaca, New York has hosted four meet-

---

[1] In fact, three. See the table on page 19.

ings, as has the University of Minnesota, Minneapolis, Minnesota. Three meetings have been held in Canada, one at Toronto and two at Ottawa. No other schools have been the hosts for more than two such meetings except the Massachusetts Institute of Technology, which entertained three such sessions. Appended[1] is a list of the forty meetings with the year and place where each was held. When possible the name of the foreign guest of honor is given. In conclusion one should point out that the first four symposia were held under the sponsorship of the Colloid Committee of the National Research Council. Beginning with the fifth symposium the Division of Colloid and Surface Chemistry of the American Chemical Society has planned and directed the meetings. The papers given at the first seven symposia were published in book form. Volumes I to VI appeared as Monographs and Volume VII was published by Wiley. This method of giving to the scientific world the important findings presented at the symposia succumbed to the Depression. In recent years, probably a better method has prevailed in that preprints of the papers go to members of the division prior to the symposia. Much better discussion of each paper is thus possible. These preprints were then published for a good many years as one issue of the *Journal of Physical Chemistry*. Now the *Journal of Colloid and Interface Science* undertakes this publication. The method of preprinting and then publishing in a journal involves refereeing of each paper, thus making certain of the high quality of the work presented.

The National Colloid Symposia, on this the fortieth birthday, demonstrate that they are mature and that they have a permanent place in American science. These meetings have given leadership to colloid and surface chemistry because of the character of the science presented and the scientific eminence of many of the participants. In fact there are several Nobel Prize winners in the list. Our younger scientists in this field of science owe much to these leaders and to the symposia in which they took such an important part.

Lloyd H. Reyerson
June 13, 1966

[1] See page 19.

Forces at Interfaces

Surface Forces and Surface Interactions

D. TABOR

Physics and Chemistry of Solids, Cavendish Laboratory, Cambridge, England

Received June 3, 1976; accepted August 12, 1976

This paper discusses briefly some of the theoretical problems relating to surface problems as seen by an experimental physicist. In particular, questions are raised concerning the calculation of interaction energies between condensed phases in atomic contact. The main experimental part of the paper deals with the adhesion between soft elastic solids and shows the importance of combining concepts of surface forces (or surface energies) with the principles of contact mechanics. It is shown that surface roughness can greatly reduce the adhesion between such solids since the higher asperities can prize the surfaces apart and break the adhesions occurring at the lower asperities. If these concepts are applied to very hard, elastic solids very small adhesions may be observed for surface roughnesses scarcely greater than atomic dimensions even if the surfaces are extremely clean. If, however, the surfaces are ductile the junctions can accommodate the prizing action of the higher asperities and strong adhesions may be observed. The overall conclusion is that adhesion between solids depends not only on surface forces but also on surface roughness and the degree of ductility of the solids themselves.

I. INTRODUCTION: SURFACE FORCES

On an occasion which celebrates the 50th Anniversary of the Colloid and Surface Chemistry Division of the American Chemical Society it is natural that a session devoted to "Forces at Interfaces" should deal with those forces most familiar to colloid and surface chemists. These forces are associated with the name of van der Waals and the most general type is that which arises from the fluctuations in the electronic charge around individual atoms. These dispersion forces were first explained by London over 40 years ago and soon after they were applied by Bradley, Derjaguin, de Boer, and Hamaker (1–4) to the problem of the forces between macroscopic bodies by pairwise addition of the forces between individual atoms.

Pairwise addition can be used to compute the lattice energy of a van der Waals solid, for example, solid argon, and although the calculation may not give the correct crystal structure, the lattice energy is correct to within less than 10%. Thus pairwise addition is a good

first approximation for certain types of calculation. The macroscopic force between two bodies separated by air or vacuum can also be calculated satisfactorily on this basis. However, in the presence of an intervening dielectric the results can be very misleading (even in the absence of electrical double layers) for reasons which I presume will be dealt with in the paper presented by Dr. Israelachvili and Professor Ninham. It is then customary to treat the bodies and the intervening medium as continua using the powerful methods developed especially by Lifshitz. By dealing with the materials in this way all the interactions which can occur can be expressed in terms of their bulk dielectric properties.

Although there has been a remarkable development in the theory of van der Waals forces using the Lifshitz approach there is very little direct experimental data by which the theoretical computations can be checked. For example, apart from the early classical work of the Russian and Dutch schools, the most direct experimental measurements of the forces

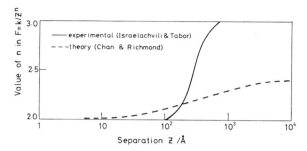

Fɪɢ. 1. Attractive forces between a sphere and a flat or between crossed cylinders as a function of separation Z. A power law $F = k/Z^2$ corresponds to predominantly nonretarded van der Waals forces; a power law $F = k/Z^3$ corresponds to predominantly retarded van der Waals forces.

between solid surfaces, especially at separations below 1000 Å, are those of Tabor and Winterton (5) and Israelachvili and Tabor (6). They used cylindrical sheets of mica arranged with their axes mutually at right angles so that the geometry is equivalent to that of a sphere on a flat. If R is the radius of each cylinder and the distance of closest approach is Z, the attractive force F_u for unretarded and F_r for retarded van der Waals forces should be

$$F_u = AR/6Z^2, \qquad [1a]$$

$$F_r = 2\pi BR/3Z^3, \qquad [1b]$$

where A is the unretarded and B the retarded Hamaker constant. From measurements of the force between the mica surfaces the change from unretarded to retarded forces could be plotted as a function of separation Z. The results are given in Fig. 1. These may be compared with a detailed computation of Chan and Richmond (7) using the Lifshitz equations and the known absorption characteristics of mica. The discrepancy is rather large and is attributed to the dominating nonretarded zero frequency contribution at large separations.

Again Chan and Richmond have shown that the mica results may be compared with the interaction energy to be expected for two parallel mica half-spaces in terms of the separation Z. If the energy is written as

$$G = -A/12\pi Z^2, \qquad [2]$$

where the Hamaker energy parameter A is a function of Z, the calculated results may be compared with the experimental. This is shown in Fig. 2. The discrepancy is large, particularly at the smaller separations. Of course, the mica surfaces are covered with an adsorbed monolayer of water. There may also be residual surface charges, although all the evidence is against this. The discrepancy, reflecting the gap between sophisticated calculations and a very limited series of experimental results, presents a challenge both to the theoretician and to the experimentalist. I must confess that on this issue I am not entirely neutral and I recall the classical comment of Coulomb in relation to another physical problem "only experiment can decide." My own gloss on this comment is "the more direct the experiment, the better."

Although as an experimental physicist my understanding of the full subtleties of the Lifshitz theory is limited, it seems to me that certain complications must arise as the condensed phases come closer and closer together. First, it is generally assumed that the dielectric properties of a continuum are constant throughout the material. This cannot be true of the surface layers where dielectric characteristics must be different from those of the bulk. These surface properties will be unimportant for separations large compared with the thickness of the modified surface layer, but for small separations they could play a significant part.

4

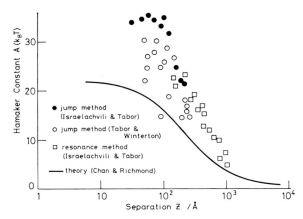

Fɪɢ. 2. Hamaker energy function A in units of k_BT as function of separation for a sphere on flat or for crossed cylinders.

Second, when the surfaces come within atomic contact the graininess of the material becomes important and atom–atom interactions must once again be given serious considerations. Both of these factors must be relevant when liquids are in contact with other liquids or with solids but we do not know how important the effect is. I hope that Dr. Haydon's paper will stimulate some discussion on this point.

Third, if solids are brought into atomic contact other short-range forces may develop which may well swamp the van der Waals interactions. For example, with insulators or semiconductors charge separation may occur in the contact zone and the Coulombic forces involved can be considerable. There has been a fair body of dispute on the relative importance of the electrostatic contribution and I believe that we shall hear something about this in Professor Derjaguin's paper. Again, it is now generally accepted that with simple metals the contribution of van der Waals forces to the surface energy is only about one-tenth of that due to the metallic bond itself. Thus the adhesion of metals in atomic contact is dominated by the metallic bond. I have no doubt that Dr. Buckley will describe some of the features of metallic adhesion in his paper.

The main point I wish to make in these introductory remarks is that even for purely van der Waals materials the interfacial forces between condensed phases in close proximity cannot be handled satisfactorily by unthinking application of the classical Lifshitz equations. The situation is even more involved if other strong short-range forces are involved.

II. THE INTERACTION OF VAN DER WAALS SOLIDS IN ATOMIC CONTACT

In this section I should like to discuss the attractive force exerted in air or vacuum between a sphere of radius R and a flat surface; in the first instance I shall deal only with van der Waals solids. In spite of all that I have said before, I shall treat the properties of the solids as those of a continuum right up to atomic contact. Consequently, we can use either the Hamaker equations or the Lifshitz equations, the only difference being the way in which the Hamaker constants are computed. We shall also assume that both surfaces are molecularly smooth.

Undeformable Solids

For separations large compared with the equilibrium size Z_0 of the atoms but still within

5

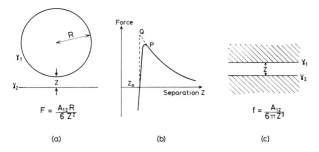

(a)   (b)   (c)

FIG. 3. Nonretarded forces between solid surfaces. (a) A sphere on a flat, (b) sketch indicating the small effect of allowing for repulsive forces when equilibrium is reached at separation Z_0, (c) arrangement of two parallel half-spaces.

the nonretarded region, the attractive force between the sphere [1] and the flat [2] is

$$F = A_{12}R/6Z^2, \qquad [3]$$

where Z is the nearest distance of approach and A_{12} is the appropriate Hamaker constant (see Fig. 3a). If the bodies come into atomic contact Z reaches the equilibrium separation Z_0 of the atoms. The attractive force, and hence the force required to pull the surfaces apart, will be

$$F_0 = A_{12}R/6Z_0^2. \qquad [4]$$

We assume here that the short-range repulsive forces do not appreciably affect the value of

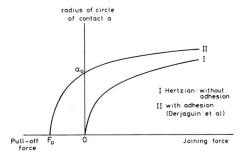

FIG. 4. Contact between an elastic sphere and a hard flat surface as a function of the load between them (schematic). Because of surface forces the circle of contact is larger than that given by the classical equations of Hertz which assume zero adhesion between the surfaces. According to Derjaguin et al. (10), the pull-off force $F_0 = 4\pi R\gamma$.

F_0 so that (see Fig. 3b) the observed force at P is essentially the same as that at Q. Now for two parallel flat half-spaces (of the same materials as the sphere and the flat) the force per unit area f for small separations Z is (see Fig. 3c)

$$f = A_{12}/6\pi Z^3. \qquad [5]$$

If we carry out a "thought-experiment" separating two bodies from separation Z_0 to infinity we can calculate the work done to create two new surfaces in place of the interface. Hence

$$\int_{Z_0}^{\infty} f \, dz = (\gamma_1 + \gamma_2 - \gamma_{12}) = A_{12}/12\pi Z_0^2, \quad [6]$$

where γ_1, γ_2, and γ_{12} are the surface energies of the sphere, the flat, and the interface, respectively. We can insert the value of A_{12}/Z_0^2 from Eq. [6] into Eq. [4] to give

$$F_0 = 2\pi R(\gamma_1 + \gamma_2 - \gamma_{12}) \qquad [7]$$

or, if both the sphere and the flat are of the same material,

$$\gamma_1 = \gamma_2 = \gamma \quad \text{and} \quad \gamma_{12} \simeq 0,$$
$$F_0 = 4\pi R\gamma. \qquad [8]$$

This was first derived by Bradley (1) using pairwise summation. As he showed, if one allows for the short-range repulsive forces, this factor enters into the calculation of the γ's (Eq. [7]) in exactly the same way as it does

6

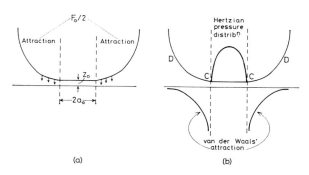

(a)   (b)

FIG. 5. Interaction between an elastic sphere and a hard flat surface according to Derjaguin *et al.*
(10). (a) When the applied normal load is zero, the attractive forces produce a finite circle of contact
of diameter $2a_0$. There is repulsion over this area and attraction in the annular region outside. (b) The
pressure distribution implied in the Derjaguin analysis (schematic).

into the calculation of F_0. Equation [8] remains unchanged.

Deformable Solids: The Approach of Derjaguin et al.

Derjaguin (2) was the first to point out that surfaces brought together by attractive forces must deform at the region of contact. The shape of the deformed surface ultimately determines the attractive forces and it is this shape that has been the subject of some dispute (8–10). In order to bring out the main physical concepts involved we shall assume (i) that the flat is undeformable, (ii) that the van der Waals properties of the sphere and the flat are the same so that A can be written in place of A_{12} and 2γ in place of $\gamma_1 + \gamma_2 - \gamma_{12}$, (iii) that the sphere has a Young's modulus E and a Poisson's ratio ν.

Derjaguin *et al.* assume that under the influence of the surface forces the sphere will deform in the contact region according to the elastic equations of Hertz. If the force pressing the sphere onto the flat is F, Hertz shows that a circle of contact is formed of radius a where

$$a = \left(\tfrac{3}{4}FR\,\frac{1-\nu^2}{E}\right)^{\frac{1}{3}}. \qquad [9]$$

If there were no adhesion, the variation of a with the external applied load F would be as

shown in Curve I in Fig. 4. Because of adhesive forces the circle of contact is larger, as shown in Curve II, and a finite force F_0 is required to pull the surfaces apart. The pull-off force at the instant of separation is found to be

$$F_0 = 4\pi R\gamma, \qquad [10]$$

so that it is equal to the value that would apply to nondeformable solids.

Of particular interest is the behavior when no external load is applied. There is a finite circle of contact of radius a_0. According to Derjaguin *et al.* (10) there is an attractive force equal to $F_0/2$ outside the region of contact which is responsible for the deformation within the region of contact, as sketched in Fig. 5. Consequently, from Eq. [9] we have

$$a_0 = \left(\frac{3\pi R^2\gamma}{2}\,\frac{1-\nu^2}{E}\right)^{\frac{1}{3}} \qquad [11]$$

(see Ref. (10, Eq. [63])). If we now attempt to pull the sphere away from the flat by applying a small pull-off force, the circle of contact diminishes and the annular region becomes correspondingly larger. With increasing pull-off force a continuously decreases until, when $a \rightarrow 0$, the pull-off force is at its maximum given by F_0 in Eq. [10]. This behavior and the values of a_0 and F_0 are indicated in Fig. 4.

7

This analysis is interesting for its attempt to link up surface forces with the use of the Hertzian elastic equations. Unfortunately, the shape of the deformed zone ignores the deformation due to the attractive forces close to the edge of, but outside, the Hertzian circle. These forces are comparable with the compression forces *within* the Hertzian circle. For a self-consistent treatment it is necessary to make the deformed shape compatible with the *overall* distribution of surface forces in accordance with the principles of contact mechanics.

Deformable Solids: The Approach of Contact Mechanics (8)

The attractive forces between the sphere and the flat are concentrated primarily in the first couple of atomic layers; to a first approximation they may indeed be regarded as being pure surface forces. In that case the interaction between the sphere and the flat may be visualized as the stitching together of the bodies with elastic thread of very limited extensibility. If Z_0 is the equilibrium separation between atoms, if the thread is stretched much beyond Z_0 the thread virtually snaps. Consequently, all the important forces are to be found within the immediate contact area; outside this they become, by comparison, negligible. Contact mechanics allows only one

stress distribution for this if we assume that the separation over the whole contact zone is constant at Z_0. The result is shown in Fig. 6a for the case where there is no external load. There are two main features. First, in the central region of the contact zone there is compression (this corresponds to a region of interfacial repulsion) and in the outer regions there is tension (this corresponds to interfacial attraction). The product of pressure times area under the compressive region exactly equals that under the tension region since there is no external load and the internal forces must balance. This means that in the central region the separation is a little less than Z_0, in the outer regions a little more; but as mentioned above we assume the separation to be constant over the whole interface at its equilibrium value Z_0. Such an assumption is also involved in the calculation of the surface energy change.

The second feature is the sharp discontinuity at the edge of the contact zone (see Fig. 6b). This clearly involves large elastic strains so that classical elasticity cannot be strictly applied. In practice there will be some rounding off at these regions so that the stresses at the edge do not exceed the limiting attractive forces that the interface can withstand.

The radius of the circle of contact when the applied load is zero is nearly double that

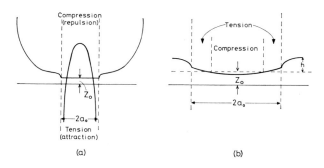

(a)   (b)

FIG. 6. Interaction between an elastic sphere and a hard flat surface according to Johnson *et al.*; (a) when the applied load is zero, the attractive forces produce a finite circle of contact of diameter $2a_0$. The central region is under compression, the peripheral region under tension. It is assumed that outside the circle of contact the interfacial forces are negligible. (b) Exaggerated features of the contact zone.

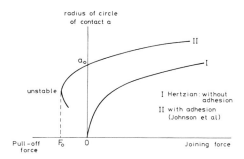

FIG. 7a. Contact between an elastic sphere and a hard flat surface as a function of the load between them. Because of surface forces the circle of contact is greater than that given by the classical equations of Hertz which assume zero adhesion between the surfaces. The analysis of Johnson et al. shows that there is a sudden instability when the pull-off force, F_0, equals $3\pi R\gamma$.

deduced by Derjaguin et al., i.e.,

$$a_0 = \left(9\pi R^2\gamma \frac{1-\nu^2}{E}\right)^{\frac{1}{3}}. \qquad [12]$$

If now we attempt to pull the sphere away from the flat the smallest pull-off force will reduce the radius of the circle of contact. This will proceed until a stage is reached when the rate of release of mechanical energy in the contact zone is greater than the surface energy requirements. The behavior resembles exactly the propagation of a Griffith crack and the surfaces immediately separate (Fig. 7a). This occurs for a pull-off force given by

$$F_0 = 3\pi R\gamma. \qquad [13]$$

This differs from the value quoted by Derjaguin et al. by having a 3 instead of a 4. To compare the two theories simply by comparing Eqs. [10] and [13] is meaningless since absolute values of γ are not easy to come by. Further, with deformable materials there are often time-dependent factors which will affect the observed pull-off force. A more significant test is provided by (i) the shape of the contacting interface and (ii) the way in which the surfaces separate.

Experiments have been carried out with an optically smooth rubber sphere (prepared by molding in optically polished glass molds) in contact with a smooth glass or Perspex flat. A study of the shape of the interface by optical interference and even by low-power optical microscopy shows that the shape of the rubber at the interface is of the type shown in Fig. 6a rather than that shown in Fig. 5a. In this connection it may be noted that the large attractive forces assumed by Derjaguin et al. to operate in the regions CD outside their assumed contact zone (see Fig. 5a) must pull these regions into contact. It is this type of interaction which produces the "neck" shown in Fig. 6a.

The way in which the surfaces separated was also consistent with the analysis of Johnson, Kendall, and Roberts. To make the results more quantitative the experiments were carried out between a rubber sphere (radius $R = 2.2$ cm) and a flat of the same rubber. For this system we may assume that $\gamma_{12} \simeq 0$. The results are shown in Fig. 7b. It is seen that a marked instability occurs for a pull-off force of about 600 dyn. From Eq. (13) this gives a value of γ_{rubber} of ca. 30 ergs cm^{-2} (mJ m^{-2}).

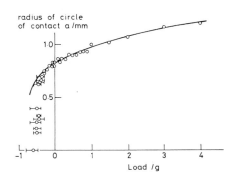

FIG. 7b. Experimental results for a rubber sphere, radius 2.2 cm, in contact with a similar rubber flat as the initial joining load of 4 g is gradually reduced and then made negative. The radius a of the circle of contact remains finite until at a critical pull-off force of about 0.6g (600 dyn) it suddenly falls to zero as the surfaces pull apart.

9

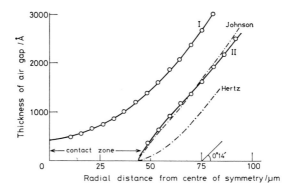

FIG. 8. Shape of air gap between crossed cylinders of mica deduced from multiple beam interferometric measurements of Israelachvili (11). The figure gives only the right-hand half about the center of symmetry. The geometry is equivalent to that of a sphere on a flat. Curve I is the shape when the surfaces are over 200 Å apart and corresponds to an equivalent sphere of radius $R = 1.25$ cm. Curve II is the shape when the surfaces are in contact at zero applied load; the equivalent radius of the circle of contact $a_0 \simeq 45\ \mu m$. The broken lines are theoretical curves for the simple Hertzian shape and for that deduced by Johnson. Although the agreement with the Johnson result is very close it should be noted that the slopes near the contact edge may be deceptive because of the enormous resolving power in the vertical, as compared with the horizontal direction.

The Behavior of Mica Surfaces in Contact

We quote here some results on contact geometry obtained in the course of a study of the van der Waals forces between mica surfaces in air. Mica sheets, a few micrometers thick, were glued on to cylindrical glass supports; these were arranged with their axes at right angles so that the geometry resembles that of a sphere on a flat. The back surfaces of the mica were silvered so that multiple-beam interferometry could be used to measure the gap between the exposed surfaces of the mica using fringes of equal chromatic order (FECO). The resolution is of the order of a few angstroms. The lateral resolution is that of a simple optical system and is therefore not better than 1 or 2 μm. Further, the sandwich consisting of a hard glass cylinder, a layer of glue, and a thin mica sheet is not an ideal representation of a homogeneous semiinfinite elastic continuum. However, provided an "effective" elastic modulus is chosen the analysis described in the previous section should apply.

Mr. D. C. B. Evans has kindly analyzed some typical interferograms obtained by Israelachvili (11) and the results are shown in Fig. 8. The gap when the surfaces are not in contact corresponds to a sphere of radius $R = 1.25$ cm. The shape of the contact zone when the surfaces are in contact under zero load is seen to correspond to a circle of radius $a_0 \simeq 45\ \mu m$. On this figure two other curves are drawn. One is the shape of the interface if the deformation outside the contact zone were Hertzian. At a distance r from the center of the contact zone (for values of $r < 2a$) the Hertzian gap h is given by

$$h = \frac{1}{\pi} \frac{a_0^2}{R} \frac{8(2)^{\frac{1}{2}}}{3} \left(\frac{r}{a} - 1 \right)^{\frac{3}{2}}. \quad [14]$$

This is the shape postulated by Derjaguin et al. According to Dr. Johnson the adhesive forces which produce the "neck" in the contact zone give an additional separation

$$h^1 = \tfrac{4}{3}(a_0^2/R)(1 - 2\pi \sin^{-1} a/r), \quad [15]$$

so that the total separation expected is $h + h^1$. The agreement between the gap calculated in this way and experiment is surprisingly good.

10

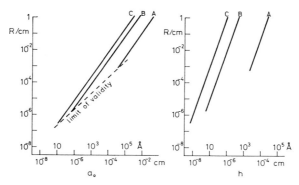

FIG. 9. Interaction between a sphere and a flat for three typical materials A, B, and C described in the text for the model described by Johnson *et al.* (12). (a) Variation of the radius a_0 of the zero-load contact area as the radius R of the sphere is varied. (b) Height of the neck h (see Fig. 6b) as a function of R.

Deformable Solids–Rigid Solids: The Limiting Case

In the analysis described in the previous section there are two basic assumptions. The first is that linear elasticity holds over the whole range of deformations involved and, as has already been mentioned, this cannot be true at the sharp edge of the contact zone. We shall not deal with this further. The second assumption is that the interfacial forces are negligible outside the contact region. We may examine this in somewhat greater detail.

It is not difficult to show that the height of the "neck" around the contact zone is of order

$$h \simeq (R\gamma^2/E^2)^{\frac{1}{3}}. \qquad [16]$$

Consequently, if h becomes comparable with Z_0 the forces between material outside the contact zone can no longer be neglected.[1] In

[1] Clearly the relevant dimensionless parameter is $h/Z_0 = (R\gamma^2/E^2Z_0^3)^{\frac{1}{3}}$.

Fig. 9 we have plotted a_0 and h as a function of sphere radius for three types of material.

A: a soft rubber,

$\gamma = \quad 40 \text{ mJ m}^{-2}, \quad E = 10^7 \text{ dyn cm}^{-2},$

B: a thermoplastic,

$\gamma = \quad 100 \text{ mJ m}^{-2}, \quad E = 10^{10} \text{ dyn cm}^{-2},$

C: a hard solid,

$\gamma = 1000 \text{ mJ m}^{-2}, \quad E = 10^{12} \text{ dyn cm}^{-2}.$

The plot of a_0 is discontinued for values of $a_0 > R/2$. We shall not discuss this further except to note that as R diminishes (other parameters remaining constant) the ratio of a_0/R actually increases; i.e., smaller particles tend to undergo proportionately greater deformation at the contact zone.

The plot of h is of greater interest. As is to be expected h is smaller, the smaller the radius of the sphere. For the soft rubber considering the smallest sphere for which the analysis is meaningful ($R \simeq 10^{-5}$ cm) h is of order 100 Å so that the second assumption is valid. For the thermoplastic the smallest R permissible is about 10^{-6} cm (100 Å) and for this h is of order 5 Å. This is just approaching the critical distance. For the hard solid a sphere of radius $R = 100$ Å gives a value of h of order 1 Å. Here, clearly the contact–mechanics analysis must be modified to allow for the attractive forces in the region outside the predicted contact zone (see Fig. 10). The solution is not known but it seems reasonable to assume that this will lead to a larger area of contact than that predicted by the simple theory outlined above. This does not necessarily imply that the final pull-off force will be larger since ultimately this will depend on the details of the energy balance as the interface peels away. It is possible that the final relation for the pull-off force will be of the form

$$F_0 = 3\pi R\gamma\phi(R\gamma^2/E^2Z_0^3), \qquad [17]$$

where the function $\phi(R\gamma^2/E^2Z_0^3)$ tends to $\frac{4}{3}$ as $(R\gamma^2/E^2Z_0^3)^{\frac{1}{3}}$ approaches unity. But whether this is or is not correct must await a full and detailed analysis.

The Effect of Surface Roughness

In all the preceding discussions we have assumed that the surfaces are "molecularly smooth" so that the geometric area of contact may be taken as being identical with the molecular area of contact. Such a situation applies to the contact between mica surfaces. It also applies approximately to the contact between optically smooth rubber surfaces of low modulus. With such a material the surfaces are deformed by very small forces so that local imperfections do not prevent intimate contact over the major portion of the interface. Recently we carried out a very simple study of the effect of surface roughness on adhesion. The experiments were between an optically smooth rubber hemisphere and a hard smooth flat surface of Perspex (PMMA). The adhesion was initially high and agreed well with Eq. [13]. The flat surface was then roughened and the adhesion was found to diminish as the roughness was increased (Fig. 11). Simple considerations show that if the roughnesses

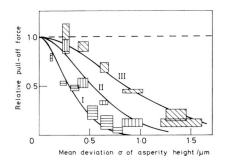

FIG. 11. Relative pull-off force for smooth rubber spheres in contact with a flat Perspex (PMMA) surface as a function of surface roughness of the Perspex. Continuous lines = theory; blocks = experiment. Effect of modulus E of the rubber: Curve I, 2.4×10^7 dyn cm^{-2}; Curve II, 6.8×10^6 dyn cm^{-2}; Curve III, 2.2×10^6 dyn cm^{-2}.

were all of uniform height (Fig. 12a) the effect of increasing roughness would be to reduce the local value of the radius of curvature R. If the horizontal pitch of the roughnesses is assumed constant this would lead to a fall-off in adhesion with increasing roughness which is very different from that observed experimentally. As Fig. 11 shows, for extremely small roughnesses the adhesion is hardly changed (it might even increase) but thereafter it falls very rapidly indeed.

As we shall see below, the main factor is the presence of a few high roughnesses which can exert a large elastic force in penetrating the rubber, the crucial quantities being the mean deviation σ of the asperity heights and the average radius R of the asperity tips. A detailed analysis (12, 13) shows that basically the adhesion depends on the ratio of the mean deviation of the asperity heights to the extension which an asperity junction can sustain before the adhesion fails. For an elastic solid of modulus E, Poisson's ratio $\frac{1}{2}$, such as rubber, this ratio is given by

$$\psi = (4\sigma/3)(4E/3\pi R^{\frac{1}{2}}\Delta\gamma)^{\frac{2}{3}}, \qquad [18]$$

where

$$\Delta\gamma = \gamma_1 + \gamma_2 - \gamma_{12},$$

as described in Eq. [6] for the two contacting

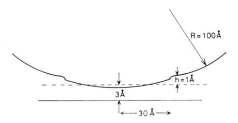

FIG. 10. Limiting case in which the neck height h (≈ 1 Å) is comparable with the equilibrium separation Z_0 ($\simeq 3$ Å) for a hard sphere of radius $R \simeq 100$ Å, giving a theoretical zero-load circle of contact radius a_0 of about 30 Å. In this situation interfacial attractions outside the theoretical circle of contact can no longer be neglected.

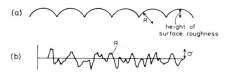

FIG. 12. Two types of surface irregularities. (a) Uniform rugosities, (b) more realistic surface defined by the mean deviation σ of the asperity heights and the average tip radius R of the asperities.

surfaces. The physical significance of ψ can be seen in a different way if we recognize that, since ψ is dimensionless, any power of it is also dimensionless. If we consider $\psi^{\frac{1}{3}}$ and omit the numerical terms we obtain a modified adhesion parameter, which we may call θ, given by

$$\theta = E\sigma^{\frac{3}{2}}/R^{\frac{1}{2}}\Delta\gamma = E\sigma^{\frac{3}{2}}R^{\frac{1}{2}}/R\Delta\gamma. \quad [19]$$

We see from Eq. [13] that the denominator of Eq. [19] is a measure of the adhesive force experienced by a sphere of radius R. The numerator (apart from a small numerical factor) turns out to be the elastic force required to push a sphere of radius R to a depth of σ into an elastic solid of modulus E. Clearly these adhesion parameters represent the statistical average of a competition between the compressive forces exerted by the higher asperities which are trying to separate the surfaces and the adhesion forces between the lower asperities which are trying to hold the surfaces together. When the adhesion parameter is small the adhesive factor dominates and the adhesion is high; as the surface roughness and hence the adhesion parameter increases the high asperities prize the surfaces apart and the adhesion falls to a low value.

It is interesting to note that roughnesses quite small compared with the overall bulk deformation of the rubber can produce an extremely large decrease in adhesion. For example, for a rubber hemisphere, $R = 2.2$ cm, $E = 2.4 \times 10^7$ dyn cm^{-2}, $\Delta\gamma = 35$ mJ m^{-2} for rubber on Perspex, the overall contact region corresponds to a depression of the rubber at the center of the region of contact of order $20\,\mu$m. Yet the adhesion on a surface of roughness $\sigma = 1\,\mu$m has fallen to almost zero

(Curve I, Fig. 11). For this surface, profilometer measurements give an asperity radius of order $100\,\mu$m. Inserting this value and the appropriate values of E and $\Delta\gamma$ in Eq. [19] gives a value of $\theta \approx 10$. We conclude that, for elastic solids, if θ exceeds a value of about 10 the adhesion will be very small.

Adhesion of Clean Solids in a High Vacuum: Effect of Roughness and Ductility

These conclusions may be applied to a completely different series of experiments which were carried out on the adhesion of hard solids carefully cleaned in a high vacuum of order 10^{-10} Torr (14). The adhesion was measured between crossed cylinders so that the geometry resembles that of a sphere on a flat and the normal loading was so small that the deformation at the contact regions was essentially elastic. We found that for clean metals the adhesion was high. For hard solids such as sapphire, diamond, and TiC, the adhesion was very small, however scrupulously we cleaned the surfaces. For clean TiC, $\Delta\gamma$ is of order 1000 ergs cm^{-2} while E is about 5×10^{12} dyn cm^{-2}. From Eq. [19] we see that θ will be greater than 10 for a value of σ of order 50 Å (assuming β remains at the value of about $100\,\mu$m). Consequently, if the results obtained with rubber apply to other elastic solids such as TiC, the presence of roughnesses on almost an atomic scale is sufficient to reduce the adhesion to very low values indeed.

Here again the lack of strong adhesion is a result of the fact that for hard elastic solids the extensibility of an adhered junction before it fails in tension is very small. If, however, one of the surfaces is ductile the junctions may survive considerable extension. In that case the adhesion should be appreciable even if the surfaces are not molecularly smooth. This is observed. For example, we obtained strong adhesion between clean copper and clean TiC although the surface roughnesses were of order 500 Å. Again with clean metals in contact with one another, the adhesion was always strong, though metals of more limited ductility, such as cobalt, adhered less strongly than ductile fcc metals.

III. SUMMARY

In this introductory paper I have attempted to outline some of the theoretical problems which seem to me, as an experimental physicist, to merit some attention in relation to surface forces. We would like to know how our current approach needs to be modified for surfaces in close proximity; in particular, how to calculate the interaction energy when condensed phases are in atomic contact.

In the second part, I have dealt with the adhesion between elastic solids and have argued that the problem can be solved theoretically only by combining surface forces with the principles of contact mechanics. Experiments show that both the shape of the deformed interface and the way in which the surfaces pull apart are in agreement with such a theory, indicating that the contact–mechanics approach has a convincing apodictic quality.

Surface roughnesses can greatly reduce the adhesion between solids. This is due to the higher surface asperities, which can prize the surfaces apart and break the adhesions occurring at the lower asperities. The critical quantity is the ratio of the extensibility before rupture of an adhered region to the mean deviation of asperity heights. Thus with materials of very high modulus, even if they are prepared with extremely clean surfaces, very small adhesions may be observed for surface roughnesses scarcely greater than atomic dimensions. If, however, one or both of the surfaces are ductile the junctions may survive much larger extensions and in those cases appreciable adhesion is observed. It is evident that adhesion depends not only on surface forces but also on surface roughness and on the mechanical properties of the solids.

Note added in proof: A more recent calculation by the same authors (*Proc. Roy. Soc. Ser. A*, in press) shows an appreciably smaller discrepancy between theory and experiment than that indicated in Fig. 1.

ACKNOWLEDGMENTS

I wish to express my sincere thanks to Dr. K. L. Johnson and Dr. J. A. Greenwood for helpful and stimulating discussions, particularly on the subtleties of contact mechanics.

REFERENCES

1. BRADLEY, R. S., *Phil. Mag.* **13**, 583 (1932).
2. DERJAGUIN, B. V., *Kolloid Z.* **69**, 155 (1934).
3. DE BOER, J. H., *Trans. Faraday Soc.* **32**, 10 (1936).
4. HAMAKER, H. C., *Physica* **4**, 1058 (1937).
5. TABOR, D. AND WINTERTON, R. H. S., *Proc. Roy. Soc. Ser. A* **312**, 435 (1969).
6. ISRAELACHVILI, J. N. AND TABOR, D., *Proc. Roy. Soc. Ser. A* **331**, 19 (1972).
7. CHAN, D. AND RICHMOND, P., private communication.
8. JOHNSON, K. L., KENDALL, K., AND ROBERTS, A. D., *Proc. Roy. Soc. Ser. A* **324**, 301 (1971).
9. DAHNEKE, B., *J. Colloid Interface Sci.* **40**, 1 (1972).
10. DERJAGUIN, B. V., MULLER, V. M., AND TOPOROV, YU. P., *J. Colloid Interface Sci.* **53**, 314 (1975).
11. ISRAELACHVILI, J. N., Ph.D. Dissertation, Cambridge University, Cambridge, Mass., 1971.
12. JOHNSON, K. L., *in* "Proceedings of the IUTAM Symposium on Deformable Bodies" (A. D. de Pater and J. J. Kalker, Eds.), Delft Univ. Press, Netherlands, 1975.
13. FULLER, K. N. G. AND TABOR, D., *Proc. Roy. Soc. Ser. A* **345**, 327 (1975).
14. GANE, N., PFAELZER, P. F. AND TABOR, D., *Proc. Roy. Soc. Ser. A* **340**, 495 (1974).

Intermolecular Forces—the Long and Short of It

J. N. ISRAELACHVILI[1] AND B. W. NINHAM

Department of Applied Mathematics, Research School of Physical Sciences, Institute of Advanced Studies, Australian National University, Canberra, A.C.T. 2600, Australia

Received June 28, 1976; accepted July 21, 1976

To a sufficient approximation, this year marks the 50th anniversary of the first correct quantum mechanical treatment of the van der Waals forces between atoms (1). Since that time we have come a long way in understanding the subtleties of these forces, culminating in the appearance of the Lifshitz theory 21 years ago (2). The impetus given to colloid science by the appearance of the Lifshitz theory, and the subsequent developments that it led to, can hardly be overestimated. Though part of this paper is devoted to describing the subtle and complex nature of van der Waals forces as revealed by modern theories, our main theme is that these advances cannot be meaningfully assessed in isolation but rather they must be viewed within the broader context of colloid science. In an attempt to put these advances into perspective, we shall try to present an overview of these areas of physics, chemistry, and biology in which intermolecular forces play a central role. As we retrace the historical development of intermolecular force theory from Maxwell to Langmuir, through London, Debye, Deryaguin, Lifshitz, and so on up to the present, we find that there have evolved three distinct approaches for interpreting the role of intermolecular forces in different phenomena. Even the title of this Symposium, "Colloid and Surface Science," is indicative of a dichotomy. We shall try to illustrate how a trichotomy arose and why it did so, and attempt a reconciliation.

[1] Also, Department of Neurobiology, Research School of Biological Sciences, I.A.S., Australian National University, Canberra.

Figure 1 shows pictorially the authors' view of the growth of intermolecular force theory over the last 50 years. The trichotomy of views is shown by the three different road systems. We follow historically the route of colloid science since the days of Young, Maxwell, and others, and see how three strands developed after Langmuir, and why. Two strands are major, and until recently have appeared irreconcilable: Are order, structure, stability, and self-assembly determined by short-range forces or long-range forces? The third strand deals with the role played by statistical mechanics. This was certainly foremost in the minds of Langmuir and Onsager, but appears to have been largely forgotten in recent years. Let us now run through these three strands.

THE ROAD OF SHORT-RANGE FORCES

The beginnings of our subject go back to the last century; the noted contributions made by such people as Young and Plateau were encapsulated in Maxwell's famous article on capillary action which appeared almost exactly 100 years ago in the Encyclopaedia Britannica (3). In this review, Maxwell laid out in his characteristically lucid style all that was then known of intermolecular forces. At that time, this knowledge was confined solely to "surface tension forces"—forces which are effective only over 1 or 2 atomic distances. The concept of such a short-range force field around atoms and molecules which is intimately related to molecular shape and geometry is axiomatic to chemists and molecular biologists. This emphasis was further espoused by Langmuir early

The Great Colloid Schism

FIGURE 1

in this century. His work on monolayers, lipid assemblies, and adsorption seemed to further support the case of the overriding importance of short-range forces. His views on this matter were delivered at the Third Colloid Symposium almost exactly 50 years ago, and were propagated with vigor in a series of important articles (4). Historically, there is no doubt that Langmuir was strongly influenced by Lord Rayleigh's work, who in turn was strongly influenced by Maxwell's work which preceded his. Maxwell in turn confesses his debt to Plateau, who together with others showed how macroscopic shapes and structures can arise solely from these short-range surface tension forces.

Langmuir made little distinction between chemical and physical forces. His ideas on adsorption, for example, as embodied in his adsorption isotherm, and further extended in the BET theory, considered an atom as sensing the presence of a surface not at a distance but only once it is in contact with that surface. Langmuir felt no need to invoke long-range attractive forces to explain any of the phenomena that he studied, and considering the systems that he was working with at the time he was probably right. Later, Langmuir believed that equilibrium colloidal structures could be accounted for by repulsive forces alone, together with a correct statistical mechanical treatment of the problem (5). He envisaged his repulsive forces as being either electrostatic in origin or as arising from a propagation of ordered water layers away from an interface, these essentially being due to short-range forces passed on from molecule to molecule. He persisted in the view that long-range attractive forces have no role to play in colloid stability. We shall return to this matter later.

Langmuir's ideas on the importance of short-range forces, steric effects, molecular geometry, and packing were readily picked up by the biologists working on protein structure, DNA, and more recently, membranes (6); and the language of the present-day molecular biologists is full of terms such as hydrophobic and hydrophilic interactions, hydrogen bonds, molecular packing, all of which are essentially short-range. Langmuir's ideas are now firmly established in the biological and surface sciences, and there has been renewed interest in theories of interfacial forces generated by the work of Fowkes, Good and others (7).

THE ROAD OF LONG-RANGE FORCES

We now come to a consideration of long-range forces as shown in our second route on the map. This idea has its origins with the work of van der Waals on the equation of state for gases and liquids; but theory had to await the advent of quantum theory before a theoretical understanding could be obtained. The classic works of Debye, Keesom, and London on long-range van der Waals forces during the 1920s formed the basis of all future development in this field (1). At about the same time Debye and Hückel, Deryaguin, Langmuir, Onsager, and others were working on bulk and surface properties of electrolytes, and derived expressions for the long-range electrostatic interactions between large bodies (5, 8). A little later de Boer and Hamaker summed the dispersion force between the atoms of condensed bodies and obtained expressions for the long-range dispersion interactions between large bodies (9). They were able to demonstrate an important result that, whereas the dispersion force between atoms and molecules is of short-range, having an inverse seventh-power distance dependence, the dispersion force between large condensed bodies is of much longer range and decays much more slowly with distance.

At that period, one of the major unexplained phenomena of colloid science was the Schulz–Hardy rule, which described the electrolyte dependence of flocculation of lyophilic sols. This was explained first by Deryaguin and Landau, and independently by Verwey and Overbeek, who reviewed the long-range electrostatic and dispersion interaction forces between colloidal particles and arrived at the famous DLVO theory of flocculation which has endured as an unchallenged pillar of colloid science (10, 11). Deryaguin and Landau stress in their original 1941 paper that they were dealing with a kinetic nonequilibrium problem. The appearance of a secondary potential minimum as sometimes arises from the balance of attractive van der Waals and repulsive electrostatic forces in the DLVO theory is, unfortunately, often taken too literally to imply that this is the sole factor in determining structure and order in equilibrium colloidal systems. The status of the DLVO theory to 1975 has recently been reviewed (12).

The DLVO theory was probably the most important contribution to colloid science; and most theoretical developments since then, such as the Lifshitz theory, or the general theory of van der Waals forces by Dzyaloshinskii, Lifshitz and Pitaevskii, and others, have served to improve and refine our understanding of the van der Waals and electrostatic forces that form the basis of that theory.

THE ROAD OF STATISTICAL MECHANICS

The third part of the trilogy has to do with the role played by statistical mechanics in determining equilibrium structure and ordering of colloidal systems and biological assemblies. Langmuir and Onsager were acutely aware of the vital role of statistical mechanics in all colloidal systems where two phases coexist in equilibrium (5, 13).

A point that is sometimes missed is that no description of a system is ever complete without a proper thermodynamic or statistical treatment. Thus a knowledge of the interaction free energy between two atoms or particles, taken in isolation, may tell us little of the properties of an ensemble of such particles. The importance of statistical mechanics is well illustrated in systems where a two-phase equilibrium exists. Such equilibria are fairly common in colloidal systems. The formation of micelles in aqueous solutions represents such a two-phase equilibrium (strictly, a multiphase equilibrium) in which large micelles and individual lipid monomers are at equilibrium in solution; at low concentrations no micelles

are formed and only monomers exist, whereas at higher concentrations—above the critical micelle concentration—a two-phase system exists. In this example, it is clear that even though in coming together to form micelles the lipid molecules can minimize their individual free energy, this in itself does not guarantee that the lipids would actually form into micelles at any specified concentration. A thorough statistical mechanical treatment is essential for an understanding of micellar stability and structure.

Thus, the first moral to be learned from statistical mechanics is that *the existence of a minimum in the two-particle interaction free energy in the associated or ordered state does not guarantee the formation of this state.* Conversely, *the existence of an associated or ordered state does not necessarily imply that the particles are sitting at a separation where there is a minimum in the two-particle interaction free energy.* This is beautifully exemplified in clays, latex particles, and assemblies of tobacco mosaic viruses, where the long-range order cannot always be explained by a DLVO-type secondary minimum, but is adequately explained by long-range repulsive forces and a correct statistical mechanical treatment (13, 14). In such systems, order is obtained through repulsive forces in which the particles attempt to get as far away from each other as possible but, being constrained in a finite solution, are forced to order themselves into either a one-phase or a two-phase system (where the gain in entropy of the more dilute disordered phase more than compensates the loss of entropy and gain in energy of the more concentrated ordered phase). The quantitative theoretical understanding of the interplay of repulsive forces and statistical mechanics has been the subject of much recent work in liquid-state physics.

AN EXAMPLE FROM BIOLOGY

We have asserted that all problems of colloid stability and structure require correct statistical mechanical treatment, but we have yet

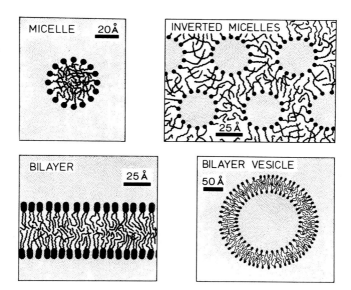

FIG. 2. The forces that govern the shapes of lipid structures are mainly short-range interfacial forces, intimately coupled with the geometric properties of these molecules.

to establish the relative importance of short-range forces and long-range forces in various colloidal systems. While there are some general principles on which to draw, there is no short-cut universal answer, and each system must be viewed on its own merits. The conventional wisdom in colloid science is that the long-range forces are the more important. An example which aptly illustrates the obverse side of the coin comes to us from the world of membrane biophysics.

Consider those biological phospholipids, such as anionic phosphatidyl serine, which naturally form into stable spherical bilayer shells known as vesicles (Fig. 2). Such vesicles have diameters typically of the order of a few hundred angstroms. Let us analyze the forces that lead to these unusual structures. These may be of long-range or of short-range. If long-range, then the equilibrium radius of a vesicle would be determined by a DLVO-type balance of van der Waals and electrostatic forces within the spherical aqueous compartment. If this were so, we should expect smaller vesicles at higher ionic strengths, where the electrostatic

forces are screened more. On the other hand, if the determining forces are short-range interfacial forces, we should expect that the surface area per lipid head group to decrease at higher ionic strengths, due to the increased screening of the electrostatic repulsion between adjacent head groups. Simple geometric packing considerations now show that this would force the lipids to pack into a bilayer shell of smaller curvature than before, thus resulting in larger vesicles in direct contrast to the smaller vesicles predicted by a long-range force analysis. As it turns out, larger vesicles are observed at higher ionic strengths, clearly pointing to the short-range interfacial forces as the main structure determining forces in vesicles. Long-range forces play a minor role in these structures. A treatment of this problem which (implicitly) includes the statistical mechanical aspects has recently been attempted by us, and this successfully explains a host of different physical properties of micelles, vesicles, and other lipid structures (6).

The lesson to be learned from this example is that, whereas the shapes of hard colloidal particles (e.g., latex spheres) are unaffected by changes in short-range interfacial forces, those particles that are composed of lipids are not rigid and their shapes are very sensitive to changes in the interfacial forces induced, for example, by changes in ionic strength, pH, etc. Thus the stability and phase equilibria of hard colloidal particles are determined by long-range forces. But for the "soft" lipid structures, such as micelles, vesicles and biological membranes, the short-range forces—intimately coupled to molecular geometry—are at least equally important in determining the overall structure and stability that such assemblages will take.

The origins of our trichotomy are now clear: Short-range forces and long-range forces could each dominate in certain systems. And as to statistical mechanics—it was just too hard to understand.

Let us now review the current state of intermolecular forces.

VAN DER WAALS FORCES

The impetus given to long-range forces by the theoretical work of Casimir on retarded van der Waals forces and the experimental measurements of these forces by Abrikosova and Deryaguin in 1951 led to the macroscopic theory of Lifshitz in 1954 (2). The Lifshitz theory encompassed all the earlier theories of Keesom, Debye, London, and Casimir, and showed up the deficiencies of the additivity approach, especially for polar media. Its most powerful aspect is that it is not model-dependent, and relies entirely on measured dielectric properties.

The generalization of the Lifshitz theory by Dzyaloshinskii, Lifshitz, and Pitaevskii (DLP) to the inclusion of a third medium between two interacting bodies was a major event in both the theory of intermolecular forces as well as a tour de force in the application of quantum field theory (2). The theory showed how the presence of an intervening medium can drastically alter the interaction between two bodies: the effects of retardation are now much more important and the force may not be described in terms of a simple power law; the forces may be attractive or repulsive—a circumstance which has often been difficult to understand but which is no more than a manifestation of the displacement principle of Archimedes in intermolecular forces; also it was found that there is a temperature-dependent (entropic) contribution to the total force that is especially strong in hydrocarbon–water systems (15). The DLP theory did more than simply give a rigorous account of the

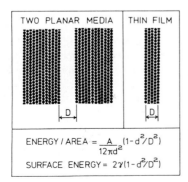
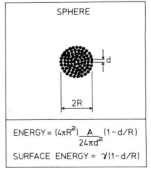

$$\text{ENERGY / AREA} = \frac{A}{12\pi d^2}(1 - d^2/D^2)$$

$$\text{SURFACE ENERGY} = 2\gamma(1 - d^2/D^2)$$

$$\text{ENERGY} = (4\pi R^2)\frac{A}{24\pi d^2}(1 - d/R)$$

$$\text{SURFACE ENERGY} = \gamma(1 - d/R)$$

FIG. 3. Approximate results of pairwise summation of London dispersion energies between all atoms for the three configurations shown. A is the Hamaker constant, and d is the intermolecular spacing. The first term in each case may be equated with the surface energy per unit area $\gamma = A/24\pi d^2$ of an isolated surface (7, 18). The long-range dispersion energy can be seen to be no more than a perturbation of the surface energy γ.

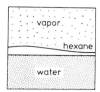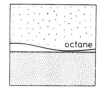

FIG. 4. When a hydrocarbon liquid spreads as a thin film on water, the system corresponds to that of a vapor and a water phase interacting across the hydrocarbon film. If the dispersion force across the film is repulsive the hydrocarbon will wet the water surface. If the force is attractive the thickness of the film will tend to decrease and the hydrocarbon liquid will not spread but collect as isolated lenses on the water surface. Viewed from the point of long-range dispersion forces, the criterion of whether the film will wet the surface depends on whether the Hamaker constant A (vapor/hydrocarbon/water) is negative (for wetting) or positive (for no wetting). It can be shown that this purely long-range force criterion is the same as that resulting from the Young–Dupré equation.

long-range van der Waals force between two bodies in a medium; their classic paper of 1961 gave a thorough treatment of thick wetting films on surfaces.

The DLP theory has since been tested experimentally in a number of ways (16). For example, the forces have been directly measured in air and in vacuum as well as indirectly across liquids, and various theoretical analyses of wetting and other phenomena have so far led to good agreement between theory and experiment (16, 17).

Unfortunately, the DLP theory was extremely difficult to apply, and it is probably due to this that it was largely ignored by most colloid scientists who persisted with the old additivity methods of Hamaker. Future progress had to await the advent of semiclassical, heuristic approaches as pioneered by van Kampen and co-workers, Langbein, Ninham, Parsegian, and others, who managed to re-derive the DLP theory using much simpler theoretical techniques (2). The semiclassical approach has proved extremely powerful and eminently suitable for extending the theory to more complex geometrical shapes, anisotropic and inhomogeneous media, and to interactions across electrolyte solutions. Simple prescriptions for computing van der Waals forces have also been developed (17).

More recent work by Ninham, Mahanty, and others has further extended the Lifshitz theory to the short-range distance regime,

thereby introducing the theory into the domain of interfacial phenomena in a respectable way (18). Emerging from this analysis is the proof that for nonpolar media, where only dispersion forces are effective, the Lifshitz interaction energy is the same as the change in the interfacial energies of the bodies. A similar treatment of surface and interfacial problems was originally given by Fowkes (7), who used London theory and the concepts of additivity to obtain a satisfactory description of the surface tension and works of adhesion of hydrocarbons, which interact only through dispersion forces.

As a simple example of the way dispersion force theory naturally links up with interfacial and surface tension forces we go back to the pairwise summation of energies between the atoms in the three shapes shown in Fig. 3. In one case we have two media interacting across a vacuum; in the second we have a thin film, and in the third we have a sphere. In each case a pairwise summation of the interaction energies between *all* atoms leads to the results shown in Fig. 3. It can be seen that for each configuration, the first term, which was ignored in Hamaker's original work, is always the same as the surface energy of the system. A further, and more instructive, example concerns the problem of the wetting of liquid hydrocarbon films on water (Fig. 4). In this case it may be shown that (assuming that the gradient of the force is monotonically decreasing with film thickness) if the long-range dispersion force across the film is attractive there will be no wetting, whereas if it is repulsive the liquid will wet the surface. By use of certain combining laws, we find that this purely long-range criterion reduces to the same result as that given by the Young–Dupré equation. Although this example is somewhat oversimplified, it nevertheless illustrates the unity of long-range and short-range dispersion forces. However, for the inverse case of a polar liquid such as water on a surface, the problem may not be treated simply in terms of dispersion forces alone. A further anomaly that has recently been resolved is the reconciliation of the Lifshitz theory of thick films with the BET theory of multilayer adsorption (19).

Another interesting theoretical finding is that the van der Waals forces between two bodies across an electrolyte solution are partially screened in a similar way to the screening of electrostatic forces between two charged surfaces. This screening manifests itself in a reduced attraction between two like bodies at increased ionic strengths (20).

The new concept of the dispersion "self-energy" of a molecule in a medium, analogous to the Born self-energy of an ion, has found a

useful application in polymer phase transitions (20). There are still many areas not yet adequately explored: thus it appears that theory sometimes predicts very strong, very long-range, attractive forces, especially between long, thin cylindrical molecules (20, 21). The whole area of the interactions involving spatially dispersive media such as metals is still not well understood but promises to be a growth area in view of its relevance to adsorption and catalysis on metal surfaces (22).

ELECTROSTATIC AND STRUCTURAL FORCES

We now turn to a quick review of the state of electrostatic force theory. By general consensus (anything else has so far proved impossibly difficult), one has to solve the Poisson–Boltzmann equation in order to obtain the free energy of a double layer as well as that of two overlapping or interacting double layers. Unlike dispersion force theory, all solutions to the problem are essentially model-dependent. The earlier practice of assuming a constant surface potential or a constant surface charge density (11) as two surfaces approach each other has been refined to allow for surface charge regulation (23). In biological cell–

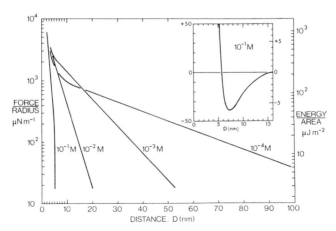

FIG. 5. Experimental results of direct measurement of force as a function of separation between two crossed mica cylinders of radius R in various KNO_3 solutions. The right-hand ordinates give the interaction energy per unit area for two parallel plates, calculated according to the "Deryaguin approximation." The ordinates of the inset are linear. [Reproduced with permission from *Nature* 262 (1976) 775.]

membrane interactions, where these are composed of proteins and lipids capable of rapid lateral diffusion, it is likely that phase separations may be induced through long-range interactions. The theories are all difficult to test and are further complicated by the existence of Stern layers and specific ion effects.

The problem of understanding double-layer forces appears to be one of the most pressing problems in colloid science at the moment. However, all theories predict an exponential repulsion with separation whose characteristic decay length is the same as the Debye length. Very recently, in Canberra, G. Adams and one of us (JNI) have started to measure directly the long-range forces between two crossed mica cylinders immersed in aqueous electrolyte solutions. This experiment is a natural extension of the earlier measurements of van der Waals forces between two mica sheets in air and in vacuum, which were carried out in Cambridge, U.K., by Tabor, Winterton, and JNI, a few years ago. The present apparatus is a new and upgraded version of the earlier models, refined to allow for the measurements of the very different types of forces under conditions that occur in liquids. We are pleased to report some preliminary results of the forces between molecularly smooth mica surfaces in various KNO_3 solutions. Figure 5 shows some of these results in which the forces are plotted as a function of separation. (Note that according to the "Deryaguin approximation," these curves for the "force" as a function of separation have exactly the same shape as the "interaction energy" as a function of separation between two plates, differing by a constant geometric factor.) The repulsive forces were all found to be exponential, with a decay length close to the Debye length. (However, below a certain separation the repulsion increased even more rapidly. The onset of this enhanced repulsion occurred very roughly at about half a Debye length separation.) In 0.1 M and 1 M KNO_3 solutions, the onset of repulsion was preceded by an attractive region, thereby indicating the existence of a secondary minimum at these concentrations. It is difficult to say whether at very small separations, below 10 Å, there is also a primary minimum, since at these separations the surfaces have appreciably deformed under the influence of the strong forces, and a more detailed analysis is needed here.

At large separations the results are reversible and reproducible, and clearly exhibit all the essential features characteristic of a DLVO-type interaction. In particular, we note that even at 0.1 M KNO_3, the decay length of the exponential repulsive region was equal to the Debye length, showing that the Poisson–Boltzmann equation appears to remain valid even at this high concentration, at least for KNO_3. The observation that at separations smaller than the Debye length, the repulsive forces increased even more rapidly appears to be consistent with calculations of double-layer forces based on the nonlinear Poisson–Boltzmann equation. The magnitudes of the attractive forces in the region of the secondary minimum are of the order to be expected from the DLP theory.

Finally, in this section it is worth mentioning that long-range repulsion may also arise from adsorbed polymers and from water structure effects at interfaces. In particular, recent findings that lecithin multilayers exhibit strong exponential repulsion at separations below 20 Å, which do not appear to be accounted for by long-range electrostatic forces of the type already discussed, probably arise from the structuring of the water molecules near the interface which manifests itself as a repulsion (24). It is satisfying that the van der Waals forces needed to give the observed equilibrium spacing in this lecithin experiment are in agreement with those predicted by the Lifshitz theory for a multilayered structure. Thus we have here a situation similar to the DLVO theory where, however, the repulsion arises from a structure-propagated ordering effect as foreseen by Langmuir 50 years ago.

As theoreticians, we believe that at the moment there are no outstanding problems in

23

colloid science beyond the realm of comprehension. We do not require any anomalous or third force, but only that the future will see an increased appreciation of the interplay between the three concepts of long-range forces, short-range forces, and statistical mechanics in all colloidal and biological studies.

REFERENCES

1. WANG, S. C., *Phys. Z.* **28**, 663 (1927).

    The work of F. London [*Z. Phys.* **63**, 245 (1930); *Z. Phys. Chem.* B **11**, 222 (1930)] and R. Eisenschitz and F. London [*Z. Phys.* **60**, 491 (1930)] followed a little later. For earlier work on the forces between dipolar molecules (Keesom forces) and on dipole-induced dipole (Debye) forces see W. M. Keesom [*Phys. Z.* **22**, 129 (1921); **22**, 643 (1921)]; P. Debye [*Phys. Z.* **21**, 178 (1920); **22**, 302 (1921); and *Phys. Z.* **13**, 97 (1912)]. Debye's theory, which gave an r^{-9} force law between nonpolar molecules gave excellent agreement with measured virial coefficients and the temperature dependence of viscosity of gases, was strongly criticized by Langmuir in the first volume of "Colloid Chemistry" in 1926. For an excellent account of the forces between atoms see H. Margenau and N. R. Kestner, "Theory of Intermolecular Forces," 2nd ed., Pergamon, Oxford, 1971.

2. LIFSHITZ, E. M., *J. Exp. Theor. Phys. USSR* **29**, 94 (1955); *Sov. Phys. JETP* **2**, 73 (1956).

    In 1894 the discoverer of light radiation pressure [P. N. Lebedev, *Wied. Ann.* **52**, 621 (1894)] wrote ". . . of special interest and difficulty is the process which takes place in a physical body when many molecules interact simultaneously, the oscillations of the latter being interdependent owing to their proximity. If the solution of this problem ever becomes possible we shall be able to calculate in advance the values of the intermolecular forces due to intermolecular inter-radiation, deduce the laws of their temperature dependence, solve the fundamental problem of molecular physics whether all so-called 'molecular forces' are confined to the already known mechanical action of light radiation, to electromagnetic forces, or whether some forces of hitherto unknown origin are involved . . . ." It seems especially fitting that his speculations and grand vision concerning the electromagnetic origin of molecular forces should have been confirmed by the Russians in the dramatic simultaneous advance in theory by Lifshitz (1955) and in experiment by B. V. Deryaguin and A. A. Abrikosova [*J. Exp. Theor. Phys.* **30**, 933 (1956); **31**, 3 (1956); *Sov. Phys. JETP* **3**, 819 (1956); **4**, 2 (1957)]. The first experiments on retarded forces were actually carried out in 1951 by these authors and their colleagues. For a review of the status of theory vs experiment to 1956, see B. V. Deryaguin, I. I. Abrikosova, and E. M. Lifshitz [*Quart. Rev. Chem. Soc.* **10**, 295 (1956)]. Lifshitz' work was inspired by Casimir's work on retardation following a suggestion of Overbeek [H. B. G. Casimir, *Proc. Kon. Ned. Akad. Wetensk.* **51**, 793 (1948); *J. Chim. Phys.* **46**, 407 (1949); and H. B. G. Casimir and D. Polder, *Nature (London)* **158**, 787 (1946); *Phys. Rev.* **73**, 360 (1948)], and by the experiments of Deryaguin and his colleagues.

    For the effect of intervening media see the general theory of I. E. Dzyaloshinskii, E. M. Lifshitz, and L. P. Pitaevskii [*Advan. Phys.* **10**, 165 (1961)]. For more recent theoretical reviews with differing emphasis, but which together give a complete survey of what has become a large literature, see J. N. Israelachvili and D. Tabor [*Progr. Surface Membr. Sci.* **7**, 1 (1973)], H. Krupp [*Advan. Colloid Interface Sci.* **1**, 111 (1967)], V. A. Parsegian [*Ann. Rev. Biophys. Bioeng.* **2**, 221 (1973)], J. N. Israelachvili [*Quart. Rev. Biophys.* **6**, 341 (1974)], D. Langbein ("Theory of van der Waals Attraction," Springer Tracts in Modern Physics, Vol. 72, Springer–Verlag, Berlin, 1974), P. Richmond [Specialist Periodical Reports, "Colloid Science" (D. H. Everett, Ed.), Vol. 2, Chemical Society, Washington, D. C., 1975], J. Mahanty and B. W. Ninham ("Dispersion Forces," Academic Press, New York, 1976), and Yu. S. Barash and V. L. Ginzburg [*Sov. Phys. Usp.* **18**, 305 (1975)].

3. MAXWELL, J. C., "Capillary Action" (1875), *in* "Encyclopaedia Britannica," 9th ed., updated by Lord Rayleigh in 11th ed. (1911).

    Lord Rayleigh [*Phil. Mag.* **30**, 285, 456 (1890)]. See also "Scientific Papers," Vol. 3, articles 176 and 186, Dover, New York, 1964. The entire state of knowledge on surface forces to that time and references to earlier work are summarized in Maxwell's article, except for the controversy between Young and Laplace. Young accused Laplace of stealing his theory verbatim!

4. LANGMUIR, I., "The Distribution and Orientation of Molecules," Colloid Symposium Monograph, Vol. III, p. 48 (1925).

    In the context of the subject matter discussed here, Volumes 6–9 of "The Collected Works of Irving Langmuir" [(C. Guy Suits, Ed.), Pergamon, Oxford, 1961] are essential reading. Especially important in understanding his ideas and their impact are the articles "The Effects of Molecular Dissymmetry on Some Properties of Matter" [*Colloid Chem.* I, 525 (1926)] and "Forces Near the Surfaces of Molecules" [*Chem. Rev.* **6**, 451 (1929)] contained therein.

5. LANGMUIR, I., *J. Chem. Phys.* **6**, 873 (1938).

While the attempt to make detailed predictions on colloid stability made here is certainly erroneous, his ideas on the importance of statistical mechanics and repulsive forces were sound and deserve special attention, especially today. He was also aware of the importance of surface charge regulation.

6. KAUZMANN, W., *Advan. Protein Chem.* **14**, 1 (1959); EISENBERG, D. AND KAUZMANN, W., "Structure and Properties of Water," Oxford, Clarendon, 1969); PAULING, L., "The Nature of the Chemical Bond," 3rd. ed., Cornell Univ. Press, Ithaca, N. Y., 1960; TANFORD, C., "The Hydrophobic Effect: Formation of Micelles and Biological Membranes," Wiley, New York, 1973; see also ISRAELACHVILI, J. N. AND MITCHELL, D. J., *Biochim. Biophys. Acta* **389**, 13 (1975); ISRAELACHVILI, J. N., MITCHELL, D. J., AND NINHAM, B. W., *J. Chem. Soc. Faraday II* **72**, 1525 (1976).

The early founders of the cell theory of biology were well aware of the importance of forces, especially short-range (surface tension) forces in determining structure. D'Arcy Thompson ("On Growth and Form," Cambridge Univ. Press, London/New York, 1917, 1936), drawing particularly on the beautiful work of Plateau, underlined this point in a treatise of extraordinary erudition. Thereafter his views were promptly buried by the biophysicists.

7. FOWKES, F. M., *Ind. Eng. Chem.* **12**, 40 (1964); *J. Colloid Interface Sci.* **28**, 493 (1968). See also articles by R. J. Good and others in "Contact Angle, Wettability and Adhesion," *Adv. Chem. Series, No. 43, Am. Chem. Soc.*, 1964. Washington, D. C.

8. DEBYE, P. AND HÜCKEL, E., *Phys. Z.* **24**, 185 (1923); DERYAGUIN, B. V. AND KUSSAKOV, M., *Bull. Acad. Sci. USSR, Ser. Chim.* No. 5, 1119 (1937).

Quoted in Ref. 9, Deryaguin appears to have carried out the first calculation of double-layer forces between both charged spheres and planes. ONSAGER, L. AND SAMARAS, N. T., *J. Chem. Phys.* **2**, 528 (1934).

The surface tension of aqueous solutions of electrolytes is obviously a subject of the highest importance to colloid science which has not progressed beyond Onsager's limiting law.

9. HAMAKER, H. C., *Physica* **4**, 1058 (1937).

10. DERYAGUIN, B. V. AND LANDAU, L., *Acta Phys. Chim. URSS* **14**, 633 (1941).

11. VERWEY, E. J. W. AND OVERBEEK, J. TH. G., "Theory of the Stability of Lyophobic Colloids," Elsevier, Amsterdam, 1948.

12. NAPPER, D. H. AND HUNTER, R. J., *in* "Surface Chemistry and Colloids" (M. Kerker, Ed.), Vol. 7, p. 161. Butterworths, London, 1975.

This article contains an exhaustive review of the status of DLVO theory to the present time.

13. ONSAGER, L., *Ann. N. Y. Acad. Sci.* **51**, 627 (1949).

Onsager's work on colloid stability follows the route of statistical mechanics and electrostatic forces, i.e., van der Waals forces have no part in the theoretical framework that accounts for the stability and ordering of charged anisometric particles. Even at rather high salt concentrations van der Waals forces play a secondary role which only affect the "fine tuning" of the ordered phase. The substantial predictions of Onsager theory for rods (tobacco mosaic virus) have been confirmed by the experiments of G. J. Oster [*J. Gen. Physiol.* **33**, 445; *J. Cell Comp. Physiol.* **49**, Suppl. 1, 129 (1957)], and the situation has been reviewed by V. A. Parsegian and S. L. Brenner [*Nature* **259**, 632 (1976)]. Inexplicably the Onsager theory has not been applied to the isotropic–nematic phase transition in dilute clay dispersions [I. Langmuir, *Ann. N. Y. Acad. Sci.* **12**, 873 (1938)].

14. HACHISU, S. AND KOBAYASHI, Y., *J. Colloid Interface Sci.* **42**, 342 (1973); HACHISU, S., KOBAYASHI, Y. AND KOSE, A., *J. Colloid Interface Sci.*, **46** 470 (1974); KOSE, A., OZAKI, M., TAKANO, K., KOBAYASHI, Y., AND HACHISU, S., *J. Colloid Interface Sci.* **44**, 330 (1973).

At very low salt concentrations the observed order–disorder transition of latexes is governed entirely by long-range repulsive *many-body* forces (see S. Marčelja, D. J. Mitchell, and B. W. Ninham, *Chem. Phys. Lett.*, **43**, 353 (1976) and at high salt by effective hard sphere interactions.

15. PARSEGIAN, V. A. AND NINHAM, B. W., *Biophys. J.* **10**, 664 (1970).

16. A detailed account of measurements of van der Waals forces in vacuum up to 1971 is given by J. N. Israelachvili and D. Tabor [*Progr. Surface Membr. Sci.* **7**, 1 (1973)]. For experiments dealing with thin films, e.g., those of Lyklema and Mysels on soap films, of Scheludko, and of Haydon and his co-workers, see Ref. 12; also J. Requena, D. F. Billett, and D. A. Haydon [*Proc. Roy. Soc. Ser. A* **347**, 141 (1975)]; and E. S. Sabisky and C. H. Anderson [*Phys. Rev. A* **7**, 790 (1973)].

17. Various theoretical analyses of experiments will be found in, e.g., Ref. 15 [P. Richmond and B. W. Ninham, *J. Low Temp. Phys.* **5**, 177 (1971); P. Richmond, B. W. Ninham, and R. H. Ottewill, *J. Colloid Interface Sci.* **45**, 69 (1973); L. R. White, J. N. Israelachvili, and B. W. Ninham, *J. Chem. Soc. Faraday II*, **72**, 2526 (1976); D. Chan and P. Richmond, *Proc. Roy. Soc. Ser. A*, to appear]. A

completely different experiment which measured the force between atoms and a metallic surface is that of A. Shih and V. A. Parsegian, [*Phys. Rev. A* **12**, 835 (1975)].

18. MAHANTY, J. AND NINHAM, B. W., *J. Chem. Soc. Faraday II* **70**, 637 (1974); MITCHELL, D. J. AND RICHMOND, P., *Chem. Phys. Lett.* **21**, 113 (1973); *J. Colloid Interface Sci.* **46**, 118 (1974), **46**, 128 (1974); ISRAELACHVILI, J. N., *J. Chem. Soc. Faraday II*, **69**, 1729 (1973); MAHANTY, J. AND NINHAM, B. W., *J. Chem. Phys.* **59**, 6157 (1973).

19. BRUNAUER, S., EMMETT, P. H., AND TELLER, E., *J. Amer. Chem. Soc.* **60**, 309 (1938).

The theory of multilayer adsorption as developed by these authors which has seen such useful service is in direct conflict with the predictions of Lifshitz theory. In the former, long-range forces are omitted, while in the latter, short-range forces do not appear. Both are in fact correct, and the BET theory emerges as a special limiting case of a more refined treatment; see, e.g., J. Mahanty and B. W. Ninham [*J. Chem. Soc. Faraday II* **70**, 637 (1974)], and for a different approach, P. Richmond (*Phys. Chem. Liquids*, to appear).

20. MAHANTY, J. AND NINHAM, B. W., "Dispersion Forces," Academic Press, London–New York, 1976.

21. CHANG, B. D., COOPER, R. L., DRUMMOND, J. E., AND YOUNG, A. C., *Phys. Lett. A*, **37**, 311 (1971);

DAVIES, B., NINHAM, B. W., AND RICHMOND, P., *J. Chem. Phys.* **58**, 744 (1973).

The existence of such very long-range forces had been suspected by F. London [*J. Phys. Chem.* **46**, 305 (1942)], and predicted by C. A. Coulson and P. L. Davies [*Trans. Faraday Soc.* **48**, 777 (1952)], who used a quantum mechanical perturbation approach.

22. There is a great deal of recent work in the physics literature; see, e.g., D. Chan and P. Richmond [*J. Phys. C: Solid State Phys.* **8**, 2509 (1974), **8**, 3221 (1974); **9**, 153 (1976); **9**, 163 (1976)]; and E. Wikborg and J. E. Inglesfield [*Solid State Commun.* **16**, 335 (1975); *J. Phys. F* **5**, 1475 (1975), **5**, 1706 (1975)].

23. NINHAM, B. W. AND PARSEGIAN, V. A., *J. Theor. Biol.* **31**, 405 (1971).

For further work on charge regulation see D. Chan, T. W. Healy, and L. R. White in several articles to appear in *J. Chem. Soc. Faraday II*.

24. LeNEVEU, D. M., RAND, R. P., AND PARSEGIAN, V. A., *Nature* **259**, 601 (1976); LeNEVEU, D. M., RAND, R. P., GINGELL, D., AND PARSEGIAN, V. A., *Science* **191**, 399 (1975); MARČELJA, S. AND RADIĆ, N., *Chem. Phys. Lett.*, **42**, 129 (1976). This is the first (phenomenological) theory of repulsive forces between surfaces due to ordering of water.

van der Waals Forces in Oil/Water Systems

J. REQUENA,[1] D. E. BROOKS,[2] AND D. A. HAYDON

Physiological Laboratory, Downing Street, Cambridge, United Kingdom

Received June 1, 1976; accepted July 21, 1976

Free energies of formation of "black" lipid films have been determined from their contact angles. In these films the repulsive forces are of such short range that it is possible to estimate relatively accurately that part of the free-energy change that originates from the van der Waals forces. It is shown that if the van der Waals free energies are interpreted on the assumption that the films are isotropic layers of hydrocarbon bounded by semi-infinite aqueous phases, the Hamaker coefficients vary considerably from one film to another, contrary to the predictions of the Lifshitz theory. If, on the other hand, it is recognized that the hydrocarbon region of a film is, in fact, a layered structure and that there are differences, albeit small, between the dielectric properties of the chains of the lipid stabilizer and the alkane solvent (some of which is retained in the film), the conflict between theory and experiment is largely removed. Thus, Hamaker coefficients calculated from the Lifshitz theory for multilayered systems agree well in nearly all instances with the corresponding experimental coefficients.

INTRODUCTION

The direct measurement of van der Waals forces in liquid/liquid systems presents considerable technical difficulties. The methods so successfully developed by Derjaguin, Abrikosova, and Lifshitz (1) and others (2–5) to study the interaction between solids in gases may perhaps be used for solids in liquids, but it is less obvious that they will succeed in purely liquid systems. For the present, therefore, it is of interest to consider less direct approaches, such as the examination of thin liquid films. A fundamental objection to this kind of approach is that it is usually very difficult to make adequate allowance for the influence of forces other than those of van der Waals. There are nevertheless certain types of systems in which these additional forces are of such short range relative to the van der Waals forces that they do not need to be known with any great accuracy. One such system is the

thin lipid film formed in aqueous solution (6). In this instance it has been shown that the contribution of the electrodynamic forces to the free energy of formation of the film may be estimated to within ca. 10% (7, 8).

The lipid films which have been investigated were formed from monoglycerides or phospholipids in alkane solution. Their drainage (under aqueous solutions) is comparable to that of aqueous soap films and consists of the flow of a dilute solution of the lipid in the alkane from between monolayers of the lipid adsorbed (or spread) at oil/water interfaces (6). The thinning process ceases when the chains of the juxtaposed adsorbed monolayers interact. The final equilibrium films have thicknesses of the order of twice the chain length of the lipid used, the precise thickness depending on the alkane solvent (6).

The change, ΔA^*, in the Helmholtz free energy accompanying the thinning of unit area of a film is given by (6, 9, 10)

$$\Delta A^* = (\sigma - 2\gamma), \qquad [1]$$

where σ is the tension of the film and γ the

[1] Present address: Centro de Biofisica, Instituto Venezolano de Investigaciones Cientificas, Caracas, Venezuela.

[2] Present address: Department of Pathology, University of British Columbia, Vancouver 8, Canada.

27

tension of the adjacent bulk phase interfaces. To å good approximation, Eq. [1] may be rewritten in terms of the contact angle θ, where θ is defined by

$$2\gamma(\cos\theta - 1) = \sigma - 2\gamma. \qquad [2]$$

Equation [1] thus becomes

$$\Delta A^* = 2\gamma(\cos\theta - 1). \qquad [3]$$

Implicit in the use of this equation is the assumption that the van der Waals and other forces acting across the thin film cease abruptly at a (hypothetical) contact line between the film and the bulk lipid solution. That there is, in fact, a transition region and that this, in principle, affects the validity of Eq. [3] is well recognized (10). It will be shown in a

later section, however, that the resulting error in ΔA^* is, in the present systems, negligible.

The only components of ΔA^* which seem at all likely to be important are those arising from the van der Waals (ΔA_{vW}), electrical double layer (ΔA_e), and steric (ΔA_s) interaction free energies.

Thus

$$\Delta A^* = \Delta A_{vW} + \Delta A_e + \Delta A_s. \qquad [4]$$

It will be argued below that ΔA_e and ΔA_s are small compared with ΔA_{vW}.

EXPERIMENTAL DATA

For the calculation of the free energy of thinning of the films it is necessary to know the film area, the interfacial tension of the bulk oil

TABLE I

The Film Thickness, Interfacial Tension, Contact Angle, and Free Energy of Thinning for Some Black Lipid Film Systems at 20°C[a]

System	Thickness of hydrocarbon region, h ((Å) $\pm$ 2%)	Interfacial tension, γ (dyn cm^{-1})	$\cos\theta$		Free energy of formation per unit area, $\Delta A^* \times 10^3$ (ergs cm^{-2})
Monoolein + n-heptane	47.1	4.70	0.999256	$\pm$ 0.000007	-6.99 ± 0.14
Monoolein + n-octane	47.8	4.38	0.999377	$\pm$ 0.000005	-5.46 ± 0.10
Monoolein + n-decane	48.1	3.84	0.999415	$\pm$ 0.000004	-4.49 ± 0.09
Monoolein + n-dodecane	45.3	3.45	0.99940(1)	$\pm$ 0.00002	-4.13 ± 0.19
Monoolein + n-tetradecane	40.7	3.04	0.99913(6)	$\pm$ 0.00002	-5.25 ± 0.20
Monoolein + n-pentadecane	36.6	2.88	0.99855(2)	$\pm$ 0.00002	-8.34 ± 0.26
Monoolein + n-hexadecane	32.7	2.51	0.995503	$\pm$ 0.000053	-22.57 ± 0.72
Monoolein + cis-5-decene	40.5	12.75	0.999712	$\pm$ 0.000006	-7.34 ± 0.18
Monomyristolein + n-decane	44.4	1.68	0.998632	$\pm$ 0.000026	-4.60 ± 0.22
Monopalmitolein + n-decane	46.3	2.70	0.999193	$\pm$ 0.000018	-4.36 ± 0.18
Monoeicosenoin + n-decane	53.1	4.26	0.999586	$\pm$ 0.000004	-3.53 ± 0.07
Monoerucin + n-decane	57.0	11.2 $\pm$ 1.0	0.999832	$\pm$ 0.000005	-3.76 ± 0.45
Mononervonin + n-decane	64.6	12.6 $\pm$ 1.0	0.999902	$\pm$ 0.000002	-2.46 ± 0.32
Monolinolein + n-decane	48.5	3.30	0.999207	$\pm$ 0.000006	-5.23 ± 0.05
Monolinolenin + n-decane	35.9	2.86	0.996402	$\pm$ 0.000065	-10.00 ± 0.75
Monolinolein + n-hexadecane	30.2	2.36	0.991079	$\pm$ 0.000050	-14.48 ± 0.61
Monooleyl glyceryl ether + n-hexadecane	35.4	3.74	0.997869	$\pm$ 0.000025	-15.94 ± 0.40
Monooleyl glyceryl ether + n-decane	48.9	4.89	0.999521	$\pm$ 0.000006	-4.68 ± 0.11
1,2 dioleyl phosphatidyl choline + n-decane	48.3	3.18	0.999228	$\pm$ 0.00001	-4.90 ± 0.30
Brain phosphatidyl serine + n-decane	50.2	2.03	0.999036	$\pm$ 0.000045	-3.91 ± 0.36

[a] The aqueous solution was 0.1 M NaCl.

VAN DER WAALS FORCES IN OIL/WATER SYSTEMS

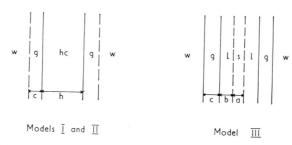

FIG. 1. Models for the black lipid film. In Model I the film is assumed to be simply a layer of liquid hydrocarbon (hc) sandwiched between two aqueous phases (w). Model II takes account of the fact that there are polar groups (g) (glyceryl esters in the present systems) between the hydrocarbon and the aqueous regions. Model III is similar to Model II except in that the hydrocarbon layer is assumed to be stratified into lipid chain (l) and solvent (s) regions. In Models I and II the whole of the hydrocarbon layer is regarded as displaceable whereas in Model III only the central solvent layer can vary in thickness.

and aqueous phases with which the film is in equilibrium, and the contact angle. In order to compare the free energies with those predicted theoretically, it is also necessary to know the thickness of the films. The determination of the film area was a trivial matter and the measurement of the tension was achieved by two independent methods, an orthodox drop-volume technique (11) and by the application of the Lippmann equation (12). For systems in which both methods were used the agreement between the two results was to within the experimental error (12). The determination of the contact angle and the film thickness require more detailed description.

The contact angle. On viewing either the Plateau–Gibbs border, or a lens of the lipid phase trapped in the black film in reflected monochromatic light, fringes can be seen from which the contours of the interfaces may be calculated (7, 8). For both the border and the lens it is possible to extrapolate to the hypothetical contact line and so obtain the contact angle. The two methods give the same angle in spite of the difference in pressure between the lens and the border. Estimates of the pressure difference and of the van der Waals forces, however, show that this was to be expected. The question also arises as to whether the extrapolated angle can legitimately be used in Eqs. [2] and [3]. This can be answered particularly easily in the case of the lens. The

interference fringes show that to within 1 μm of their edges the lenses, which are some 100 μm in diameter, consist very precisely of two identical segments of a sphere, one on either side of the plane of the black film. At 1 μm from the edge, the angle of the surface with the plane of the film is within 2% of θ, the angle obtained by extrapolation to the contact line. At the same point, the lens is between 500 and 1000 Å in thickness and the van der Waals forces should be negligibly small. The error in ΔA^* through using Eq. [3] should therefore be, at most, about 4%.

The film thickness. For lipid films in aqueous media the film thickness is, in most instances, more readily and accurately accessible from electrical capacitance than from optical reflectance measurements (6). The electrical method yields the thickness of the hydrocarbon region only, however, and that of the polar group layers has to be estimated from the consideration of molecular models. Fortunately, the nature and dimensions of the polar group seem, from experiment, to be of only slight importance in the interpretation of the free energies (see below). Furthermore, since only alkyl chain molecules are involved, the dielectric constant, which is needed to calculate the thickness from the capacitance, is very insensitive as to which lipid and alkane solvent are used to form the film. Confirmation of the validity of the electrical method of thickness measurement has come from the examination of "solventless" phospholipid membranes, where capacitance data, and X-ray diffraction studies of the stacked membranes in liquid crystalline material, have given very similar results (13, 14).

Experimental data are displayed in Table I. The film capacitances and compositions from which the thicknesses were calculated have been given elsewhere (8, 15).

DISCUSSION

The Calculation of the van der Waals Component of the Free Energy of Thinning

In order to determine ΔA_{vw} from Eq. [4] it is necessary either to know ΔA_e and ΔA_s or to be able to show that these quantities may be neglected. There are several reasons for thinking that ΔA_e is very small, one being that neither the adsorption of ions to the film, nor large variations in the ionic composition of the aqueous solutions appear to have appreciable effects on ΔA^* (8, 15). A more useful demonstration of the likely magnitudes of ΔA_e and ΔA_s comes from the recognition that they are both "repulsive" terms and can be studied by

compressing the films under applied electrical potentials. A detailed description of such experiments and their interpretation is too long to be included here, but the outcome for the two systems that were fully examined (monoolein + n-decane and monoolein + n-hexadecane) was that $\Delta A_e + \Delta A_s$ amounted to approximately 9% of ΔA^* (8). The compressibility, and other properties, of the remaining systems were largely intermediate to those mentioned above and it has been assumed that a similar relationship between the repulsive and total free energy exists also in these instances. Except in Fig. 3, the difference between ΔA_{vw} and ΔA^* will be ignored, so that it should be borne in mind that, elsewhere, the absolute values of the experimental Hamaker coefficients could be a few percent low.

Three models were explored in interpreting the free energies.

The Three-Layer Model

The experimental Hamaker coefficients. The essential features of the three-layer model are shown in Fig. 1. The film is considered to be

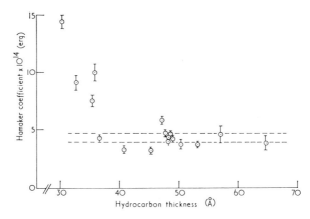

FIG. 2. Hamaker coefficients as a function of the hydrocarbon thickness of the lipid film. The experimental values are indicated by the points and were calculated from the free energies using Eq. [5]. No correction for the repulsive forces has been attempted, so that the true experimental values should be ca. 9% higher. The theoretical Hamaker coefficients were calculated assuming Model I (Fig. 1) and all fell within the range indicated by the dashed lines. If Model II had been used, the values would have been fairly consistently ca. 10% lower.

simply a layer of liquid hydrocarbon between two semi-infinite aqueous phases. The polar groups of the lipids are disregarded because they constitute a layer which is thin relative to the hydrocarbon layer and because they are interspersed by and, in their dielectric properties, similar to water. No distinction is made between the chains of the hydrocarbon and of the alkane solvent which remains in the film, and hence the formation of the film amounts merely to the drainage of a layer of bulk aliphatic hydrocarbon from between the two aqueous solutions. The van der Waals free energy may then be written (16),

$$\Delta A^* \simeq A_{\mathrm{vW}} = -A/12\pi h^2, \qquad [5]$$

where A is the Hamaker coefficient. In this equation the Hamaker coefficient should be independent of the film thickness, h. However, a plot of the data (Fig. 2) shows that this is not so. Although A appears to approach a constant value at large h, it rises increasingly rapidly as h decreases and reaches some four times this value in the thinnest film. The explanation of the nonconstancy of A will be considered below, but first it is of interest to compare the present data with Hamaker coefficients calculated from Lifshitz theory.

The theoretical Hamaker coefficients. As has been pointed out by Ninham and Parsegian (17), the van der Waals forces in the present systems are unlikely to be appreciably retarded. For the three-layer model (Fig. 1), Ninham and Parsegian (17) give the following expression for the Hamaker coefficient:

$$A = -\tfrac{3}{2}kT \sum_{n=0}^{\infty} {}' \int_0^{\infty} x \ln \ (1 - \Delta_{\mathrm{hc-w}}{}^2 e^{-x})dx, \qquad [6]$$

where

$$\Delta_{\mathrm{hc-w}} = \frac{\epsilon_{\mathrm{hc}}(i\xi_n) - \epsilon_{\mathrm{w}}(i\xi_n)}{\epsilon_{\mathrm{hc}}(i\xi_n) + \epsilon_{\mathrm{w}}(i\xi_n)} \qquad [7]$$

and

$$\xi_n = (2\pi kT/h)n. \qquad [8]$$

In these expressions ϵ_{w} and ϵ_{hc} are the dielectric permeabilities of water and hydrocarbon, respectively, and both are functions of the imaginary frequency ω $(=i\xi_n)$. $2\pi h$ is Planck's constant and n is an integer. In the summation in Eq. [6] the prime indicates that the term for $n = 0$ should be multiplied by $\tfrac{1}{2}$.

The dielectric permeabilities may be written in the form

$$\epsilon_j(i\xi_n) = 1 + \frac{C_{\mathrm{mw}}}{1 + \xi/\omega_{\mathrm{mw}}} + \sum \frac{C_{\mathrm{ir}}}{1 + (\xi/\omega_{\mathrm{ir}})^2}$$
$$+ \sum \frac{C_{\mathrm{uv}}}{1 + (\xi/\omega_{\mathrm{uv}})^2}. \qquad [9]$$

The assignment of values to the constants for hydrocarbon/water systems has been discussed generally by Ninham and Parsegian (17) and Gingell and Parsegian (18) and, for the present thin film systems by Brooks *et al.* (19).

The calculated Hamaker coefficients vary slightly over the range of systems in Table I because the composition of the hydrocarbon in the film varies. But this effect is very small and since, from the nature of the model, the theoretical A cannot be a function of h, the results all lie in a narrow band, as shown in Fig. 2. It is notable that there is excellent agreement between the theoretical Hamaker coefficients and the experimental values for the thicker films. In absolute magnitude, the coefficients are not much greater than the zero frequency term in Eq. [6], i.e., approximately $3kT/4$. It is for this reason, and because the films are relatively thin, that retardation can be disregarded.

Despite the apparent success of the theory as applied to the thicker films, there is no way of accounting for the variation in A with h, and it is necessary to examine more carefully the validity of the simple three-layer model. An obvious deficiency is the neglect of the polar groups of the lipids. It is possible that the interactions of these groups with similar groups, and with water, on the far side of the film is significant and that, as the films become thinner, these interactions become dominant. With this in mind, films were examined in which the head groups differed. The last three sets of data in Table I are for phospholipids and a glyceryl ether. As can be seen, however, the free energies for these films are very similar to those for monoglyceride films of comparable thickness. Further evidence for the unimportance of the polar groups may be found in the results for monoglyceride films formed in glycerol (15). In such systems the distinction between the polar groups and the solvent was largely removed, but the Hamaker coefficient still increased markedly as the film thickness decreased. Finally, calculations of the theoretical Hamaker coefficient were carried out on the basis of Model II of Fig. 1, in which the polar groups are considered to constitute two additional homogeneous layers. The equations appropriate to this model have been derived by Ninham and Parsegian (20) and are

$$A = -\tfrac{3}{2}kT \sum_{n=0}^{\infty}{}' \int_0^{\infty} x \ln \left\{ 1 \right.$$

$$\left. - \left[\frac{\Delta_{hc-g} + \Delta_{g-w}e^{-xc/h}}{1 + \Delta_{hc-g}\Delta_{g-w}e^{-xc/h}} \right]^2 e^{-x} \right\} dx, \quad [10]$$

where

$$\Delta_{hc-g} = \frac{\epsilon_{hc}(i\xi_n) - \epsilon_g(i\xi_n)}{\epsilon_{hc}(i\xi_n) + \epsilon_g(i\xi_n)} \quad [11]$$

and

$$\Delta_{g-w} = \frac{\epsilon_g(i\xi_n) - \epsilon_w(i\xi_n)}{\epsilon_g(i\xi_n) + \epsilon_w(i\xi_n)}. \quad [12]$$

In these expressions h and c are the thicknesses of the hydrocarbon and polar group layers, respectively, and ϵ_g is the dielectric permeability of the polar groups. The polar group layer was assumed to be made up of pure glycerol and the spectroscopic data were estimated as described in an earlier paper (19). The resulting Hamaker coefficients were ca. 10% smaller than those for the three-layer model (I) but showed a very similar lack of variation with film thickness. They were also insensitive to uncertainties in the spectroscopic data for glycerol.

It must be concluded that the major part of the variation of the Hamaker coefficient with film thickness is unlikely to originate from the presence of the polar groups of the lipids. The three-layer model could, however, be more seriously inadequate in the way in which it depicts the hydrocarbon region. Notwithstanding the fact that this region consists of two distinct species, lipid chains and alkane solvent, Model I presupposes a homogeneous layer of the appropriate average composition. There are reasons, discussed in the next section, for thinking that this model is too simple.

The Seven-Layer Model

It has been shown elsewhere (21) from consideration of the compressibility of the lipid films, that the lipid chains and the solvent are almost certainly not perfectly mixed. Instead, the lipid chains appear to be concentrated at the edges and the solvent in the middle of the films. The precise state of affairs seems to depend strongly on the chain length of the alkane used. Consider the lipid monoolein. In decane and shorter chain alkanes the film thicknesses are close to twice the extended length of the glyceride chains. The relatively high compressibility of these films is, however, consistent with the idea that only a very small proportion of the glyceride chains reach the middle of the film at any one time (21). The large majority of the chains thus appear to spend most of their time packed close to the sides while the central region of the film consists largely of alkane solvent. In other words, the decane does not penetrate completely into the glyceride chain region. When hexadecane is used, this lack of penetration is still more pronounced, to the extent that in this instance the tendency of the lipid chains to be fully extended is so slight that the film thickness is well under its maximum value, and the amount of solvent retained in the film is proportionately reduced.

From the above consideration it seems that it would be reasonable to consider a model in which the films thin through the drainage of the bulk lipid solution from between two close-packed arrays of lipid chains. This model is the opposite extreme of the three-layer model (I) in that it assumes no mixing of the solvent with the lipid chains, while Model I assumes complete mixing. Obviously a layer of nearly pure solvent in the middle of the film is not consistent with the notion that the film is stabilized by the interaction of the lipid chains but, as an approximation to the internal structure of the film, this model is not obviously less satisfactory than Model I. In the calculation of the effective Hamaker coefficient from Eq. [5] the distance variable is now a (Fig. 1, Model III). This may be found from the composition of the film (8, 15). Thus the value of a is proportional to the volume fraction of alkane solvent present. If ϕ is the volume fraction of the lipid chains,

$$2b = \phi h \qquad [13]$$

and

$$a = h - 2b. \qquad [14]$$

The parameter b is the thickness of the close-packed lipid chains (Fig. 1) and is found from a knowledge of the adsorption of the lipid and an assumption as to the partial molar volumes of the lipid chains in the film. Values of ϕ and a for some of the systems of Table I are shown in Table II.

The experimental Hamaker coefficients. As can be seen from Table II, the magnitudes of the Hamaker coefficients according to Model III are considerably less than those for Model I. This is merely because the distance a is always less than h. Thus

$$A_{III}{}^{exp} = A_{I(II)}{}^{exp}(a^2/h^2). \qquad [15]$$

Physically, the smaller A arises from a change in the reference energy state, i.e., in the infinitely thick film. Instead of being directly in contact with the displaceable hydrocarbon, as in Models I or II, the polar phases are now always screened by the layers of close-packed lipid chains. For comparable changes in the distance variable, therefore, the changes in the van der Waals interactions as the two polar phases approach each other are less in Model III than in Models I or II.

The theoretical Hamaker coefficients. The appropriate Lifshitz equations for the seven-layer model (III) may be derived from a relativistic expression given by Parsegian and

TABLE II

Theoretical and Experimental Hamaker Coefficients for the Seven-Layer Model (III)

	Volume fraction ϕ of lipid chains	a (Å)	$A^{exp} \times 10^{15}$ (ergs)	$A^{theor} \times 10^{15}$ (ergs)
Monoolein + n-heptane	0.52	22.6	13.5	9.8
Monoolein + n-octane	0.51	23.4	10.9	9.7
Monoolein + n-decane	0.51	23.6	9.4	9.5
Monoolein + n-dodecane	0.54	21.0	6.9	8.5
Monoolein + n-tetradecane	0.60	16.0	5.1	6.9
Monoolein + n-pentadecane	0.67	12.0	4.5	5.2
Monoolein + n-hexadecane	0.74	8.5	6.2	3.7
Monoolein + cis-5-decene	0.60	16.2	7.3	7.0
Monomyristolein + n-decane	0.42	25.8	11.5	12.8
Monopalmitolein + n-decane	0.47	24.5	9.9	10.9
Monolinolein + n-decane	0.49	24.7	12.0	10.5
Monolinolein + n-hexadecane	0.79	6.3	6.4	3.1
Monolinolenin + n-decane	0.65	12.6	12.3	7.1
Monoeicosenoin + n-decane	0.51	26.0	9.0	9.5
Monoerucin + n-decane	0.52	27.4	10.6	9.2
Mononervonin + n-decane	0.51	31.7	9.3	9.5

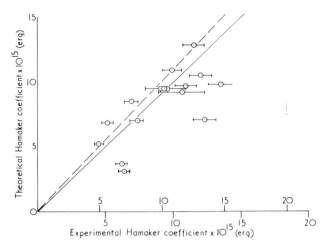

FIG. 3. A comparison of experimental and theoretical Hamaker coefficients according to Model III (Fig. 1). The full and dashed lines represent the equality of the theoretical coefficients with, respectively, the experimental values without (lower scale) and with (upper scale) a 9% correction for repulsion.

Ninham (22). The Hamaker coefficient is then

$$A = -\tfrac{3}{2}kT \sum_{n=0}^{\infty}{}' \int_{0}^{\infty} x \ln \left\{ 1 \right.$$

$$\left. - \left[\frac{\Delta_{s-1} + \Delta e^{-xb/a}}{1 + \Delta_{s-1}\Delta e^{-xb/a}} \right]^{2} e^{-x} \right\} dx, \quad [16]$$

where

$$\Delta = \frac{\Delta_{l-g} + \Delta_{g-w}e^{-xc/a}}{1 + \Delta_{l-g}\Delta_{g-w}e^{-xc/a}}, \quad [17]$$

and Δ_{s-1} and Δ_{1-g} are given by relationships similar to those of Eqs. [7], [11], and [12]. The differences between the dielectric permeabilities of the lipid chain and solvent layers are obviously very small but may be estimated as shown previously (15, 19). The Hamaker coefficients are listed in Table II. They still originate largely from the zero frequency terms, but they now vary from one system to another considerably more than did the theoretical values for Model I. For completeness, the contributions of the polar group layers have been included but, as for Models I and II, they affect the results by at most a few percent.

The experimental and theoretical Hamaker coefficients for Model III are compared graphically in Fig. 3. It is immediately obvious that this model is more successful than I or II in that the thinner films no longer appear to be anomalous. In fact, for more than half of the systems examined the agreement is to within 20% and in no instance is there a discrepancy of more than about a factor of 2. Moreover, such discrepancies as do remain may be qualitatively accounted for by recognizing that the true condition of the films will be intermediate between that assumed in the Models I (or II) and III, i.e., there will be some intermixing of lipid chains and solvent. This point is illustrated in Fig. 4 where, in the monoolein–heptane system, results are shown for various degrees of intermixing. In order to obtain these curves it has been supposed that given amounts of the solvent from the central region have been transferred successively to the lipid chain regions, the homogeneity of the latter being preserved and the overall dimensions and composition of the film remaining constant. The experimental Hamaker coefficient decreases simply because, as the solvent layer is made thinner (at con-

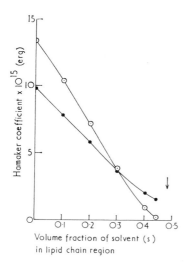

FIG. 4. The effect of intermixing solvent and lipid chains on the Hamaker coefficients according to Model III. The film is of monoolein and n-heptane, and the abscissa indicates the amount of solvent assumed to be present in the chain region. The values of A at zero mixing are thus as given in Table II. The arrow indicates the point at which all the solvent has been intermixed and the distance a reduced to zero. O, experimental; and ●, theoretical Hamaker coefficients.

stant free energy) the distance variable a decreases. The theoretical Hamaker coefficient changes because the transfer of the solvent alters the thickness and dielectric permeability of the screening layer of lipid chains. The fact that the two curves intersect is encouraging in that it suggests that with a more detailed knowledge of the film structure it should be possible to improve still further the agreement between the experimental and theoretical Hamaker coefficients.

CONCLUSIONS

Two of the more interesting findings which have emerged from the above investigation are

(1) that despite the very close similarity of the chains of the stabilizing lipids to the alkane solvent, the partial segregation of the two species in a thin film can be of crucial

importance for the calculation of the van der Waals forces; and

(2) that the agreement between the theoretical (Lifshitz) and experimental Hamaker coefficients is exceedingly good.

Taken together these two conclusions suggest that in the thicker lipid films, the interactions are predominantly those of the aqueous phases across an effectively homogeneous layer of aliphatic hydrocarbon whereas, in the thinner films, it is the interaction of the screening layers of lipid chains across the solvent layer which becomes predominant. This last observation has obvious implications for the understanding of thin liquid films generally.

REFERENCES

1. DERJAGUIN, B. V., ABRIKOSOVA, I. I., AND LIFSHITZ, E. M., *Quart. Rev. Chem. Soc.* **10**, 295 (1956).
2. KITCHENER, J. A., AND PROSSER, A. P., *Proc. Roy. Soc. London Ser. A* **242**, 403 (1957).
3. BLACK, W., DE JONGH, J. G. V., OVERBEEK, J. TH. G., AND SPARNAAY, M. J., *Trans. Faraday Soc.* **56**, 1597 (1960).
4. ROUWELER, G. C. J., AND OVERBEEK, J. TH. G., *Trans. Faraday Soc.* **67**, 2117 (1971).
5. ISRAELACHVILI, J. N., AND TABOR, D., *Proc. Roy. Soc. London Ser. A* **331**, 19 (1972).
6. FETTIPLACE, R., GORDON, L. G. M., HLADKY, S. B., REQUENA, J., ZINGSHEIM, H. P., AND HAYDON, D. A., *in* "Methods in Membrane Biology" (E. D. Korn, Ed.), Vol. 4, p. 1. Plenum Press, New York, 1974.
7. HAYDON, D. A., AND TAYLOR, J. L., *Nature (London)* **217**, 739 (1968).
8. REQUENA, J., BILLETT, D. F., AND HAYDON, D. A., *Proc. Roy. Soc. London Ser. A* **347**, 141 (1975).
9. SCHELUDKO, A., *Advan. Colloid Interface Sci.* **1**, 391 (1976).
10. DE FEIJTER, J. A., AND VRIJ, A., *J. Electroanal. Chem.* **37**, 9 (1972).
11. AVEYARD, R., AND HAYDON, D. A., *Trans. Faraday Soc.* **61**, 2255 (1965).
12. REQUENA, J., AND HAYDON, D. A., *J. Colloid Interface Sci.* **51**, 315 (1975).
13. HAYDON, D. A., *Ann. N. Y. Acad. Sci.* **264**, 1 (1975).
14. BENZ, R., FRÖHLICH, O., LÄUGER, P., AND MONTAL, M., *Biochim. Biophys. Acta* **394**, 323 (1975).
15. REQUENA, J., AND HAYDON, D. A., *Proc. Roy. Soc. London Ser. A.* **347**, 161 (1975).
16. PARSEGIAN, V. A., *Ann. Rev. Biophys. Bioeng.* **2**, 221 (1973).

17. NINHAM, B. W., AND PARSEGIAN, V. A., *Biophys. J.* **10**, 646 (1970).

18. GINGELL, D., AND PARSEGIAN, V. A., *J. Theor. Biol.* **36**, 41 (1972).

19. BROOKS, D. E., LEVINE, Y. K., REQUENA, J., AND HAYDON, D. A., *Proc. Roy. Soc. London Ser. A* **347**, 179 (1975).

20. NINHAM, B. W., AND PARSEGIAN, V. A., *J. Chem. Phys.* **52**, 4578 (1970).

21. ANDREWS, D. M., MANEV, E. D., AND HAYDON, D. A., *Spec. Discuss. Faraday Soc.* No. 1, 46 (1970).

22. PARSEGIAN, V. A., AND NINHAM, B. W., *J. Theor. Biol.* **38**, 101 (1973).

The Metal-to-Metal Interface and Its Effect on Adhesion and Friction

DONALD H. BUCKLEY

Lewis Research Center, Cleveland, Ohio 44135

Received March 31, 1976; accepted July 1, 1976

The interface established by two metal surfaces brought into solid state contact is much more rigidly predetermined than is the interface for the other states of matter contacting themselves or solids. Thus, solid state structural factors at the surface such as orientation, lattice registry, crystal lattice defects, and structure are shown to have an effect on the character of the resulting interface established for two metals in contact. The interfacial structural character affects the adhesion or bonding forces of one solid to another. This in turn influences the forces necessary for tangential displacements of one solid surface relative to the other. Because the nature of the metal-to-metal interface is determined to an extent by the solid surficial layers' surface tools such as LEED, Auger emission spectroscopy and field ion microscopy are used to characterize the solid surfaces prior to contacts and after the establishment of an interface. In addition to the foregoing structural considerations, many of the properties of matter which influence the nature of the interface of the various states of matter with metals are shown to effect the metal-to-metal interface. These include metal surface chemistry and the influence of alloying on surface chemistry and bulk chemical behavior. The nature of the interface, adhesion and friction properties of noble metals, platinum metals, Group IVB metals, and transition metals are considered. Surface chemical activity of the noble and platinum metals are shown to effect metal-to-metal interfaces as does valence bonding in the transition metals. With the Group IVB metals the degree of metallic nature of the elements is shown to effect interfacial behavior. The effect of surface segregation of alloy constituents such as silicon in iron and its influence on the metal-to-metal interface is discussed. In addition, the effect of alloy constituents on changes in bulk properties such as transformations in tin is shown to effect interfacial adhesion and friction behavior.

INTRODUCTION

Considerable research efforts have been expended in the studies of various types of interfaces. These have included the gas–liquid, liquid–liquid, gas–solid, and liquid–solid (1). However, it has not been until recently that the solid–solid and particularly the metal–metal interface has been examined in any detail. Prior to the advent of field ion microscopy, Auger emission spectroscopy, low-energy electron diffraction and scanning electron microscopy much of the information relative to the metal–metal interface was derived from grain boundary studies (2). Currently a wealth of information relative to such interfaces is emerging (3, 4).

It has become obvious that surfaces can no longer be thought of as extensions of the bulk properties of materials and studies are addressed to the thermodynamics of surfaces (5), surface structure (6), and the stresses and strains associated with surfaces (7). A good deal of metal–metal interface information is being derived from the deposition of films on substrates and examining the resulting interface (3, 4, 8, 9). A new terminology is evolving with consideration of the metal–metal interface. Lattice disregistry, misfit dislocations, coherency strain, and anisotropic interfaces are terms currently in use.

Most of the metal-to-metal interfacial studies have been conducted with thin metal films deposited by ion plating, sputtering, or vapor deposition onto the surface of another metal. The generation of an interface by vapor phase transport of one metal to the surface of the other can, because of the mobility of the incoming species, result in such conditions as interfacial epitaxy (3, 4). These conditions are less likely to occur in those situations where both of the metal surfaces are solids with "frozen" lattices. An important interface is that developed between solid metals contacting in technological mechanisms such as electrical contacts, bearings, gears, and seals.

The objective of this paper is to consider the interface that develops between two bulk metals in contact and the effect of that interface on adhesive bonding, resistance to tangential displacements on friction, and the interfacial transport of metal from one surface to another.

METALS IN CONTACT WITH THEMSELVES

If it were possible to bring two metal single crystals of the same surface together with a perfect match of atomic planes and crystallographic directions and the two surfaces were defect-free on near touch contact the two single crystals would join to form one continuous interface free metal single crystal. As a practical matter such matching is not experimentally possible and the result is that when two metal crystals of the same orientation are brought into contact the equivalent of defect-ladened interfaces develop which are analogous at best

to grain boundaries. Such interfaces will contain voids and misfit dislocations.

When two metal single crystals of the same orientation are brought into contact the bonding forces at the interface will depend very heavily on the degree of lattice mismatch across the interface. The greater the degree of lattice mismatch, the greater the concentration of misfit dislocations and the greater the interfacial energy (10). Where the misfit is slight the two metal crystal lattices will be pulled into registry at the interface so that very low interfacial energy will exist, although because of the mismatch long range elastic distortions into the bulk metals will occur. The minimum or zero energy condition exists in the complete absence of lattice mismatch and an interface. Adhesion with copper crystals in contact indicate that mismatch of crystallographic directions along a common crystal axis results in a decrease in the force required to pull the interface in tension to fracture. Further, where different crystallographic planes of copper are brought into contact the force required for tensile fracture of the interface is less than where planes of the same orientation are brought in contact (11).

With respect to the adhesion and bonding of various matched crystallographic planes and directions in general, the high atomic

TABLE I. - SOME PROPERTIES OF THREE PLANES OF COPPER TOGETHER

WITH MEASURED ADHESIVE FORCES TO THOSE PLANES

Copper surface plane	Coordination number of surface	Atomic arrangement of surface unit mesh	Number of surface, atoms /cm^2	Elastic modulus, dynes/cm^2	Surface energy, ergs/cm^2	Force of adhesion to gold,[a] mg
(111)	9		$1.7 \cdot 10^{15}$	$19.4 \cdot 10^{11}$	2499	80
(100)	8		$1.5 \cdot 10^{15}$	$6.67 \cdot 10^{11}$	2892	185
(110)	7		$1.1 \cdot 10^{15}$	$13.1 \cdot 10^{11}$	----	390

[a]Applied load, 20 mg; Au (100) surface; contact time, 10 seconds.

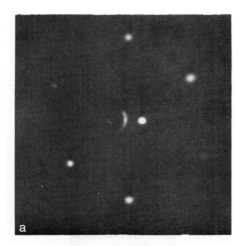

FIG. 1. LEED patterns of copper (111) surface before and after adhesive contact with gold. (a) clean (111), (b) after 20-ml contact with gold.

Polycrystalline metals when brought into contact with themselves present at the interface in addition to a variety of mismatches in crystal lattice, grain boundaries. These boundaries have their own characteristic energy. It will vary with the mismatch in the orientation of the adjacent grains generating the boundary. In general, stronger adhesion bonding forces will be developed between such surfaces than is observed with the high atomic density single crystal surfaces (11).

DISSIMILAR METAL INTERFACE

The case of a metal contacting itself and the generation of an interface can be compared to the grain boundary in a polycrystalline metal sample. With dissimilar metals in contact such an analogy cannot be made. The species generating the interface differ in atomic size, lattice spacing, binding energy, and other properties. Surface orientation does exert an influence on the interface formed between the dissimilar metals and the bond strength of that interface.

In Table I various properties for three atomic planes of copper are presented together with adhesion data for a gold (100) surface to those planes. As with copper in contact with itself bonding forces are least to the (111) copper surface. If the copper surface is examined by LEED and AES after the interface is pulled to tensile fracture gold is found to have transferred to the copper surface. AES

density planes exhibit the weakest interfacial bonding and the low atomic density planes exhibit the greatest interfacial adhesion. Thus, for the face-centered cubic metals such as copper the (111) orientation in contact with itself yields the minimum in bonding force, while for the body-centered cubic metals it is the (110) orientation and for the hexagonal close-packed metals it is the (0001) orientation (12).

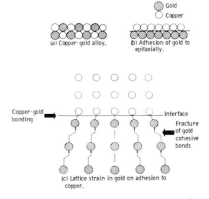

FIG. 2. Atomic arrangement and lattice bonding.

analysis of the surface indicates the presence of gold in addition to copper peaks in the Auger spectrum. Figure 1 is a LEED pattern for the copper (111) surface before and after contact by gold.

The LEED pattern of Fig. 1 together with AES analysis indicate pseudomorphic transfer of gold to the copper. The gold lattice is contracted from that observed for bulk gold. The gold accommodates itself to the copper lattice as indicated in the schematic of Fig. 2.

In Fig. 2a the atomic arrangement of copper to gold atoms in a bulk alloy is indicated. For alloy systems in the solid solution region, Vegard's Law can be used to predict lattice spacing.

With two dissimilar metals contacting across an interface as indicated in Fig. 2b, the atomic arrangement is different than that seen for the alloy in Fig. 2a. After an interface is established between the copper and gold as indicated in Fig. 2b, strong bonds are formed between the two elements. Lattice strain occurs in the gold in order that the gold may accommodate itself to the copper lattice as indicated in Fig. 2c.

When a tensile force is exerted normal to the copper–gold interface fracture will occur in the weakest region. Since on pulling the specimen to fracture, gold was found remaining on the copper surface in a pseudomorphic manner, fracture had to occur in the gold. Thus, the interfacial bonds developed between the copper and the gold were stronger than the cohesive bonds in the gold.

The observation that the interfacial bonds formed between two metals in contact are stronger than the cohesive bonds in the weaker of the two metals is a general observation and occurs in other dissimilar metal systems as well (13). The transfer of metal across the interface does not generally occur in a pseudomorphic manner but will normally occur in accordance with the general rule. This will be discussed further when reference is made to the iron contacting noble metal data.

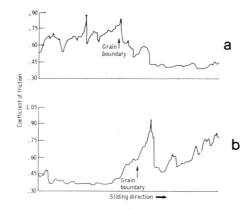

FIG. 3. Recorder tracings of friction force for copper slider sliding across grain boundary on copper bicrystal. Load, 100 g; sliding speed, 1.4 mm/min. (a) From (210) to (111), (b) from (111) to (210).

GRAIN BOUNDARY AND ORIENTATION EFFECTS ON INTERFACIAL TRANSFER

Orientation not only affects the nature and strength of an interface between metals in contact but also exerts an influence on tangential displacements such as those associated with sliding. Studies with a polycrystalline slider moving across a copper bicrystal (one grain the (111) and the other the (210) orientation) resulted in differences in friction not only on the surface of the grains but also in the grain boundary region. This effect is manifested in the data of Fig. 3.

In Fig. 3a in sliding from the (210) grain to the (111) grain, friction is higher on the (210) plane and in the grain boundary region than it is on the (111) plane. Grain boundary effects can be seen much more readily when sliding is initiated on the (111) surface as indicated in Fig. 3b. There is a pronounced increase in the friction for the slider–grain boundary interface. The grain boundary is atomically less dense than the grain surfaces on either side of that boundary.

Examination with scanning electron microscopy of the (111) and (210) grain surfaces after sliding and a single pass of the slider across the surface revealed severe surface

40

disturbance as a result of the contact as indicated in the micrographs of Fig. 4. The micrographs for the contacted surface area on both grains are at the same magnification.

On both grain surfaces in Fig. 4 fracture cracks are observed. These cracks are surface initiated. The wear face of the cracks is extremely smooth indicating crack initiation

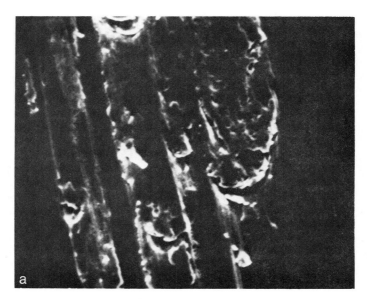

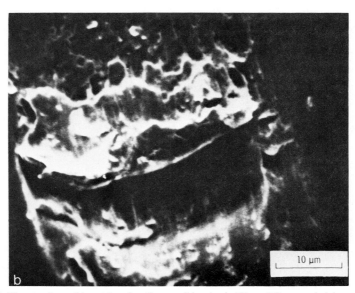

FIG. 4. Wear tracks on bicrystal grains. Copper slider; load, 100 g; sliding speed, 1.4 mm/min. (a) (111) grain, (b) (210) grain.

along slip bands. As indicated in the micrographs, the size of the cracks is much larger on the (210) surface than on the (111) surface. Sectioning of the wear track and measurement of the crack angle of orientation relative to the surface orientation indicate that the fracture cracks form along slip bands in the copper grains.

A wake of metal just ahead of the fracture crack stands above the surface of the grain itself. This occurs for both grain surfaces, but again, the amount of metal standing above the surface is greater for the (210) than for the (111) grain.

The metal-to-metal interfacial mechanism responsible for the manifested friction behavior of Fig. 3 and the surface conditions of Fig. 4 can best be explained with the aid of Fig. 5.

When the copper slider is first brought into touch contact with either grain surface and a load is applied, deformation of surface asperities results in penetration of surface contaminating films and metal-to-metal interface formation with strong adhesive bonding. As tangential motion is commenced fracture must occur in the weakest interfacial region. The weakest region is not at the interface but rather in the cohesive bonds between adjacent slip planes in the individual copper grains. Thus,

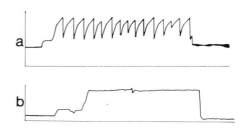

FIG. 6. Friction traces for iron (110) sliding on a tin (110) single-crystal surface at 24° and −46°C. Sliding velocity, 0.7 mm/min; load, 10 g; pressure, 10^{-8} N/m^2 (10^{-10} Torr). (a) Temperature, 24°C; white tin, (b) temperature, −46°C; gray tin.

with tangential force atomic bonds along the copper slip fracture with the formation of a surface initiated crack, as indicated in Fig. 5.

With a continued application of a tangential force, at some point the applied force will be sufficient to exceed that of the interfacial slider to grain bond and fracture of that interfacial bonding will occur. The slider will move on until adhesion again occurs.

After the interfacial slider-to-grain bond has fractured a wake or curl of metal will remain above the plane of the grain surface, as indicated in Fig. 5. Subsequent passes result in shear of test surface protuberance of metal with the resulting formation of a wear particle. Thus, for polycrystalline copper in contact with a single crystal (grain) of copper the interface develops bonds which offer greater resistance to fracture than cohesive bonds along slip planes in the copper single crystal (grain).

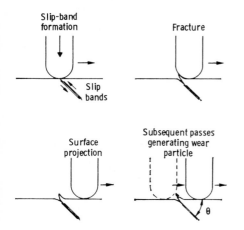

FIG. 5. Origin of surface fracture and formation of wear particle.

EFFECT OF CRYSTAL STRUCTURE ON THE METAL–TO–METAL INTERFACE

A number of metals in the periodic table of elements are polymorphic; that is, they exist in more than one crystalline form. This ability to exist in more than one crystalline form raises the question as to the effect of the various crystalline forms of a single metal on the nature of the interface formed with itself and with other metals. Tin is polymorphic, existing

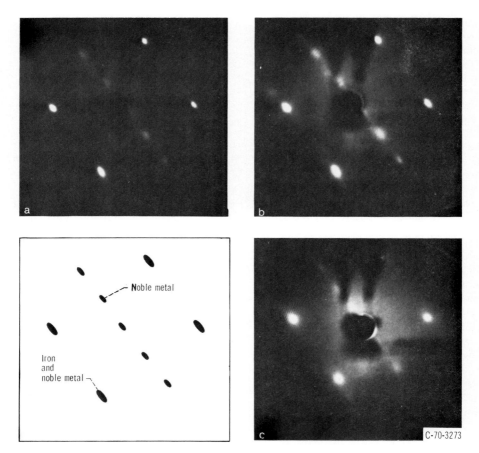

FIG. 7. LEED photographs of iron (011) surface after adhesion to noble metals. (a) Silver, (b) copper, (c) gold.

as gray tin below 13°C and white tin above this temperature (14).

Gray tin has a diamond type of crystal structure, with each tin atom tetrahedrally coordinated by four other tin atoms. White tin has a body-centered tetragonal structure and appears as a distorted diamond structure.

When an interface is formed on contact of the two forms of tin with iron and tangential motion initiated differences in friction behavior are observed. These differences are reflected in the data of Fig. 6.

White tin in Fig. 6a has a stick-slip or saw-tooth type of friction trace indicating the formation of strong adhesive junctions at the interface between the iron and white tin. With the continued application of a tangential force the interfacial bonds are broken and slip occurs. This is manifested in Fig. 6a by the sharp drop in the friction force occurring at regular intervals. After slip adhesion again occurs and the applied tangential force continues to increase until the interfacial bond strength is once again exceeded and slip occurs once again. The

43

process continues to repeat itself. It is a commonly observed behavior pattern for metal in contact with metal where strong interfacial bonds form.

The gray or diamond form of tin exhibited a continuous smooth friction trace (Fig. 6b). There is an absence in the trace of the stick-slip or sawtooth behavior seen in Fig. 6a.

Examination of the iron surface after contact with the two forms of tin revealed that with white tin random islands of tin remained on the iron surface as a result of adhesion and fracture of tin bonds in the bulk tin. With gray tin a smooth, continuous, uniform interfacial transport of tin to iron was observed.

The data of Fig. 6 indicate that the nature of the metal-to-metal interface for tin to iron is different for the two forms of tin. It is not, therefore, just the atomic character of the two elemental metals which form the interface but also the crystal lattice in which they find themselves.

In gray tin the atoms are in stacked sheets of continuously linked hexagonal rings parallel to the (111) planes with shear taking place along these planes. This then may account for the smooth uniform transfer film of tin observed on iron.

With white tin the tetragonal structure permits slip in two systems, on the (110) planes of atoms in the [001] direction at low temperatures and (110) planes of atoms in the [111] direction at higher temperatures (15). This multiple slip behavior allows more readily for the type of transfer observed for white tin to iron.

THE NOBLE METAL–TO–IRON INTERFACE

In attempting to understand the nature of the interface between metals, a seemingly logical consideration is the similarity that might exist for those metals having like properties based upon their classification in the periodic table. The noble metals, silver, copper, and gold have many properties in common. When these metals are brought into

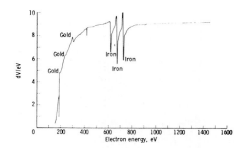

Fig. 8. Auger emission spectrometer trace of iron (011) surface with adhered gold.

contact with an iron (001) surface interfacial adhesion occurs for each of these metals to iron.

Separation of the noble metals from the iron and subsequent examination of the iron surface with LEED and Auger emission spectroscopy indicate marked similarity in the adhesive behavior of all three noble metals.

The LEED patterns obtained with all three noble metals are presented in Fig. 7. The basic LEED structures are identical for all three noble metals on iron as indicated in the diagrammatic sketch of Fig. 7. All three noble metals were found transferred to the iron with Auger emission spectroscopy analysis. An Auger spectrum for gold on iron is presented in Fig. 8.

From the LEED and Auger analyses of the iron surface, interfacial adhesion of the noble metals to iron occurred. Since the basic LEED structures are the same, the iron dictated the structural interfacial arrangement of the noble metals on iron. With the application of tensile forces to fracture the interface, the cohesively weaker noble metals were found to have transferred to the cohesively stronger iron. Again, as was noted previously, the adhesion bonds at the interface are stronger than the cohesive bonds in the weaker noble metals.

It is of interest that all three of the noble metals behaved and transferred to the iron in a similar manner. Such results indicate that basic similarities in the properties of noble metals are reflected in like similarities in their interfacial adhesive behavior.

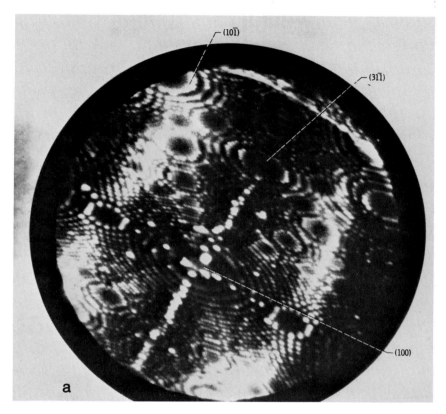

Fɪɢ. 9. Field ion micrographs of iridium–gold contact. Image gas, helium. (a) Iridium prior to contact. Voltage, 13.0 kV; liquid-nitrogen cooling. (b) Iridium after contact at 12.0 kV; liquid-helium cooling. (c) Iridium after gold contact at 13.0 kV with voltage raised to 15.2 kV for 30 sec; liquid-helium cooling.

INTERFACES OF MEMBERS OF THE PLATINUM METALS GROUP

The field ion microscope has been very useful in the study of the metal-to-metal interface (16, 17). Metals can be brought into solid state contact and the interfacial results of that contact examined at the atomic level. A number of the members of the platinum metals family have been examined in this manner.

Figure 9a is a photomicrograph of an iridium surface as seen in the field ion microscope, each individual white spot indicating an atom site with some of the atomic planes called out. This is an asperity free surface.

When gold is brought into contact with the iridium, adhesion of the gold to the iridium occurs. On tensile fracture of the specimen bonding, gold is found to have adhered to the iridium surface as shown in Fig. 9b. The white spots are now due to the presence of the adhered gold. There appears to be an ordered distribution of the gold on the iridium surface but without any preference of the gold for a specific atomic plane of iridium. Again, the cohesively weaker metal has transferred to the cohesively stronger.

The adhered gold of Fig. 9b can be removed by field evaporation. Where the gold has been

45

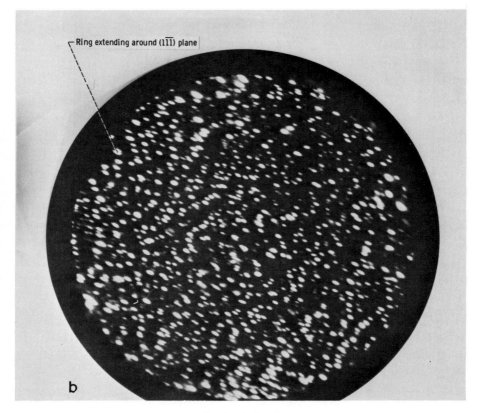

FIG. 9—*Continued*

removed the original iridium surface is seen. This is shown in the micrograph of Fig. 9c. The atomic planes of iridium can be identified once again.

The iridium-to-platinum interface was also studied with field ion microscopy. A field ion micrograph of the iridium surface prior to contact is presented in Fig. 10a and after contact with platinum in Fig. 10b.

Platinum transferred to the iridium surface with a fairly high degree of order. There are two nonimaging areas between the (100) and (311) planes of Fig. 10b. If field evaporation of the platinum was conducted the last region to lose platinum was the (100) region. Examination during field evaporation of the platinum covered iridium surface indicated that the

platinum had adhered to the iridium in a near-epitaxial manner.

The near epitaxial transfer of platinum to iridium, a sister element in the metals of the platinum family, is analogous to the ordered transfer of gold to copper, a sister noble metal. In each case the cohesively weaker of the two metals comes into atomic registry with the cohesively stronger fracture occurring in the cohesively weaker metal. Adhesive interfacial bonding is again stronger than cohesive bonding in the weaker of the two contacting metals.

The interfacial behavior of the platinum metals was examined in contact with a single metal to determine the relative differences in behavior. Loads ranging from 1 to 10 g were applied to the surfaces in contact with tangen-

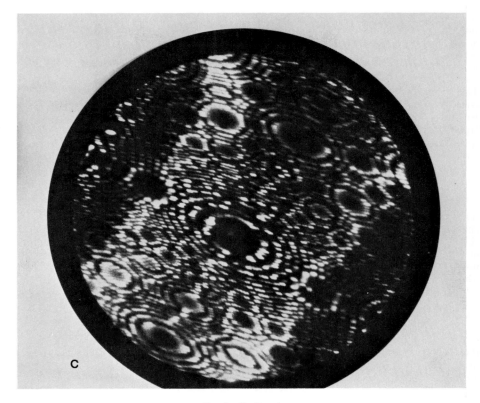

FIG. 9—*Continued*

tial motion then initiated and friction forces measured.

Figure 11 is a summary figure of the friction data obtained for a gold (111) surface in sliding contact with the various metals except osmium. The figure indicates that even though the pin was identical for all platinum metals and the transfer of gold occurred to all of the platinum metal surfaces as indicated by Auger emission spectroscopy, differences in friction behavior existed (18).

The highest friction was obtained with the metals platinum and palladium and the lowest with ruthenium, rhodium, and iridium. All metals were in single crystal form with the highest atomic density, lowest surface energy plane exposed to contact with gold. The

atomic planes have the same atomic packing. These orientations were selected to eliminate crystallographic orientation as a variable.

Both platinum and palladium are chemically more active than ruthenium, rhodium, and iridium. As the atomic number in period 5 containing the elements ruthenium, rhodium, and palladium is increased, the contribution to bonding of d electrons is increased. Likewise, a similar behavior is observed in period 6 with the elements osmium, iridium, and platinum (19). Thus, stronger bonding of gold to platinum and palladium would be anticipated from the valence–bond model when that model is applied to metallic systems. As indicated in Ref. (19) there is no reason not to apply it to metal systems since it involves the same basic electronic bonding as is involved in other systems for which the model was originally developed.

In considering the transition elements a knowledge of the contribution of d electrons to metallic bonding is necessary. An examination of the heats of atomization of the elements in the periodic table clearly indicates the importance of the d electrons to bonding. The most stable metallic structures are those which use as many d electrons as possible in bonding. The contribution to d-electron bonding increases with increasing atomic number; thus, Ru < Rh < Pd and Os < Ir < Pt. With increasing d-electron bonding there is a corresponding decrease in the sp-electron contribution to bonding.

GROUP IV ELEMENTS

The Group IV elements silicon, germanium, tin, and lead exhibit many common properties. One very interesting property of the elements as a group is the increase in metallic character in moving through the Group from silicon to lead. Germanium for example, is very brittle while white tin has good ductility. Tin itself exhibits differences in metallic character. White tin is very ductile, while gray tin is much less so.

Because silicon and germanium are very friable it is difficult to prepare flat surfaces of these elements which readily lend themselves to interface, adhesion, and friction studies. Thin films (800 Å in depth) of silicon, germanium, tin, and lead were therefore deposited by ion plating on a common substrate, namely a (011) nickel crystal surface for interfacial bonding studies. A gold (111) surface was brought into contact with these various films, loads applied, the surfaces separated and the interfaces examined for interfacial transport.

By examining a very thin film, 800 Å, which is just sufficient to form a continuous film over the nickel substrate, the effect of the basic chemistry of the elements can be compared without too much concern for differences in the mechanical deformation behavior of the film. Insight into the more fundamental effects of the electronic nature of these elements can thereby be achieved.

Adhesion was greater for the tin and lead in contact with gold than it was for silicon and germanium. Initiation of tangential motion and recording of frictional force indicated markedly different interfacial behavior as reflected in the frictional force data of Fig. 12 for the germanium and tin films.

In Fig. 12a high initial friction was noted as indicated by the spike to the left in the friction trace. This reflects very strong interfacial bonding between the gold and the germanium. Once tangential motion has begun the force drops to a very low value. If the movement is stopped and the specimens allowed to stand in contact under load for a period of time and then tangential motion reinitiated, high friction is again obtained as indicated in the spikes to the right side of Fig. 12a.

Identical experiments with tin films yielded the friction results of Fig. 12b. The stick-slip behavior observed earlier with iron in contact with tin is again seen in Fig. 12b with films of tin. As discussed earlier, this behavior is characteristic of strong metal-to-metal interfacial bonding. Observations similar to those presented in Fig. 12 for germanium and tin were observed when silicon and lead were compared.

The experimental results herein indicate that for films of the Group IV elements on a common substrate adhesion and friction are less for the covalently bonded elements than for the metallic bonded metals in contact with a metal. The cohesive binding energies for silicon and germanium are greater than those for tin, lead, and the gold pin. The stronger the interatomic bonding within the element the more closely the valence electrons are held to the nucleus. The covalent bond character of the Group IV elements is due to the sp^3 hybrid formation. In this study, the electron pair bonds are strongest in silicon and become weaker with the other elements. The electrons become less and less of a valence type and tend

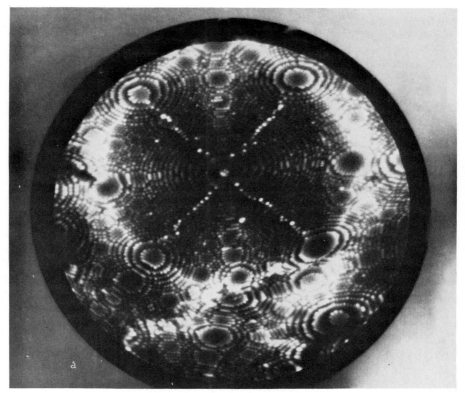

Fig. 10. Field ion micrographs of iridium–platinum contact. Image gas, helium; liquid-nitrogen cooling. (a) Iridium prior to contact. (b) Iridium after platinum contact at 18.0 kV.

to resemble free electrons more and more when moving from silicon to germanium to tin and finally to lead.

Valence electrons require a greater degree of specificity in interfacial electron compounds than is required with free electrons. Thus, bonding can be expected to occur more readily with free electron elements.

The good adhesion resistance of germanium was recognized in early engineering studies (20). These early observations were, however, not related to bonding.

THE RELATION BETWEEN METAL BOND CHARACTER AND INTERFACIAL BONDING

Pauling in 1948 formulated a resonating-valence–bond theory of metals and inter-metallic compounds in which numerical values could be placed on the bonding character of the various transition elements (21). While there have been critics of the theory it appears to be the most plausible in explaining the interfacial interactions of transition metals in contact with themselves and other metals.

When two metal surfaces are placed into contact in the atomically clean state the intermetallic bonds that form are going to depend heavily on the character of the bonding in each of the metals. One might predict from Pauling's theory that those metals which have strong d character would be less likely to interact, forming strong interfacial bonds with other metals, than those metals which do not have this strong character.

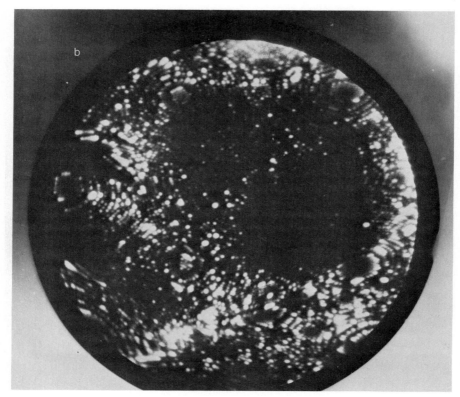

FIG. 10—*Continued*

Adhesion and friction experiments have been conducted with transition metals both in bulk and thin film form. Results for bulk metal friction measurements are presented in Fig. 13. The surface in contact with each of the transition metals in Fig. 13 was a gold (111) surface. The data of Fig. 13 indicate a decrease in friction with an increase in d character of the metallic bond. Similar results were obtained in adhesion experiments.

When thin films (2000 Å) of some of the transition metals examined in Fig. 13 were placed on a quartz substrate by sputter deposition and examined in adhesion and friction experiments, adhesion and friction decreased with increasing d bond character to iron. With iron and those metals having stronger d character (e.g., platinum) the interface between the transition metal and the quartz substrate was weaker than that between the gold and the transition metal and with tangential motion the metal film separated from the quartz substrate. With iron an abrupt decrease occurred in friction and with all the metals which separated from the quartz the friction was essentially the same as that for gold in contact with quartz.

ALLOYING AND ITS EFFECT ON THE
METAL–TO–METAL INTERFACE

Small amounts of alloying elements can markedly alter the character of metal surfaces via such mechanisms as equilibrium surface segregation (22–25). The segregation of alloy

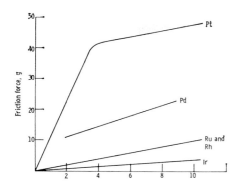

Fig. 11. Friction force as function of applied load for gold (111) single crystal sliding on various members of platinum metals group. Sliding velocity, 0.7 mm/min; ambient pressure, 1.33×10^{-8} N/m^2 (10^{-10} Torr); temperature, 23°C.

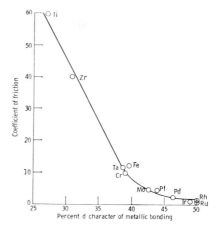

Fig. 13. Coefficient of friction as a function of the percent d bond character for various metals. Sliding velocity, 0.7 mm/min, load, 1 g, 23°C and 10^{-8} N/m^2.

constituents to the surface has been found to result in concentrations of alloy constituents on the surface far in excess of the bulk. With copper–aluminum alloys an alloy containing 10% aluminum had in its surface layer pure aluminum atoms with lateral packing equivalent to one-third bulk atomic packing along (111) planes (24). When two metals are brought into solid state contact, the presence

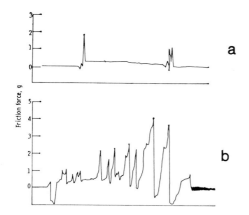

Fig. 12. Friction traces for 8×10^{-8} m (800 Å) films of geranium and tin films ion plated onto nickel (011) surface. Load, 1 g; sliding velocity, 7.0 m/min; temperature, 23°C; pressure, 1.33×10^{-8} N/m^2 (10^{-10} Torr). (a) Geranium, (b) tin.

of these segregated species can and does alter the nature of the metal-to-metal interface.

An example of the effect of surface segregation is seen with copper–aluminum alloys contacting a gold surface. The adhesive bonding of gold to a copper and one atomic aluminum alloy resulted in measured adhesive forces five times those for elemental copper in identical experiments. The adhesive bonding forces are identical to those measured for a gold (111) surface contacting an elemental aluminum (111) surface. The aluminum had segregated out of the alloy matrix on the surface of the alloy such that the interface upon contact with gold was one of gold to aluminum rather than one of gold to an alloy of aluminum in copper. It is for this reason that great care must be taken in using bulk metal properties to predict surface behavior as surfaces may not always be reflections of the bulk.

With the copper–aluminum alloys just described the aluminum, once it has segregated to the alloy surface due to heating or strain, remains on the surface. Another type of surface segregation and one which is more elusive to study is that of silicon in iron.

When an iron and 6.55 at.% silicon alloy is heated the silicon segregates from the matrix to the surface. Auger emission spectroscopy analysis of the alloy while being heated indicates growth of the concentration of silicon at the surface. This growth is indicated in the data of Fig. 14.

An examination of Fig. 14 indicates that at temperatures above about 300°C the amount of silicon on the surface due to segregation from the bulk increases. It continues to increase with increases in specimen temperature to 700°C. When the specimen is cooled the silicon returns to the alloy matrix, as indicated in Fig. 14. Thus, the segregation is reversible.

The adhesion behavior of gold to the iron and 6.55 at.% silicon alloy of Fig. 14 was studied over the same temperature range. The results obtained are presented in Fig. 15 together with data for elemental iron.

The data of Fig. 15 indicate a decrease in the interfacial adhesive bonding of the gold to the iron–silicon alloy as the temperature of 300°C is approached. It appears from the data that with silicon segregation adhesion forces decrease. There is still strong adhesion but the binding force of gold to silicon is less than it is to iron.

Beyond 300°C the adhesive binding force of the gold to the alloy remained relatively constant, as reflected in the adhesion data of Fig. 15. This is an observation similar to that made with copper–aluminum alloys (22).

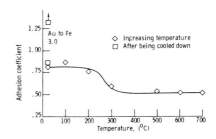

Fig. 15. Adhesion of gold (111) surface to an iron–6.5% silicon alloy crystal (110) at various temperatures. Load 1.0 cm and 10^{-8} N/m² environmental pressure.

When the specimens of Fig. 15 were cooled to room temperature the adhesion coefficient returned to near the original room temperature value. This is indicated in the single data point of Fig. 15.

The observations of Fig. 15 for iron–silicon are contrary to those normally observed for clean metals in contact. Generally with clean metals in contact the interfacial adhesive binding forces increase with temperature increases. In Ref. (26) strong adhesion for gold was observed to commence at about 247°C. A similar observation was made for gold to iron. When iron was contacted by gold at temperature above 250°C, binding forces were so strong that separation of the specimens constituted

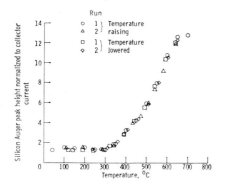

Fig. 14. AES silicon peak-to-peak height normalized to collector current for iron–6.55 at.% silicon (single crystal) against temperature.

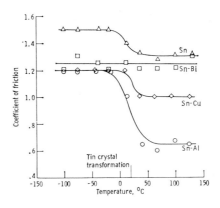

Fig. 16. Coefficient of friction for polycrystalline tin and tin alloys at various temperatures. Sliding velocity, 0.7 mm/min; load, 10 g; pressure, 1.33×10^{-8} N/m² (10^{-10} Torr).

tensile fracture experiments rather than adhesive bond force studies. It must be concluded from Fig. 15 that silicon segregation to the surface of iron–silicon alloys reduced interfacial adhesive bonding.

The interesting aspect of the data of Fig. 15 is that the segregation of silicon is reversible and so is the adhesive behavior. With increasing temperature, silicon segregates and adhesion goes down, while with a return to room temperature silicon returns to the matrix and adhesion goes back up.

In addition to the segregation of alloy constituents at the surface of alloys which influences interfacial metal-to-metal behavior there are other effects of alloy constituents which bear upon interfacial behavior. One such effect is that of alloying elements on the kinetics of crystal transformations. As was discussed earlier in reference to Fig. 6, tin transforms from one crystalline form to another at 13°C. Some alloying elements accelerate the kinetics of transformation, while others retard it or arrest it completely.

Changes in interfacial friction properties of tin occur with an alteration of the kinetics of crystal transformation. This is demonstrated by the data of Fig. 16 when various elements are added to tin in a concentration of one-at.% of the alloying element.

The data of Fig. 16 indicate that a decrease in friction coefficient occurs with the transformation of gray tin to white tin. Bismuth arrests the transformation and the data of Fig. 16 indicate that a change in friction is also arrested. Both copper and aluminum accelerate the kinetics of transformation and the data of Fig. 16 indicate marked changes in friction with accelerated transformation, the greatest being noticed with aluminum. Thus, these data indicate that bulk as well as surface effects with alloying will influence metal-to-metal interfacial behavior.

CONCLUDING REMARKS

There are many properties of metals which influence the nature of the interface developed when two metals are brought into contact. These include surface orientation, lattice spacing, grain boundaries, crystal structure, nature of bond character and alloying elements. The effect of alloying elements can alter interfacial behavior by segregation to the surface of metals or by altering bulk properties through crystal transformation kinetics.

With dissimilar metals in contact pseudomorphic and epitaxial transfer from one metal to another has been observed. The bonding at the interface between dissimilar metals is, as a general rule, stronger than the bonding in the cohesively weaker of the two metals with the result that on separation of the metals transport of the cohesively weaker to the cohesively stronger is observed.

REFERENCES

1. ADAMSON, A. W., "Physical Chemistry of Surfaces," 2nd ed., Interscience, New York, 1967.
2. McLEAN, D., "Grain Boundaries in Metals," Oxford, London, 1957.
3. Proceedings of the International Conference on the Structure and Properties of Grain Boundaries and Interfaces, *Surface Sci.* **31**, 1 (1972).
4. Proceedings of the Sixth International Vacuum Congress, 1974, *Japan J. Appl. Phys., Suppl. 2*, Pt. 1, 525–700 (1974).
5. COUCHMAN, P. R. AND JESSER, W. A., *Surface Sci.* **34**, 212 (1973).
6. GOMER, R., "Topics in Applied Physics," Springer–Verlag, New York, 1975.
7. COUCHMAN, P. R., JESSER, W. A. AND KUHLMANN–WILSDORF, O., *Surface Sci.* **33**, 429 (1972).
8. VAN DER MERWE, J. H., *Surface Sci.* **31**, 198 (1972).
9. VAN DER MERWE, J. H. AND VAN DER BERG, N. G., *Surface Sci.* **32**, 1 (1972).
10. DUNN, C. G. AND LIONETTI, F., *J. Metals* **1**, 125 (1949).
11. BUCKLEY, D. H., "Adhesion or Cold Welding of Materials in Space Environment," STP-431, pp. 248–271. Amer. Soc. for Testing and Materials, Philadelphia, 1967.
12. BUCKLEY, D. H. AND JOHNSON, R. L., *Wear* **11**, 405 (1968).
13. BUCKLEY, D. H., *Wear* **20**, 89 (1971).
14. AZAROFF, L. V., "Introduction to Solids," McGraw-Hill, New York, 1960.
15. SCHMID, E. AND BOAS, W., "Kristallplastizitat, mit Besonderer Berucksichtigung der Metalle," Springer, Berlin, 1935.

16. MULLER, E. W. AND NISHIKAWA, Z., "Adhesion and Cold Welding Materials in Space Environment," STP-431, pp. 67–87. Amer. Soc. for Testing and Materials, Philadelphia, 1967.

17. BRAINARD, W. A. AND BUCKLEY, D. H., NASA TN D-6492 (1971).

18. BUCKLEY, D. H., NASA TN D-7896 (1975).

19. BREWER, L., "Electronic Structure and Alloy Chemistry of the Transition Elements" (P. A. Beck, Ed.), pp. 221–235. Interscience, New York. (1963).

20. ROACH, A. E., GOODZEIT, C. L., AND HUNNICUTT, R. P., *Trans. ASME* **78**, 1659 (1956).

21. PAULING, L., *Proc. Roy. Soc., Ser. A* **196**, 343 (1949).

22. FERRANTE, J. AND BUCKLEY, D. H., *ASLE Trans.* **15**, 18 (1972).

23. JOSHI, A. AND STEIN, D. F., *J. Inst. Metals* **99**, 178 (1971).

24. FERRANTE, J., *Acta Met.* **19**, 743 (1971).

25. STEIN, D. F., WEBER, R. E., AND PALMBERG, P. W., *J. Metals* **23**, 39 (1971).

26. BUCKLEY, D. H., NASA TN D-2985, 1965.

Catalysis

Adsorption and Catalytic Reaction of Hydrogen on Chromia at Low Temperatures

ROBERT L. BURWELL, JR., AND KENNETH S. STEC

Ipatieff Laboratory, Department of Chemistry, Northwestern University, Evanston, Illinois 60201

Received May 19, 1976; accepted July 19, 1976

Some of the H_2 adsorbed on chromia (microcrystalline α-chromia, amorphous chromia, and Cr_2O_3/SiO_2) at temperatures below $-130°C$ is not removed by flowing helium at the temperature of adsorption, T. Such adsorbed hydrogen, $H_{irr}(T)$, increases monotonically as T is lowered from -130 to $-205°C$. $H_{irr}(-196)$, $CO_{irr}(-78)$, and activity for $H_2 + D_2 = 2HD$ at $-130°C$ increase in parallel as the temperature of activation in helium of chromia hydrogel is increased from 300 to 500°C. The hydrogel was pretreated with H_2 at 300°C. $CO_{irr}(-78)$ completely poisons $H_{irr}(-196)$. Pulses of D_2 at $-196°C$ completely and rapidly displace $H_2(g)$ from $H_{irr}(-196)$ without formation of HD. At $-163°C$ substantial HD is formed. The weak chemisorption of H_2 at lower temperatures appears to be a molecular polarization adsorption at $Cr^{3+}(cus)$ $O^{2-}(cus)$ ion pairs. Displacement by D_2 may involve adsorption to form H_2*D_2 followed by preferential release of H_2. An isotope effect of about 50 favors adsorption of D_2 vs H_2 at $-196°C$. The isotopic exchange at temperatures somewhat above $-196°C$ appears to involve rearrangement of H_2*D_2 to $HD*HD$ perhaps via $H^-*(HD_2)^+$.

INTRODUCTION

Studies of the adsorption of hydrogen on oxides started in the 1920's (1) and have continued to the present. One may say, as is conventional, that much has been learned but that many areas remain unresolved. Adsorption of hydrogen on chromia has been of interest throughout this period. The doctoral dissertation of one of the authors represented one of the earlier investigations on this subject and that of the other author, one of the latest. In the earlier investigation (2) which involved amorphous chromia activated *in vacuo* at 400°C, it was reported that adsorption of hydrogen at 0°C was very small, that physisorption appeared at lower temperatures and became large by $-185°C$, that beginning at about 75°C an adsorption of hydrogen became evident for which E_a and ΔH_{ads} were both about 10 kcal, and that another form of adsorption began at about 130°C with $E_a = 21$ kcal, independent of coverage, and $\Delta H_{ads} = 27$ kcal (3). The discovery that $H_2 + D_2 = 2HD$ proceeded on chromia at $-185°C$ (4) aroused considerable interest at that time. The later observation that chromia catalyzed the hydrogenation of ethylene at $-78°C$ also seemed surprising (5, 6).

More recently, there have been many investigations of adsorption and catalytic reactions of hydrogen on a variety of oxides at lower temperatures (meaning, in this article, at temperatures below 0°C). These investigations are too voluminous to review fully here, but mention may be made of more recent studies on ZnO (7, 8, 9), MgO (10, 11), Co_3O_4 (12), Al_2O_3 (13, 14, 15, 16), and V_2O_3 and MnO (9). The reaction, $H_2 + D_2 = 2HD$, was observed on all of these materials at lower temperatures although, for the latter two, only at about 0°C. In the first long period of the periodic system, activity for both $H_2 + D_2 = 2HD$ and $H_2 + C_2H_4 = C_2H_6$ peaks at Cr_2O_3 and Co_3O_4 (17).

Although the general idea that surface atoms are in some way unsaturated is old (1), it has recently become useful specifically to treat many types of catalytic sites as being coordinatively unsaturated surface (cus) ions or atoms (18). An electrically neutral particle of oxide constituted exclusively of M^{n+} and O^{2-} ions should contain both ions in its surface layer with coordination numbers lower than those in the bulk, i.e., both ions are cus. Exposure to water will hydroxylate such surfaces by the reaction, $O_s^{2-} + H_2O = 2O_sH^-$ (where the subscript s designates an atom in the surface layer). This reaction, carried to completion, forms a surface layer composed solely of hydroxide ions; it covers all M^{n+} ions, and it restores their coordination number to that in bulk. Complete hydroxylation may be slow even when the oxide is immersed in liquid water at 25°C (19). Oxides prepared in aqueous solution should be completely hydroxylated.

Heating an oxide whose surface is hydroxylated will generate $M^{n+}(cus)$ and $O^{2-}(cus)$ pairs by the reverse of the hydroxylation reaction, $2O_sH^- = O_s^{2-} + H_2O(g)$. All of the oxides mentioned above required pretreatment at higher temperatures to develop sites for adsorption or catalytic action and all are poisoned by exposure to water vapor. Carbon monoxide and oxygen chemisorb on activated chromia (18, 21) at −78°C to equal extents and they cannot be desorbed at that temperature. On amorphous chromia the quantity of such irreversible adsorption increases with the temperature of activation and it measures the amount of $Cr^{3+}(cus)$ (18).

As one might expect, probably generally, but best characterized on chromia, dehydroxylation appears to produce more than one kind of surface site and sites of different degrees of coordinative unsaturation (19, 22). Sites of different catalytic activity are also formed (18).

Increase in the oxidation number of surface metal ions of altervalent metals can permit the surface of an oxide to become covered with oxygen rather than with hydroxide, without formation of (cus) ions of metals, and yet with preservation of electrical neutrality. For example, either amorphous chromia or α-Cr_2O_3 exposed to oxygen at 350–400°C becomes covered with a monolayer of chromate (20).

This surface layer is reduced by hydrogen at 300°C.

As already mentioned for chromia, more than one form of adsorbed hydrogen is often observed. Kokes has observed two forms of adsorbed hydrogen on ZnO at room temperatures, one forming more rapidly (type I) and the other more slowly (type II). Amenomiya (15) has reported that five different types of adsorbed hydrogen appear on alumina and that some of these types involve the same sites. On zinc oxide, however, types I and II appear to involve different sites (8).

There has been considerable discussion of the nature of adsorbed hydrogen on metal oxides. Various structures have been proposed.

(1) Heterolytic dissociative adsorption (18).

$$H_2(g)$$
$$ H^- H^-$$
$$M^{n+}(cus) \quad O^{2-}(cus) \rightarrow M^{n+} \quad O$$

Consequent to infrared spectroscopic studies of adsorbed H_2, HD, and D_2 (7, 8), type I chemisorption on zinc oxide (an adsorption which occurs readily even at −78°C) provides the clearest example of heterolytic dissociative adsorption. Type I H_{ads} appears to be involved in the hydrogenation of olefins at room temperatures (7).

(2) Reductive adsorption. This might occur thus (18):

$$H_2(g)$$
$$ H^- H^-$$
$$O_s^{2-} \, M^{n+} \, O_s^{2-} \, M^{n+} \rightarrow O_s \, M^{(n-1)+} \, O_s \, M^{(n-1)+}$$

Here, H_2 is adsorbed as two protons and the electrons reduce one, or as shown above, two M^{n+} ions. Such adsorption is probably involved in reduction of oxides particularly in the reduction of higher to lower oxides. It might be followed by dehydroxylation. Further, it should be noted that reductive and heterolytic dissociative adsorption are isomeric.

(3) Oxidative adsorption. This can be represented as

$$H^- H^- H^- H^-$$
$$M^{(n+2)+} \quad \text{or} \quad M^{(n+1)+} \, M^{(n+1)+}.$$

The form at the left is analogous to *oxidative addition* of hydrogen to a number of coordination complexes, for example (24),

$$L_4Rh(I) + H_2 \rightarrow L_4Rh(III)H_2.$$

However, this reaction seems to be characteristic of d^8 and d^{10} species which are unlikely to be present in oxides of current catalytic importance. In general, oxidative chemisorption should be favored by oxides of altervalent metals in which the metal is in a low oxidation number, perhaps, for example, Cr(II)O. No formal evidence for oxidative adsorption has been reported.

(4) *Homolytic dissociative adsorption at* $2O_s^-$.

$$H_2$$
$$H^- \quad H^-$$
$$O_s^- \quad O_s^- \rightarrow O_s \quad O_s$$

This reaction has been proposed as occurring on activated MgO at temperatures above 25°C (23). The reverse reaction is reported at high temperatures (23) and unlike dehydroxylation it does not generate (*cus*) cations. It may be noted that $O^{2-} = O^- + e^-$ is strongly exothermic in the gas phase and that O^{2-} is stable in solid oxides because of the Madelung energy of the crystal. O^- should, therefore, be more stable at the surface than in bulk.

(5) *Molecular chemisorption at* −196°C. The best-analyzed example is a weak, readily reversible adsorption on zinc oxide (7) at −196°C. Infrared spectroscopy using H_2, HD, and D_2 establishes that the adsorption is molecular and presumably at Zn^{2+}(*cus*). A chemisorption on Co_3O_4 at −196°C which does not desorb at that temperature has been reported (12) and might turn out to be molecular. Most or all of the oxides mentioned above give a "magnetic" interconversion of ortho- and para-hydrogen at temperatures of liquid nitrogen by some kind of molecular adsorption. It may be noted that, on alumina, sites for the interconversion and for exchange between H_2 and D_2 at temperatures above −140°C develop in a parallel fashion during activation and the sites are similarly poisoned by carbon dioxide (16). It was suggested that the sites which give

molecular adsorption at −196°C are those which give some form of dissociative adsorption at higher temperatures. As measured by the poisoning, the site concentration is rather low, 0.1 site/nm² (per 100 Å).

The objective of the present investigation was the further elucidation of the nature of the adsorption of hydrogen on chromia at low temperatures and the correlation (we hoped) of this with the H_2–D_2 exchange reaction and the hydrogenation of olefins at low temperatures.

EXPERIMENTAL TECHNIQUES

Unless otherwise specified, our experiments started with amorphous chromia hydrogel which had been dried in air at 85°C, washed with water, and then dried at 135°C. It had been prepared by McDaniel, who showed that it contained 500 meq/mole Cr of excess oxygen beyond stoichiometric Cr_2O_3 (20). A chromia which had merely been dried at 85°C and which contained 127 meq of excess oxygen and a chromia originally prepared by G. J. Antos, which had been dried at room temperature and which was devoid of excess oxygen (20), gave similar results. The content in Cr_2O_3 of these hydrogels was 69.9, 65.0, and 60.3%, respectively, as determined by ignition. Ordinarily 0.2 g samples of the hydrogels were employed. The samples with excess oxygen were first reduced to Cr(III) by treatment for 6 hr at 300°C with flowing hydrogen diffused through palladium–silver alloy, swept with helium at a temperature rising to a final value, T_{act}, at which there was a hold for 12 hr, and cooled in helium. The resulting material is symbolized $Cr_2O_3(T_{act}$. Flow rates were 40 cm³/min unless otherwise stated and heating rates were 5°/min. The helium was diffused through the walls of fused silica capillaries in the helium purifier of Electron Technology, East Kearny, N. J., Model SLM-L. As shown by X ray diffraction, $Cr_2O_3(500)$ is microcrystalline α-Cr_2O_3, but samples of $T_{act} < 450°C$ are amorphous.

Only helium, hydrogen, and deuterium (which was purified like hydrogen) were passed continuously over the activated chromia. These gases were additionally purified by passage through a trap of Davison Grade 59 silica gel at $-196°C$. It was essential to keep the content in possible poisons such as oxygen and water below 0.1 ppm and to monitor impurities regularly to ensure that this limit was not exceeded. Content in oxygen was determined by passing helium, H_2 or D_2 through a trap of silica gel at $-196°C$ for about 15 min and then releasing the adsorbed oxygen for catharometric measurement. The released pulse could be separated into N_2 and O_2 on a column of Linde molecular sieves 5A. With carrier gases of adequate purity, the fractional loss of surface sites was negligible during the experiments to be described.

No greased stopcocks were used upstream of the chromia samples, only Nupro bellows valves. The carrier gas from the silica gel trap was passed through a Carle injection valve which was purged with helium to eliminate the effects of any leaks and then over the sample of chromia held in one arm of a U-tube of fused silica and attached to the apparatus by Swagelock fittings with Teflon and Zytel ferrules. H_2, D_2, HD, CO, O_2, or Ne could be introduced into the 0.104 cm³ injection loop. Carbon monoxide and hydrogen could be removed from the flowing helium beyond the chromia by passage through a trap of molecular sieve, Linde 5A, at $-196°C$ and released for measurement by warming the trap to $90°C$ for carbon monoxide and to room temperature for hydrogen. Pulses of gas which passed the chromia or which were liberated from traps were measured in a Gow–Mac four filament catharometer, Model 10-292-6, potted for immersion in water. H_2, HD, and D_2 in helium were converted to water before entering the catharometer by copper oxide at $550°C$. Analysis for H_2, HD, and D_2 in gas released from chromia was effected by trapping on Linde 5A followed by release of the mixture as a pulse which was passed over $FeCl_3/Al_2O_3$ for chromatographic separation (25). Samples from mixtures of $H_2 + D_2$ passed over chromia

were injected into helium carrier for similar analysis. The reaction, $H_2 + D_2 = 2HD$, was also investigated by passage of pulses of D_2 over the catalyst in H_2 carrier. The resulting pulse of HD + D_2 was separated by passage through a 60 cm column of Linde 5A (26) and passed directly to the catharometer.

EXPERIMENTAL RESULTS

Irreversibly Adsorbed Hydrogen

Initial pulses of H_2 injected into helium passing over activated chromia at $-196°C$ are completely adsorbed and no H_2 appears beyond the catalyst in 1.5 hr. Later pulses break through. If chromia is exposed to H_2 at $-196°C$, swept with helium at $-196°C$, and warmed, the released pulse of H_2 can be trapped and later measured. It decreases in magnitude by about 20% as the sweep time by helium increases from 3 to 15 min but further increase in sweep time to 4 hr produces negligible further decrease. The amount of H_2 held after a sweep time of 30 min will be designated by $H_{irr}(T)$ measured in micromoles per gram of starting hydrogel where T is the temperature of the adsorption experiment. $H_{irr}(-196)$ was the same for times of passage of H_2 over chromia varying from 5 to 240 min, for p_{H_2} between 0.1 and 0.8 atm, and for passage of 10 pulses of H_2. The latter procedure for exposure to H_2 was experimentally most convenient and was the one usually used.

Mere removal of the low temperature bath effected complete release of H_{irr} as was shown by mole balance between the moles of H_2 in the pulses, the moles of H_2 which passed through the bed of chromia unadsorbed, and the moles of H_2 released upon warming. No further H_2 was released upon warming the chromia sample from room temperature to $300°C$. In temperature programmed desorption at $5°/min$ from $-196°C$, H_2 was released from $Cr_2O_3(340)$ and $Cr_2O_3(440)$ as broad pulses without structure and with maximum intensities at about $-170°C$. The pulse from $Cr_2O_3(340)$ reached zero at $-120°C$; the

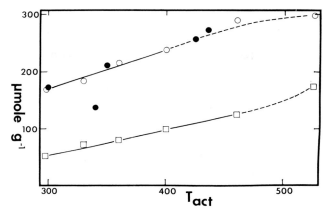

FIG. 1. $CO_{irr}(-78)$ (empty circles, this work; full circles, Ref. (18)) and $H_{irr}(-196)$ (squares) in moles per gram of chromia hydrogel vs T_{act}. The sample of this work was successively activated in helium to higher temperatures and measurements of adsorption made after each activation.

larger pulse from $Cr_2O_3(440)$ reached zero at $-90°C$.

The sample of chromia devoid of excess oxygen gave the same value of $H_{irr}(-196)$ whether or not it was exposed to hydrogen at 300°C during activation.

Activated chromias were exposed to carbon monoxide or oxygen at $-78°C$ for 1 hr and purged with helium for 30 min. $H_{irr}(-196)$ was zero on such samples. The value of $CO_{irr}(-78)$ was measured by exposure to CO at $-78°C$ followed by warming to 200°C and measuring the evolved pulse. Mass balance showed that all CO_{irr} was released. $CO_{irr}(-78)$ was zero on silica gel.

Figure 1 exhibits $CO_{irr}(-78)$ and $H_{irr}(-196)$ as $f(T_{act})$. Figure 2 exhibits the values of $H_{irr}(T)$ and $D_{irr}(T)$ as $f(T)$ for $Cr_2O_3(351)$ and $Cr_2O_3(501)$.

If 4.58% CrO_3 is deposited on Davison Grade 62 silica gel, 60–80 mesh, and the material is reduced in H_2 and activated in helium in the same fashion as chromia, a Cr_2O_3/SiO_2 gel results which behaves like chromia except that the value of $H_{irr}(-196)$ increases with time of exposure. However, silica gel itself holds H_2 irreversibly at $-196°C$ but adsorption, presumably in small micro-

pores, is slow. Thus, per gram of silica gel alone, irreversible adsorption of H_2 amounts to about $3 \mu mole/g$ after an exposure to hydrogen for 5 min. This increases linearly to 7.5 at an exposure for 30 min. The rate of increase of irreversibly adsorbed H_2 with time

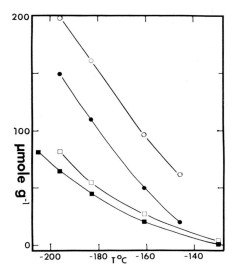

FIG. 2. Values of $H_{irr}(T)$ (full symbols) and $D_{irr}(T)$ (empty symbols) in micromoles per gram of hydrogel for $Cr_2O_3(351)$ (circles) and α-$Cr_2O_3(501)$ (squares).

61

TABLE I

Composition of Exit Pulses from Passage over $Cr_2O_3(350)$

Pulse no.	D_2 pulses on $H_{irr}(-196)$			D_2 pulses on $H_{irr}(-161)$		
	H_2 (%)	HD (%)	D_2 (%)	H_2 (%)	HD (%)	D_2 (%)
1	100	0	0	100	0	0
2	100	0	0	32	20	48
3	82	0	18	0	3	97
4	0	0	100	0	0	100
Desorb[a]	0	0	100	0	0	100
	H_2 pulses on $D_{irr}(-196)$			H_2 pulses on $D_{irr}(-161)$		
1	80	0	20	68	8	25
2	87	0	13	72	12	17
3	94	0	6	78	21	1
4	100	0	0	83	16	1
5	100	0	0	80	20	0
6	100	0	0	87	13	0
7	100	0	0	93	7	0
8	100	0	0	100	0	0

[a] Composition of gas liberated from the chromia on warming.

of exposure is the same for Cr_2O_3/SiO_2, i.e., subtraction of the curve for adsorption on silica gel from that for Cr_2O_3/SiO_2 gives a line independent of time. For $Cr_2O_3/SiO_2(408)$ this time independent quantity is 17 μmole/g. Exposure to H_2 via 10 pulses also gives 17 mole/g. Thus, adsorption of H_2 is slow enough that the adsorption from several pulses is negligible.

At $-196°C$ neon pulses passed over $Cr_2O_3(310)$ and $Cr_2O_3(398)$ exhibited no irreversible adsorption and no significant broadening of the pulses. Neon pulses liberated no hydrogen when passed over these catalysts treated to contain $H_{irr}(-196)$.

Reaction of $D_2(g)$ with H_{irr}

Four pulses of D_2 were passed at $-196°C$ over $Cr_2O_3(352)$ containing $H_{irr}(-196)$ directly to the catharometer. There was no structure in the exit pulses. The experiment was repeated but the pulses were trapped, released, and passed through the $FeCl_3/Al_2O_3$ column. Results appear in Table I. The chromia which had been dried at 85°C behaved similarly. In a similar experiment with

$H_{irr}(-161)$, HD was present in the exit pulses. Pulses of HD were similarly passed over $Cr_2O_3(352)$ containing $H_{irr}(-196)$ at $-196°C$. The first two exit pulses contained only H_2, the third pulse was mixed, and the fourth pulse consisted only of HD. Pulses of D_2 were then injected. The exit pulses consisted of HD, then HD + D_2, and then D_2 only. Upon warming, only D_2 was released. On this chromia, $H_{irr}(-196)$ was 83.1, $HD_{irr}(-196)$ was 90.0, and $D_{irr}(-196)$ was 100 μmole/g.

Pulses of H_2 were passed over $Cr_2O_3(348)$ containing $D_{irr}(-196)$ and $D_{irr}(-161)$ with results shown in Table I. Upon warming the sample at $-196°C$, 37% of the original deuterium was released as a mixture of D_2 and HD. However, the HD was formed during the warming since if, in a similar experiment, the final desorption step was replaced by passage of pulses of D_2, only H_2 and D_2 were observed in the exit pulses. Sensitivity for small amounts of H_2, HD, and D_2 was 1–2% although sensitivity for small amounts of HD was somewhat less when H_2 was large.

Pulses of D_2 were passed over $Cr_2O_3(525)$ and $Cr_2O_3/SiO_2(375)$ containing $H_{irr}(-196)$.

Behavior like that in Table I was observed. Silica gel, Davison Grade 62, was exposed to flowing helium at 350°C for 12 hr, to H_2 at −196°C for 30 min, and then swept with helium at that temperature for 30 min. Pulses of D_2 displaced no H_2.

At −196°C pulses of an equimolar mixture of H_2 and D_2 were passed over $Cr_2O_3(369)$. The first three pulses were completely adsorbed. The next five passed the chromia and contained only H_2. The eighth pulse consisted of $H_2/D_2 = 0.2/0.8$ and the ninth of H_2/D_2 = unity. Warming liberated D_2 only. In similar experiments, 60 pulses of a mixture of H_2 and D_2 richer in H_2 were passed over $Cr_2O_3(369)$ at −196°C, swept with helium for 30 min, warmed, and the evolved pulse analyzed. In a typical experiment, the composition of the inlet pulse was about H_2/D_2 = 0.95/0.05 and that of the gas evolved on warming was about 0.3/0.7. The ratio $[D_{irr}/H_{irr}]/[D_2(g)/H_2(g)]$ is, thus, about 50, but the experiment is not one of high precision.

Pulses of D_2 were passed over Cr_2O_3 at various increasing temperatures and the exit pulse was analyzed. HD was barely detectable at 0°C and undetectable below 0°C. The content in HD was 10% at 40°C and 22% at 100°C (see also Ref. (9)). The experiment was also run on $Cr_2O_3(302)$ for which the reduction by hydrogen at 300°C had been omitted. At the same temperature formation of HD was about one-fifth as large as on the sample activated normally. On the other hand, no HD was observed at 100°C upon passage of a pulse of deuterium over $Cr_2O_3/SiO_2(347)$. A pulse of D_2 was trapped on a sample of $Cr_2O_3(347)$ at −183°C for 6 hr. No HD was evolved on warming.

A pulse of HD was passed at −196°C over $Cr_2O_3(348)$ and $Cr_2O_3/SiO_2(385)$ containing $HD_{irr}(−196)$ and the exit pulse was analyzed. It contained only HD. The sample of $Cr_2O_3(348)$ was isolated at −196°C with an amount of HD in excess of HD_{irr}. After 5 hr, helium flow was restored and pulses of D_2 were injected. Trapping and analysis showed that 7% of the HD had been converted to H_2.

Isotopic Exchange between H_2 and D_2

The exchange reaction was examined by two methods. In method A an equimolar mixture of H_2 and D_2 was passed continuously through a bed of activated chromia usually at 40 cm³/min and the product was sampled periodically. No HD could be detected at −183°C, conversion to HD at −130°C was useful, and equilibrium was reached at −78°C. In method B pulses of D_2 were injected into H_2 flowing continuously over the bed and usually at 40 cm³/min. The effluent pulse was analyzed for HD and D_2. At −196°C no HD could be detected, the conversion to HD at −183°C was useful, and the conversion was nearly the equilibrium one at −130°C.

In method A the data were converted into a rate constant by the equation

$$k_1 = L \ln [HD_\infty/(HD_\infty - HD)],$$

where L is the flow rate in cubic centimeters (STP) of hydrogen per minute per gram, HD represents the fraction HD in the product, and HD_∞, the equilibrium fraction. In 11 runs with $Cr_2O_3(298)$, k_1 was 95 cm³ min⁻¹ g⁻¹ over values of L ranging from 100 to 275 cm³ min⁻¹ g⁻¹. The variation in k_1 was ±10% and there was no correlation with the values of L. In these experiments on 0.2 g of hydrogel, at 40 cm³/min the conversion was about 20%.

In method B, the data followed the equation

$$k_0 = L(n_{HD}),$$

where n_{HD} is the number of moles of HD in the product pulse, over a range of L from 140 to 300 cm³ min⁻¹ g⁻¹ on $Cr_2O_3(298)$. The conversion was about 10% at $L = 200$ cm³ min⁻¹ g⁻¹. Further, k_0 was the same with a 0.260 cm³ pulse as with the usual 0.104 cm³ pulse. The value of k_0 was measured as a function of T_{act} by successively activating a sample of chromia from 300 to 525°C. The value of k_0 increased from an initial value of 290 mole cm³ min⁻¹ g⁻¹ to 750 at 365°C following which k_0 decreased to 75 at 525°. A duplicate run checked well and a sample activated directly to 525°C gave the same value of k_0 as the first sample.

The effect of adsorbed carbon monoxide upon k_0 measured by method B and upon $H_{irr}(-196)$ was studied by covering $Cr_2O_3(362)$ with $CO_{irr}(-78)$. Both k_0 and $H_{irr}(-196)$ were zero on this material. In measuring H_{irr}, the adsorbent was warmed only to $-78°C$. The catalyst was then swept with helium for 30 min at $-40°C$, the evolved carbon monoxide was trapped and measured, and k_0 and H_{irr} were determined. This process was repeated after sweeping with helium at 0, 25, 60, 90, 150, and 200°C. The total quantity of carbon monoxide evolved up to 200°C was 216 μmole/g. After the sweep at 200°C, $H_{irr}^0(-196)$ was 92.6 μmole and k_0^0 was 952 mole cm³ min⁻¹ g⁻¹. Results are presented in Fig. 3. Results were similar on $Cr_2O_3(450)$ except that the slope of the straight line segment of H_{irr} vs $(1 - \theta_{CO})$ near $1 - \theta_{CO} = 1$ was 2.0 rather than 2.5 and the curve for k_0/k_0^0 bowed above the thick dashed line of unit slope.

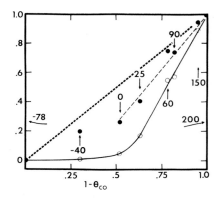

FIG. 3. $H_{irr}(-196)/H_{irr}^0(-196)$ (full line, empty circles) and k/k_0^0 (dashed line, full circles) vs $1 - \theta_{CO}$. The numbers are the temperatures (in °C) to which the sample was heated successively after initial saturation with carbon monoxide at $-78°C$. The chromia is $Cr_2O_3(362)$.

DISCUSSION

Molecular Chemisorption of Hydrogen on Chromia

On activated chromia (microporous, amorphous of about 250 m² per g of anhydrous gel; mesoporous, microcrystalline α-Cr_2O_3 of about 80 m² per g; and Cr_2O_3/SiO_2) some of the hydrogen adsorbed at temperatures below $-130°C$ is not removed by flowing ultrapure helium in a period of several hours. Adsorption is initially fast at $-196°C$ but complete saturation requires about one minute. If activated chromia at a temperature T below $-130°C$ is exposed to hydrogen, swept with helium until the rate of desorption of hydrogen has become negligible, and then warmed, hydrogen is desorbed into the flowing helium and desorption is complete by about $-100°C$. The evolved H_2 measures what will be called $H_{irr}(T)$.

At $-196°C$, $D_2(g)$ but not neon displaces $H_2(g)$ from $H_{irr}(-196)$. Presaturation of the surface at $-78°C$ by carbon monoxide or oxygen reduces $H_{irr}(-196)$ to zero. The formation of H_{irr} is, thus, site selective and the

efficacy of displacement by pulses of another gas does not follow a physisorption sequence. The adsorption would appear to be best classified as a weak chemisorption.

Since $H_{irr}(-196)$ and $CO_{irr}(-78)$ increase in parallel with increasing T_{act} as shown in Fig. 1, H_{irr} involves the same sites as CO_{irr}. $H_{irr}(T)$ decreases monotonically as T increases (Fig. 2) but even at $-205°C$, the lowest temperature investigated, $H_{irr}(T)/CO_{irr}(-78)$ is only about 0.5. Presumably, during the first step in preparing $H_{irr}(-196)$, the sites for adsorption of carbon monoxide would be fully covered which requires, for $Cr_2O_3(400)$, 8 cm³ of hydrogen, a substantial fraction but less than all of the adsorbed hydrogen reported at higher pressures of hydrogen at $-196°C$ (2, 21). If the pressure of hydrogen is reduced at $-196°C$ and the temperature is then raised during passage of helium, the adsorption sites would be progressively depopulated in the sequence weak to strong in capacity to form *CO. This is demonstrated by the linear section of slope 2.5 in $H_{irr}(-196)/H_{irr}^0(-196)$ vs $1 - \theta_{CO}$ near $\theta_{CO} = 0$ in Fig. 3. Since $CO_{irr}(-78)/H_{irr}(-196)$ is 2.5 for $Cr_2O_3(362)$ (Fig. 1), the most strongly held carbon monoxide eliminates adsorption of hydrogen molecule for molecule until nearly all $H_{irr}(-196)$ has been eliminated. Although adsorbed carbon monoxide is held more strongly than adsorbed hydrogen, $CO_{irr}(T)$ similarly decreases with T from $\theta_{CO} = 1$ at $-78°C$ to $\theta_{CO} = 0$ at $200°C$ (Fig. 3).

The rate of desorption of hydrogen or carbon monoxide into flowing helium must follow an equation of the general character of

$$d\theta/dt = (\text{constant}) \exp(-E_{des}^0/RT) \exp(-a/\theta),$$

where θ is fractional coverage, a is a constant, and E_{des}^0 is the activation energy at some fixed value of θ. At any given T, $d\theta/dt$ decreases with time until it reaches a value of m which is indistinguishable from zero in our technique. The residual value of θ corresponds to $H_{irr}(T)$ or $CO_{irr}(T)$. If the temperature is now increased, $d\theta/dt$ increases and then sinks to m as θ falls to a new and lower value. A series of nonequilibrium steady states results.

Adsorption of Deuterium

As shown in Fig. 2, $D_{irr}(T)/H_{irr}(T)$ is greater than unity, it increases with T, and it is larger on α-$Cr_2O_3(501)$ than on $Cr_2O_3(351)$.

As shown in Table I, at $-196°C$, pulses of D_2 passed over $H_{irr}(-196)$ liberate H_2 devoid of HD and form $D_{irr}(-196)$. The reaction is rapid and complete. Also $HD(g)$ displaces $H_{irr}(-196)$ and D_2 displaces $HD_{irr}(-196)$. Pulses of H_2 displace D_2 from $D_{irr}(-196)$ again without formation of HD, but the reaction is slow and incomplete (Table I). However, passage of H_2 over $D_{irr}(-196)$ for 10 min leads to the complete displacement of D_{irr}. The isotope effect upon absorption was measured by passing a mixture of $H_2/D_2 = 19$ over activated chromia, sweeping with helium and warming. The adsorption of D_2 vs H_2 is favored by a factor of about 50. Such a large isotope effect must result from being able to run a chemical reaction at $-196°C$ (27).

The complete absence of HD during displacement reactions at $-196°C$ suggests that the chemisorption of hydrogen is molecular

and that the displacement can be formulated

$$\begin{array}{cccc}
\text{H}_2(\text{g}) & \text{D}_2(\text{g}) & & + \text{H}_2(\text{g}) \\
\longrightarrow & \longrightarrow & \longrightarrow & \\
& \text{H} & \text{D H} & \text{D} \\
* & * \mid & \mid * \mid & \mid * \\
& \text{H} & \text{D H} & \text{D}
\end{array} \quad [1]$$

Displacement of H_2 by D_2 under conditions where helium does not elute H_2 and the higher boiling neon does not displace hydrogen would seem to require some specific interaction between D_2 and the site to which $H_2(ads)$ is attached. In the formulation above, the binding of H_2 in D_2*H_2 should be weaker than in $*H_2$. D_2*H_2 decomposes selectively to $H_2(g)$ and D_2*. There are parallels to such displacement in coordination chemistry. For example, $(acac)Rh(C_2H_4)_2$ (acac is the conjugate base of acetylacetone), a 16-electron complex, exchanges rapidly with C_2D_4 and almost surely by the intermediate, $(acac)Rh(C_2H_4)_2(C_2D_4)$. NMR observations indicate that formation and decomposition of the tris–ethylene complex in chloroform in the presence of ethylene is rapid at $-58°C$ (28). There is an alternate mechanism for displacement which, although it cannot be excluded, appears less likely on the basis of present data: $H_2* \rightleftarrows * + H_2(g)$ followed by $* + D_2(g) \rightarrow D_2*$. At the end of the helium sweep, P_{H_2} would be less than 10^{-6} atm yet adsorption–desorption would have to be assumed so fast that D_2 rapidly replaces H_2. Should, however, this mechanism prove correct, a relation for $d\theta/dt$ of the same form as that given earlier in this article would still be required, but E_{des}^0 would be replaced by ΔH_{des}^0, the enthalpy of desorption.

Isotopic Exchange between H_2 and D_2

As shown in Table I, substantial but much smaller than equilibrium amounts of HD result from passage of pulses of D_2 over H_{irr} at $-161°C$. Larger amounts of HD appear when pulses of H_2 are passed over D_{irr} at $-161°C$. At $-181°C$ about one-half so much HD is formed from $D_2 + H_{irr}$ as at $-161°C$.

When an equimolar mixture of H_2 and D_2 is passed over chromia (method A), convenient conversions to HD are obtained at $-130°C$. When pulses of D_2 are injected into H_2 carrier passing over chromia (method B), convenient conversions occur at $-183°C$. In method A about 10 cm³ of HD is made per minute; in B, only 0.01 cm³.

In method A the rate follows the conventional first-order approach to equilibrium and k_1 increases monotonically by a factor of 2 rather closely paralleling $CO_{irr}(-78)$ (Fig. 1) as T_{act} increases from 300 to 500°C. Method B must be complicated in detail. D_2 moves slowly through the bed by successive adsorptions and desorptions by displacement. Thus, the effect of T_{act} is difficult to interpret. However, interpretation of the poisoning experiments by carbon monoxide seems straightforward.

On both $Cr_2O_3(362)$ (Fig. 3) and $Cr_2O_3(450)$, $H_{irr}(-196)$ is nearly zero at $\theta_{CO} = 0.5$, but $k_0/k_0^0 = 0.25$ for $Cr_2O_3(362)$ as against 0.75 for $Cr_2O_3(450)$. Thus, on $Cr_2O_3(362)$ isotopic exchange by method B goes primarily on the sites which adsorb carbon monoxide most strongly, but on $Cr_2O_3(450)$, primarily on those which adsorb carbon monoxide most weakly. On the latter catalyst, even at $\theta_{CO} = 0.8$, $k_0/k_0^0 = 0.45$. As judged by $H_{irr}(-196)/CO_{irr}(-78)$, the sites for adsorption of hydrogen are relatively stronger at higher T_{act} (Fig. 1).

The Nature of H_{irr} and of the Isotopic Exchange Reaction

The reaction, $H_2 + D_2 \rightarrow 2HD$, must proceed at low temperatures without formation of any intermediate of energy greater or less than that of $H_2 + D_2$ by more than a few kilocalories (if greater, the reaction would be too slow; if less, the surface would become poisoned). This consideration places a serious constraint upon mechanism.

On the basis of the available evidence, we suggest that hydrogen is adsorbed molecularly at lower temperatures by some form of polarization interaction (7, 16, 29) and probably between $Cr^{3+}(cus)$ and $O^{2-}(cus)$ as shown by A

below, where $\oplus$ is $Cr^{3+}(cus)$ and $\ominus$ is $O^{2-}(cus)$.

$$
\begin{array}{cccc}
\oplus\quad\ominus & \overset{D-D}{\oplus\quad\ominus} & \oplus H \overset{\cdot\cdot D+}{\underset{\cdot H'}{\ominus D}} & \overset{H-D}{\oplus\quad\ominus} \\
H-H & H-H & & H-D \\
\mathbf{A} & \mathbf{B} & \mathbf{C} & \mathbf{D}
\end{array}
$$

B is the intermediate or transition state in the displacement, $D_2(g) + H_{irr}$. At temperatures of $-183°C$ and above, it is assumed to rearrange slowly to D. This results in isotopic exchange via some species like C, here shown as hydrogen bonded.

An attempt to introduce dissociatively adsorbed forms of hydrogen incurs difficulties. Except for chromias activated at the highest temperatures, the chromias of this work were relatively slightly dehydroxylated. For example, $Cr_2O_3(400)$ has only one (cus) pair per nm² out of a potential number of ten (18). Dissociative adsorption of deuterium in either the heterolytic or the reductive form would generate O_sD^- amid many O_sH^- yet no isotopic exchange of O_sH^- occurs below 0°C. The assumption that only a small fraction of the O_sD^- formed are equivalent to any adjacent O_sH^- so that desorption can only regenerate the molecule which originally adsorbed appears rather artificial. It might be assumed that formation of HD proceeds by dissociative heterolytic or reductive adsorption on such a small fraction of the surface that isotopic exchange with surface O_sH^- is invisible. Indeed, it has been proposed that isotopic exchange on magnesium oxide occurs on a very small fraction of the surface (11). However, it seems unlikely that sites of such highly special properties would be distributed over the wide range of sites for adsorption of carbon monoxide and hydrogen shown by Fig. 3.

One could imagine that displacement adsorption in $D_2(g) + H_{irr}(-198)$ results from a very large isotope effect favoring desorption of H_2 from $H(ads) + D(ads)$. Were such the case, however, $HD(g) + HD_{irr}$ and $H_2 + D_{irr}$ would give H_2 and HD respectively at $-196°C$. Neither is observed.

Oxidative addition on close $Cu^{3+}(cus)$ ion pairs (31) might be imagined. However, displacement adsorption would necessitate the addition of a second molecule of $H_2(D_2)$ to form $Cr^{5+}(H^-)_2$ for which there is no precedent in inorganic chemistry. Further, $CO_{irr}(-78)$ should occur on isolated $Cr^{3+}(cus)$ as well as on $Cr^{3+}(cus)$ close pairs. One would expect that $Cr^{3+}(cus)$ pairs would be generated at higher temperatures than isolated $Cr^{3+}(cus)$, and yet $CO_{irr}(-78)$, $H_{irr}(-196)$, and k_1 by method A increase in a nearly parallel fashion with increasing T_{act}.

Similar problems appear with the proposed mechanism involving heterolytic dissociative adsorption of hydrogen in the hydrogenation of ethylene at -100 to $0°C$ (18). Since activity for hydrogenation rises much more rapidly with T_{act} than does $CO_{irr}(-78)$ (18), only some of the sites for adsorption of carbon monoxide are active in hydrogenation and possibly the fraction is so small that exchange of O_sH^- is invisible. Work by Antos (32) showed that hydrogenation of ethylene and propylene at low temperatures is accompanied by extensive formation of polymer which deactivates the catalyst rather rapidly. This suggests that $C_2H_5(ads)$ is formed and undergoes insertion by ethylene to form polymer. This could result from the presence of some form of $H(ads)$ or of a reaction like

$$C_2H_4(ads) + H_2(ads) \rightarrow C_2H_5(ads) + H(ads).$$

ACKNOWLEDGMENT

Mr. Edson L. Blackman constructed some of the apparatus used in this research and ran some useful experiments preliminary to those described in this paper.

REFERENCES

1. TAYLOR, H. S., Chem. Rev. 9, 1 (1931).
2. BURWELL, R. L., JR., AND TAYLOR, H. S., J. Amer. Chem. Soc. 58, 697 (1936).
3. HOWARD, J., AND TAYLOR, H. S., J. Amer. Chem. Soc. 56, 2259 (1934).
4. GOULD, A. J., BLEAKNEY, W., AND TAYLOR, H. S., J. Chem. Phys. 2, 362 (1934).
5. WELLER, S. W. AND VOLTZ, S. E., J. Amer. Chem. Soc. 76, 4695 (1954).
6. BURWELL, R. L., JR., LITTLEWOOD, A. B., CARDEW, M., PASS, G., AND STODDART, C. T. H., J. Amer. Chem. Soc. 82, 6272 (1960).
7. KOKES, R. J., Accounts Chem. Res. 6, 226 (1973).
8. NAITO, S., SHIMIZU, H., HAGIWARA, E., ONISHI, T., AND TAMARU, K., Trans. Faraday Soc. 67, 1519 (1971).
9. SHIGEHARA, Y. AND OZAKI, A., Bull. Chem. Soc. Japan 45, 634 (1972).
10. ELEY, D. D. AND ZAMMITT, M. A., J. Catal. 21, 377 (1971).
11. BOUDART, M., DELBOUILLE, A., DEROUANE, E. G., INDOVINA, V., AND WALTERS, A. B., J. Amer. Chem. Soc. 94, 6622 (1972).
12. SHIGEHARA, Y. AND OZAKI, A., J. Catal. 21, 78 (1971).
13. ACRES, G. J. K., ELEY, D. D., AND TULLO, J. M., J. Catal. 4, 12 (1965).
14. OHNO, S. AND YASUMORI, I., Bull. Chem. Soc. Japan 41, 2227 (1968).
15. AMENOMIYA, Y., J. Catal. 22, 109 (1971).
16. HALL, W. K., Accounts Chem. Res. 8, 257 (1975).
17. HARRISON, D. L., NICHOLLS, D., AND STEINER, H., J. Catal. 7, 359 (1967); DOWDEN, D. A., McKENZIE, N., AND TRAPNELL, B. M. W., Proc. Roy. Soc. London Ser. A 237, 245 (1956).
18. BURWELL, R. L., JR., HALLER, G. L., TAYLOR, K. C., AND READ, J. F., Advan. Catal. 20, 1 (1969).
19. ZECCHINA, A., COLUCCIA, S., GUGLIELMINOTTI, E., AND GHIOTTI, G., J. Phys. Chem. 75, 2774 (1971).
20. McDANIEL, M. P. AND BURWELL, R. L., JR., J. Catal. 36, 394 (1975).
21. MacIVER, D. S. AND TOBIN, H. H., J. Phys. Chem. 64, 451 (1960).
22. PERI, J. B., J. Phys. Chem. 78, 588 (1974).
23. GIESEKE, W., NÄGERL, H., AND FREUND, F., J. Phys. Chem. 78, 758 (1974).
24. COTTON, F. A. AND WILKINSON, G., "Advanced Inorganic Chemistry," p. 773. Interscience, New York, 1972.
25. SHIPMAN, G., Anal. Chem. 34, 877 (1962).
26. OHKOSH, S., FUJITA, Y., AND KWAN, T., Bull. Chem. Soc. Japan 31, 770 (1958).
27. BURWELL, R. L., JR. AND STEC, K. S., J. C. S. Chem. Comm. 577 (1976).
28. CRAMER, R., J. Amer. Chem. Soc. 86, 217 (1964); CRAMER, R. AND SEWELL, L. P., J. Organometal. Chem. 92, 245 (1975).
29. CHANG, C. C., DIXON, L. T., AND KOKES, R. J., J. Phys. Chem. 77, 2634 (1973).
30. DOWDEN, D. A. AND WELLS, D., in "Actes du Deuxiéme Congrès International de Catalyse," Vol. 2, p. 1499. Technip, Paris, 1961.
31. STONE, F. S., Chimia 23, 490 (1969).
32. ANTOS, G. J., Doctoral Dissertation, Northwestern University, 1973.

The Nature of Weak Adsorption on Transition Metals

B. E. NIEUWENHUYS AND W. M. H. SACHTLER

Gorlaeus Laboratoria, Rijksuniversiteit Leiden, P.O. Box 75, Leiden, The Netherlands

Received March 8, 1976; accepted August 17, 1976

Results on the adsorption of nitrogen and xenon on various crystal faces of platinum and iridium, as obtained by the field emission–probe hole technique, are reviewed in conjunction with photoelectron emission data on xenon adsorption on films of various fcc metals. These data are analyzed in order to discriminate between different models on the nature of weak adsorption on metals. The popular model which attributes this adsorption solely to the presence of B_5 sites is disproved. Xenon adsorption on individual crystal faces of Ir remains limited to certain sites for the atomically rough faces, but a close-packed Xe layer is formed on the smooth faces at high coverage. The nature of the adsorption bond of xenon on fcc metals appears to be a combination of a chemical "charge transfer–no bond" interaction with a bonding by dispersion forces.

I. INTRODUCTION

In processes of heterogeneous catalysis, molecules are chemisorbed on a surface and the adsorbed atoms are subsequently desorbed in different combinations. It is, therefore, clear that the interest of catalytic chemists has been focused on the nature of strong chemical adsorption, where strong chemical bonds are formed and broken. Much less interest has been devoted to weak adsorption such as, for instance, rare gases.

For the former case of strong chemisorption a number of conclusions have been drawn: The strength of the chemisorption bond depends on the chemical nature of the absorbent and it is often very specific for the nature of the adsorbing crystal face. The heat of adsorption is particularly large if the surface atoms have a high degree of coordinative unsaturation. Strong chemisorption is confined to sites and the rules of chemical valency determine the stoichiometry of the adsorbent/adsorbate complexes. The number of molecules which can be adsorbed on a given surface without corrosion is, hence, a consequence of both geometry and stoichiometry. The polarity of the adsorption bond is determined by the electronegativity of the atoms involved;

oxygen, nitrogen, hydrogen, and halogens when strongly chemisorbed on metals are usually charged negatively with respect to the metal surface atoms. In consequence, the electronic work function Φ is increased by this type of chemisorption, i.e., the surface potential sp,

$$sp = -\Delta\Phi,$$

is usually negative.

By contrast, weak adsorption has enjoyed the interest of catalytic chemists only as a practical tool to determine the surface area of a solid. This important application is based on the belief that this adsorption is virtually independent of the chemical nature of the adsorbent and is not confined to certain sites, but a close-packed monolayer can be defined where the surface area per adsorbed molecule is constant, i.e., ignoring both the chemical and the crystallographic nature of the adsorbing surface. To emphasize the independence of any chemical parameters, the term "physical adsorption" is often used for adsorptions with a heat release of less than 10 kcal/mole of adsorbate. It has often been assumed that Van der Waals–London forces are responsible for this type of adsorption. A striking phenomenological aspect is the large positive surface potential first discovered by Mignolet (1).

However, with the exception of this aspect, most of the mentioned criteria of physical adsorption are based more on intuition than on unambiguous facts. Indeed, serious doubts have been brought forward with respect to some of the above statements. Field emission data by Ehrlich *et al.* (2, 3) and Sachtler *et al.* (4) provided evidence that the adsorption of Kr and Xe on tungsten is highly face specific and on some faces presumably confined to sites. On films, Ponec *et al.* (5) and Brennan *et al.* (6) found equal adsorbate/metal atom ratios for Kr and Xe in spite of their widely different sizes. For the weak adsorption of nitrogen on metals of Group VIII of the periodic table, van Hardeveld and van Montfoort (7) even postulated that this adsorption is confined to specific sites on the surface called B_5-sites which are present on (110) and rougher faces but are absent on (111) and (100) faces.

That Van der Waals–London forces cannot be the sole cause of "physical" adsorption was stated by various authors. Mignolet (8, 9) argued that Mulliken's charge transfer–no bond model (10) provided a better description, and Ehrlich (11) reasoned that for Van der Waals–London bonds the relative heat of adsorption of nitrogen and krypton should be given by their relative polarizabilities, which is in contrast to the experimental facts. Hall (12), Klemperer, and Snaith (13) and Müller (14, 15) argued that polarization by a hypothetical surface field was responsible for both the high positive surface potential and the heat of adsorption, exceeding the value attributable to Van der Waals–London forces. van Oirschot and Sachtler (16) drew attention to the strong dependence of Xe adsorption on the chemical nature of the adsorbent; on Ru (17) it was very difficult to remove Xe at $T > 100°K$ by pumping, but on potassium no detectable adsorption took place at 78°K since on this metal the heat of adsorption is apparently inferior to the heat of sublimation of Xe.

These doubts concerning the simple physical nature of weak adsorption justify a reinvesti-gation of this phenomenon with modern methods. The present authors have therefore studied the adsorption of nitrogen on Pt and Ir by the field emission–probe hole technique and the adsorption of Xe on various group VIII and group Ib metals using either the same technique or photoelectric emission. With the former technique the desorption of adsorbed Xe was monitored while the temperature was increased in a programmed manner. From the rate of desorption thus determined,

$$r_{des} = \frac{kT}{h} \theta \exp\left(-\frac{\Delta F^{\ddagger}_{des}}{RT}\right)$$

(k and h are Boltzmann and Planck's constants, respectively, R is the gas constant, T is the absolute temperature, θ the degree of coverage, and $\Delta F^{\ddagger}_{des}$ the free activation energy of desorption), the free activation energy $\Delta F^{\ddagger}_{des}$ is calculated, which for the system under discussion may be assumed roughly equal to the heat of adsorption Q_{ads} because the activation entropy for desorption $\Delta S^{\ddagger}_{des}$ and the activation enthalpy for adsorption $\Delta H^{\ddagger}_{ads}$ are presumably very small:

$$-\Delta F^{\ddagger}_{des} = -\Delta H^{\ddagger}_{des} + T\Delta S^{\ddagger} \approx -\Delta H^{\ddagger}_{des}$$
$$= Q_{ads} + \Delta H^{\ddagger}_{ads} \approx Q_{ads}.$$

In this way the heat of adsorption was determined on various crystal faces on field emission tips of the metals Ir and Pt. By means of photoelectric work function measurements, a number of adsorption isotherms were determined at different temperatures on films of various metals. The heats of adsorption were then obtained from adsorption isosteres. The metals examined were Ni, Rh, Pt, Cu, and Au.

In this way it was possible to characterize the systems of interest by the average surface potentials on films, the specific surface potentials on individual crystal faces, and the average and individual heats of adsorption. In the case of Xe on Ir and Pt it was also possible to follow qualitatively the dependence of the heat of adsorption on the coverage for a number of crystal faces.

THE NATURE OF WEAK ADSORPTION ON TRANSITION METALS

A discussion of the totality of the data leads to conclusions on the face specificity of adsorption, the role played by surface geometry, coordinative unsaturation and work function on the strength of the adsorption bond, and the circumstances under which adsorption remains confined to certain a priori sites and those where close-packed monolayers are formed. By facing the data on surface potentials with those on the heats of adsorption measured *in situ* on the same surfaces, the theoretical models under discussion on the nature of weak adsorption are critically examined and the model which is able to describe the data consistently is identified. Since the individual results have been published elsewhere (18–25), the present paper tries to give a review of some of the more general conclusions.

II. EXPERIMENTAL TECHNIQUES

(a) Principle of Field Emission–Probe Hole Technique

The field emission–probe hole technique is based on the conventional field emission microscope invented in 1937 by Müller (26). The most essential parts are a sharp needle tip with a radius of curvature of the order of 1000 Å and an electrically conducting fluorescent screen opposite to the tip. Field emission of electrons, i.e., tunneling of electrons through the energy barrier at a metal surface, can be obtained by means of a high electric field of 0.2–0.5 V/Å with the negative potential on the tip and the screen serving as the anode. Such a high electric field is already attained at a relatively low voltage of the order of 1000 V due to the small radius of curvature of the tip. The emitted electrons diverge radially outward from the tip to the anode, i.e., the screen. Upon striking the screen the electrons produce a highly magnified image of the emitting tip surface, approximately according to a stereographic projection.

Because of the small radius the tip is almost always part of a single crystal. The emitting region usually has a nearly hemispherical shape

exposing a large number of different crystal planes. Since each individual crystal plane is characterized by its own work function, the magnified pattern will exhibit emission anisotropy which corresponds to the crystal symmetry. The different crystal planes on the emitting surface can thus easily be recognized from the symmetry of the pattern and by means of stereographic projection maps.

In the field emission–probe hole technique the local current density of an individual tip region is measured by means of an electron collector placed behind a small hole in the center of the screen. The electrons from the desired region are directed into the hole by means of deflection of the electron beam by an adjustable magnetic or electric field or by mechanical shifting of the tip with the aid of a manipulator.

For a thorough discussion of the applications of field emission–probe hole microscopy the reader is referred to a number of review articles (27–30). We mention here that the technique enables us to determine:

(a) the change in work function due to adsorption on each crystal plane,

(b) the heat of adsorption of a gas on each crystal face,

(c) the activation energy for diffusion of an adsorbed gas on the individual tip areas,

(d) the energy distribution of the field emitted electrons from each tip region. This yields information on the electron band structure of the metal. Specific surface states have been identified on metals. More recently it has been realized that the phenomenon of field emission resonance tunneling can provide information on the atomic states of molecules adsorbed on metal surfaces.

The strong power of the field emission–probe hole technique for adsorption studies is that adsorption can be studied on various well-defined crystal faces under exactly identical experimental conditions allowing the determination of the crystal face specificity in adsorption processes.

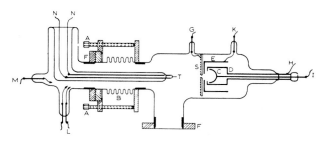

Fig. 1. Field emission–probe hole tube: (A) adjusting screws, (B) bellows, (C) collector, (D) shield, (E) lens electrode, (F) flange, (G) electric lead to screen, (H) electric lead to shield, (I) Kovar rod connected with collector, (K) electric lead to lens electrode, (L) connection with thermocouple wires, (M) extra electric lead, (N) lead to tip assembly, (S) fluorescent conductive screen, (T) tip assembly.

The probe-hole tube constructed by us for the studies described in the present paper is shown schematically in Fig. 1. A complete description of the tube can be found in Ref. (18). The tip has been spotwelded on a metal loop. The tip assembly is mounted on tungsten rods which are led through a dewar filled with liquid nitrogen. In this way the tip can be cooled to 78°K and heated to any desired temperature. A hole of 1.0 mm diameter in the center of the screen S serves as a diaphragm to the electrons emitted by the tip. The tip can be positioned by means of a manipulator A. The electrons of the desired tip region pass through the hole and arrive at a hemispherical collector C which is in direct contact with an electrometer in I.

(b) Determination of the Surface Potential and the Heat of Adsorption of Xe on Films of Various Metals

The technique used was photoelectron emission. We shall give here only a brief description of the procedure followed. For experimental details we refer the reader to Ref. (24).

The films were deposited on a glass substrate cooled at 78°K and were subsequently annealed at room temperature. At a fixed temperature T a number of Xe surface potentials were determined at stepwise increasing Xe pressures. This procedure was then repeated at different temperatures. The curves obtained by plotting surface potential versus p_{Xe} at constant T are essentially adsorption isotherms since the surface potential is a unique function of the coverage. From each set of isotherms $\ln p$ versus $1/T$ plots were constructed for several constant values of the surface potential and, hence, of the degree of coverage. The isosteric heats of adsorption were evaluated from the slope of the isosteres thus constructed.

III. NITROGEN ADSORPTION ON Pt AND Ir

In the periodic system of elements Pt and Ir are neighbors in group VIII. They crystallize in the same cubic face-centered lattice $O_h{}^5$, and their atomic radii differ only by 2%. Both metals have very high work functions, $\Phi_{Ir} = 5.0$ eV, $\Phi_{Pt} = 5.32$ eV. In spite of these close similarities the weak adsorption of nitrogen on the two metals shows marked differences. The data in Table I, collected from Refs. (18, 19), where the results are described in greater detail, show that nitrogen adsorption is face specific on either metal, but this specificity is much more marked on Ir. Here, even a negative surface potential is found in a surface region close to the one where for Pt the ubiquitous positive surface potential has its highest value. The heat of adsorption, which is nearly constant on all faces for Pt, differs by almost a factor of 2 between different areas on Ir.

During adsorption of nitrogen the emission current from the (100) and (511) faces of Ir

72

passes through a minimum while for Pt it increases monotonously on all crystal faces under the same circumstances. These results show that the adsorption of nitrogen on these metals depends in a very subtle manner on their chemical properties. The higher heat of adsorption on Ir is in line with the higher heat of formation of other Ir compounds in comparison to corresponding compounds of the more noble element Pt. It can be visualized that a chemical adsorption bond is formed by donation of electrons of nitrogen into vacant d-orbitals of the metal and stabilized by donation from filled d-orbitals into the vacant antibonding orbitals of the nitrogen molecule. It is not inconceivable that such a bond would be stronger for Ir than for Pt.

The crystal face specificity of the heat of adsorption of nitrogen cannot be correlated with the concentration of B_5 sites on these faces. This concept of the "demanding" adsorption requiring certain specific surface sites is, hence, insufficient.

IV. ADSORPTION OF Xe ON METALS

(a) Crystal Face Specificity and Its Causes

Using the adsorption of Xe on a single crystal Ir tip we have tried to tackle the following questions:

Can the initial (i.e., at low coverage) heat of adsorption on different crystal faces of a given metal be correlated with:

(1) a *geometrical* parameter such as the number of metal surface atoms in contact with

TABLE I

Crystal Face Specificity of Nitrogen Adsorption on Pt and Ir

Surface region	sp (eV)		Q_{ads} (kcal/mole)	
	Pt	Ir	Pt	Ir
Total surface	0.65	0.08	9	–
(111) face	0.3	0.2	9	–
(100) face	0.3	0.7	9	7–8
Around (110)	–	−0.1	–	10–11
(210) ~ (320)	0.8	0.3	9	13–14

TABLE II

Adsorption of Xe on Various Crystal Faces of Ir

Surface region	Φ (eV)	sp (eV)	Q_{ads} (kcal/mole)		
			Maximum	Average	Minimum
Total surface	5.0	+1.18	–	–	–
(111) face	5.79	+1.8	7.5	7.0	6.5
(100) face	5.67	+1.6	7.5	6.9	6.5
(110) vicinals	5.0	+0.8	7.0	6.5	6.3
(210) face	5.0	+1.3	7.2	7.0	6.7
(321) face	5.4	+1.0	7.8	7.6	7.0
(511) face	–	–	7.2	6.8	6.5
(531)–(731)	4.9	–	7.2	7.0	6.7

the Xe atom adsorbed in an appropriate dimple on the particular crystal face?

(2) the *chemical* parameter of the coordinative unsaturation of the metal atoms containing an adsorbed Xe atom?

(3) the *electronic* parameter of the work function Φ_{hkl} of the particular face?

A careful analysis of the experimental data is published in Ref. (22), where the possible influence of Ir atoms slightly further away than the ideal Xe–Ir distance was also considered. In addition, the possible influence of the electrical field present in the technique used was analyzed and was found to be negligible.

It appears that the geometrical parameter, if considered as the sole cause for differences in heats of adsorption on different crystallographic regions, would predict the following sequence of decreasing initial heats of adsorption on different crystal faces:

$(321) > (531) > (331) \sim (210) \sim (320)$
$\sim (310) > (110) > (100) > (111).$

When this is compared with the sequence actually observed (for numbers see Table II, column 3),

$(321) > (100) \sim (111) > (531)$
$\sim (210) > (110)$ vicinals.

It is seen that the (111) and (100) regions have initial heats of adsorption higher than predicted; for all other faces the measured data follow the order predicted by the geometrical

TABLE III

The surface Potential (sp), the Maximum Heat of Adsorption (Q_{max}), and the Difference in Maximum and Minimum Heat of Adsorption (ΔQ) of Xe on Various Crystal Faces of Pt and Ir

Tip region	Pt			Ir		
	sp (eV)	Q_{max} (kcal/mole)	ΔQ (kcal/mole)	sp (eV)	Q_{max} (kcal/mole)	ΔQ (kcal/mole)
(111)	–	7.6	1.4	1.8	7.5	1.0
(100)	1.0	7.5	1.1	1.6	7.5	1.0
Around (110)	–	7.1	0.7	0.8	7.0	0.7
(210)	1.1	7.2	0.4	1.3	7.2	0.5
(321)	0.9	7.8	0.9	1.0	7.8	0.8
(531)	–	–	–	–	7.2	0.5

parameter which would be decisive if the bond were purely of the Van der Waals–London type. Since the two deviating faces (111) and (100) do not possess particularly coordinatively unsaturated atoms, this chemical parameter cannot be invoked to explain the deviation. The two deviating faces do, however, excel by particularly high work functions and this parameter is important in the charge transfer–no bond type of chemical bonding. The data therefore show that both a good geometrical ligancy of an ad-atom *and* a high work function of the adsorbing face favor a high initial heat of adsorption. This result is consistent with the idea that the actual bond type is a combination of Van der Waals–London dispersion bonding with a bonding of the charge transfer–no bond type.

(b) Site Adsorption or Close-packed Monolayer

It was mentioned in the Introduction that weak "physical" adsorption is utilized in determinations of the surface areas of solids, in particular of catalysts. The underlying assumption that a monolayer coverage can be defined in which the surface area per adsorbed molecule is independent of the chemical and the crystallographic nature of the adsorbing surface has been questioned, and the problem arises whether weak adsorption is confined to certain sites; the spacing in ad-molecules at saturation would then reflect the spacing of these sites and consequently would not be independent of the nature of the adsorbing surface.

The present authors have attempted to obtain relevant information on this problem by determining the heat of adsorption of Xe on Ir and Pt not only on different crystal faces but also for each face at different coverages (22, 23). Since the atomic topography of each adsorbing crystal face is known, one can easily define which dimples in that face are available as attractive adsorption sites. If, indeed, adsorption in the initial stages is confined to these sites, the heat of adsorption should change very little until these sites are filled. It should decrease strongly when, after the filling of the attractive sites, additional Xe atoms are adsorbed at less favorable places, possibly also including a reshuffle of the previously adsorbed atoms. Further, not only adsorbate/metal but also adsorbate/adsorbate interaction will play a role in the arrangement of the ad-atoms at high coverage. This is because the adsorbate/adsorbate interaction may be attractive due to dispersion attraction or repulsive as a consequence of the mutual repulsion between electric dipoles on the surface. Clearly, the former favors the formation of a close-packed layer. Most data obtained on single crystal planes indicate that the latter is more important than the former for Xe on crystal planes of Pd (31), W (32), Pt (23), and Ir (22), since the heat of adsorption decreases with increasing coverage. A relatively large decrease in heat of adsorption with coverage may, therefore, be expected on faces where a close-packed layer is formed.

By analyzing in this way the measured data on Q_{ads} at initial adsorption, medium coverage, and for the loosest bonded portion (for numbers, see Table III), the following conclusions were drawn.

On those planes which possess attractive sites, adsorption remains confined to these sites. This is particularly true for the (321), (531), and (210) faces. Filling of all these sites will not occur on (210) because they are too densely spaced. In consequence, the calculated surface area per adsorbed atom remains very large. On (531) all sites can easily be filled because they are widely spaced for an Xe atom. The heat of adsorption, consequently, decreases very little on this plane. On (321) the available favorable sites are spaced in one direction at a distance compatible with the size of an Xe atom. In the other directions the favorable sites are widely spaced. Once the favorable sites are filled, adsorption stops under the conditions of our experiments and no less favorable sites become occupied. The packing in one dimension is then reasonably dense. The heat of adsorption therefore decreases more on (321) than on (210) and (531).

Quite differently, on the smooth faces such as (111) and (100) the data show that the heat of adsorption does decrease markedly with coverage indicating that a close-packed layer is formed at high coverage, the lateral interaction between neighboring Xe atoms being mainly repulsive. A close-packed ad-layer on several smooth planes of various metals had also been detected by LEED (31, 33), which confirms our results. These show, however, that on solids exposing atomically rough crystal faces, surface area measurements by Xe adsorption are unreliable since the monolayer coverage reflects the spacing of the attractive sites.

(c) The Nature of the Adsorption Bond

Considering the relatively high values found for the heat of adsorption, Q_{ads}, and the order in the maximum heat of adsorption on the different crystal faces (see Section IVa), it follows that additional interactions are operative. Two types of interaction are under discussion:

(1) a polarization of the adsorbed atom by a surface electric field, and

(2) an interaction according to the "charge transfer–no bond" model.

If, in a crude approximation, the actual heat of adsorption is considered as a linear combination of the contributions due to dispersion forces, polarization and charge transfer–no bond interaction, respectively, i.e.,

$$Q_{ads} = Q_{dis} + Q_{pol} + Q_{CTNB},$$

we have to decide how large the relative contribution is of each of the three types of interaction. Before we analyze our experimental data for this question a few words on the nature of each of the three interaction types may be appropriate.

(*i*) *Van der Waals–London dispersion forces.* In terms of Bohr's atom model negative electrons rotate around a positive nucleus which is equivalent to a rotating dipole. It induces, on a neighboring atom, a dipole of opposite sign. The two dipoles attract each other. In terms of quantum chemistry, the interaction between two atoms would be zero only if the ground states are exclusively considered, but a positive (= attractive) interaction results when the excited states are included in the description. For the case of a gas, interacting with a metal surface, calculations have been published by Lennard-Jones (34), Margenau and Pollard (35), Prosen and Sachs (36), Bardeen (37), and, more recently, Mavroyannis (38). Using the formulas derived by the latter author, a value of 3.1 ± 0.3 kcal/g atom is calculated for Q_{dis} of Xe for the following metals: Au, Cu, Ni, Pt, and Rh.

(*ii*) *Polarization by surface field.* Electrostatic forces near the surface are due to the presence of an electrical double layer. It arises from two features, called hereafter the "spreading effect" and the "smoothing effect. Conduction electrons are not only fairly mobile within the metal lattice but they can also spill over the outermost row of nuclei. According to Bardeen (39) the resulting spreading of negative charge above the surface can, for many purposes, be simulated by a plane parallel to the surface with charge density $-\delta q$. Electroneutrality requires that a parallel plane with positive charge density $+\delta q$ is located below the negative plane. The distance d between the planes is of the order of angstroms; the system of both planes represents a plate condenser of elemental dipoles $\mu_{spr} = -qd$. The negative sign indicates that the negative pole of the elemental dipole points away from the metal.

Spreading of electrons can also occur parallel to the surface. In particular, for crystal faces with widely spaced atoms, there is a high probability that electrons will be found in the space between the nuclei. These troughs, hence, acquire a net negative charge while the atom cores become positive because the charge of the nucleus is incompletely compensated. An equivalent way of describing this situation is to say that the equipotential lines near the surface do not follow precisely the contours of the outermost atoms as given, e.g., by a marble model, but the surface is "smoothed." It was Smoluchowski (40) who first realized that the dipoles between the negative troughs and the positive atom cores will also have a small vertical component μ_{sm}, pointing with the positive end outward. To the extent that both models are valid the actual surface double layer is composed by the moments due to the spreading and the smoothing effects:

$$\mu_{actual} = \mu_{spr} + \mu_{sm},$$

with

$$\mu_{spr} < 0 \text{ and } \mu_{sm} > 0.$$

The presence of the double layer causes a potential jump at the metal surface which forms an important contribution to the electronic work function $\Phi = A - 4\pi\sigma\mu_{actual}$, where A is the contribution due to the interior of the metal. The experimental fact that the work function for a given metal depends on the crystal face and that it is smaller for the rough than for the smooth faces is easily understood in terms of the above model, as μ_{spr} and μ_{sm} have opposite signs and μ_{sm} is largest for the atomically rough faces of high Miller indices while μ_{spr} is nearly isotropic. In the present context the question how such a double layer can induce a dipole moment on an adsorbed atom or molecule is relevant. According to classical physics no electric field exists outside a plate condenser with plates infinitely large and perfectly parallel with the charge spread homogeneously over the plates. For a condenser of finite size, edge effects will arise which we shall, however, ignore in the present context. A homogeneous electric field exists, however, in the space between the two plates. By consequence, an atom approaching a large metal surface will become polarized only if it can squeeze itself in the space between the two charged planes mentioned above. This explanation for the positive surface potential of inert gases on transition metals was first proposed by Mignolet (1). Later, Mignolet (9) and other authors (2, 41) argued that this picture is probably too simple. When the electron distribution near a metal surface could indeed be represented by a double layer in the form of an infinite parallel plate condenser it is rather improbable that a big Xe atom will be able to penetrate deeply in this double layer. Let us for the moment assume that this could occur. An Xe ad-atom experiences, then, a field strength which will be proportional to μ_{actual}. The atom will, hence, be polarized with the positive charge pointing away from the surface if $|\mu_{spr}| > |\mu_{sm}|$. This polarization should be largest on the smooth crystal planes but smaller or even of opposite sign on the atomically rough faces. For the family of crystal faces with $|\mu_{spr}| > |\mu_{sm}|$ Q_p should be largest on the smooth faces with high work function. The relatively large surface potential and heat of adsorption on Ir(111) and Ir(100) and on W(110) (32, 42) would be in agreement with the model. More in general, an at least qualitative relationship between the surface potential and the work function of the crystal

THE NATURE OF WEAK ADSORPTION ON TRANSITION METALS

TABLE IV

Adsorption of Xenon on Metal Films

Metal	Φ (eV)	sp (eV)	Q_{ads} (kcal/mole)	ΔS°_{ads} (cal deg^{-1} mole^{-1})	Q_{disp} (kcal/mole)
Au	5.28	0.52	4.6	−14	2.9
Cu	4.52	0.63	5.2	−23	2.8
Ni	4.90	0.82	6.4	−20	2.9
Pt	5.62	0.95	7.6	−20	3.4
Rh	4.98	1.08	8.7	−30	3.2

plane of a metal would be expected. Table II shows, however, that such a relationship does not exist.

More recently, it has been argued (2, 41) that the surface electric field will be very inhomogeneous over the metal surface due to the nonuniform charge distribution at the surface, in particular on stepped faces. Here the outermost atoms having a deficit of electrons are separated by troughs with an excess of negative charge. An adsorbed Xe atom cannot come into contact with the troughs but is bound to the outermost atoms with an electron deficit. Polarization will, hence, result in a layer with a net positive charge on the outside in accord with the observed positive surface potential. A similar suggestion has been brought forward recently by Müller (15). He has also suggested that according to this induced polarization model, a quantitative correlation may be expected between the Xe surface potential and the work function of the bare face. The surface potential has to decrease with increasing work function. This prediction is obviously in contrast to the results presented in Table II. However, the uncertainties in the value and even in the sign of the electric field and Müller's view that $Q_p > Q_{CTNB}$ appears to justify a search for further experimental evidence to decide on this question.

A fairly general consequence of the polarization model is that the surface potentials should be proportional to the surface field, while Q_p is proportional to the square of this field. In consequence, a quadratic relationship between Q_p and sp would be expected if $Q_p > Q_{CTNB}$ and if the influence of dispersion forces on the

surface potential is small. This consequence of the model can be examined by experiment.

(iii) *The charge transfer–no bond interaction.* This type of interaction was originally suggested by Mulliken (10) for the interaction of halogen molecules with benzene, e.g., while Mignolet (8) was the first to realize that it provides on appropriate description for the weak adsorption of rare gases by metals. The model assumes an electron transfer from the gas atom to the metal followed by resonance stabilization. The wavefunction for the adsorption system is formed by a linear superposition of the wavefunctions for the "no bond state" (Xe, metal) and for the charge transfer state (Xe$^+$, metal$^-$). The published calculations (24) provide formulas for the surface potential sp and for the interaction energy Q_{CTNB}. For those metals where only the d-band is involved in the interaction, the formulas show that Q_{CTNB} and sp are interrelated by

$$Q_{CTNB} = Csp.$$

Here, C depends very little on the metal and has the average value given by log $C = 13.7$.

The result that a linear relation is postulated between the interaction energy and the surface potential for this type of bonding, while a quadratic relationship was predicted for the polarization energy, provides an important clue for deciding experimentally which of the two types of interaction predominates in reality. What is needed for such a decision is a set of data on heat of adsorption and surface potential, measured under identical conditions for Xe adsorbed on a number of metals.

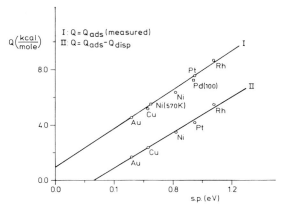

FIG. 2. Correlation with surface potentials of (I) measured heats of adsorption, and (II) heats of adsorption corrected for dispersion forces.

(iv) *Experimental results.* In order to decide between the models discussed, we have studied the adsorption of Xe on metal films prepared under ultrahigh vacuum using photoelectric emission to determine the surface potential as a function of the Xe equilibrium pressure in the cell for various values for the film temperature. In this way, the work functions of the pure metals and the surface potentials, the isosteric heats and standard entropies of adsorption were determined for Xe on films of Au, Cu, Ni, Pt, and Rh. The results are listed in Table IV; the surface potentials correspond to saturation of the surface with Xe at 78°K; the heats of adsorption correspond to a coverage where the surface potential has a value equal to half of the maximum surface potential. In the last column of Table IV the contribution of the interaction by dispersion forces to the heat of adsorption as calculated by the formulas given by Mavroyannis (38) is listed. The experimental values for the heat of adsorption vary strongly with the metal; on Rh, e.g., it is nearly twice that on Au. On the other hand, the values for the dispersion force interaction as emerging from the equation of Mavroyannis (38) and from the other dispersion force theories are nearly equal on the five metals used in this study. Hence, it may be concluded that the variation on heat of adsorp-

tion with the metal is largely determined by $Q_{pol} + Q_{CTNB}$. In Fig. 2 the measured heats of adsorption are plotted versus the surface potentials. In this graph are also included data of the (100) face of Pd (31) and data of a Ni film sintered at a higher temperature (570°K) than the other films. All points are, evidently, located on or very near a straight line. A linear relation is also obtained if the calculated value of Q_{disp} is first substracted from the measured heats of adsorption. This is in agreement with expectation if

$$Q_{CTNB} \gg Q_{pol}, \quad \text{i.e.,} \quad Q_{ads} - Q_{disp} \sim Q_{CTNB}.$$

The question can be asked whether the straight lines when extrapolated to $Q_{ads} - Q_{disp} = 0$, should pass through the origin or whether a finite surface potential should be attributed to a purely Van der Waals–London type of adsorption as predicted by Antoniewicz (43). We do not think that our data can decide this question, as a good fit is also obtained by assuming a smaller value, viz., $Q_{disp} = 1$ kcal/mole than the value calculated according to Mavroyannis for the dispersion interaction. In that case the points of $Q_{ads} - Q_{disp} = f(sp)$ are located on a perfectly straight line passing through the origin with a slope near the theoretical value predicted for a charge transfer–no bond interaction, i.e., log $C = 13.7$.

78

None of the graphs can be fitted, however, to a quadratic relationship postulated for the case $Q_{pol} \gg Q_{CTNB}$.

With the criteria available at present, we therefore feel that it is safe to conclude that the bond type of Xe adsorption on metals is predominantly of the charge transfer–no bond type of additive to a Van der Waals–London dispersion.

REFERENCES

1. MIGNOLET, J. C. P., *Discuss. Faraday Soc.* **8**, 105 (1950).
2. EHRLICH, G., AND HUDDA, F. G., *J. Chem. Phys.* **30**, 493 (1959).
3. EHRLICH, G., *Advan. Catal.* **14**, 255 (1963).
4. ROOTSAERT, W. J. M., VAN REIJEN, L. L., AND SACHTLER, W. M. H., *J. Catal.* **1**, 416 (1962).
5. PONEC, V., AND KNOR, Z., *Collect. Czech. Chem. Commun.* **27**, 1091 (1962).
6. BRENNAN, D., AND GRAHAM, M. J., *Phil. Trans. Roy. Soc. London Ser. A* **258**, 325 (1965).
7. VAN HARDEVELD, R., AND VAN MONTFOORT, A., *Surface Sci.* **4**, 396 (1966).
8. MIGNOLET, J. C. P., *J. Chem. Phys.* **21**, 1298 (1953).
9. MIGNOLET, J. C. P., *in* "Chemisorption" (W. E. Garner, Ed.), p. 118. Butterworths, London, 1957.
10. MULLIKEN, J., *J. Amer. Chem. Soc.* **74**, 811 (1952).
11. EHRLICH, G., HEYNE, H., AND KIRK, C. F., *in* "The Structure and Chemistry of Surfaces" (G. A. Somorjai, Ed.), p. 49-1. Wiley, New York, 1969.
12. HALL, P. G., *Chem. Commun.*, 877 (1966).
13. KLEMPERER, D. F., AND SNAITH, J. C., *Surface Sci.* **28**, 209 (1971).
14. MÜLLER, J., *Chem. Commun.*, 1173 (1970).
15. MÜLLER, J., *Surface Sci.* **42**, 525 (1974).
16. VAN OIRSCHOT, TH. G. J., AND SACHTLER, W. M. H., *Ned. Tijdschr. Vacuumtech.* **8**, 96 (1970).
17. BOUWMAN, R., AND SACHTLER, W. M. H., *Ber. Bunsenges. Physik. Chem.* **74**, 1273 (1970).
18. NIEUWENHUYS, B. E., AND SACHTLER, W. M. H., *Surface Sci.* **34**, 317 (1973).
19. NIEUWENHUYS, B. E., MEIJER, D. TH., AND SACHTLER, W. M. H., *Surface Sci.* **40**, 12 (1973).
20. NIEUWENHUYS, B. E., VAN AARDENNE, O. G., AND SACHTLER, W. M. H., *Thin Solid Films* **17**, S7 (1973).
21. NIEUWENHUYS, B. E., BOUWMAN, R., AND SACHTLER, W. M. H., *Thin Solid Films* **21**, 51 (1974).
22. NIEUWENHUYS, B. E., AND SACHTLER, W. M. H., *Surface Sci.* **45**, 513 (1974).
23. NIEUWENHUYS, B. E., MEIJER, D. TH., AND SACHTLER, W. M. H., *Phys. Status. Solidi A* **24**, 115 (1974).
24. NIEUWENHUYS, B. E., VAN AARDENNE, O. G., AND SACHTLER, W. M. H., *Chem. Phys.* **5**, 418 (1974).
25. NIEUWENHUYS, B. E., *Ned. Tijdschr. Vacuumtech.* **13**, 41 (1975).
26. MÜLLER, E. W., *Z. Phys.* **106**, 541 (1937).
27. GOMER, R., "Field Emission and Field Ionization," Harvard Univ. Press, Cambridge, Mass., 1961.
28. SACHTLER, W. M. H., *Angew. Chem.* **7**, 668 (1968).
29. GADZUK, J. W., AND PLUMMER, E. W., *Rev. Mod. Phys.* **45**, 487 (1973).
30. MÜLLER, E. W., *in* "Modern Diffraction and Imaging Techn. in Materials Sci." (S. Amelinckx *et al.*, Eds.), p. 683. North-Holland, Amsterdam, 1970.
31. PALMBERG, P. W., *Surface Sci.* **25**, 598 (1971).
32. ENGEL, T., AND GOMER, R., *J. Chem. Phys.* **52**, 5572 (1970).
33. CHESTERS, M. A., HUSSAIN, M., AND PRITCHARD, J., *Surface Sci.*, **35**, 161 (1973).
34. LENNARD-JONES, J. E., *Trans. Faraday Soc.* **28**, 334 (1932).
35. MARGENAU, H., AND POLLARD, W. G., *Phys. Rev.* **60**, 128 (1941).
36. PROSEN, E. J. R., AND SACHS, R. G., *Phys. Rev.* **61**, 65 (1942).
37. BARDEEN, J., *Phys. Rev.* **58**, 727 (1940).
38. MAVROYANNIS, C., *Mol. Phys.* **6**, 593 (1963).
39. BARDEEN, J., *Phys. Rev.* **49**, 653 (1936).
40. SMOLUCHOWSKI, R., *Phys. Rev.* **60**, 661 (1941).
41. TOMPKINS, F. C., *in* "The Solid–Gas Interface" (E. A. Flood, Ed.), p. 765. Dekker, New York, 1967.
42. NIKLIBORC, J., AND DWORECKI, Z., *Acta Phys. Polon.* **32**, 1023 (1967).
43. ANTONIEWICZ, P. R., *Phys. Rev. Lett.* **32**, 1424 (1974).

Catalysis and Surface Chemistry

I. Cyclopropane Reactions over Reduced Molybdena–Alumina

MARIANO LO JACONO[1] AND W. KEITH HALL[2]

Department of Chemistry, University of Wisconsin, Milwaukee, Wisconsin 53201

Received May 25, 1976; accepted August 12, 1976

The surface chemistry of an 8% Mo on alumina catalyst was varied by a previously described method which determines the extent of reduction, the hydroxyl concentration introduced into the molybdena layer, the number of anion vacancies formed by removal of H_2O as the catalyst is reduced with H_2, and the amount of reversibly adsorbed hydrogen, H_R. The latter was found to correlate linearly with the vacancy concentration. The isomerization of cyclopropane was studied using a microcatalytic pulse technique at temperatures between 0 and 109°C. At a very low extent of reduction (H_2 consumed as $H_C/Mo = 0.10$ atom/atom) the mass balance was good when He-carrying gas was used and the only products observed were propene and a trace of ethene. As the extent of reduction was increased, however, nearly equal quantities of ethene and n-butenes appeared in the products in increasing amounts together with propene. Also, increasing amounts of carbonaceous material were retained by the catalyst. The isomerization reaction correlated linearly with the sum of the concentrations of vacancies and hydroxyls introduced as Mo–OH during the reduction with H_2. Olefin metathesis and loss from the mass balance, on the other hand, could be correlated with vacancies, suggesting that these occur on coordinatively unsaturated Mo^{+4} centers. When H_2-carrying gas was used, the olefins were completely hydrogenated under all conditions tested and the fraction of the cyclopropane reacted was increased considerably. At a low extent of reduction, the metathesis rate was about the same as in He as judged by the amount of *ethane* produced. As the extent of reduction was increased, however, both metathesis and formation of carbonaceous residues were repressed by H_2 until propane was the only product. Hydrogenolysis of either cyclopropane or propane (to form methane) was not appreciable below 390°C. Both isomerization and metathesis were poisoned by adsorption of H_2O; both activities could be restored by evacuation of the catalyst at elevated temperatures, the isomerization reaction recovering faster with removal of H_2O. Carbon monoxide was more weakly chemisorbed, but also exhibited poisoning effects.

INTRODUCTION

Molybdena–alumina catalysts are used extensively for hydrodesulfurization of petroleum feed stocks. Because coal also contains substantial quantities of chemically bound sulfur, which would poison or modify metallic catalysts under hydrogenating conditions, molybdena–alumina catalysts have become prime candidates for use in coal hydrogenation. Thus, an objective of our research has been to develop basic understanding of these catalysts with a view toward future process development.

Recently, Hall and Massoth (1) discovered that a substantial quantity of hydrogen remained associated with molybdena–alumina catalysts when they were reduced with H_2. Moreover, since it could be removed as H_2O but not as H_2 on raising the temperature, the results suggested that it was held as OH in the epitaxial molybdena monolayer. If so, it may then be conjectured that these OH groups act as Brönsted acid sites in catalytic reactions.

Anion vacancies were also introduced into the molybdena layer as evidenced by formation of H_2O during the reduction. Coordinatively

[1] On leave from the Centro di studio su struttura ed attivita catalitica di sistemi di ossidi, Istituto Chimica Generale, Citta Universitaria, Piazzale delle Scienze, 00100 Rome, Italy.

[2] To whom all correspondence should be addressed.

unsaturated molybdenum ions should then be present at these sites. Hall and Lo Jacono (2) reasoned that these ions should be Mo^{+4} while the OH should count the Mo^{+5}. Since their data agreed quantitatively with the earlier results (1), it was possible to seek relationships between this continuously varying surface chemistry and catalyst function.

The role of surface hydroxyl groups as catalytic sites has been described in a recent review (3). If such sites function as Brönsted acids in their reactions with nitrogen bases such as pyridine, it is virtually certain that they will catalyze the isomerization of cyclopropane via the cyclopropyl carbonium ion (4–6). Whereas the uncatalyzed thermal reaction does not occur much below 500°C, the acid-catalyzed reaction proceeds at an easily measurable rate below 100° over silica–alumina and above 200° over alumina. The homogeneous gas phase isomerization has been characterized (8–14) as a unimolecular reaction having an activation energy of about 65 kcal/mole; hence, it is immeasurably slow below about 425°C. The heterogeneous proton-catalyzed reaction, on the other hand, is quite facile. Much of the known chemistry of cyclopropane in homogeneous media is centered around its ability to become protonated as if it were an olefin. Kiviat and Petrakis (7) have shown that pyridinium ions form readily on the surfaces of molybdena-alumina catalysts whereas they do not on the alumina support. Cyclopropane, besides being uniquely suited to test the acidity of such groups, affords the possibility of simultaneously testing the activity of the catalysts for hydrogenation of the olefin formed by isomerization, and for hydrogenolysis of the paraffin by simply adding H_2 to the reactant gas.

Over alumina and silica–alumina, the catalytic activity parallels the hydroxyl content of the catalyst (15) and over the latter at least, isotope exchange occurs stepwise via the catalyst protons (4–5). The possibility that coordinatively unsaturated cations function as catalytic sites has been suggested many times, particularly in connection with reactions where hydrogenation or deuterium exchange was involved (16). The anion vacancies introduced as H_2O is removed should expose Mo^{+4} ions (d^2). The object of the present paper is to show the progress made in understanding the reaction system by identifying the several different reactions with particular kinds of sites.

MATERIALS AND METHODS

Equipment and Procedures

The reactions of cyclopropane were studied using a microcatalytic pulse reactor. The weight of raw catalyst was 1.00 g. It was contained in a quartz reactor which was connected by 10/30 standard taper glass joints to the reaction system. The reactor could be surrounded by a furnace and pretreated or used at temperatures regulated to within about 0.1°C by a thyratron circuit. A temperature-controlled water bath was substituted for the furnace in the experiments carried out below 100°C, but above room temperature. Pulses of reactant were picked up by the helium or hydrogen carrying gas stream from a dosing device made from stopcocks similar to that used by Hall and Emmett (17); its real volume was about 1.15 cm^3. The pulse was carried over the catalyst at 1 atm total pressure. The hydrocarbon content of the pulse could be varied by altering its pressure in the doser, i.e., by diluting with carrying gas. The products emerging from the catalyst were collected in a liquid nitrogen trap for 15 to 40 min; a period as long as practical was needed because some of the hydrocarbon desorbed very slowly from the catalyst. Then after removal of the liquid nitrogen, the hydrocarbon was flashed onto a 13′ × 0.25″ (o.d.) column packed with 40% dimethylsulfolane on anichrome C22A (80–100 mesh) which was thermostated at 0°C. The chromatogram was developed using the stream of He which was passing over the catalyst. Provision was made to recover samples of the products for mass spectrometric analyses as they were eluted from the column. Deficits from the mass balance represent

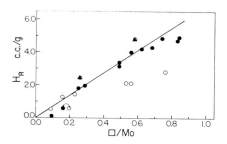

Fig. 1. Relationship between the reversibly chemisorbed hydrogen, H_R, and the amount of H_2O produced, calculated as oxygen vacancy concentration where $\bigcirc$ = points obtained in an earlier study at a lower ambient H_2-pressure, $\bullet$ = points obtained in present work, $\blacktriangle$ = points obtained by extending evacuation time to 150 min from usual 1 hr. $\bullet\!\bigcirc$ indicates that the open point falls on the same position as the solid point.

material held up on the catalyst or which desorbed too slowly to be caught in the liquid nitrogen trap.

Reagents

All gases were obtained from commerical sources. The helium and hydrogen were passed through glass traps containing anhydrous $CaCl_2$ and $MgClO_4$ to remove most of the water before passing them over high-area activated charcoal thermostated at $-195°C$. Propane, propene, and cyclopropane were supplied by the Union Carbide Corporation, Linde Division. They were twice distilled from -78 to $-195°C$ and outgassed at the lower temperature. No impurities were detected by chromatography.

Catalysts

All the experiments were made over a single sample of the same 8% Mo on $\gamma\text{-}Al_2O_3$ catalyst studied in our earlier work (2). This was made by impregnating a sample of Ketjen CK-300 alumina with an ammonium paramolybdate solution, followed by drying at 120°C before a final air calcination at 540°C for 16 hr. Surface areas and pore volumes for the support and catalyst were 192 and 185 m^2/g; and 0.5 and

0.42 cm^3/g, respectively. The chief impurities (Ca, Cr, Cu, Mg, and Si), as determined by spectrographic analysis, were all less than 100 ppm.

The surface chemistry of the catalyst could be varied systematically in a predetermined way by controlling the extent of reduction (2). Moreover, following an experiment it could be restored to its initial condition by reoxidation. Thus, the standard procedure was to subject the catalyst to an overnight pretreatment in a flow of dry O_2 at 500°C, then to evacuate it for 40 min at the reduction temperature (400° or 500°C) before contacting it with a measured volume of H_2 contained in the all-glass circulating loop. The excellent reproducibility found in our earlier work (2) made it possible to estimate accurately all of the parameters defining the surface chemistry from the pressure drop during reduction. These included the extent of reduction in terms of H_2 consumed [defined as H_c/Mo (atom/atom)], the amount of hydrogen held irreversibly by the catalyst, H_I/Mo, and the number of surface sites of low coordination (or vacancy concentration) measured as H_2O (oxygen anions) removed, $\square/Mo$. Following the reduction step, the catalyst was cooled to the reaction temperature where it was contacted with flowing helium or hydrogen-carrying gas. An

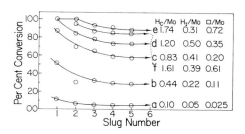

Fig. 2. Microcatalytic results for decomposition of cyclopropane over a molybdena–alumina catalyst reduced to various extents with H_2. He-carrying gas was used with a flow rate of 65 cm^3/min and the reaction temperature was 109°C. The extents of reduction and resulting surface chemistry are indicated numerically on the figure. Δ^0 and Δ represent moles of cyclopropane initially contained in the pulse and emerging from the reactor, respectively.

TABLE I

Cyclopropane Isomerization Data at 109°C Using He-Carrying Gas[a]

Pulse no.	Product analysis (molecules $\times 10^{-18}$)					Fraction of initial reactant converted				
	$E_e =$ C_2H_4	$P_e =$ C_3H_6	$\Delta = c\text{-}$ C_3H_6	$trans\text{-}$ C_4H_8	$cis\text{-}$ C_4H_8	$\dfrac{(\Delta^0 - \Delta)}{\Delta^0}$	$\dfrac{P_e + 2E_e}{\Delta^0}$	$\dfrac{L}{\Delta^0}$	$\dfrac{L^*}{\Delta^0}$	$\dfrac{2E_e}{\Delta^0}$
1	0.1	1.6	12.9	—	—	0.108	0.12	0 ± 0.03	−0.014	0.014
2	0.1	1.4	13.5	—	—	0.066	0.10		−0.040	
3	0.1	1.3	13.7	—	—	0.053	0.10		−0.048	
4	0.1	1.1	13.8	—	—	0.046	0.08		−0.040	
5	0.1	1.1	13.8	—	—	0.046	0.08		−0.040	
6	0.1	1.05	14.0	—	—	0.032	0.08		−0.048	
	$H_e/Mo = 0.10$	$H_I/Mo = 0.05$						$\square/Mo = 0.025$		
1	0.8	3.9	7.1	0.2	—	0.51	0.38	0.24	+0.13	
2	0.5	4.4	10.1	0.2	—	0.30	0.37	0 ± 0.035	−0.06	
3	0.5	4.2	9.7	0.2	—	0.32	0.35		−0.03	
4	0.5	4.3	10.2	0.2	—	0.29	0.36		−0.06	
5	0.5	4.0	10.2	0.2	—	0.28	0.34		−0.05	0.07
	$H_e/Mo = 0.44$	$H_I/Mo = 0.22$						$\square/Mo = 0.11$		
1	2.5	5.4	1.9			0.87	0.71	0.49	+0.15	0.35
2	1.5	6.0	4.5	0.7	0.2	0.69	0.62	0.27	+0.07	0.20
3	1.4	6.4	5.1	1.1	0.5	0.65	0.63	0.20	+0.014	0.19
4	0.9	6.1	5.9	0.9	0.2	0.59	0.54	0.17	+0.05	0.13
5	1.2	6.5	6.2	1.1	0.2	0.57	0.61	0.12	−0.04	0.17
	$H_e/Mo = 0.83$	$H_I/Mo = 0.415$						$\square/Mo = 0.205$		
1	0.7	0.0	0.0			1.00	0.10	1.00	+0.90	
2	3.3	5.15	2.35	0.1		0.84	0.81	0.48	+0.03	0.46
3	2.0	5.55	3.20	0.9		0.78	0.66	0.39	+0.12	0.27
4	1.6	5.70	3.85	0.7		0.73	0.61	0.34	+0.12	0.22
5	1.5	6.10	3.85	0.7		0.73	0.63	0.31	−0.11	0.20
	$H_e/Mo = 1.20$	$H_I/Mo = 0.50$						$\square/Mo = 0.35$		
1		0.0	0.0			1.00	0.00	1.00	+1.00	
2	4.2	0.3	0.0			1.00	0.60	0.98	+0.40	0.58
3	3.3	4.0	1.05	0.2	0.3	0.93	0.73	0.65	+0.20	0.46
4	2.5	5.8	1.30	1.2	0.2	0.91	0.74	0.51	+0.17	0.35
5	2.3	5.5	1.70	1.1	0.3	0.88	0.70	0.50	+0.19	0.32
6	2.3	6.5	1.25	1.5	0.4	0.91	0.76	0.46	+0.15	0.32
	$H_e/Mo = 1.74$	$H_I/Mo = 0.31$						$\square/Mo = 0.72$		
1	1.25	0.0	0.0			1.00	0.17	1.00	0.83	0.17
2	3.9	3.8	0.8			0.95	0.80	0.68	0.15	0.53
3	2.2	6.1	1.85	1.85	0.7	0.87	0.72	0.45	0.15	0.30
4	2.1	6.25	2.05	2.2	0.8	0.86	0.72	0.43	0.14	0.29
5	1.9	6.25	2.2	2.0	0.8	0.85	0.69	0.42	0.16	0.26
	$H_e/Mo = 0.83$	$H_I/Mo = 0.415$								

[a] Catalyst weight = 1.00 g; doser volume = 1.15 ml; initial cyclopropane pressure = 388 Torr; reaction temperature = 109°C; reactant cyclopropane in each pulse $(\Delta^0) = 14.5 \times 10^{18}$ molecules; Δ = cyclopropane in product; $L/\Delta^0 = [\Delta^0 - (P_e + \Delta)]/\Delta^0$; $L^*/\Delta^0 = [\Delta^0 - (P_e + 2E + \Delta)]/\Delta^0$.

experiment consisted of introducing a series of about five pulses into the carrying gas stream as described above. Following the experiment, the catalyst was then reoxidized overnight. The catalyst could be repeatedly used and gave reproducible results.

RESULTS

At the moment when the reduction of the catalyst is terminated, a certain amount of chemisorbed hydrogen is present in equilibrium with the gaseous H_2. This reversible chemisorption, H_R, is removed when the catalyst is evacuated prior to testing or to reoxidation (2). The data of Fig. 1 show that H_R correlated with vacancy concentration. The relationship appears linear and the slope indicates that about one in three vacancies is occupied by a reversibly adsorbed H_2. The points shown as triangles falling above the curve were obtained

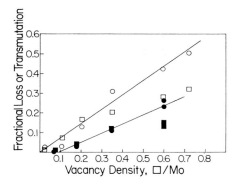

FIG. 3. Correlation of the fractional loss and olefin metathesis reactions with vacancy concentration using He-carrying gas where $\bigcirc$ and $\bullet$ represent L/Δ^0 at 109°C and 55°C respectively; $\square$ and $\blacksquare$ are values of $2E_e/\Delta^0$ for these temperatures, and $L = \Delta^0 - (P_e + \Delta)$; E_e and P_e are the number of ethylene and propylene molecules produced in the reaction and Δ^0 and Δ represent the cyclopropane contained in the pulse before contacting the catalyst and in the product, respectively.

in experiments where the evacuation time was extended to 150 min (from the usual 1 hr) following the reduction step. Higher values would result if some additional H_2 was removed during the additional evacuation. Several points (shown as open circles) from an earlier study (1) fall badly below the curve. The corresponding ambient H_2 pressures for these experiments were lower than those used in the present work. The similar plot published earlier (1) was also linear but did not pass through the origin because of the improper definition of the percent reduction. These data suggest that vacancies act as sites for hydrogen chemisorption.

The microcatalytic results for the decomposition of cyclopropane over the catalyst reduced to various extents are shown in Fig. 2. The catalyst activity, [defined as the fraction of the cyclopropane pulse consumed, $(\Delta^0 - \Delta)/\Delta^0$], increased with the extent of reduction and fell with the first couple of pulses to a steady lined-out value. Data obtained from the fifth pulse corresponding to these lined-out values have been plotted in the subsequent figures. The data of Table I show that the product

propylene underwent an increasing amount of olefin metathesis as the depth of the reduction increased. The accuracy of the data is indicated by the mass balances obtained with the samples having the lowest extents of reduction. Similar tabular data are available from the authors for all runs.

A blank run was made without reducing the catalyst following pretreatment in oxygen. The results at 109°C were quite similar to those shown in Table I for the preparation reduced to the smallest extent ($H_e/Mo = 0.10$). Thus, the unreduced catalyst had sufficient acidity to effect isomerization of from 5 to 10% of the cyclopropane contained in the pulse.

Data for a reaction temperature of 55°C were similar to those for 109°C, but the conversions were correspondingly lower. This is demonstrated by the data of Fig. 3 where the total loss, L, and the extent of the metathesis reaction, $2E_e$, are plotted against the vacancy concentration which increased with the extent of reduction. The parameter, L, was obtained from the mass balance of the C_3 species alone, i.e., it represents everything but propene and cyclopropane and includes molecules retained by the catalyst surface. The extent of the metathesis reaction was calculated as twice the number of ethylene mole-

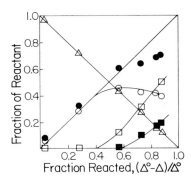

FIG. 4. Correlation of product distribution with cyclopropane conversion at 109°C over catalyst with increasing extent of reduction where $\triangle = \Delta/\Delta^0$, $\bigcirc = P_e/\Delta^0$, $\bullet = (P_e + 2E_e)/\Delta^0$, $\square = L/\Delta^0$, and $\blacksquare = L^*/\Delta^0$, and $L^* = \Delta^0 - (P_e + \Delta + 2E_e)$; $L = \Delta^0 - (P + \Delta)$. L, Δ^0, Δ, P_e, E_e are defined in legend of Fig. 3.

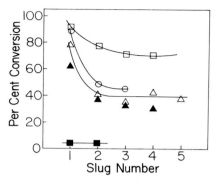

FIG. 5. Effect of adsorbed water on the catalytic activity for cyclopropane decomposition at 109°C. Reduced catalyst ($H_e/Mo = 1.43$; $H_I/Mo = 0.50$; $\square/Mo = 0.47$) exposed to 25 Torr H_2O vapor for 3 min prior to starting flow of He-carrying gas. No reaction or loss of cyclopropane was apparent until H_2O had been desorbed for 30 min at 110°C, ■; then evacuated overnight at 210°C, △; treated in flowing He at 109°C overnight following the preceding step, ▲; evacuated for an additional 1.5 hr at 305°C, ○; and then evacuated overnight at 500°C, □.

cules, E_e, formed because a butene molecule is produced with each ethylene molecule from two propylene molecules; thus two propylene molecules disappeared for each ethylene produced. The fractional loss, L/Δ^0, correlated fairly well with the vacancy concentration although at the lower temperature the curve intercepted the abscissa, indicating that the first few vacancies formed are not effective in generating a loss at this temperature. Moreover, below $\square/Mo = 0.5$, an equally good correlation exists with the fractional olefin metathesis, $2E_e/\Delta^0$; above this value the points fell substantially below the curve, perhaps because here desorption of the products is very slow. The same may be said for the real fraction loss, L^*/Δ^0, shown in Table I. These values represent cyclopropane molecules which cannot be accounted for by either isomerization or olefin metathesis; presumably they remain retained by the catalyst. Although the real fractional loss tended to increase with the ex-

TABLE II

Cyclopropane Isomerization Data at 109°C Using H_2-Carrying Gas[a]

Pulse no.	Product analysis (molecules $\times 10^{-18}$)				Fraction of initial reactant converted			
	$E_a = C_2H_6$	$P_a = C_3H_8$	$B_a = C_4H_{10}$	$\Delta = c\text{-}C_3H_6$	$(\Delta^0 - \Delta)/\Delta^0$	$(P_a + 2E_a)/\Delta^0$	L/Δ^0	L^*/Δ^0
1	0.5	12.3	0.6	1.1	0.92	0.92	0.07	0.00
2	0.5	12.0	0.5	1.4	0.90	0.90	0.07	0.00
3	0.6	11.4	0.8	2.0	0.86	0.87	0.07	+0.01
4	0.5	10.4	0.5	2.3	0.84	0.79	0.12	0.05
5	0.5	9.8	0.6	3.2	0.78	0.75	0.10	0.03
R. T. 6	—	2.5	—	10.8	0.25	0.17	0.08	0.08
	$H_e/Mo = 0.47$	$H_I/Mo = 0.23$			$\square/Mo = 0.12$			
1	0.4	13.0	0.3	—	1.00	0.96	0.10	0.04
2	0.4	12.9	0.35	0.1	0.99	0.95	0.10	0.04
3	0.45	13.5	0.45	0.2	0.99	0.99	0.05	0.00
4	0.45	13.3	0.45	0.5	0.96	0.98	0.05	+0.02
5	0.45	12.8	0.60	0.8	0.94	0.95	0.06	+0.01
R. T. 6	—	4.9	—	9.3	0.36	0.34	—	0.02
	$H_e/Mo = 0.62$	$H_I/Mo = 0.31$			$\square/Mo = 0.15$			
1	0.2	14.2		—	1.00	1.01	0.02	+0.01
2	0.2	14.2		—	1.00	1.01	0.02	+0.01
3	0.2	14.2	traces	—	1.00	1.01	0.02	+0.01
4	0.2	14.2	traces	—	1.00	1.01	0.02	+0.01
	$H_e/Mo = 1.04$	$H_I/Mo = 0.50$			$\square/Mo = 0.26$			
1	0.2	14.0	—	—	1.00	1.00	0.03	0.00
2	0.15	14.0	—	—	1.00	0.99	0.03	0.01
3	0.2	13.8	—	—	1.00	0.98	0.04	0.02
4	0.15	14.3	—	—	1.00	1.01	0.01	+0.01
R. T. 5	0.25	14.3	—	—	1.00	1.02	0.01	+0.02
	$H_e/Mo = 1.53$	$H_I/Mo = 0.46$			$\square/Mo = 0.53$			

[a] Reaction temperature was 109°C except where indicated otherwise (R.T. = room temperature). See footnotes on Table I for other definitions.

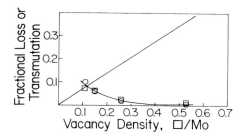

FIG. 6. Variation of olefin metathesis with vacancy concentration during hydrogenation at 109°C. H_2-carrying gas was used. The curve of Fig. 3 is shown for comparison; $\bigcirc = L/\Delta^0$ and $\square = 2E_a/\Delta^0$ where E_a = ethane. The other symbols have been defined in previous figures.

tent of reduction, it was relatively insensitive to temperature within the range studied. It is, of course, included in L/Δ^0.

Some additional aspects of the data are delineated in Fig. 4, where product formation is related to the cyclopropane reacted. The fraction of the initial cyclopropane which remains (Δ/Δ^0) is necessarily inversely proportional to the fraction reacted $[(\Delta^0 - \Delta)/\Delta^0]$. If propene were the only product, a plot of the fraction of the initial cyclopropane appearing as propene (P_e/Δ^0) would fall along the 45° proportionality line. Indeed, up to about 40% conversion, this was found to be true $(\bigcirc)$. Above this level, however, the extent of conversion to propene remained about constant while increasing amounts of olefin metathesis

occurred. Thus the sum of the fraction appearing as propene and twice the fraction appearing as ethylene $(\bullet)$ followed the theoretical curve up to about 80% conversion. These data are consistent with the notion that the olefin metathesis reaction is consecutive to the isomerization. The fractional total loss $(\square)$ and the real loss $(\blacksquare)$ represent the deviations of the open and filled circles from the theoretical curve.

Reference to Table I shows that the amounts of butene recovered were frequently less than those of ethylene, although the difference in some cases was not large. The chromatographic peaks for ethylene were quite sharp whereas those for the butenes were broad due to their long retention times. Moreover, in the earlier experiments, the products were collected in a liquid nitrogen trap for only 10 to 15 min before flashing onto the chromatographic column; in later work, where 30 to 40 min were allowed, the agreement between the amounts of ethylene and 2-butene produced was much improved (see, for example, the last three pulses of the preparation with $H_e/Mo = 0.83$). Thus, with preparations having low and medium extents of reduction, the real loss not recovered as identifiable products, L^*/Δ^0, was zero after the first pulse. Consequently, $[(\Delta^0 - \Delta)/\Delta^0]$ could be taken as equal to $[(P_e + 2E_e)/\Delta^0]$. At higher extents of reduction, the real fractional loss was 100% of the first pulse and from 10 to 20% of later ones. Consequently, the material lost from the mass

TABLE III

Hydrogenation of Propene[a]

Temp (°C)	Pulse no.	Product analysis (molecules × 10⁻¹⁸)			Fraction of initial reactant converted		
		$E_a = C_2H_6$	$P_a = C_3H_8$	$B_a = C_4H_{10}$	$(P_e^0 - P_e)/P_e^0$	$(P_a + 2E_a)/P_e^0$	L^*/P_e^0
27	1	0.3	13.2		1.00	0.95	0.05
27	2	0.3	13.4		1.00	0.97	0.03
27	3	0.3	13.5		1.00	0.98	0.02
27	4	0.5	14.0	traces	1.00	1.03	−0.03
27	5	0.5	13.9	traces	1.00	1.03	−0.03
0	6	0.35	13.2	—	1.00	0.96	0.04
0	7	0.35	13.2	—	1.00	0.96	0.04
110	8	0.45	13.2	0.5	1.00	0.98	0.02
		$H_e/Mo = 0.45$	$H_I/Mo = 0.22$			$\square/Mo = 0.11$	

[a] Reaction temperature as indicated; $P_e^0 = 14.45 \times 10^{18}$ molecules; E_a = ethane, P_a = propane, B_a = butane; see footnote on Table I for other definitions. $L^*/P_e^0 = P_e^0 - (P_a + 2E_a)/P_e^0$.

TABLE IV

Hydrogenolysis of Cyclopropane[a] and Propane[b]

Temp (°C)	Pulse no.	Product analysis (molecules $\times 10^{-18}$)				Fraction of initial reactant converted		
		$M_a = CH_4$	$E_a = C_2H_6$	$P_a = C_3H_8$	$\Delta = c\text{-}C_3H_6$	$(\Delta^0-\Delta)/\Delta^0$	$(P_a+E_a)/\Delta^0$	Loss[a]
391	1	0.95	1.2	12.6	—	1.00	0.96	0.04
391	2	0.50	0.65	13.0	—	1.00	0.95	0.05
391	3	0.55	0.50	14.0	—	1.00	1.00	0
391	4	0.70	0.65	13.7	—	1.00	0.99	0.01
391	5	0.60	0.55	13.9	—	1.00	1.00	0
391	6	0.60	0.60	13.9	—	1.00	1.00	0
111	7	0.15	0.20	13.2	—	1.00	0.93	0.07
111	8	0.15	0.20	13.5	—	1.00	0.95	0.05
		$H_c/Mo = 1.71$		$H_I/Mo = 0.32$		$\square/Mo = 0.7$		

Temp (°C)	Pulse no.	Product analysis (molecules $\times 10^{-18}$)			Fraction of initial reactant converted			
		$M_a = CH_4$	$E_a = C_2H_6$	$P_a = C_3H_8$	$(P_a^0-P_a)/P_a^0$	M_a/P_a^0	E_a/P_a^0	Loss
344	1	0.7	0.8	13.6	0.06	0.05	0.055	0.00
360	2	0.25	0.4	14.1	0.024	0.017	0.028	0.00
365	3	0.4	0.45	14.1	0.024	0.028	0.028	0.00
505	4	3.0	2.7	10.3	0.29	0.21	0.19	0.09
506	5	9.5	4.0	6.3	0.56	0.66	0.28	0.15
502	6	11.3	5.2	3.7	0.74	0.78	0.36	0.24
405	7	2.7	2.0	12.1	0.16	0.19	0.14	0.00
410	8	2.7	2.1	12.1	0.16	0.19	0.14	0.00
		$H_c/Mo = 1.47$		$H_I/Mo = 0.5$		$\square/Mo = 0.48$		

[a] Catalyst weight = 1.00 g; doser volume = 1.15 ml; sample pressure = 388 Torr; reaction temperature = as indicated; pulse size (Δ^0) = 14.45 $\times 10^{18}$ molecules; E_a = ethane; Loss* = $1 - (P_a + E_a)/\Delta^0$.

[b] Reaction temperature as indicated; P_a^0 = propane inlet = 14.45 $\times 10^{18}$ molecules; P_a = propane recovered as product; E_a = ethane; M_a = methane; Loss = $1 - (P_a/P_a^0) - 2/3(E_a/P_a^0) = 1/3(M_a/P_a^0)$.

balance accumulated continuously on the catalyst as carbonaceous residue.

The effect of added back H_2O on the catalytic activity of the reduced catalyst was tested and the data are plotted in Fig. 5. The reaction conditions were identical with those used to obtain the data of Fig. 2. The reduced catalyst $(H_c/Mo = 1.43)$ was nearly completely poisoned by exposure to H_2O at its vapor pressure at room temperature for 3 min. When the H_2O was allowed to desorb for about 30 min at 110°C, however, a conversion of about 5% was obtained. As indicated, the conversion level could be further increased by desorbing water at higher temperatures and after evacuation overnight at 500°C, the initial activity could be completely restored ($\square$). Evacuation at 300°C only partially restored the activity; vide infra.

Results for the reactions of cyclopropane with H_2 are given in Table II. The ratio of H_2 to cyclopropane in the pulse is not accurately known because dilution occurs by diffusion;

however, it is estimated to be about 1:1. Inspection of the data contained in the table reveals the following points: (1) for catalyst samples with about the same extents of reduction, the conversion of cyclopropane $[(\Delta^0 - \Delta)/\Delta^0]$ was about twice as great with H_2-carrying gas as with He at the same temperature; (2) losses from the mass balance were in all cases relatively small; (3) the main reaction product was found to be propane, but small amounts of ethane and butane also appeared; (4) olefins were entirely missing. Figure 6 shows that there was no real loss; deviations from the mass balance can be accounted for entirely by metathesis with the olefins formed being hydrogenated to the corresponding paraffins. The extent of the metathesis reaction decreased with increasing vacancy density rather than increasing as it did in the absence of H_2. (The straight line drawn to the origin in Fig. 6 is the line for the reaction at the same temperature taken from Fig. 3.) Evidently, hydrogenation is much faster than metathesis and, since both are consecutive to isomerization, competes favorably with it. Almost identical results were obtained when propene was substituted for cyclopropane over a catalyst where $H_c/Mo = 0.45$ (Table III). The mass balance was good and the only products were propane containing traces of ethane and butane.

Table IV contains data concerning hydrogenolysis of cyclopropane or propane. In these experiments, catalysts having high extents of reduction ($H_c/Mo = 1.71$ and 1.47) were used because these had the highest hydrogenating activity. Only small amounts of methane and ethane were produced, even at temperatures in excess of 500°C.

DISCUSSION

The chemistry of the reduction process has been discussed in detail elsewhere (2). It has been suggested that sites of the kind

$$[1]$$

appear as the valence state of molybdenum is reduced from $+6$ to $+4$. This idealized model may be used to explain the data of Fig. 1. The linear correlation of reversibly adsorbed hydrogen, H_R, with vacancies may reflect the heterolytic cleavage of H_2 according to

$$[2]$$

This is identical to the process described by Kokes and Dent (18) for ZnO. However, other possibilities are not excluded.

In Fig. 7, the catalytic data are compared with the surface chemistry parameters determined in our earlier study (2). It is clear that the total loss, ($\square$), follows generally the vacancy concentration curve (long dashes) and that the propylene produced ($\bigcirc$) and the extent of isomerization ($\bullet$) compare favorably with H_I, the irreversibly held hydrogen (short dashes). The former relationship is described more quantitatively in Fig. 3.

The percent conversion (lined-out values of Fig. 2) are plotted versus $\square/Mo$ and H_I/Mo in Figs. 8 and 9, respectively. Recalling that over the range $0 < H_c < 1.0$, both H_I/Mo and $\square/Mo$ increase linearly (see dashed lines of Fig. 7), the first three points at low extent reduction would be expected to correlate in the same way on both curves; both are linear. As shown in Fig. 8, however, a further increase in vacancy concentration effects only a small increase in the percent conversion, whereas this quantity seems to correlate better with H_I/Mo (Fig. 9). These data suggest that the conversion of cyclopropane is primarily related to the introduction of Brönsted acid sites onto the surface during the reduction. This is in accordance with the generally accepted idea that the facile catalytic isomerization of cyclopropane is acid-catalyzed (3–6, 19, 20). It is further supported by the results of Kiviat and Petrakis (7) which showed that molybdena–alumina catalysts have the ability to convert pyridine to pyridinium ions, whereas the corresponding alumina supports do not, and that the ratio of Lewis to Brönsted sites decrease substantially with reduction.

The two flagged points on Fig. 9 lie well above the straight line; these higher isomerization rates correspond to H_I/Mo values lower than 0.5 by virtue of the fact that they were reduced beyond $H_e/Mo = 1.5$ into the range where the Mo^{+5} sites of Eq. [1] are being converted to Mo^{+4} (see Fig. 7). This increase in activity *per acid site* may be explained as due to an increase in the strength of the site caused by the increasing ratio of vacancies to Brönsted sites. At lower extents of reduction, the ratio $\square/H_I$ has values: 0.5, 0.5, 0.5, and 0.7; for the two points which deviate from the line, it has values of 1.6 and 2.3 corresponding to $H_e/Mo = 1.61$ and 1.7, respectively. A similar correlation of the increase in polarization of the OH-bond with increasing acid strength has been noted previously in zeolite catalysis; an increase in activity for acid catalyzed reactions invariably results from dehydroxylation of a hydrogen zeolite.

Significantly, the two flagged points of Fig. 9 discussed above can be brought onto a straight line if the percent conversion is plotted against the sum of the Brönsted sites and vacancies. This is shown on Fig. 10, where the same points are flagged.

The percent conversion represents the cyclopropane which has disappeared. At high extents of reduction, a substantial fraction of this is retained irreversibly by the catalyst (possibly on multiple vacancies). If it is

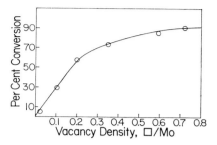

Fig. 8. Correlation of conversion of cyclopropane to all products at 109°C, $(\Delta^0 - \Delta)/\Delta^0$, with vacancy concentration.

assumed that vacancies react directly with cyclopropane to form a carbonaceous material, it may not be necessary to invoke the argument concerning the increased acidity of the Brönsted sites. This uncertainty is the result of the experimental technique used in this work; it may be resolved by future experiments with a conventional flow reactor.

The presence of ethylene and 2-butene can readily be explained by the olefin metathesis reaction with the propene formed by cyclopropane isomerization. The data of Figs. 3 and 4 indicate that this reaction is subsequent to the isomerization and that it is catalyzed by the anion vacancies introduced as H_2O is removed. Olefin disproportionation is catalyzed by a variety of supported heterogeneous catalysts including the oxides, carbonyls and sulfides of molybdenum and tungsten (21–29). Factors controlling the activity and selectivity of the catalyst (22, 29) have been reported and the kinetic data interpreted as a Rideal mechanism (30) or a Langmuir–Hinshelwood mechanism (27, 31, 32). The main characteristics of the reaction have been firmly established by tracer data (26, 36). It has been found that C–H bonds are not broken as the double bonds are taken apart and reassembled.

An attractive theory of olefin metathesis involving a transition state having a four-membered ring centered on a transition metal ion has been advanced by Mango and Schachtschneider (34). According to them this reaction, which is forbidden in the Woodward–

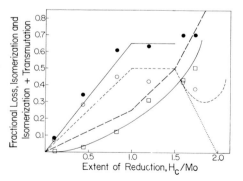

Fig. 7. Comparison of catalytic data with surface chemistry. Solid lines and points represent catalytic data; dashed curves are taken from Ref. (2), where they represent H_I/Mo and $\square/Mo$, (the latter being the longer dashes); dotted curve is an idealized extrapolation. $\bigcirc = P_e/\Delta^0$, $\square = L/\Delta^0$ and $\bullet = (P_e + 2E_e)/\Delta^0$.

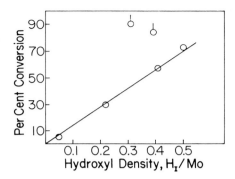

FIG. 9. Correlation of conversion of cyclopropane to all products at 109°C, $(\Delta^0 - \Delta)/\Delta^0$, with irreversibly held hydrogen introduced on reduction. Flagged points are for catalysts reduced to $H_e/Mo > 1.5$.

Hoffman sense (35), is made facile by a transition metal ion having degenerate d-orbitals matching in symmetry those required for the concerted conversion. According to this theory, the coordinatively unsaturated Mo^{+4} site would be ideally suited for this purpose; it is a d^2 ion having a geometry that could simultaneously hold two olefin molecules in adjacent ligand positions. These ideas will not be pursued further here, however, because it has been recently shown that they may be inapplicable. Grubbs *et al.* (36) and Luckner and Wills (37) have provided substantive evidence that olefin metathesis may, in fact, occur by a carbene mechanism. Distinction between these two possible reaction paths can be made only with mixtures of olefins more complicated than those usually investigated. Whatever the pathway, considerable evidence exists (26, 29, 37) that the electronic configuration of the metal ion, its oxidation number, and its symmetry are all important mechanistic requirements.

For catalysts reduced extensively, the amounts of butene in products from the first couple of pulses were invariably less than those of ethylene (see *Note added in proof*). A similar observation was made by Davie *et al.* (27). Probably the 2-butene from these pulses remained irreversibly adsorbed and thus contributed to the real loss. Of course, some hydrocarbon may have desorbed very very slowly

between pulses and became lost from the mass balance. For this reason, the extent of the metathesis reaction has been calculated as twice the number of ethylene molecules obtained; this hydrocarbon is much more weakly adsorbed.

One of the more surprising features (to us) of the present work was the very high ability of reduced molybdena–alumina to hydrogenate olefins. The olefins produced by cyclopropane isomerization were completely hydrogenated to the corresponding paraffins, both at 109°C and room temperature. When the hydrogenation of propene was studied, the results were even more impressive. Now all of the molecules contained in the pulse were hydrogenated, even at 0°C. (As shown in Table II, at low extents of reduction less than 25% of the cyclopropane was isomerized and consequently hydrogenated at room temperature.) Another interesting feature was that the ethene and butene formed by olefin metathesis were also hydrogenated to the corresponding paraffins. Moreover, metathesis was very greatly repressed when hydrogen carrying gas was used, and in fact, became less the higher the extent of reduction (Fig. 6). Since the reversibly held hydrogen appears to be chemisorbed on vacancies and since these vacancies also appear to catalyze the olefin metathesis reaction, these

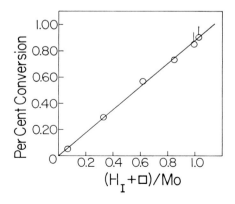

FIG. 10. Relationship of conversion of cyclopropane to all products at 109°C, $(\Delta^0 - \Delta)/\Delta^0$, with the sum of the concentrations of vacancies and acid sites. Flagged points are for $H_e/Mo > 1.5$.

data strongly suggest that hydrogenation and metathesis are competitive; they take place on the same sites, with hydrogenation being the faster. Interestingly, this catalyst which is so very active for olefin hydrogenation is quite inactive for paraffin hydrogenolysis (see Table IV).

One other feature needs to be noted. The extent of isomerization of cyclopropane was also increased when hydrogen-carrying gas was used with catalysts having equal extents of reduction. (Compare the data of Tables I and II.) Since this reaction is believed to be acid-catalyzed, it may be that the chemisorbed hydrogen, H_R, can contribute Brönsted sites in the form of the OH group of Eq. [2]. Alternatively, this observation may only reflect a cleaner surface maintained in H_2.

Note added in proof. Very recently, Gasman and Johnson (*J. Amer. Chem. Soc.* **98**, 6057 (1976)) have shown that olefin metathesis catalysts react with cyclopropane to form the methylene carbene, freeing C_2H_4 in the process. It is postulated that the metal ion inserts itself into the cyclopropane ring to form the metallocyclobutane which decomposes to the carbene and olefin. Possibly, the excess ethylene over butene observed in the present experiments stems from carbene formation.

ACKNOWLEDGMENTS

The support of this work by the National Science Foundation under Grant CHE74–11539 is gratefully acknowledged. Thanks are also due to the Gulf Oil Foundation for a gift of equipment and financial support to initiate it. One of us, MLJ, is grateful to the Consiglio Nazionale delle Ricerche, Italy, for a leave of absence to study at UWM.

REFERENCES

1. HALL, W. K. AND MASSOTH, F. E., *J. Catal.* **34**, 41 (1974).
2. HALL, W. K. AND LO JACONO, M., in "Preprints of Papers," Sixth International Congress on Catalysis, London, July 1976.
3. HALL, W. K., *Accounts Chem. Res.* **8**, 257 (1975).
4. HIGHTOWER J. W. AND HALL, W. K., *J. Amer. Chem. Soc.* **90**, 851 (1968).
5. LARSON, J. G., GERBERICH, H. R., AND HALL, W. K., *J. Amer. Chem. Soc.* **87**, 1880 (1965).
6. HIGHTOWER, J. W. AND HALL, W. K., *J. Phys. Chem.* **72**, 4555 (1968).
7. KIVIAT, F. E. AND PETRAKIS, L., *J. Phys. Chem.* **77**, 1232 (1973).
8. CHAMBERS, T. S. AND KISTIAKOWSKI, G. B., *J. Amer. Chem. Soc.* **56**, 399 (1934).
9. PRITCHARD, H. O., SNOWDEN, R. G., AND TROTMAN-DICKENSON, A. F., *Proc. Roy. Soc. (London) Ser. A* **217**, 563 (1953).
10. FALCONER, W. E., HUNTER, T. P., AND TROTMAN-DICKENSON, A. F., *J. Chem. Soc.* **1**, 609 (1961).
11. RABINOVITCH, B. S., SCHLAG, E. W., AND WIBERG, K. B., *J. Chem. Phys.* **28**, 504 (1958).
12. SCHLAG, E. W. AND RABINOVITCH, B. S., *J. Amer. Chem. Soc.* **82**, 5996 (1960).
13. BLADES, A. T., *Canad. J. Chem.* **39**, 1401 (1961).
14. BENSON, S. W., *J. Chem. Phys.* **34**, 521 (1961).
15. HALL, W. K., LUTINSKI, F. E., AND GERBERICH, H. R., *J. Catal.* **3**, 512 (1964).
16. BURWELL, R. L., JR., *Chem. Eng. News*, p. 56. August 22, 1966.
17. HALL, W. K. AND EMMETT, P. H., *J. Amer. Chem. Soc.* **79**, 2091 (1957).
18. KOKES, R. J. AND DENT, A. L., *Advan. Catal. Relat. Subj.* **22**, 1 (1972).
19. ROBERTS, R. M., *J. Phys. Chem.* **63**, 1400 (1959).
20. BASSETT, D. W. AND HABGOOD, H. W., *J. Phys. Chem.* **64**, 769 (1960).
21. BIELANSKI AND DATKA, J., *J. Catal.* **37**, 383 (1975).
22. BAILEY, G. C., *Catal. Rev.* **3**(1), 37 (1969).
23. KHIDAKEL, M. L., SHEBALDOVA, A. D., AND KALECHITS, I. V., *Russ. Chem. Rev.* **40**, 669 (1971).
24. BANKS, R. L. AND BAILEY, G. C., *Ind. Eng. Chem. Prod. Res. Dev.* **3**, 170 (1964).
25. HECKELSBERG, L. F., BANKS, R. L., AND BAILEY, G. C., *Ind. Eng. Chem. Prod. Res. Dev.* **8**, 259 (1969).
26. BRADSHAW, C. P. C., HOWMAN, E. J., AND TURNER, L., *J. Catal.* **7**, 269 (1967).
27. DAVIE, E. S., WHAN, D. A., AND KEMBALL, C., *J. Catal.* **24**, 272 (1972).
28. SMITH, J., HOWE, R. F., AND WHAN, D. A., *J. Catal.* **34**, 191 (1974).
29. KOBYLINSKI, T. P. AND SWIFT, H. E., *J. Catal.* **33**, 83 (1974).
30. BEGLEV, J. W. AND WILSON, R. T., *J. Catal.* **9**, 375 (1967).
31. LEWIS, M. J. AND WILLS, G. B., *J. Catal.* **15**, 140 (1969); **20**, 182 (1971).
32. MOFFAT, A. J. AND CLARK, A., *J. Catal.* **17**, 264 (1970).
33. MOL, J. C., MOULYN, J. A., AND BOELHOUWER, C. J., *J. Catal.* **11**, 87 (1968).
34. MANGO, F. D. AND SCHACHTSCHNEIDER, J. H., *J. Amer. Chem. Soc.* **93**, 1123 (1971); MANGO, F. D., *Advan. Catal.* **20**, 291 (1969).
35. WOODWARD, R. B. AND HOFFMAN, R., *J. Amer. Chem. Soc.* **87**, 2045 (1965).
36. GRUBBS, R. H., BURK, P. L., AND CARR, D. D., *J. Amer. Chem. Soc.* **97**, 3265 (1975).
37. LUCKNER, R. C. AND WILLS, G. B., *J. Catal.* **28**, 83 (1973).

Hydrogen Spillover in Catalytic Reactions

I. Activation of Alumina

STANISLAS JEAN TEICHNER, ALAIN RENÉ MAZABRARD, GÉRARD PAJONK,
GÉRARD E. E. GARDES, AND CAN HOANG–VAN

*Laboratoire de Thermodynamique et Cinétique Chimiques (L. A. N° 231 CNRS), Université
Claude Bernard (Lyon I) 69621, Villeurbanne, France*

Received June 2, 1976; accepted September 8, 1976

Hydrogen, adsorbed and dissociated by the metal (Pt or Ni) supported on alumina, migrates on admixed alumina and, depending on the nature of the metal, of alumina and on the temperature (i) participates in some reactions such as hydrogenation of ethylene, or (ii) activates alumina for hydrogenation (ethylene, butenes, benzene) or isomerization (methylcyclopropane) reactions, occurring at fairly low temperatures with high selectivity for some products. This new behavior of alumina is easily observed because the catalyst (metal on alumina) is removed from admixed alumina in a special design reactor, provided with a lifting device. These selective reactions on alumina are masked by the metal part of the oxide-supported catalyst in the conventional reactors where the catalyst cannot be separated from admixed oxide.

A. INTRODUCTION

Molecular hydrogen is dissociatively chemisorbed on metals and participates in various hydrogenation reactions. However, most metal catalysts also contain a carrier, such as silica or alumina, in order to preserve the dispersion of the metal, to contribute the mechanical properties of the catalyst, and to provide, in some cases, a second catalytic function (such as isomerization) due to the carrier. It was shown by various experiments (1) that hydrogen atoms can migrate from a metal such as Pt or Pd to another substance, in contact with the metal. This phenomenon was termed "hydrogen spillover." In the case of a catalyst, a carrier such as alumina is the first substance which accepts hydrogen. However, this migration may extend further to a second hydrogen acceptor, like tungsten trioxide, in mechanical contact with the catalyst. In this way, hydrogen bronzes of WO_3, of a blue color, were formed at ambient temperature (2), exhibiting a composition H_xWO_3 ($x < 0.46$) (3). With molecular hydrogen, in the absence of a catalyst (Pt or Pd) this reaction, occurs

only above 500°C (4). Finally, migrating hydrogen atoms may reduce the acceptor, partially or completely. This reduction occurs when Pt or Pd is placed directly on the oxide to be reduced or even when the metal supported on alumina or silica carrier is physically admixed with the oxide to be reduced. The temperature of the reduction of the oxide is either markedly decreased (V_2O_5, CrO_3, MoO_3, WO_3, UO_3, Re_2O_7, CO_3O_4) or the catalyst has only a small effect on the reduction (CuO, Cu_2O, NiO, ZnO, SnO_2). The requirement for a pronounced effect is the availability of an oxidation state one below that in the oxide being reduced (5).

Now, "hydrogen spillover" was also investigated in the case of an organic acceptor reagent to be hydrogenated. By admixing supported metal catalyst with further carrier (mainly alumina), an enhancement of the catalytic activity, in comparison to a non-diluted catalyst, was observed. A Pt (0.05%) supported on silica catalyst, diluted with alumina (1:9), was seven times more active in the hydrogenation of ethylene than the undiluted

93

catalyst (6). Similarly, the activity in the hydrogenation of benzene was increased by dilution (1:200) with further alumina of a Pd (2.2%) supported on alumina catalyst. The enhancement of the activity was explained by the reaction of the hydrocarbon, adsorbed on the diluent surface, with the hydrogen spilled over.

These results were questioned, however, by some authors (7, 8) who were unable to reproduce these experiments. If alumina diluent may act as a scavenger for any contaminant of the catalyst, an enhancement of the catalytic activity may be observed, but it is not correlated with hydrogen spillover (8).

It was therefore uncertain whether the increase of the activity of supported metal, after dilution with the support, was due to hydrogen spillover or to some other phenomenon. This problem was difficult to solve because the effect of the catalyst in a reaction may be a major one and the added contribution of the admixed carrier could escape the observation, as for the negative experiments previously quoted (7, 8). To yield unambiguous results, we imagined a reactor in which the catalyst, first in contact with the extra carrier (acceptor), may be withdrawn and isolated after hydrogen spillover (9, 10). The interaction between the hydrocarbon to be hydrogenated and the hydrogen, molecular and spilled over, is therefore observed in the absence of the supported metal, with only the carrier present in the reactor. The previous results concerning the hydrogenation of ethylene, after hydrogen spillover from an Ni/Al$_2$O$_3$ catalyst to alumina, will be briefly recalled and the new effect, observed on alumina after hydrogen spillover from a Pt/Al$_2$O$_3$ catalyst, will be described in detail.

B. EXPERIMENTAL METHODS AND MATERIALS

(a) Reactor

The reactor shown in Fig. 1 was described in detail heretofore (9, 10). The catalyst (10–30 mg) is placed on the bottom of the

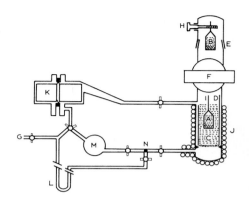

Fig. 1. Reactor allowing the activation of alumina by a metal on alumina catalyst.

suspended pan A and is covered by alumina C (about 1 g) on which the effect of hydrogen spillover is to be studied. The holder D has a porous glass bottom in this constant volume recirculation reactor. The lifting device H allows the removal of the catalyst (with some oxide) in the pan suspended by an inert wire and which takes the position B. The stopcock F then isolates the catalyst in the pan from the reactor during the catalytic run. The vacuum line and the gas supplies are connected through the stopcock G. A recirculation pump K is provided for homogenization of the gas mixture. The reactor may be heated by the oven J. The liquid nitrogen (or dry ice) trap L is included in the circuit agitated by the pump. A calibrated volume M allows the introduction of a known amount of reagent into the reactor. Sampling for gas chromatographic analysis is performed through the device N.

(b) Catalysts

Nickel (28%) on amorphous alumina catalyst is of an aerogel type (11) and has a surface area of 650 m^2/g with nickel surface area of 9.2 m^2 Ni/g catalyst. Platinum (2.7%) on amorphous alumina catalyst was described previously (12). Its surface area is 110 m^2/g, with platinum surface area of 3.65 m^2 Pt/g catalyst.

(c) Aluminas

Aluminas used as acceptors of hydrogen spilled over from the catalyst belong to three types:

(i) δ-alumina, prepared in the laboratory by hydrolysis of aluminum chloride vapor in the hydrogen–oxygen flame (13), is very pure (impurity content <5 ppm per element) and of a nonporous type, with a surface area of 110 m^2/g (14). Also, commercial (Degussa P 110) δ-alumina was used, with or without a further purification ("standardization") in an oxygen flow at 400°C for 15 hr.

(ii) Amorphous alumina aerogel (porous) was prepared in a non-aqueous medium in the autoclave (11, 15) and has a surface area of 500 m^2/g and an impurity content <35 ppm per element.

(iii) Amorphous alumina xerogel (porous) was prepared in an aqueous medium (16) and has a surface area of 120 m^2/g and an impurity content <10 ppm per element.

Hydrogen and organic reagents were of a purity better than 99.95%.

C. HYDROGEN SPILLOVER FROM NICKEL CATALYST

In these experiments, Ni/Al$_2$O$_3$ catalyst (30 mg) is covered by amorphous alumina aerogel (1 g) in the reactor of Fig. 1 and the system is heated in H$_2$ (760 Torr) for 8–13 hr, then cooled in H$_2$ to 110°C, still in the presence of the catalyst, and maintained at this temperature for 8 hr. Finally, the catalyst (and some alumina) is withdrawn from the reactor by the lifting device and isolated by the stopcock F. These conditions were found to be the best for the hydrogenation at 110°C of 50 cm^3 of ethylene introduced in the reactor from the volume M. Indeed, 45 cm^3 of ethane are formed in 20 hr (10). An evacuation of this system after the first test followed by the introduction of the reaction mixture (H$_2$ + C$_2$H$_4$), at 110°C, leads only to traces of ethane after 20 hr. The catalytic properties of alumina activated by hydrogen spillover are therefore exhausted after the first test. If the migration of nickel from catalyst to alumina during the activation was responsible for the high catalytic activity in the first test, an activity of the same order of magnitude should be observed in the second test. Moreover, if in the activation treatment the cooling in hydrogen in the presence of the catalyst is performed down to 25°C and then only the catalyst is removed, only 0.5 to 2.5 cm^3 of ethane are formed (from 50 cm^3 of C$_2$H$_4$) at 110°C in 20 hr. These unfavorable results encountered when the temperature of the system is decreased to 25°C are also observed for platinum catalyst activation of alumina and will be discussed later. They also show that the migration of nickel from catalyst to alumina cannot be responsible for the reaction, which would proceed to the same extent (at 110°C) with or without intermediate cooling of alumina to 25°C.

The amount of ethane formed (45 cm^3 in 20 hr) after the standard activation procedure is barely related to the amount of hydrogen (1.3 cm^3) which is spilled over, at 110°C, on alumina (1 g) in contact with nickel catalyst (30 mg). Molecular hydrogen present in the

TABLE I

Hydrogenation at 110°C of Ethylene (50 cm^3) by Aluminas Activated in H$_2$
at 300°C in the Presence of Ni/Al$_2$O$_3$ Catalyst

Alumina	% of C$_2$H$_4$ transformed	Influence of NO	% of C$_2$H$_4$ transformed in a second test	Influence of NO
δ (Lab.)	100% in 20 hr	Inhibition	37% in 2¼ hr	Inhibition
δ (Degussa)	30% in 40 hr	Inhibition	—	Inhibition
Amorphous xerogel	5% in 75 hr	—	2% in 50 hr	—
Amorphous aerogel	90% in 20 hr	Inhibition	3% in 50 hr	—

system (around 1000 cm³) must also be involved in the hydrogenation of ethylene (45 cm³). But if the activation treatment of alumina in H_2 at 300°C is performed without the catalyst, or if after the standard activation procedure (with the catalyst), hydrogen is evacuated at 110°C prior to the reaction $H_2 + C_2H_4$, only negligible amounts of ethane are formed after 20 hr. These experiments show that the spilled-over hydrogen is also required for the catalytic activity. The evacuation of hydrogen prior to the test shows, moreover, that the migration of nickel from catalyst to alumina is excluded. If not, alumina should be active even after evacuation of hydrogen at 110°C.

Because of the simultaneous need for adsorbed (spilled-over) hydrogen and molecular hydrogen, in order to hydrogenate ethylene, a chain reaction in the adsorbed phase was envisaged (10) where the propagation steps are

$$C_2H_4 + H \rightarrow C_2H_5,$$
$$C_2H_5 + H_2 \rightarrow C_2H_6 + H.$$

The inhibiting effect of NO injected (0.5 cm³) during the reaction is in agreement with this scheme, NO acting as radical scavenger.

In conclusion, these experiments have shown that hydrogen, spilled over alumina, participates, together with molecular hydrogen, in the hydrogenation of ethylene. This phenomenon would not be detected in a conventional

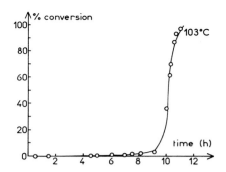

FIG. 3. Second test of conversion of ethylene at 103°C on δ-alumina after evacuation at 430°C.

reactor in which the catalyst, admixed with an extra alumina, remains in the system and participates (more effectively than Al_2O_3) in the reaction.

It is shown below that the activation of δ-alumina in hydrogen by platinum catalyst requires a temperature of 430°C. Nickel catalyst on amorphous alumina aerogel activated at 430°C and on two other types of alumina gives negative results for the hydrogenation of ethylene at 110°C. The results concerning the activation by nickel catalyst at 300°C of δ-aluminas and amorphous aluminas are summarized in Table I.

The main difference from the previous results, concerning amorphous alumina aerogel, is the permanent catalytic behavior of δ-aluminas (Table I). Indeed, after evacuation of the system at 110°C (10⁻⁵ Torr) after the first test, the hydrogenation of 50 cm³ of C_2H_4 in a second test gave similar results, whereas the activity of amorphous alumina aerogel decreased enormously. The spilled-over hydrogen must be exhausted for this alumina after the first test (chain reaction mechanism) and in the second test only a few sites of a different type (permanent) are active in the reaction, whereas δ-alumina behavior tends to show that its sites are able to activate molecular hydrogen (perhaps as H) in a permanent way. This property involves only a few sites of amorphous alumina aerogel and the spilled-over hydrogen is required for high

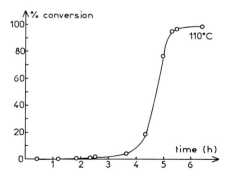

FIG. 2. Conversion at 110°C of ethylene into ethane on δ-alumina activated by Pt/Al_2O_3 catalyst.

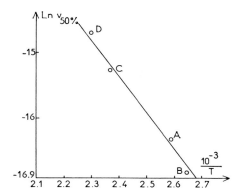

FIG. 4. Activation energy determination for the fast reaction in successive tests. A, first test; B, C, D, successive tests.

activity. Amorphous alumina xerogel does not retain spilled-over hydrogen and its initial activity is comparable to that of amorphous alumina aerogel after the first test.

The permanent catalytic behavior, for all aluminas, is encountered after the activation by Pt/Al_2O_3 catalyst. The nature of the alumina as well as the nature of the activating catalyst determine, therefore, the modifications produced on the acceptor oxide by the spilled-over hydrogen.

D. HYDROGEN SPILLOVER FROM PLATINUM CATALYST

1. Hydrogenation of Ethylene

The experiments described below were performed on commercial δ-alumina (Degussa P 110) but a comparison was also made with three other types of alumina.

(a) The catalyst is separated from alumina at the reaction temperature (above 100°C). As the preliminary results have shown that δ-alumina is practically inactive in the hydrogenation of ethylene after activation at 300°C in the presence of Pt/Al_2O_3 catalyst and hydrogen, the temperature of activation was raised to 430°C. The standard activation procedure is then composed of the following treatments: (i) evacuation (10^{-5} Torr) of the system (δ-alumina with Pt/Al_2O_3 catalyst in position A of Fig. 1)

at 430°C for 8 hr, (ii) interaction with 760 Torr of H_2 at 430°C for 12 hr, (iii) cooling (for 1 hr 30 min) to the reaction temperature (100°C and above), (iv) removal and isolation of the catalyst (with some alumina), (v) introduction of 50 cm³ of ethylene. In the second procedure, described below, the catalyst is removed at 430°C (instead of 100°C and above) and isolated. The kinetics of the hydrogenation of ethylene are different for these two procedures of activation.

For the first activation procedure the results are shown in Fig. 2, giving the conversion of ethylene into ethane at 110°C as a function of time. An induction period (about 4 hr) precedes a fast reaction period (about 2 hr). If the same sample of alumina is then evacuated (10^{-5} Torr) at 430°C (8 hr) and hydrogen (760 Torr) is introduced at this temperature (12 hr), which is then decreased (1 hr 30 min) to 103°C and 50 cm³ of ethylene is introduced, the reaction proceeds in a similar way according to Fig. 3. The catalytic properties of δ-alumina activated by Pt/Al_2O_3 catalyst are therefore of a permanent character. This behavior allows the determination of the activation energy of reaction for successive tests at various temperatures, separated by evacuation at 430°C. The activation energy of the fast reaction is 11.5 kcal/mole (Fig. 4), whereas the activation energy of the induction period is 18 kcal/mole (Fig. 5). Point A (reaction

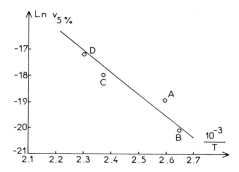

FIG. 5. Activation energy determination for the induction period in successive tests. A, first test; B, C, D, successive tests.

after the activation by the catalyst in H_2) lies on the same straight line as points B, C, and D corresponding to successive tests separated by evacuation at 430°C (without further contact with the catalyst). The sites created after the activation conserve their character after many reactions and evacuations. This behavior is therefore different from that shown by amorphous alumina aerogel activated in H_2 at 300°C by Ni/Al_2O_3 catalyst where this permanent character was not observed. Also, there is no inhibiting influence of NO detected nor is the evacuation of hydrogen, prior to the reaction, detrimental to the conversion of ethylene. It is inferred from these results that the spilled-over hydrogen is no longer required (through the chain mechanism) for the reaction to proceed. δ-alumina is modified by the activation treatment in such a way as to develop some permanent active sites on its surface. The reaction here would be of a conventional, associative, type. But in this case, several successive doses (50 cm³) of ethylene should be converted, with an excess of hydrogen (1000 cm³) present in the reactor volume.

Figure 6 shows two successive conversions at 120°C of ethylene (without evacuation at 430°C) during which the absence of the inhibiting effect of NO was registered. The induction period is no longer observed for the second and third doses of ethylene. This result tends to show that the formation of

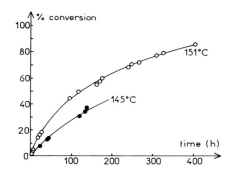

FIG. 7. Conversion of ethylene into ethane on δ-alumina activated by Pt/Al_2O_3 catalyst according to the second procedure and cooled to the reaction temperature (see text).

active sites, which requires an induction period, results from the first interaction around 100°C between the activated alumina, ethylene, and hydrogen. The second dose of ethylene reacts on the same sites without the induction period. If after the first test the system is evacuated at 430°C (Fig. 3), the induction period reappears because the active sites, resulting from the first interaction between alumina and the reagents, are destroyed by evacuation. The active sites should therefore be formed by some adsorbed species resulting from this interaction and desorbed during evacuation.

The conditions of the formation of these species, active in the hydrogenation of ethylene around 100°C, are sensitive to the temperature of contact with hydrogen. Indeed, if after the first test δ-alumina is evacuated at 430°C, subjected to H_2 (760 Torr), and cooled down (1 hr 30 min) to 25°C instead of 110°C, the test, performed after the temperature is raised to 110°C and C_2H_4 introduced, is negative. Hydrogen adsorbed on δ-alumina between 110°C and 25°C is therefore unable to form the active adsorbed species. This detrimental influence of cooling in hydrogen to 25°C was previously mentioned in the case of amorphous alumina aerogel activated by Ni/Al_2O_3 catalyst.

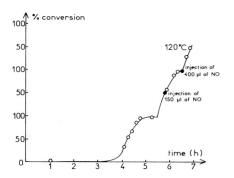

FIG. 6. Three successive conversions of ethylene at 120°C without evacuation between the tests. Influence of NO.

These results also show that platinum does not migrate from catalyst to alumina.

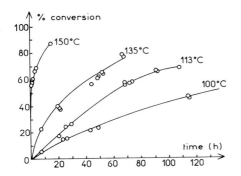

FIG. 8. Conversion of ethylene into ethane on δ-alumina activated by Pt/Al₂O₃ catalyst according to the second procedure and cooled to 25°C (see text).

Indeed, it would be immaterial, if platinum was present, to decrease the temperature of the system from 430°C to 110°C, with or without a plateau at 25°C, to observe the reaction.

If after this unsuccessful attempt with δ-alumina, cooled down to 25°C, the sample is evacuated at 430°C and cooled in H_2 from this temperature to 110°C only, the new test shows that the reaction proceeds in the usual way (induction period and fast reaction). Hydrogen adsorbed at 25°C in an unfavorable form is therefore eliminated, the sites on alumina conserving their properties towards a new formation of adsorbed species at 110°C, active in the reaction.

Finally, the sites created on δ-alumina during activation are not destroyed by the contact with air at 25°C. Indeed an evacuation of such a sample at 430°C followed by cooling in H_2 to 110°C produces results in the hydrogenation of ethylene similar to those of Fig. 3. It was also checked that the catalytic aptitude of δ-alumina is not created by activation at 430°C in the absence of Pt/Al₂O₃ catalyst or of H_2. No traces of Pt were detected (by neutron activation) on activated δ-alumina.

(b) *The catalyst is separated from alumina at 430°C.* If this second procedure of activation is applied and δ-alumina is cooled (1 hr 30 min) in H_2 from 430°C to the reaction temperature in the absence of the catalyst, the kinetics of the reaction are different, as shown in Fig. 7.

At 150°C, the reaction requires 600 hr to reach completion and the induction period is no longer observed. A second test, performed after evacuation at 430°C, give similar results. This activation procedure tends, therefore, to create different types of sites on δ-alumina which are less active than those formed in the first activation procedure. The absence of the induction period may indicate the absence of interaction between alumina, hydrogen, and ethylene forming the final active site.

If after the second type of activation procedure (removal of the catalyst at 430°C), δ-alumina is cooled (1 hr 30 min) in H_2 from 430° to 25°C (15 min) and then reheated (30 min) to 150°C, ethylene (50 cm³) is entirely converted into ethane after only 18 hr instead of 600 hr and the induction period is again unobserved (Fig. 8). It is recalled that for the first activation procedure (removal of the catalyst at 110°C) cooling in H_2 down to 25°C is detrimental to the reaction performed at 110°C.

If after the first test (Fig. 8), δ-alumina is evacuated at 430°C and then cooled in H_2 down to 25°C and finally re-treated to various reaction temperatures, for similar tests the corresponding curves (Fig. 8) exhibit similar trends. The activation energy calculated from these plots is 33 kcal/mole. The same order of magnitude of the activation energy is calculated from the plots of Fig. 7 (low activity). Once the catalyst is removed at 430°C, the cooling in hydrogen down to 25°C after this activation procedure, or between the successive tests separated by an evacuation at 430°C, develops more active sites than cooling in H_2 to 110°C.

Summing up, the removal of the catalyst at 430°C creates permanent sites in δ-alumina of type S_1 which are moderately active, around 110°C, in the hydrogenation of ethylene and do not exhibit an induction period. These sites are very active if, prior to the reaction, δ-alumina is cooled in H_2 down to 25°C.

If the catalyst is conserved in δ-alumina during cooling from 430°C to around 110°C, a second type of site S_2 is created. These sites

TABLE II

Hydrogenation at 110°C of Ethylene (50 cm³) by Aluminas Activated in H_2 at 430°C or at 300°C in the Presence of Pt/Al_2O_3 Catalyst

Alumina	Activation at			
	430°C		300°C	
	% of C_2H_4 transformed	Influence of NO	% of C_2H_4 transformed	Influence of NO
δ (Degussa)	100% in 8–10 hr	No effect	23% in 150 hr	—
δ (Lab.)	100% in 30 min.	No effect		
Amorphous xerogel	50% in 35 hr	No effect	25% in 25 hr	No effect
Amorphous aerogel	Undetectable	—	95% in 24 hr	No effect

become active after interaction with hydrogen and ethylene (induction period) at the reaction temperature. But if δ-alumina having sites S_2 is cooled to 25°C prior to the reaction (around 110°C) the catalytic activity is destroyed, probably by a different interaction between S_2 and H_2 at 25°C compared to that at 110°C. The species formed with C_2H_4 after this interaction at 25°C do not form active sites for the reaction at 110°C, whereas the species formed with C_2H_4 after the interaction with H_2 of the same sites S_2, but at 110°C instead of 25°C, are active sites for the reaction. It has been reported in a few cases for alumina that an active site for an isomerization reaction is formed by the interaction of a site on an alumina surface with the olefin (17).

(c) *Activation of other types of alumina.* The activation in H_2 in the presence of Pt/Al_2O_3 catalyst was performed at 430°C or at 300°C. For both temperatures, the catalyst was removed after cooling in H_2 to the reaction temperature (110°C). The results are summarized in Table II.

It can be seen that the purest δ-alumina (prepared in the laboratory) is more active than commercial (Degussa) "standardized" δ-alumina which, in turn, is more active than the commercial "unstandardized" δ-alumina still containing some chlorine. In all cases an induction period is observed which characterizes the type of activation procedure used for the experiments of Table II. A second test after the evacuation of alumina samples (at 430°C or at 300°C) gives comparable results

showing the permanent character of created sites.

The results of Table II allow the following classification of the catalytic activity of alumina samples with respect to the activation temperature in H_2 in the presence of Pt/Al_2O_3 catalyst:

at 430°C: δ > amorphous xerogel
> amorphous aerogel,

at 300°C: amorphous aerogel
> amorphous xerogel > δ.

The sequence is reversed by changing the temperature of activation. A first conclusion deduced from this classification concerns the absence of a migration of platinum from the catalyst to alumina, during activation. Indeed, in the opposite case, all aluminas would exhibit comparable catalytic properties, irrespective of their nature and of the temperature of activation. The second conclusion comes on recalling the sequence of activity established after activation in the presence of Ni/Al_2O_3 catalyst at 300°C (the activation at 430°C gives negative results for all samples). This sequence is

δ ∼ amorphous aerogel > amorphous xerogel.

The metal–hydrogen bond strength may be involved in hydrogen spillover from metal to acceptor, because the temperature is the criterion of the activation, at least for nickel (no activation at 430°C). But for platinum, where activation is possible at both temperatures, and also for nickel, active at 300°C, the nature of alumina is the second property to be concerned with in this phenomenon. As long as the

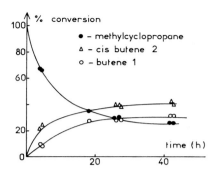

FIG. 9. Isomerization of methylcyclopropane at 25°C on amorphous alumina xerogel activated by Pt/Al₂O₃ catalyst.

surface properties of aluminas used in this work, after activation in H_2 in the presence of the catalyst, remain unknown (possibility of surface reduction of alumina, formation of surface lattice defects, exposed cations, basic or acid sites, surface stabilization of protons, etc.), the nature of catalytic sites created by activation will remain unsolved.

2. *Isomerization of Methylcyclopropane and Hydrogenation of Butenes*

If the interaction between the spilled-over hydrogen and the surface of alumina leads to some reduction of the oxide, an eventual formation of acid sites may be envisaged as it was shown previously for the amorphous alumina xerogel (18). These sites should be active for some isomerization reactions. As the hydrogen spillover seems to be involved in hydrogenolysis of methylcyclopropane on Pt/Al₂O₃ catalyst admixed with extra alumina (amorphous xerogel) or silica–alumina (19), an attempt was made at the isomerization of this cycloparaffine on amorphous alumina xerogel activated by Pt/Al₂O₃ catalyst. The activation in H_2 (760 Torr) at 430°C was performed in the way already described (14 hr). The system was cooled to 25°C (1 hr 30 min) and the catalyst was withdrawn and isolated. A mixture of 50 cm³ of methylcyclopropane (7.6%) and helium was then introduced into the reactor. After about 40 hr at 25°C a

pseudoequilibrium was established (Fig. 9) corresponding to the formation of *cis*-butene-2 and butene-1, without detectable traces of *trans*-butene-2 (20). Neither butane nor isobutane was found. This last negative result gives an argument against the migration of platinum to alumina, because otherwise the only products of reaction would be butane and isobutane (19). The isomerizing activity of alumina, characterized by a high stereoselectivity (no *trans*-butene-2), is probably correlated with a formation of peculiar sites on alumina activated in H_2 by Pt/Al₂O₃ catalyst.

Indeed, if the activation procedure of alumina is performed under H_2 but without Pt/Al₂O₃ catalyst, or with this catalyst but under helium, no isomerization of methylcyclopropane is recorded at 25°C. The reaction becomes possible on nonactivated alumina only above 200°C and without stereospecificity.

The active sites created by hydrogen spillover have a permanent character. A second test with 50 cm³ of the previous mixture, after a preliminary evacuation of alumina at 430°C (10^{-5} Torr) and introduction of helium at this temperature followed by cooling down to 25°C, gives results very similar to those shown in Fig. 9. Also it may be inferred that hydrogen is not involved in the mechanism of the skeletal isomerization of methylcyclopropane because in the second test only helium is present.

If after the activation of amorphous alumina xerogel and after the isomerization of methylcyclopropane in the presence of hydrogen at 25°C (Fig. 9), the temperature is increased to 151°C, it is observed that within 80 hr the hydrocarbon mixture is composed of 50% butane, 10% *cis*-butene-2, 20% butene-1, and 20% methylcyclopropane. *Cis*-butene-2 is converted to butane more rapidly than butene-1. Isobutane was not found. If the platinum was present the remaining methylcyclopropane would be converted, in part, into isobutane. The same sample of alumina is therefore able to isomerize methylcyclopropane at 25°C into olefins at 25°C and hydrogenate these products around 150°C in the presence of hydrogen. The hydrogenating activity, which was previously demonstrated in the case of ethylene, is thus conserved. Also, the same sample of alumina, after evacuation, hydrogenates 50 cm³ of ethylene in 20 hr at around 140°C, with an activation energy for the fast reaction of 16 kcal/mole.

The skeletal isomerization activity is usually associated with the presence of acid sites on the oxide catalyst surface. Hence, an ammonia poisoning test was performed by injecting NH_3 (3.5 Torr) into the system during isomerization at 25°C of methylcyclopropane on amorphous alumina xerogel. The formation of butenes was immediately stopped. If, however, after evacuation at 430°C of the catalyst, ammonia is injected during the standard test of hydrogenation of ethylene at 125°C, no poisoning effect toward the formation of ethane is observed. Two types of sites seem, therefore, to be developed on alumina by the activation treatment. The sites responsible for the skeletal isomerization of methylcyclopropane at 25°C are inhibited by ammonia, whereas those responsible for the hydrogenation of ethylene above 100°C are not sensitive to this poison.

The sites active in the isomerization at 25°C are not developed by a mere dehydroxylation of the alumina surface at high temperature (430°C here), as is the case for sites active above 200°C (21). If they *were* developed, an isomerization activity would be observed after activation in hydrogen at 430°C in the absence of Pt/Al_2O_3 catalyst, or with this catalyst but in the presence of helium instead of hydrogen.

A second explanation of the isomerization—the presence of protons (Brönsted acidity) resulting from the migration of hydrogen (as H^+) from the catalyst to alumina during activation—may be rejected. This hypothesis was formulated by Khoobiar (2) in order to explain the migration of hydrogen on an oxide. However, in the second test of isomerization, made after evacuation at 430°C, only helium (and methylcyclopropane) is present (no H_2). The

sites formed on alumina by activation and which exhibit a permanent character in isomerization are, therefore, rather of a Lewis type.

Finally, the isomerization activity is not restricted to amorphous alumina xerogel. Indeed δ-alumina (Lab.) activated by the same process as that mentioned previously for amorphous alumina, gives a positive test for the isomerization of methylcyclopropane at 25°C. The composition of the hydrocarbon mixture after 30 hr of reaction is: methylcyclopropane, 60%; *cis*-butene-2, 25%; butene-1, 15%. Again, no isobutane and butane are formed and hence the possibility of migration of platinum may be rejected. Similarly, the high stereoselectivity is observed (no *trans*-butene-2).

3. Hydrogenation of Benzene

After the demonstration of the hydrogenation activity toward olefins (ethylene and butenes) of alumina activated by hydrogen spillover, it seemed worthwhile to investigate an eventual aptitude in the hydrogenation of aromatics. A blank showed that δ-alumina (Degussa or prepared in the laboratory), heated in H_2 at 430°C (without the catalyst), exhibits no catalytic property toward hydrogenation of benzene.

In the test performed in a differential dynamic reactor (22) under a flow (1.1 liters/hr) of hydrogen (708 Torr) and of vapor of benzene (52 Torr) on 60 to 80 mg of alumina, which was previously activated in H_2 by Pt/Al_2O_3 catalyst at 430°C (the catalyst being withdrawn and isolated at 110°C), the conversion of benzene at the stationary state, at 144°C, is of the order of 1%, which is a value usually achieved in a differential reactor. For δ-alumina (Degussa), the products of reaction are cyclohexane and cyclohexene with a ratio of 2:1. This high selectivity (33%) into cyclohexene again gives an argument against the migration of platinum, from the catalyst to alumina, the platinum exhibiting an aptitude for the total hydrogenation of benzene.

The activation in H_2 of alumina (δ and amorphous xerogel) by Ni/Al_2O_3 catalyst at 300°C gives negative results for the hydrogenation of benzene between 144° and 170°C.

In the static test performed in the reactor of Fig. 1 on δ-alumina (Lab.) activated at 430°C under H_2 with Pt/Al_2O_3 catalyst (withdrawn at 110°C), the hydrogenation of benzene (8 Torr) at 115°C shows a conversion of 2.5% after 70 hr, with a composition: cyclohexane—1.5% and cyclohexadiene—1%. If the temperature is then raised to 180°C, the conversion after 40 hr is of 30%, with a composition: cyclohexane, 22%; cyclohexadiene, 8%; and no traces of cyclohexene. The relatively high proportion of cyclohexadiene in the static conditions again gives an argument against the migration of platinum onto alumina. The cyclohexene is found only at the beginning of the static test at 115°C.

A more detailed examination of the hydrogenation of benzene is in progress.

SUMMARY AND CONCLUSIONS

Some new and unforeseen catalytic properties of alumina, resulting from the hydrogen spillover initiated by a nickel or platinum on alumina catalyst, are described. These properties can be observed because after the spillover the catalyst is separated from alumina to be activated and isolated. Only alumina remains in the reactor and analysis by atomic absorption spectroscopy for nickel content of alumina (10) or by neutron activation for platinum is negative. The catalytic properties so far investigated are of two types. The hydrogenation of olefins (ethylene, butenes) which is observed at around 100°C and the hydrogenation of benzene at around 140°C belong to the first type. The second type of catalysis, observed at 25°C, is the skeletal isomerization (methylcyclopropane) and the double bond shift (butenes). In this way, amorphous alumina aerogel which has been submitted at 300°C to a hydrogen treatment in the presence of Ni/Al_2O_3 catalyst is, after removal of the catalyst, an efficient contact for the hydrogen-

ation of ethylene around 110°C. Here this reaction implies spilled-over hydrogen atoms and molecular hydrogen from the gas phase through the chain reaction process.

For Pt/Al_2O_3 catalyst, at 300° or 430°C, depending on the nature of alumina (amorphous or δ), the hydrogen spillover from platinum modifies alumina in such a way that this oxide remains active in hydrogenation even if the spilled-over hydrogen is removed. The sites created on alumina are active in the hydrogenation of olefins (ethylene, butenes) and benzene above 100°C and also are able to isomerize methylcyclopropane at 25°C.

Because of such novel behavior of alumina, the hypothesis of a migration of the metal, from the catalyst to alumina, was envisaged, but many cross checks did not finally give support to this possibility. For instance, if nickel migrates to alumina at 300°C (in an amount which cannot be detected by analysis) and is responsible for the hydrogenation of ethylene, this migration should still be more important at 430°C. Now, whereas the activation of alumina in H_2 in the presence of Ni/Al_2O_3 catalyst at 300°C gives positive results for the hydrogenation of ethylene, the activation at 430°C results in a complete lack of catalytic activity. On the other hand, various types of alumina (δ, amorphous xerogel, amorphous aerogel) are activated (or are not) to various extents for given activation conditions, in H_2 in the presence of the catalyst (Pt or Ni). This behavior is difficult to understand if the metal migrates to alumina; rather, this migration depends on the physical conditions (23).

Also, cooling in H_2 down to 25°C of aluminas activated by Pt/Al_2O_3 catalyst results either in a lack of activity at 110°C, or in an increase of activity at 110°C, depending on the activation procedure (catalyst withdrawn at 110°C or at 430°C). If platinum migrates on alumina and is responsible for the hydrogenation activity it is difficult to understand why in the first case it loses its activity upon being cooled down to 25°C whereas in the second case this activity is enhanced.

Finally, the high selectivity for some reaction products is in contradiction with the presence of the metal. For instance, the isomerization of methylcyclopropane gives butene-1 and *cis*-butene-2 and no traces of *trans*-butene-2 and of butanes. The hydrogenation of benzene gives cyclohexadiene (static conditions) or cyclohexene (dynamic conditions) in addition to cyclohexane.

The activation by hydrogen spillover onto an oxide, such as alumina, for unusual (as far as alumina is concerned) catalytic reactions probably results from a modification of surface properties of the oxide. This peculiar activation procedure is probably not restricted to different varieties of alumina but may be exercised on other materials. In each case the explanation of the catalytic behavior should be based on the modification of surface properties by hydrogen spillover.

REFERENCES

1. BOND, G. C., AND SERMON, P. A., *Catal. Rev.* **8**, 211 (1973).
2. KHOOBIAR, S. J., *J. Phys. Chem.* **68**, 411 (1964).
3. SERMON, P. A., AND BOND, G. C., *J. Chem. Soc. Faraday I* **72**, 730 (1976).
4. TANSTER, S. J. AND SINFELT, J. H., *J. Phys. Chem.* **74**, 3831 (1970).
5. BOND, G. C., AND TRIPATHI, JAI B. P., *J. Chem. Soc. Faraday I* **72**, 933 (1976).
6. CARTER, J. L., LUCCHESI, P. J., SINFELT, J. H., AND YATES, D. J. C., *Int. Congr. Catal. 3rd* **1**, 644 (1965).
7. VANNICE, M. A., AND NEIKAM, W. C., *J. Catal.* **23**, 401 (1971).
8. SCHLATTER, J. C., AND BOUDART, M., *J. Catal.* **24**, 482 (1972).
9. GARDES, G. E. E., PAJONK, G. M., AND TEICHNER, S. J., *J. Catal.* **33**, 145 (1974).
10. BIANCHI, D., GARDES, G. E. E., PAJONK, G. M., AND TEICHNER, S. J., *J. Catal.* **38**, 135 (1975).
11. ASTIER, M., BERTRAND, A., BIANCHI, D., CHENARD, A., GARDES, G. E. E., PAJONK, G., TAGHAVI, M. B., TEICHNER, S. J., AND VILLEMIN, B. L., "Preparation of Catalysts," p. 315. Elsevier, Amsterdam, 1976.
12. HOANG-VAN, C., COMPAGNON, P. A., AND TEICHNER, S. J., *Bull. Soc. Chim. Fr.* 1226 (1974).
13. CUER, J. P., ELSTON, J., AND TEICHNER, S. J., *Bull. Soc. Chim. Fr.* 81 (1961).
14. JUILLET, F., LECOMTE, F., MOZZANEGA, H., TEICHNER, S. J., THEVENET, A., AND VERGNON, P., *J. Chem. Soc. Faraday I* **7**, 57 (1973).
15. VICARINI, M., NICOLAON, G. A., AND TEICHNER, S. J., *Bull. Soc. Chim. Fr.* 1966 (1969).
16. HOANG-VAN, C., AND TEICHNER, S. J., *Bull. Soc. Chim. Fr.* 1498, 1504 (1969).
17. OZAKI, A., AND KIMURA, K., *J. Catal.* **3**, 395 (1964).
18. GHORBEL, A., HOANG-VAN, C., AND TEICHNER, S. J., *J. Catal.* **30**, 298, (1973); *J. Catal.* **33**, 123 (1974).
19. COMPAGNON, P. A., HOANG-VAN, C., AND TEICHNER, S. J., The Sixth International Congress on Catalysis, London, paper AH, 1976
20. HOANG-VAN, C., MAZABRARD, A. R., MICHEL, C., PAJONK, G., AND TEICHNER, S. J., *C. R. Acad. Sci.* **281C**, 211 (1975).
21. PERI, J. B., *J. Phys. Chem.* **69**, 220 (1965).
22. PAJONK, G., TAGHAVI, M. B., AND TEICHNER, S. J., *Bull. Soc. Chim. Fr.* 983 (1975).
23. KUCZYNSKI, G. C., AND CARBERRY, J. J., in press.

Aerosols

Some Recent Reflections on Light Scattering [1]

MILTON KERKER

Clarkson College of Technology, Potsdam, New York 13676

Received July 9, 1976; accepted September 9, 1976

Three recently discovered unusual light scattering effects are described in a qualitative way. These effects are concerned with: (1) nearly invisible particles, (2) internal heating of particles by radiation, (3) Raman and fluorescent scattering by molecules embedded in particles.

INTRODUCTION

It has been evident, at least since the Lord first displayed the rainbow to Noah, that the scattering of light by small particles can result in spectacular effects.

Consider the matter of visibility in the atmosphere as presented in Table I. The second and third columns give typical values of the radii and concentrations of the molecules, particles, and raindrops comprising clear air, cigarette smoke, and a light rain. The fourth column gives the densities of each and the last column the turbidity, a quantity inversely proportional to the visibility. The degradation of visibility by two orders of magnitude in light rain and by six orders of magnitude in smoke is not unexpected. However, it may come as a surprise that the density of aerosol which causes this effect is less than 1% of the density of air and that the density of rain is even smaller. Considered on a mass basis, smoke degrades visibility more effectively than air by nearly nine orders of magnitude.

In this paper, I would like to describe, in only a qualitative way, some recently discovered light scattering effects, which although not as striking, are still unusual. These effects are concerned with: (1) nearly invisible particles, (2) internal heating of particles by radiation, (3) Raman and fluorescent scattering by molecules embedded in particles.

[1] Supported by NSF Grant CHE 74-03910 and the Paint Research Institute.

It will be helpful at first to touch upon some fundamental aspects of light scattering (1). Figure 1 gives the physical picture. The radiation incident on the particle is described by two vectors $\mathbf{E}^i$ and $\mathbf{H}^i$ which obey Maxwell's equations; the radiation within the particle is $\mathbf{E}^r$, $\mathbf{H}^r$; the radiation scattered by the particle is $\mathbf{E}^s$, $\mathbf{H}^s$. Solutions for the internal and scattered fields are obtained by consideration of the boundary conditions which require that the tangential components of the vectors be continuous across the particle boundary.

The total power scattered by a particle can be represented by the scattering cross section

$$C_{\text{SCA}} = \frac{\lambda^2}{2\pi} \sum_{n=1}^{\infty} (2n + 1)\{|a_n|^2 + |b_n|^2\}, \quad [1]$$

where λ is the wavelength and a_n and b_n are functions which depend upon the size, shape, and optical properties of the particle.

This series converges more or less rapidly depending upon the particle size. For small particles (relative to the wavelength) convergence occurs after only a small number of terms; indeed for particles several times smaller than the wavelength, the leading term of the above series

$$C_{\text{SCA}} = (3\lambda^2/2\pi) |a_1|^2 \quad [1a]$$

suffices. This limit is called Rayleigh scattering.

The general equation can be given a simple physical interpretation. Scattering can be conceived as arising from synchronous oscillations of the electrons in the particle induced by the oscillating electromagnetic field and subsequent reradiation. The resulting radiation field can be envisaged as arising from the superposition of fields due to oscillating electric and magnetic multipoles located at the center of the particle. The electric multipoles depicted in Fig. 2 are the electric dipole, quadrupole, and octupole. Higher-order multipoles can be described mathematically, but of course cannot be represented graphically.

Each term in the above expression for the scattering cross section describes the partial field associated with a particular oscillating multipole, i.e., a_1, the electric dipole; b_1, the magnetic dipole; a_2 the electric quadrupole, etc. For small particles, as Lord Rayleigh had correctly surmised, the scattering can be adequately described as if it were due only to an oscillating dipole. His equation can be obtained either as the limiting case of the general theory given by Eq. [1] or by a relatively simple quasi-static analysis.

2. NEARLY INVISIBLE PARTICLES

We are able to see nonluminous objects because they scatter or absorb light. A commonplace example is the sky, whose brightness results from the preferential scattering of blue light by air molecules. Indeed, Lord Rayleigh developed his dipolar theory of scattering in order to account for the particular blue hue He demonstrated that the color arose from the

TABLE I

Visibility through the Atmosphere

Atmospheric condition	Radius (cm)	Number (cm^{-3})	Density (g/cm^3)	Turbidity[a] (cm^{-1})
Clear air	7×10^{-9}	1×10^{19}	1×10^{-3}	1×10^{-7}
Cigarette smoke	5×10^{-5}	1×10^{7}	5×10^{-6}	2×10^{-1}
Light rain (3 mm/hr)	5×10^{-2}	1×10^{-3}	5×10^{-7}	1×10^{-5}

[a] The visible range is inversely proportional to the turbidity.

dependence of the scattering upon the inverse fourth power of the wavelength.

Invisible particles neither scatter nor absorb light; nearly invisible particles hardly scatter or absorb, and they have other interesting optical properties. The effects upon which we will comment are based entirely upon theoretical analysis, but experimental systems have been observed which exhibit these effects.

We were looking for something quite different. It was in connection with the inclusion of microvoids in paints. We sought to account for the enhancement of hiding power that is observed when colloidal air spaces called microvoids are incorporated into paint films (2). In some cases, there are synergistic effects such that the hiding power of the microvoid coating is greater than what might be expected from the sum of the light scattering properties of the pigment and the air spaces. We proposed that models A and B, shown in Fig. 3, might lead to such synergistic effects. Model A consists of a spherical pigment particle surrounded by a void concentric shell, all encased in a resin.

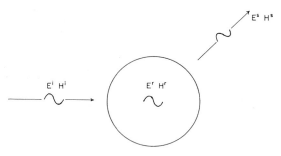

FIG. 1. Scattering theory schematicized. The incident wave is depicted by $\mathbf{E}^i$, $\mathbf{H}^i$; the scattered wave by $\mathbf{E}^s$, $\mathbf{H}^s$; the internal wave by $\mathbf{E}^r$, $\mathbf{H}^r$.

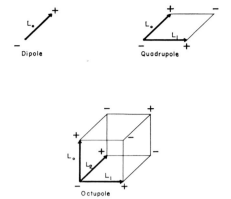

FIG. 2. First three electric multipoles.

Of course this is not realistic since it depicts the pigment sphere suspended at the center of a void. The justification for this is that we had already developed a scattering theory for such a pair of concentric spheres (3). Hopefully, any interesting effects might carry over into the more realistic case where the pigment particle sits on the floor of the void. In model B, the pigment material forms a concentric shell enclosing the microvoid.

The results are quite striking. Consider Fig. 4, which corresponds to model A, in which a titania sphere is surrounded by a concentric spherical shell of void all encased in a resin (4).

The ordinate represents the hiding power of a film containing such particles. The abcissa represents the volume fraction of the microvoid which is occupied by a pigment core. Thus the left side, $f = 1$, corresponds to pure pigment particles, and the right side represents pure microvoids.

The curves are for two sizes, in which the outer circumferences relative to the wavelength are $\nu = 0.3$ and 1.60. The larger parameter is approximately that of commercial titania pigment. The smaller is a Rayleigh particle.

Contrary to expectation, the scattering decreases very sharply for intermediate values of f. For $\nu = 0.3$ there is a decrease of four orders of magnitude when a titania sphere with $f = 0.42$ is inserted into a microvoid sphere. Instead of becoming more opaque, a film of particles having such a configuration becomes virtually transparent. The particles become nearly invisible. This effect becomes greater, as the size of the microvoid becomes smaller.

We have been able to develop a theory of this effect for a pair of confocal ellipsoids such as that illustrated in Fig. 5. The theory is an extension of Rayleigh's dipolar theory for particles which are small compared to the wavelength. Of course, this model reduces to concentric spheres for equal axial ratios. Also, the more general theory for concentric spheres reduces to the same form in the limit of small particles. When such an ellipsoidal particle is oriented so that one of its principal axes is aligned parallel to the electric vector of the polarized light beam, it scatters as if the radiation were emanating from an oscillating dipole

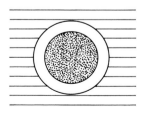

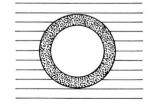

A

B

FIG. 3. Models for pigmented microvoid coating. (A) Pigment sphere within microvoid spherical shell; (B) microvoid sphere within spherical shell of pigment.

whose dipole moment is given by (5)

$$p = \frac{8\pi}{3}\epsilon_3 E_0 \frac{\epsilon_2 S - \epsilon_3 R}{\epsilon_2 SB - \epsilon_3 RB + 2\epsilon_3 R/abc}, \quad [2]$$

where a, b, c are the lengths of the semiaxes of the inner ellipsoid, ϵ_1, ϵ_2, ϵ_3 are the dielectric constants of the inner ellipsoid, the intermediate ellipsoidal shell, and the outer medium. The incident radiation is described by E_0 and B, S, and R are functions of elliptic integrals which depend upon the various optical and geometrical parameters.

The requirement for zero scattering is that the dipole moment be zero, viz.,

$$p = \epsilon_2 S - \epsilon_3 R = 0, \quad [2a]$$

and this leads to the condition that the dielectric constant of the external medium be intermediate between the values for the two regions in the compound particle; more specifically:

$$0 < \epsilon_1 < \epsilon_3, \quad \epsilon_2' > \epsilon_2 > \epsilon_3,$$
$$\epsilon_3 < \epsilon_1 < \infty, \quad \epsilon_3 > \epsilon_2 > \epsilon_2''. \quad [3]$$

The limits to which the dielectric constant of the intermediate region is constrained (ϵ_2' and ϵ_2'') depend upon the geometry. Since negative values of the dielectric constant may be admitted, this also allows for the invisibility of dielectric particles encased in plasma sheaths, or of confocal plasma ellipsoids.

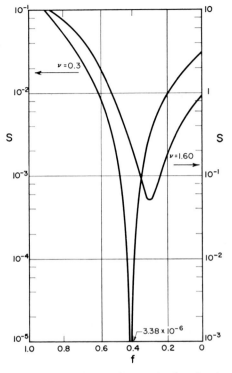

FIG. 4. Scattering function S versus f, volume fraction of core for model A where $m_1 = 2.97$ (pigment core), $m_2 = 1.00$ (microvoid spherical shell), $m_3 = 1.51$ (resin), wavelength $\lambda = 0.546\ \mu m$ for two sizes, $\nu = 0.3$ and 1.60 (ν is ratio of outer circumference to wavelength).

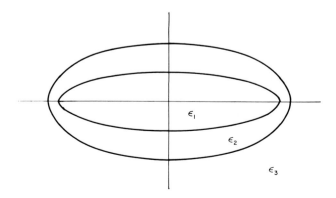

FIG. 5. Confocal ellipsoids with dielectric constants ϵ_1, ϵ_2, ϵ_3.

FIG. 6. Procedure for incrementing microvoid coating. Volume fraction of cores $f = 1$, 0.1, and 0.01; a is inner radius, b is outer radius.

In retrospect, it is now apparent why such configurations become invisible. In the Rayleigh approximation, the equivalent dipole from which the scattering emanates is due to the polarization of the media. When the dielectric constant of a particle is greater than that of the external medium, the polarization is positive; when the dielectric constant is less

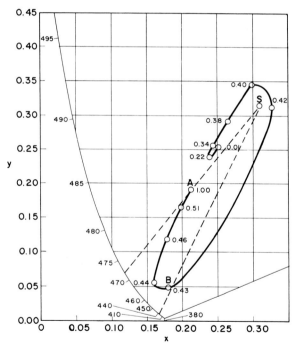

FIG. 7. Chromaticity diagram for core size distribution with modal radius $a_M = 0.03$ μm and standard deviation $\sigma_0 = 0.01$. Refractive index of dispersion corresponds to the weighted average of the ordinary and extraordinary rays for rutile. The elliptical curve is the locus of the chromaticity as the cores are encased in increasingly larger microvoid shells. Numbers alongside this curve are values of f, S represents the chromaticity of the tungsten light source, A is the chromaticity for the bare rutile particles ($f = 1$), and B is the chromaticity for that configuration for which the luminance is minimum ($f = 0.43$, the nearly invisible particle). Numbers along the outer curve indicate the wavelengths (nm) of spectrally pure colors.

111

than that of the external medium it is negative. These nearly invisible particles have been so contrived that part of the particle is positively polarized; part is negatively polarized; the net polarization, and hence the scattering, is zero.

We have so far considered the Rayleigh limit where the particles are sufficiently small that the dipole contribution dominates the scattering. However, when the particle morphology corresponds to what we have euphemistically called an invisible particle, there will still be some scattering from the higher multipoles and because the dipole term has been suppressed, these residual contributions will dominate.

Some insight into the residual scattering can be obtained by expressing the scattering cross section in the following way.

$$C_{\text{SCA}} = 2\pi k^4 \{3(\alpha^2 + \alpha'^2) + 6(\alpha\beta + \alpha'\beta')k^2 + (3\beta^2 + 3\beta'^2 + 6\alpha'\beta + 5f^2 + 5f'^2)k^4 + \cdots\}, \quad [4]$$

where the wavenumber $k = 2\pi/\lambda$. In this expression only the dipole and quadrupole terms in the multipole expansion have been used and these have each been expanded in a power series in k. The series is truncated at k^8.

Now a fascinating thing occurs. The condition which has already been specified for invisibility leads to zero values for both α and α'. But zero values of α and α' also cause the disappearance of the second term so that the leading term in the above expansion is proportional to k^8 or to λ^{-8}. Accordingly, the residual scattering of a nearly invisible particle should

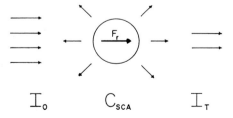

FIG. 8. Origin of radiation pressure. I_0 is incident irradiance, I_t is transmitted irradiance, C_{SCA} is scattering cross section. Momentum removed from incident direction as a result of scattering is restored to the particle, resulting in the radiation force F_r.

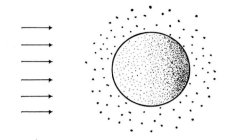

FIG. 9. Temperature distribution within an illuminated partially absorbing sphere. Darker shading indicates higher temperature. Dots outside the sphere represent gas molecules.

exhibit an inverse eight-power wavelength dependence.

This λ^{-8} dependence was actually observed quite a number of years ago in optical scattering by phase-separated glass (7–10) and also in small-angle X-ray scattering from precipitated alloys (11). In addition, the glass exhibited preferential backscattering. Both of these effects are contrary to conventional wisdom, which anticipates an inverse fourth-power wavelength dependence and preferential forward scattering.

The concentric sphere model in which the external medium has a refractive index or dielectric constant intermediate between the core and the concentric shell is quite appropriate for these systems. For example, in the glass, one can imagine that a precipitate is formed by diffusion of a particular component of the glass to a nucleation center. Then the refractive index of the bulk of the glass is intermediate between that of the highly refractive precipitate and the surrounding spherical shell. This shell will have a lower refractive index than the bulk because it has been depleted of the highly refractive component. This model corresponds to our nearly invisible particles, which also scatter proportionally to the inverse eighth power of the wavelength and scatter preferentially into the backward directions.

The particular hue of the sky, which is a consequence of the inverse fourth-power wavelength dependence of Rayleigh's scattering

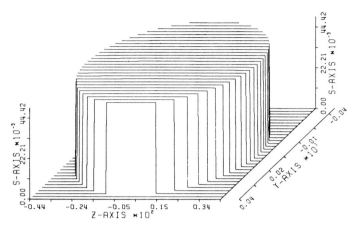

FIG. 10. Projection of the source function for the central plane of the sphere as viewed from an elevation of 45° along the diagonal from the forward to the backward half-space. Incident light propagates from the left. Size parameter $\alpha = 0.05$ refractive, index $m = 1.5-0.1i$, wavelength $\lambda = 0.5$ μm.

law, is a thing of beauty. One can ask how a dispersion of nearly invisible particles, which follow the inverse eighth-power law, might appear to a human observer?

We have made a number of calculations for dispersions of titania surrounded by a microvoid shell all encased in a transparent resin (12). One such calculation starts with a distribution of pigment particles which are then surrounded by increasingly large concentric microvoid shells. The incrementing procedure is illustrated in Fig. 6, which shows a lone pigment particle and also configurations in which the pigment particle comprises 10 and 1% of the total volume. Figure 7 shows how the color changes as a microvoid shell grows around a distribution of titania spheres. The modal radius is 0.03 μm with a 1% standard deviation. This calculation takes into account the dispersion of the refractive index of rutile over the visible spectrum, the particular spectral characteristics of the light source, and the psychophysical response of a human observer. It is an arduous calculation.

Any particular color sensation is represented by an x–y position on this chromaticity diagram. The locus of pure spectral colors is given by the outer boundary. The point designated by S is the chromaticity of the light source. The hue of a particular point is obtained by the extension of the line from the source through a point to the spectral curve. Thus, the small titania particle ($a = 0.03$ μm) at point A is sky-blue. It has a hue of 475 nm and a purity of 50%, corresponding to a mixture of equal parts of spectral pure blue light and of the white source. The purity is given by the ratio of the line segment AS to the length of the extended line.

The oval-shaped curve is the locus of the chromaticity as a microvoid of increasing thickness surrounds each pigment particle. Then the color becomes more violet and much purer as the microvoid shell surrounding the rutile particles is expanded. Indeed, for volume fraction $f = 0.43$ the color has shifted into the violet ($\lambda = 440$ nm) and the purity has increased to 88%. But this is the point at which the luminance or brightness goes through a deep minimum (12). This is the configuration of a nearly invisible particle for which the residual scattering is proportional to λ^{-8}. A film containing such particles would be nearly transparent, unless it were sufficiently thick or contained a sufficiently high particle density,

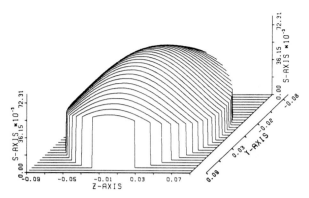

Fig. 11. Same as Fig. 10 for $\alpha = 1.0$

in which case it would take on a very pure and a very beautiful violet coloration.

3. INTERNAL HEATING OF PARTICLES BY RADIATION

The internal heating of small absorbing particles depends in a sensitive way upon the particle size and complex refractive index, and this influences phenomena as diverse as the heating of plasma spheres by lasers in order to ignite nuclear fusion, physiological damage due to microwave radiation, and the effect of particulates upon the energy balance in the planetary atmosphere.

Radiation pressure and photophoresis describe more subtle phenomena. Radiation pressure influences the movement of particles in space. It originates in the change of momentum suffered by a light beam upon being scattered as illustrated in Fig. 8. Some of the momentum carried by the incident beam is transferred to the particle, which then experiences a radiation force or radiation pressure. Peter Debye elucidated the relationship between light scattering and radiation pressure in his 1909 Ph.D. thesis in order to account for the manner in which the tails of comets are pointed away from the sun (13). His equation can be written

$$F_r = (1/c)[C_{SCA}(1 - \langle \cos \theta \rangle_{SCA}) + C_{ABS}], \quad [5]$$

where C_{SCA} and C_{ABS} are the cross sections for scattering and adsorption and c is the velocity of light. The factor $\langle \cos \theta \rangle_{SCA}$, which arises from the asymmetry of the scattering, accounts for that component of the momentum abstracted from the incident beam as scattered radiation which is returned into the forward direction.

Debye failed to account for an additional contribution to the radiation force for absorbing particles which arises because the internal heating of the particle is nonuniform. This gives rise to asymmetric thermal radiation. We have proposed the equation (14)

$$F_r = (1/c)[C_{SCA}(1 - \langle \cos \theta \rangle_{SCA}) + C_{ABS}(1 - \langle \cos \theta \rangle_{ABS})], \quad [5a]$$

where $\langle \cos \theta \rangle_{ABS}$ accounts for the asymmetry in the thermal radiation.

This effect can be visualized with the aid of Fig. 9, which schematically depicts the uneven internal heating of a particle. In this case the back part of the particle is heated more intensely than elsewhere and because the back surface would have a higher steady-state temperature, it would radiate thermal energy more intensely. In this particular case there would be a net contribution to the momentum of the particle back into the direction of the light beam, thereby reducing the radiation force.

When such a particle is in a gas rather than in a vacuum, the uneven temperature distribution results in a radiometer effect in which

114

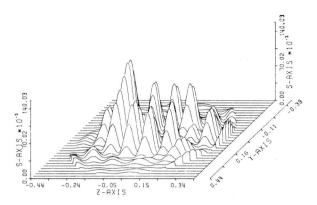

Fig. 12. Same as Fig. 10 for $\alpha = 5.0$ and $m = 3.0 - 0.1i$.

molecules rebounding from the hotter side carry excess momentum. The radiometer effect predominates over the radiation pressure, even in gases at reduced pressures. This phenomenon, called photophoresis, may play a significant role in the distribution of aerosol particles in the upper atmosphere.

The local distribution of heat sources within the particle is the necessary starting point for an analysis of the effect of heating upon either radiation pressure or photophoresis. Heat progresses through the particle in the following steps:

1. Electromagnetic radiation is absorbed and reappears as thermal energy. The local distribution of these thermal sources is the *source function*.

2. Heat flows to the surface of the sphere by thermal conduction.

3. Thermal energy and its associated momentum is transported from the surface by radiation and, in the presence of a fluid, by conduction and convection.

4. For volatile media there will also be evaporation with its attendent mass, heat, and momentum transfers.

If we look at Fig. 1, which outlined the model for scattering, it is apparent that the same boundary value problem which leads to the scattered field can also provide the local elec-tromagnetic field within the particle, and in turn this leads to the source function (15).

We have only very recently obtained some typical results for the source function.[2] For example, Fig. 10 is a topographical projection of the source function within a Rayleigh particle with size parameter $\alpha = 0.05$, $m = 1.5 - 0.1i$, and $\lambda = 0.5$ μm. The incident light propagates along the positive z-axis. This diagram depicts the equatorial slice through the sphere which is parallel to the incident radiation. The center of the sphere is at the origin of the z–y plane. The source function is plotted along the S-coordinate, perpendicular to the z–y plane. This projection gives a topographical view from an elevation angle of 45° looking along the z–y diagonal. In this case the source function is uniform throughout the equatorial plane, in agreement with the expectation for a Rayleigh particle.

For the large particle ($\alpha = 1.0$) shown in Fig. 11, more heat is generated in the interior than along the edge. Also, somewhat more heat is generated in the front half (left side of diagram) of this equatorial slice.

Figure 12 provides a spectacular panorama. This is the source function in the equatorial slice for $\alpha = 5.0$ and $m = 3.0 - 0.1i$. Most of

[2] We are indebted to Mr. Peter Dusel for these results, which will appear in more detail in his forthcoming M. S. thesis.

the heat is generated well away from the surface of the particle. The source function is highly asymmetric and highly irregular. The hottest spots are in the front half.

A somewhat different view of this same terrain can be glimpsed if one looks along the incident beam direction (still from a 45° elevation) rather than along the diagonal. This is shown in Fig. 13.

Figure 14 is for a "carbon" particle, $m = 1.95 - 0.66i$ with $\alpha = 5.0$. In this case the back surface is the hottest part of the particle, quite contrary to previous expectations based upon geometrical optics considerations. Such a spherical particle should experience a strong photophoretic force back into the direction of the light beam.

We are planning a systematic study of photophoresis starting with such source functions. This is a complex phenomenon. Consider the following variables: intensity and spectral distribution of the radiation, dispersion of the real and imaginary parts of the refractive index, thermal conductivity and emissivity, thermal and viscous properties of the gas, and finally, the particle size.

4. RAMAN AND FLUORESCENT SCATTERING BY MOLECULES EMBEDDED IN SMALL PARTICLES

A number of problems of practical importance involve fluorescent and Raman scattering by molecules embedded in a particle whose dimensions are comparable to the wavelength. For example, the particle may be a biological cell which has been tagged with fluorescent molecules that attach specifically to either the nucleus, the cytoplasm, or the cell membrane. This has made possible very high speed enumeration and separation of component cells of a mixed cell population (16).

In another area, concern with air and water pollution and with inhalation toxicology has led to an increasing demand to monitor and measure particulates. Electron probe microanalysis has provided a powerful tool for obtaining the elemental composition of single particles in the micrometer range but unfortunately this does not provide a basis for molecular or crystal identification. At present, Raman spectroscopy and fluorescence analysis appear to offer a distinct possibility for effective application in this direction. At least one instrument has been described recently which demonstrates the feasibility of obtaining highly specific analyses of chemical species by Raman spectra from single particles with dimensions in the micrometer domain (17).

Since neither the Raman frequency and lineshape nor the fluorescence emission spectra are affected by the fact that the molecules are a constituent within a small particle, Raman and fluorescent scattering can be used for qualitative chemical analysis on a straightforward

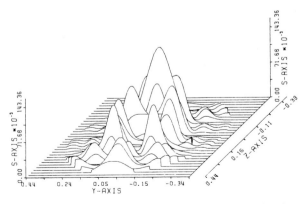

FIG. 13. Same as Fig. 12 as viewed into the incident beam direction.

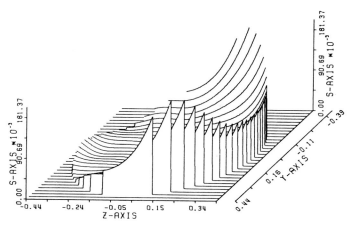

FIG. 14. Same as Fig. 10 for $\alpha = 5.0$ and $m = 1.95-0.66i$.

basis. However, it has not been recognized heretofore that quantitative analysis is considerably more complicated, because particle size and refractive index strongly influence the intensity of Raman and fluorescent scattering, as well as the angular distribution of the scattering. The manner in which this occurs can be visualized with the aid of Fig. 15.

The incident electric field $\mathbf{E}_{\omega_0}{}^i$ at frequency ω_0 induces an internal field $\mathbf{E}_{\omega_0}{}^r$ within the particle (medium) and a scattered field $\mathbf{E}_{\omega_0}{}^s$ outside the particle (medium).

We now consider the excitation of molecular transitions by the internal field, which ultimately results in Raman or fluorescent emission at some shifted frequency ω. The quantum processes are not affected by the fact that the molecules are contained within the small particle, so that the effects which we consider are classical. They occur for two reasons.

In the first place, each molecule undergoing a transition sees only the local electromagnetic field. We have already seen that this is highly nonuniform within a small particle and that it depends in a sensitive way upon the particle size and refractive index. It is this field which leads to the highly irregular source function in the case where the electromagnetic radiation is degraded to heat. In the present instance, the rate of molecular transitions, and hence the rate of luminous emission at any location within the particle, is directly proportional to the local electric field intensity. The emitted field, in turn, is proportional to the product of the local field and the molecular polarizability.

In the second place, it is also necessary to consider how the emitted radiation propagates out of the particle. In the simplest instance, the molecules within the particle act as dipole radiators. Interestingly, the problem of a dipole radiator arbitrarily located within a dielectric sphere had not heretofore been solved. We approached this problem by assuming that the internal field $\mathbf{E}_{\omega}{}^r$ arising from emission at a particular location within the particle can be constructed by superposition of the field of the dipole radiator plus an unknown induced field due to the boundary. This leads to the construction of the shifted field $\mathbf{E}_{\omega}{}^s$ outside the particle, which can now be obtained as the solution of a standard boundary value problem. However, there is a considerable amount of mathematical complexity involved because the arbitrarily located dipole radiator is not necessarily located at the center of the particle. The detailed analysis has been published elsewhere (18).

The result of the analysis is an expression for the electromagnetic field at some position outside of the particle which arises from a single source within the particle. This may now be averaged over the particle volume in the case of coherent emission, such as in stimulated Raman scattering. In the more usual case, where the emission from the various locations is incoherent, it is the time average power appropriate to each location which must be averaged. Because of this averaging procedure there is very little added complexity if there is some arbitrary variation of concentration of the Raman or fluorescent species within the particle.

The theory has been extended to particles consisting of concentric shells having different optical properties (19). For example, the three-layer case can serve as a model for a biological cell consisting of a nucleus, cytoplasm, and a membrane.

It seems reasonable to expect that the angular distribution and polarization would differ quite significantly from the unshifted Lorenz–Mie scattering. Some preliminary experiments with a monodisperse latex of methylmethacrylate copolymerized with 9-vinyl anthracene indicate that this is the case (20). Computer calculations and further experiments are presently under way in order to articulate these phenomena more precisely.

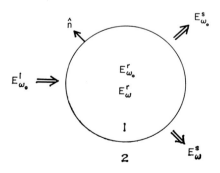

FIG. 15. Model for fluorescent or Raman scattering. $E_{\omega_0}{}^i$, $E_{\omega_0}{}^r$, and $E_{\omega_0}{}^s$ represent the field vectors of the incident wave, the wave within the particle, and the scattered waves at the incident frequency ω_0. The corresponding quantities at the shifted frequency ω are $E_\omega{}^r$ and $E_\omega{}^s$, where the former quantity is the sum of the field due to the dipole emitter and an induced field.

TABLE II

Particle Absorption Factor for Rayleigh Scatters

m	η'	$1 - R$	η
1.00–0.01i	1.00	1.00	1.00
1.33–0.01i	0.84	0.98	0.86
1.50–0.01i	0.75	0.96	0.78
2.00–0.01i	0.50	0.89	0.56
2.50–0.01i	0.33	0.82	0.40
3.00–0.01i	0.22	0.75	0.29

Yet, a "back of the envelope" calculation for Rayleigh particles does verify that substantial effects can be anticipated. Consider that in fluorescence some energy is absorbed and is later reemitted at a longer wavelength. Accordingly, the intensity of the fluorescence will be proportional to the rate at which energy is absorbed at the incident frequency and this absorption cross section in turn depends upon the particle size and complex refractive index. For a material with refractive index $m = n(1 - \kappa i)$, the ratio of the absorption cross section when the fluorescent molecules are dispersed in Rayleigh particles to the absorption cross section for the same mass of fluorescent material dispersed in the bulk medium is (21)

$$\eta = \frac{(-1.5/n\kappa)\ \mathrm{Im}\left[(m^2 - 1)/(m^2 + 2)\right]}{1 - |(m - 1)/(m + 1)|^2}. \quad [6]$$

The denominator corrects for the reflection of part of the incident beam at the front surface of the bulk medium (perpendicular incidence). The effect of the real part of the refractive index is shown in Table II. In this table η' and R represent the numerator and denominator, respectively, in Eq. [6].

This result demonstrates that fluorescent molecules embedded in a highly refractive material such as titania will fluoresce only 29% as strongly when the titania is dispersed as Rayleigh-size particles compared with the fluorescence in the bulk. We anticipate much more striking effects, especially in the angular distribution of the radiance, for particles whose dimensions are comparable to the wavelength of light.

It is clear that these particle-size effects must be considered in order to interpret properly fluorescent and Raman scattering signals. We have already indicated how such signals may figure importantly in analytical techniques useful in research in air pollution and cytology.

REFERENCES

1. KERKER M., "The Scattering of Light and Other Electromagnetic Radiation," Academic Press, New York, 1969
2. SEINER, J. A. AND GERHART, H. L., *Ind. Eng. Chem. Prod. Res. Develop.* **12**, 98 (1973).
3. ADEN, A. L. AND KERKER, M., *J. Appl. Phys.* **22**, 1242 (1951).
4. KERKER, M., COOKE, D. D., AND ROSS, W. D., *J. Paint Technol.* (*No. 603*), **47**, 33 (1975).
5. KERKER, M., *J. Opt. Soc. Amer.* **65**, 376 (1975).
6. CHEW, H. AND KERKER, M., *J. Opt. Soc. Amer.* **66**, 445 (1976).
7. KOLYADIN, A. I., *in Proceedings of the Third All-Union Conference on the Glassy State, Leningrad, 1959*, p. 202.
8. VISHVILLO, N. A., *Opt. Spectrosk.* **12**, 412 (1962); English translation: *Opt. Spectrosc.* **12**, 225 (1962).
9. MAURO, R. D., *J. Appl. Phys.* **33**, 2132 (1962).
10. HAMMEL, J. J. AND OHLBERG, S. M., *J. Appl. Phys.* **36**, 1442 (1965).
11. WALKER, C. AND GUINIER, A., *Acta. Met.* **1**, 568 (1963).
12. KERKER, M. AND COOKE, D. D., *J. Coatings Technol.* (No. 621) **48**, 35 (1976).
13. DEBYE, P., *Am. Physik* **30**, 57 (1909).
14. KERKER, M., *Amer. Sci.* **62**, 92 (1974).
15. KERKER, M. AND COOKE, D. D., *Appl. Opt.* **12**, 1378 (1973).
16. KAMINSKY, L. A., *Advan. Biol. Med. Phys.* **14**, 93 (1973); STEINKAMP, J. A., HANSEN, K. M., AND CRISSMAN, H. A., *J. Histochem. Cytochem.* **24**, 292 (1976).
17. ROSASCO, G. J., ETZ, E. S., AND CASSATT, W. A., *Appl. Spectrosc.* **29**, 396 (1975).
18. CHEW, H., McNULTY, P. J., AND KERKER, M., *Phys. Rev. A* **13**, 396 (1976).
19. CHEW, H., KERKER, M., AND McNULTY, P. J., *J. Opt. Soc. Amer.* **66**, 440 (1976).
20. LEE, P. AND KRATOHVIL, J. P., private communication.
21. KERKER, M., *Appl. Opt.* **12**, 2787 (1973).

Formation of Atmospheric Ultrafine Particles and Ions from Trace Gases

J. BRICARD, M. CABANE, AND G. MADELAINE

University of Paris VI and Center of Nuclear Studies of Fontenay-aux-Roses, Paris, France

Received June 10, 1976; accepted September 9, 1976

Firstly, the importance of ultrafine particles in the atmospheric aerosol is reviewed. Based on conventional nucleation theory, it is shown that water vapor alone cannot initiate the formation of such particles. We consider a mixture of water vapor and sulfuric acid and, pursuant to a discussion of theoretical results obtained by various writers, we consider the case under atmospheric conditions. We demonstrate that this leads to orders of magnitude which compare with the results of simulation experiments now under way. Secondly, we study the mobility spectrum of atmospheric ions, based on mobility mass correlations, and we demonstrate that Thomson's conventional theory, calling upon only the presence of water vapor in the atmosphere, does not allow for the interpretation of the atmospheric ions' mobility spectrum observed in our experiment. This spectrum is qualitatively interpreted by generalizing the theory of Thomson and by introducing in it the simultaneous presence of water vapor and sulfuric acid.

INTRODUCTION

Many air-suspended particles may be considered as having been created from gaseous-state matter that has been transformed into particles. This is specifically the case of H_2SO_4 vapor which is easily condensable, issuing from SO_2 oxidation. These gases reside for only a limited time in the air—a few days to 2 weeks at a maximum, possibly only some hours in heavily polluted atmospheres. These particles, whose study is relatively recent, have at the start of their formation dimensions of tens of angstroms (1). Under the action of Brownian motion, they coagulate among themselves or with larger particles that are already in suspension in the atmosphere, and they establish a steady state between their formation and disappearance through coagulation. The atmospheric aerosol may result from such a balance.

In this paper, we propose to summarize the formation processes of aerosol particles and also of ions in the lower troposphere and then to show the fundamental role played by the presence of sulfuric acid in the atmosphere on the formation of atmospheric particles.

Let us recall here that there are normally present in tropospheric air some gaseous impurities to the extent of 10^{-2} to 10^{-4} ppm, while water vapor, depending on meteorological conditions, is present to the extent of some 10^4 ppm.

A. NEUTRAL PARTICLE FORMATION

1. Formation of Particles from Pure Vapor

The classical approach to the theory of homogeneous nucleation consists primarily in determining the variation of the free enthalpy ΔG corresponding to the formation of an embryo of radius R such as

$$\Delta G = n_l(\mu_l - \mu_g) + 4\pi R^2\sigma, \qquad [1]$$

where n_l is the total number of liquid molecules in the embryo, μ_l and μ_g are the chemical potentials for a macroscopic amount of the liquid phase and in the gas phase; σ is the surface tension.

The variation of ΔG as a function of the radius, when the vapor is supersaturated (i.e., $S > 1$, S is the relative humidity in the case of water vapor and is defined by the ratio of the vapor pressure to the pressure of saturated vapor at the same temperature), shows that above a certain critical dimension, the embryos develop, while below they are unstable.

It would seem pertinent to point out the very approximate nature of the above reasoning, the numerous assumptions on which the theory is based, and in particular, the significance of the surface tension in the case of particles containing less than hundreds of molecules. In this respect, we suggest the detailed discussion found in the book of Zettlemoyer (2). This type of reasoning has been generalized by Reiss (3) in the case of a system with several components. The application of this generalization to the water–sulfuric acid system has been done by Dole (4), Kiang et al. (5), and Mirabel and Katz (6).

(a) By generalizing the relation [1], one finds that the free enthalpy of formation of an embryo of arbitrary radius R and composition is

$$\Delta G = n_1(\mu_{1l} - \mu_{1g})$$
$$+ n_2(\mu_{2l} - \mu_{2g}) + 4\pi R^2\sigma, \quad [2]$$

where n_1 and n_2 are the numbers of molecules of components 1 and 2 in the embryo, μ_{1l} and μ_{2l} are the chemical potentials of components 1 and 2 taken for a microscopic amount of a liquid phase of the same composition, μ_{1g} and μ_{2g} are the chemical potentials of components 1 and 2 in the gas phase, and σ is the surface tension of the mixture. Instead of representing ΔG in two-dimensional space, as in the case of relation [1], we will represent it in three-dimensional space, or ΔG (n_1, n_2). One finds that the surface has a saddle point directed toward the increasing ΔG, and its peak ΔG^* corresponds to the point of nucleation, i.e., to the formation of the critical embryo.

In the case of interest to us, we represent the constituent 1 by water and 2 by sulfuric acid, and we designate the relative humidity by $S = P_1/P_{1\infty}$ and the activity by $a = P_2/P_{2\infty}$, P_1 and $P_{1\infty}$, P_2 and $P_{2\infty}$ being respectively the pressure and saturated pressure at the same temperature for water vapor and sulfuric acid.

Critical characteristics are defined for each given activity and relative humidity. However, the saturated vapor pressure $P_{2\infty}$ is rather poorly known at the present time. In line with other authors, it would have the value between 3.5×10^{-4} Torr (7) and 10^{-6} Torr (4, 9). We may therefore assume that the vapor concentration of H_2SO_4 is of the order of 4×10^{-4} at 1.3×10^{-2} ppm. Calculations show (6) that in assuming a value $P_{2\infty} = 3.5 \times 10^{-4}$ Torr, the critical radius is weakly dependent on humidity and is of the order of 7 Å, and this would correspond to a content in acid of some tens of molecules. We will find a value of the same order of magnitude by taking $P_{2\infty} = 10^{-6}$ Torr (9).

Experimentally, the determination of the value of the critical radius of atmospheric aerosols is a delicate operation due to the fact that such small particles cannot be detected directly, but only by condensation nuclei counters. An order of magnitude of 5 to 8 Å has been suggested by extrapolation of measurements of the aerosols produced by radiolysis of gaseous impurities of air by means of a diffusion battery.

(b) Following this type of reasoning, one finds that the nucleation frequency for this type of binary mixture takes the form of

$$I = C \exp(-\Delta G^*/KT), \quad [3]$$

where ΔG^* as defined earlier represents free enthalpy for the formation of the critical embryo. The proportionality constant depends upon the concentration of vapor molecules present in the medium and the rate at which new molecules are incorporated in the critical embryo.

Following Kiang et al. (5), we write Eq. [3] as

$$I = 4\pi R^{*2} \frac{N_2kT}{(2\pi m_2kT)^{\frac{1}{2}}} N_1 \exp\left(-\frac{\Delta G^*}{kT}\right), \quad [4]$$

where $4\pi R^{*2}$ is the surface of the embryo, N_1 is the concentration of compound 1, and the remainder of the preexponential factor is the flux of active gas molecules.

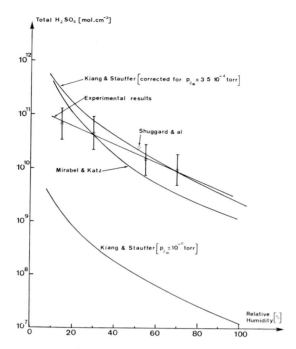

FIG. 1. H_2SO_4 vapor concentration as a function of relative humidity RH in percent corresponding to $I = 1$ nucleus cm^{-3} sec^{-1} from calculations of Mirabel and Katz, and those of Kiang and Stauffer corrected for $p_2 = 3.5 \times 10^{-4}$ Torr. Marked points represent experimental points and calculations of Shugard, Heist, and Reiss. Sulfuric acid and vapor are introduced directly into the experimental chamber.

The most recent numerical computations were carried out by Mirabel and Katz (6) with the same data as those used for the case of the critical radius given above; therefore $P_{2\infty} = 3.5 \times 10^{-4}$ Torr (see also Ref. (9) with $P_\infty = 10^{-6}$ Torr).

Figure 1 shows the results obtained during our preliminary experiments, obtained from measurements of nucleation frequency in a simulation chamber. Shown as ordinates are sulfuric acid vapor concentration and on the abscissa the relative humidity. In Full lines are the theoretical curves calculated by Mirabel and Katz and also Kiang et al. from relation [4] by taking $P_{2\infty} = 3.5 \times 10^{-4}$ Torr, corresponding to a nucleation frequency of 1 particle/cm³/sec. Due to the precision inaccuracy, it is not possible to draw a definitive conclusion from these curves. However, it would seem

that the above calculations are not in unacceptable disagreement with measurements when the value $P = 3.5 \times 10^{-4}$ Torr is adopted, owing to the uncertainty of the calculation.

In fact, the calculations given above use the mixture of H_2SO_4 vapor and pure H_2O, but do not take into account the formation of H_2SO_4, H_2O hydrates. This has been studied recently by Reiss et al. (10) and introduced in the calculation of nucleation frequency by Shugard et al. (11) covering relative humidities of 50–100%, and 200%. Results at 50% are shown in Fig. 1. One sees that the corresponding points are much closer to each other.

2. Further Development

(a) Due to the high value of the ratio N_1/N_2, 10^5 to 10^9 molecules of water will strike

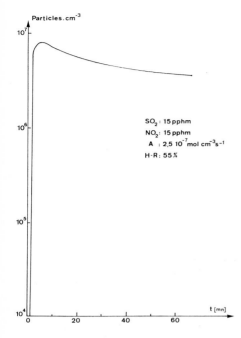

SO$_2$: 15 pphm
NO$_2$: 15 pphm
A : 2,5 10^{-7}mol cm^{-3}s^{-1}
H-R: 55%

Fig. 2. Variations of the number of particles per cubic centimeter as a function of time, measured by Boulaud *et al.*, for given activity and relative humidity. Sulfuric acid is produced in the experimental chamber by radiolysis of the mixing SO$_2$, NO$_2$, water vapor system.

the droplet which has just appeared on a critical embryo during the interval of time between the fixing of two successives molecules of sulfuric acid. Let us suppose that all the acid molecules striking a droplet are incorporated into the droplet. For a critical embryo containing a total of 50 molecules, the fixing of such a high number of water molecules is impossible and most water molecules will have reevaporated. Thus, between the fixing of two successives acid molecules (that is, when n_2 remains constant), an equilibrium relative to the dimension and composition of a droplet occurs such that $(\partial \Delta G/\partial n_1) = 0$. Let us designate by N_p the concentration of already formed particles at a given moment, including embryos, and let us assume that we are working

in a closed chamber which allows control of the experimental conditions.

Assuming a monodispersed aerosol, we will write (9)

$$dN_2/dt = A - (\alpha + N_p)N_2,$$
$$dN_p/dt = (\alpha/n_2^*)N_2 - KN_p^2. \qquad [5]$$

Here A represents the rate of production in H$_2$SO$_4$ molecules, α is the conversion rate of sulfuric acid to embryos defined by

$$\alpha = In_2^*/N_2,$$

n_2^* is the number of acid molecules per critical embryo, I is the frequency of nucleation, γ is the rate of fixing of sulfuric acid vapor on already formed particles which can be evaluated (18) from the condensation heat period, and K is the coagulation constant. The size of particles depends on time, α, and K.

(b) Taking into account the heterogeneity of particles, we assume a mean value for $K = 10^{-8}$ cm^3 sec^{-1}. For $A = 10^6$ cm^{-3} sec^{-1}, we find that N_p increases, reaches a maximum, and then decreases in a manner in line with the experimental results obtained by Bricard *et al.* (19); see Fig. 2. This maximum follows from relation [5] by setting $dN_p/dt = dN_2/dt = 0$. The result is that

$$(N_p)_{\max} = \left(\frac{\alpha}{n_2^*K}\right)^{\frac{1}{2}} \left(\frac{A}{\alpha + \gamma N_p}\right)^{\frac{1}{2}}. \qquad [6]$$

In addition let us assume that $\alpha \gg N_p$. This means that later vapor condensation by embryos which have crossed the critical radius is negligible in comparison to direct nucleation. The result is that

$$(N_p)_{\max} = (A/n_2^*K)^{\frac{1}{2}}. \qquad [7]$$

Figure 3 represents experimental results of Boulaud *et al.* (12, 13). We have shown on ordinates the maximum number of particles formed per cubic centimeter as a function of relative humidity. The points indicated give the theoretical data of Takahashi for a given value of A. We can therefore conclude that as long as relation [7] remains valid—that is, for

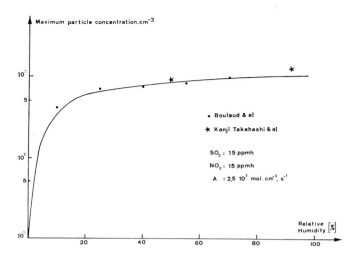

FIG. 3. Variations of maximum number of particles per cubic centimeter as a function of relative humidity for a given activity, measured by Boulaud *et al.* The marked points correspond to calculations by Takahashi *et al.* Sulfuric acid is produced in the experimental chamber by radiolysis of the mixing SO_2, NO_2, water vapor system.

the first phase of formation—later condensation of sulfuric acid is negligible and the development of particles is achieved through coagulation. On the other hand, this is not true when particles are sufficiently developed. Embryos can then play an important role as centers of condensation (14, 17).

COMMENTS

Let us note here that these results cannot be directly related in generalized terms with atmospheric conditions, and that they represent only a first approximation because we must take into account the presence of other gaseous impurities in air. This is particularly true for ammonia since it has been well established that an important fraction of atmospheric aerosol includes ammonium sulfate (Megaw (15)). Yet, it is quite evident that the above conclusions are valid if sulfuric acid vapors appear during the first stage and then are transformed to give particles which would then react with other impurities in the air.

Coffey and Mohnen (20) have shown that the direct action of NH_3 on SO_2 gives a polymer in the form $(NH_3)_2 (SO_2)_n$. However, by working in natural filtered air, Bricard *et al.* (22) and Friend (24) have seen that the addition of ammonia does not lead to the formation of new particles. From what Friend has shown under these conditions, the embryos $NH_4^+ HSO_4^-$ would form saline droplets which constitute an ideal medium for the catalytic oxidation of SO_2. The development of particles continues as long as NH_3 is fixed and can neutralize the acid resulting from the catalytic oxidation of SO_2 fixed by the particle. This process could explain the development of sulfate particles whose radii exceed 0.1 μm, even if the humidity has not reached saturation.

Let us also mention here that we have seen that in a certain number of cases atmospheric particles contain nitrosylsulfuric acid SO_4HNO (16), with the latter absent when working in the presence of an excess of ammonia. The formation of particles containing SO_4HNO could be attributed to the action of NO_2 on SO_2 in the presence of water and to the nucleation of SO_4HNO. This could lead us to invalidate the nucleation mechanism that we have described above so that the question is still unresolved. In fact, we could be led to think that the origin of particles through the nucleation of the mixture $H_2SO_4 + H_2O$ is a first approximation, and that we should consider the nucleation of the ternary mixture H_2SO_4, NO_3H, H_2O. Relation [4] is just a first approximation of the problem of atmospheric nucleation. This problem has been brought up by Kiang et al. (21) in the case of a multicomponent gaseous mixture, but at the present time we have no specific data on this problem.

B. NUCLEATION BY IONS

1. Composition of Ions

Near the ground, α, β, γ radiation from radon and thoron and their daughters, and cosmic rays are the main ionizing agents leading to the separating of charges between electrons and positive ions bearing a single basic charge. At high altitudes, up to 60 km, cosmic rays represent the main sources of atmospheric ionization, and above this altitude photo-ionization becomes primary.

Free electrons, after thermalizing, fix on neutral molecules, and positive as well as negative ions are subjected to a series of molecular reactions (23). As long as studies *in situ* have not indicated the composition of atmospheric ions we will be forced to remain satisfied with chemical models based on laboratory measurements of constants of formation for all possible interactions between ions and the atmospheric constituents. In this paper, we will limit ourselves to the study of positive ions. Let us recall that the evolution of oxonium ions $H^+(H_2O)_n$ (with $n = 4$ or 5) takes place within a simple gaseous system containing N_2, O_2, CO_2, and H_2O, and that it is the most important ion in the atmosphere above 85 km in altitude. A currently accepted hypothesis is that the evolution of oxonium ions begins at the base of strato-

spheric and tropospheric ions, but that this ion is certainly not the last of the series.

Loeb (25) has shown that below water vapor saturation, the chemical bond energies of hydrated ion are too weak to enable the development of critical embryos (see later). The dominant forces generating stable small agglomerates about simple ions are electrostatic resulting from the interaction between the ionic charge and the dipole moment of the fixed molecule. Other forces also get involved, e.g., dipole–dipole, dipole–quadrupole, etc.

A more complete analysis of atmospheric ion composition can be found in (23).

2. Ion Mobility

All ions in equilibrium with neighboring gaseous molecules are subjected to thermal motion. This motion, when the medium is not subjected to an electric field, is carried out with a velocity v_i such that $\frac{1}{2}Mv_i^2 = \frac{3}{2}kT$, where M represents the ion mass, k is the constant of Boltzmann ($k = 1.38 \times 10^{-16}$ CGS), and T is the absolute temperature. Under the action of electric field E the ion covers a drift motion along the field force lines with a velocity V_i, which is superimposed over the motion. Let us note that $V_i \ll v_i$. For lower values of the ratio E/p, p representing pressure, the energy of the ion whose mass is not much higher than that of neighboring molecules does not increase substantially over $\frac{3}{2}kT$. The ionic mobility is defined by the ratio

$$K_i = V_i/E. \qquad [8]$$

We will not attempt here to discuss the various methods for measuring the ionic mobility. Due to their inaccuracy we can feel content to represent their results by means of the Langevin theory, which enables us to express the ion mobility in function of its mass M, of its collision cross section Ω_i with molecules of the carrier gas of mass m, and we can therefore write (26)

$$K_i = \frac{3e}{16n} \frac{m + M^{\frac{1}{2}}}{mM} \left(\frac{2\pi}{kT}\right) \frac{1}{\Omega_i}, \qquad [9]$$

where n represents the concentration of molecules in the carrier gas and e is the basic charge. The collision cross section depends also on the assumed ion forms (spheric, stretched, etc.) and the nature of interactions generated between the ion and the carrier gas components (elastic collision, molecule polarization, etc.).

A change in gas density ρ in which the ion travels leads to a variation in ion/molecule collision frequency, such that when ρ increases, the drift velocity V_i and, therefore the mobility K_i diminishes. This lead us to introduce the concept of reduced mobility K:

$$K = K_i(\rho/\rho_0), \qquad [10]$$

where $\rho_0 =$ gas density under experimental conditions, and $\rho =$ gas density under normal temperature and pressure conditions. Thus,

$$K = K_i(p/760)(273/T), \qquad [11]$$

where p is the pressure of the gas expressed in Torr and T is its temperature in degrees Kelvin.

We have used the last relation to express mobility values described in this paper. Figure 4 shows what Kilpatrick (27) calls a correlation of mass mobility of ion obtained by means of the simultaneous measurement of two parameters, the mass and the mobility. These are the first measurements of this type obtained at atmospheric pressure. Unlike mass, the study of ion mobility could be achieved directly on atmospheric ions, but the accuracy of these two types of measurements cannot be compared. In particular, it is not possible to identify the type of ion from a measurement of its mobility, as such a measurement could only lead to qualitative indications.

3. Results of Mobility Measurements

We will not describe in detail the measuring method used by Bricard et al. (22) which consists in producing ions with a radioactive source, in separating the ions of each sign under the action of an electric field, then letting them travel under the action of an adequate electric field within an aging chamber for a more or less lengthy time (1 to 500 msec), and

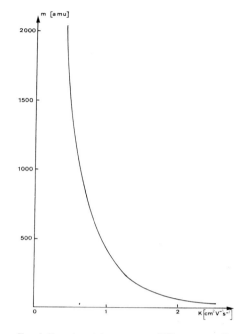

FIG. 4. Experimental curve $m = f(K)$, corresponding to Kilpatrick.

finally measuring their mobility within a measuring space.

The measuring space and the aging chamber are separated by a screen consisting of two coplanar parallel wire systems between which a repetitive voltage difference can be established. Ions move in the measuring space and their mobility can be derived from τ_d (transit time) as they move to cross this space under an auxiliary electric field. Table I shows results obtained in laboratory air for various values of ions age in the case of positive ions.

The results obtained also apply to negative ions, although their mobility is a little higher for the same aging time.

(a) Table I shows that two categories of ions exist: The first corresponds to mobilities above 1 cm² V⁻¹ sec⁻¹ and the second represents mobilities between 1 and 0.1 cm² V⁻¹ sec⁻¹ when the ion/air contact time is between 1 and 500 msec. We will call the first line "spectra

TABLE I

TABLE I

t (msec)	K (cm² V⁻¹ sec⁻¹)			
1	2.1	1.9	1.5	1.35
8		1.9	1.5	0.65
20		1.9	1.5	0.5
40			1.5	0.4
120				0.3
500				0.2

ions," and the second "band spectra ions." In fact, the second can be made of discrete mobility bands that the method does not enable us to separate for lack of sensibility, since the concentration of ions in this area is smaller than in the first.

(b) For a given aging, using a laboratory air sample, if we progressively add water vapor or SO₂ groups 2.1 and 1.9 disappear in succession and the remainder of the spectrum moves toward weaker mobilities.

Always, for a given aging, if we work within a pure gas (nitrogen, argon) that is humidified, we find in our experiments that the mobility of each ion category is linked to the content in water vapor p expressed in parts per million by a relation of the form

$$K^{-\frac{1}{2}} = A + B(c + p10^{-4})^{\frac{1}{3}}, \quad [12]$$

the parameter B depending on aging time. In the case of ions with a mobility that is lower than 1 cm² V^{-1} sec⁻¹, the constant $c = 0.22$, while $c = 0$ is that of ions with higher mobility.

4. Study of Mass/Mobility Correlation

Mobility measurements are not accurate enough to deduce the composition of a given mobility for an ion by using the Kilpatrick curve (27) (Fig. 4). Nevertheless, we believe that atmospheric ions are responsible for the sequence of oxonium ion intervention, of form H_3O^+ $(H_2O)_n$, mixed motions that call upon impurities derived from SO₂ and N₂ in their composition.

The systematic study (work under way) of ion evolution with a mobility that is higher than 1 cm² V^{-1} sec⁻¹ is carried out with a system

that is similar to that of McDaniel (34), so that ion mobility and the mass of ions created within a gaseous mixture may be measured simultaneously. In our case, the mobility was measured at atmospheric pressure, and ions of given mass were extracted from the reaction chamber and focused over a quadrupole. This device has enabled us to show:

In nitrogen with water vapor (total pressure = 760 Torr, temperature of 20°C and for an ion-carrier gas contact time of the order of a millisecond, the mass spectrum of positive ions consists essentially of ions of the series $H^+(H_2O)_n$ (n being less than or equal to 7 for a water content of less than 1000 ppm in volume).

Where contact time increases, peak amplitude corresponding to ions of the series $H^+(H_2O)_n$ diminishes, while new ions of the type X^+ or $H^+(H_2O)_nX_m$ appear. The existence of these ions can be attributed to reactions between hydrooxonium ions and organic or mineral impurities which may emerge from degassing products of the chamber or from the actual carrier gas. Some of these impurities have been identified as being acetone (CH₃–CO–CH₃) or ethanol (CH₃–CH₂OH).

The determination of rate and equilibrium constants shows that the production of $H^+(H_2O)_n$ ions is more likely at an increased pressure.

5. Apparent Mobility of Ions Belonging to the series $H^+(H_2O)_n$

We are currently studying the mobility of ions $H^+(H_2O)_4$ and $H^+(H_2O)_5$ under conditions such that the equilibrium between ions of series $H^+(H_2O)_n$ is attained (the residence time in the aging space of the order of 5 msec), and that the mass spectrum be essentially composed of ions $H^+(H_2O)_4$ (73 amu), and $H^+(H_2O)_5$ (91 amu) (the content in water vapor of the vector gas being held at 20 ppm in volume).

The experimental system was so devised that we only measured the mobility of ions with a given mass at the outlet of the reaction

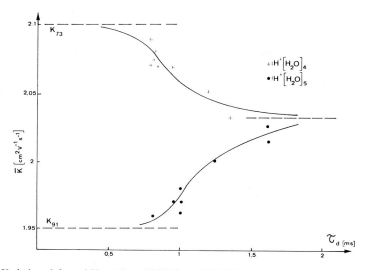

Fig. 5. Variation of the mobility of ions $H^+(H_2O)_4$ and $H^+(H_2O)_5$ as a function of their transit time.

chamber (we have checked to see that there are no associative or dissociative collisions for the ions within the ion focusing space and within the mass measurement space).

Measurements of mobility were made for various values of the electric field, i.e., for various values of the transit time in the measurement space. This mobility was then the apparent mean value $\bar{K}$ of the mobility of the different states of the ion in the measurement space. The values of the field were always less than 600 V/cm in order to comply with the "thermalizing" condition on ions.

Experimental results are shown in Fig. 5 with, on the abscissa, the transit time τ_d of ions in the measurement space, and on the ordinate, the mean mobility $\bar{K}$. In the same figure, we have shown by a dotted line the values of mobility of ions $H^+(H_2O)_4$ and $H^+(H_2O)_5$ obtained by Huertas (28) from measurements obtained at a pressure of nearly 20 Torr: $K_{73} = 2.1$ cm^2 V^{-1} sec^{-1} and K_{91} $= 1.95$ cm^2 V^{-1} sec^{-1}, as well as a mobility K_{eq} which represents the equilibrium value of the mean mobility of ions. For large values of the transit time, $H^+(H_2O)_4$ and $H^+(H_2O)_5$ ions reach an equilibrium between themselves ac-

cording to the reaction

$$H^+(H_2O)_4 + H_2O \rightleftarrows H^+(H_2O)_5.$$

If $H^+(H_2O)_4$ eq and $H^+(H_2O)_5$ eq represent the concentrations of the two ions at equilibrium measured under experimental conditions, we have

$$K_{eq} = \frac{K_{73}H^+(H_2O)_4 \text{ eq} + K_{91}H^+(H_2O)_5 \text{ eq}}{H^+(H_2O)_4 \text{ eq} + H^+(H_2O)_5 \text{ eq}}.$$
$$[13]$$

It is therefore impossible to say if mobilities shown in Table I correspond to individual ions or represent a group different mobilities corresponding to relation [13].

6. Modified Thomson's Theory (30)

Thomson's theory (30) permits one to calculate the free energy of an ion as a function of the number and type of molecules composing it (8, 29). In a closed system at temperature T, in which ions and vapor that are free of foreign particles are enclosed, the formation of droplets with a radius R about these ions corresponds to a variation of free energy of the

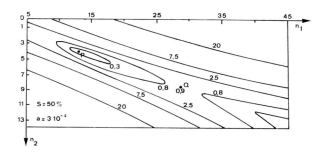

FIG. 6. Number of water (n_1) and sulfuric acid (n_2) molecules in an embryo as a function of the ratio $\Delta G/kT$.

system after the relation [2]

$$\Delta G = n_l(\mu_l - \mu_g) + 4\pi R^2 \sigma$$
$$+ \frac{Q^2}{2}\left(1 - \frac{1}{\epsilon}\right)\left(\frac{1}{R} - \frac{1}{R_0}\right), \quad [14]$$

where R_0 represents the ion radius, Q its charge, ϵ the dielectric constant of the liquid of surface tension σ. To write this relation, we assume on the one hand that $R_0{}^3$ is negligible in front of R^3, and on the other hand that the notion of surface energy of a primary ion of a radius R_0, or what could correspond to it, makes no sense.

It is possible to summarize for our case the results of Thomson theory as follows (22): The nucleation of water vapor in the presence of a charge has the effect of creating a stable state yielding ions of the line spectral type as reported in this paper. Furthermore, for the values of $P/P_\infty > 1$ there appears to be an unstable state corresponding to ions that grow in time or with age. But the fact that ions of this type appear for values of $P/P_\infty < 1$ seems to indicate that the charge and the water vapor alone are not sufficient to form a stable particle.[1]

Castelman *et al.* (33) have shown that ions emerging from an electric discharge in the presence of SO_2 have a complex composition and could be composed, in particular, of $H_3O^+(H_2O)(SO_2)_m$.

[1] P is the vapor pressure around the ion and P_∞ the saturated pressure at the same temperature.

Moreover, relation [14] must involve, in addition to humidity, an impurity that has not yet been identified, but this impurity can be detected by its influence on the shaping of the mobility/humidity curves for ions of given age.

(a) We have seen that in addition to water vapor, natural air contains sulfuric acid vapors. Let us adopt the notations used in relation [2]. By generalizing relation [14] to a binary vapor mixture, we can write the variation in free energy, corresponding to the passage from radius R_0 to radius R of the droplet bearing the basic charge, in the form (31)

$$\Delta G^c = \Delta G^0 + \frac{e^2}{2}\left(\frac{1}{R} - \frac{1}{R_0}\right), \quad [15]$$

by assuming that the permittivity of the liquid is much higher than unity. ΔG^0 represents the variation in energy corresponding to a neutral particle whose radius has been subjected to the same variations, and corresponds to relation [2]; since this relation is only valid with the same restrictions as those for relation [14], we can derive only very approximate indications. As in the case of relation [2], knowing that $S = P_1/P_{1\infty}$ and $a = P_2/P_{2\infty}$, as a function of n_1 and n_2, it is possible to determine ΔG^0 and ΔG^c. As a function of n_1 and n_2, we can show level curves in $\Delta G^c/kT$. A typical case is shown in Fig. 6 for droplets carrying a basic charge. The reference $\Delta G^c = 0$ has been taken at point P and this corresponds to $R_0 = 3$ Å. At P we have a stable condition, and if a droplet shifts away from this condition, it will tend to return to it, as the energy there is a minimum (32).

At Q, there is an unstable condition (saddle point). Any droplet leaving this unstable condition could evolve toward weaker energy states either by losing molecules and returning to the stable condition P, or by fixing molecules and developing in a continuous fashion. For a given value of relative humidity, it is possible to calculate the values n_1, n_2, and R corresponding to stable and unstable states by varying the activity. We obtain the curve in Fig. 7, which also shows, as a dotted line, the curve of unstable conditions corresponding to neutral particles for a relative humidity of 50%. The portion on the left of the maximum corresponds to stable states; the part at the right, to unstable states. The maximum corresponds to a radius $R_M = 6$ Å, therefore to a mobility $K = 1$ cm^2 V^{-1} sec^{-1} if we are to assume the validity of Langevin's relation. All of this is in agreement with experimental results represented in Table I. Furthermore, for $a_1 = 4.5 \times 10^{-4}$ and 50% relative humidity, there is a spontaneous development of droplets.

(b) Knowing the characteristics of both stable and unstable states, it is possible to calculate the rate of nucleation, which is the number of droplets to cross over the saddle point (unstable condition) per unit of volume and time. In the case of neutral particles, by designating R_0^* and ΔG^0 as the radius and energy of an electrically neutral droplet at the saddle point, and N_1 and N_2 as the concentrations of air in molecules 1 and 2 (water and sulfuric acid), the rate of nucleation is given in a first approximation by relation [4], which, if $N_1 \gg N_2$, can be written as

$$I_0 = 4\pi R_0^{*2} \beta_2 N_1 \exp\left(-\frac{\Delta G^{*0}}{kT}\right), \quad [16]$$

where β_2 represents the flux of molecules of H$_2$SO$_4$ which arrives on the droplet and governs the fixing of water molecules.

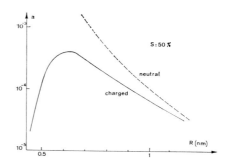

FIG. 7. Values of activity (ordinates) corresponding to particles of radius R (abscissa) bearing a basic charge for a 50% relative humidity. Stable states are at left of the maximum, unstable states at right. The dotted curve corresponds to electrically neutral particles.

In the case of electrically charged particles, droplets forming on ions (N_i ions per volume unit), nucleation centers are no longer the N_2 neutral molecules but the N_i ions. Moreover, ΔG^{*0} must be replaced by ΔG^{*c}, the variation of energy at the unstable state. The nucleation frequency is then written by defining the droplet radius R_c at the saddle point:

$$I_c = 4\pi R_c^{*2} \beta_2 N_i \exp\left(-\frac{\Delta G^{*c}}{kT}\right). \quad [17]$$

As above, let us take $P_{2\infty} = 3.5 \times 10^{-4}$ Torr. For a given activity that is lower than $a = 4.5 \times 10^{-4}$, we find that

$$\frac{I_c}{I_0} = \frac{R_c^{*2}}{R_0^{*2}} \frac{N_i}{N_1} \exp\left(\frac{\Delta G^{*0} - \Delta G^{*c}}{kT}\right), \quad [18]$$

is higher than 10^4 N_i. Nucleation is therefore favored by the presence of the charge. Moreover, when a is higher than a_1, there is spontaneous development of particles. Therefore, ions quickly reach the first stage $n_2 = 0$, then develop by attaching H$_2$SO$_4$ and H$_2$O molecules. It should be noted that particle/molecule collisions in these calculations were viewed as elastic and that the collision cross section increases for the particles with an electric charge, thus increasing the frequency of particle/molecule collisions.

(c) Let us designate by N_p and N_Q the concentrations of ions at points P (minimum) and Q (saddle point) of Fig. 6. By using the notations given above, we write the following, while realizing that

$$\Delta G_P = 0: \quad \frac{N_Q}{N_P} = \exp\left(-\frac{\Delta G_Q}{kT}\right) \simeq 0.5.$$

The concentration of ions that have crossed over the saddle point, with a mobility that is lower than the critical mobility, is thus always lower than that of small ions. This is actually what was found experimentally.

ACKNOWLEDGMENT

This work has been supported in part by the French Ministry of Life Quality.

REFERENCES

1. "Inadvertent Climate Modification," Report on the study of Man's Impact on Climate (SMIC), MIT Press, Cambridge, Mass. London, 1971.
2. ZETTLEMOYER, A. C., "Nucleation." Deker, New York, 1969.
3. REISS, H., *J. Chem. Phys.* **18**, 840 (1950); **19**, 253 (1951).
4. DOYLE G., *J. Chem. Phys.* **35**, 795 (1961).
5. KIANG, C. S., STAUFFER, D., MOHNEN, V. A., BRICARD, J., AND VIGLA, D., *Atmos. Environ.* **7**, 1279 (1973).
6. MIRABEL, P., AND KATZ, J. L., *J. Chem. Phys.* **60**, 1138 (1974).
7. ABEL, J., GMITRO, J. I., AND VERMEULEN, T., *AICHE* **10**, 741 (1964).
8. WOOD, W. P., CASTELMAN, A. W., AND TANG, I. N., *J. Aerosol Sci.* **6**, 267 (1975).
9. TAKAHASHI, K., KASAHARA, J., AND ITOH M. I., *J. Aerosol Sci.* **6**, 45 (1975).
10. HEIST, R. H., AND REISS, H., *J. Chem. Phys.* **61**, 573 (1974).
11. SHUGARD, W. J., HEIST, R. H., AND REISS, H., *J. Chem. Phys.* **61**, 5298 (1974).
12. BOULAUD, D., MADELAINE, G., VIGLA, D., AND BRICARD, J., "Air, Water, and Soil Pollution." **4**, 635 (1975).
13. BOULAUD, D., BRICARD, J., AND MADELAINE, G., A paraître dans *J. Aerosol Sci.*, 1976.
14. WHITBY, K. T., HUSAR, R. B., AND LIU-DANS HIDY, B. Y. H., "Aerosol and Atmospheric Chemistry." Academic Press, New York, 1972.
15. MEGAW, G. W., "Research Progress Report." Health, Physics, and Medical Division I.K.A.E.A., 1966.
16. BOURBIGOT, Y., BRICARD, J., MADELAINE, G., AND VIGLA, D., *C. R. Acad. Sci. Ser. T* **276**, 547 (1973).
17. GOETZ, A., AND PUESCHEL, R., *Atmos. Environ.* **1**, 287 (1967).
18. FUCHS, N. A., "Evaporation and Droplet Growth in Gaseous Media." Pergamon, Oxford, 1959.
19. BRICARD, J., BILLARD, F., MADELAINE, G., AND VIGLA, D., *J. Geophys. Res.* **73**, 4487 (1967).
20. COFFEY, P. E., AND MOHNEN, V. A., *Bull. Amer. Phys. Soc.* **17**, 392 (1972).
21. KIANG, C. S. P., CADLE, R. D., HAMIL, P., MOHNEN, G. W., AND YVE, G. K., *J. Aerosol Sci.* **6**, 465 (1975).
22. BRICARD, J., CABANE, M., MADELAINE, G., AND VIGLA, D., *J. Colloid Interface Sci.* **39**, 42 (1972).
23. MOHNEN, V., Fifth International Conference on Atmospheric Electricity, Garmisch Partenkirchen, 1974.
24. FRIEND, J. P., A paraître dans *Atmos. Environ.*
25. LOEB, L. B., *J. Aerosol Sci.* **2**, 133 (1971).
26. LANGEVIN, P., *Ann. Phys. Chim.* **5**, 245 (1905).
27. KILPATRICK, W. D., Proceedings of the 19th Conference on Mass Spectrometry, Atlanta, 1971.
28. HUERTAS, M. L., Thèse de Doctorat, Toulouse, 1972.
29. CASTELMAN, A. W., AND TANG, I. N., *J. Chem. Phys.* **57**, 3629 (1972).
30. THOMSON, J. J., "Conduction of Electricity through Gases," Cambridge University Press, London, 1906.
31. WIENDTL, E., Fifth International Conference on Atmospheric Electricity, Garmisch Partenkirchen, September, 1974.
32. CABANE, M., FAUSSOT, C., MADELAINE, G., AND BRICARD, J., *C. R. Acad. Sci. Paris Ser. B Q.* **281**, 209 (1975).
33. CASTELMAN, A. W., TANG, I. N., AND MENKELWITZ, H., *Science* **173**, 1025 (1971).
34. ALBRITTON, D. L., MILLER, T. M., MARTIN, D. W., AND McDANIELL E. W., *Phys. Rev.* **173**, 115 (1968).

The Use of Nucleation and Growth as a Tool in Chemical Physics

HOWARD REISS,[1] DEAN C. MARVIN, AND RICHARD H. HEIST[2]

Department of Chemistry, University of California, Los Angeles, California 90024

Received June 7, 1976; accepted July 21, 1976

The possibility of employing nucleation and growth for the amplification and detection of a variety of chemical physical phenomena is discussed. Preliminary experiments, employing a diffusion cloud chamber, and aimed at the studies of (1) the gas phase photooxidation of SO_2 and (2) gas phase photopolymerization are described. In the first case an interesting reaction, periodic in time, has been discovered. It appears unlikely that single neutral monomeric molecules can be detected by gas phase nucleation, although very small traces of H_2SO_4 are detectable. Preliminary cloud chamber studies confirm the order of magnitude of previous estimates of the pseudo unimolecular rate constant for the SO_2 photooxidation. Specialized design problems for a cloud chamber to be used in connection with these various studies are discussed.

I. INTRODUCTION

It has been known for many years that nucleation and growth constitute a process capable of detecting and amplifying events occurring at the atomic and molecular level. The photographic process is a case in point in which, in effect, single electronic defects (in solids) can be detected. The application of the Wilson cloud chamber by high-energy physicists to the study of single nuclear events, and sometimes to the detection of single ions, furnishes another example. Given these examples, it is surprising that nucleation has not been used for detection and amplification in connection with a wide variety of more conventional chemical physical phenomena. The reasons seem to lie with the difficulty of detecting neutral molecules as opposed to ions.

On the other hand radiation-induced nucleation, especially photoinduced nucleation in vapors, has been observed for many years. It appears to be a ubiquitous effect in photochemical smog, where it has been variously attributed to the photoformation of sulfinic acids, to the photooxidation of SO_2, to the photopolymerization of organic components, or to the effects of several of these phenomena acting in concert (1, 2). In all instances, the mechanisms involved are at best poorly understood. Photoinduced particulate formation has even been observed in fairly pure mixtures of helium and unsaturated water vapor, the experiments having been performed in a reaction chamber (3).

Photoinduced nucleation has also been observed in cloud chamber studies, where it usually has been regarded as a nuisance (4), and steps have been taken to avoid rather than understand it. The common view attributes it to uncontrollable photochemistry associated with trace impurities, leading to products of very low volatility. It has even been suggested that certain single neutral molecules could act as condensation nuclei (5, 6), and that the merest trace of impurity presents a problem unless light of the relevant frequency is excluded.

During the past few years we have been trying to adapt nucleation and growth processes to the problem of detection and amplification

[1] All correspondence should be addressed to Dr. Reiss at the address given above.
[2] Present address: Department of Chemical Engineering, University of Rochester, Rochester, N.Y. 14627.

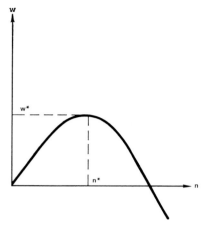

FIG. 1. Free energy of cluster formation plotted versus *n*, the number of molecules in the cluster.

geneous nucleation in a one-component system is usually expressed as J, the number of drops nucleated per cubic centimeter per second, and has the form (8–18)

$$J = K \exp[-W^*/k_B T], \qquad [1]$$

where K is a slowly varying prefactor depending on pressure p, temperature T, density of liquid, and several other parameters, while k_B is the Boltzmann constant. W^*, the free energy of formation of the critical nucleus, has the form

$$W^* = \text{const } v^2 \sigma^3 / [T \ln (p/p_s)]^2, \qquad [2]$$

where σ is the surface tension and v is the volume per molecule in the liquid. J proves to be a very sensitive function of the supersaturation, p/p_s, where p is the actual pressure and p_s, the saturation pressure. For the one-component system, the work of formation, W, of a cluster of n molecules gives the curve shown in Fig. 1 when plotted versus n for supersaturated liquids. The maximum corresponds to W^*. Beyond the maximum, drops

(as suggested above) in connection with several chemical physical phenomena involving neutral molecules. Our primary instruments in these studies have been one or another variant of the diffusion cloud chamber developed to a high level of precision during the past decade by Katz (4, 7). Our efforts have uncovered many specialized experimental difficulties in adapting the diffusion cloud chamber in this manner, and we have still not achieved *quantitative* success. Nevertheless, at this juncture quite a few dramatic qualitative and semiquantitative observations have been made, most of which have not been reported in the open literature, and the anniversary symposium of the colloid division seems a particularly appropriate vehicle for their first publication. In this paper, therefore, we report on these phenomena as well as on the technical progress (including the solution of certain instrumental problems) achieved in adapting the diffusion cloud chamber to the detection and amplification of various chemical processes.

II. BRIEF ACCOUNT OF THE THEORY OF NUCLEATION

It is convenient to provide an abbreviated account of the theory of the nucleation of drops from the vapor. The rate of homo-

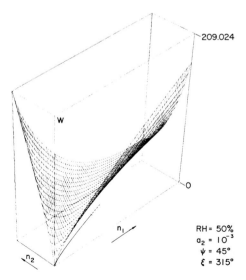

FIG. 2. Portion of the free-energy surface for nucleation in H_2O–H_2SO_4 mixtures at 50% relative humidity, and H_2SO_4 activity of 10^{-3} ($\sim 10^{-7}$ Torr, H_2SO_4).

may grow with a decrease in free energy. The free-energy barrier is produced by the work which must be expended in forming the interface between drop and liquid. This accounts for the strategic position of σ in W^*.

For a binary system, it may be shown that the free-energy *hill* in Fig. 1 becomes a free-energy *surface* (17), since now the free energy of formation of a cluster depends upon the numbers of molecules, n_1 and n_2, of *both* species in the drop. Portions of two such surfaces are shown in Figs. 2 and 3, taken from a paper dealing with the binary system H_2O–H_2SO_4 (18). The clusters, on the way to becoming drops, flow over the surface and through a mountain pass. The saddle point marking the location of this pass provides some measure of W^* for the binary system.

Under certain conditions in a supersaturated vapor (one-component or binary), the flux J can be increased dramatically (because W^* is lowered) by the insertion of even a single molecule other than the molecules normally incorporated into the nucleus. This possibility has been exploited by high-energy physicist

FIG. 4. Partial pressures and supersaturation in diffusion cloud chamber.

for many years in cases where the "other" species of molecule is an *ion* which, because of dielectric polarization, is able to lower W^* to the point where the single ion signalizes its presence by the formation of a macroscopic liquid drop. Thus, a high-energy event producing ion trails can be observed in a cloud chamber containing supersaturated vapor. This represents almost the ultimate in amplification and detection, although in a certain sense the cascading feature of an ion gauge is a similar but not exactly comparable phenomenon.

III. CLOUD CHAMBER EXPERIMENTS

The earliest cloud chamber, and the one most commonly discussed, is the Wilson expansion chamber (19), a piston device in which cooling and supersaturation are induced by adiabatic expansion. This chamber is cleared of foreign or "heterogeneous" condensation nuclei by repeated expansion and the settling of the drops condensed on these nuclei. It can be cleared of ions by maintaining an electric field across its dimension. The critical limit of supersaturation at which the vapor forces the issue itself (homogeneous or self-nucleation) is determined by identifying that degree of expansion beyond which the supersaturated state can no longer be sustained. At this point J in Eq. [1] is considered to lie in the neighborhood of unity, and the value of the supersaturation obtained, as a rule, by setting J equal to unity is usually defined as the *critical* supersaturation. The expansion chamber has been developed to a high state of refinement (6).

As indicated earlier, we shall be interested, mainly, in the diffusion cloud chamber, which

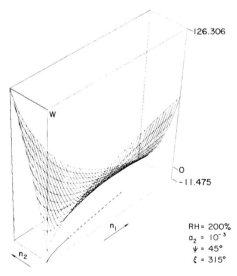

FIG. 3. Portion of the free-energy surface for nucleation in H_2O–H_2SO_4 mixtures at 200% relative humidity, and H_2SO_4 activity of 10^{-3} ($\sim 10^{-7}$ Torr, H_2SO_4).

TABLE I

Critical Supersaturations for H_2O Vapor in the Presence of H_2SO

Temp °C (initial)	Temp °C (final)	H_2O pressure (Torr)	H_2SO_4 pressure (Torr)	Theoretical critical supersaturation for pure H_2O vapor	Observed critical supersaturation
21.8	−28.5	4.29	4×10^{-10}	7.38	5.56
24.2	−29.6	3.63	2.2×10^{-9}	7.61	4.46
21.5	−35.6	1.24	2.5×10^{-8}	9.27	2.60
21.5	−38.7	0.32	4.3×10^{-7}	10.5	0.87
21.8	−44.0	0.09	3.5×10^{-6}	13.3	0.39
22.5	−51.7	0.019	2.0×10^{-5}	20.9	0.17

has experienced considerable development during the last decade (4, 7, 20, 21, 22). It maintains a steady state of supersaturation.

In describing the operation of the chamber, it is best to consider a definite case, namely that of operation with H_2O. Figure 4 will be helpful in this connection. As shown in the figure, the chamber consists of two parallel metal plates maintained at temperatures T_1 and T_2, respectively, with $T_2 < T_1$. A pool of "working fluid" (e.g., water) rests on the lower plate. The space above the pool is filled with a light, inert gas such as helium. The walls of the chamber form a cylinder, and can be constructed of glass, quartz, or some other material containing quartz windows through which light of optical and other frequencies can be transmitted.

At temperature T_1, water evaporates and diffuses through the helium to the upper plate where it condenses at the lower temperature T_2. The upper plate is slightly beveled and the water flows to the sides where it returns to the lower pool. A steady state of reflux is set up in which the relative humidity, or supersaturation, p/p_s, as shown in the figure, passes through a maximum at a point relatively distant from either plate. This sort of chamber was first used by Frank and Hertz (20) and refined by Katz (4, 7, 21) for the study of homogeneous nucleation. The dependence of supersaturation on elevation in the chamber must be computed by solving the associated transport problem. It is obvious that data on transport coefficients and vapor pressures are mandatory.

Under normal operation, the experiment begins with $T_2 = T_1$, and then T_2 is gradually reduced so that the curve for p/p_s develops the bulge shown in the figure. If heterogeneous condensation nuclei are present, they cause the formation of drops which fall to the pool on the lower plate. Thus, the chamber is self-cleaning, and when it is thus cleaned, T_2 is lowered so as to increase the bulge in p/p_s to the point where homogeneous nucleation just begins, as evidenced by a steady average rate of drop formation (1–10 cm^{-3} sec^{-1}) which takes place near the maximum in p/p_s. In this way the critical supersaturation for homogeneous nucleation can be determined as a function of temperature.

IV. TRACE EFFECTS IN CLOUD CHAMBERS

Physicists working with expansion chambers have, for many years, noticed that the presence of light of certain frequencies seemed to produce nucleation. Since these frequencies have not been high enough to cause ionization, some assumed that single neutral molecules could lead to condensation (5, 6), i.e., to the formation of a drop in much the same manner that an ion inserted into a cluster could convert it to a nucleus. Aerosols have even been observed to form in mixtures of helium and water vapor under the action of UV light, the water vapor being *undersaturated* (3). Although these various authors have laid the blame on photochemistry or radiation chemistry, the situation after nearly a century remains obscure. Yet as far back as 1939 (5) it was suggested that this phenomenon, properly understood, and the cloud chamber, might provide an exciting new tool for the study of several aspects of chemical physics, based primarily on its assumed ability to detect microscopic consequences of single atomic and molecular events.

However, recent theoretical work suggests that it may be impossible to detect single neutral molecules by means of nucleation. This work involved the theory of nucleation in vapor mixtures of H_2SO_4 and H_2O. H_2SO_4, because of its known high binding energy for water, should be one of the most promising (if not the most promising) monomeric neutral molecules for detection by nucleation. The theory has been developed through several stages of refinement by several authors (23, 24, 25), all of whom arrived at essentially the same conclusion. As mentioned earlier, for the binary system under consideration, the growth "fluxes" are now two-dimensional since the clusters are characterized by the numbers of molecules of both species, and the "flow" toward condensation occurs over a free-energy "surface" rather than a "hill" and through a "saddle point" on that surface.

From these theories it is very clear that even under the best of conditions, a single H_2SO_4 molecule, unlike a single ion, is not *guaranteed* to produce a macroscopic drop if present in supersaturated water vapor. However, very small concentrations of H_2SO_4 molecules can nevertheless induce nucleation. For example, in water vapor of relative humidity, about 300%, the presence of 10^7 H_2SO_4 molecules per cubic centimeter should, at room temperature, produce nucleation (25). These predictions have been semiquantitatively confirmed by recent experiments employing an expansion cloud chamber.

Table I illustrates some results from this work (26). The first column lists the initial temperature of the vapor mixture in the expansion chamber while the second column

shows the temperatures at the point of condensation during the expansion. The third and fourth columns list the initial partial pressures of H_2O and H_2SO_4, respectively, in the vapor mixtures. The fifth column shows the critical supersaturations expected on the basis of theory (for pure water vapor) if H_2SO_4 were absent from the initial mixture. Finally, the last column indicates the observed critical supersaturations. These are substantially lower than those expected for pure water in the fifth column, and indicate the nucleating effect of H_2SO_4 at very low partial pressures. The error limits are not shown in Table I. Errors are sufficiently large so that the data must be regarded as semiquantitative. The original paper should be consulted for further information in this respect.

Even though theory shows that one cannot *detect* a single impurity molecule, it shows that there may be cases in which a nucleus *contains* only a single H_2SO_4 molecule. But this is not the same as having *every* impurity molecule serve as a nucleus. Thus, out of the 10^7 H_2SO_4 molecules present in a cubic centimeter of water at 300% relative humidity, only one may act as a nucleus, so all 10^7 molecules have to be present before the nucleation event has a reasonable chance to occur.

In terms of the scheme in which a growing cluster adds molecules one at a time, we may analyze this problem as follows. Suppose we denote an impurity molecule by B. Then we may write

$$B + A = AB,$$
$$AB + A = A_2B, \qquad [3]$$
$$A_2B + A = A_3B,$$
$$\text{etc.}$$

This is a chain reaction similar to the normal nucleation process (in which case the B molecule would be replaced by A). The addition of B to a cluster may occur at a more advanced level of growth than is indicated in Eq. [3], e.g.,

$$A_n + B = A_nB, \qquad [4]$$

but the results are eventually the same. In any event, each cubic centimeter of supersaturated water vapor contains about 10^{18}–10^{19} H_2O molecules (A molecules). Thus, the single chain in Eq. [3] is competing with 10^{18}–10^{19} chains of the sort corresponding to the one-component process in which B is replaced by A. If we denote the flux for Eq. [3], similar to the J of Eq. [1], by J_B we may write

$$J_B = K_B \exp[-W_B^*/k_B T]. \qquad [5]$$

In essence the multiplicity of chains, for Eq. [1], shows up in K and

$$K/K_B \approx 10^{19}. \qquad [6]$$

Thus, for J_B to be competitive with J, we require

$$\exp[-(W_B^* - W^*)/k_B T] \approx 10^{19} \qquad [7]$$

and W^* would have to exceed W_B^* by about 1.2 eV per nucleus. This is a very unlikely figure for a neutral molecule.

Since H_2SO_4 is the most likely monomeric molecule to act invariably as a condensation nucleus (and it does not), it is even less likely that nitric acid, derived from oxides of nitrogen, should be effective, in this respect. In fact, estimates (27) indicate that concentrations of nitric acid about 10^5 times greater than H_2SO_4 are required for a comparable nucleating effect.

V. EXAMPLE OF THE CLOUD CHAMBER AS A DETECTION DEVICE; THE PHOTOOXIDATION OF SO₂

To illustrate some of the chemical physical applications of the cloud chamber in the role of detector it is convenient to focus on a particular case. For this purpose we consider a chemical process which we have already investigated with the aid of the cloud chamber, and for which some experimental results are available. The process in question is the gas phase photooxidation of SO_2.

Because it has been regarded as an important atmospheric reaction the photooxidation of SO_2 has, at this juncture, received considerable attention from many competent investigators. The understanding of the mechanism of this reaction, however, remains incomplete. On both theoretical and experimental grounds, the reaction appears to be very slow. In air, in strong sunlight (3000–4000 Å) with a photon flux of about 10^{13} cm^{-2} sec^{-1}, SO_2 seems to be *consumed* at the rate of about 0.25% per hour by photooxidation. It is assumed that the product is SO_3.

Because of the extreme slowness of the reaction relative to diffusion of the reactants one cannot, by the usual methods of detection and analysis, be sure that the observed loss of SO_2 is entirely due to a homogeneous process. Catalysis at the walls of the containing vessel is an ever present possibility.

Basically, two methods have been applied to the study of the gas-phase process. The first depends on the observation of the *disappearance* of reactant SO_2. The second focuses attention on the photophysical processes involved in the primary steps of photoexcitation and has concentrated heavily on the spectroscopy of triplet SO_2, i.e., $^3SO_2$.

The latest studies (28, 29) in both of these categories conclude by postulating the same mechanism. This is

$$\begin{aligned} SO_2 + h\nu &\rightarrow {}^1SO_2, &\text{(a)}\\ {}^1SO_2 &\rightarrow {}^3SO_2, &\text{(b)}\\ {}^3SO_2 + O_2 &\rightarrow SO_4, &\text{(c)}\\ SO_4 + SO_2 &\rightarrow 2SO_3, &\text{(d)} \end{aligned} \qquad [8]$$

In this series of reactions, $^1SO_2$ refers to the first excited state while $^3SO_2$ is the triplet. Reaction (b) involves an intersystem crossing. To the best of our knowledge the species SO_4 has never been observed in the reacting mixture, nevertheless it is postulated. The latest work on this system based on the study of the disappearance of SO_2 is due to Smith and Urone (28). These authors also studied the effects of NO_2, C_3H_6, and H_2O on the process. For example, small additions of NO_2 accelerate the reaction while larger additions retard it. These studies are of course complicated by the slowness of the disappearance of SO_2, and possible associated wall effects, as well as by

TABLE II

Summary of Quenching Rate Constant Data for Sulfur Dioxide Triplet Molecules with Various Atmospheric Components and Common Atmospheric Contaminants at 25°C

Compound	Parameters of Stern–Volmer plots[a]		k_1 liter/mole-sec ($\times 10^{-8}$)
	Intercept sec^{-1} ($\times 10^{-4}$)	Slope (mm sec)$^{-2}$ ($\times 10^{-4}$)	
Nitrogen	4.19 ± 0.36	0.458 ± 0.051	0.85 ± 0.10
Oxygen[b]	(1) 0.611 ± 0.020	(1) 0.516 ± 0.036	(1) 0.96 ± 0.07
	(2) 0.955 ± 0.061	(2) 0.622 ± 0.135	(2) 1.16 ± 0.25
	(3) 4.14 ± 0.38	(3) 0.515 ± 0.059	(3) 0.96 ± 0.11
	(4) 9.18 ± 0.57	(4) 0.463 ± 0.089	(4) 0.86 ± 0.16
			wtd. av: 0.96 ± 0.05
Water	4.34 ± 0.49	4.78 ± 0.63	8.9 ± 1.2
Argon	4.30 ± 0.37	0.280 ± 0.029	0.52 ± 0.05
Helium	3.92 ± 0.63	0.365 ± 0.035	0.68 ± 0.07
Xenon	3.83 ± 0.86	0.381 ± 0.058	0.71 ± 0.11
Carbon monoxide	—	—	0.84 ± 0.04[c]
Carbon dioxide	4.53 ± 0.29	0.614 ± 0.039	1.14 ± 0.07
Nitric oxide	4.19 ± 0.33	399 ± 18	741 ± 33
Ozone	3.77 ± 0.43	5.93 ± 0.62	11.0 ± 1.2
Sulfur dioxide	—	—	3.9 ± 0.1[d]
Methane	—	—	1.16 ± 0.16[e]
Propane	—	—	5.11 ± 0.58[e]
Propylene	—	—	850 ± 87[f]
Cis-2-butene	—	—	1340 ± 98[f]

[a] Derived from the least-squares treatment of the $1/\tau$ vs pressure of reactant plots in Figs. 2–4; the error limits reported are the 95% confidence limits (twice the standard deviation).

[b] The four independent estimates of the rate constant for the $^3SO_2$-quenching by O_2 are from runs at various SO_2 and Ar pressures: (1) $P_{SO_2} = 0.197$ Torr, (2) $P_{SO_2} = 0.197$, $P_{Ar} = 1.30$ Torr, (3) $P_{SO_2} = 1.55$ Torr, (4) $P_{SO_2} = 4.58$ Torr.

[c] Data from Jackson and Calvert.

[d] Data from Sidebottom et al.

[e] Data from Badcock et al.

[f] Data from Sidebottom et al.

uncertainties regarding the products of reaction since the concentrations of these are not determined.

A great deal of work on the primary photophysical processes in SO_2 photooxidation has been performed by Calvert and his associates at Ohio State. In one of the most recent papers in this area due to Sidebottom et al. (29), a frequency doubled laser was used to pump, at 3828.8 Å, directly through the "forbidden" band corresponding to the transition $SO_2 \rightarrow$ $^3SO_2$ where SO_2 is the ground-state singlet. It was then possible for them to study, in the presence of various quenching agents (N_2, O_2, H_2O, He, NO, CH_3CHCH_2, etc.), the decay of the triplet by monitoring the phosphores-

cence accompanying its return to the ground state. In this way it was possible to measure the rate constant for quenching corresponding to the respective agents. Table II (taken from their paper) lists these quenching constants as k_1 in the last column.

Quenching could be *physical* corresponding to

$$^3SO_2 \rightarrow SO_2 + h\nu_p, \qquad [9]$$

where ν_p is the frequency of phosphorescence, or it could be *chemical* corresponding to reaction c in Eq. [8] (in the case of O_2). Such chemical quenching leads, of course, in the scheme of Eq. [8], to the production of SO_3.

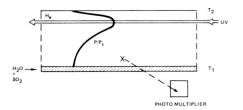

FIG. 5. Adaptation of cloud chamber to the study of the photooxidation of SO_2.

The table shows that O_2, N_2, and CO have, on a relative basis, a quenching power of about unity, while H_2O and O_3 have quenching powers of about 10. In contrast, still speaking in relative terms, molecules like NO and CH_3CHCH_2 are about 1000 times as effective. Since the quenching coefficients listed in the table include *both* the physical and chemical effects, this has given rise to speculation that NO and CH_3CHCH_2 are largely chemical quenchers. On the other hand, the work of Smith and Urone (28) indicates that NO and NO_2 can retard the reaction responsible for the consumption of SO_2.

The missing step in all of these studies concerns the rate of *arrival* of SO_3 molecules. If measurements on the rate of production of SO_3 were possible, we could, for example, compare them with the quenching data of Sidebottom, *et al.*, to see where a positive (chemical quenching) or negative (physical quenching) correlation exists. Abundant other information concerning the mechanism would also be produced. It should be reemphasized that the measurement of the rate at which SO_2 disappears is somewhat insufficient since the product of reaction may not be SO_3 alone. For example, it may under some circumstances be sulfur, and, in the presence of traces of olefin, sulfinic acid could be produced.

The diffusion cloud chamber represents an ideal device for completing these various studies on the photooxidation of SO_2 because, in view of the powerful nucleating ability of SO_3 or H_2SO_4, it permits us to observe the arrival of SO_3 via the photooxidation process. Furthermore it is possible to observe this ar-

rival in a manner guaranteed to be free of the effects of walls. In order to see how this may be accomplished we refer to Fig. 5, which is a variant of Fig. 4. The chamber is brought to a steady condition such that a given small p/p_s (as shown in Fig. 5) is established. The working fluid is H_2O containing SO_2, so that besides water vapor (pressure p) SO_2 exists in the vapor phase at steady state. Since the Henry's law constant is known for SO_2 in H_2O, the distribution of SO_2 with elevation is also known.

In Fig. 5, the ✗ marks the position of an He–Ne laser beam in a direction emerging from the page. Both T_2 and T_1 are adjusted so that homogeneous nucleation does not occur. No drops fall through the laser beam to scatter light into the photomultiplier tube denoted by P. The output of the photomultiplier is monitored by a counting circuit, and the tube is focused on an optically defined volume of observation. When drops fall through this volume they are counted automatically. The chamber is of course allowed to self-clean, as explained earlier.

At this point, a UV beam (about 1 mm in vertical thickness and 1 cm in horizontal width) is allowed to pass through the chamber above the laser beam and at right angles to it, as shown in the figure. The beam is positioned at the level of the maximum in the small p/p_s curve. Usually after a certain delay, nucleation takes place in the beam as a result of the production of SO_3 and the drops fall through the volume of observation and are counted. The nucleation rate is adjusted by fixing T_2 so that it is a Poisson-like process amounting to a few drops per cubic centimeter per second. This rate is a measure of the rate of photooxidation in the beam. Because of the capability of detecting the merest trace of H_2SO_4, by its nucleation of water vapor, the rate of oxidation is measured *as the oxidation occurs*, and not only after an accumulation of product is available, as in most previous methods. Furthermore, it is clear from the configuration of the experiment that reactions *on the chamber wall* are *not* being measured; any SO_3 formed in the layers of water which cover the plates would, because of its high solubility, stay there. Thus, two stubborn problems, mentioned earlier, are immediately eliminated.

In describing the steady chemical state in the chamber, it is helpful to have in mind a particular reaction, even one which may not conform in all its particulars to the actual process occurring in the chamber. Assume (although it may not be true if very long-lived intermediates are involved in the photochemical process) that SO_3 is produced only within the UV beam. A negligible fraction of this product is consumed by the nucleation process. The remainder diffuses to the upper and lower plates where it is absorbed essentially completely by the pools of water on those plates. Thus, a steady state is set up in which, according to the model, it can be shown that the concentration C_{SO_3} of SO_3 in the beam is given by

$$C_{SO_3} = k^*C_{SO_2}(hl/D) \qquad [10]$$

where C_{SO_2} is the concentration of SO_2, hl is a known parameter having the dimensions of area (h is the height of the cloud chamber and l the thickness of the UV beam), and D is the diffusion coefficient of H_2SO_4 while k^* is the pseudo-first-order rate constant for the reaction. The steady-state concentration of SO_3 (or H_2SO_4) determines the rate of nucleation, and thus the rate of nucleation provides information on k^* as it depends on the concentration of possible oxidants, light intensity, frequency, etc. The best estimate of k^* for the photooxidation of SO_2 in sunlit air yields values of the order of 10^{-6} sec^{-1} (24). Under the conditions of our experiments which yield values for all of the variables on the right of Eq. [10], this value of k^* requires a steady-state concentration C_{SO_3} of SO_3 equal to 5×10^6 cm^{-3}. Since the reaction occurs in water vapor having a relative humidity of approximately 300% and the observed nucleation rate is of the order of 1 drop per cubic centimeter per second, this is in accord with theory which predicts unit nucleation rate at 300% relative humidity if the concentration

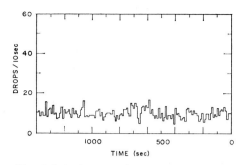

FIG. 6. Strip-chart recording of nucleation induced by the photooxidation of SO_2 in H_2O vapor (steady state).

of H_2SO_4 is in the neighborhood of 10^7 cm^{-3} (25). Thus from an order-of-magnitude standpoint the cloud chamber confirms previous estimates of the pseudo-first-order rate constant, but by the important alternative of observing the rate of arrival of product.

Figure 6 is a typical, steady-state, strip-chart recording showing the output of the photomultiplying tube. The ordinate represents the number of drops falling through the volume of observation in a 10 sec period while the abscissa represents time in seconds, every interval corresponding to 100 sec. It will be observed that the output shows good Poisson statistics.

It is easy to demonstrate that the nucleation is due to some photoreaction of SO_2, but more difficult to establish that the particular product, SO_3, is the nucleating agent. One can prove that SO_2 is involved by adding a small amount of NaOH to the pool of working fluid. Any SO_2 in the vapor is immediately absorbed by the formation of bisulfite ion, and concomitantly the nucleation ceases. Another test involves passing the UV radiation through an absorption cell containing SO_2, prior to its admission to the chamber. Again nucleation ceases.

Finally, by the use of narrow (100 Å) band-pass interference filters, it has been observed that the highest rate of nucleation occurs in a band centered at 2900 Å, the center of the band leading to the first excited singlet state of SO_2.

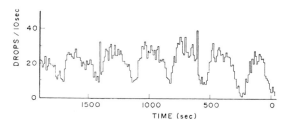

F<small>IG</small>. 7. Strip-chart recording of nucleation induced by the photooxidation of SO_2 in H_2O vapor (process, periodic in time).

The identity of the actual product causing nucleation is another matter. The two most prominent possible alternatives to SO_3 are the formation of molecular sulfur or the photoformation when traces of olefin are present of small amounts of highly involatile sulfinic acid. However, if the steady-state concentration of sulfur were 5×10^6 cm^{-3}, as we estimate for SO_3, above, the sulfur vapor would be highly undersaturated since its vapor pressure at room temperature is such that a concentration of about 10^{13} cm^{-3} would be required merely to establish *saturation*. Since the k^* for the production of sulfur is not likely to exceed that for the production of SO_3 (for then much higher rates of *disappearance* of SO_2 would be observed via this alternative process), sulfur particles acting as heterogeneous nuclei can pretty well be ruled out. Sulfinic acids are also unlikely because even if the steady concentrations attained were to correspond to the instant conversion of all the olefin present at the maximum possible concentration, the resulting vapor pressure would be barely enough to saturate the system with respect to this compound. It is obviously unlikely that such a maximum steady-state concentration could be achieved. If it were achieved, the olefin would be rapidly consumed and nucleation would cease. The evidence is therefore quite strong in favor of SO_3 as the nucleating agent.

Some preliminary experiments have been performed on the effects of deliberate addition of O_2 on the rate of nucleation. If O_2 acts primarily as a physical quenching agent this would retard reaction (c) of Eq. [8] and O_2 should reduce the rate of nucleation. On the other hand if its action is primarily through reaction (c) of Eq. [8], i.e., as a chemical quenching agent, then the rate of nucleation should be accelerated. We have observed that small additions of O_2 accelerate the process while larger additions retard it. Although these experiments are not yet conclusive, because of difficulties of control described below, they do indicate that oxygen can, not surprisingly, act in both modes. Since oxygen is itself a triplet molecule ($^3\Sigma_g$) it could be very well quench triplet SO_2 effectively.

Perhaps the most dramatic effect observed during the course of photooxidation in the cloud chamber has been the occurrence of a reaction periodic in time, associated with the composite process photooxidation of SO_2-nucleation. The periodic process seems to occur when a particularly high rate of photooxidation is maintained. Figure 7 is a strip-chart recording similar to Fig. 6, but for a higher rate of photooxidation, achieved by increasing the UV intensity. The oscillations in the rate of nucleation are quite evident. The period of oscillation is constant at about 260 sec. The remarkable thing about this process is the fact that the noise which constituted the Poisson fluctuation in the steady state is superposed upon the steady fluctuation which constitutes the low-frequency oscillation. This noise is not completely disconnected from the main oscillation, as is evident from the increase in its amplitude in the presence of the periodic process. Reactions having periodic structure in space and time are currently the subject of much investigation. A famous example is the Zhabotinski reaction (30) involving the oxidation of malonic acid in the presence of cerium sulfate and potassium bromate. The oscillations occur in the concentrations of Ce^{3+} and Ce^{4+} ions. However, the Zhabotinski reaction shows only the principal mode and not the noise shown in Fig. 7.

Periodic processes are of importance in nonphysical as well as physical phenomena. In fact, one of the earliest studies of such a process involves the famous Volterra equations (frequently called the Lotka–Volterra problem) (31) which were formulated to explain competition between species. The well-known problem deals with the situation in which big fish feed on little fish and flourish but eventually deplete their food supply so that they themselves diminish, whereupon the food supply (small fish) makes a comeback so that the big fish can once more feed and flourish. The process continues with alternating periods of prosperity for big fish and little fish. Other examples of nonphysical periodic processes can be found in the cyclic phenomena of economics. The situation is well summarized by Glansdorf and Prigogine in a recent book (32). These authors suggest that the establishment of such periodic structures in space and time enables nature to establish dynamic stability far from equilibrium. Our case obviously presents an interesting example.

It is likely that the periodicity is due to coupled processes involving both chemical reaction and transport. In fact a possible explanation is suggested immediately. The following four-step mechanism is self-explanatory:

Step (1): production of $SO_3 \rightarrow$ diffusion to
$\downarrow$ the walls
nucleation of H_2O drops,
Step (2): SO_3 + drops → dissolved SO_3 [11]
(removal from vapor),
Step (3): cessation of nucleation,
Step (4): return to Step (1).

The basic idea in this mechanism is that SO_3 produced by photooxidation at a high rate causes abundant nucleation. The resulting

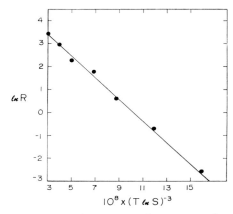

FIG. 8. Logarithm of observed rate, R, of photoinduced nucleation (at constant UV intensity) plotted versus $(T \ln S)^{-3}$.

drops of water contiguous to the high concentration of H_2SO_4 serve as sinks which by dissolving the SO_3 remove it from the vapor and stop the nucleation. Then the photooxidation, proceeding continuously, replenishes the supply of SO_3 and initiates nucleation once more.

The kinetics of this process can be modeled to yield the following pair of rate equations:

$$dC_{SO_3}/dt = G - \alpha C_{SO_3} - \gamma C_{SO_3} C_D, \quad [12]$$

$$dC_D/dt = a + bC_{SO_3} - gC_D. \quad [13]$$

In these equations C_{SO_3} represents the SO_3 concentration while C_D is the drop concentration in the UV beam. G is the steady rate of production of SO_3, the term αC_{SO_3}, with α constant, indicates diffusion to the walls, and γC_{SO_3}, with γ constant, is the rate of sorption of SO_3 by the drop. The quantity $a + bC_{SO_3}$ is the nucleation rate and gC_D, with g constant, is the rate at which drops fall out of the UV beam under the influence of gravity. It is possible to provide reasonable order-of-magnitude values for all of the constants appearing in Eqs. [12] and [13], although there is not room to provide the details in this paper. Unfortunately, although Eqs. [12] and [13], because of the nonlinear nature of Eq. [12], are capable, in principle, of sustaining oscillations, no such steady oscillations are possible

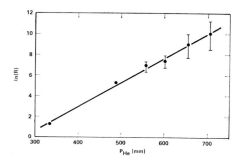

FIG. 9. The dependence of the logarithm of observed rate, R, of photoinduced nucleation on helium partial pressure, P_{He}, in the diffusion cloud chamber.

with constants of the estimated proper orders of magnitude. The main difficulty is with effects due to the acceleration of gravity summarized by the constant g. This is far too large, and as a result the drops do not remain in the UV beam long enough to produce oscillations. Gravity would have to be reduced by many orders of magnitude to make the above mechanism work.

Thus, as things now stand, it is necessary to look for another mechanism. Only the chemical and not the transport process may be involved, but this is unlikely. In any event, it is important to establish the correct mechanism because both the amplitude and period of the oscillations and their dependences on controllable parameters must yield considerable information on both chemical reaction and transport processes in the chamber.

VI. ANOTHER EXAMPLE OF THE CLOUD CHAMBER AS A DETECTOR; PHOTOPOLYMERIZATION

If, in the cloud chamber, mixtures of relatively pure supersaturated water vapor and helium are exposed to UV light of wavelength less than 2500 Å, droplets are observed to form at supersaturations well below that required for self-nucleation in water vapor. The rate of nucleation is measured using the arrangement in Fig. 5. Figure 8 is a plot of the logarithm of this measured rate versus $10^8/(T \ln S)^3$, where S is the maximum supersaturation in the chamber arranged to occur at the level of the

UV beam in Fig. 5. The UV intensity is maintained constant. The plot is an excellent straight line. Furthermore, nucleation is clearly photoinduced since it can be initiated or stopped by turning the UV on or off. An electric field applied vertically across the chamber has no effect, so ions are not involved. Furthermore, the UV frequency is not high enough to produce ions in a single photon process. If the helium pressure in the chamber is increased the rate of nucleation is increased. This is evident in Fig. 9, where the logarithm of the nucleation rate is plotted versus helium pressure. This strongly suggests that the nucleation is due to a trace impurity in the helium.

One possible explanation of Fig. 8 is free-radical addition polymerization (33–50). As an example consider the following well-known reaction involving vinyl monomers and R·, an initiating free radical:

$$\text{R} \cdot + \text{CH}_2 = \text{CHX} \rightarrow \text{RCH}_2 - \overset{\overset{\displaystyle X}{|}}{\text{CH}} \cdot,$$

$$\text{RCH}_2 - \overset{\overset{\displaystyle X}{|}}{\text{CH}} \cdot + \text{CH}_2 = \text{CHX} \rightarrow \text{RCH}_2 \qquad [14]$$

$$- \text{CHX} - \text{CH}_2 - \overset{\overset{\displaystyle X}{|}}{\text{CH}} \cdot, \quad \text{etc.}$$

If this process occurs in the presence of highly supersaturated water vapor, as might be found in the cloud chamber, and a high degree of polymerization is achieved, the single polymer molecule might eventually look like an oil drop to the water vapor, and a macroscopic drop of water would nucleate and grow on it. Thus, in this case we would be detecting a single neutral molecule, albeit a polymer molecule!

An early example of such gas phase addition polymerization is the polymerization of methylacrylate studied by Melville as early as the late 1930's (45). Melville was able to show that, although the principal reaction occurred on the walls of the reaction vessel, gas-phase polymers with degrees of polymerization as high as 100 were produced. It is an easy matter to show that Eq. [14], proceeding in the steady state, would have, for the rate R_g at which polymers of degree of polymerization g were produced, the expression

$$R_g = k_0 \left(\frac{k}{k + \beta} \right)^{g-1}. \quad [15]$$

In this equation k_0 is the rate constant for initiation, k the rate constant for propagation, and β the rate constant for the escape of polymer from the region of high supersaturation marked by the UV beam. The ratio $k/(k + \beta)$ is the probability of propagation in the face of the possibility of loss by escape from the region of high supersaturation. Since, in order to achieve a polymer of size g, propagation must occur g times in succession it is not surprising that the probability of propagation raised to the g power appears in Eq. [15].

Now we can proceed to the possible explanation of Fig. 8. Suppose that the nucleus required for the self-nucleation of water, at the supersaturation S, contains ω^* water molecules. Furthermore, assume that for a polymer molecule to act as a nucleus its degree of polymerization g^* must be proportional or roughly proportional to ω^*. Thus we have

$$g^* = \kappa\omega^*, \quad [16]$$

where κ is a constant.

The radius of the condensation nucleus (for self-nucleation) is prescribed by the Gibbs–Thomson relation as

$$r = 2\sigma v/k_B T \ln S, \quad [17]$$

where σ is the surface tension, v the volume per molecule in the liquid, and S the supersaturation. The volume of the nucleus goes as r^3 and the number of molecules, ω^*, in it, is proportional to the volume. Thus using Eq. [17] we have

$$\omega^* = (4\pi/3v)(2\sigma v/k_B)^3 [1/(T \ln S)^3]$$
$$= \Gamma/(T \ln S)^3, \quad [18]$$

where Γ is a constant only slightly dependent on temperature.

Employing Eq. [16] we may write

$$g^* = \kappa\Gamma/(T \ln S)^3 = \Omega/(T \ln S)^3, \quad [19]$$

where Ω is another constant. Substituting Eq. [19] into Eq. [15] with $g = g^*$ and taking the logarithm of both sides of the resulting equation gives

$$\ln R_{g^*} = \frac{k_0(\beta + k)}{k} + \left\{ \Omega \ln \frac{k}{k + \beta} \right\}$$
$$\times (T \ln S)^{-3}. \quad [20]$$

Since $\ln (k/\beta + k)$ is negative and only slightly dependent on temperature, Eq. [20] predicts that the logarithm of the photoinduced nucleation rate will decrease linearly with $(T \ln S)^{-3}$. This is just what is observed in Fig. 8.

Although no vinyl monomer was deliberately added to the system involved in Fig. 8, such very small concentrations of monomer could be effective in the process. For example, in the parts per million range, just above the limit (10^{-4} Torr) of mass spectrometric detection, assuming the reactive free-radical head of the growing polymer to have a collision cross section of 5×10^{-15} cm^2 and the relative thermal velocity of the monomer to be approximately 10^5 cm sec^{-1}, while its concentration remains at 10^{-4} Torr or at a concentration of about 10^{12} cm^{-3}, the number of collisions per second will be

$$Z = (5 \times 10^{-15})(10^5)(10^{12}) = 500 \text{ sec}^{-1}. \quad [21]$$

Thus, if every collision is fruitful, it requires only 1 sec to reach a degree of polymerization of 500. Since this probably exceeds the size of a nucleus, nucleation may easily occur.

If such low concentrations of monomer are effective a special problem of purity is posed. At water vapor supersaturations typical of our experiments, the nucleus for homogeneous nucleation contains several hundred molecules, i.e., ω^* is of the order of several hundred. Thus, as indicated earlier, a degree of polymerization of, say $g^* = 500$, should be sufficient if the polymer is to serve as a condensation nucleus. Since observed nucleation rates are of the order of 1 cm^{-3} sec^{-1}, this implies that about one polymer of degree of polymerization, 500, is produced each second, and that 500 monomer units are consumed each second in each cubic centimeter. Even at the mass spectrometer limit of 10^{12} molecules cm^{-3}, it would require about

$$10^{12}/500 = 2 \times 10^9 \text{ sec} \approx 100 \text{ yr}$$

to consume all the monomer in a cubic centimeter. This leads to an obvious problem with initial purity. Even substances with vapor pressures of the order of 10^{-6} Torr, if adsorbed on the walls of the chamber, will evaporate about 10^{14} molecules per square centimeter each second. Obviously there are many ways to achieve enough impurity to support the described mechanism insofar as the concentration of reactant is concerned.

The observed photonucleation could still be due to an entirely different cause besides polymer formation. For example, if a photochemical reaction converted some trace impurity into a steady concentration (in the UV beam) equivalent to about 10^{-5} Torr of some involatile (vapor pressure, approximately 10^{-6} Torr) monomeric product it would be quite possible for this product to undergo self-nucleation and produce particles which subsequently serve as heterogeneous nuclei for the condensation of water. Recent theoretical studies (51) show that such self-nucleation could occur even in the nonuniform environment of the cloud chamber, provided that a steady concentration of product at the stated level were achieved.

VII. CLOUD CHAMBER DESIGN PROBLEMS

We conclude this report with a discussion of special problems associated with the design of a cloud chamber for the study of phenomena like those described in the preceding sections. Katz (4, 7) has already given exhaustive accounts of the special problems associated with cloud chambers aimed at the study of nucleation itself. However, new problems arise when one wishes to conduct photochemical and spectroscopic studies in the chamber. In the course of our research we have solved a number of these and noted several others. It is our hope

that other investigators will turn their attention to this field, and a brief account of problems of design and their solutions should be of value in this respect.

As has been known for some time, tangential temperature gradients (parallel to the cloud chamber plates) are to be avoided. If these gradients become too large, Bénard convection (52) is liable to occur within the pool of working fluid, leading eventually to unstable transport throughout the chamber. The chamber cannot function in this nonquiescent mode. Tangential gradients are usually avoided by constructing the plates out of highly conductive metal. On the other hand, if corrosive reagents (e.g., H_2SO_4, SO_2, stearic acid, etc.) are to be used one cannot place the reactive solution in direct contact with metal. Protecting the plates with a layer of inert metal (e.g., gold) proves infeasible because the best plating invariably possesses microscopic weak spots where corrosion eventually begins. Covering the plates with a uniform layer of Teflon is impractical because then the working fluid will not wet the upper plate. Incidentally, even if the plate surfaces are metal it proves practically impossible to keep them clean enough during operations that they will be wet by water. Thus water, a very important substance, cannot be used as a working fluid in conjunction with metallic plate surfaces. A partial but far from satisfactory solution is obtained by placing a layer of glass wool on the upper plate, to be held there by capillary action. This same partial solution can be used with Teflon-coated plates, but in this case the layer of glass wool must be literally held in place by mechanical action, i.e., by the establishment of a vacuum behind each fiber. In any event, whether one is concerned about corrosion or with wetting by water, some nonmetallic layer must be deposited on the plate.

Our best solution to date has been to lay down a thin (10 mil) and very uniform layer of compatible glass on the metal substrate. This process has been perfected by the Pfaudler Company of Rochester, New York. However, the plates must be fashioned out of special low-carbon steel. As such they are not as thermally conducting as they would be if made out of aluminum or copper, but if the layer of glass is sufficiently uniform tangential gradients can be avoided.

Even with glass-coated plates it is no simple matter to wet the upper plate with water in a uniform film. The plates must be scrupulously clean, since even a monolayer of grease can be detrimental. We have found the most satisfactory cleaning procedure to be one in which the cloud chamber is assembled and filled with cleaning solution (NOCHROMIX, Godax Labs, New York, N. Y.) and then flushed with highly purified water. Since the chamber is relatively massive and at the same time has fragile components it is necessary to construct a special jig so that the assembled chamber can be swiveled around a horizontal axis to facilitate the cleaning and flushing process which utilizes ports in the chamber walls. Once the chamber is clean it can only be exposed to the most highly purified gases and liquids.

Glass or Teflon layers introduce still another problem. They make it impossible to maintain an electric field across the chamber for the purpose of sweeping out ions. This problem could conceivably be solved by introducing a slightly conducting glass, but this solution is not available at the moment. The same problem occurs when vapors, capable of dissolving in water as electrolytes (e.g., SO_2), are used in the chamber. Then conducting streams of refluxing liquid run down the chamber wall so that the two plates are short-circuited.

The chamber wall constitutes another set of problems. If one wishes to perform photochemical or spectroscopic experiments at ultraviolet frequencies provision must be made for the entry of such radiation. Simple glass walls are obviously unsuitable for this purpose. An optically polished quartz cylinder can and has been used, but it is very expensive, and it is not easy to equip it with the several ports necessary for cleaning and for the accurate introduction of small amounts of gases and liquids. Furthermore internal reflection makes quantitative spectroscopy difficult. A metal

ring equipped with quartz ports has many desirable features but it also poses a problem because of its high thermal conductivity; uniform thermal differentials between the plates are difficult to maintain. Our best solution, to date, involves the use of a Pyrex glass ring with screw-on quartz ports sealed by Viton O-rings. Other ports admit various fluids, and accurate metering is performed by means of a gas syringe. Except for Viton O-rings which seal the glass cylinder to both plates the entire interior of the chamber is now glass. This constitutes a distinct advantage in cleaning and degassing. The various optical ports must be kept defogged by streams of hot air.

Because of the obvious problems associated with purity all fluids entering the chamber must be scrupulously purified in advance. Water, for example, must be distilled several times in the presence of oxidizing agents. Then it must be distilled "within the system" under highly purified helium. It may have to be irradiated with UV during the course of distillation to convert any residual SO_2 to H_2SO_4, which will then have no volatility in the chamber. Helium must be passed through molecular sieves and charcoal at liquid nitrogen temperatures. Teflon stopcocks must be used and the manifold for treating gases, as well as the connections to the cloud chamber, must contain no plastic joints. All gases admitted to the chamber must be purified by distillation. Water must be degassed, finally, by freezing under vacuum. It is desirable to use a mercury-free vacuum system.

There are many other highly specialized problems. For example, when SO_2 is used in the chamber its concentration in the vapor (as a function of elevation) can only be known if its concentrations in both the pool on the lower plate and the aqueous layer on the upper plate are known. Even though the Henry's law constant for SO_2 in H_2O is known the two plates are at different temperatures. For this purpose it is not sufficient merely to know the temperatures of the plates, and the total amounts of water and SO_2 in the system. The actual steady-state partitioning of the water between the two plates must be known. Many other specialized problems arise but there is not space enough to discuss them.

VIII. CONCLUSION

We have provided qualitative and semiquantitative evidence that the diffusion cloud chamber represents a promising tool as a detector for a variety of interesting chemical physical phenomena. It is unlikely, however, that single neutral molecules can be detected, although the possibility remains for detecting individual polymer molecules of a substantial degree of polymerization. A variety of specialized design problems have been partially solved or remain to be solved before the diffusion chamber can be applied with maximum efficiency in this mode. However, as the number of workers in the field increases much of the potential of these kinds of experiments should be realized.

ACKNOWLEDGMENT

The authors wish to acknowledge the support of the National Science Foundation, Grant MPS74-22922.

REFERENCES

1. BRICARD, J., CABANE, M., MADELAINE, G., AND VIGLA, D., *in* "Aerosols and Atmospheric Chemistry" (G. M. Hidy, Ed.), p. 27. Academic Press, New York, 1972.
2. MILLER, M. S., FRIEDLANDER, S. K., AND HIDY, G. M., *in* "Aerosols and Atmospheric Chemistry" (G. M. Hidy, Ed.), p. 301. Academic Press, New York, 1972.
3. CLARK, I. D., AND NOXON, J. F., *Science* **174**, 941 (1971).
4. KATZ, J. L., *J. Chem. Phys.* **62**, 448 (1975).
5. CRANE, H. R., AND HALPERN, J., *Phys. Rev.* **56**, 232 (1951).
6. ALLEN, L. B., AND KASSNER, J. L., JR., *J. Colloid Interface Sci.* **30**, 81 (1969).
7. KATZ, J. L., *J. Chem. Phys.* **52**, 4733 (1970).
8. BECKER, R., AND DORING, W., *Ann. Phys. (Leipzig)* **24**, 719 (1935).
9. VOLMER, M., AND WEBER, A., *Z. Phys. Chem. (Leipzig)* **119**, 277 (1926).
10. FARKAS, L., *Z. Phys. Chem. (Leipzig) A* **125**, 236 (1927).
11. VOLMER, M., *Z. Phys. Chem. (Leipzig)* **25**, 555 (1929); "Kinetik der Phasenbildung," p. 122. Steinkopff, Darmstadt, Germany, 1939.
12. ZELDOVICH, J. B., *Acta Physicochem.*, U.R.S.S., **18**, 1 (1943).
13. McDONALD, J. E., *Amer. J. Phys.*, **30**, 870 (1962).
14. FRENKEL, J., "Kinetic Theory of Liquids," Chap. VII. Dover, New York, 1946.
15. DUNNING, W. J., *in* "Nucleation" (A. C. Zettlemoyer, Ed.), Chap. 1. Dekker, New York, 1969.
16. REISS, H., *Ind. Eng. Chem.* **44**, 1284 (1952).
17. REISS, H., *J. Chem. Phys.* **18**, 840 (1950).
18. HEIST, R. H., AND REISS, H., *J. Chem. Phys.* **61**, 573 (1974).
19. WILSON, C. T. R., *Phil. Trans. Royal Soc. London Ser. A* **190**, 403 (1898).
20. FRANK, J. R., AND HERTZ, H. G., *Z. Phys.* **143**, 559 (1956).
21. KATZ, J. L., AND OSTERMIER, B. J., *J. Chem. Phys.* **47**, 478 (1967).
22. HEIST, R. H., AND REISS, H., *J. Chem. Phys.* **59**, 665 (1973).
23. DOYLE, G. J., *J. Chem. Phys.* **35**, 795 (1961).
24. MIRABEL, P., AND KATZ, J. L., *J. Chem. Phys.* **60**, 1138 (1974).
25. SHUGARD, W. J., HEIST, R. H., AND REISS, H., *J. Chem. Phys.* **61**, 5298 (1974).
26. REISS, H., MARGOLESE, D. I., AND SCHELLING, F. J., *J. Colloid Interface Sci.*, **56**, 511 (1976).
27. KIANG, C. S., AND STAUFER, D., *Discuss. Faraday Soc.* **57** (1974).
28. SMITH, J. P., AND URONE, P., *Environ. Sci. Technol.* **8**, 742 (1974).
29. SIDEBOTTOM, H. W., BADCOCK, C. C., JACKSON, G. E., CALVERT, J. G., REINHARDT, G. W., AND DAMON, E. K., *Environ. Sci. Technol.* **6**, 73 (1972).
30. ZHABOTINSKI, A. M., *Biofizika* **2**, 306 (1964).
31. LOTKA, A. J., *J. Amer. Chem. Soc.* **42**, 1595 (1920); VOLTERRA, V., "Theorie Mathématique de la Lutte pour la Vie." Gauthier–Villars, Paris, 1931.
32. GANSDORFF, P., AND PRIGOGINE, I., "Structure, Stability, and Fluctuations." Wiley, New York, 1971.
33. FLORY, P. J., "Principles of Polymer Chemistry," Chap. IV. Cornell Univ. Press, Ithica, N.Y., 1953.
34. NORTH, A. M., "The Kinetics of Free Radical Polymerization." Pergamon, Elmsford, N.Y., 1966.
35. BAGDASAR'YAN, KH. S., "Theory of Free Radical Polymerization" (translated by J. Schmarak). Israel Program for Scientific Translations, 1968.
36. HUYSER, E. S., "Free Radical Chain Reactions," Chap. 12. Wiley, New York, 1970.
37. HAM, G. E., *in* "Kinetics and Mechanisms of Polymerization–Vinyl Polymerization (G. E. Ham, Ed.), Vol. 1. Dekker, New York, 1967.
38. TAYLOR, H. S., AND BATES, J. R., *Proc. Nat. Acad. Sci.* **12**, 714 (1926).

39. OLSON, A. R., AND MEYERS, C. H., *J. Amer. Chem. Soc.* **48**, 389 (1926).

40. BATES, J. R., AND TAYLOR, H. S., *J. Amer. Chem. Soc.* **49**, 2439 (1927).

41. TAYLOR, H. S., AND HILL, D. G., *J. Amer. Chem. Soc.* **51**, 2922 (1929).

42. TAYLOR, H. S., AND EMELEUS, H. J., *J. Amer. Chem. Soc.* **52**, 2150 (1930).

43. STORCH, H. H., *J. Amer. Chem. Soc.* **56**, 374 (1934).

44. RICE, O. K., AND SICKMAN, D. V., *J. Amer. Chem. Soc.* **57**, 1384 (1935).

45. MELVILLE, H. W., *Proc. Roy. Soc. Ser. A* **163**, 511 (1937); **167**, 99 (1938).

46. DANBY, C. J., AND HINSHELWOOD, C. N., *Proc Roy. Soc. Ser. A* **179**, 169 (1941).

47. ARVIA, A. J., AYMONINO, P. J., AND SCHUMACHER H. J., *Z. Phys. Chem. (Frankfurt)* **28**, 393 (1961).

48. CORDISCHI, D., DELLE SITE, A., LENZI, M., ANI MELE, A., *Chem. Ind. (Milan)* **44**, 1101 (1962).

49. HEICKLEN, J., *J. Amer. Chem. Soc.* **87**, 445 (1965).

50. MARSH, D. C., AND HEICKLEN, J., *J. Amer. Chem. Soc.* **88**, 269 (1966).

51. BECKER, C., AND REISS, H., *J. Chem. Phys.*, **65** 2066 (1976).

52. BENARD, H., *Rev. Gen. Sci. Pures. Appl.* **11**, 1261; 1309 (1900); *Ann. Chim. Phys.* **23**, 62 (1901).

On Unipolar Diffusion Charging of Aerosols in the Continuum Regime

BENJAMIN Y. H. LIU AND DAVID Y. H. PUI

Particle Technology Laboratory, Mechanical Engineering Department, University of Minnesota, Minneapolis, Minnesota 55455

Received July 23, 1976; accepted August 12, 1976

Some recently completed unipolar diffusion charging experiments in the continuum regime are described. Monodisperse aerosols generated by the vibrating orifice generator and the mobility classification method were exposed to unipolar positive ions produced by a corona discharge under conditions approximating pure ionic diffusion, and the resulting particle charge was measured as a function of the charging parameter, $N_0 t$, where N_0 is the unipolar ion concentration and t is the charging time. The particle diameter was varied between 0.1 and 5.04 μm and the $N_0 t$ product, between 2.56×10^6 and 5.1×10^7 (ions/cm^3) (sec).

A comparison of the data with the available theories shows that the best agreement is obtained with the theory of Fuchs and Bricard, whereas the agreement with the theory of White is less satisfactory. It is concluded that for unipolar charging in the continuum regime, the theory of Fuchs and Bricard should be used.

INTRODUCTION

Charging of aerosol particles by unipolar ions is an important phenomenon in aerosol physics with many practical applications. Charging of particles can occur by ionic diffusion, or by the action of an applied electric field. The former is used in the Electrical Aerosol Analyzer (1) for measuring the size distribution of aerosols by electrical mobility, while the latter is used in electrostatic precipitators for removing suspended particles for gas cleaning or other purposes.

Although there is considerable interest in the phenomenon of unipolar diffusion charging, detailed experimental study on the process is lacking. There is much confusion in the literature regarding the applicable laws in unipolar diffusion charging in the continuum regime. For instance, in literature dealing with electrostatic precipitation (see Ref. (2), for example), White's equation (3) is often used, whereas in literature dealing with atmospheric ions and particles the equation due to Fuchs (4), Bricard (5–7), and Pluvinage (8–10) is more commonly used. There has been no attempt to reconcile the difference between these equations. In this paper, a detailed experimental study on the unipolar diffusion charging process in the continuum regime is described and the results are compared with the available theories.

THEORETICAL CONSIDERATIONS

Consider a spherical particle of radius a, carrying n_p elementary unit of charge placed in a gaseous medium containing N_0 unipolar ions/cm^3. In the continuum regime, the flux of ions crossing any spherical surface of radius r around the particle is given by

$$j = -4\pi r^2 [D_i (dN/dr) - Z_i E N], \quad [1]$$

where D_i and Z_i are the diffusion coefficient and the electrical mobility of the ions, E is the local electric field, and N, the local ion concentration at r. Using Einstein's relation

$$D_i = (kT/e)Z_i, \quad [2]$$

we have

$$j = -4\pi r^2 D_i \left[\frac{dN}{dr} - \frac{e}{kT} EN \right], \quad [3]$$

where k is Boltzmann's constant, T is the absolute temperature, and e is the elementary unit of charge. In the steady state, the flux j is a constant independent of r. With the boundary condition, $N = N_0$ as $r \to \infty$, Eq. [3] can be integrated to give

$$N = \exp(-e\phi/kT) \left[N_0 + \left(\frac{j}{4\pi D_i} \right) \right.$$

$$\left. \times \int_r^\infty \left(\frac{1}{r^2} \right) \exp(e\phi/kT) dr \right], \quad [4]$$

where ϕ is the electric potential. Taking into account both the Coulombic and image forces, we have

$$E = \frac{n_p e}{r^2} - \frac{K-1}{K+1} \frac{ea^3(2r^2 - a^2)}{r^3(r^2 - a^2)^2} \quad [5]$$

and

$$\phi = -\int_r^\infty E dr = \frac{n_p e}{r}$$

$$- \frac{K-1}{K+1} \frac{ea^3}{2r^2(r^2 - a^2)}, \quad [6]$$

where K is the dielectric constant of the particle.

On account of the image force term (the second term in Eq. [6]) the electric potential $\phi \to -\infty$ as $r \to a$. Therefore, in order for the ion concentration, N, to remain finite as $r \to a$, the quantity in the bracket in Eq. [4] must approach zero as $r \to a$, or

$$N_0 + \left(\frac{j}{4\pi D_i} \right) \int_a^\infty \left(\frac{1}{r^2} \right) \exp(e\phi/kT) dr = 0, \quad [7]$$

from which it follows that

$$j = -4\pi D_i a N_0 / I, \quad [8]$$

where I is the definite integral,

$$I = \int_1^\infty \left(\frac{1}{x^2} \right)$$

$$\times \exp \left[\frac{e^2}{akT} \left[\frac{n_p}{x} - \frac{K-1}{K+1} \frac{1}{2x^2(x^2-1)} \right] \right] dx.$$

$$[9]$$

Equations [8] and [9] were first derived by Fuchs (4) and independently (7) by Bricard (5, 6) for the case of a conducting sphere ($K = \infty$). If the image force term is dropped by letting $K = 1$, we have

$$I = \int_1^\infty \left(\frac{1}{x^2} \right) \exp \left(\frac{n_p e^2}{akTx} \right) dx$$

$$= \frac{\exp(n_p e^2/akT) - 1}{(n_p e^2/akT)}. \quad [10]$$

This equation was first derived by Fuchs (4) and by Pluvinage (8–10).

In contrast to the above, White (3) suggested that for a charged particle placed in a gaseous medium containing unipolar ions, the distribution of ions around the particle should be described by Boltzmann's law,

$$N = N_0 \exp(-n_p e^2/rkT). \quad [11]$$

He then calculated the ionic flux, j, using the kinetic theory expression

$$j = -(4\pi a^2)(\tfrac{1}{4} N_{r=a} \cdot \bar{c}_i), \quad [12]$$

where $\bar{c}_i$ is the mean thermal speed of the ions and $N_{r=a}$ is the ion concentration at $r = a$. Thus, according to White,

$$j = -\pi a^2 \bar{c}_i N_0 \exp(-n_p e^2/akT). \quad [13]$$

In unipolar diffusion charging, a neutral aerosol is usually exposed to unipolar ions, and the resultant particle charge is then measured. Assuming a quasi-steady charging process, we have

$$dn_p/dt = -j, \quad [14]$$

from which the time, t, needed to increase the particle charge from 0 to n_p units can be

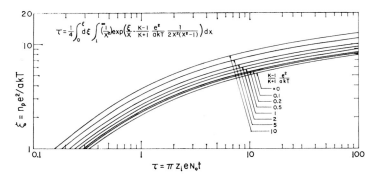

FIG. 1. Relationship between dimensionless charge and dimensionless time for unipolar diffusion charging according to the theory of Fuchs and Bricard.

determined:

$$t = \int_0^{n_p} (dn_p/-j).$$ [15]

Defining a dimensionless charge

$$\xi = n_p e^2 / akT$$ [16]

and a dimensionless time

$$\tau = \pi Z_i e N_0 t,$$ [17]

we can integrate Eq. [15] with the Fuchs–Bricard equation for the ionic flux (Eqs. [8] and [9]) to give

$$\tau = \frac{1}{4} \int_0^\xi d\xi \int_1^\infty \left(\frac{1}{x^2}\right)$$

$$\times \exp\left(\frac{\xi}{x} - \frac{K-1}{K+1} \frac{e^2}{akT} \frac{1}{2x^2(x^2-1)}\right) dx.$$ [18]

If the image force is neglected,

$$\tau = \frac{1}{4} \int_0^\xi [(e^\xi - 1)/\xi] d\xi.$$ [19]

Neither Eq. [18] nor Eq. [19] can be integrated analytically. However, both can be integrated numerically. The result of this integration is shown in Fig. 1. Similarly, defining a dimensionless time $\tau' = \pi a e^2 \bar{c}_i N_0 t / kT$ and using White's equation (Eq. [13]), Eq.

[15] can be integrated to give

$$\xi = \ln (1 + \tau').$$ [20]

Equations [18], [19], and [20] all predict that the particle charge, n_p, is a function of the particle radius a, and the $N_0 t$ product. In the experiments described below, this functional relationship is established empirically, and the results are compared with the theoretical predictions.

In order to satisfy the continuum requirement in the experiments, the particle radius, a, must be large compared to the mean free path of the ions, λ_i. The mean free path of ions can be calculated by the Maxwell–Chapmann–Enskog theory of diffusion,

$$D_i = (1 + \epsilon_{ij}) \frac{3\pi^{\frac{1}{2}}}{8} \lambda_i \left(\frac{m_i + m_j}{m_j}\right)^{\frac{1}{2}} \frac{\pi^{\frac{1}{2}}}{8^{\frac{1}{2}}} \bar{c}_i.$$ [21]

The calculated ionic mean free path is $\lambda_i = 1.45 \times 10^{-6}$ cm using the following constants:

$D_i = 0.0357$ cm²/sec (corresponding to $Z_i = 1.4$ cm²/V-sec),

$m_i =$ molecular mass of ions $= 109$ amu (corresponding to the hydrated proton $H^+ (H_2O)_6$),

$m_j =$ molecular mass of air $= 28.97$,

$\bar{c}_i =$ mean thermal speed of ions, $= (8kT/\pi m_i)^{\frac{1}{2}} = 2.39 \times 10^4$ cm/sec for $m_i = 109$ amu,

153

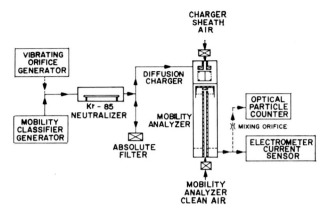

FIG. 2. Experimental system for studying unipolar diffusion charging of particles.

$\epsilon_{ij} = 0.132$ (a constant having the value 0.016 for $m_i = m_j$ and 0.132 for extremely unequal molecular masses, m_i and m_j).

In the experiments described below, the particle diameter was varied between 0.1 and 5.04 μm. At the lower particle size limit of 0.1 μm, Kn $= 0.0145/0.05 = 0.29$, and the continuum condition is therefore not completely satisfied. However, the data are included here and the departure of data from theory will be noted.

EXPERIMENTAL

Generation of Monodisperse Aerosols

Because of the wide range of particle sizes covered, two different techniques had to be used in order to generate the monodisperse aerosols needed in the experiments.

For generating monodisperse aerosols in the 0.75 to 5.06 μm diameter range, the vibrating orifice generator (11) was used. The aerosol was generated by dissolving oleic acid in isopropyl alcohol and spraying the solution through the vibrating orifice. The size of the oleic acid aerosol generated was calculated by the equation

$$D_p = (6Q_1C/\pi f)^{\frac{1}{3}}, \qquad [22]$$

where Q_1 is the liquid flow rate through the orifice, f is the vibrating frequency of the orifice, and C is the volumetric concentration of oleic acid in the solution. The particle size was calculated to an accuracy of $\pm 2\%$ based on estimated accuracies for Q_1, C, and f. The particles were very monodisperse, with estimated geometrical standard deviations, σ_g, of about 1.02.

For generating aerosols in the 0.1 to 0.75 μm diameter range, a modified apparatus based on the mobility classification principle described by Liu and Pui (12) was used. The modification consists of replacing the Collison atomizer used in the original apparatus by an atomization–condensation aerosol generator (13) in order to improve the monodispersity of the particles produced for mobility classification. While the original apparatus was limited to an upper particle size limit of about 0.1 μm, the modified apparatus was capable of producing monodisperse aerosols up to about 1 μm in diameter. More details concerning this modified apparatus can be found in Ref. (14). In the present experiments, aerosols of DOP (di-octyl phthalate) were used and the equivalent σ_g of the aerosol was estimated to be about 1.03. The particle size was calculated to an accuracy of about 2% using the measured electric mobility of the singly charged particles as previously described.

154

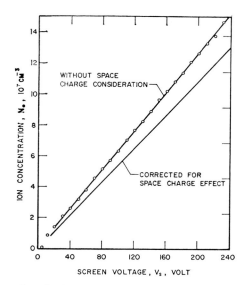

FIG. 3. Operating characteristics of the aerosol charger.

Experimental System and Procedure

Figure 2 is a schematic diagram of the experimental system used. Monodisperse aerosols produced by the procedures described above were introduced into the aerosol charger and mobility analyzer as shown. The particles downstream of the mobility analyzer were then detected by an electrometer current sensor or by an optical particle counter. The electrometer current sensor was used with the DOP aerosols generated by the mobility classification method while the optical particle counter (Royco Model 220) was used when oleic acid aerosols produced by the vibrating orifice generator were used.

The charger and mobility analyzer used in the present study were those used in the Electrical Aerosol Analyzer previously described (1, 15, 16). While the Electrical Aerosol Analyzer was intended primarily as a size distribution analyzer for aerosols, it contained the necessary components and was used in the present study.

The operating principle of the aerosol charger is briefly as follows. A fine, 25 μm

diameter tungsten wire is placed along the axis of two coaxial metal cylinders to produce unipolar positive ions by the application of a high voltage. A section of the inner cylinder is covered with a metal screen to allow the ions to pass through. By maintaining the outer cylinder at a negative potential relative to the inner cylinder, the ions are drawn through the screen openings into the annular space between the cylinders. The aerosol particles flowing through this annular space are exposed to unipolar ions and become electrically charged. The residence time of the particles in the charging region is determined by the flow rate and the dimensions of the devices,

$$t = \pi(r_{2c}^2 - r_{1c}^2)w/q_{tc}. \qquad [23]$$

where $r_{1c} = 1.670$ cm and $r_{2c} = 2.972$ cm are the inner and outer radii of the annular charging region, $w = 0.953$ is the width of the screen opening on the inner cylinder, and q_{tc} (cm^3/sec) is the total charger flow. The total charger flow consists of an aerosol flow of q_{ac} and a clean sheath air flow q_{sc}; the latter is used in order to displace the aerosol stream away from the inner cylinder and to prevent particles from entering the high-intensity corona discharge region within the inner cylinder. In these experiments, the charger sheath air flow was kept at 25% of the total charger flow, the latter being varied between 1.25 and 5 liters/min. The corresponding residence time of the particle in the charging region is between 0.217 and 0.868 sec.

As is shown in Refs. (1, 14), the nominal ion concentration in the charging region can be calculated by the following equation if the space charge of ions is ignored:

$$N_0 = I_c \ln (r_{2c}/r_{1c})/V_s Z_i \pi 2we, \qquad [24]$$

where I_c is the charging current collected by the outer cylinder, V_s is the potential of the screen relative to the outer cylinder, and Z_i is the mobility of the ions. To correct for the effect of the space charge of the ions, Poisson's equation can be used. In cylindrical coordinates, Poisson's equation for the electric field E

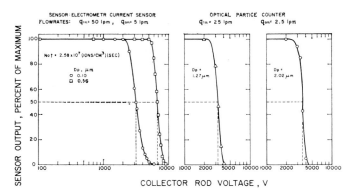

FIG. 4. Typical experimental data showing the penetration of particles through the mobility analyzer.

in the annular charging space is

$$(1/r)(d(ER)/dr) = 4\pi Ne. \qquad [25]$$

With the boundary conditions, $E = E_1$ and $N = N_1$ at $r = r_1$ and $E = E_2$, $N = N_2$ at $r = r_2$, Eq. [25] can be solved to give (17)

$$\left(\frac{E_2}{E_1}\right)^2 = \left(\frac{r_1}{r_2}\right)^2 \left(1 - 4\pi e \frac{r_1 N_1}{E_1}\right)$$

$$+ 4\pi e \frac{r_1 N_1}{E_1}. \qquad [26]$$

Under steady-state conditions, the ion flux through the annular gap space is a constant, so that

$$E_1 r_1 N_1 = E_2 r_2 N_2. \qquad [27]$$

Equations [26] and [27] are sufficient to determine the mean ion concentration in the annular charging region. In Fig. 3, we show the calculated ion concentration as a function of the screen voltage with and without taking into account the space charge effect of the ions. It is seen that the effect of the space charge of ions is to reduce the effective ion concentration by 15% from the calculated value using Eq. [24]. The value of N_0 given by the lower curve in Fig. 3 is used in reporting the data given below.

The operating procedure for each experiment consists of measuring the particle penetration through the mobility analyzer with the elec-

trometer current sensor or the optical particle counter as a function of the applied voltage on the collector rod in the mobility analyzer. Some sample data are shown in Fig. 4, where the percent penetration of the particles is shown as a function of the collector rod voltage. The median collector rod voltage, V, is then used in the following equation to calculate the mobility of the charged particles,

$$Z_p = (q_{tm} - \tfrac{1}{2}q_{am}) \ln (r_{2m}/r_{1m})/2VL\pi, \quad [28]$$

where q_{tm} and q_{am} are the total flow and aerosol flow in the mobility analyzer, $r_{1m} = 1.151$ cm and $r_{2m} = 2.985$ cm are the inner and outer radii of the annular precipitating region in the mobility analyzer, and $L = 30.48$ cm is the length of the collector rod. Knowing the mobility of the particles, the particle charge, n_p, can then be calculated by means of the equation

$$Z_p = n_p eC/6\pi\mu a, \qquad [29]$$

where e is the elementary unit of charge, C is the slip correction, and μ is the gas viscosity.

RESULTS

Unipolar diffusion charging experiments were performed with the apparatus described above using particles in the 0.1 to 5.04 μm diameter range. A total of 15 different particle sizes were used. The ion concentration in the charger was varied between 2.95×10^6 and 1.19×10^8

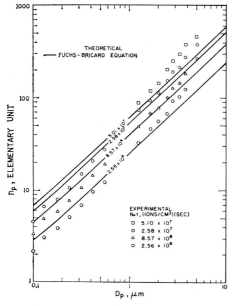

FIG. 5. Comparison of data with the theory of Fuchs and Bricard.

is decidedly better for the theory of Fuchs and Bricard than it is for the theory of White. Further, it can be shown that the discrepancy between the Fuchs–Bricard theory and the experimental data for $D_p = 0.1$ and 0.24 μm is explainable by the finite mean free path of the ions. For these particle sizes, the Knudsen numbers are 0.29 and 0.12, respectively. Thus, the effect of the finite ionic mean free path is not entirely negligible for these small particles.

A further comparison of the data with the theory of Fuchs and Bricard is shown in Fig. 7, in which the dimensionless particle charge, ξ, is plotted against the dimensionless time, τ, for various values of the image force parameter, $[(K - 1)/(K + 1)](e^2/akT)$. The dielectric constant $K = 5.1$ is used for DOP. For oleic acid, $K = 2.46$ is used. Only data in the 0.32 to 2.02 μm diameter range are included in this comparison, since these are the data for which the finite mean free path of the ions and the applied electric field both have negligible effects. The agreement between theory and data is again seen to be good.

cm^{-3}, and the $N_0 t$ product, between 2.56×10^6 and 5.1×10^7 (sec/cm^3).

In Fig. 5, the experimental results are compared with the theory of Fuchs and Bricard (Eq. [18]). It is seen that the agreement is good over much of the particle size range covered. The discrepancy between theory and experiment for $D_p \geq 2.5$ μm and for $N_0 t = 5.10 \times 10^7$ (ion/cm^3)(sec) can be attributed to the finite electric field used in the aerosol charger. The effect of this electric field, which was between 12 and 225 V/cm, is considered in more detail in Ref. (14).

A detailed comparison of the data for $D_p < 2.5$ μm with the theory of Fuchs and Bricard and the theory of White (Eq. [20]) is shown in Fig. 6. It is interesting to note that in the case of small particles ($D_p = 0.10$ and 0.24 μm), the discrepancy between theory and experiment is about the same for both theories. However, in the case of the larger particle sizes, viz., for $D_p = 0.56$, 1.27, and 2.02 μm, the agreement between theory and experiment

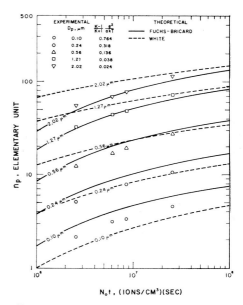

FIG. 6. Comparison of data with the theory of Fuchs and Bricard and with the theory of White.

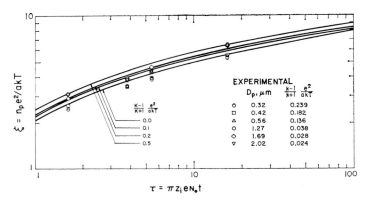

FIG. 7. Theoretical (Fuchs and Bricard) and experimental relationship between the dimensionless charge and the dimensionless time.

On the basis of the above comparisons, it is concluded that the theory of Fuchs and Bricard provides an accurate description of the unipolar diffusion charging process in the continuum regime and should be used in practical applications.

ACKNOWLEDGMENT

The research is supported by the Energy Research and Development Administration under Contract E (11-1)-1248.

REFERENCES

1. Liu, B. Y. H., and Pui, D. Y. H., *J. Aerosol Sci.* **6**, 249 (1975).
2. Smith, W. B., and McDonald, J. R., *J. Aerosol Sci.* **7**, 151 (1976).
3. White, H. J., *AIEE Trans.* **70**, 1186 (1951).
4. Fuchs, N. A., *Izv. Akad. Nauk. SSSR, Ser. Geogr. Geofiz.* **11**, 341 (1947).
5. Bricard, J., *C.R. Acad. Sci. Paris* **226**, 1536 (1948).
6. Bricard, J., *J. Geophys. Res.* **54**, 39 (1949).
7. Bricard, J., "Problems of Atmospheric and Space Electricity" (S. C. Coroniti, Ed.) pp. 82–117, Elsevier, Amsterdam, 1965.
8. Pluvinage, P., "Etude theorique et experimentale de la conductibilite electrique dans les nuages non orageux," Theses, Faculte des Sciences de l'Universite de Paris, 1945.
9. Pluvinage, P., *Ann. Geophys.* **2**, 31 (1946).
10. Pluvinage, P., *Ann. Geophys.* **3**, 2 (1947).
11. Berglund, R. N., and Liu B. Y. H., *Environ. Sci. Technol.* **7**, 147 (1973).
12. Liu, B. Y. H., and Pui, D. Y. H., *J. Colloid Interface Sci.* **47**, 155 (1974).
13. Liu, B. Y. H., and Lee, K. W., *Amer. Ind. Hyg. Assoc. J.* **36**, 861 (1975).
14. Pui, D. Y. H., "Experimental Study of Diffusion Charging of Aerosols," Ph.D. thesis, Mechanical Engineering Department, University of Minnesota, Minneapolis, 1976.
15. Liu, B. Y. H., Pui, D. Y. H., and Kapadia, A., "Electrical Aerosol Analyzer: History, Principle and Data Reduction," presented at the Aerosol Measurement Workshop, University of Florida, Gainesville, March 24–26, 1976.
16. Pui, D. Y. H., and Liu, B. Y. H., "Electrical Aerosol Analyzer: Calibration and Performance," presented at the Aerosol Measurement Work shop, University of Florida, Gainesville, March 24–26, 1976.
17. Hewitt, G. W., *AIEE Trans.* **76**, 300 (1957).

Solid Surfaces

Atomic and Molecular Processes at Solid Surfaces

G. A. SOMORJAI

Materials and Molecular Research Division, Lawrence Berkeley Laboratory, and Department of Chemistry, University of California, Berkeley, California 94720

Received June 3, 1976; accepted July 19, 1976

Development of new techniques over the past 15 years permits the investigation of the structure, the composition, and reactivity of solid surfaces on the atomic scale. Our present knowledge of the structure of clean surfaces and of adsorbed monolayers as studied by low-energy electron diffraction is reviewed. Surface crystallography provides information on the nature of the surface chemical bond.

Electron spectroscopy studies of the surface phase diagrams (surface composition vs bulk composition and temperature) of binary alloy systems are reviewed. The surface enrichment of one of the alloy constituents is commonly observed. The various thermodynamic and chemical variables that control the surface composition are discussed.

The reactive scattering of molecular beams from surfaces reveals the kinetics and mechanisms of surface reactions. From the data the minimum surface residence times of the reactants and products and the partitioning of energy between the molecules and the surface are determined. The available experimental data are reviewed and the utility of this technique is demonstrated by discussion of the H_2–D_2 exchange on platinum surfaces.

INTRODUCTION

In recent years a very large number of new techniques have become available that permit the investigation of surface phenomena on the atomic scale. Most of these involve the scattering of electrons, atoms, molecules or ions or photons (especially in the vacuum ultraviolet or higher energy range). Particles such as electrons and atoms are scattered very strongly by the topmost monolayer; thus the scattered species provide information primarily about the surface. Elastically scattered (diffracted) electrons in the 15–200 eV range are the probes utilized in low-energy electron diffraction to determine the structure of an ordered surface or of an adsorbate (surface crystallography). Electrons that are emitted from the valence shell or from the inner shells of surface atoms as a result of bombardment by electrons or photons of higher energy (10–10⁴ eV) are the probes of ultraviolet photoelectron spectroscopy, Auger electron spectroscopy, and X-ray photoelectron spectroscopy. These techniques are employed to determine the chemical composition of the surface and the oxidation states of surface atoms with a sensitivity of about 1% of a monolayer ($\sim 10^{13}$ atoms/cm²). Once we know the arrangement and composition of atoms at the surface we may study a variety of surface phenomena involving atom transport, such as surface reactions, and establish how the reaction rate and reaction path leading to observed product distribution depend on the surface structure and composition. Reactive scattering of a beam of molecules that takes place upon their collision with the surface has been used to unravel the elementary surface reaction steps. Studies of surface reactions by molecular beam scattering reveal the minimum surface residence time of the reactant that leads to a chemical reaction, the reaction probability (number of product molecules/number of incident reactant molecules), and the partitioning of energy (translation, vibration, and rotation) among the reaction partners, the reactants, surface atoms, and the desorbing product molecules. Other methods of studying chemical surface reactions at higher surface coverages than those used in molecular beam–surface scattering experiments can also profit from relating the reactivity of the surface with its atomic structure and composition.

This paper is divided into three parts. First we discuss our present knowledge of the atomic surface structure. Much of the atomic scale information has been obtained by low-energy electron diffraction studies over the past 15 years and these will be reviewed. Then we discuss what is known of the surface composition of multicomponent systems, primarily metal alloys since these have been studied in most detail. Finally, we review our present understanding of energy transfer during surface reactions and how the reactivity depends on the structure and composition of solid surfaces. We neglect discussions of the techniques that have been utilized. Reviews on the use of low-energy electron diffraction (1–3), electron spectroscopy (4–6), and molecular beam scattering (7, 8) are available and should be consulted by the interested reader.

THE STRUCTURE OF SOLID SURFACES

Figure 1 depicts schematically a solid surface. The surface is structurally hetero-geneous; there are several different atomic sites which are distinguishable by their number of nearest neighbors or coordination number. There are atoms in terraces where they are surrounded by the largest number of neighbors. There are atoms in steps, and at kinks in the steps. These sites have fewer neighbors than the atoms in the terraces. Adatoms that are located on top of the terrace have, of course, the smallest coordination number. This model of the solid surface has been utilized to develop the theory of the growth of crystals and to explain the vaporization mechanisms of solids and other surface

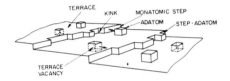

FIG. 1. Model of heterogeneous solid surface depicting different surface sites. These sites are distinguishable by their number of nearest neighbors.

phenomena. New information that has been emerging from our studies, primarily of single-crystal surfaces, is that each of the surface sites may have a different chemical reactivity (9). Our studies of transition metal surfaces indicate that chemical bond breaking (H–H, C–H, and C–C bonds) is facilitated at sites of low coordination number such as atomic steps and kinks that can be present on suitably prepared surfaces with high concentrations (10, 11).

If each crystal plane, step, and kink site at the surface has a different chemical reactivity, much of the complexity revealed in studies of heterogeneous catalysis or corrosion can readily be rationalized. Small particles used in heterogeneous catalysis, for example, are bound by different atomic planes and must have large concentrations of atoms in steps and in kink positions. A crystallographic model of one such particle is shown in Fig. 2. Thus the experimentally determined reaction rate and product distribution is that obtained by the sum total of the rates and products obtained at each surface site. It is very difficult, if not impossible, therefore, to decipher the reactivity of each active surface site from the experimental data. It is possible, however, to prepare surfaces of single crystals where one or more surface sites are present in large concentrations and study them by low-energy electron diffraction (LEED), a technique that can identify these various surface sites. A scheme of the LEED experiment is shown in Fig. 3a. Low-energy electrons (10–200 eV) are back-scattered from a crystal surface and the elastically scattered fraction (those electrons that did not lose energy in

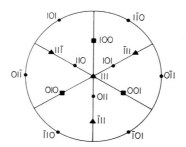

Standard Cubic (III) Projection

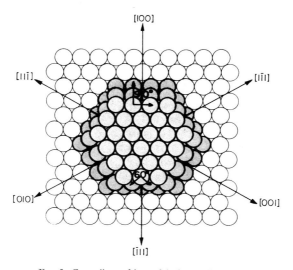

FIG. 2. Crystallographic model of a small particle.

the collision with the surface) is accelerated and impinges on a fluorescent screen where the diffraction spots are displayed (1). The diffraction patterns and schematic representations of the three types of surfaces of platinum that give large differences in relative concentrations of terrace, step, and kink sites are shown in Fig. 3b. Surfaces where most of the atoms are in terrace positions can be easily prepared and these exhibit high thermal and chemical stability. Alternately, stable surfaces with over 20% of the surface atoms in step positions have been prepared. By appropriate cutting of a single crystal, one can introduce

a high density of kinks into the steps (11). One may explore the chemistry of each of these surface sites by studying the reactivity as a function of step or kink concentration on a series of single-crystal surfaces and compare the results with those obtained on small particles. This, then, is the approach we took in many catalytic studies using single-crystal platinum surfaces to explore the mechanism of hydrocarbon reactions (11).

A. The Structure of Clean Solid Surfaces

There are perhaps two major findings that emerge from low-energy electron diffraction studies over the past 15 years that were previously not known.

(1) Surface reconstruction. Many surfaces have atomic structures that are different from that expected from the projection of the X-ray bulk unit cell. The surface atoms assume new equilibrium positions, by out-of-plane buckling or by "relaxing" inward (contraction), that often result in an entirely different ordered surface structure.

(2) Surfaces that exhibit ordered surface irregularities (steps and kinks) are the stable structure of crystal planes of high Miller index. A large number of crystal planes are characterized by an atomic structure in which terraces a few atoms wide are separated by periodic steps of one atom in height. This ordered step–terrace arrangement persists up to the melting point of many solids. We shall give examples of both of these properties of solid surface structure.

Figure 4 displays the diffraction patterns and schematic representation of the structures of the (100) crystal face of platinum. This surface exhibits the so-called (5×1) surface structure shown in the figure (Fig. 4a). There are two perpendicular domains of this structure and there are $\frac{1}{5}$, $\frac{2}{5}$, $\frac{3}{5}$, or $\frac{4}{5}$ order spots between the (00) and (10) diffraction beams. The surface structure appears to be stable at all temperatures from 25°C to the melting point although at elevated temperatures impurities from the bulk can come to the surface and cause a transformation of this structure to the impurity stabilized (1×1) surface structure (Fig. 4c). Preliminary calculations by Clark *et al.* (12) and in this laboratory indicate that a model for Pt(100) in which the surface atoms assume a disordered hexagonal configuration by out-of-plane buckling is favored. The apparent (5×1) unit cell is then the result of the coincidence of the atomic position of atoms in the surface, i.e., in the distorted hexagonal layer, with atoms of the undistorted second layer below.

The surface atoms in any crystal face are in an anisotropic environment which is very

different from that around the bulk atoms. The crystal symmetry that is experienced by each bulk atom is markedly higher than for atoms placed on the surface. The change of symmetry and the lack of neighbors in the direction perpendicular to the surface permit displacement of the surface atoms in ways that are not allowed in the bulk. Surface relaxation can give rise to a multitude of surface structures depending on the electronic structure of a given substance. The (100) crystals of gold and iridium that are neighbors of platinum in the periodic table exhibit the same surface reconstruction and the same surface structure as those of platinum that are shown in Fig. 4. The (110) crystal faces of these three elements are also restructured and exhibit a different unit cell than that expected from the bulk X-ray structure. The (111) crystal face of these three metals has the same surface structure as that indicated by the bulk unit cell. Other metals such as tin, antimony, and bismuth also exhibit surface reconstruction. For semiconductors, however, most crystal planes that have been studied show reconstruction (1, 3). Monatomic (Si, Ge) and diatomic (GaAs, InSb, CdSb, CdS, ZnO, etc.) semiconductor surfaces have been investigated in large numbers and surface reordering has been observed for most of them (1, 3). Frequently there are changes of surface structure with temperature (order–order transformation) that are often irreversible.

The mechanisms of surface reconstruction is not well understood. There are models invoking changes of electronic structure at the surface or proposing the presence of surface relaxation or ordered vacancies to explain these findings (13). Further studies, both experimental and theoretical, are clearly necessary to understand the mechanism and nature of surface restructuring.

The excess free energy introduced by the creation of the surface induces other types of chemical changes in addition to surface reconstruction. Alkali–halide surfaces have unit cells not visibly different from what is expected from the bulk X-ray unit cells.

However, the surface composition appears to be markedly different. Auger electron spectroscopy studies of alkali–halides revealed that when excess vacancies are introduced (13), the surface becomes enriched in either cations or in anions depending on the relative sizes of the positive ions. Oxide surfaces, such as the surface of aluminum oxide and vanadium pentoxide, exhibit both nonstoichiometry and change of surface structure (14). It appears that under appropriate conditions the cation is reduced at the surface and a surface oxide with a lower oxidation state forms (Al_2O or V_3O_5) with surface structure and composition that are different from those of the bulk oxide. Still another type of surface restructuring is exhibited by the surface of crystals of large organic molecules. For example, phthalocyanine crystals can be epitaxially grown on ordered metal surfaces (15). When copper phthalocyanine was grown on the copper (111) surface the surface structure of the growing organic single-crystal layer did not resemble the structure of any of the simpler crystal planes in the bulk structure of the organic crystal. It appears that the ordered metal substrate predetermined the orientation and packing of the phthalocyanine monolayer, which in turn control the orientation and packing of organic layers deposited on top of it. For large molecules such as phthalocyanine, restructuring into a more stable crystallographic arrangement requires molecular rotation and diffusion processes that are too slow to occur under conditions of crystal growth. Thus the molecules are frozen into a surface structure that is predetermined by the structure of the substrate and the first adsorbed organic monolayer.

The preparation of surfaces with large concentrations of stable and ordered irregularities (steps and kinks) can be carried out by cutting crystal surfaces along high Miller index directions. Figures 5a and 5b show the cutting angles for obtaining surfaces with given step and kink densities and also show the schematic representations of these surfaces. Stepped surfaces of several metals, semi-

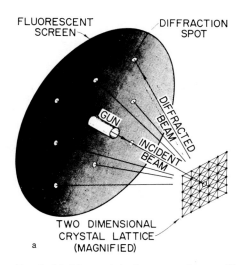

FIG. 3. (a) Scheme of the low-energy electron diffraction (LEED) experiment. (b) Low-energy electron diffraction patterns and schematic representations of the surface configurations of platinum single crystal surfaces: (a) Pt(111) containing less than 10^{12} defect/cm^2; (b) Pt(557) surface containing 2.5×10^{14} atoms/cm^2 with an average spacing between steps of 6 atoms and (c) Pt(679) surface containing 2.3×10^{14} step atoms/cm^2 and 7×10^{14} kink atoms/cm^2 with an average spacing between steps of 7 atoms and between kinks of 3 atoms.

conductors, and oxide surfaces were prepared this way (13). It appears that ordered steps one atom in height separated by terraces of low Miller index orientation that are of the same width, on the average, are the stable surface structures of many high Miller index surfaces of solids regardless of their chemical bonding. Not all stepped surfaces are stable, however (9). Figure 6 shows a stereographic triangle for platinum with all the crystal faces that were cut and studied in our laboratory indicated by the points on the triangle. With the exception of crystal faces in the regions indicated by arrows, the ordered step structure proved to be the stable surface structure on all these crystal faces. However, when a crystal face was cut that was located in the region indicated by one of the arrows (the (510) surface plane, for example) this face was

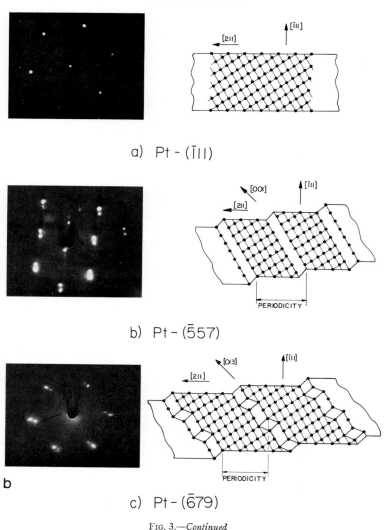

a) Pt - (Ī11)

b) Pt - (5̄57)

c) Pt - (6̄79)

Fɪɢ. 3.—*Continued*

unstable and on heating it undergoes facetting into the two surfaces shown at the ends of the arrows (the (100) and (210) crystal faces). Thus studies of the relative structural stabilities of the various surfaces permits determination of their relative surface free energies (the more stable surfaces having the lower surface free energy).

In several studies of adsorption and surface reaction on platinum surfaces we have found

that atomic steps and kinks are the active sites for breaking strong H–H, C–H, and C–C chemical bonds. In the absence of high concentration of these sites (for example on the (111) surface), thermal activation or high adsorbate pressures were necessary to obtain bond breaking activity comparable to that detectable on stepped surfaces. Theoretical calculations by Falicov et al. (16–18) revealed two major reasons for this enhanced chemical

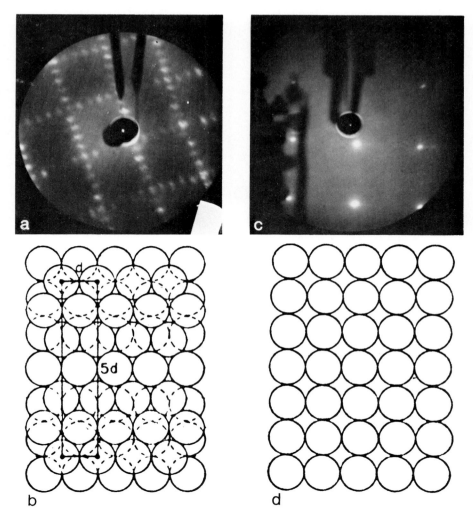

FIG. 4. (a) Diffraction pattern from the Pt(100)–(5 × 1) structure. (b) Schematic representation of the (100) surface with hexagonal overlayer. (c) Diffraction pattern from the Pt(100)–(1 × 1) structure. (d) Schematic representation of the (100) surface.

activity of surface sites of low coordination number. One of the reasons is that at a corner site there is enhanced amplitude of charge density fluctuations that leads to an increased potential energy ($\Delta\phi$) of free electrons on the corner atom. The magnitude of this potential energy difference for electrons at the corner site and away from it depends on the local atomic structure at the surface irregularity. As a result, some of the free electrons are displaced away from the corner site, leaving behind a net positive charge. The number of electrons ΔN that are removed from the corner site is proportional to $\Delta N \sim -\Delta\phi \times D(E_f)$ where $D(E_f)$ is the density of states at the Fermi level. The magnitude

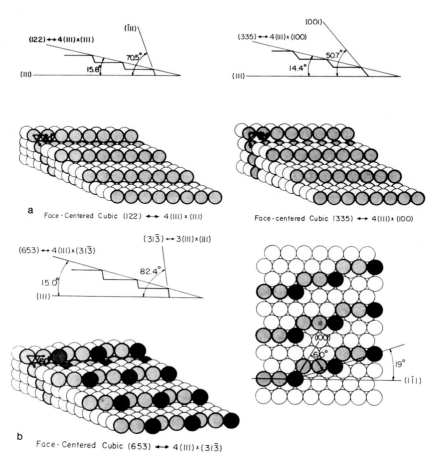

FIG. 5. The cutting angles necessary to obtain stepped and kinked surfaces whose schematic diagrams are shown.

of $\Delta\phi$ is determined by the local step structure. Thus there is a large local electric field present at the corner sites of the order of 0.3–0.4 V/Å that should help to polarize further the incoming molecules that have well-defined polarizabilities and to break them apart. The higher the density of states at the Fermi level, $D(E_f)$, the larger the positive charge in the corner site. For transition metals the density of states at the Fermi level is very large indeed, while for nontransition metals $D(E_f)$ can be quite small. Some of the values of $D(E_f)$ that have been determined are: plati-

num, 2.1 electrons/volt/atom; nickel, 1.1; tungsten, 0.7; copper, 0.4; and gold, 0.3. These values yield ΔN of -0.6, -0.3, and -0.2 for platinum, nickel, and tungsten, taking $\Delta\phi$ equal to 0.3 V. Thus for metals with large values of $D(E_f)$ there are large variations of charge density at surface irregularities, i.e., sites of low coordination number. For gold, on the other hand, ΔN is about -0.09. The surface irregularities for gold do not show much charge density variation with respect to atoms at surface sites away from steps. Therefore the gold surface is likely to present uniform

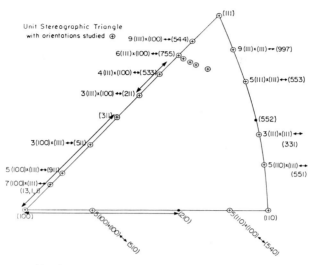

FIG. 6. A stereographic triangle of a platinum crystal depicting the various high Miller index surfaces of platinum that were studied.

charge density to the incoming reactants and is homogeneous regardless of variations of surface sites. These conclusions are certainly supported by our experiments of chemisorption and chemical reactions on platinum (11), gold (19), and iridium (20, 21) surfaces. While surface irregularities like atomic steps have different chemistries for platinum and iridium, both metals with a high density of states, gold surfaces show the same chemical behavior regardless of surface atomic structure.

The second reason for the different chemistry with changing coordination number is due to the local field experienced by the transition metal ion at various surface sites. This local field has two major effects: (1) causing a sizable electron transfer from the stepped corner to the bulk or vice versa; (2) producing different d-orbital level splittings and different d-orbital occupations depending on the position of the metal ion. Tsang and Falicov (17) have calculated the energy levels of localized d-electrons at different surface sites when the crystal field is turned on. These calculations were made across the periodic table for ions with different number of d-electrons. The d-electron wavefunction is a product of the radial and the angular part. Assuming that the radial part is constant, they display the resultant constant contour of the angular part for these various ions. These contours clearly indicate that the stereochemistry of the corner atoms can be quite different from that of a terrace atom; therefore, a different kind of bonding is favored at different sites. The result of their calculations has been applied to explain the large differences between the catalytic activities of various transition metal surfaces. Studies of the charge density, the energy, and spatial arrangement of localized electronic orbitals at surface sites of low coordination number, i.e., steps and kinks, appear to be an important and challenging area of theoretical surface chemistry.

B. The Structure of Adsorbed Monolayers

Over the past 15 years, low-energy electron diffraction studies of monolayers of adsorbates on crystal surface have shown that (1) adsorbate monolayers are largely ordered under proper conditions of the experiment, (2) the surface crystallography reveals the nature of the surface chemical bond, and (3) there is a marked crystal plane sensitivity of the surface chemical bond. Surface irregularities and kinks have different bond-breaking abilities as compared to atoms in terraces.

It is well established that atoms or small diatomic molecules order readily on low Miller index surfaces. Over 200 surface structures that were detected have been tabulated (1, 22). In fact, the tabulated data permitted us to propose rules of ordering for these small molecules that allow prediction (within certain limits) of their surface structure (23).

The surface structures of over 50 hydrocarbon molecules have been studied on the platinum (111) and (100) surfaces. Again, ordering in the adsorbed monolayer was more of a rule than an exception. However, caution should be excerised in establishing the best experimental conditions for studying ordering of the adsorbed layers. The same hydrocarbon molecule, benzene for example, that orders readily at 300°K on the Pt(111) surface may decompose readily on the Ir(111) surface or fail to adsorb altogether on the (111) surface of gold under identical experimental conditions. The metal–adsorbate interaction can be either too strong or too weak and thus ordering is prevented in the monolayer. Ordering can be facilitated by lowering the surface temperature since most bond-breaking processes at the surface require activation energy, or, in the case of weak surface bonding, a lowered surface temperature results in increased surface coverage of the adsorbate.

Ordering of large molecules is generally best on high-symmetry surfaces (Pt(111) rather than Pt(100), for example). Aromatic molecules with high rotational symmetry order the best with small substituent groups and at low incident vapor flux. These conditions allow maximum opportunity for a molecule, once adsorbed, to reorient itself for incorporation into the growing ordered region. Thus ordering of large molecules can be seen to be somewhat different from site adsorption

TABLE I

Adsorbate–Substrate Bond Lengths Determined by LEED (15)

Substrate	Adsorbate	Bond length[a] (experimental)	Bond length[a] (predicted)
Ni(001)	O	1.97	1.90
	S	2.18	2.28
	Se	2.27	2.41
	Te	2.58	2.61
	Na	3.37	3.10
Ni(110)	O	1.91	1.90
	S	2.17	2.28
Ni(111)	S	2.02	2.28
Ag(001)	Se	2.80	2.61
Ag(111)	I	2.75	2.77
Al(100)	Na	3.52	3.32
Mo(001)	N	2.02	2.08
W(110)	O	2.08	2.05

[a] In angstroms.

for smaller molecules. In the former case the molecule may overlap many surface-bonding sites and, besides requiring sufficient translational mobility, the molecule must also have sufficient rotational mobility.

The location of the adsorbed atom or molecule, its bond distances, and bond angles with respect to other atoms in the surface can be determined uniquely by low-energy electron diffraction, from analysis of the diffraction beam intensities. Over the past 10 years the theory of low-energy electron diffraction has been developed to the point where we can determine the surface structure of simple adsorbates. The field of surface crystallography has been born.

In Table I we list the experimentally determined bond lengths that were obtained from surface crystallography studies of several adsorbed atoms on metal surfaces. These results are compared with the bond length obtained by summing the covalent radii. In most cases the differences are within 0.1 Å that is claimed for the uncertainty of the experimental determination, and in no case is the discrepancy greater than 10%. This result suggests that the chemisorption bond is basically covalent for these atoms, which

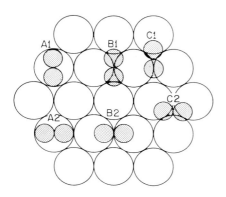

Fig. 7. Trial geometries of C_2H_2 on the Pt(111) surface. Divisions a, b and c refer to coordination to 1, 2 or 3 neighboring platinum atoms. The labels 1 and 2 distinguish molecules rotated 90°.

means that theoretical treatments in terms of localized surface complexes and clusters should be applicable to chemisorption. Recently we have analyzed the surface structure of the first adsorbed molecule, that of acetylene (C_2H_2), on the (111) crystal face of platinum (25). The nature of bonding of this molecule to a metal surface has important catalytic implications. Figure 7 shows the various molecular positions that were considered for C_2H_2 as possible bonding sites. It was concluded that acetylene is chemisorbed on platinum (111) in one of two possible local bonding modes at a distance of 1.95 ± 0.1 Å above the topmost plane of platinum atoms. In the most likely bonding mode (C_2) the molecule is centered on a triangular site, the carbon atoms are equivalent by symmetry, and relevant C–Pt distances are 2.25 and 2.259 Å. In the other possible bonding mode (B_1) the molecule is in an approximately twofold position with each carbon coordinating to three platinum atoms, the C–Pt distances being 2.47 and 2.65 Å. The C_2 mode of bonding is found to occur in various trimetallic metal–alkine complexes, whereas the bridging structure analogous to B_1 occurs in bimetallic complexes. These findings and structure analysis contadict popular notions

of acetylene adsorption on transition metal surfaces that were thought to involve acetylene π-complex coordinated to a single metal atom (model geometry A_1 or A_2) or a di-sigma model has often been cited in connection with the mechanism of dehydrogenation of ethylene, C_2H_4, to an acetylenic species upon adsorption with the ligand molecular orbitals in an sp^2 hybrid configuration. Although these bonding modes could be operable in some other bonding situations, they are not the favored bonding arrangements of acetylene on the Pt(111) surfaces.

Surface crystallography by low-energy electron diffraction permits us to determine the chemical bonding of atoms or simple molecules on solid surfaces. This information in turn will aid in determining the nature of the surface chemical bond. This may also aid our learning about adhesion and lubrication and other surface phenomena in the molecular level.

Surface irregularities, steps and kinks, play important roles in breaking chemical bonds that otherwise would not easily be broken in the absence of these sites. There are many reports of the variation of heats of adsorption and the types of binding states that are available, from crystal face to crystal face. A polycrystalline surface or the surfaces of small particles provide several bonding sites each with its own chemical activity. It is difficult to compare the reactivities of these structurally and therefore chemically heterogeneous surfaces that are prepared in different laboratories since minor changes in surface preparation change their surface structure and thus may markedly alter their reactivity. It is therefore essential to correlate the atomic surface structure and reactivity of model systems such as single-crystal surfaces and, using the similarity in chemical reactivity between these crystal surfaces and polydispersed systems, to learn about the atomic surface structure of dispersed small particles that are utilized frequently as heterogeneous catalyst systems.

THE SURFACE COMPOSITION OF BINARY ALLOYS

The composition of alloys in the topmost monolayer is likely to be very different from the bulk composition for several reasons, and we briefly review each of them.

(a) For systems that exhibit regular solution behavior, the surface free energy (which is always positive) is minimized by the segregation of that constituent with the lower surface tension. For metals that have high surface tensions as compared to organic surfaces or most oxide surfaces, such surface enrichment is likely to be very marked indeed. For a system obeying regular solution theory, it has been shown (26, 27) that the composition of the surface monolayer is given by

$$\frac{x_1^A}{x_1^B} = \frac{x_b^A}{x_b^B} \exp\left(\frac{(\sigma^B - \sigma^A)a}{RT}\right)$$

$$\times \exp\left[\frac{\Omega(l+m)}{RT}((x_b^B)^2 - (x_b^A)^2)\right.$$

$$\left. + \frac{\Omega l}{RT}((x_1^A)^2 - (x_1^B)^2)\right], \quad [1]$$

where x_1 and x_b are the atom fractions in the surface monolayer and bulk, respectively. In this expression σ is the surface energy, a is the molar surface area, T is the absolute temperature, R is the ideal gas constant, and Ω is the regular solution parameter, which is given from the heat of mixing ΔH_m by

$$\Omega = \Delta H_m/(x_b^B x_b^A). \quad [2]$$

The packing parameter l gives the fraction of an atom's nearest neighbors which are in the same plane parallel to the surface as the atom. Similarly m is the fraction of nearest neighbors which are in an adjacent parallel plane. For instance, in an fcc lattice, an atom has 12 nearest neighbors so a (111) plane has 6 nearest neighbors in the plane of an atom and 3 nearest neighbors in the plane below the the atom. Thus in this case $l = 6/12$ and $m = 3/12$.

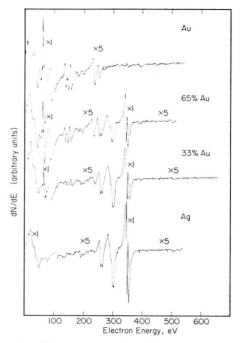

FIG. 8. Auger spectra of Au–Ag alloys and of pure Au and Ag.

Electron spectroscopy techniques, especially Auger electron spectroscopy and photoelectron spectroscopy, have been eminently successful in determining the surface phase diagrams, i.e., the surface composition as a function of bulk composition and temperature, for several binary metal systems. A typical Auger spectrum of an alloy system, that of silver–gold is shown in Fig. 8 (28). From the Auger peak intensities one may determine surface compositions for a series of bulk compositions and from these data the surface phase diagram is deduced. The most serious experimental difficulty in converting the Auger peak intensities to surface concentrations lies in estimating the depth near the surface from which the Auger electrons of different kinetic energy are emitted and the relative scattering power of the two types of atoms in this volume sampled by the electrons. Since part of the Auger signal is due to atoms below the surface,

172

TABLE II

Surface Enrichment of Binary Metal Alloys[a]

System	Object
Ag–Au (28)	Ag
Ag–Pd (31)	Ag
Al–Cu (32)	Al
Au–Cu (33)	Au
Au–In (34)	In
Au–Ni (35)	Au
Cu–Ni (36, 37)	Cu
In–Pb (38)	Pb
Ni–Pd (39)	Pd
Fe–Cr (40)	Cr
Pt–Sn (41)	Sn

[a] Numbers in parentheses indicate references.

it is essential to subtract this part from the signal intensities in order to extract the surface contribution. Much of the discrepancy between results reported from different laboratories on the same alloy system can be attributed to difficulties in estimating relative contributions of surface atoms to the various Auger peak intensities.

(b) Adsorbed gases or impurities segregating at the surface can markedly change the surface composition. If one of the constituents forms strong chemical bonds with the adsorbate (oxygen, CO, carbon, sulfur, or calcium, most commonly), it will be accumulating on the surface even though that it may not be at the surface in the absence of the adsorbate. CO, for example, which forms strong bonds with many transition metal atoms, may aid the segregation of these atoms at the surface at the expense of other constituents that form weak CO bonds. In essence an adsorbate converts a two-component system to a three-component system, thereby altering the surface composition at equilibrium.

(c) Solute atoms with large or small radii with respect to the size of the solvent (majority) atoms may be "excluded" from the bulk because they introduce excessive strain in the crystal lattice. As a result, surface segregation of one component that is misfit in the crystal lattice can be observed. The strain energy contribution may enhance the

surface concentration of that component that would segregate on the surface on account of its effect of decreasing the total surface free energy. Conversely this strain energy contribution may oppose the effect of surface segregation of one constituent in some cases.

(d) There are complex phase diagrams that exhibit the formation of stable compounds of high lattice energy. In this circumstance the energy necessary to exchange the two atoms, A and B, by removing them from their equilibrium lattice site is so large that it overrides the influence of surface forces that would induce surface segregation. Thus the bulk composition pins the surface composition, i.e., the surface and bulk compositions remain identical. We have found in studying the complex gold–tin phase diagram that this phenomenon occurs at the bulk composition of AuSn that forms a compound with high cohesive energy.

In Table II we list those metal systems that have been studied by electron spectroscopy and indicate which element is found in excess at the surface.

Since the topmost surface layer is the first line of defense of the solid against external chemical attack such as corrosion, it is essential that we determine and learn how to control the surface composition. Perhaps alloys with unique corrosion resistance under high-temperature and high-pressure conditions can be developed by the addition of small quantities of appropriately chosen constituents that passivate the surface layer by segregating there.

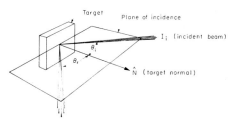

Fig. 9. Scheme of the molecular beam–surface scattering experiment.

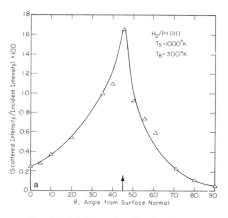
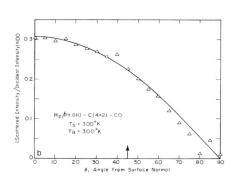

Fig. 10. (a) Scattering distribution of H_2 from a clean surface. (b) Scattering distribution of H_2 from a CO-covered surface.

MOLECULAR BEAM–SURFACE SCATTERING

One of the important questions we must answer in order to unravel the nature of surface reactions is how the energy partitions in a chemical reaction between the reactant, product molecules and the surfacea toms (7, 8). Also, it is useful to know the necessary minimum residence time of molecules on the surface for the chemical reaction to take place. To answer these questions we are studying the interaction of well-defined beams of molecules with crystal surfaces. The scheme of the molecular beam–surface scattering experiment is shown in Fig. 9. A beam of molecules impinges on the surface and the scattered product molecules are detected as a function of angle (angular distribution) by a rotatable mass spectrometer detector. By chopping the incident beams periodically, the flight time of the molecules can be determined. This experimental information yields the surface residence time of the impinging molecules before readmission to the gas phase, and the velocity and the momentum of the incident molecules and of the scattered products. The surface structure and composition of the crystal surfaces are also determined by simultaneously utilizing low-energy electron diffraction and Auger electron spectroscopy.

Using this sensitive molecular beam probe we have determined that the amount and type of energy transfer between the incident beam and surface depend very much on the surface structure and composition. Molecules scattered from ordered clean metal surfaces undergo little loss of their kinetic energy (velocity). If the surface is covered with a layer of adsorbed molecules, for example CO, the impinging atoms and molecules lose all their kinetic energy to the surface and are trapped long enough that complete energy equilibration occurs between the gas and the surface before the reemission of the molecule (29).

This effect is demonstrated in Figs. 10a and 10b. Here the linear signal intensity due to the scattered beam is plotted as a function of the scattering angle. The peaked angular distribution of hydrogen and the fact that the angle at which the scattered molecular beam has maximum intensity equals the angle of incidence of the incoming beam (which is indicated by the arrow at the abscissa) are characteristics of specular scattering (Fig. 10a). This type of ping-pong-ball-like scattering is associated with poor energy exchange between the incident gas molecules and the surface. In the case of particles emitted from the surface that have completely equilibrated

TABLE III

Preexponential Factors, Activation Energies, and Reaction Probabilities for Several Surface
Reactions Studied by Molecular Beam Scattering

Reaction	A^a	E_a (kcal/mol)	Reaction probability	Reference
$H + D_2 \xrightarrow{Pt} HD$ (<600°K)	2×10^5 (sec⁻¹)	4.5	~10^{-1}	30
$H + D_2 \xrightarrow{Pt} HD$ (>600°K)	1×10^2 (sec⁻¹)	0.6	~10^{-1}	30
$D + O_2 \xrightarrow{Pt} D_2O$ (700°K)	—	12	—	42
$CO + O \xrightarrow{Pt} CO_2$ (700°K)	—	20	~10^{-3}	43
$C_2H_4 + O_2 \xrightarrow{Ag} CO_2$ (800°K)	—	8	<10^{-2}	44
$2H \xrightarrow{graphite} H_2$ (800–1000°K)	1.06×10^{-2} cm²/atom sec	15.9	10^{-3}–10^{-2}	45
$H_2 \xrightarrow{Ta} 2H$ (1100–2600°K)	—	75	4×10^{-1}	46
$HCOOH \xrightarrow[decomp]{Ni} CO_2$ (<455°K)	10^{12} (sec⁻¹)	20.7	—	47
$HCOOH \xrightarrow[decomp]{Ni} CO_2$ (>455°K)	5.8×10^{13} (sec⁻¹)	2.5	~0.9	47
$C + O_2 \longrightarrow CO$ (1000–2000°K)	2.5×10^7 (sec⁻¹)	30	10^{-3}–10^{-2}	49
$C + O_2 \longrightarrow CO$ (1000–2000°K)	3×10^{12} (sec⁻¹)	50	10^{-3}–10^{-2}	49
$C + 4H \longrightarrow CH_4$ (500–800°K)	1.27×10^{-18} cm⁴/atom sec	3.3	10^{-3}–10^{-2}	45
$2C + 2H \longrightarrow C_2H_2$ (>1000°K)	1.59 cm²/atom sec	32.5	10^{-3}–10^{-2}	45
$Ge + O_2 \longrightarrow GeO$ (750–1100°K)	10^{16} (sec⁻¹)	55	2×10^{-2}	50
$Ge + O \longrightarrow GeO$ (750–1100°K)	10^{16} (sec⁻¹)	55	3×10^{-1}	51
$Ge + O_3 \longrightarrow GeO$ (750–1100°K)	10^{16} (sec⁻¹)	55	5×10^{-1}	52
$Ge + Cl_2 \longrightarrow GeCl_2$ (750–1100°K)	10^7 (sec⁻¹)	25	3×10^{-1}	47
$Ge + Br_2 \longrightarrow GeBr_2$ (750–1100°K) $\searrow GeBr_4$	10^7 (sec⁻¹)	20	3×10^{-1}	53
$Si + Cl_2 \longrightarrow SiCl_2$ (1100–1500°K)	10^8 (sec⁻¹)	40	3×10^{-1}	47
$Ni + Cl_2 \longrightarrow NiCl$ (900–1400°K)	10^7 (sec⁻¹)	30	8×10^{-1}	48

^{a} For bimolecular reactions, the preexponential factor also includes the surface concentration of one of the reactants that are held constant during the experiments.

with the surface, one obtains a cosine-like angular distribution, as shown in Fig. 10b. When hydrogen is scattered from a monolayer of adsorbed carbon monoxide, complete energy equilibration occurs between the incident molecules and the surface. Much of the energy accommodation between incident particles and the surface depends on the nature of energy transfer between the translational and rotational modes of the gas molecules (and vibrational modes if these can be excited or de-excited during the collision) and the vibrational modes of the surface atoms. It appears that the localized low-frequency bending modes of adsorbed carbon monoxide are very effective in adsorbing much of the translational energy of the incident hydrogen molecules. A recent review summarizes the results of energy transfer studies during nonreactive molecular beam–surface scattering (7).

Reactive scattering studies can be divided into two groups: (1) the reaction takes place on the surface under the catalytic influence

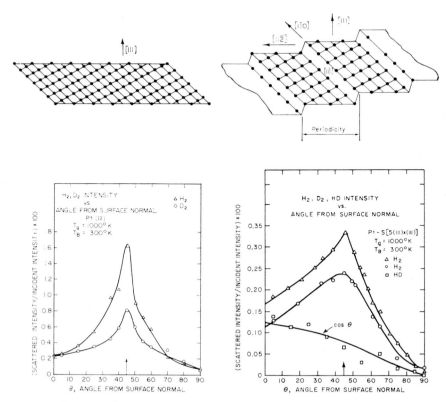

FIG. 11. (a) Angular distribution of H_2 and D_2 scattered from Pt(111) single crystal (HD signal is undetectable). (b) Angular distribution of H_2–D_2 and HD scattered from a stepped platinum single crystal. Schematic diagrams of the surfaces are shown above the figures.

of the surface atoms that, however are not present among the reactants or the products (for example,

$$A_{2(gas)} + B_{2(gas)} \xrightarrow{\text{surface}} 2AB_{(gas)});$$

(2) the surface atom is one of the reactants

$$(A_{2(gas)} + S \rightarrow SA_{2(gas)}).$$

Using a mass spectrometer the product distribution and the rates of formation of the product molecules (reaction probabilities) are determined as a function of the system variables. From the dependence of the reaction rate on the incident beam velocity, the activation energy for adsorption (if any) is determined. From the surface temperature de-

pendence of the rate, the activation energy of the surface reaction is obtained.

In Table III we list all of the preexponential factors, activation energies and reaction probabilities for the surface reactions that were studied by molecular beam scattering techniques and where such data were reported. From this information a great deal can be learned about the kinetics and mechanisms of surface reactions.

Let us consider, as an example, the mechanism of H–H bond breaking on platinum surfaces that can be investigated by studying the exchange of hydrogen and deuterium to form HD using molecular beam scattering (30). One of the fundamental questions of heterogeneous catalysis is how surfaces lower the activation energy for simple reactions on the atomic scale so that they proceed readily on the surface while the same reaction in the gas phase is improbable. The reaction of hydrogen and deuterium molecules to form hydrogen deuteride is one of the simple reactions that take place readily on metal surfaces, even at temperatures below 100°K. The same reaction is completely inhibited in the gas phase by the large dissociation energy of H_2 or D_2 (103 kcal/mol). Once the H_2 molecule is dissociated, the successive atom–molecule reaction $(H + D_2 = HD + D)$ in the gas phase still has a potential energy barrier of about 10 kcal/mole. If we were to study this reaction at high pressures (about 10^3 Torr), the exchange takes place so rapidly on most transition metal surfaces that equilibrium is readily established among the various hydrogen molecules, H_2, D_2, and HD. In order to study the kinetics of this reaction one has to carry out the experiments far from equilibrium. Molecular beam–surface scattering experiments work the best far from equilibrium at low incident beam pressures (about 10^{-7}–10^{-6} Torr) and at low surface coverages. The H_2–D_2 exchange reaction was studied by Bernasek *et al.* (30) using platinum single-crystal surfaces of low and high Miller index. Under the conditions of the experiments, which put strict limitations on the residence time of the detected molecules, the reaction product, HD, could not be detected from the (111) crystal face. However, the reaction product was readily detectable from the high Miller index stepped surface. The integrated reaction probability defined as total desorbed HD flux/H_2 flux incident on the surface is approximately 10^{-1} while HD formation was below the limit of detectability on the platinum (111) surface (reaction probability less than 10^{-5}). High Miller index single-crystal surfaces are characterized by ordered arrangements of atomic steps of one atom height separated by terraces of low Miller index orientation. Thus atomic steps at the platinum surface must play controlling

roles in dissociating the diatomic molecules. Figure 11 shows the scattering distributions from both the (111) and the stepped platinum surfaces. Varying the chopping frequency of the incident molecular beams has yielded HD residence times of about 25 msec and longer up to 1 sec on a stepped platinum surface in the 700–1000°K surface temperature range. Such long residence times should result in complete thermal equilibration between the surface and the reactant products. Indeed, it was found by experiments that the desorbing HD exhibits cosine angular distribution, as seen in Fig. 11.

The reaction probability was also dependent on the angle of H_2 incidence. The maximum reaction probability was obtained upon incidence normal to the step edges. The experiments have also revealed that the exchange reaction occurs by a two-branch mechanism. The faster reaction takes place at the step, while a slower reaction (with surface residence times of seconds) occurs on the atomic terraces. The absence of beam kinetic energy dependence of the rate indicates that adsorption and dissociation of hydrogen do not require activation energy. The surface is able to store a sufficiently large concentration of atoms which react with other surface atoms by a two-branch mechanism. The rate constants for the H_2–D_2 reaction were also determined under conditions of constant hydrogen atom coverage and are given in Table III. The rate-determining step appears to be the diffusion of D atoms on the surface to a site where HD is formed by a two-centered reaction subsequent to H_2 or D_2 dissociation at the step. At higher temperatures the reaction between an adsorbed H atom and an incident D_2 gas molecule may also become an important reaction channel. The catalyst action of the platinum surface for the exchange reaction is due to its ability to adsorb and dissociate hydrogen molecules with near zero activation energy and to store atomic hydrogen on the surface, thereby converting the gas phase molecule–molecule reaction to an atom–atom reaction of low activation energy.

It would be of great importance to measure the velocity of the scattered products from the time of flight measurements using a mass spectrometer, in addition to their angular distribution, in order to determine the nature of energy transfer between the HD product molecules and the surface prior to desorption. It is likely that during certain exothermic surface reactions (for example, hydrogen or oxygen atom recombination) much of the chemical energy is converted into internal energy and translational energy of the product molecules. Studies of energy transfer during exothermic surface chemical reactions are in progress in several laboratories.

ACKNOWLEDGMENT

This work was supported by the U. S. Energy Research and Development Administration.

REFERENCES

1. KESMODEL, L. L. AND SOMORJAI, G. A., *MPT Int. Rev. Sci., Phys. Chem. Ser. 2* **7**, 1 (1975).
2. PENDRY, J. B., "Low Energy Electron Diffraction." Academic Press, London, 1974.
3. STROZIER, J. A., JEPSEN, D. W., AND JONA F., *in* "Surface Physics of Materials" (J. M. Blakely, Ed.), Vol. 1. Academic Press, New York, 1975.
4. SIEGBAHN, K., *et al.*, "ESCA-Atomic, Molecular and Solid State Structure Studied by Means of Electron Spectroscopy." Almqvist and Wiksells, Uppsala, 1967.
5. BRUNDLE, C. R. *in* "Surface and Defect Properties of Solids," Vol. 1, p. 171. The Chemical Society, Washington, D. C., 1972.
6. SZALKOWSKI, F. J. AND AND SOMORJAI, G. A., *Advan. High Temp. Chem.* **4**, 137 (1971).
7. BERNASEK, S. L. AND SOMORJAI, G. A., *Progr. Surface Sci.* **5**, 377 (1975).
8. BRUMBACH, S. B. AND SOMORJAI, G. A., *Crit. Rev. Solid State Sci.* **4**, 429 (1974).
9. SOMORJAI, G. A., *Advan. Catalysis*, to appear (1976).
10. BLAKELY, D. W. AND SOMORJAI, G. A., *Nature (London)* **258**, 580 (1975).
11. BLAKELY, D. W. AND SOMORJAI, G. A., *J. Catalysis*, **42**, 181 (1976).
12. CLARKE, T. A., MASON, R., AND TESCARI, M., *Surface Sci.* **30**, 553 (1972).
13. KESMODEL, L. L. AND SOMORJAI, G. A., *Accounts Chem. Res.* (1976).
14. CHESTERS, M. A. AND SOMORJAI, G. A., *Ann. Rev. Mat. Sci.* **5**, 99 (1975).
15. BUCHHOLZ, J. C. AND SOMORJAI, G. A., *Accounts Chem. Res.*, **9**, 333 (1976).
16. KESMODEL, L. L. AND FALICOV, L. M., *Solid State Commun.* **16**, 1201 (1975).
17. TSANG, Y. W. AND FALICOV, L. M., *J. Phys. C. Solid State Phys.* **9**, 51 (1976).
18. TSANG, Y. W. AND FALICOV, L. M., *Phys. Rev. B* **12**, 2441 (1975).
19. CHESTERS, M. A. AND SOMORJAI, G. A., *Surface Sci.* **52**, 21 (1975).
20. HAGEN, D. I., NIEUWENHUYS, B. E., ROVIDA G., AND SOMORJAI, G. A., *Surface Sci.*, **59**, 155 (1976).
21. NIEUWENHUYS, B. E., HAGEN, D. I., ROVIDA, G., AND SOMORJAI, G. A., *Surface Sci.*, **57**, 632 (1976).
22. SOMORJAI, G. A., *Surface Sci.* **34**, 156 (1973).
23. SZALKOWSKI, F. J. AND SOMORJAI, G. A., *J. Chem. Phys.* **54**, 389 (1971).
24. GLAND, J. L. AND SAMORJAI, G. A., *Advan. Colloid Interface Sci.* **5**, 203 (1976).
25. KESMODEL, L. L., STAIR, P. C., BAETZOLD, R. C., AND SOMORJAI, G. A., *Phys. Rev. Lett.*, **36**, 1316 (1976).
26. OVERBURY, S. H., BERTRAND, P. A., AND SOMORJAI, G. A., *Chem. Rev.* **75**, 547 (1975).
27. OVERBURY, S. H. AND SOMORJAI, G. A., *Discuss. Faraday Soc.* **60**, 279 (1975).
28. OVERBURY, S. H. AND SOMORJAI, G. A., *Surface Sci.* **55**, 209 (1976).
29. BERNASEK, S. L. AND SOMORJAI, G. A., *J. Chem. Phys.* **60**, 4552 (1974).
30. BERNASEK, S. L. AND SOMORJAI, G. A., *J. Chem. Phys.* **62**, 3149 (1975).
31. WOOD, B. J. AND WISE, H., *Surface Sci.* **52**, 151 (1975).
32. FERRANTE, J., *Acta Met.* **19**, 743 (1971).
33. McDAVID, J. M. AND FAIN, S. C., *Surface Sci.* **52**, 161 (1975).
34. THOMAS, S., *Appl. Phys. Lett.* **24**, 1 (1974).
35. BURTON, J. J., HELMS, C. R., AND POLIZZOTTI, R. S., *J. Vac. Sci. Technol.* **13**, 204 (1976).
36. HELMS, C. R. AND YU, K. Y., *J. Vac. Sci. Technol.* **12**, 276 (1975).
37. HELMS, C. R., *J. Catalysis* **36**, 114 (1975).
38. BERGLUND, S. AND SOMORJAI, G. A., *J. Chem. Phys.* **59**, 5537 (1973).
39. STODDART, C. T. H., MOSS, R. L., AND POPE, D., *Surface Sci.* **53**, 241 (1975).
40. LEYGRAF, C., HULTQUIST, G., EKELUND, S., AND ERIKSSON, J. C., *Surface Sci.* **46**, 157 (1974).
41. BOUWMAN, R., TONEMAN, L. H., AND HOLSCHER, A. A., *Surface Sci.* **35**, 8 (1973).
42. SMITH, J. N. AND PALMER, R. L., *J. Chem. Phys.* **56**, 13 (1972).
43. BONZEL, H. P. AND KU, R., *Surface Sci.* **33**, 91 (1975).

44. SMITH, J. N., PALMER, R. L., AND VROOM, D. A., *J. Vac. Sci. Technol.* **10**, 5027 (1968).

45. BALOOCH, M. AND OLANDER, D. R., *J. Chem. Phys.* **63**, 4772 (1975).

46. KRAKOWSKI, R. A. AND OLANDER, D. R., *J. Chem. Phys.* **49**, 5027 (1968).

47. MADIX, R. J., *in* "Physical Chemistry of Fast Reactions" (D. O. Hayward, Ed.), Vol. 2. Plenum, New York, 1976.

48. MCKINLEY, J. D., *J. Chem. Phys.* **40**, 120 (1964).

49. OLANDER, D. R., JONES, R. H., SCHWARTZ, J. A., AND SIEKHAUS, W. J., *J. Chem. Phys.* **57**, 421 (1972).

50. ANDERSON, J. B. AND BOUDART, M., *J. Catalysis* **3**, 216 (1964); MALIX, R. J. AND BOUDART, M., *J. Catalysis* **7**, 240 (1967).

51. MADIX, R. J. AND SUSU, A. A., *Surface Sci.* **20**, 377 (1970).

52. MADIX, R. J., PARKS, R., SUSU, A. A., AND SCHWARTZ, J. A., *Surface Sci.* **24**, 288 (1971).

53. MADIX, R. J. AND SUSU, A. A., *J. Vac. Sci. Technol.* **9**, 915 (1972).

Heterogeneous Chemical Kinetics by Modulated Molecular Beam Mass Spectrometry

Limitations of the Technique

DONALD R. OLANDER

Materials and Molecular Research Division of the Lawrence Berkeley Laboratory and the Department of Nuclear Engineering of the University of California, Berkeley, California 94720

Received June 7, 1976; accepted August 12, 1976

The advantages and limitations of modulated molecular beam, mass spectrometry as applied to the study of heterogeneous chemical kinetics are reviewed. The process of deducing a model of the surface reaction from experimental data is illustrated by analysis of the hydrogen reduction of uranium dioxide.

I. INTRODUCTION

The study of gas–solid chemical reactions by the combination of modulated molecular beams, high vacuum, mass spectrometry, and phase-lock or digital processing techniques has, since its inception by Smith and Fite (1) 13 years ago, generated more analysis than data. The literature in this field is unique in the annals of experimental chemistry in having produced, in its relatively short life, nearly as many review papers as experimental reports. These reviews (2–7) have thoroughly detailed the advantages of using modulated molecular beam–mass spectrometric methods over conventional methods of studying heterogeneous reactions. They have also rather completely summarized all of the molecular beam surface chemistry experiments which have been reported. Rather than repeat what has been admirably reviewed by the practitioners of this field, the present paper concentrates on the difficulties and limitations of the technique, with emphasis on the ability to deduce a chemically and physically acceptable mechanism from the massive amount of data which a modulated molecular beam experiment can produce. The precise control of reaction conditions and the variety and high informa-tion content of the data which are obtained are the singular advantages claimed for the new method. By the use of a case study, we examine the extent to which this potential can be realized.

II. THE MODULATED BEAM METHOD AND ITS ADVANTAGES

The general features of a molecular beam apparatus for heterogeneous chemical kinetic studies are shown in Fig. 1. The species detection, signal processing, and data analysis methods have been described in detail elsewhere (3, 6). The reactant gas emanating from an effusive or nozzle source is modulated by a rotating slotted disk and collimated into a collision-free beam which is ∼ 3 mm in diameter as it strikes the solid target. Reflected reactant gas and any reaction products emitted from the target are sampled by another collimator separating the target chamber and the chamber housing the mass spectrometer. The mass spectrometer output signal is fed into a lock-in amplifier where it is combined with a reference signal derived from the chopper motor. Each experimental point consists of the amplitude and phase angle of a selected reaction product and of the reflected reactant gas. The ratio of

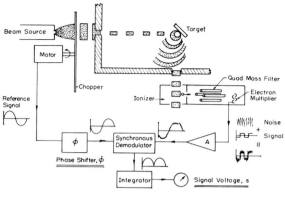

XBL 717- 6985

FIG. 1. Schematic of a modulated molecular beam apparatus for surface chemical kinetic studies.

the signal amplitude of the reaction product to that of the reflected reactant is denoted as the apparent reaction probability, ϵ. The phase difference between these two signals is called the reaction phase lag, ϕ. These two experimental quantities are corrected for various parasitic effects (3), before comparison with theory. Apparent reaction probabilities and reaction phase lags are measured as functions of the modulation frequency, ω, the target temperature, T_s, and the reactant beam intensity, I_0.

As an alternative to the synchronous modulation/lock-in amplifier as a means of signal processing, the entire waveform of the reflected reactant and the desorbed product can be obtained by pulse counting techniques combined with data accumulation in a multi-channel analyzer. Recent advances in nuclear radiation detection make this method accessible although expensive. In principle, the pulse counting method can achieve any level of signal-to-noise ratio in the output simply by using sufficiently long counting times. The signal-to-noise improvement attainable by a lock-in amplifier, on the other hand, is limited by the bandpass of this instrument. In addition, lock-in amplifiers are generally restricted to modulation frequencies greater than a few Hertz. Although complete waveform accumu-

lation has its adherents (4), most experimenters in the field continue to use lock-in amplification as the principal data processing tool. In our laboratory, both methods have been tried and no dramatic advantage of the pulse counting technique has been found. However, the analysis of very complicated surface mechanisms which give rise to peculiar shapes of product waveforms may benefit from being able to see the entire waveform rather than merely to record the amplitude and phase of the first Fourier component.

A third method of obtaining data in a molecular beam experiment is called the correlation technique. This method is similar to the lock-in amplifier technique in that it utilizes analog signals, but instead of being synchronously modulated (i.e., equal off and on times), the primary beam is modulated in a pseudorandom manner. Cross correlation of the primary beam modulation with the detected product signal produces information equivalent to that obtainable by a series of experiments using lock-in amplifiers tuned to many frequencies. The correlation technique is described by Visser et al. (8) and has been applied to a nonreactive scattering system by Hurkmans et al. (9).

Both the inherent advantages and the limitations of the molecular beam method lie in the way in which the reactant gas is delivered to the target. A molecular (or atomic) beam is a narrow, nearly parallel ray of particles moving in straight lines at the velocities imparted to them by the flow through the source. The primary beam molecules collide with nothing (including each other) until they strike the solid reactant. The reactant delivery system permits control of the angle at which the molecules strike the surface, a property which has been exploited in chemisorption studies (7). The collimated nature of the incident beam makes it possible to study particular crystallographic faces of the target solid, which, in some instances, reveals spectacular differences in reactivity (10).

Similarly, the products desorbing from the surface are detected in collision-free flight, which avoids quenching errors inherent in conventional sampling of hot reactive gases. Collisionless sampling also permits free radical reaction intermediates to be detected should they be emitted from the solid surface. However, the modulated beam experiments reported to date have observed the same sort of reaction products that one would expect in conventional heterogeneous kinetic experiments (e.g., CO and CO_2 in the case of graphite oxidation (10); NO and N_2 for N_2O decomposition (11)). In the case of reactions involving an element with multiple valence states, the low equivalent pressure of the reactant gas tends to favor products which have a small number of atoms of the gaseous reactant; thus the sole products of the reaction of Cl_2 with Ge and Si are the dichlorides rather than the tetrachlorides which would result in experiments at higher gas pressures (12). It was an early expectation that the modulated beam method would reveal unusual products of heterogeneous reactions which had remained hidden by gas phase and wall collisions in conventional, normal pressure kinetic experiments. This expectation has not been confirmed, and theories have been provided (13, 14) to explain why; under conditions where desorption from the surface is rate controlling and all preceding reactions on the surface are at equilibrium, the products of the hetero-geneous reaction should be those which thermodynamics indicates are the stable gas phase equilibrium species. This theory works well in some metal oxidation systems (13) but it is less successful when nonmetallics such as graphite are involved (10). Nonetheless, the quasi-equilibrium theory has at least considerably dimmed the prospects of discovering unusual products of gas–solid reactions by molecular beam studies.

The collimated nature of the incident reactant gas beam has also provided the experimentalist with a new means of exploiting isotope exchange as a technique of unraveling surface processes. In addition to the common technique of utilizing mixtures of enriched isotopes as reactants, one isotope may be contained in the modulated reactant beam and the other caused to impinge on the surface from a nonmodulated source. The latter method was applied to the interaction of water vapor with graphite (15) and of hydrogen on Pt (16).

Perhaps the most valuable aspect of introducing the gaseous reactant as a collimated beam is that the reactant supply can be modulated on a time scale which nature has fortunately made the characteristic time of many interesting surface reactions. This point can be demonstrated by considering the simple reaction of gas A with solid B to produce the chemisorbed intermediate AB which slowly desorbs to produce the reaction product signal detected by the mass spectrometer. This surface process can be described by the elementary reactions of sticking (and reaction) of A on the surface of B:

$$A\,(\mathrm{g}) + B\,(\mathrm{s}) \xrightarrow{\eta} AB\,(\mathrm{ads}) \qquad [1]$$

followed by

$$AB\,(\mathrm{ads}) \xrightarrow{k} AB\,(\mathrm{g}). \qquad [2]$$

Here η is the sticking probability of A on the surface of B and k is the desorption rate constant for the product AB. As noted by Madix (6), the characteristic time of the first reaction is of the order of picoseconds and thus appears as an instantaneous event in the modulated beam experiment. The reciprocal of the rate constant of reaction [2], however, is the mean lifetime of AB on the surface. It is this lifetime which must be of the same order of magnitude as the modulation frequency to be quantitatively measurable by the modulated beam method. Analysis of the elementary mechanism represented by reactions [1] and [2] leads to the predicted apparent reaction probability and reaction phase lag (as defined in the first paragraph of this section) given by (6)

$$\epsilon = \eta/[1 + (\omega/k)^2]^{\frac{1}{2}}, \qquad [3]$$

$$\phi = \tan^{-1}(\omega/k). \qquad [4]$$

The utility of beam modulation is illustrated by these equations; a steady-state experiment (whether or not molecular beams are used) would produce data which are equivalent to setting $\omega = 0$ in Eq. [3] and disregarding Eq. [4] entirely. That is, the steady-state experiment would provide only the sticking probability but not the desorption rate constant. In the modulated beam method, a single simultaneous measurement of ϵ and ϕ permits both η and k to be measured. Moreover, Eqs. [3] and [4] prescribe the functional dependence of ϵ and ϕ upon modulation frequency, and systematic variation of this variable provides a means of verifying the validity of the assumed surface mechanism (reactions [1] and [2] in this example).

Equations [3] and [4] also show that the range of rate constants which can be quantitatively measured is limited on the fast side by the restriction $\omega/k \gtrsim 0.05$ (which assumes a minimum detectable phase lag of 3°) and on the slow side by $\omega/k \lesssim 100$ (for a sticking probability of 0.01 and a minimum detectable apparent reaction probability of 10^{-4}). Readily attainable modulation frequencies range from 10 to 1000 Hz (60 to 6000 rad/sec) which means that rate constants between ~ 1 and 10^5 sec^{-1} are obtainable—a very respectable range. The lower limit to k generally cannot be attained because the slow desorption of AB relative to the rate of supply of A from the

beam yields surface coverages comparable to unity, and the sticking coefficient becomes less than the bare surface value assumed in the example.

III. LIMITATIONS OF THE MOLECULAR BEAM METHOD

The brief summary in the preceding section and the more detailed discussions in Refs. (2–6) more than adequately present the affirmative aspects of the modulated molecular beam method for surface chemistry studies. The limitations of the method, on the other hand, have never been discussed with much relish. Several of the obvious restrictions are not especially critical; the reaction product must be volatile; one cannot study the corrosion of iron by oxidizing gases by this method, for example. The reactants and products must all have different masses to be differentiated by the mass spectrometer detector. This restriction effectively precludes study of isomerization reactions and reactions which produce species of the same mass (N_2O and CO_2 or CH_4 and O, for example). However, in the latter case, the cracking patterns of the equal mass species may be sufficiently different to enable the fragment ions rather than the parent ions to be used for product detection. A less obvious limitation is the requirement that the reaction product should not be one of the fragmentation products of the reactant molecules which are ionized in the mass spectrometer. If this is the case (and it is for the reaction of H_2O with graphite which produces H_2 (Ref. 15)), the measured apparent reaction probability and phase lag of H_2 are dominated by the reflected reactant contribution, not by the true product of the surface reaction. For the same reason, it would be very difficult to study atom recombination by modulated molecular beam methods unless the molecular contaminant of the primary beam were very low.

One of the first questions that an impartial observer would ask is the following: If the modulated beam method has so many advantages over the conventional methods of heterogeneous chemical kinetics, why has it not been more fully exploited by the scientific community? Compared to the explosive rise in the use of physical surface techniques of comparable sophistication (low energy electron diffraction (LEED) and Auger electron spectroscopy (AES)), modulated molecular beam–mass spectrometery has relatively few practitioners. The number of laboratories throughout the world which consistently publish in this area can be counted on one hand. Part of the answer to the question may be the inherent difficulty of the experiment proper. A modulated beam system for surface chemical kinetic studies cannot be purchased commercially as can a LEED or AES apparatus. Each chemical system selected for study requires extensive experimental and analytical work, with the result that even laboratories which devote a substantial portion of their time to this field produce no more than one experimental paper every 2 or 3 years.

There are, however, more fundamental reasons why the molecular beam method is limited in its scope. The principal limit is in the nature of the molecular beam reactant itself. The strength of the primary molecular beam impinging on the solid reactant is limited by the available pumping speed and geometrical constraints of the apparatus. Most experiments produce primary beams which have an intensity of $\sim 10^{14}$ molecules/cm²-sec striking the target. This impingement rate corresponds to a random gas pressure of $\sim 10^{-7}$ Torr. The highest reactant beam intensities which have been used in surface chemical studies approach 10^{-4} Torr equivalent pressure. In principle, one could obtain somewhat stronger beams by using nozzle beam sources, but the cost of the vacuum pumps needed to exhaust the flow from the source producing the beam quickly becomes prohibitive. Supersonic beams are more useful in nonreactive scattering and gas–gas scattering experiments than they are in reactive surface beam experiments. In the first two cases, the use of the monoenergetic beams produced

by a nozzle source makes comparison of theory with experiment simpler than if a Maxwellian beam had been used. In reactive surface scattering, however, variation of beam temperature (provided the reactant is not thermally dissociated) has little effect on the reaction (10), so that at room temperature, Maxwellian beams are sufficient, provided that they are of high intensity. High primary beam intensities are attainable only by making the distance between the source and the target and between the target and the mass spectrometer as small as possible (because the beam intensity decreases as the inverse square of the distance travelled). Given the necessity for a mechanical chopper, vacuum walls with collimators for differential pumping, and a reasonable flow area for exhausting the gas load, flight paths less than a few centimeters do not appear to be feasible.

Since gas–solid reactions of commercial interest are carried out at pressures about seven orders of magnitude higher than those characteristic of molecular beam experiments, the benefits of the modulated-beam method are generally not translatable to chemical systems of technological interest. An extrapolation of a reaction mechanism determined by a low-pressure molecular beam experiment by a factor of 10^7 in reactant gas pressure is justifiably viewed with some suspicion by those interested in the application of gas–solid kinetics to practical situations. However, problems generated in newly developed technologies, including space vehicle reentry into the upper atmosphere and chemical attack of fusion reactor structures by the confined plasma, can directly benefit from the detailed sort of low-pressure chemical kinetic data which the molecular beam method is well suited to provide (17, 18).

The problem of sensitivity severely limits the class of reactions amenable to examination by the molecular beam method. Experience has shown that with the state-of-the-art in signal-to-noise enhancement and vacuum system cleanliness, the lowest apparent reaction probability which can be measured with some confidence in the molecular beam method is $\sim 10^{-4}$. With very strong primary beams and a product species at a mass number where there is little background gas, this limit might be reduced by an order of magnitude. However, most gas–solid reactions at high pressures proceed with very much lower probabilities and are detectable in conventional kinetic experiments only because each reactant molecule is permitted very many collisions with the solid reactant. In a molecular beam experiment, on the other hand, each reactant molecule has only one chance to react with the surface. In order to render low-probability reactions detectable by the molecular beam technique, part of the reactant gas energy must be supplied in the molecular beam source rather than by the solid surface where the reaction takes place. In a study of hydrogen interaction with copper by Balooch et al., (19) reactant adsorption was strongly enhanced by increasing the energy of the incident beam. Madix et al. (20) could not observe sufficient reaction of O_2 with Ge or Si and were forced to use O_3 or O as the reactant form of oxygen. Molecular hydrogen interacts with graphite as does a rare gas, and the graphite–hydrogen reaction can only be made visible in a beam experiment by predissociating the hydrogen in an oven source (21).

Despite the critical need for strong signals, no gas–surface reaction system has exploited the scintillation-type detector of very high efficiency (22, 23) or electron bombardment ionizers which are commonly installed in gas–gas scattering apparatus (24). However, it should be noted that the primary objective of gas–gas scattering experiments is to measure the intensity of a product species at a particular angle in a specified energy range. Modulation of the beam is used primarily to permit distinction between background gases and the desired signal. In a gas–surface reaction experiment, on the other hand, beam modulation is the means of imparting phase information to the reaction and product signals. Although it may be possible to measure the amplitude of a weak signal quantitatively (say one yielding

a few ions per second striking the electron multiplier), the phase of such a signal cannot be measured with any reasonable accuracy.

Because of the inherently transient (although periodic) and spatially collimated character of the reactant gas supply to the surface, modulated beam experiments are often considerably complicated by diffusional processes on the surface or in the bulk of the solid target. These transport processes, although interesting in their own right, must be properly identified and accounted for in data analysis if an accurate picture of the surface chemistry is to be had. In a modulated molecular beam study of H_2 dissociation of H_2 on tantalum (25), surface migration of H atoms to the cool target holder where recombination was favored considerably reduced the dissociation rate from the rate that would have occurred if the system had not exhibited strong spatial nonuniformities in temperature and reactant pressure. In a similar vein, the study of CO oxidation on Pt by Palmer and Smith (26) showed evidence of CO migration to the perimeter of the beam spot.

Nonmetallic solid targets appear to be more susceptible to bulk solution–diffusion processes competing with surface steps than are metallic substrates. The diffusion of hydrogen (21), oxygen (10), and water (15) into and out of pyrolytic graphite in response to the pulsed incident beam presents difficult data interpretation problems in graphite surface chemistry. In the next section, we show that bulk solution and diffusion of either product or reactant also appear important in reactions involving ceramic oxides and gases.

Long-range surface and bulk diffusion on or in the solid caused by spatial and temporal nonuniformities in the reactant gas supply are deduced only indirectly by the manner in which the phase lag varies with frequency. These data cannot be rationalized solely by the chemical elementary surface steps with which heterogeneous kineticists have usually interpreted experimental results (e.g., Langmuir–Hinschelwood or Ely–Rideal mechanisms). Until recently, there was no direct experimental evidence that long-range transport processes were in fact taking place along with the surface reactions. This state of affairs naturally leaves the experimentalist with the nagging suspicion that he may have fabricated a physical process to explain an experimental artifact. The study of molybdenum oxidation by Ullman and Madix (27) has provided convincing molecular beam evidence of surface diffusion of chemisorbed oxygen to grain boundaries where oxidation occurs. In addition, however, Ullman and Madix took the next step of examining the reacted surface visually and found that indeed exposure of molybdenum to oxygen resulted in preferential etching of the grain boundaries in the region of the target illuminated by the primary beam. This work is a good illustration of the need to make all possible tests on the system in addition to simply measuring reaction probabilities and phase lags as functions of temperature, modulation frequency, and primary beam intensity.

In the study of the reaction of water vapor with graphite (15), H_2O continued to ooze out of the sample for tens of seconds after the reactant beam had been shut off. Such behavior can best be attributed to the storage of water from the primary beam in the bulk of the target, especially when this interpretation is consistent with the molecular beam results which represent a much smaller time scale.

It is important to note that, save for porous solids (which have never been used as targets in molecular beam studies) or the grain boundary role in molybdenum oxidation (27), none of the long-range diffusional processes would ever have occurred, much less have been detected, in conventional steady-state kinetic experiments. When the reactant gas supply is time-independent and is uniformly distributed over the entire solid specimen, diffusional effects disappear after the startup transient. The surface and the bulk become saturated with reactant gas and the surface chemical reactions proceed unimpeded by transport processes. In the modulated molecular beam experiment, on the other hand, there

is no experimental way of avoiding long-range transport of products or reactants; the technique eliminates one type of diffusional resistance (external gas phase mass transfer) only at the expense of introducing another transport resistance.

Short-range diffusional processes, usually involving migration of species over the solid surface to reactive sites, have been uncovered in many modulated beam chemical studies (see Ref. (6) for a good summary). In contrast to long-range diffusion, short-range migrational steps of this type represent integral parts of the surface chemistry and are not merely experimental artifacts of the molecular beam method. It is to the credit of this technique that it is the only one which has thus far been capable of elucidating the fundamental importance of surface transport in many gas–solid reactions. These steps are made visible in modulated molecular beam experiments only by virtue of phase lag information; surface migration to active sites produces no new products nor does it affect the steady-state reaction rate; it simply introduces a time lag in the release of products which could not possibly be detected by any steady-state experiment.

A limitation of the modulated molecular beam method as a tool for surface chemical studies lies in the complexity of the data interpretation aspect of the experiments. Heterogeneous chemical kineticists are used to dealing with reaction rates whose temperature dependence can be understood as one or a combination of familiar exponentially activated terms. A straight line Arrhenius plot is very comforting but is rarely found in modulated beam experimental data (or in most low-pressure chemical kinetic studies, for that matter). In fact, at high temperatures where all surface processes have become rapid, the reaction probability becomes temperature-independent. This result simply means that the entire process is limited by the rate at which reactant gas can be supplied to the surface; the reaction probability is identical to the sticking probability. High-pressure kinetic experiments are rarely supply-limited in this manner because some gas phase mass transfer step invariably intervenes before the kinetic theory supply limit is reached.

The notion that a phase lag can contain valuable surface kinetic information is not one with which chemists are comfortable. Yet in many recent experimental studies by modulated beam methods, the phase lag data have proved to be more valuable in mechanistic interpretation than has the corresponding reaction probability information. This state of affairs arises because the phase lag does not contain the sticking probability (at least for linear reaction mechanisms) and, in contrast to the reaction probability, does not require calibration or estimation of the relative sensitivity of the mass spectrometer for the reactant and product species. Mass spectrometer sensitivity involves the electron impact ionization cross section and the secondary electron emission coefficient of the electron multiplier, neither of which is readily available. The sensitivity of the phase lag to the surface chemical processes means that intuitive interpretation of such data is risky. Only after a detailed analysis of the proposed mechanism and comparison of the data with the theory can one hope to identify the chemical steps responsible for the observed phase lag behavior. However, Schwarz and Madix (5) have made an attempt to codify phase lag and reaction probability behavior characteristic of certain classes of prototypical surface reactions. Application of these rules of thumb to complex surface processes in place of detailed analysis should be avoided, although they may be useful in suggesting likely directions in which to start the hunt for the correct mechanism.

While interpretation of data from conventional steady-state heterogeneous kinetic experiments in terms of nonlinear processes is not particularly forbidding, nonlinearities create considerable interpretation problems in modulated beam experiments. The exact method for analyzing surface mechanisms which include a nonlinear elementary step (28) (such as atom recombination) are so laborious that they have

been used only once (10). To avoid this stumbling block, Schwarz and Madix (5) and Foxon *et al.* (4) suggest an experimental linearization technique which renders nonlinear processes pseudo first order. This is accomplished by dosing the surface with a steady flux of reactant gas so that the modulated beam (consisting of the same reactant gas) presents only a small perturbation in the total reactant impingement rate. This strategem, although it definitely alleviates the data interpretation difficulties, also reduces the sensitivity of the experiment (which may already be poor) by casting the bulk of the reaction product signal into a dc rather than an ac mode. The dc signals are rejected by the lock-in amplifier but contribute instead to the background from which the information-bearing modulated signal must be extracted. As an alternative to this approach, Olander and Ullman (29) have suggested that the theory should be linearized, not the experiment. Although this procedure renders the theory approximate, the errors incurred appear to be acceptably low, especially in view of the tremendous time saving afforded by the ability to apply rapid Fourier expansion techniques to nonlinear systems.

IV. A CASE STUDY—REDUCTION OF UO_2 BY ATOMIC HYDROGEN

The points discussed in the previous section can best be illustrated by following the process by which molecular beam data are utilized to deduce a reaction mechanism. Such considerations are usually not found in a finished report on a modulated beam study but nonetheless are critical to the final choice of a reaction model. Basically, the requirements of a satisfactory reaction model are: (1) It must fit the data better than its competitors. (2) It should have as few adjustable parameters as possible. (3) It should consist solely of elementary steps. (4) The rate constants (e.g., preexponential factors and activation energies) which are obtained from the data must be physically and chemically acceptable.

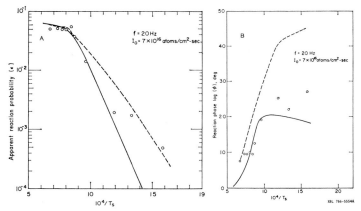

FIG. 2. Effect of solid temperature. (A) Reaction probability; (B) phase lag.

We present here tentative data and a mechanistic interpretation relating to the reduction of uranium dioxide by atomic hydrogen (30). The data will be refined and extended and other mechanisms will be tested before the work is completed.

The reaction of molecular hydrogen and UO_2 is not detectable by molecular beam methods. Hence the reactant gas was partially dissociated thermally in the source prior to impingement on the solid (the dissociation source is described in Ref. (21)). Reduction of UO_2 either to the metal or to an oxide of an oxygen-to-metal ratio less than 2 is accompanied by production of water vapor, which is the volatile product detected by the mass spectrometer. The apparent reaction probability and phase lag of H_2O (relative to reflected H) have been measured as functions of modulation frequency, solid temperature, and H atom beam intensity. The problem is to find a reaction mechanism which best reproduces the observed functional dependence of ϵ and ϕ on ω, T_s, and I_0.

The data are shown in Figs. 2–4. The solid and dashed lines on these plots represent the predictions of two different reaction models which are described below. The data suggest several aspects of the mechanism. First, the supply-limited plateau in Fig. 2A is attained at $T_s \simeq 1300°K$. The decrease in reaction probability at lower temperatures is possibly due to increasing coverage of the surface by adsorbed reactant (or else by product prior to desorption), which decreases the sticking probability of the incident reactant. To account for this behavior, the sticking probability in the theory is permitted to vary as the fraction of open surface, as in the Langmuir model of adsorption–desorption. The presence of a coverage-dependent sticking probability should introduce a nonlinearity into the surface processes only at temperatures where the coverage is comparable to unity. Figures 4A and 4B substantiate this expectation. ϵ and ϕ are independent of I_0 at 1300°K, which is an indication of a linear mechanism at this temperature. At 1000°K, however, the nonlinearity is manifested by distinct dependence of ϵ and ϕ on I_0.

Increasing coverage as temperature is decreased is not the only mechanism of representing a reaction which smoothly moves from nonlinearity to linearity as the temperature is raised. The same effect could be obtained if there were two chemical reactions competing for the adsorbed H atoms. If the H_2O-producing reaction were first order and the competing reaction were higher than first order and possessed a higher activation energy than the water-producing reaction, the same general behavior of the reaction probability and phase

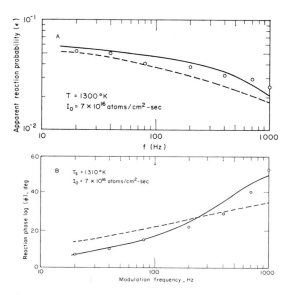

FIG. 3. Effect of modulation frequency. (A) Reaction probability; (B) phase lag.

lag on solid temperature and beam intensity would be observed. The most likely "hidden" reaction is recombination of surface H atoms to form H_2 which cannot be detected because of the large quantity of H_2 in the primary reactant beam.

The slow variation of the phase lag with frequency shown in Fig. 3B has been found in the past (10, 15, 21) to be due to bulk solution–diffusion of reactants or products in the solid. Had a true surface reaction been responsible for the shape of the phase–frequency variation in Fig. 3B, a much different curve would have been found. For example, it is impossible to fit the data on Fig. 3B with the theoretical phase lag formula of Eq. [4]. In the absence of a long-range transport process, phase–frequency data of the type shown in Fig. 3B are also exhibited by branched reaction mechanisms in which the reaction proceeds by two parallel channels with different rate constants (3, 6). This possibility has not been explored for urania reduction.

Having selected bulk solution–diffusion as a part of the mechanism, there remains the

question of the nature of the diffusing species. In the hydrogen–UO_2 system, H, OH, H_2O, or lattice oxygen ions may migrate. There is no a priori way of deciding which of these to choose.

Last, one must select the slow chemical step ultimately responsible for production of water vapor and the location of this step. Following Grabke's (31) interpretation of the reaction of H_2O with FeO, a reasonable mechanism would involve slow reaction of adsorbed hydrogen atoms with lattice oxygen atoms on the surface to produce surface hydroxyl radicals. This slow reaction is followed by rapid attack of the hydroxyl radical by another adsorbed hydrogen atom to produce H_2O, which immediately desorbs.

A different reaction scheme is suggested by the recent conventional kinetic study of UO_2 reduction by H_2 reported by Woodley (32). He found that the rate was proportional to the square root of the H_2 pressure, which suggests that atomic hydrogen is involved in the rate-limiting step. His data also indicate that the rate is independent of the surface-to-volume ratio of the solid. Most of the surface area in the sintered polycrystalline specimens used by Woodley consisted of internal porosity. This peculiar observation implies either that the internal surfaces are ineffectual because of buildup of the reaction product in long pores or that the reaction takes place in the bulk solid rather than on the surface.

We have selected for detailed analysis one reaction model based on a rate-limiting surface reaction and another model involving a rate-limiting bulk reaction. The two models are shown in Table I. The first step is adsorption of atomic hydrogen from the beam. The sticking probability η is assumed to vary linearly

with the fraction of the surface not covered by adsorbed hydrogen. The second step is surface recombination of H atoms to return H_2 to the gas. The third step represents dissolution of surface hydrogen into the bulk followed by diffusion in the UO_2 lattice. The following steps represent the slow chemical process to produce OH. In the bulk reaction model, the OH formed must diffuse to the surface. Step 5, the rapid removal of OH from the surface, is common to both models. In the analysis of these models, a surface mass balance on adsorbed hydrogen atoms (surface concentration denoted by n) provides the boundary condition for the bulk diffusion equation for

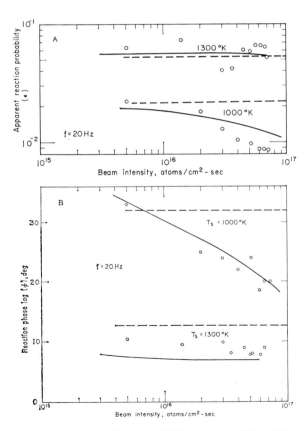

FIG. 4. Effect of primary beam intensity. (A) Reaction probability; (B) phase lag.

TABLE I

Models for UO_2 Reduction by Atomic Hydrogen

1. $H(g) \xrightarrow{\eta} H(ads)$	Coverage-dependent sticking
2. $H(ads) + H(ads) \xrightarrow{k_e} H_2(g)$	Recombination
3. $H(ads) \underset{H, D_H}{\rightleftharpoons} H(soln)$	Bulk solution–diffusion of H
4a. $H(ads) + O(surf) \xrightarrow{k_1} OH(surf)$ or:	Slow step (in surface reaction model)
4b. $H(soln) + O(bulk) \xrightarrow{k_1} OH(soln)$ and: $OH(soln) \xrightarrow{D_{OH}} OH(surf)$	Slow step (in bulk reaction model)
5. $H(ads) + OH(surf) \xrightarrow{fast} H_2O(g)$	Rapid product formation

hydrogen in the UO_2 lattice. A similar diffusion equation is needed for OH in the bulk reaction model.

Surface Reaction Model

The surface mass balance is

$$\frac{dn}{dt} = \eta_0 I_0 \left(1 - \frac{n}{N_s}\right) g(t) - 2k_1 n$$
$$- 2k_e n^2 + D_H \left(\frac{\partial C_H}{\partial z}\right)_{z=0}, \quad [5]$$

where η_0 is the bare surface sticking probability of H on UO_2, I_0 is the amplitude of the modulated hydrogen atom beam and $g(t)$ is the gating function (any periodic function of frequency ω rad/sec). N_s is the density of sites on the UO_2 surface and n/N_s is the fractional coverage of these sites by atomic hydrogen. The first term on the right represents the supply term from the beam. The second and third terms represent removal steps according to reactions 4a and 2, respectively, of Table I. The last term is the Fick's law flux of atomic hydrogen into the bulk UO_2.

The hydrogen concentration in the solid just beneath the surface is assumed to be related to the surface concentration by an equilibrium solubility relation analogous to Henry's law:

$$C_H(0, t) = Hn(t), \quad [6]$$

where H is the solubility constant and $C_H(z, t)$

is the time-dependent concentration distribution of hydrogen in the UO_2. Diffusion of hydrogen in the solid is governed by Fick's second law:

$$\partial C_H/\partial t = D_H(\partial^2 C_H/\partial z^2), \quad [7]$$

where D_H is the diffusion coefficient of hydrogen in UO_2. The boundary condition for Eq. [7] at $z = 0$ is given by Eqs. [5] and [6] and at $z = \infty$:

$$C_H(\infty, t) = \text{finite}. \quad [8]$$

These equations are solved for a periodic gating function of frequency ω by the one-term Fourier expansion method (3, 6, 29). This procedure yields the predicted apparent reaction probability ϵ and phase lag ϕ as functions of the modulation frequency, the beam intensity and the physical parameters of the model. The latter include the bare surface sticking probability η_0 and the chemical rate constants k_1 and k_e. The solution–diffusion process introduces the combined parameter $H^2 D_H$. The site density N_s is considered known and is assigned a value of 10^{15} cm^{-2}. If the three temperature-dependent quantities k_1, k_e, and $H^2 D_H$ follow Arrhenius behavior, each is characterized by a preexponential factor and an activation energy. In all, the theoretical functions $\epsilon(\omega, T_s, I_0)$ and $\phi(\omega, T_s, I_0)$ contain seven adjustable parameters which are obtained by fitting the model to the data of Figs. 2–4 with a digital computer. The solid curves on these plots represent the best fit

TABLE II

Parameters of the UO₂ Reduction Models

Parameter (units)	Surface reaction		Bulk reaction	
	Preexponential	Activation[a] energy	Preexponential	Activation[a] energy
η_0	0.13	—	0.12	
k_1 (sec^{-1})	1.4×10^7	96	4×10^{10}	170
k_e (cm^2/atom-sec)	3×10^{-11}	0	—	—
$H^2 D_H$ (sec^{-1})	7200	0	3×10^6	23
$H^2 D_{OH}$ (sec^{-1})	—	—	2×10^4	−18

[a] kJ/mole.

(obtained by a multivariable least-squares-fitting subroutine) of the data to the surface reaction model. The parameters of the model which produce the best fit are given in Table II.

Bulk Reaction Model

For this case, the surface balance on H(ads) takes the form

$$\frac{dn}{dt} = \eta_0 I_0 \left(1 - \frac{n}{N_s}\right) g(t)$$

$$- D_{OH}\left(\frac{\partial C_{OH}}{\partial z}\right)_{z=0} + D_H \left(\frac{\partial C_H}{\partial z}\right)_{z=0} \quad [9]$$

The next-to-the-last term in this balance is the flux of OH to the surface, which is equal to the rate of removal of H(ads) by reaction 5 in Table I. To keep the number of adjustable parameters the same as in the surface reaction model, the recombination step was neglected. The diffusion equations in the bulk are

$$\partial C_H/\partial t = D_H(\partial^2 C_H/\partial z^2) - k_1 C_H \quad [10]$$

and

$$\partial C_{OH}/\partial t = D_{OH}(\partial^2 C_{OH}/\partial z^2) + k_1 C_H. \quad [11]$$

Because reaction 5 in Table I is assumed to be fast, all OH reaching the surface by diffusion from the interior is immediately removed, and a boundary condition for Eq. [11] is

$$C_{OH}(0, t) = 0. \quad [12]$$

Another boundary condition on C_{OH} far from the surface is similar to Eq. [8] for hydrogen.

The model is treated analytically by the method used for the surface reaction mechanism and the best-fit parameters produce the dashed curves in Figs. 2–4. These parameters are also shown in Table II.

Discussion of the Models

The high reactivity of atomic hydrogen with UO₂ is seen from Table II and Fig. 2A. At high temperatures, about one out of every seven incident H atoms sticks to the surface and returns to the gas phase in an H₂O molecule. of the two reaction models, the surface reaction mechanism fits the data best. The recombination rate constant k_e is so small that it could have been disregarded in the data fitting process.

In addition to fitting the data less well, the bulk reaction fails to reproduce the nonlinearity of the data at $T_s = 1000°K$ (Fig. 4) even though there is a nonlinear term in the mechanism. This occurs because the fitting computation converged on a highly activated OH bulk solution/diffusion parameter which provides the temperature dependence of Fig. 1A. Slow reaction in the bulk controls the overall process in this model.

Although the surface reaction model appears to be superior to the bulk reaction model, the question of uniqueness persists. The theoretical functions $\epsilon(\omega, T_s, I_0)$ and $\phi(\omega, T_s, I_0)$, each containing seven adjustable parameters, may be viewed as surfaces in multidimensional space. The fitting computation tries to find the extremum on this surface which represents the minimum disagreement with the ensemble of the data. Depending upon the goodness of the initial guess of the adjustable parameters of the model, the extremum eventually found by the computation may only be a local extremum. In other words, there may be other sets of seven numbers which provide better fits to the data than those shown in Table II.

Finally, a basic physical process may have been omitted from both of the models tried thus far. In particular, we suspect that the correct assignment of the bulk diffusion species may be neither H nor OH but lattice oxygen. Other variants of the models described above include (1) permitting step 5 in Table I to occur in the bulk rather than on the surface (in this case, H_2O bulk diffusion occurs rather than OH migration); and (2) allowing the coverage dependence in the H atom supply term to be governed by a slow H_2O desorption step. The first of these modifications would probably not provide a much better fit to the data than the original bulk OH diffusion model. The second mechanism adds two more adjustable parameters (to characterize the H_2O desorption rate constant) to be fitted to the data.

V. CONCLUSION

The present state-of-the-art of modulated molecular beam surface chemistry has been described and a case study illustrating the application of the method to a particular system has been presented. The question of the best data acquisition system (i.e., lock-in amplification or pulse counting) has not been settled. The present experimental devices could benefit from ion detectors of higher efficiency than the standard electron multipliers currently employed. Contrary to gas–gas scattering experiments, modern, on-line data processing methods do not appear to be particularly applicable to gas–surface studies. A cross section fitted to polynomials in velocity and angle is still a basic physical property of a gas–gas molecular interaction. A curve fit to a phase lag vs frequency plot in a gas–surface inter-

action study has no predictive or extrapolable qualities whatsoever. The primary interest in gas–surface chemical studies by modulated molecular beams is to elucidate the combination of elementary steps which together constitute the mechanism of a surface reaction. Deduction of a mechanism from the data obtained in a molecular beam experiment often requires considerable chemical ingenuity and is not likely to be aided by automating the experiment.

ACKNOWLEDGMENT

This work was supported by the United States Energy Research and Development Administration.

REFERENCES

1. SMITH, J. N., JR. AND FITE, W. L., *in* "Proceedings of the Third International Symposium on Rarefied Gas Dynamics," Vol. 1, p. 430, 1963.
2. MERRILL, R. P., *Catal. Rev.* 4, 115 (1970).
3. JONES, R. H., OLANDER, D. R., SIEKHAUS, W. J., AND SCHWARZ, J. A., *J. Vac. Sci. Technol.* 9, 1429 (1972).
4. FOXON, C. T., BOUDRY, M. R., AND JOYCE, B. A., *Surface Sci.* 44, 69 (1974).
5. SCHWARZ J. A. AND MADIX, R. J., *Surface Sci.* 46, 317 (1974).
6. MADIX, R. J., *in* "Physical Chemistry of Fast Reactions" (D. O. Hayward, Ed.), Vol. 2, Plenum, New York, 1975.
7. PALMER, R. L. AND SMITH, J. N., JR., *Catal. Rev.* 12, 279 (1975).
8. VISSER, C. A., WOLLESWINKEL, J., AND LOS, J., *J. Phys. E* 3, 483 (1970).
9. HURKMANS, A., OVERBOSCH, E. G., OLANDER, D. R., AND LOS, J., *Surface Sci.* 54, 154 (1976).
10. OLANDER, D. R., SIEKHAUS, W., JONES, R. H., AND SCHWARZ, J. A., *J. Chem. Phys.* 57, 408, 421 (1972).
11. WEST, L. A. AND SOMORJAI, G. A., *J. Vac. Sci. Technol.* 9, 71 (1968).
12. MADIX, R. J. AND SCHWARZ, J. A., *Surface Sci.* 24, 264 (1971).
13. BATTY, J. C. AND STICKNEY, R. E., *J. Chem. Phys.* 51, 4475 (1970).
14. ULLMAN, A. Z. AND MADIX, R. J., *in* "Rarefied Gas Dynamics," p. 381. Academic Press, New York, 1974.
15. ULLMAN, A. Z., ACHARYA, T. R., AND OLANDER, D. R., to be published.

16. BERNASEK, S. A. AND SOMORJAI, G. A., *Surface Sci.* **48**, 204 (1975).

17. ROSNER, D. E. AND ALLENDORF, H. D., *AIAA J.* **6**, 650 (1968).

18. BALOOCH, M. AND OLANDER, D. R., *Trans. Amer. Nucl. Soc.* **22**, 164 (1975).

19. BALOOCH, M., CARDILLO, M. J., MILLER, D. R., AND STICKNEY, R. E., *Surface Sci.* **46**, 358 (1974).

20. MADIX, R. J., PARKS, R., SUSU, A. A., AND SCHWARZ, J. A., *Surface Sci.* **24**, 288 (1971).

21. BALOOCH, M. AND OLANDER, D. R., *J. Chem. Phys.* **53**, 4772 (1975).

22. DALY, N. R., *Rev. Sci. Instrum.* **31**, 264 (1960).

23. GIBBS, H. M. AND COMMINS, E. D., *Rev. Sci. Instrum.* **37**, 1385 (1966).

24. LEE, Y. T., MCDONALD, J. D., LEBRETON, P. R., AND HERSCHBACH, D. R., *Rev. Sci. Instrum.* **40**, 1402 (1969).

25. KRAKOWSKI, R. A. AND OLANDER, D. R., *J. Chem. Phys.* **49**, 5027 (1968).

26. PALMER, R. L. AND SMITH, J. N., JR., *J. Chem. Phys.* **60**, 1453 (1974).

27. ULLMAN, A. Z. AND MADIX, R. J., *High Temp. Sci.* **6**, 342 (1974).

28. OLANDER, D. R., *in* "Structure and Chemistry of Solid Surfaces" (G. A. Somorjai, Ed.), Wiley, New York, 1969.

29. OLANDER, D. R. AND ULLMAN, A. Z., *Int. J. Chem. Kinet.*, VII, 625 (1976).

30. Data obtained by M. Balooch. Data analysis performed by D. Dooley.

31. GRABKE, H. J. AND GALA, A., *in* "Symposium on High Temperature Gas–Metal Reactions in Mixed Environments" (S. A. Jansson and Z. A. Foroulis, Eds.), p. 348. The Metallurgical Society of AIME, 1972.

32. WOODLEY, R. E., "Hydrogen Reduction of Mixed-Oxide Fuel: Reaction Kinetics and Their Application to Fuel Processing," HEDL-TME 75-136, 1976.

The Crystallography of Clean Surfaces and Chemisorbed Species as Determined by Low-Energy Electron Diffraction

J. E. DEMUTH

IBM Thomas J. Watson Research Center, Yorktown Heights, New York 10598

Received July 12, 1976; accepted August 12, 1976

Low-energy electron diffraction (LEED) is a valuable technique for determining not only two-dimensional translational periodicities of surface atoms but also their detailed locations via an analysis of LEED intensities. We outline the experimental diffraction intensity data and various methods of theoretical analysis including dynamical calculations, data averaging, and data inversion methods needed for such structural determinations. The relative merits of the dynamical analysis of either azimuthal intensity distributions at fixed energy, $I(\phi)$, or the energy dependence of the intensity, $I(E)$, to determine surface crystallography are examined via model calculations. Dynamical analysis of the $I(E)$ spectra of the oxygen (2×1) overlayer and the hydrogen (1×2) overlayer structures on Ni(110) are presented. We determine that the oxygen atoms in the (2×1) structure reside in the twofold bridge sites on top of $\langle 1\bar{1}0 \rangle$ rows of the Ni(110) surface and are located 1.92 ± 0.04 Å from the two adjacent Ni atoms. For the case of hydrogen on Ni(110) we find that little agreement is found between calculated and experimental $I(E)$ spectra for simple hydrogen adsorption models. A surface distortion model is found to best describe the experimental data where adjacent $\langle 1\bar{1}0 \rangle$ rows of Ni atoms are alternately attracted or repelled 0.1 ± 0.05 Å together or apart and compressed into the bulk by 0.1 ± 0.05 Å. The chemical trends of these and other recent LEED determinations as well as the future directions of surface crystallography are briefly discussed.

An understanding of gaseous or solid matter to date has largely come from an understanding of its geometric and electronic structure. Although gases and solids have been the object of nearly 100 years of scientific study, only for about the last 15 years have the properties of clean surfaces been carefully studied. This has been largely due to the technology necessary to prepare and maintain surfaces for such studies. In this paper, we discuss the application of low-energy electron diffraction (LEED) to the study of the geometric structure of clean single crystal surfaces and such surfaces with adsorbed atoms or molecules. We hope to provide an overview of the application of this technique to obtain structural information, to provide some examples of its application and finally, to provide a framework for interpreting what such structural information can tell us regarding the surface and for the properties of adsorbed species. Since there have been several recent reviews regarding the topic of this paper, we will present only a rather brief phenomenological discussion of the relevant theory, experimental methods, and the present status of the field.

I. BACKGROUND AND METHODS OF ANALYSIS

A. The Diffraction Phenomena

When electrons with energies, E, between 20 and 1000 eV interact with matter, they are very strongly attenuated. This attenuation restricts the elastically (or inelastically) scattered electrons emitted from the surface to be those nearest the topmost layers of the material. If the wavelength λ ($\lambda \sim (150)^{\frac{1}{2}}/E$ for λ in angstroms and E in electron volts) of a monochromatic, coherent incident electron beam is comparable to the atomic spacings, then the elastic scattering of electrons by each atom in a

197

periodic material gives rise to diffraction via the conservation of energy and momentum within the crystal. Diffracted beams will be observed if the domain size of the crystal is of the same size of larger than the coherence zone of the incident electron beam—typically 500–1000 Å. The directions of the backscattered diffracted beams are determined by the two-dimensional periodicity in the surface region and the electron wavelength via Bragg's Law for a two-dimensional grating. These elastically backscattered electrons can be displayed directly on a spherical fluorescent screen, as in commercially available LEED optics, or they can be measured directly with a Faraday collector. The diffraction patterns so observed are commonly referred to as LEED patterns.

The study of LEED patterns gives direct information regarding the periodicities of the arrangement of surface atoms in the plane of a single crystal surface. Not only can clean crystalline surfaces be studied but also the periodicities of adsorbed atoms or periodic irregularities (steps) on crystalline surfaces can be observed. LEED patterns, however, do not allow detailed determination as to the number of atoms within the two-dimensional surface unit cell or the locations of these atoms relative to the bulk unit cell or "substrate" atoms. To obtain such information the LEED intensities must be analyzed.

The intensity of each diffraction beam is modulated from what one would expect in X-ray diffraction as a result of "weak" momentum conservation perpendicular to the surface as well as by multiple-scattering contributions, both the result of the strong electron scattering. Such multiple scattering renders a direct single-scattering or kinematic analysis of limited value. It is to a detailed analysis of these intensities that much attention has been given.

B. Analysis of LEED Intensities

The intensity variations of each diffraction beam as a function of an experimentally variable parameter, usually energy, contain structural information. Three approaches have been taken to obtain information from the diffraction intensities. First, one can simply attempt to get rid of the nonkinematic features in the diffraction intensities and perform a simple kinematic analysis. Lagally and Webb (6) have shown that the averaging of the intensities of one particular beam over all polar and azimuthal angles for constant-momentum transfer largely gets rid of the nonkinematic features. This allows an analysis of the intensity as a function of electron energy, denoted as $I(E)$ spectra hereafter, in terms of simple kinematic theory. However, as found experimentally (6–11) and also in theoretical calculations (12), one cannot average over a large enough range of angles to remove all the nonkinematic features. Thus some nonkinematic "noise" in the averaged spectra complicates detailed analysis. To surmount the problems of these nonkinematic features, Burkstrand and Kleiman (10) have devised an optimization scheme to compare averaged results of constant-momentum to kinematic theory so as to minimize the subjective nature of such comparisons.

Another approach as proposed by Clark, Mason, and Tescari (13) and others (14–16) is to Fourier invert the experimental intensities to obtain atomic geometries as done in present X-ray analysis. For LEED, numerical problems exist in transforming real experimental data without obtaining spurious noise, not only because of nonkinematic scattering but also because of the small energy range (i.e., reciprocal space) usually accessible experimentally (14–16). Such numerical problems have been largely surmounted for clean surfaces by Adams and Landman (14) by the use of a window function which accounts for many kinematic features, e.g., scattering factors and surface vibrations.

The final technique for analysis of LEED intensity spectra consists of actually calculating an intensity spectrum $I(E)$ or $I(\Theta, \phi)$ within a physically realistic LEED model (1–3), which includes *all* scattering properties and multiple scattering expected in a real surface, and comparing these to experimental data for several assumed model geometries. Although this technique is indirect and somewhat cumbersome, i.e., a structure is proposed and an intensity spectrum is calculated and compared to experiment, it makes use of the great amount of detailed structure in the intensity spectra imposed by the nonkinematic contributions to the scattering intensities.

To summarize the status of these methods, the data reduction or inversion schemes are being developed and tested. The inversion method has been successfully applied to clean surfaces but has yet to be applied to surfaces with chemisorbed species. Results for the averaging approach to date suggest it may be most useful to obtain information regarding clean surfaces (6, 7, 10) and the contents of the unit cell for adsorbed atoms (7). This same approach, if more quantitatively applied as done by Burkstrand and Kleiman, appears in one case to have enabled a determination of the precise location of adsorbed atoms (11). Further tests are under way on other systems to check the viability of this approach in determining adsorbed atom locations. The final and last approach, perhaps the most detailed of all, is the one approach that has been successfully tried and tested for a large number of clean surfaces and surfaces with adsorbed atoms. These clean surfaces include simple metals such as Al or Ti; transition metals such as Ni, Fe, Mo, W, and Pt (17); layered compounds such as MoS_2 (18) and $NbSe_2$ (19); oxides such as ZnO (20) and MgO (21); and III–IV semiconductors such as GaAs (22). The locations of chemisorbed atoms such as O, N, S, Se, or Te have also been determined on much of these same surfaces (17). The application of this method also to the determination of the location of an adsorbed molecule on a Pt surface has been recently completed (23).

Our prognosis for the future application of these techniques to surface structure determination is that although full dynamical calculations have been the most successful to date, and will be exemplified in this paper, the data reduction schemes if perfected, will be the most practical method. This is true not only because of the large computations needed for dynamical calculation but by necessity. The complexity of the many interesting surface structures may prohibit such dynamical calculations, e.g., molecules and/or surfaces with several atoms per unit cell or with a surface unit cell several times larger than that of the bulk. It is possible that computational efficiencies as utilized by Pendry (2) and Tong and Van Hove (3) or a new calculation approach could render more complicated systems tractable. While the accomplishments of dynamical calculations cannot be overlooked, and will likely continue to play an important role in surface structural analysis for many years to come, the merits of such data reduction schemes cannot be overlooked.

C. Experimental Measurements

LEED intensities have been measured by two methods: via the measurement of the intensity of the fluorescent spots on a LEED display system with an optical "spot" photometer (24), or directly with a Faraday collector (6, 9). Photographic recording of the spot intensities has also been used (25). All measurement methods appear to give results similar enough to be useful for structural analysis (9, 25). The systematics of the measurements largely determine the reliability of the measurements in that either stray magnetic fields, surface contaminants, or poorly defined angles can render the data meaningless (24). Another important aspect of the measurements is that the incident electron beam can change or dissociate what one is trying to measure (23, 24). A detailed accounting of experimental procedures is discussed in some detail elsewhere (24).

III. APPLICATION OF THE DYNAMICAL METHOD TO SURFACE CRYSTALLOGRAPHY

The application of dynamical calculations to determine a structure requires that the calculation be correct. This is usually accomplished by checking the calculation against the clean

1×1 surface and perhaps modifying certain parameters used to describe physical processes, such as electron attenuation or Debye temperatures, so as to provide optimal agreement. The scattering potentials used in the intensity calculation are treated as fixed quantities predetermined from either bulk band-structure calculations or from the superposition of atomic charge densities. An important point to be noted is that for LEED structure work at sufficiently high energies ($\gtrsim 30$ eV), the scattering from these potentials is dominated by the ion core electrons and not the valence electrons, thereby permitting surface structure determination without a priori knowledge of the state of valency or the

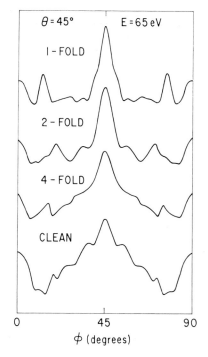

FIG. 2. The (00) beam rotational LEED spectra, $I(\phi)$, at 65 eV for different Na bond sites where the Ni–Na bond length remains constant. The clean surface $I(\phi)$ spectra is also shown. The same scale is used for all plots.

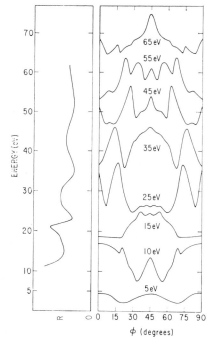

FIG. 1. The (00) beam rotational LEED spectra, $I(\phi)$, for the $c(2 \times 2)$ Na overlayer on Ni(100) for different incident energies. The Na location is that determined in Ref. (37), the fourfold hollow site and $d_\perp \sim 2.23$ Å. The same scale is used for all plots. The $I(E)$ spectrum for the (00) beam for the same angular conditions and $\phi = 0$ is shown at the left.

nature of bonding of the atoms/molecules involved.

In the case of calculations for ordered adsorbed layers, much controversy has arisen as a result of different model calculations giving different results (26–28). The reasons for these discrepancies are now understood and the disagreements have been resolved (28). (We note that an error in computer coding occurred in one case (29).) Other dynamical calculations on adsorbed species, which used different computational methods and model parameters, have given identical structural results (30, 31) to demonstrate that the location of the adsorbed atom is the most crucial parameter in this approach. The calculation method used here is chiefly the layer–KKR method de-

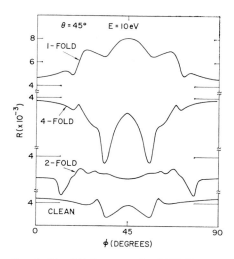

FIG. 3. The (00) beam rotational LEED spectra, $I(\phi)$, at 10 eV for different Na bond sites as in Fig. 2 and the clean surface. The same scales are used for all plots and shown are the actual reflectivities.

for clean Ni(100) and a $c(2 \times 2)$ adsorbed layer of Na on Ni(100) to provide some insight as to these questions.

We consider an incident beam of electrons at an angle of 45° to the sample and calculate the diffraction intensities as a function of azimuthal rotation, ϕ, using the model parameters which gave optimal agreement in the clean Ni and Na overlayer work (37). In Fig. 1 we show the calculated $I(\phi)$ spectra for the (00) beam from 5–65 eV for the $c(2 \times 2)$ Na overlayer on Ni(100). Although there are many

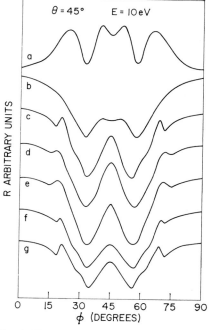

FIG. 4. The sensitivity of the (00) beam rotational spectra at 10 eV to use of various model parameters (see Ref. (37). (a) No reflection condition at the surface with a matching of muffin-tin zeros of Na and Ni potentials; (b) same as (a) but with the Fermi levels characteristic of the Na and Ni potentials set equal; (c) same as (b) but with inclusion of reflections at the surface barrier—the optimal parameters as found in Ref. (37); (d) same as (c) but with barrier plane 0.5 Å closer to surface; (e) same as (c) but with barrier 30% narrower; (f) same as (c) but with the truncation of the Na muffin-tin radius; (g) same as (e) but with less electron adsorption ($\beta = 2.0$ versus 2.5 eV).

veloped by Marcus and Jepsen (32). We shall not discuss the calculational procedures or how various physical effects such as electron attenuation or surface vibrations are included in the LEED model but refer the reader elsewhere (1–3, 32, 33). New results and possible insights to LEED dynamical structural analysis will now be presented.

A. $I(\phi)$ Spectra for Surface Structure Determination

An important question in LEED analysis is what form of intensity data is the most useful, i.e., as a function of energy, $I(E)$, or as a function of incident, $I(\Theta)$, or aximuthal angle, $I(\phi)$, for fixed energy. Rotational spectra $I(\phi)$ have been analyzed (34, 35) but a fundamental question regarding whether this form of the data is more useful or practical than $I(E)$ data (36) remains unanswered. It can be argued that $I(E)$ spectra allow one to see wavelength-related variations of intensity while in rotational spectra more dynamical features are observed; the latter is less readily interpreted. We have performed calculations of $I(\phi)$

interesting features in these spectra which are modulated at different energies we see no obvious relation of these modulations to the original clean surface, the location of the Na atom on Ni, or to the $I(E)$ spectra for the (00) beam also shown in Fig. 1. We note that the notches in many of the $I(\phi)$ spectra can be correlated with the emergence of diffraction beams out of the surface. The sensitivity of the $I(\phi)$ spectra to adsorbate bond site is shown again for the (00) beam for incident electron energies of 65 and 10 eV in Figs. 2 and 3, respectively. Here, we have used the determined Na–Ni bond distance (37) for each bond site. We observe that at 65 eV the general shape of the spectra is not strongly dependent on the bond site, while at 10 eV the spectra are very dependent on the bond site. In examining the sensitivity of the calculated rotational spectra to the details of the matching conditions at the surface (the inner potential of the overlayer versus the bulk, or the surface barrier and its shape), we observe some striking modifications in $I(\phi)$ at 10 eV as shown in Fig. 4. At 65 eV these same matching conditions produce weaker modifications than those shown in Fig. 4 and are comparable to the modifications observed for different site geometries or displacements from the surface. We find similar results for other beams and we expect the same general effects (Figs. 1–4) to be observed at other angles of incidence.

Our general observations regarding $I(\phi)$ spectra for surface structure analysis is that a conflict exists between the sensitivity of the spectra to adsorbate location and the choice of model parameters. That is, at low energies where strong site and displacement sensitivity exists the $I(\phi)$ spectra are most sensitive to the choice of model parameters. At higher energies where the choice of model parameters does not as strongly affect $I(\phi)$ spectra, weak site and displacement sensitivity occurs. Such results certainly suggest uncertainties regarding the usefulness and reliability of the analysis of $I(\phi)$ spectra in determining surface structure.

One useful aspect of such $I(\phi)$ studies, once we have obtained the structure from $I(E)$

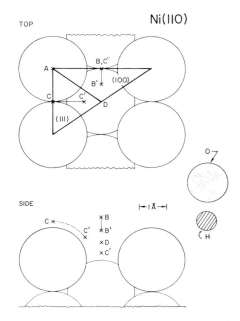

FIG. 5. Top and side views of the crystallography of the Ni(110) surface with high- and low-symmetry sites denoted as described in text. The relative sizes of O, H, and Ni atoms are shown. The Ni(110) surface contains atomic arrangements characteristic of (100) and (111) planes (see text).

analysis, is to use the sensitivity to model parameters as a probe of the detailed properties of the surface boundary conditions. An area of related interest is the angular dependence of the emission of photoexcited and, to a lesser extent, electron emission via the Auger process, where similar angular distributions are being investigated (38–42). Clearly, such largely unknown surface boundary conditions will also influence low kinetic energy photoemission (PE) angular distribution spectra. Such $I(\phi)$ studies would be relevant and useful to PE angular studies and model calculations of the latter. We also expect that the use of PE or Auger angular distributions to obtain bonding geometries may have many of the same limitations as $I(\phi)$ spectra.

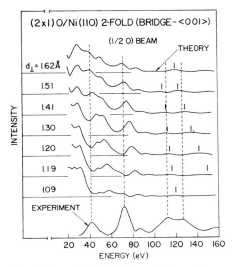

FIG. 6. The calculated $(\frac{1}{2} 0)$ beam $I(E)$ spectra as a function of $d_\perp$ for the bridge site along the $\langle 100 \rangle$ direction and the experimental $I(E)$ spectra of Demuth and Rhodin (34). Both are for normal incidence.

B. The Determination of the Structure of (2 × 1) Oxygen and (1 × 2) Hydrogen on Ni(110)

(2 × 1) Oxygen on Ni(110). The (2 × 1) O LEED pattern is part of an interesting sequence of LEED patterns, i.e., two-dimensional structures, that leads to oxide formation on the Ni(110) surface (24). The experimental $I(E)$ measurements for the (2 × 1) O pattern to which we compare our calculated results were reported and documented thoroughly by Demuth and Rhodin (24). The model parameters used were similar to those reported previously in our structural determination of the $c(2 \times 2)$ O on the Ni(100) system. However, we have used the oxygen potential found by Batra and Robaux (43) for an Ni_4O cluster. The difference between this potential and that used previously (26, 27) is insignificant above ~ 30 eV.

The surface geometry of the (110) surface, in particular its openness, in conjunction with the small size of the oxygen atom, makes it necessary to explore many adsorption sites lower of symmetry than those previously

studied for the Ni(100) surface. For such low-symmetry sites possible domain structures must be taken into account, where necessary, by averaging appropriate beams. In Fig. 5 we illustrate these high- and low-symmetry sites modeled in our calculations. The high-symmetry positions are A (top), B and C (bridged), and D (the fourfold hollows of the Ni(110) surface). Lower-symmetry bonding positions may occur if the adsorbed atom rotates about the twofold positions C → C′, where C′ is a threefold hollow site similar to that on Ni(111) and B → B′, where B′ is similar to the fourfold hollow site of an Ni(100) surface. In addition, below position B another twofold position C″ occurs, similar to C, while position D is analogous to position A, but on the second layer of the crystal. We have examined all these models from $d_\perp \sim 0.2$ Å to 3 Å (the range being site dependent), where $d_\perp$ is defined as the distance between the ion core centers of the first substrate layer and the adsorbate layer. We note that in examining displacements smaller than ~ 0.5 Å, in particular for positions C″ and D, we have collaborated

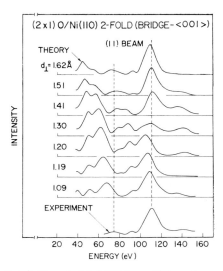

FIG. 7. The calculated (11) beam $I(E)$ spectra as a function of $d_\perp$ for the bridge site along the $\langle 100 \rangle$ direction and the experimental $I(E)$ spectra of Demuth and Rhodin (34). Both are for normal incidence.

with Tong and Van Hove and used their layer-doubling method (3) so as to include 157 beams and allow accurate intensity calculations at the smaller spacings (44).

In Fig. 6 we show the sensitivity of the calculated spectra for the (1/2 0) beam, associated with the (2 × 1) structure, to $d_\perp$ for the bridged bond site C (Fig. 5) and compare with experiment. Such sensitivities to $d_\perp$ and adsorption site are typical of such LEED calculations. We note that if the scattering were kinematic then the spectra shown in Fig. 6 would consist of a rather smooth decreasing curve which would basically reflect the atomic scattering factor of O. The extra structural features clearly visible in the experimental spectra are due to multiple scattering interferences and are the features which we attempt to reproduce in the calculated spectra. For that reason we indicate in Fig. 6 with the dashed vertical lines the experimental peak positions and observe similar positions occurring in the calculation for $d_\perp \sim 1.41$–1.51 Å. It should be noted that in Fig. 6 the decrease in the experimental $I(E)$ spectra below 40 eV (versus the increase in the calculated spectra) could be the result of problems in measuring the $I(E)$ spectra, as this particular diffraction beam is on the fringe of the LEED display optics at these energies. Because of such experimental difficulties, as well as sensitivities of the model calculation to details of the surface boundary conditions and scattering potentials at low energies ($\lesssim 30$ eV), we place more emphasis on the agreement achieved between theory and experiment at higher energies.

In Fig. 7 we show a similar comparison for the same O binding site for the (11) integral order beam which can also be associated with the clean substrate. Here, we find agreement between the calculated spectra for $d_\perp \sim 1.51$ Å consistent with the ($\frac{1}{2}$ 0) beam. In general, the fractional order beams are the most useful for such an analysis, whereas the integral order beams may be rendered less useful by the strong inherent substrate features which they nominally contain. In many cases where the ordering of adsorbed atoms on the surface does not pro-

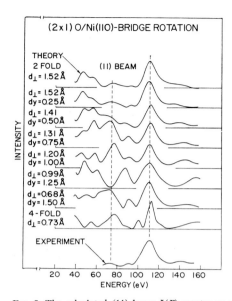

FIG. 8. The calculated (11) beam $I(E)$ spectra as a function of rotation of the bridge-bonded O about the adjacent Ni atoms into the fourfold hollow site. The experimental results of Demuth and Rhodin (34) are also shown. Both are for normal incidence.

vide fractional order beams, i.e., when the unit cell of the adsorbed layer is equal to the substrate, the integral order beams are the only beams available for analysis.

In Fig. 8 we show the changes in the calculated $I(E)$ spectra as the adsorbed atom is rotated about the twofold bridge position C to position C' and also for position D of Fig. 5. Actually we have calculated a series of $d_\perp$ as in Figs. 6 and 7 for many positions along C → C' and here show only the spectra corresponding to the loci of positions with nearly the same Ni–O bond length as found for the twofold bridge site.

In summary, we find that oxygen adsorbed in the (2 × 1) array on Ni(110) provides by far the best agreement with experiment for oxygen atoms lying in the twofold bridge site C for $d_\perp$ between 1.41 and 1.51 Å. We have also examined reconstruction models, i.e., mixed O and Ni atoms in the first layer, but find little or no agreement with experiment for

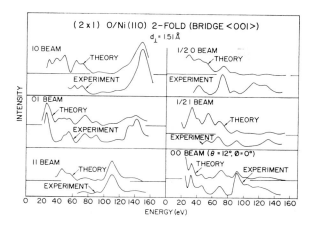

FIG. 9. Summary of agreement between the experimental results of Demuth and Rhodin (34) and calculated $I(E)$ spectra for oxygen in the bridged position along the $\langle 100 \rangle$ direction and $d_\perp \sim 1.51$ Å for normal incidence and for the (00) beam at $\theta = 12°$, $\phi = 0°$.

such models. In Fig. 9 we summarize the agreement achieved between experiment and calculated results for $d_\perp = 1.51$ Å and the bridge bonding site. The overall agreement is sufficiently good, although not quite as good as that found for other cases (26, 27), that we can confidently say the structure has been determined. Efforts have been under way to quantify the level of "agreement" so as to place a confidence scale on structural determinations by this method (27, 45, 46), but they will not be discussed here. An interesting aspect of this determined structure is that the oxygen atom resides atop the row of Ni atoms and not in the hollow (or troughs) as found for O on Ni(100) or (111) surfaces. This particular result will be discussed later.

(1 × 2) Hydrogen on Ni(110). Hydrogen exposure to Ni at room temperature produces intense new diffraction spots characteristic of a (1 × 2) arrangement of atoms on the surface. The nature of the surface structure associated with this LEED pattern has been the subject of much controversy. Namely, as a result of the strong intensity of these new diffraction features, Germer and McRae (47) postulated that H induces Ni atoms to "reconstruct" the surface and give a new (1 × 2) pattern. Bauer

(48) justifiably argued that multiple scattering of electrons between H and the substrate could account for the strong intensities observed. Finally, Tucker (49) suggested that a small distortion in the location of $\langle 1\bar{1}0 \rangle$ rows of substrate atoms, induced by the presence of hydrogen, could also account for the observed diffraction features. Long after these initial interpretations we are now able to calculate diffraction intensities and possibly discriminate between these models. Even though the present computations are restricted to two atoms per surface unit cell, we can treat all models with the limitation that for the Tucker model we cannot include hydrogen.

There is more uncertainty in the choice of model parameters for the case of an adsorbed layer of hydrogen than for the O on Ni case. In particular, we can superimpose atomic hydrogen charge densities to arrive at a suitable scattering potential as done previously, but we cannot take into account the effect of chemisorption on the H(1s) potential. Recent realistic calculations for H on a simple metal (50) suggest that these changes are small and that such an atomic superposition potential for H should be adequate for structure determination. Another, perhaps more serious, question exists regarding the surface vibrations of chemisorbed hydrogen at room temperature: If they are sufficiently large, can these vibrations be adequately represented within the framework of our rigid lattice formalism without the inclusion of more phase shifts (1–3, 33)? Recent inelastic neutron scattering work for hydrogen on Raney Ni suggests a mean square vibrational amplitude $\langle u^2 \rangle$ of $\sim 0.04 \pm 0.02$ Å² (51), and we thus estimate a $\langle u^2 \rangle$ for hydrogen chemisorbed on Ni of 0.02 Å². (We note that LEED work for clean Ni uses $\langle u^2 \rangle \sim 0.0026$ Å², while for adsorbed atoms of sulfur and oxygen $\langle u^2 \rangle$ values of ~ 0.004 Å² are typically used.) Fortunately, changes in the $\langle u^2 \rangle$ only cause overall intensity changes in the $I(E)$ spectra and not changes in the peak position. This poses no problems to a structural analysis with this method as long as geometric sensitivity exists in the calculated $I(E)$ spectra. Interestingly we find that use of our estimated $\langle u^2 \rangle$ of ~ 0.02 Å² provides $I(E)$ spectra with only a weak dependence on $d_\perp$, but it shows clear differences between high-symmetry adsorption sites. (This weak dependence might be expected, as the strong vibrational motion of H "averages out" its scattering contributions to the diffraction intensities over distance related to the mean square vibrational amplitude.) In view of the uncertainties for $\langle u^2 \rangle$ for H on Ni, we have also performed intensity calculations for adsorbed H with $\langle u^2 \rangle \sim 0.004$ Å².

For the simple adsorption models we find that no high-symmetry site provides calculated $I(E)$ spectra which are in even general agreement with the experimental $I(E)$ spectra. As for O on Ni(110) we again have utilized the layer-doubling method and 157 diffraction beams so as to explore $d_\perp$ to a value as small as 0.2 Å. We note that we have not completely explored all the lower-symmetry sites possible for H on Ni(110), as done for O on Ni(110).

In considering reconstruction models we have modeled a (1×2) Ni overlayer on the Ni(110) surface for the Ni atoms on top, bridged, and in the hollow locations on the (110) surface. Again little agreement is found between our calculated spectra and experiment. We do

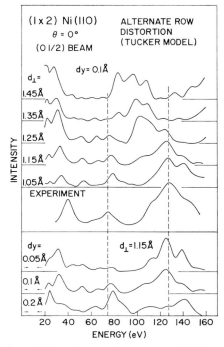

FIG. 10. Comparison of the experimental $(0\frac{1}{2})$ beam $I(E)$ spectra of Demuth and Rhodin (34) for the (1×2) hydrogen structure on Ni(110) with model calculations for an alternate row distortion dy of $\langle 1\bar{1}0 \rangle$ rows as a function of $d_\perp$ (top), and for a fixed $d_\perp$ of 1.15 Å as a function of alternate row distortions dy (bottom).

find that although a (1×2) Ni overlayer on Ni(110) alone does not provide agreement, the addition of a coincident (1×2) hydrogen overlayer slightly improves agreement. We find that for Ni atoms in the hollow position D (with $d_\perp = 1.17$ Å) and with H in the bridge position C (with $d_\perp = 1.92$ Å) some agreement is obtained between theoretical and experimental peak positions. Similarly, if hydrogen is located 0.75 Å directly above the (2×1) Ni overlayer on Ni(110) some agreement is also achieved. However, better agreement exists between yet another structural model and experiment, suggesting that these models do not have the correct structure.

The final surface structure models examined consist of distorting the Ni surface atoms so as

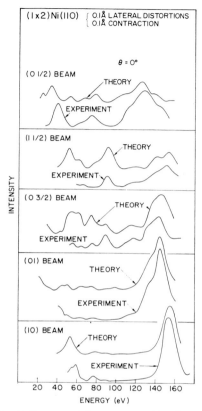

FIG. 11. Summary of the agreement achieved between the experimental results of Demuth and Rhodin (34) for the (1×2) hydrogen structure on Ni(110) and calculated $I(E)$ spectra for an alternate distortion dy of $\langle 1\bar{1}0 \rangle$ rows by 0.1 Å and a 0.1 Å inward contraction of the topmost layer of Ni atoms. Both results are for normal incidence.

to produce a (1×2) periodicity. Models considered were a simple depression or raising of every alternate row of Ni atoms in the $\langle 110 \rangle$ directions, a pairwise distortion of every alternate row in the plane of the surface (the row distortions suggested by Tucker (49)), or a modified Tucker distortion involving rotation as proposed by Taylor and Estrup (52). As a result of the strong scattering of Ni we find that such distortion models produce very distinct and structure-sensitive changes in the $I(E)$ spectra.

We find that of all the models studied the type of model proposed by Tucker (49) describes the experimental data the most satisfactorily. In Fig. 10, we show the typical sensitivity of the calculated $I(E)$ spectra to either a row distortion, dy, for a fixed $d_\perp$ of the top surface layer, or an expansion or contraction of the Ni(110) surface layer from its equilibrium value of ~ 1.25 Å for a fixed row distortion. (The row distortions dy shown are for each row so that the distance between alternate pairs of rows changes by twice this amount.) The optimal agreement occurs for $d_\perp \sim 1.15 \pm 0.05$ Å, i.e., compression of the surface layer of ~ 0.1 Å and a 0.1 ± 0.05 Å alternate displacement of rows in the $\langle 1\bar{1}0 \rangle$ direction. The overall agreement of the calculated $I(E)$ spectra for this structure with experiment is shown in Fig. 11. The agreement, particularly at lower electron energies, is not as good as that found in several previous cases, e.g., O or S on Ni(100) or Ni(110), and it is not clear that a better structural model could be found. Hydrogen has not been included in this model, but from our experience with the effects of an adsorbed hydrogen layer, we do not believe the addition of hydrogen will severely modify the results of Fig. 11, particularly if hydrogen has a large mean square vibrational amplitude on the surface at room temperature. The merits of the agreement shown in Fig. 11, however, cannot be overlooked and we can certainly rule out the simple adsorption or reconstruction models we have examined in favor of the Tucker distortion model.

IV. DISCUSSION AND CONCLUSIONS

The structural determinations presented for O and H on Ni(110) indicate two features of the dynamic approach which deserve comment: first, that the calculated spectra show great sensitivity to the structure or location of the adsorbed atomic specie and, second, that there always exists the question as to whether better agreement can be achieved with a different model. Also, for simple chemisorbed atoms such as O, it is fairly direct to select possible models, whereas as shown for H on Ni(110) more complex models may exist which at present are not possible to calculate $I(E)$ spectra for. It is for such cases that more efficient calculations and data inversion or reduction methods may prove very worthwhile in the future.

As a result of determining the structure we are in a position to examine bond lengths and local geometries to understand better the surface chemistry involved in chemisorption bonding. For (2×1) O on Ni(110) our determination of $d_\perp$ of ~ 1.41 to 1.51 Å gives a bond length of 1.92 ± 0.04 as opposed to 1.88 ± 0.06 for O on Ni(111) and approximately 1.98±0.05 on Ni(100) (26, 27). We note that the summation of metallic radii for the substrate and covalent radii for the adsorbate or a more detailed application of Pauling's rules (53) as carried out by Madukar (54) produces bond lengths in agreement with those found. The differences in bond lengths for O on the (110) versus the (100) and (111) crystal faces as well as similar differences for S (Ref. (26), third paper), can be accounted for by local coordination and valency saturation effects (54).

The most significant aspect of the bonding geometry found for O on Ni(110), as opposed to that found for O on the (100) or (111) surfaces, is that a bridge-bonded specie occurs. All adsorbates (O, S, Se, Te, etc.) studied to date have resided in the highest coordination site on the surface—the site where another substrate atom would sit if the lattice were continued. A minor exception to this rule is O on W(110) (45) where the hollow site is preferred but it is not the hollow site which would be filled if the lattic were to be extended. The fact that bridge bonding is preferred suggests that O bonds to two adjacent Ni atoms via O p_x and p_y atomic orbitals. Substrate t_{2g} d-orbitals (55) protruding upward from the surface (through site A in Fig. 5) may form bonds to these p_x and p_y orbitals as Ni d-orbitals are generally believed to play an important role in bonding. In fact, if we consider that the maximum in d-orbital charge density is located about 0.35 Å (56) in t_{2g} orbitals above each Ni atom, these "lobes" prescribes a bridge

bond angle of $\sim 96°$, very close to what one expects on the basis of X_2O compounds (53). Although such modeling of bonding is crude, Since we do not know the precise wavefunctions at the surface or the role of more diffuse Ni s-electrons, a very simple and consistent chemical viewpoint of the geometry arises. We note that the local bonding geometry of O in NiO, as well as in a metastable Ni–O phase observed by Alessandrini and Freedman (57), is completely consistent with our determined bridge-bonding site. We also note that the results of ion-scattering work by Heiland and Taglauer (58) on the (2×1) O on the Ni(110) system is also consistent with our determined structure.

Such a bridge-bonding mode on Ni(100) and (111) surfaces may also exist even though bridge-bonding positions are not found. This postulate is more plausible if one considers that a rotation of a bridge-bonded species toward the high-symmetry sites (analogous to $C \rightarrow C'$ in Fig. 5) places the atom to within ~ 0.06 Å of the determined locations on (100) and (111) surfaces. One might then envision an additional bonding interaction after such a rotation—perhaps even the formation of resonance bridge bonds between the oxygen atom and pairs of Ni atoms. Thus, for O on Ni(110) we find an exception to the general trends found on other Ni surfaces that may provide further insight into the chemistry involved in oxygen bonding.

From our analysis of the hydrogen-induced features occurring on Ni(110) we cannot determine the location of chemisorbed hydrogen, but we find that a distortion in the locations of surface Ni atoms is likely. This apparent distortion of the lattice may be significant in that the attractive or repulsive forces involved in this surface distortion may play a role in bulk hydrogen diffusion into metals, hydride formation and the hydrogen embrittlement problem. Similar substrate distortion for other adsorbates may be of importance for corrosive processes at surfaces. For O on Ni(110), distortions of the top layer of Ni atom positions were considered so as to improve the agreement between theory and experiment shown in Fig. 9. However, in examining the simple distortion of moving the top Ni layer inward or away from the bulk by 0.05 and 0.1 Å for O on Ni(110), we have not found any significant improvement in the agreement. Similarly, we also find no evidence of such a subsurface expansion for the case of O in Ni(100). An exception to the lack of substrate distortion observed for O on Ni, is the case recently found for a (1×1) O structure on Fe(100) by Legg and Jona (59). Here, they find the same bond site as that on Ni(100) and a similar bond length, *but* a 0.1 Å displacement of the top layer of Fe atoms away from the bulk. Such distortions are likely related to oxide formation and corrosive oxidation problems in this same material.

Finally, another notable case worth mentioning is a recent LEED structural determination of the locations of the carbon atoms for acetylene molecules adsorbed on Pt(111) by Kesmodel et al. (23). They find two preferred locations for the carbon atoms on the surface. In one, the carbon atoms reside near adjacent threefold hollows (i.e., the molecule strattles the bridge position), and in the other both carbon atoms are equally near *one* threefold hollow site (i.e., the molecule sits over one of the hollows). Although the carbon–carbon bond distance could not be clearly resolved to be either 1.2 or 1.4 Å and the hydrogen atoms and locations were neglected, the location of the carbon atoms of the molecule relative to the surface rules out many suggested bonding modes of acetylene to Pt(111).

In conclusion, LEED intensity analyses are currently providing a broad range of fundamental information regarding the crystallography of the clean surfaces as well as surfaces with chemisorbed species. A general conclusion of such structural work to date is that chemical bonding of foreign atoms to the surface appears similar to that occurring in the bulk, differing only in terms of the local coordination. That is, the reduced coordination on the surface produces bond lengths slightly smaller than those found in bulk compounds.

To reiterate an introductory remark, surface crystallographic information is complemented by valence orbital electronic structure studies (60, 61) such that the determination of both crystallographic and electronic structures will greatly further our understanding of the chemistry of surfaces.

ACKNOWLEDGMENT

The author wishes to acknowledge many useful discussions with P. M. Marcus and D. W. Jepsen as well as their collaboration in determining many of the structures cited.

REFERENCES

1. STROZIER, J. A., JR., JEPSEN, D. W., AND JONA, F., *in* "Surface Physics of Materials" J. M. Blakely, Ed.), pp. 1–77. Academic Press, New York, 1975.
2. PENDRY, J. B., "Low-Energy Electron Diffraction Theory." Academic Press, London, 1974.
3. TONG, S. Y., "Progress in Surface Science" (S. G. Davison, Ed.), Vol. 7, Part 2, Pergamon, Oxford, 1975.
4. BAUER, E., *in* "Techniques of Metals Research" (R. F. Bunshah, Ed.), Vol. II, Part 2, Interscience, New York, 1969.
5. ESTRUP, P. J., AND McRAE, E. G., *Surface Sci.* 25, 1 (1971).
6. LAGALLY, M. G., NGOC, T. C., AND WEBB, M. B., *Physics. Rev. Lett.* 26, 321 (1971); NGOC, T. C., LAGALLY, M. G., AND WEBB, M. B., *Surface Sci.* 35, 117 (1973); WEBB, M. B., AND LAGALLY, M. G., *Solid State Phys.* 28, 301 (1973).
7. LAGALLY, M. G., BUCHHOLZ, J. C., AND WANG, G. C., *J. Vac. Sci. Technol.* 12, 213 (1975).
8. McDONNELL, L., WOODRUFF, D. P., AND MITCHELL, K. A. R., *Surface Sci.* 45, 1 (1975).
9. BURKSTRAND, J. M., KLEIMAN, G. G., ARLINGHAUS, F. J., *Surface Sci.* 46, 43 (1974).
10. KLEIMAN, G. G., AND BURKSTRAND, J. M., *Surface Sci.* 50, 493 (1975).
11. BURKSTRAND, J. M., KLEIMAN, G. G., TIBBETTS, G. G., AND TRACY, J. C., *J. Vac. Sci. Technol.* 13, 291 (1976).
12. DUKE, C. B., AND SMITH, D. L., *Phys. Rev. B* 5, 4730 (1972).
13. CLARK, T. A., MASON, R., AND TESCARI, M., *Surface Sci.* 30, 553 (1972).
14. LANDMAN, U., AND ADAMS, D. L., *J. Vac. Sci. Technol.* 11, 195 (1974); ADAMS, D. L., AND LANDMAN, U., *Phys. Rev. Lett.* 33, 585 (1974); *J. Vac. Sci. Technol.* 12, 260 (1975); *Surface Sci.* 51, 149 (1975).

15. Buchholz, J. C., Lagally, M. G., and Webb, M. B., *Surface Sci.* **41**, 248 (1974).

16. Woodruff, D. P., Mitchell, K. A. R., and McDonnell, L., *Surface Sci.* **42**, 355 (1974).

17. See Ref. (1) for a compilation of systems and further references.

18. Mrstik, B. J., Tong, S. Y., Kaplan, R., and Gangzely, A. K., *Solid State Commun.*, to appear.

19. Mrstik, B. J., Kaplan, R., Reinecke, T. L., Tong, S. Y., and Van Hove, M. A., *Bull. Amer. Phys. Soc.* **20**, 406 (1975).

20. Lubinsky, A. R., Duke, C. B., Chang, S. C., Lee, B. W., and Mark, P., *J. Vac. Sci. Technol.* **13**, 189 (1976); *Surface Sci.*, to appear.

21. Leg, K. O., Kinniburgh, C. G., and Prutton M., *Bull. Amer. Phys. Soc.* (*II*) **19**, 233 (1974).

22. Lubinsky, A. R., Duke, G. B., Lee, B. W., and Mark, P., *Phys. Rev. Lett.* **36**, 1058 (1976).

23. Kesmodel, L. L., Stair, P. C., Baetzold, R. C., and Somorjai, G. A., *Phys. Rev. Lett.* **36**, 1316 (1976).

24. Demuth, J. E., and Rhodin, T. N., *Surface Sci.* **42**, 261 (1974); **45**, 249 (1974).

25. Stair, P. C., Kaminska, T. J., Kesmodel, L. L., and Somorjai, G. A., *Phys. Rev. B* **11**, 623 (1975),

26. Demuth, J. E., Jepsen, D. W., and Marcus, P. M.. *Solid State Commun.* **13**, 1311 (1973); *Phys. Rev. Lett.* **31**, 540 (1973); **32**, 1182 (1974); Van Hove, M., and Tong, S. Y., *J. Vac. Sci. Technol.* **21**, 230 (1975).

27. Marcus, P. M., Demuth, J. E., and Jepsen, D. W., *Surface Sci.* **53**, 501 (1975).

28. See, for example, Rhodin, T. N., and Tong, S. Y., *Physics Today*, p. 23, October 1975.

29. C. B. Duke recently informed the author of a coding error in one part of their program. This has been corrected and now provides agreement with our O and S on Ni(100) (Ref. (26)) results. These corrections have been included in all subsequent work.

30. Hutchins, B. A., Rhodin, T. N., and Demuth, J. E., *Surface Sci.* **54**, 419 (1976).

31. Van Hove, M., Tong, S. Y., and Stoner, A., *Surface Sci.* **54**, 259 (1976).

32. Jepsen, D. W., Marcus, P. M., and Jona, F., *Phys. Rev. B* **6**, 3864 (1972).

33. Demuth, J. E., Marcus, P. M., and Jepsen, D. W., *Phys. Rev. B* **11**, 1460 (1975).

34. Bandoing, R., Debersuder, L., Ganbert, G., Hoffstein, V., Lauzier, J., Taub, H., *J. Vac. Sci. Technol.* **9**, 634 (1972).

35. Laramore, G. E., *Phys. Rev. B* **6**, 2950 (1972).

36. Group D'etude des Surfaces, *Surface Sci.* **48**, 497 (1975).

37. Demuth, J. E., Jepsen, D. W., and Marcus, P. M., *J. Phys. C: Solid State Phys.* **8**, L25 (1975).

38. Leibsch, A., *Phys. Rev. Lett.* **32**, 1203 (1974).

39. Rogers, D. L., and Fong, C. Y., *Phys. Rev. Lett.* **34**, 660 ((1975).

40. Rowe, J. E., Traum, M. M., and Smith, N. V., *Phys. Rev. Lett.* **33**, 1333 (1974).

41. Noonan, J. R., Zehner, D. M., and Jenkins, L. H., *J. Vac. Sci. and Technol.* **13**, 183 (1976).

42. Woodruff, D. P., *Surface Sci.* **53**, 538 (1975).

43. Batra, I. P., and Robaux, O., *Surface Sci.* **49**, 653 (1973).

44. At smaller spacings more beams must be considered to obtain accurate $I(E)$ spectra. With the maximum number of beams (i.e., 116 beams) that could be used in the Layer–KKR program (Ref. (32)) we could only examine $d_\perp$ to 0.5 Å without uncertainties. We have thus used the layer doubling method and more beams to check the KKR results and go to $d_\perp \sim 0.2$ Å.

45. Van Hove, M., and Tong, S. Y., *Phys. Rev. Lett.* **35**, 1092 (1975).

46. Zanazzi, E., and Jona, F., to be published.

47. Germer, L. H., and McRae, A. U., *J. Chem. Phys.* **37**, 1382 (1962).

48. Bauer, E., *Surface Sci.* **5**, 142 (1966).

49. Tucker, C. W., quoted as a private communication by G. Ehrlich, *Annu. Rev. Phys. Chem.* **17**, 295 (1966); *Surface Sci.* **26**, 311 (1971).

50. Land, N. L., and Williams, A. R., *Phys. Rev. Lett.* **34**, 531 (1976).

51. Stockmeyer, R. Conrad, H. M., Renouprez, A., and Fouilloux, P., *Surface Sci.* **49**, 549 (1975).

52. Taylor, T. N., and Estrup, P. J., *J. Vac. Sci. Technol.* **11**, 244 (1974).

53. Pauling, L., *in* "The Nature of the Chemical Bond," 3rd ed., Cornell Univ. Press, Ithaca, N. Y., 1960.

54. Madhukar, A., *Solid State Commun.* **16**, 461 (1975).

55. See for example, Bond, G. C., *Faraday Soc. Discuss.* **41**, 200 (1966).

56. Williams, A. R., private communication; see, for example, Janak, J. F., Moruzzi, V. L., and Williams, A. R., *Phys. Rev. B* **12**, 1257 (1975).

57. Alessandrini, E. I., and Freedman, J. F., *Acta Crystallogr.* **16**, 54 (1963).

58. Heiland, W., and Taglauer, E., *J. Vac. Sci. Technol.* **9**, 620 (1972).

59. Legg, K. O., and Jona, F., *J. Phys. C: Solid State Phys.* **8**, L492 (1975).

60. Feuerbacher, B., and Willis, R. F., *J. Phys. C: Solid State Phys.* **9**, 169 (1976).

61. Demuth, J. E., and Eastman, D. E., *J. Vac. Sci. Technol.* **13**, 283 (1976), and references therein.

210

The Characterization of Surfaces by Electron Spectroscopy

F. J. SZALKOWSKI

Science Center, Rockwell International, Thousand Oaks, California 91360

Received July 6, 1976; accepted October 25, 1976

An introduction to the surface-sensitive electron spectroscopic techniques of X-ray photoelectron spectroscopy (XPS or ESCA) and Auger electron spectroscopy (AES) is presented. These techniques provide information concerning the chemical composition and electronic properties existing within the first few monolayers of a free surface. A brief description is given of the energetics of the physical processes involved, the experimental techniques used, and the degree of surface sensitivity attainable. The extent to which XPS and AES can provide qualitative analysis, quantitative analysis, and chemical bonding information is discussed, XPS proving to be generally superior because of the greater simplicity of the process and the higher energy resolution attainable. AES possesses the advantages of higher two-dimensional resolution and greater speed of analysis than XPS.

INTRODUCTION

The meteoric resurgence of interest in surface studies during the past decade has, in no small way, been spurred on by the development of electron spectroscopic techniques to investigate the chemical and physical properties of surfaces. This overview of electron spectroscopy will concern itself with two of the most widely used and powerful of these techniques, X-ray photoelectron spectroscopy (XPS, also known as ESCA: electron spectroscopy for chemical analysis) and Auger electron spectroscopy (AES). Of course the observation of X-ray photoemission and of Auger electron emission is not recent—pioneering experiments were performed back in the 1910's and 1920's, respectively (1, 2)—but their effective application to the study of solid surfaces awaited the development of an ultrahigh vacuum ($< 10^{-8}$ Torr) technology.

Completely definitive studies of surface phenomena require that we obtained detailed information about the atomic surface structure and the chemical composition of a thin surface layer, ideally one not greater than two monolayers thick. Provided that single crystals are available, low-energy electron diffraction (LEED) is well suited to addressing the structural question (3). The chemical compositional problem may be attacked by electron spectroscopic methods. Implicitly included in the term chemical composition is a knowledge of the qualitative composition of the surface layer, of the concentration of each atomic species present, and of the chemical bonding state(s) that each is in. These methods are not restricted by the single-crystal boundary condition and are able to provide information on atomically heterogeneous systems. In addition to providing this basic compositional information, it is desirable that the techniques (1) produce little or no perturbation of the surface conditions during the course of the experiment, (2) acquire information rapidly enough that the experiment may be concluded before appreciable surface contamination from residual ambient gases occurs, and (3) achieve a reasonable degree of resolution in the two-dimensional plane of the surface.

Of course no single technique is superior in meeting each of these requirements. Qualitative data is approximately equally accessible by both XPS and AES. Due to the greater simplicity of the mechanisms involved and its inherently greater energy resolution, XPS yields somewhat more quantitative information than AES and mirrors chemical bonding effects

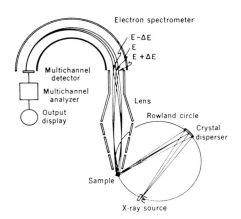

FIG. 1. An XPS spectrometer system employing monochromatization of the incident beam.

more readily. Also, it disturbs the system which it is probing to a lesser degree than AES largely because of the smaller incident beam fluxes involved. On the other hand, an AES spectrum can be recorded more rapidly and with greater spacial resolution, thereby facilitating the observation of transient and of localized phenomena.

The following discussion will limit itself to a brief description of the methodology and energetics involved in these two techniques, an indication of their surface sensitivity, and the presentation of a number of examples demonstrating their application to surface-related problems. A number of more extensive reviews (4–12) are available on the subject of electron spectroscopy and can provide more detailed information on the subject to the interested reader.

TECHNIQUES AND ENERGY RELATIONSHIPS

The technique of XPS is illustrated in Fig. 1. The sample is exposed to a beam of soft (<2 keV) X-radiation, preferably monochromatic in order to eliminate satellite peaks and the Bremsstrahlung, and those electrons which absorb a photon and are emitted from the sample are velocity-analyzed in a magnetically shielded, high-resolution spectrometer,

usually of the hemispherical variety. Because of the low photon flux densities involved ($\approx 10^7$–10^8 photons/cm² sec), electron counting techniques must be employed. The photoelectric process is schematically shown on the left-hand side of Fig. 2 and the kinetic energies of the photoemitted electrons in the vacuum are determined by the energy conservation relation

$$\hbar\omega = E_{W_0} + \Phi_C + E'_{KIN}, \qquad [1]$$

and therefore

$$E'_{KIN} = \hbar\omega - E_{W_0} - \Phi_C, \qquad [2]$$

where $\hbar\omega$ is the energy of the incident photon, E_{W_0} is the binding energy of a given core level, Φ_C is the work function of the crystal, and E'_{KIN} is the kinetic energy of the emitted electron. Since the sample and the spectrometer are typically in electrical contact, this implies that their Fermi levels are aligned. Due to the difference in work functions between the two, a contact potential, Φ_{CP}, exists and the emitted electron undergoes a slight velocity change so that the kinetic energy actually measured turns out to be independent of the sample's work function (13). That is,

$$E_{KIN} = \hbar\omega - E_{W_0} - \Phi_A, \qquad [3]$$

where Φ_A is the work function of the analyzer.

The photoelectric process may be looked upon as a precursor to the emission of Auger electrons. In fact any process which creates an electron vacancy in an inner shell will give rise to Auger emission. Since the atom is in an excited state, it will decay to a lower energy state by means of an electronic rearrangement process. There are two major competing processes which dissipate the energy released in the de-excitation process when an electron from a higher energy level drops down to fill a vacancy. One of these is X-ray fluorescence, the energy being emitted as a photon. The second possible process is Auger electron emission. This is usually described in terms of the Wentzel two-electron interaction mechanism (14). In this case, the energy released by the electronic transition is transferred, by

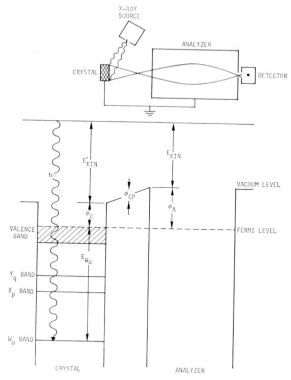

FIG. 2. Schematic diagram illustrating the modification of the kinetic energy of an emitted photo-electron by the sample crystal-analyzer contact potential, Φ_{CP}.

Coulombic interaction, to another bound electron which may be ejected into the vacuum leaving behind a doubly ionized atom. In Fig. 3, the Auger mechanism of de-excitation is illustrated in a schematic diagram of the electronic band structure of a typical metallic solid of atomic number Z. The energy bands have been designated by the generalized notation W_0, X_p, and Y_q with the respective mean energies $-\bar{E}_{W_0}(Z)$, $-\bar{E}_{X_p}(Z)$, and $-\bar{E}_{Y_q}(Z)$ relative to the Fermi level. In the figure a vacancy has been assumed to have been produced in the W_0 band and the de-excitation electron from the X_p band interacts with a Y_q-band electron to produce the Auger electron which is usually designated by the $j-j$ coupling process label $W_0 X_p Y_q$. The kinetic energy in the vacuum of the emitted electron is

given by an expression of the form

$$\bar{E}_{W_0 X_p Y_q}(Z) = [\bar{E}_{W_0}(Z) - \bar{E}_{X_p}(Z)]$$
$$- \bar{E}_{Y_q}(Z^*) - \Phi_C, \quad [4]$$

where the term $\bar{E}_{Y_q}(Z^*)$ is used to indicate that the Auger electron is emitted from a doubly ionized atom.

As with the photoemission process, electronic relaxation and correlation effects have to be taken into consideration to provide a detailed picture of the energetics involved (12) but a discussion of these lies outside the scope of this paper.

The Auger electrons are then energy analyzed using either a velocity analyzer of either the hemispherical or cylindrical mirror type or a retarding field energy analyzer, such as that shown in Fig. 4. The retarding field

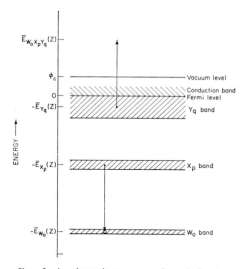

FIG. 3. A schematic representation of the Auger electron emission process from a metallic solid containing an initial electron vacancy in the W_0 energy band, an X_p-band electron undergoing de-excitation to fill the vacancy, and a Y_q-band electron being ejected from the sample.

analyzer has the advantage of doubling as a LEED optics; the velocity analyzers have a greater dynamic range so the results may be displayed on an oscilloscope and transient events observed. The core-level vacancies which give rise to Auger events are generally created by high-energy (3–10 keV) electron bombardment. This allows the use of analog data collection techniques but results in a large background so that the derivative of the energy distribution, as shown in the lower half of Fig. 5, is the quantity generally reported.

SURFACE SENSITIVITY

The question naturally arises as to how thick a surface layer is probed by electron spectroscopic techniques. The answer is indicated in the curve shown in Fig. 6, which graphs the observed variation in the mean free path of an electron in various solid materials vs its kinetic energy. Since it is the energy of the photoelectron and of the Auger electron that is the

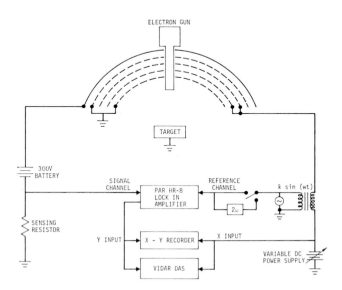

FIG. 4. Schematic diagram of the retarding field energy analyzer and its supporting electronics when used for Auger electron spectroscopy.

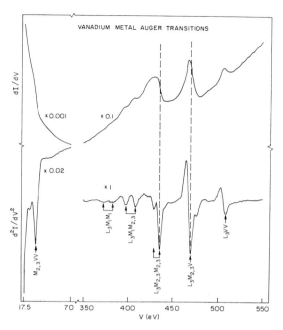

FIG. 5. The energy distribution (dI/dV) and its derivative (d^2I/dV^2) Auger spectra of a vanadium metal (100) surface.

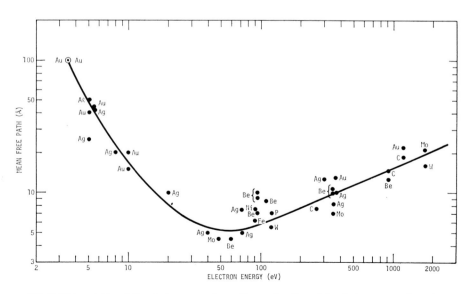

FIG. 6. The variation of the electron mean free path as a function of the electron energy in various solid materials.

215

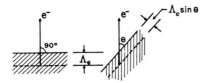

Fɪɢ. 7. Schematic illustration of the surface enhancement achieved for low angles of electron emission with respect to the surface plane. (Λ_e is the electron mean free path and θ is the emission angle relative to the surface.)

limiting factor in determining the mean escape depth (and not the penetration depth of the incident X-ray or electron beam), it is seen that for the 50–2000 eV electrons involved in XPS and AES the escape depths vary in the range of 5–30 Å.

As Fadley (15) and Frazer et al. (16) have demonstrated, it is also possible to increase the surface sensitivity by roughly an order of magnitude by selectively analyzing those electrons which are emitted at small angles with respect to the solid surface. This effect is shown in Fig. 7 and it relies on the increased path length for escape from a given depth below the surface as the exit angle is decreased. If the mean free path is defined as Λ_e, a consideration of the figure then indicates that the average depth of emission for an arbitrary exit angle θ will be $\Lambda_e \sin\theta$.

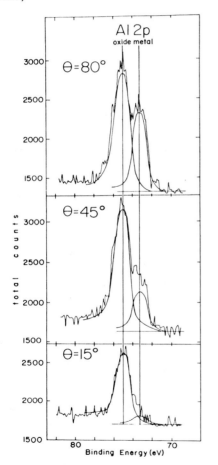

Fɪɢ. 8. Al $2p$ core-level spectra from an aluminum specimen with oxide- and carbon-containing surface layers as a function of the electron emission angle.

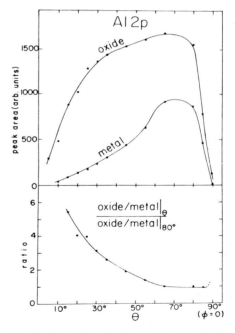

Fɪɢ. 9. Al $2p$-oxide and Al $2p$-metal angular distribution intensities and their ratio (normalized to $\theta = 80°$) from the specimen of Fig. 8 as a function of the electron emission angle.

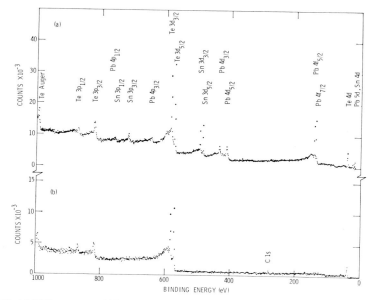

Fig. 10. (a) XPS spectrum of $Pb_{0.62}Sn_{0.38}Te$ specimen which had been cleaned by sputtering. Major photoelectron lines are identified; (b) XPS spectrum of PbTe specimen which had been etched in 20% HNO_3 for 98 min at 20°C (surface was *not* sputter cleaned).

An example of this effect is seen in Fig. 8, which shows the Al $2p$ peaks (≈ 1410 eV kinetic energy) observed on a slightly oxidized Al sample (17). The doublet peak results from the existence of both Al_2O_3 and metallic Al within the escape depth of the sample, the chemical shift between the Al in these two different oxidation states being readily observed. As the takeoff angle into the spectrometer is varied away from the surface normal, the relative enhancement of the peak associated with the oxide overlayer is evident. Figure 9 graphs the dramatic change in the oxide/metal peak intensity ratio (normalized to that observed at $\theta = 80°$) as a function of takeoff angle. A factor of 5 intensity-ratio increase is observed even for these relatively high kinetic energy electrons. In more favorable cases, therefore, one can legitimately speak of sampling only the first atomic layer of the surface. The obvious tradeoff parameter of reduced peak intensity per unit time is also shown in the figure, longer sampling times being necessary to acquire equal data statistics as the takeoff angle is reduced toward grazing.

APPLICATIONS OF XPS

(1) Qualitative Analysis and Chemical Bonding State Information

The earlier description of the XPS mechanism showed that the kinetic energy of a photoelectron is a function of the atomic energy levels; therefore, after suitable calibration with known standards, one can identify the existence of an element by correlating positions and relative intensities of the peaks in a measured energy spectrum with one's data base. Also, the energy of an electron emitted from a given core level varies slightly depending upon the chemical environment of the atom (i.e., whether the element is in its pure state or part of a compound). The size of this chemical shift changes with each particular environment and its roots lie in the variation of the atomic screening parameters and the consequent changes in the relaxation

217

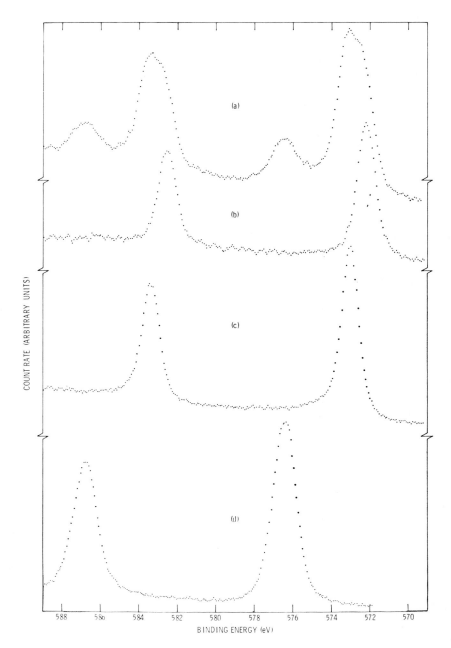

FIG. 11. XPS spectra in region of Te *3d* lines of (a) PbTe which had been etched in $\approx 17\%$ H_2SO_4 for 10 min at 35°C, (b) PbTe (sputter cleaned), (c) Te (sputter cleaned), (d) TeO_2.

energy of the system as the surrounding electron orbitals collapse toward the newly created electron vacancy. These two concepts are well illustrated in the following example.

Because its bandgap is a continuously variable function (through zero) of its alloy composition, the $Pb_{1-x}Sn_xTe$ system is useful in constructing infrared detectors. Unfortunately, such compound semiconductor materials frequently exhibit marked surface leakage currents across a heterojunction interface, thereby limiting device performance. The electrical passivation of the surface represents a substantial problem, due at least in part to the increased complexity of surface chemistry which can occur in a multicomponent system. If a correlation of surface chemistry with device performance could be obtained, one can hope to reduce surface leakage problems by a systematic method. To this end, Grant and co-workers (17) undertook an investigation of the surface leakage problem in $Pb_{1-x}Sn_xTe$ devices as a function of chemical etching procedures using XPS.

Since it is often difficult to obtain completely reproducible surface properties on $Pb_{1-x}Sn_xTe$ materials even when these materials are processed by supposedly identical procedures, electrical and surface chemical measurements were performed on a single surface with the hope of partially avoiding this problem. Because the spatial resolution of the XPS technique is poor (typically 1 mm² is analyzed), large area $Pb_{1-x}Sn_xTe/PbTe$ diode arrays were fabricated. The array contains rows of equal numbers of large and small diodes that differ in area by roughly an order of magnitude and there are two large areas to facilitate XPS measurements; one of these areas is a PbTe surface while the other is $Pb_{0.8}Sn_{0.2}Te$. The entire surface was processed simultaneously so that the diode surfaces and the large area surfaces should have the same average surface chemistry. The diode leakage currents were measured at a small reverse bias and it was assumed that this current could be expressed as

$$i = SP + BA, \qquad [5]$$

where S is the surface leakage parameter (current/length), P is the junction perimeter, B is the bulk leakage parameter (current/area), and A is the junction area. P and A are defined by the etched mesa sizes, and S and B can be obtained by solving the two simultaneous equations generated because of the two different size diodes.

Figure 10 contains low-resolution survey spectra (0–1000 eV binding energy) of a $Pb_{0.62}Sn_{0.38}Te$ alloy surface which was cleaned by argon ion sputtering (curve a) and of a PbTe surface which had been etched in a 20% HNO_3 solution (curve b). The core level photolines of Pb, Sn, and Te which fall in this energy range are readily visible in Fig. 10a. The HNO_3 etch on PbTe has obviously depleted the surface of Pb.

Figure 11 contains high-resolution spectra in the vicinity of the Te $3d$ peaks of (a) a PbTe surface which has been etched in a H_2SO_4 solution, (b) an ion-bombardment cleaned PbTe surface, (c) a clean Te surface, and (d) a TeO_2 surface. The chemical shifts between the elemental Te surface and Te in PbTe and in TeO_2 are quite evident. Also, it is clear that the H_2SO_4 etched surface contains approximately equal amounts of elemental Te^0 and Te^{2-} and a lesser amount of Te^{4+}, an excellent example of the ability of XPS to observe different oxidation states on a heterogeneous surface.

Keeping these data in mind, let us return to the surface passivation problem. Figure 12a shows another PbTe surface which was etched in a 30% HNO_3 solution, which we have seen produces a Te-rich surface. The left-hand side of the graph shows the Te $3d_{\frac{5}{2}}$ peak and the right-hand side of Pb $4f_{7/2}$ peak. Using the data presented in the previous figure, it is seen that the etched surface is completely composed of elemental Te within the detected volume. As one removes the Te^0 layer by ion sputtering the Te peak first broadens (concurrently with the appearance of the Pb photoline) as the Te^{2-} of PbTe is encountered and then centers itself about the energy value associated with pure PbTe.

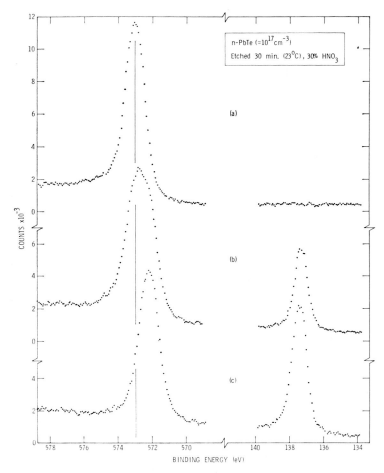

FIG. 12. XPS spectra in region of Pb $4f_{7/2}$ and Te $3d_{5/2}$ lines for a specimen of PbTe which had been etched for 30 min in 30% HNO_3 at 23°C (a) immediately after etching, (b) after sputtering surface for 30 min, (c) after sputtering surface for 46 min.

Figure 13 is a compilation of the results of their HNO_3 etching experiments. The average layer thickness was defined as the sputtering depth where the Te $3d_{\frac{5}{2}}$ intensities from Te^0 and Te^{2-} are equal. The rate at which the Te^0 layer builds up is quite dependent on the acid concentration and monotonically increases with the etching time. The mechanism involved in building up such thick Te^0 layers is unclear at this time. However, scanning electron microscopy studies of the etched surfaces revealed the existence of deep pits in the Te^0 layer, the number of which correlates well with the dislocation density present in the PbTe sample. Consequently, the suggestion has been made that chemical reaction and transport along these lattice defects may provide the source of new Te required to build up the thick layers.

The effect on the surface and bulk leakage currents of several array processing steps is shown in Table I. In addition, the chemical

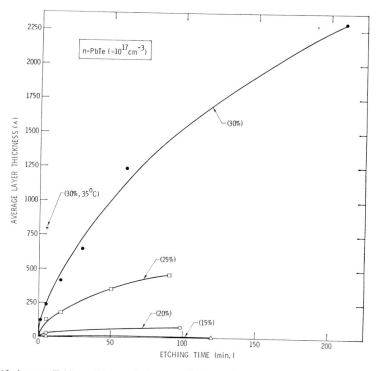

FIG. 13. Average Te^0 layer thickness built up on PbTe as a function of etching time for various concentrations of HNO_3.

nature of the Te overlayer as determined by XPS is also presented. One will note that in the case of the first two treatments the surface has a thick Te^0 layer and the surface leakage parameter is very small. In the processing procedures where this layer is not formed and the presence of Pb, Sn, and Te oxides is observed on the surface, S increases significantly, implying that Te^0 acts as a surface passivant in the case of $Pb_{1-x}Sn_xTe$ (the bulk parameter remaining virtually unchanged). One possible explanation for this effect may

TABLE I

Compositional and Leakage Characteristics of a $Pb_{1-x}Sn_xTe$ Diode Array

Treatment	Nature of surface Te as determined by XPS	S (mA/cm)	R (mA/cm²)
Norr etch; acid clean-up etch	Thick Te^0 layer, small amount of Te^{4+}	0 ± 1	1750 ± 200
Exposed to air for few hours	Thick Te^0 layer, oxide layer ≈ 3 times thicker than after Norr etch	-1 ± 1	1850 ± 350
H_2O_2 based etch	Small amount of Te^0; bulk Te^{2+} observable, several monolayers of oxide	80 ± 15	2300 ± 1300
Norr etch; 9% HNO_3	Small amount of Te^0; bulk Te^{2+} observable, small amount surface oxide	27 ± 8	2700 ± 1300

TABLE II

Surface Atomic Fractions and the Ti/Sr Ratio on (100) SrTiO₃ as a Function of Various Treatments

	Sr	Ti	O	Ti/Sr
Hydrogen-reduced	0.10	0.19	0.71	1.84
Hydrogen-reduced + etch	0.11	0.14	0.75	1.21
Mechanical polish + etch	0.15	0.19	0.66	1.26
Mechanical polish + etch + electrochemical cycling	0.18	0.22	0.61	1.24
Cleaved	0.21	0.20	0.59	0.98
Cleaved + H₂O rinse	0.17	0.24	0.59	1.37

be that surface states introduced by the oxides are removed at the $Te^0/Pb_{1-x}Sn_xTe$ interface.

(2) Quantitative Analysis

Generally speaking, unless one is able to manipulate the situation so that only the first monolayer of the surface is detected, electron spectroscopic techniques should be viewed as being only semiquantitative in nature. The basic difficulties in quantification lie in (1) the uncertainty involved in exactly determining the escape depth of the (different energy) electrons in a given material and (2) the problem of determining the depth distribution of the various species present in heterogeneous systems. However, under the favorable circumstances of chemical homogeneity, similar electron escape depths, and little surface contamination, the hope of carrying out a quantitative analysis does exist.

A case in point is some work (18) performed on the SrTiO₃ system. The d-band perovskites have recently attracted considerable interest, both theoretical and experimental, as potential catalysts. Morin and Wolfram (19) have shown that they possess high concentrations of electronic surface states of the appropriate symmetry for catalytic activity. In particular, SrTiO₃ has been shown to be a good electrode material for the photoelectrolytic decomposition of water (20). In general, the implicit assumption is made in such studies that, in the absence of contrary experimental evidence, the bulk

perovskite composition is maintained right up to the surface of the material.

During recent work in our laboratory, it was decided to monitor the surface chemistry of the (100) surface of SrTiO₃ single crystals as a function of preparative treatments and electrochemical voltage cycling. The $2p_{\frac{3}{2}}$ photolines of Sr and Ti (binding energies of 268 and 458 eV, respectively) and the $1s_{\frac{1}{2}}$ photoline of O (529 eV binding energy) were observed. The peak areas were obtained by digital integration of the data and a correction was made for inelastic energy loss-related background differential across the peaks. The relative X-ray cross sections for Sr, Ti, and O were obtained using high-purity samples of SrCO₃ and TiO₂. An analysis was then carried out which yielded the atomic fraction of each element present within the detected volume of the sample. These results, together with the Ti/Sr atom ratios, are shown in Table II.

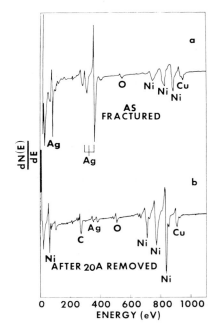

FIG. 14. Auger electron spectra from electroplated Ni–Ni bond surface (a) as fractured, (b) after 20 Å removed from surface.

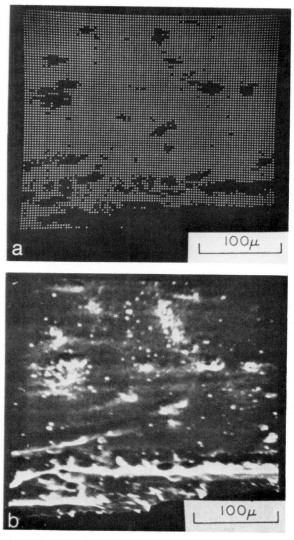

FIG. 15. Auger electron image of silver-contaminated Ni–Ni bond surface. (a) Silver Auger electron image. (b) Secondary electron micrograph.

The first four treatments were sequentially carried out on the same specimen. The quench from growth temperature in a 0.04 H_2–0.96 N_2 atmosphere produces a sample with n-type conductivity. Etching was performed using an 85% H_3PO_4 solution and the electrochemical cycling was performed in a 0.5 M K_2SO_4 solution. Few conclusions can be drawn from the atomic fraction data, the sample also generally displaying some carbon contamination which was ignored in the analysis. It is clear from the Ti/Sr ratios, however, that the hydrogen quench produces quite a Ti-rich surface. The other three treatments produce a

223

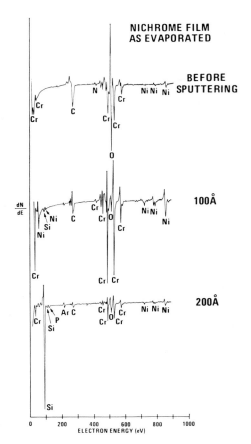

FIG. 16. Representative Auger spectra from various depths of a 150 Å nichrome film on a silicon substrate.

$Sr(OH)_2$ formation. The important point for our purposes, however, is the excellent quantitativeness of the atomic fraction data, the cleaved surfaces showing almost exactly the 1:1:3 Sr:Ti:O composition to be expected in such a case.

It might be noted that a simple dissolution model does not account for the observed Ti/Sr enhancement, predicting a ratio of approximately 1.13 (18). A plausible explanation, however, comes from the consideration of expected surface nonidealities. Surfaces of the type under consideration are not atomically flat, but contain numerous terraces and ledges related to defects in the material and the angular accuracy of the cleave. Such features give rise to substantial concentrations of high-index crystal faces and multilayer solvation of exposed SrO planes from such faces could serve as an added source of surface Sr depletion. This process would also tend to explain the greater Sr depletion observed in the (cleaved + H_2O) samples relative to the (etched) and the (etched + electrochemical cycling) data which were taken on as-grown surfaces; etching would be expected to microscopically smooth the surfaces considerably, substantially reducing the incidence of high-index faces.

APPLICATIONS OF AES

The greater complexity of the Auger emission process, its ability to be generated by electron impact (by emitted as well as incident electrons), and the lower energy resolution involved in Auger spectra make the extraction of chemical compositional information much more complicated than in XPS. In order to avoid redundancy in this paper, the reader is referred to the aforementioned reviews of the subject for an insight into the problems involved.

The major advantages of the Auger technique over XPS are (1) the high spatial resolution (≈ 500 Å) in the plane of the surface attainable by using a finely focused excitation beam of electrons and (2) the ability

surface with approximately equal Ti/Sr ratios. This fact, given the diversity of the etching and electrochemical cycling solutions, led to the suggestion that perhaps a common factor was acting in these cases, namely H_2O. This theory was tested by cleaving some crystals, performing an analysis on the freshly exposed surface, washing the samples in deionized H_2O, and repeating the analysis. The results are shown in the lower part of the table and it does indeed appear that selective removal of Sr from the interface is taking place, possibly not a surprising fact considering the aqueous solubility of SrO and its reactivity toward

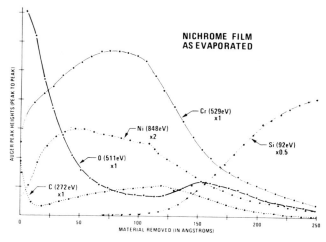

FIG. 17. Amplitudes of the various Auger peaks from a nichrome film on a silicon substrate as a function of the amount of material sputter etched from the film.

to observe transient effects and carry out time-efficient elemental profiles normal to the surface (employing inert gas ion sputtering techniques) as the result of the relatively large Auger peak currents involved.

An example of the former quality is seen in the work of Waldrop and Marcus (21), who investigated bond line failure in macroscopic multilayer electrodeposits of nickel. Samples were fractured in the ultrahigh vacuum

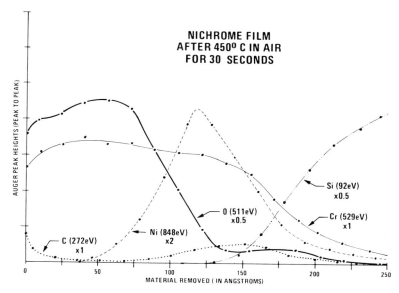

FIG. 18. Amplitude of the Auger peaks from a heat-treated nichrome film as a function of the amount of material sputter removed.

225

environment and the Auger spectrum of the fracture surface was seen to be that of Fig. 14a; Ag and Cu impurities are seen to be present in significant quantities. After the removal of approximately 20 Å of the surface by ion bombardment, the Ag signal had virtually disappeared but that of the Cu remained, suggesting that the Ag was the cause of the embrittlement at the deposit interface. They then employed their own recent development of performing AES in a scanning electron microscope (SEM) where, by using the appropriate Auger electrons to form an image, a two-dimensional map of the surface distribution of an element can be generated. The element chosen in this case was, of course, Ag. The resultant map of the Ag distribution on the fracture surface is shown in Fig. 15a, an image point being white if Ag is present above the 10% total intensity level and black otherwise. Comparing the Auger results with a conventional SEM micrograph of the sample clearly shows that Ag covers the surface except where plastic tearing during fracture has taken place.

The speed advantage of AES is seen in its capability to continuously monitor the surface elemental composition (generally using a velocity analyzer system) while simultaneously stripping away the surface, monolayer by monolayer, using the inert gas ion bombardment sputtering technique. One method used to produce resistive films for the computer industry is to evaporate nichrome upon Si substrates followed by a thermal treatment to obtain the proper electrical properties. Since the film is only a few hundred Ångstroms thick and is a multicomponent system, it is desirable to ascertain the changes which occur in the film composition during processing. Figure 16 presents full-scale Auger spectra from a 150 Å thick film; the spectra are representative of the film constituents at the surface, 100 Å into the film, and after 200 Å have been etched away. Marked changes in the relative elemental intensities are observed. Electronic multiplex techniques may be employed to sample the peak intensities and

construct a graph depicting the profile of each element in a direction normal to the surface. Typical sputtering rates are of the order of 10–20 Å/min. Such a plot is shown in Fig. 17 for an "as evaporated" nichrome film. The NiCr/Si interface is easily visible. Also note the monotonic decrease of the 0 peak intensity near the free surface and the shape of the Ni profile. Figure 18 shows the profiles obtained from an identical film which has been heated to 450°C in air for a duration of 30 sec. Clearly the effect of the heat treatment was to produce a fairly pure chromium oxide layer about 75 Å thick at the surface while the Ni has diffused to the film/substrate interface. As might be expected, changes of such a nature are paralleled by variations in the resistive properties of the film.

REFERENCES

1. ROBINSON, H., AND RAWLINSON, W. F., *Phil. Mag.* **28**, 277 (1914).
2. AUGER, P., *J. Phys. Radium* **6**, 205 (1925).
3. SOMORJAI, G. A., AND FARRELL, H. H., in "Advances in Chemical Physics" (I. Prigogine and S. A. Rice, Eds.), Vol. XX. Wiley, New York, 1971.
4. SOMORJAI, G. A., AND SZALKOWSKI, F. J., in "Advances in High Temperature Chemistry," (L. Eyring, Ed.), Vol. 4, p. 137. Academic Press, New York, 1971.
5. CHANG, C. C., *Surface Sci.* **25**, 53 (1971).
6. TODD, C. J., *Vacuum* **23**, 195 (1973).
7. RIVIERE, J. C., *Contemp. Phys.* **14**, 513 (1973).
8. BRUNDLE, C. R., *J. Vac. Sci. Technol.* **11**, 212 (1974).
9. SICKAFUS, E. N., *J. Vac. Sci. Technol.* **11**, 299 (1974).
10. TRACY, J. C., AND BURKSTRAND, J. M., *Crit. Rev. Solid State Sci.* **4**, 381 (1974).
11. SIEGBAHN, K., *J. Electron Spectrosc. Relat. Phenom.* **5**, 3 (1974).
12. SHIRLEY, D. A., *J. Vac. Sci. Technol.* **12**, 280 (1975).
13. SIEGBAHN, K., NORDLING, C., FAHLMAN, A., NORDBERG, R., HAMRIN, K., HEDMAN, J., JOHANSSON, G., BERGMARK, T., KARLSSON, S.-E., LINDGREN, I., AND LINDBERG, B., in "ESCA: Atomic, Molecular, and Solid State Structure Studied by Means of Electron Spectroscopy." Almquist and Wiksalls, Uppsala, Sweden, 1967.
14. WENTZEL, G., *Z. Phys.* **43**, 524 (1927).
15. FADLEY, C. S., AND BERGSTRÖM, S. Å. L., *Phys. Lett. A* **35**, 375 (1971).

16. FRAZER, W. A., FLORIO, J. V., DELGASS, W. N., AND ROBERTSON, W. D., *Surface Sci.* **36**, 661 (1973).

17. GRANT, R. W., PASKO, J. G., LONGO, J. T., AND ANDREWS, A. M., *J. Vac. Sci. Technol.* **13**, 940 (1976).

18. SZALKOWSKI, F. J., RALEIGH, D. O., AND TENCH, D. M., submitted to *Appl. Phys. Lett.*

19. MORIN, F. J., AND WOLFRAM, T., *Phys. Rev. Lett.* **30**, 1214 (1973).

20. MAVROIDES, J. G., KAFALAS, J. A., AND KOLESAR, D. F., *Appl. Phys. Lett.* **28**, 241 (1976).

21. WALDROP, J. R., AND MARCUS, H. L., *J. Test. Eval.* **1**, 194 (1973).

22. WEBER, R. E., *Res./Develop.* **23**, 22 (1972).

Water at Interfaces

Water at Interfaces: Molecular Structure and Dynamics

K. KLIER AND A. C. ZETTLEMOYER

Center for Surface and Coatings Research, Lehigh University, Bethlehem, Pennsylvania 18015

Received July 15, 1976; accepted October 12, 1976

Recent advances in molecular mechanics and dynamics of water on inorganic surfaces are reviewed. The structures of hydroxylated surfaces and water–hydroxyl adducts, determined by LEED–Auger studies and infrared spectroscopy, are described to a detail only recently resolved. New mechanistic concepts of nucleation and freezing have ensued from the analysis of time-correlation functions, showing that low-frequency angular perturbations dominate the different behavior of water in bulk phases and at surfaces. While the rotational motion of water molecules in the liquid phase can be interpreted as rotation modulated by making and breaking hydrogen bonds with the neighbors, water adsorbed on nucleating catalysts undergoes an irreversible and complete reorientation in a fraction of the rotational period.

1. INTRODUCTION

The interaction of water with surfaces gives rise to a great variety of phenomena ranging from a strong attack of the solid such as corrosion and dissolution to very subtle effects involving, for example, heterogeneous nucleation of liquid water and ice. Among the water–organic interactions, perhaps the most intensely studied are those which result in conformational changes of biomacromolecules. A review covering the analysis of the state of water on both organic and inorganic surfaces will appear shortly (1). The present paper is concerned with recent advances in *water–inorganic systems*, in which the practical objectives are important enough to call for analysis of the mechanisms and time scales of molecular events. The molecular mechanics and dynamics, as little as we understand them today, are illustrated in a sequence involving interactions with selected metal surfaces, oxide surfaces, silica and silicate surfaces, and water with itself in the presence of the various surfaces mentioned above.

2. INTERACTIONS WITH METAL SURFACES

We first take a clean metal surface and expose it to water vapor. Iron may be chosen as an example of a reactive metal, and in order to pursue the elementary interactions with confidence, a defined surface of a single crystal is the first experimental subject. Water absorbs on the (100) iron crystal face with a sticking probability depicted as a function of surface coverage in Fig. 1, obtained in a meticulous LEED–Auger spectroscopic study of Dwyer and Simmons (2). The "monolayer" in this figure is one in which a single water molecule (or its fragment, hydroxyl, as is discussed shortly) occupies two iron atoms. Three features are imminent from the sticking probability curves in Fig. 1 (top).

(1) The initial sticking probability is high (0.55), exceeding that in which the water molecule had lost all (0.02), or two (0.08), or even only one (0.38) rotational degree of freedom.

(2) For a range of surface coverages, the sticking probability is the same whether the

surface is occupied or free from the stable product. Therefore, prior to anchoring itself at a certain position, the water molecule tumbles over the bare *and* covered metal surface, retaining most of its rotational and translational motion.

(3) Only 80% of a monolayer monitored as an oxygen Auger signal is formed, beyond which point the adsorption ceases, despite the irreversible and stable oxygen-containing product formed.

That the product is hydroxyl (Fig. 1, bottom) rather than oxygen is supported by electron spectroscopic measurements of Kishi and Ikeda (3) in which the O(1s) signal appeared at energies close to those of hydroxyls and remote from those of surface oxide. In addition, the structure and chemical reactivity of the water-produced hydroxylated surface are different from those of the adsorbate produced by oxygen. The structural features are displayed in Fig. 2, where the observed fuzzy half-order spots (top left) can be simulated by a model involving exclusion of nearest-neighbor sites (2). The results of computer simulation are represented in Fig. 2 (bottom) as the overlay structure and in Fig. 2 (top right) as the LEED pattern generated from the simulated structure by diffraction of a laser beam. The basic feature of the structure in Fig. 2 is the existence of domains separated by boundaries which include more vacant sites than the "perfect" structure of either domain. If the computer-simulated adsorption is allowed to proceed by random fixing with the exclusion of the nearest-neighbor sites, the area of vacant sites is exactly 22% that of the monolayer, and in agreement with the experimental observation that 20% of the positions are excluded from the adsorption process. The picture of water adsorption on an iron surface is thus one of a mobile and rotating precursor water molecule which fixes itself by forming firmly bound hydroxyls and hydrogen which is found partly as a gas phase product. The vacant spaces at the boundaries may play an important role in corrosion. They leave room

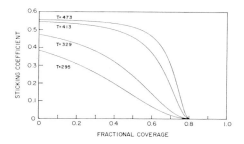

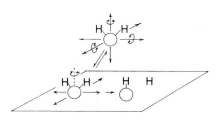

FIG. 1. Top: Sticking probabilities of water adsorption on (100) crystal face of iron. Fractional coverage is given as a fraction of a monolayer in which one hydroxyl is bound to two iron atoms. Bottom: Schematic representation of incomplete loss of rotational degrees of freedom when water adsorbs into the precursor state, and a subsequent formation of stable and immobile hydroxyls. From Ref. (2).

for oxygen chemisorption, which is stronger than that of water. One can anticipate a lateral distribution (or inhomogeneity) of work function and electric potential between the domains and the boundaries, and thereby the beginning of the formation of cathodic and anodic regions on the iron surface.

In contrast to water, hydrogen sulfide forms a perfect structure accompanied by sharp $C(2 \times 2)$ LEED spots, fills the surface to complete the monolayer without vacant boundaries, and poisons the surface completely for oxygen chemisorption (2). In view of the previously developed ideas, the fragments of hydrogen sulfide, be they sulfhydryl or sulfur atoms, must be mobile until they settle to make the perfect overlayer structure, and exclude oxygen from chemisorption. Although not based on experimental detail such as those in the water–iron studies, initial mechanisms for metal oxidation by water had been pro-

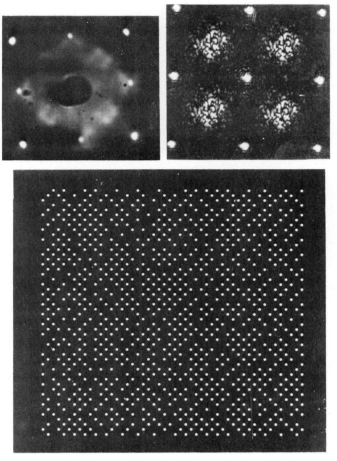

Fig. 2. Top: LEED patterns of (100) Fe surface exposed to water. Left upper figure, observed pattern showing the integral order spots of the (100) iron plane and fuzzy half-order spots of the water-generated overlayer; right upper figure, laser diffraction pattern from the array shown in the lower part of the figure, simulating the fuzzy half-order spots. Bottom: Computer-simulated overlayer structure in which each water was allowed to form one hydroxyl per two iron atoms with the exclusion of the nearest neighbor sites to occupied sites. The nonexcluded positions are chosen by a random number generating program. From Ref. (2).

posed for water–copper (4), nickel (5), tungsten (6), and platinum (7) interactions. The mechanism outlined above thus appears general for many reactive metals, and is not confined to single-crystal surfaces.

The surface hydroxylation will of course continue on a number of metals into the bulk if the water concentration or temperature is increased. To give an extreme example, Linnenbom showed as early as in 1958 (8) that iron powder immersed in oxygen-free water is totally converted, at room temperature, into ferrous hydroxide with the liberation of an equivalent amount of hydrogen.

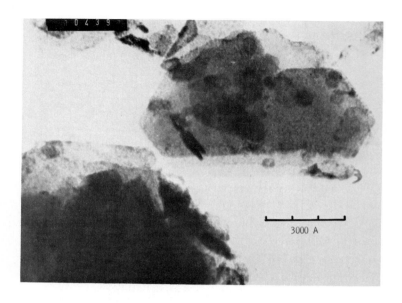

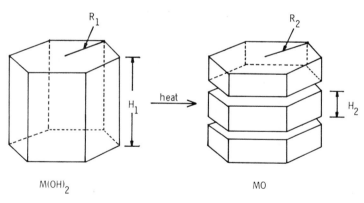

$M(OH)_2$ MO

FIG. 3. Electron micrograph of NiO layers (top) resulting from decomposition of nickel hydroxide on the microscope hot stage. The micrograph shows slipped layers schematically represented in the bottom part of the figure. The surfaces of the slipped layers remain hydroxylated. From Ref. (9).

3. INTERACTIONS WITH OXIDE SURFACES

The prevalence of hydroxyl groups has become apparent also for oxide surfaces. For a long time, the thermal decomposition of nickel hydroxide to yield high-area nickel oxide was thought to be a reaction in which particles of highly porous nickel oxide are generated and supported on an undecomposed matrix of the hydroxide. An entirely different picture holds today, as revealed primarily by the microscopic, spectroscopic, adsorption, and calorimetric studies of the Lyon, Zagreb, and Lehigh groups. As in the previous comments, the most recent observations giving an updated description of this interesting system will be discussed presently. Figure 3 demonstrates what happens

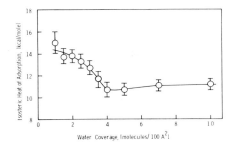

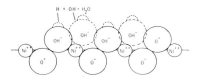

(1̄10) Cross Section Through the (111) Surface of NiO

Fig. 4. Isosteric heats of adsorption (top) and the model for partial rehydroxylation of the (111) surface of NiO (bottom). The higher heats of adsorption up to 4 molecules/100 Å² correspond primarily to completion of coordinatively unsaturated Ni^{2+} ions of the (111) plane. Approximately half of the surface is covered by hydroxyls after decomposition of $Ni(OH)_2$ at 200°C, the remaining half being replenished by water adsorption.

during the thermal decomposition of nickel hydroxide, followed directly on the hot stage of an electron microscope (9).

The platelets of the cadmium iodide structure of the hydroxide, oriented with the hexagonal axis out of the picture, retain their shape during the whole decomposition process in which nickel oxide is formed, except for the separation of layers with faces parallel to the (111) plane of NiO. Here the product NiO and its orientation are identified by electron diffraction. The (111) plane would normally be occupied by either the Ni^{2+} cations or the oxygen anions only, and therefore it would be unstable. However, near-infrared investigations of Cronan (9) show that they are actually terminated by surface hydroxyls, as illustrated in Fig. 4, when formed by the decomposition of the hydroxide at 200°C.

It may be noted that such *topotaxic* splitting was suggested as early as 1947 by Zettlemoyer and Walker (10) for thermal decomposition of $Mg(OH)_2$, but the first solid evidence was obtained only by the combination of electron microscopic and spectroscopic investigations mentioned above. The surface hydroxyls are sufficient to prevent oxidation of the nickel ions. Oxygen chemisorbs on only approximately 5% of the surface which coincides with the geometric edge surface of the NiO platelets. The edge surfaces are also likely to be involved in redox catalysis while the hydroxylated surfaces turn out to be inert.

The hydroxyls can be removed further from many oxide surfaces until a hydroxyl-free surface is formed. However, such a process is not an easy one and usually gives rise to adsorbents with highly energetic surfaces. Yao (11) and later Yao with Kummer (12) obtained calorimetric heats of water adsorption for the (0001) surface of hydroxyl-free ferric oxide and other similar structures, and assigned energies of 30 kcal/mole to water–cation interactions, and about 20 kcal/mole to water–anion interactions. From the extent of adsorption, cations and anions were found to be present in nearly equal amounts on surfaces which accommodate only one kind of ion if atomically smooth. Therefore, the conclusion of Yao *et al.* was that anions and cations form patches of nearly equal size which are separated by one atomic step. The magnitude of adsorption heats for water indicates, however, that in systems with very small amounts of water present these patches will soon be totally covered by hydroxyls.

4. INTERACTION OF WATER WITH SILICAS AND SILICATES

There are certain oxides in which the removal of *surface* hydroxyls by heat is not possible without morphological changes and sintering. Such materials are represented by silicas on which most work on water–interface interactions has been done. It is a crudely accurate statement to say that at temperatures below 600°C the amount of surface hydroxyls on silicas remains constant (13, 14) and the presence of water does not influence their formation or disappearance. However, water does interact with the hydroxyls and a fairly detailed mechanistic picture of this process is available today through combined spectroscopic and adsorption studies. Figure 5 displays water adsorption on a nonporous HiSil silica followed by infrared spectroscopy in the region of stretching overtones of free silanols (lower curves with maxima at 7300 cm^{-1}) and adsorbed water (curve progression).

Clearly, the free silanols become bound silanols as their frequency shifts by a large amount (100–150 cm^{-1}) into the spectrum of the OH–OH$_2$ adduct. The molecular water bands ($\nu + \delta$) at 5300 cm^{-1} show only small (10–20 cm^{-1}) shifts compared to the monomer and, since large shifts indicate the attack of hydrogen atoms in the OH oscillator, water clearly sits "oxygen down" in the beginning stages of adsorption. At a later stage, lower-frequency water bands appear, indicating the formation of hydrogen-bonded water clusters. Interestingly the water clusters are formed before all the free hydroxyls are taken up, as demonstrated in Fig. 6, where black circles indicate the uptake by free silanols while the white circles show the concentration of water in "oxygen down" geometry. The incomplete uptake by all the free hydroxyls is caused by the relationship between adsorption energies, energies per water molecule in water clusters, and the corresponding entropies. The calorimetric heat of a single hydroxyl–water bond is approximately 6 kcal/mole (15), less than the heat of liquefaction, 10.5 kcal/mole. Water is therefore energetically favored in the liquid compared to the bonding by a single hydroxyl. As far as *cluster stabilities* are concerned, no experimental data are available but recent quantum mechanical calculations (mainly those of Del Bene and Pople (16) and Kistenmacher, Popkie, and Clementi (17)) can be accepted with a great deal of confidence because of accurate accounts for the energies of water dimers and water–ion interactions.

Figure 7 shows examples of various cluster sizes and geometries for which the following energies per water molecule have been obtained (in kcal/mole):

−3.05	−3.09	−3.36
Dimer	C_{2v} trimer	C_{2v} closed trimer
(XXX)	(XXV)	(XXVi)
−4.9	−5.6 to −2.1	−3.0
Linear trimer	Cyclic trimer	Cyclic trimer
(XXVii)	(XXViii)	(XXiX)
−10.5(S_4)		
−9.5(C_4)–4.3(D_{2h})	−7.35	−7.92
Cyclic tetramer	Central tetramer	Pentamer
(XXXi)	(XXXii)	(XXXiii)

The open pentamer was calculated to be stabilized by 11.65 kcal/mole and a hexamer with linear hydrogen bonds, by 10.76 kcal/mole, both close to the latent heat of liquefaction. These calculations indicate that water molecules are stabilized in small clusters containing five or six molecules as much as they are stabilized in bulk liquid. The clustered molecules are, however, more translationally and rotationally restricted in their motion than those weakly bound to the silanols, and therefore have a lower entropy. Thus, at low coverages water adsorption on silanols is favored by entropy, whereas at higher coverages clustering is favored by energy stabilization in the clusters. These concepts are supported by a statistical mechanical theory the main result of which is that the adsorbate density will abruptly change before full occupation of the surface is reached, provided that the adsorption energy is smaller than the heat of liquefaction.

The earlier example was given for a nonporous silica with a relatively low concentration of surface hydroxyls. With larger hydroxyl concentrations, multiple bonding of water molecules accompanied by higher heats of adsorption has been observed (18), in which case the surface is occupied to a large degree before clustering occurs. To summarize briefly, water clusters, or nuclei of a liquid phase, are more easily formed on surfaces that appear mildly hydrophobic, which is realized on

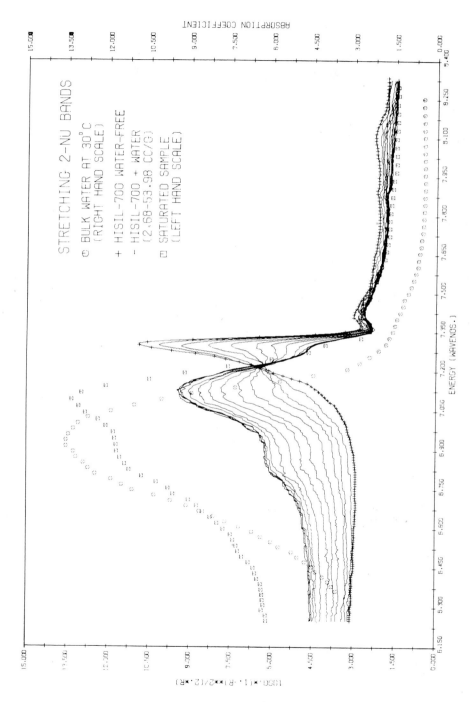

silicas by distances between hydroxyls that prevent multiple water–hydroxyl bonds from being formed.

5. CORRELATION MOTION

Having discussed the structure and the energy, and having touched upon the entropy of water in adsorbates, we are still left with little information about the *motion* of water molecules. And yet the character of motion, particularly of restricted rotations, determines the many thermodynamic and kinetic properties of water, both in bulk phases and in the adsorbed state.

It is therefore a great advantage that water has a permanent dipole as well as various transitional dipoles through which we can follow reorientations of the water molecules. In fact, we may select a particular transition dipole to follow reorientations about a selected axis. We have chosen for our investigations a combination band $(\nu_3 + \delta)$ appearing in the near infrared at 1.9 μm where no interference occurs from such surface species as hydroxyls. The presence of the deformation component δ ensures that we follow an integral water molecule; at the same time the asymmetric vibrations ν_3 give rise to large excursions compared to the deformations δ so that motion about the molecular axis will be predominantly monitored because the transition dipole of this band is nearly perpendicular to the molecular axis.

The time development of molecular motion is fundamentally connected with Fourier transforms of infrared bands exciting the above-mentioned dipoles. The connection can be demonstrated by both classical and quantum theory, the latter having the advantage of correctly accommodating the zero-point vibrations. In the quantum picture, the time Fourier

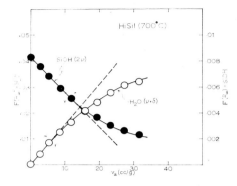

FIG. 6. Relation between intensities at band maxima of the SiOH (2ν) band (right-hand scale, full circles), of the H_2O $(\nu + \delta)$ band (left-hand scale, open circles) and the amount of water adsorbed (abscissa axis) at 25°C on HilSi heat-treated at 700°C. $F(R_\infty) \equiv (1 - R_\infty)^2/2R_\infty$, where R_∞ is the diffuse reflectance from the semi-infinite layer. $F(R_\infty)$ is proportional to the absorption coefficient. From Ref. (14).

transform relates to its original in a way Heisenberg mechanics employing moving operators relates to Schrödinger mechanics of stationary observable quantities. Gordon, in 1965, demonstrated this relationship for a transition dipole of infrared absorption (19) and the basic connection is briefly reviewed as follows.

The intensity of an infrared band is determined by the transition probability:

$$I(\omega) = \frac{\hbar c}{4\pi^2} \frac{\sigma(\omega + \omega_0)}{(\omega + \omega_0)}$$

$$\times \{1 - \exp(-\hbar[\omega + \omega_0]/kT)\}^{-1}$$

$$= \sum_{l,u} \rho_l |\langle l| \boldsymbol{\varepsilon} \cdot \mathbf{m}^v |u\rangle|^2$$

$$\times \epsilon[(E_u - E_l)/\hbar - \omega],$$

where $|l\rangle$ is the initial rotational state, $|u\rangle$ is

---

FIG. 5. The near-infrared spectra of silanols and water in the 2ν region is HiSil heat-treated at 700°C. The adsorbed amounts and intensities corresponding to the individual bands are represented in Fig. 6. The H_2O (2ν) band of bulk liquid water is shown for comparison (◯). The left-hand ordinate scale is a function of diffuse reflectance R_∞, proportional to the absorption coefficient of the powdered silicas; the right-hand ordinate scale is the absorption coefficient of bulk liquid water; the abscissas are wavenumbers (in thousands of cm⁻¹). From Ref. (14).

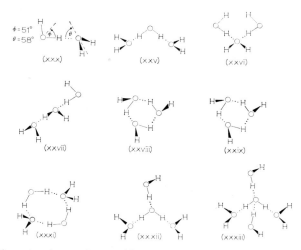

F<small>IG</small>. 7. Geometry of water clusters with corresponding binding energies listed in text.

the final rotational state, $\mathbf{m}^v$ is the vibrational transition dipole, $\boldsymbol{\varepsilon}$ is the exciting field, and σ is the absorption coefficient. If rotational–vibrational coupling is neglected, then ω is the frequency displacement from the vibrational band center at ω_0.

Representing the δ-function by its Fourier integral, we obtain:

$$I(\omega) = \frac{1}{2\pi} \int_{-\infty}^{\infty} dt \sum_{l,u} \rho_l \langle l | \boldsymbol{\varepsilon} \cdot \mathbf{m}^v | u \rangle \langle u | \boldsymbol{\varepsilon} \cdot \mathbf{m}^v | l \rangle$$

$$\times \exp[i(E_u - E_l)t/\hbar] \exp[-i\omega t]$$

$$= \frac{1}{2\pi} \int_{-\infty}^{\infty} dt \langle [\boldsymbol{\varepsilon} \cdot \mathbf{m}^v(0)][\boldsymbol{\varepsilon} \cdot \mathbf{m}^v(t)] \rangle$$

$$\times \exp[-i\omega t],$$

where $\langle \, \rangle$ stands for the ensemble average. Here

$$\mathbf{m}^v(t) = e^{iHt/\hbar} \mathbf{m}^v e^{-iHt/\hbar}$$

is the Heisenberg moving dipole operator and $\mathbf{m}^v(t)$ is a solution of Heisenberg's equation of motion:

$$(d/dt)\mathbf{m}^v(t) = [\mathbf{m}^v(t), H].$$

Upon Fourier transformation and averaging over all directions of $\boldsymbol{\varepsilon}$ in an isotropic sample,

normalizing the intensity,

$$\hat{I}(\omega) = I(\omega) \bigg/ \int_{\text{band}} I(\omega)d\omega,$$

and introducing a unit vector $\hat{\mathbf{u}}$ along the direction of the transition dipole moment,

$$\hat{\mathbf{u}} = \mathbf{m}^v / \langle (\mathbf{m}^v)^2 \rangle^{\frac{1}{2}},$$

we obtain the fundamental relation

$$\langle \hat{\mathbf{u}}(0) \cdot \hat{\mathbf{u}}(t) \rangle = \int_{\text{band}} \hat{I}(\omega) e^{i\omega t} d\omega.$$

It follows from this relation that the Fourier transform of the normalized band intensity describes the time development of the projection of the transition dipole onto its value at an arbitrarily chosen zero time. The expression $\langle \hat{\mathbf{u}}(0) \cdot \hat{\mathbf{u}}(t) \rangle$ is the transition dipole time correlation function. Since the transition dipole rotates with the molecule, the correlation function describes the progress of randomization of rotational motion over the ensemble average. The assumption of the absence of roto-vibrational coupling appears adequate for many liquids in a wide range of temperatures and pressures. The notable exception is *water*, for

239

which all experimental evidence unambiguously points to strong coupling of vibrational and rotational motion. The roto-vibrational coupling was treated to the first order of perturbations with the following principal results (20).

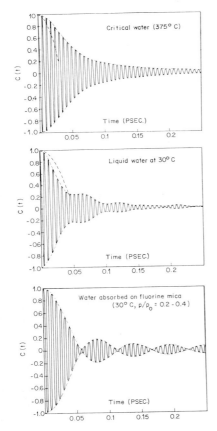

FIG. 8. Time dipole correlation functions $C(t)$ of water in critical state (top), in bulk liquid at 30°C (center), and in an effective monolayer on fluorophlogopite (bottom). The normalized total correlation functions, obtained according to Eq. [9] of Ref. (20). involve vibrations of the transition dipole of the $(\nu_3 + \delta)$ band displayed as fast oscillations, rotations, and angular perturbations appearing as contours on the vibrational correlation functions. The dashed line in the center figure represents rotation about the axis of the least inertial moment of the free water molecule, showing that the motion in this liquid is initially accelerated by intermolecular forces.

SCHEMATIC REPRESENTATION OF
ANGULAR POTENTIALS FOR WATER

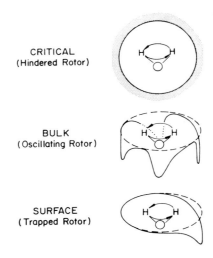

FIG. 9. The form of angular potentials for water estimated classically from correlation functions in Fig. 8.

An angularly and radially dependent perturbation

$$V = V(r, \vartheta, \varphi)$$

was introduced to act upon both the vibrational and rotational states. The Fourier transform of an infrared band is then given by

$$\int_{band} \hat{I}(\omega)e^{i\omega t}d\omega$$
$$= \sum_l \rho_l \langle l| \hat{\mathbf{u}}(0) \cdot \hat{\mathbf{u}}(t)|l\rangle \exp[i(\Delta\omega_l - \Delta\omega_\nu)t],$$

where

$$\Delta\omega_l = (1/\hbar)[\langle l| V_e|l\rangle - \langle l| V_g|l\rangle];$$

V_e is the average perturbation of the *excited vibrational* state, V_g the average perturbation of the *ground vibrational* state, and $\Delta\omega_\nu$ the shift of the band center.

Compared with the unperturbed time-correlation function, the angular perturbation V produced modulation factors $\exp[i(\Delta\omega_l - \Delta\omega_\nu)t]$ on each rotational state.

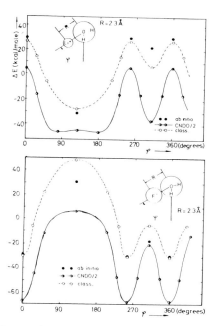

FIG. 10. Angular dependences of the binding energy ΔE for planar $[Li(H_2O)]^+$ (top) and $[F(H_2O)]^-$ (bottom) showing the spatial reversal of water binding to cations and anions.

In the ultimate case, where V has a crystal symmetry, the perturbation V executes a crystal field splitting of the rotational states, converting them into crystal vibrations. At short times, the unperturbed correlation function, e.g., for linear molecules, has the form (19)

$$C(t) \equiv \langle \hat{\mathbf{u}}(0) \cdot \hat{\mathbf{u}}(t) \rangle \cong 1 - (kT/I)t^2,$$

where I is the moment of inertia, whereas that involving rotovibrational coupling,

$$C(t) \cong 1 - (kT/I)t^2 - \tfrac{1}{2}t^2 \sum_l \rho_l (\Delta\omega_l - \Delta\omega_\nu)^2 \langle l \,|\, \hat{\mathbf{u}}(0) \cdot \hat{\mathbf{u}}(t) \,|\, l \rangle,$$

has an additional term, quadratic in time and negative, showing that the angular perturbation produces *faster* than the inertial decay of the correlation function (20).

Armed with this brief theoretical introduction, we can now discuss the correlation func-

tions of water in various states of agglomeration, both in bulk phases and at surfaces.

How such correlation functions look for the $(\nu_3 + \delta)$ transition dipole of water is shown in Fig. 8, where the motion is resolved on a time scale of 10^{-14} sec. Here the fast oscillations are caused by vibrations and the contours by perturbed rotational motion. The behavior of water in the critical state is typical for most common liquids and corresponds to very frequent and random rotational reorientations giving rise to a near-exponential time decay. Similar behavior will be displayed by a molecule rotating in a uniform viscous medium, often characterized as the *Debye liquid*. The correlation function of bulk water, however, is not that of a Debye liquid, showing (a) an oscillatory character of rotations, and (b) a rapid initial reorientation which is faster than the inertial rotation of water about its fastest rotational axis, indicating that an attractive force is operating to accelerate the water rotors. The third example is that of water motion on a surface of fluoride mica. Here the reorientation, although initially not as rapid as in bulk water, is more complete and is followed by low-frequency beats in the new orientation of the molecule. The three different motions are illustrated in Fig. 9, in which the bulk water motion shows the number of encounters with the neighbors (approximately four collisions during a rotational period) and the shapes of potential wells calculated classically from the periodic accelerations and decelerations of the correlation functions, after the removal of the dissipative "Debye" potential. It should be noted that recent computer-simulation studies of Rahman, Stillinger, and Lemberg (21) of water correlation motion display both the oscillatory character and the faster-than-inertial decay observed earlier (20).

While it is apparent that the perturbing potential in bulk water corresponds to the forming and breaking of hydrogen bonds between near-neighbor water molecules, the potential causing a fast reorientation on the mica surface requires further discussion. We first note that the fluorophlogopite surface

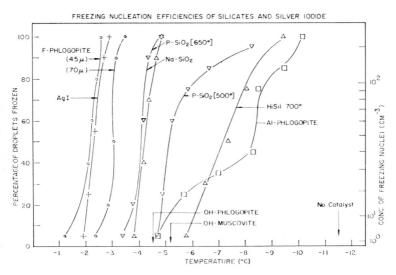

FIG. 11. Ice nucleation from liquid water in the presence of nucleation catalysts, from liquid water according to Shen, Klier, and Zettlemoyer (23). The fluorine micas show nucleation efficiencies close to that of silver iodide whereas hydroxyl micas are relatively poor nucleation catalysts, not significantly better than amorphous HiSil, heat-treated at 700°C.

consists of an aluminosilicate plane network with both the potassium cations and fluoride anions accessible to water. The binding of water to these ions is exemplified by Fig. 10, where lithium, instead of potassium, is used to demonstrate the cation–water binding. The ion–water binding energies, taken from the work of Breitschwerdt and Kistenmacher, and Lischka et al. (22) are calculated here as a function of the angle between the water–hydroxyl axis and the ion–water axis and clearly the configurations that are stable for the cation are unstable for the anion and vice versa. These angular potentials suggest that water which may first be adsorbed on the more exposed cations of the mica surface will be angularly trapped by hydrogen bonds to the fluoride. That fluoride is indeed involved in water anchoring is strongly corroborated by freezing nucleation experiments (Fig. 11) where mica surfaces containing hydroxyls in place of fluoride ions are found to be only moderate nucleation catalysts whereas fluorine micas are excellent nucleants equaling or surpassing

silver iodide. The graph in Fig. 11 shows nucleation efficiencies at a rate of cooling 0.05°C/min (23). A proposed structure taking into account the acceptor hydrogen bonding on fluorine atoms is illustrated in Fig. 12, which also shows, in a vertically expanded fashion, an ice nucleus growing on the mica surface and anchored on those fluoride ions which are exposed because one-half of the potassium ions are missing in the surface for maintaining electrostatic neutrality, as also documented by the surface analytical work of Goldsztaub et al. (24).

The graph in Fig. 11 contains information on the nucleating properties of a number of materials, gathered during a decade of effort to replace silver iodide, introduced in 1947 by Vonnegut (25) and Langmuir for the control of atmospheric precipitation through cloud seeding, by the more ecologically acceptable silicaceous materials. Many different mechanisms of the freezing catalysis are involved. However, a comparison of the nucleating efficiencies and the previously discussed corre-

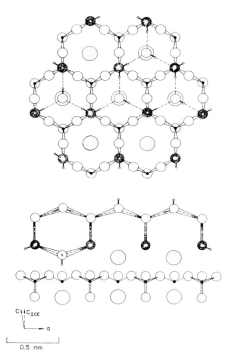

Fig. 12. Proposed structure of an ice nucleus on a fluorophlogopite basal plane. Fluorophlogopite: small black circles, Si or Al; medium open circles, O; large open circles, F$^-$; large dotted circles, K$^+$; only the surface layer with some of the K$^+$ ion missing is shown. Water molecules in ice are represented by open or shaded oxygen atoms with rodlike connections to the small circle hydrogens. The bottom part of the figure is vertically expanded to show how the hydrogen bonds to the fluoride ions are formed.

lation motion allows us to put forward a mechanism for freezing nucleation and conditions that a good nucleant should satisfy. The molecular picture of the freezing of water can be viewed as a conversion of perturbed angular motion into one in which the rotating molecules no longer escape the potential wells of the angular perturbation, i.e., into the rocking motion denoted as libration. It should be pointed out that the periodic motion observed on the water correlation functions on fluorine micas has a frequency close to the libration of ice ($\sim$800 cm^{-1}) observed in the far infrared

(26) and thus one aspect of heterogeneous nucleation is a catalysis inducing vibrations of suitable frequency in the adsorbate. At least some of the following conditions must be satisfied by a good ice nucleation catalyst: (a) the catalyst should have suitable but not too strong adsorption centers for water (F$^-$, I$^-$, OH$^-$); (b) it should locate water molecules close to the crystallographic positions of ice (epitaxy); (c) it should induce in water molecules librations having a frequency close to those of ice. This last condition is satisfied by attractive angular perturbations of moderater stength (0.5–2 kcal/mole).

In conclusion, we have discussed structures of water and its products at surfaces, ranging from strong interactions to those in which only near-thermal energies are involved such as in phase transformation. Molecular events on a time scale of 10^{-14} to 10^{-13} sec appear to be important in phase transformation and it is also this time scale which must encompass the lifetime of water clusters and agglomerates in the liquid and liquid-like adsorbates.

REFERENCES

1. TEXTER, J. A., KLIER, K., AND ZETTLEMOYER, A. C., Water at surfaces, *Progr. Surf. Membr. Sci.*, in press.
2. DWYER, D. J., AND SIMMONS, G. W., to be published.
3. KISHI, K., AND IKEDA, S., *Bull. Chem. Soc. Japan* **46**, 341 (1973).
4. SUHRMANN, R., HERAS, J. M., VISCIDO DE HERAS, L., AND WEDLER, G., *Ber. Bunsenges. Phys.* **72**, 855 (1968).
5. SUHRMANN, R., HERAS, J. M., VISCIDO DE HERAS, L., AND WEDLER, G., *Ber. Bunsenges. Phys. Chem.* **68**, 671 (1964).
6. IMAI, H., AND KEMBALL, C., *Proc. Roy. Soc. Ser. A* **302**, 399 (1968).
7. JOWETT, C. W., DOBSON, P. J., AND HOPKINS, B. J., *Surface Sci.* **17**, 474 (1969).
8. LINNENBOM, J., *Electrochem. Soc.* **105**, 322 (1958).
9. CRONAN, C. L., THESIS, Lehigh University, 1976.
10. MICALE, F. J., Topic, M., Cronan, C. L., Leidheiser, H., AND Zettlemoyer, A. C., *J. Colloid Interface Sci.* **55**, 540; (1976) CRONAN, C. L., MICALE, F. J., TOPIC, M., LEIDHEISER, H., ZETTLEMOYER, A. C., AND POPOVIC, S., *J. Colloid Interface Sci.* **55** 546 (1976).
11. YAO, Y.-F.Y., *J. Phys. Chem.* **69**, 3930 (1965).

12. KUMMER, J. T., AND YAO, Y.-F.Y., *Canad. J. Chem.* **45**, 421 (1967).

13. BASSETT, D. R., BOUCHER, E. A., AND ZETTLEMOYER, A. C., *J. Colloid Interface Sci.* **34**, 3 (1970).

14. KLIER, K., SHEN, J. H., AND ZETTLEMOYER, A. C., *J. Phys. Chem.* **77**, 1458 (1973).

15. BASSETT, D. R., Thesis, Lehigh University, 1967.

16. DEL BENE, J. E., AND POPLE, J. A., *Chem. Phys. Lett.* **4**, 426 (1969); *J. Chem. Phys.* **52**, 4858 (1970).

17. KISTENMACHER, H., POPKIE, H., AND CLEMENTI, E., *J. Chem. Phys.* **58**, 1689 (1973); 5627; 5842.

18. SHEN, J. H., Thesis, Lehigh University, 1974.

19. GORDON, R. G., *J. Chem. Phys.* **42**, 3658 (1965).

20. KLIER, K., *J. Chem. Phys.* **58**, 737 (1973).

21. RAHMAN, A., STILLINGER, F. H., AND LEMBERG, H. L., *J. Chem. Phys.* **63**, 5223 (1975).

22. BREITSCHWERDT, K. G., AND KISTENMACHER, H., *Chem. Phys. Lett.* **14**, 288 (1972); LISHKA, H., PLESSER, T. H., AND SCHUSTER, P., *Chem. Phys. Lett.* **6**, 263 (1970).

23. SHEN, J. H., KLIER, K., AND ZETTLEMOYER, A. C., U. S. Patent 3,858,805 (Jan. 7, 1975).

24. DEVILLE, J. P., EBERHARD, J. P., AND GOLDSZTAUB, S., *C. R. Acad. Sci. Ser. B* **264** 142 (1967); 289; GOLDSZTAUB, S., DAVID, G., DEVILLE, J. P., AND LANG, B., *C. R. Acad. Sci. Ser. B* **262** 1718 (1966).

25. VONNEGUT, B., *J. Appl. Phys.* **18**, 593 (1947).

26. IKAWA, S. I., AND MAEDA, S., *Spectrochim. Acta A* **24** 655 (1968).

The Influence of Hydration upon the Potential at the Shear Plane (Zeta Potential) of a Hydrophobic Surface in the Presence of Various Electrolytes

DONALD EAGLAND AND ANTHONY P. ALLEN

School of Pharmaceutical Chemistry, University of Bradford, Bradford BD7 1DP, West Yorkshire, England

Received June 9, 1976; accepted September 14, 1976

Evidence is presented from the literature for the importance of water structure for the stability of colloidal systems having hydrophilic character and the confused nature of the evidence for the involvement of hydration of hydrophobic surfaces. Streaming potential results at a polyvinylacetate/ water interface in the presence of LiCl, LiBr, LiI, Li$_2$SO$_4$, NaCl, KCl, CsCl, Me$_4$NCl, Et$_4$NCl, and Bu$_4$NCl are presented; the data suggest a cation dependent yield stress within the liquid contained in the diaphragm pores. Together with concomitant surface conductance determinations these results gave values of electrokinetic potential which indicated strong specific anion adsorption at the apolar interface which was also observed in the case of Cs$^+$, K$^+$, and the tetraalkylammonium ions; no specific adsorption was observed in the case of Li$^+$ and Na$^+$. The evidence obtained is used to support a possible explanation of the confusion with regard to the importance of hydration for the stability of hydrophobic sols.

Data published by Freundlich in 1909 (1) on the effect of different electrolytes upon the coagulation of a sulfur sol (Table I) suggest that the interaction of a hydrophobic sol with a counterion is more specific than is suggested by the DLVO theory (2, 3); indeed, Freundlich hypothesized that the order of effectiveness of the ions in promoting coagulation which follows a lyotropic or Hofmeister (4) series, strongly implicated the involvement of solvent in the interaction process. In this connection it is interesting to note the influence of the sulfate ion; much higher concentrations of Li$_2$SO$_4$ than of LiCl are required for the flocculation of a sulfur sol. In the field of hydrophilic colloids Bull and Breeze (5) have reported that the protein egg albumin is preferentially hydrated in sodium sulfate solution and the salt is negatively adsorbed at the protein/solution interface. Behavior of this kind is very similar to that reported by Payne (6) for the mercury/ solution interface, suggesting that at negative

potentials a layer of sulfate-free water exists between the solution and the mercury surface.

Evidence, in the case of hydrophilic colloids, that a correlation exists between lyotropic series and hydration properties is now very substantial (7); the observance of similar lyotropic series for hydrophobic colloids suggests that more consideration should be given to the importance of hydration in the stability of hydrophobic systems. This has not generally been the case; apart from the pioneering work of Lyklema (8) and Dirkhin and Derjaguin (9, 10) progress in determining the role of solvent water within the electrical double layer has been left largely to workers in the field of electrochemistry (11–14).

Evidence (15, 16) suggests that, in a simplified model, hydrophobic and hydrophilic particles can often be considered as being of similar shape but exposing different proportions of hydrophobic and hydrophilic surfaces; attention can therefore be given as to determining whether only water adjacent to the hydrophilic areas of the particles should be considered of importance or whether water adjacent to the hydrophobic regions also has a role to play. Is hydration similar in kind for all hydrophilic sites or does the sign of the site charge influence the nature of the hydration? One must also consider whether the role of water may be more or less important in the diffuse region of the double layer vis-à-vis the Stern region.

Theoretical studies in the case of the diffuse region suggest that the nature of the hydrophilic site is not of importance, but only the overall sign of the charge and the magnitude of the potential at the outer Helmholtz plane (17, 18). Within the Stern region Levine and Matijevic (19) have shown that the linear relationship of the integral capacity to ψ_0 observed experimentally (20) can be reproduced by variation of the relative proportions of partially hydrated counterions and water molecules within the Stern layer, suggesting again that the sign of the charge is the prime factor.

Trasatti (21), however, has reported a quantitative relationship between the capacity of the Stern region of the double layer and the interaction of water dipoles, depending upon the degree of water orientation for metal surfaces in the order

$$\text{Hg, Bi, Sb} < \text{Pb} < \text{Zn} < \text{Cd} < \text{Ga,}$$

strikingly similar to the hydrophilicity sequence of metals published by Frumkin and co-workers (22).

Clavilier and Huong (23) suggest that the F^- is not specifically adsorbed into the dense region of the double layer but that specific adsorption increases in the sequence $Cl^- < Br^- < I^-$; comparison of adsorption of a given ion on Au, Hg, Bi, and Ga indicates a decrease of adsorption related to the passage from a hydrophobic metal to one of a more hydrophilic character.

Specific adsorption of anions at the negatively charged AgI surface follows the sequence already outlined for Hg and other metals according to Berube and de Bruyn (24) but the

TABLE I

Critical Coagulation Concentration of Electrolytes for a Sulfur Sol

Electrolyte	CCC (mM)
Li$_2$SO$_4$/2	1500
LiCl	750
NaCl	190
KCl	85
RbCl	80
CsCl	95

reverse sequence

$$Cl^- \simeq ClO_4^- \simeq NO_3^- > I^-$$

was observed with the strongly hydrophilic negatively charged hydroxylated TiO$_2$ surface; such a reversal of the sequence would be expected from the variation of anion adsorption with hydrophilicity of the (metal) surface reported by Frumkin (22). The direct relationship of the specific adsorption sequence of anions to the nature of the hydration layer at the solid/liquid interface is confirmed by the flocculation studies of Freundlich (25) on a positively charged Fe$_2$O$_3$ sol and the data of Gann (26) and Ishizaka (27), on positively charged Al$_2$O$_3$ sols, the sequences following the order

$$Cl > Br \simeq NO_3^- \simeq CNS^- > I^-,$$

similar to that observed by Berube and de Bruyn.

Berube and de Bruyn suggest the development of a highly ordered water atmosphere around the TiO$_2$ particle of appreciable thickness; this is supported by the evidence of Webb et al. (28). It is interesting, therefore, that water adsorption isotherm studies by Parfitt and co-workers on TiO$_2$ surfaces indicate that only slightly more than monolayer coverage by highly ordered water molecules occurs. Similar discrepancies appear to occur in studies of hydroxylated SiO$_2$ surfaces; Tadros and Lyklema (29) have suggested a porous gellike state for water adjacent to an SiO$_2$ surface, whereas NMR studies by Pearson and Derbyshire (30) suggest that of the order of three

246

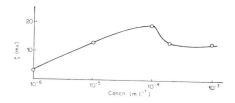

FIG. 3. Zeta potential of the poly(vinylacetate) surface in the presence of NaCl, uncorrected for surface conductance effects.

for forming molecular spatial structures; no yield stress was observed in non-hydrogen-bonding liquids such as acetone, chloroform, and carbon tetrachloride, whereas mixtures of acetone and chloroform which are thought to produce hydrogen bonding exhibit yield stress behavior.

Within the alkali metal halide sequence shown in Table III, no dependence of yield value upon the nature of the anion is apparent with the exception of the sulfate ion and the cation-specific nature of the dependence only becomes apparent at approximately 5×10^{-4} M when the rapid increase in the pressure difference is observed; a similar dependence of charge density at the δ surface and in uncorrected zeta-potential values in this and other work (61) is also observed, implying some form of limiting condition before interaction occurs, notwithstanding the low number of ions present in 5×10^{-4} M (approximately 1×10^{-6} ions/nm^{-2} in the plane of shear). The hydrated crystal radii of the alkali metal cations (nm) are (62)

Li$^+$	Na$^+$	K$^+$	$^+$Cs
0.34	0.276	0.23	0.228

From these results the resistence to flow would be expected to follow the sequence

$$Li^+ > Na^+ > K^+ > Cs^+,$$

the reverse of the experimental findings. These values, however, vary with the amount of water associated with the ions as determined by dynamic measurement; since we are concerned here with yield stress, the destruction of a static condition, the ionic cosphere radii (63) might

be considered more appropriate:

Li$^+$	Na$^+$	K$^+$	Cs$^+$
0.34	0.375	0.413	0.449

The sequence is now in agreement with the experimental data, suggesting that electrostrictive hydration structure is responsible for the appearance of a yield stress.

The phenomenon of surface conductance also shows a marked increase in magnitude over the same concentration range of electrolyte; this trend has also been remarked upon by Morimoto and Kittaka (64) in their study of the effects of surface conductance of metal oxides in electrolyte solution, with the exception of Na$_2$SO$_4$, where no marked increase in surface conductance occurred.

The pressure difference values obtained with the tetraalkylammonium chlorides seem to confirm the picture previously outlined; considerable structuring of the solvent water by the tetraalkylammonium cations, particularly the higher members, has been widely reported (65) but is of a fundamentally different character from that surrounding the alkali metal cations —hydrophobic hydration which can be thought of as stabilized normal water structure. It might be expected, therefore, that the yield stresses should be only slightly different from that of pure water, which is indeed the case.

(b) Surface Conductance and ζ Potential

In many studies of electrokinetic phenomena deviations from the classical theory of Smoluchowski for a single capillary (66) occur and many reports have been published (52) employing modifications designed to overcome the limitations of the original theory. In the work reported here, because, by the nature of the surface being considered, the requirement of low surface potentials (≤ 25 mV) is complied with in the large majority of the studies reported; hence it was very much more straightforward to ensure that the second main requirement, $Ka \gg 1$, be met, thus enabling the classical theory to be applied, albeit with the modification of surface conductance. The particle diameter of 304 μm, giving an effective pore radius of 62 μm (52), satisfies this condition.

The Smoluchowski relationship of streaming potential to applied pressure allowing for the phenomenon of surface conductance is

$$E_s = \frac{\epsilon P \zeta}{4\pi\eta(K_0 + [2K_s/a])}, \qquad [1]$$

where E_s is the developed streaming potential for the applied pressure P, ϵ is the permittivity of the medium and η its viscosity, K_0 is the specific conductance; ζ is the electrokinetic potential which, according to the criteria suggested by Dukhin (70), is coincident with ψ_d in this work.

The values of the ζ potential, uncorrected for the effect of surface conductance, indicated the presence of the maximum value at an electrolyte concentration of approximately 5×10^{-4} M; similar maxima have been reported by Rutgers (67, 68) and James and Humphreys (69); Fig. 3 illustrates one particular example for NaCl.

The concept of K_s may be extended to diaphragms, providing the condition of $Ka \gg 1$ remains satisfied; however, because of the complex geometry involved, large errors arise. It is, therefore, an advantage to utilize the associated variable of the specific conductivity of the liquid within the diaphragm pores, K_p. The experimentally determined resistance of the diaphragm R_d is inversely proportional to K_p, the proportionality coefficient C_d^0 being dependent upon the pore space, and not directly determinable.

Utilizing the method proposed by Briggs (70) C_d^0 may be determined and the corrected zeta potential can then be found from the equation

$$\frac{\zeta}{\zeta_{sm}} = 1 + \frac{K_p - K_0}{K_0} = 1 + \frac{K_s}{K_0},$$

where ζ_{sm} is the zeta potential obtained from the classical Smolukowski equation. The de-

TABLE IV

Specific Surface Conductance (K_s) of the Electrolyte Solution within the Diaphragm (ohm^{-1} m^{-1})

Concentration (M)	LiCl	LiBr	LiI	Li$_2$SO$_4$	NaCl	KCl	CsCl	(CH$_3$)$_4$NCl	(C$_2$H$_5$)$_4$NCl	(C$_4$H$_9$)$_4$NCl
1×10^{-3}	1.61×10^{-2}	2.50×10^{-2}	1.60×10^{-2}	1.87×10^{-2}	0.63×10^{-2}	0.89×10^{-2}	1.2×10^{-2}	0.92×10^{-2}	0.64×10^{-2}	0.63×10^{-3}
5×10^{-4}	3.08×10^{-3}	—	—	—	0.21×10^{-3}	—	—	—	—	—
1×10^{-4}	—	0.93×10^{-3}	1.49×10^{-3}	—	—	—	—	—	—	—
1×10^{-5}	4.50×10^{-4}	—	—	—	2.55×10^{-4}	1.42×10^{-4}	0.32×10^{-4}	1.10×10^{-4}	0.29×10^{-4}	1.06×10^{-4}
1×10^{-6}	4.97×10^{-4}	6.48×10^{-4}	5.98×10^{-4}	8.16×10^{-4}	3.0×10^{-4}	1.21×10^{-4}	0.92×10^{-4}	1.62×10^{-4}	0.14×10^{-4}	1.36×10^{-4}

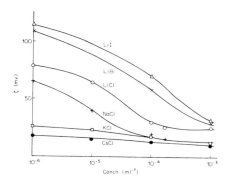

FIG. 4. Zeta potential of the poly(vinylacetate) surface in the presence of various alkali halides, corrected for surface conductance effects.

rived values of specific surface conductance are given in Table IV. It is seen that K_s decreases with decreasing concentration; however, the effects of K_s become more important in the more dilute solutions as can be seen from examining the ratio of K_s/K_0 for LiCl at 10^{-3} M where the value is approximately 1.5, while at 10^{-6} M it is the order of 50. The general trend appears to be that of a leveling-out of the ratio K_s/K_0 at concentrations below 5×10^{-4} M.

No dependence of the surface conductance upon the nature of the anion is apparent in the case of LiCl, LiBr, and LiI but, significantly, the value for Li_2SO_4 at 10^{-6} M is much higher than those observed for the halides. A clear cationic sequence, indicating a lowering of surface conductance in going from Li^+ to Cs^+, is apparent but no clear dependence upon the nature of the cation is apparent within the tetraalkylammonium sequence. The absolute values for the tetraalkylammonium ion sequence are in general much lower than those for most of the alkali metal ions, approximating, at 10^{-6} M, to that of the Cs^+ ion.

Surface conduction effects could be manifested in any or all of the three regions of the electrical double layer:

(a) the inner Helmholtz plane,
(b) the outer Helmholtz plane,
(c) the diffuse region.

Morimoto ane Kittaka consider that in the case of Al_2O_3, Cr_2O_3, and Fe_2O_3 surface conductance originates from ion transfer, not only within the diffuse region but also within the Stern layer. As noted earlier oxide surfaces appear to carry considerable surface hydration of a hydrophilic character; it is therefore unlikely that similar mechanisms of surface conductance will operate in such a dissimilar system as the hydrophobic apolar interface considered in this work. Astrakhamtseva and Us'Yarov (71) have reported surface conductivity in aqueous electrolyte wetting films on Pyrex glass; the authors report that studies on LiCl, NaCl, KCl, and RbCl solutions were undertaken, and the sequence is in the order

$$LiCl \simeq NaCl < KCl < RbCl,$$

regarded by the authors as evidence for increasing adsorbability and mobility of the cations in going from Li^+ to Rb^+. Large differences were obtained between experimentally determined values of surface conductance and those calculated assuming that ζ and ψ_d are identical; the conclusion is, therefore, drawn that the plane of electrokinetic potential lies some distance from the outer side of the Helmholtz plane.

In the data reported here the reverse sequence of ions is apparent in the magnitude of surface conductance. The effect of the Li^+ ion is much greater than that of the Cs^+ ion; no clear trend is discernible in the values of the different lithium halides but the sulfate ion produces a much higher value of surface conductance.

The conditions for equating the electrokinetic potential with ψ_d are satisfied in the data reported here. The corrected values of ζ (ψ_d) shown in Figs. 4 and 5 are, therefore, of interest; in the case of the anions strong specific adsorption is apparent with a slight trend in the sequence of $I^- > Br^- > Cl^-$. The value for the Li_2SO_4 solution, at 1×10^{-6} M concentration, not shown in the diagram, is 36 mV, much lower than values observed in the case of the halides suggesting much less specific adsorption of the sulfate ion. This result is in agreement with the evidence reported by Payne for the Hg/electrolyte interface. With regard to the alkali metal cations the large negative value of ζ associated with the lithium salts suggests that little, if any, transfer of Li^+ ions from the diffuse region of the double layer into the outer Helmholtz plane occurs; it appears that some transfer occurs in the case of the Na^+ ion but considerable transfer is observed in the case of the K^+ and Cs^+ ions. The evidence thus suggests that surface conductance occurs to a large extent within the diffuse region of the double layer and cations transferred into the outer Helmholtz plane contribute little to the phenomenon: this seems particularly clear in the case of the sulfate ion.

The lyotropic sequence observed in the values of ζ are in the sequence that would be expected from the critical coagulation studies of hydrophobic colloids reported; the high concentrations of lithium salts required for flocculation are reflected in the inability of the Li^+ ion to approach the outer Helmholtz plane of a hydrophobic surface. Such a situation is unlikely to be determined by the polarizability of the ion but from the entropic term arising from the interaction of the hydrophobic hydration sheath of the poly(vinylacetate) surface and the hydration B shell (72) of the cation.

Support for the viewpoint that adsorption of the Li^+ and Na^+ into the Stern region of the double layer does not occur arises from the suggestion of Benton and Elton (73) that for a 1:1 electrolyte in water the rate of change of ζ with concentration when no specific adsorption occurs is close to 20 mV per decade of change concentration—the data for Li^+ and Na^+ give changes of this order of magnitude.

The strong specific adsorption of the halide anions follows the known hydration characteristics of the ions and gives rise to the appreciable negative surface potentials at low electrolyte concentrations on what is otherwise a nonpolar surface with no bound charge sites.

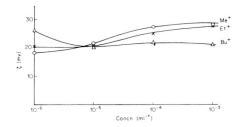

Fig. 5. Zeta potential of the poly(vinylacetate) surface in the presence of various tetraalkylammonium chlorides, corrected for surface conductance effects.

Similar specific adsorption of halide anions at an apolar or a relatively apolar hydrocarbon surface has been reported by Kovylov and Krasikov (74) for polyacrylonitrite suspensions and by Benes and Paulenova (75) for polyethylene surfaces.

The very low values of the ζ potential obtained in the case of the tetraalkylammonium ions reveal that considerable transfer of cations into the Stern region of the double layer occurs; the trend in the magnitude of the ζ potential at $10^{-6}\ M$ of $Bu_4N^+ > Me_4N^+$ reflects the increasing size of the alkly group. The data suggest that the basic mechanism of transfer of the tetraalkylammonium ions into the Stern region is through hydrophobic interaction between the alkyl groups of the ion and the apolar poly(vinylacetate) surface; this implies that specific adsorption of the organic cations into the inner Helmholtz plane of the interface occurs.

We may recall that in the case of the yield stress for capillary flow of these electrolyte solutions against the poly(vinylacetate) surface the lyotropic sequence is reversed compared to the sequence observed for zeta potentials. The sequence observed with regard to zeta potential is similar to that obtained for critical coagulation concentrations of these salts for hydrophobic colloids, including AgI, the sequence observed in the yield stress behavior is similar to that reported for surface potential studies on the double layer of AgI.

It seems to us that an explanation of this reversal of behavior lies in considering the hydration of the ion when in motion or stationary; in the stationary state Gurney cosphere hydration behavior appears to dominate as in the case of yield stress behavior for poly-(vinylacetate) and surface-potential studies of AgI. In the mobile state Jones–Dole "B coefficient" behavior is dominant, as in the case of zeta-potential studies of poly(vinylacetate) and critical-coagulation studies of hydrophobic colloids.

REFERENCES

1. VERWEY, E. J. W., AND OVERBEEK, J. TH. G., "Theory of the Stability of Lyophobic Colloids." Elsevier, Amsterdam, 1948.
2. DERJAGUIN, B. V., AND LANDAU, L., *Acta Physicochim. URSS* 14, 633 (1941).
3. FREUNDLICH, H., "Kolloid und Capillairchemie," 3rd ed. (English translation). Methuen, London, 1926.
4. HOFMEISTER, F., *Arch. Exp. Pathol. Pharmakol.* 24, 247 (1888).
5. BULL, H. B., AND BREEZE, K., *Arch. Biochem. Biophys.* 128, 497 (1968).
6. PAYNE, R., *Electroanal. Chem. Interfac. Electrochem.* 60, 183 (1975).
7. EAGLAND, D., *in* "Water–A Comprehensive Treatise" (F. Franks, Ed.), Vol. 4. Plenum, New York, 1975.
8. LYKLEMA, J., *Discus. Faraday Soc.* 42, 81 (1966).
9. DUKHIN, S. S., AND DERJAGUIN, B. V., *Dokl. Akad. Nauk. SSSR* 159, 636 (1964).
10. DERJAGUIN, B. V., DUKHIN, S. S., AND KOPTELOVA, M. M., *Kolloid Z.* 31, 359 (1969).
11. DAMASKIN, B. B., AND FRUMKIN, A. N., *Electrochim. Acta* 19, 173 (1974).
12. TRASATTI, S., *Electroanal Chem. Interfac. Electrochem.* 54, 437 (1974).
13. BOCKRIS, J. O.'M., AND HABIB, M. A., *J. Electroanal. Chem.* 65, 473 (1975).
14. PARSONS, R., *Electroanal Chem. Interface. Electrochem.* 59, 229 (1975).
15. SHRAKE, A., AND RUPLEY, J. A., *J. Mol. Biol.* 79, 351 (1973).
16. EAGLAND, D., PILLING, G., AND WHEELER, R. G., *Discuss. Faraday Soc.* 57, 181 (1974).
17. HASTED, J. B., RITSON, D. M., AND COLLIE, C. H., *J. Chem. Phys.* 16, 1 (1948).
18. SANFELD, A. DEVILLEZ, C., AND TERLINCK, P., *J. Colloid Interface Sci.* 32, 33 (1970).
19. LEVINE, S., AND MATIJEVIC, E., *J. Colloid Sci.* 23, 188 (1967).
20. LYKLEMA, J., *Trans. Faraday Soc.* 59, 48 (1963).
21. TRASATTI, S., *J. Electroanal Chem.* 65, 815 (1975).
22. FRUMKIN, A., DAMASKIN, B., GRIGORYEV, N., AND BAGOTSKAYA, I., *Electrochim. Acta* 19, 69 (1974).
23. CLAVILIER, J., AND HUONG, N. V., *Electroanal. Chem. Interfac. Electrochem.* 41, 259 (1973).
24. BERUBE, Y. G., AND DE BRUYN, P. L. J., *J. Colloid Interface Sci.* 28, 92 (1968).
25. FREUNDLICH, H., *Z. Phys. Chem.* 44, 151 (1903).
26. GANN J. A., *Kolloid chem. Beih.* 8, 125 (1916).
27. ISHIZAKA, N., *Z. Phys. Chem.* 83, 97 (1913).
28. WEBB, J. T., BHATNAGOR, P. D., AND WILLIAMS, D. G., *J. Colloid Interface Sci.* 49, 346 (1974).
29. TADROS, T. F., AND LYKLEMA, J., *Electroanal Chem. Interfac. Electrochem.* 17, 267 (1968).
30. PEARSON, R. T., AND DERBYSHIRE, W., *J. Colloid Interface Sci.* 46, 232 (1974).
31. GARGALLO, L., BEPULVEDA, L., AND GOLDFARB, J., *Kolloid Z. Z. Polym.* 229, 51 (1969).
32. US'YAROV, O. G., *Kolloid. Z* 37, 79 (1975).
33. BARAN, A. A., SOLOMENTZEVA, I. M., AND KURILENKO, O. D., *Kolloid Z.* 37, 219 (1975).
34. KLOMPE, M. A. N., AND KRUYT, H. R., *Kolloid Beih.* 54, 484 (1943).
35. LYKLEMA, J., *Discuss. Faraday Soc.* 42, 81 (1966).
36. DAMASKIN, B. B., *J. Electroanal. Chem.* 65, 799 (1975).
37. PAYNE, R., *Electroanal Chem. Interfac. Electrochem.* 60, 183 (1975).
38. WROBLOVA, H., KOVAC, Z., AND BOCKRIS, J. O. M., *Trans. Faraday Soc.* 61, 1523 (1965).
39. MACKOR, E. L., *Rec. Trav. Chim.* 70, 763 (1951).
40. LYKLEMA, J., *Kolloid Z.* 175, 129 (1961).
41. PIEPER, J. H. A., DE VOOYS, D. A., AND OVERBEEK, J. TH. G., *J. Electroannal. Chem.* 65, 429 (1975).
42. FRANKS, F., *Chem. Ind.*, 204 (1961).
43. UNNI, A. D. R., ELIAS, L., AND SCHIFF, H. I., *J. Phys. Chem.* 67, 1216 (1973).
44. JONES, G., AND DOLE, M., *J. Amer. Chem. Soc.* 51, 2950 (1929).
45. KAMINSKY, M., *Z. Phys. Chem. (Frankfurt)* 8, 173 (1950).
46. EAGLAND, D., AND PILLING, G., *J. Phys. Chem.* 76, 1902 (1972).
47. ONSAGER, L., AND FUOSS, R. M., *J. Phys. Chem.* 26, 2689 (1932).
48. OWEN, B. B., *J. Chem. Phys.* C-72, 49 (1952).
49. DOGGET, H. H., BAIN, E. J., AND KRAUS, C. A., *J. Amer. Chem. Soc.* 73, 799 (1951).
50. SHEDLOVSKY, Y., *J. Amer. Chem. Soc.* 54, 1411 (1932).

51. OWEN, B. B., AND ZELDES, K. A., *J. Chem. Phys.* **18**, 1083 (1950).

52. DUKHIN, S. S., AND DERJAGUIN, B. V., *in* "Surface and Colloid Science, Vol. 7" (E. Matijevic, Ed.,), p. 167. Wiley, New York, 1974.

53. BULL, H. B., *J. Amer. Chem. Soc.* **57**, 259 (1925).

54. HUNTER, R. G., AND ALEXANDER, A. E., *J. Colloid Sci.* **17**, 781 (1962).

55. PARRIERA, H. C., *J. Colloid Sci.* **20**, 1 (1965).

56. BALL, B., AND FUERSTENAU, D. W., *Miner. Sci. Eng.*, **5**, 267 (1973).

57. VON ENGLEGARDT, D., AND TUNN, W., *U. S. Geol. Surv. Circ.* **194**, 17 (1955).

58. LOW, P. F., *Advan. Agron.* **13**, 269 (1961).

59. NERPIN, S. V., AND BONDARENKO, N. F., *Dokl. Akad. Nauk. SSSR* **114**, 833 (1957).

60. BONDARENKO, N. F., *Dokl. Akad. Nauk. SSSR* **177**, 383 (1967).

61. WRIGHT, M. H., AND JAMES, A. M., *Kolloid Z. Z. Polym.* **251**, 745 (1973).

62. ROBINSON, R. A., AND STOKES, R. H., "Electrolyte Solutions", Butterworths, London 1968.

63. FRIEDMAN, H. L., AND KRISHNAN, C. V., *in*

"Water–A Comprehensive Treatise" (F. Franks, Ed.), Vol. 3. Plenum, New York, 1973.

64. MORIMOTO, T., AND KITTAKA, S., *J. Colloid Interface Sci.* **44**, 289 (1973).

65. WEN, W. Y., AND SAITO, S., *J. Phys. Chem.* **68**, 2639 (1964).

66. SMOLUCHOWSKI, M., *Krak. Anz.*, 182 (1903).

67. RUTGERS, A. J., *Trans. Faraday Soc.* **36**, 69 (1940).

68. RUTGERS, A. J., AND DE SMET, M., *Trans. Faraday Soc.* **43**, 102 (1947).

69. JAMES, A. M., AND HUMPHREYS, M. N., *J. Colloid Interface Sci.* **29**, 696 (1969).

70. BRIGGS, D. K., *J. Phys. Chem.* **32**, 641 (1928).

71. ASTRAKHANTSEVA, N. P., AND US'YAROV, O. G., *Kolloid Z.* **36**, 827 (1974).

72. FRANK, H. S., AND EVANS, M. W., *J. Chem. Phys.* **13**, 507 (1945).

73. BENTON, D. P., AND ELTON, G. A., *Proc. Int. Congr. Surface Activ. 2nd L.* **3**, 28 (1957).

74. KOVYLOV, A. E., AND KRASIKOV, N. N., *Kolloid Z.* **33**, 74 (1969).

75. BENES, P., AND PAULENOVA, M., *J. Colloid Polym. Sci.* **252**, 472 (1974).

Water at Interfaces: A Colloid-Chemical Approach

J. LYKLEMA

Laboratory for Physical and Colloid Chemistry of the Agricultural University,
De Dreijen 6, Wageningen, The Netherlands

Received May 25, 1976; accepted July 19, 1976

A systematic study with silver iodide, involving double-layer measurements, stability studies, and electrokinetics, leads to the conclusion that the electrokinetic potential ζ and the outer Helmholtz plane potential ψ_d coincide within experimental error. Consequently, there are no thick stagnant water layers at the interface. The same is concluded for flow in thin liquid films.

INTRODUCTION

It is obvious that water in the immediate surroundings of a solid surface is different from bulk water. The big questions are: How different is it and how far from the surface do these differences persist? This quantitative issue has drawn the interest of many scientists and it was not surprising that the Anniversary Meeting of the Colloid and Surface Division of the American Chemical Society paid special attention to it.

Basically, there are two reasons why any quantitative assessment is bound to be problematic. First of all, there is the problem of which property to study and how to measure it. Different properties may have different sensitivities to the presence of a nearby wall. Second, contiguous water cannot be isolated. Rather, deviations from bulk character are inferred from anomalies in systems containing the two bulk phases in addition to the interface under consideration. In other words, one is actually looking for an *excess* quantity. A given property may be assigned to an interfacial region only if the same property for the two bulk phases is very well known and properly accounted for.

Let us, by way of example, discuss two types of experiments that have currently been used to infer properties of water near surfaces.

The first is viscous flow in narrow capillaries. For ordinary flow, according to Poiseuille, the liquid flux is $Q = \pi r^4 P/8\eta l$ where r is the capillary radius, η the liquid viscosity, and P/l the pressure drop. Over and again, it has been observed that in very narrow capillaries the flux is lower than predicted by Poiseuille's law. It is tempting to attribute this to either an increased effective viscosity or a decreased effective radius. Both interpretations definitely state that the capillary wall exerts its influence over some distance into the water. The objection to this conclusion is that the validity of Poiseuille's law is presupposed, but this is not correct. In narrow capillaries, pores, and slits, the applied pressure more often than not leads to a streaming potential that, in turn, produces a counter-electro-osmotic flow. This counteracting effect reduces the flow considerably and consequently Poiseuille's law requires substantial modification. The difficult point is the quantitative one. Several serious attempts have been made to account for it. The complete solution of the problem requires, among other things, insight into the composition of the Stern layers, double-layer overlap, and surface conductance (1–9). For a judicious estimation of the thickness of stagnant water layers on the capillary wall or, for that matter, of the increase of the effective viscosity, it is necessary to have full command over the physics of surface conduction, Stern-layer adsorption, double-layer overlap and to know the relevant experimental parameters accurately. In the opinion of this author, no complete solution encompassing all of this for one system has as yet been achieved. Any conclusion concerning the properties of adjacent water must be considered with great reservation as long as it is not convincingly demonstrated that there is no defect in the underlying physical picture.

The second example concerns the thickness d of thin aqueous films on solid surfaces. In equilibrium, the sum of the forces determining d is zero. In simple cases, there are only three types of forces, an electrical one F_{el}, a dispersion force F_{vdw}, and a capillary force F_c. This enables one to consider two of them as "known" and the third as the "unknown." For example, in this way $F_{vdw}(d)$ has been found from the equilibrium thickness of liquid films. Pushing the analysis one step further, one could substitute values for all forces involved to the best of present-day knowledge. Then, as a rule, the sum of the forces will not exactly add up to zero. What is left may be looked upon as an additional force, perhaps to be interpreted as due to a water structure effect, emanating from the surface (10). It is clear that this is also a dangerous procedure. In the first place, the calculation of F_{el} suffers from inadequate knowledge of the outer Helmholtz plane potentials ψ_d and the actual distance between the two outer Helmholtz planes, a problem to be discussed below. The dispersion energy contains factors of the shape

$$[\epsilon_w(\omega) - \epsilon_s(\omega)]/[\epsilon_w(\omega) + \epsilon_s(\omega)],$$

which are to be integrated over the entire frequency range for any distance from the surface. Here, $\epsilon(\omega)$ is the dielectric dispersion, the indices w and s referring to water and solid, respectively. It is clear that F_{vdw} can be computed only if, also for the adjacent layers, $\epsilon_w(\omega)$ is fully available. In practice, this is not the case. Usually, F_{vdw} is calculated as $A/6\pi d^3$, where A is the *bulk* Hamaker con-

stant. The consequence is that again any deficiency of the calculation is assigned to the "additional force."

These two examples, selected because of their topicality, can be extended by many others. Interested readers are referred to an old review by Henniker (11), who, on the basis of often debatable analyses, arrives at the conclusion that the effect of a surface can be felt inside an adjacent liquid up to considerable distances.

The question may be asked whether there are model systems available allowing, with some reliability, the establishment of the effective thickness of surface layers. Below, this question will be approached from the colloid-chemical side; i.e., in this paper we refrain from a discussion of various spectroscopic techniques.

THE ψ_d–ζ PROBLEM

Asking for the difference between the OHP potential ψ_d and the electrokinetic potential ζ is one way to phrase the issue in colloid-chemical terms. At the same time, it is one of the major problems of colloidal stability. Both ζ and ψ_d are elusive parameters; nevertheless, it is impossible to describe colloidal stability and electrokinetics without knowing ψ_d or ζ, respectively.

In most current double-layer models, the outer Helmholtz plane is the plane where the diffuse double layer starts or, for that matter, the plane beyond which the charge distribution obeys Poisson–Boltzmann statistics. Any deviation from ideality, notably finite ionic volume and specific adsorption effects, is assigned to the Stern layer, that is, the layer between the plane of the surface charge and the OHP. It is good to realize that attributing all nonideality effects to the Stern layer only is a kind of physical abstraction. In reality, the double layer becomes progressively more ideal with increasing distance from the surface. However, experience accumulated over the past decades has shown that this model works very well, apparently because of the fact that the two nonideality effects decay very rapidly with distance. The thickness Δ of the Stern layer is a few tenths of a nanometer, say two to four water layers. Further narrowing down this thickness would be unrealistic since all defects in the diffuse theory (for instance, the neglect of surface roughness and discreteness-of-charge effects) are artificially attributed to the Stern layer and consequently are interpreted in terms of the Stern-theory parameters, among which are Δ and the specific adsorption free energy ϕ.

There is no way to measure ψ_d directly. The most viable approach seems to be assessing it indirectly from colloidal stability. For example, by analyzing stability simply in terms of critical coagulation concentrations c_c, it is possible to relate ψ_d to c_c with the following modification of an equation of the DLVO-theory (12–14):

$$c_c = f(T, \epsilon, \Delta) \left[\tanh\left(ze\psi_d/4kT\right)\right]^4/A^2 z^6, \quad [1]$$

$$c_c \sim f'(T, \epsilon, \Delta)\psi_d^4/A^2 z^2. \quad [2]$$

Here, A is the Hamaker constant (that is, the short-distance limit of the Hamaker function), z the counterion valency, e the elementary charge, and f and f' are known functions of the temperature T, relative dielectric constant ϵ, and Δ. Equation [2] is the low-potential limit of [1]. It is a good approximation in most of the cases which are encountered. This equation illustrates that, because of the fourth power of ψ_d, estimation of this potential is relatively reliable, even if Δ and A are not very precisely known. Using Eqs. [1] and [2] presupposes that diffuse double-layer overlap takes place at fixed ψ_d, a point that also requires discussion.

The electrokinetic potential ζ is also an abstraction. It is commonly interpreted as the potential of the slipping plane, tacitly presupposing that the tangential motion of a liquid involves a stagnant layer, discretely bounding ordinary bulk fluid. In other words, the viscosity η is thought to be infinite close to the surface, jumping to its bulk value at the slipping plane. However, it would be more realistic to treat $\eta(x)$ as a gradually decreasing function. In the Lyklema–Overbeek model, $\eta(x)$ is obtained from $\eta(E)$, where $E(x)$ is the double-layer field strength (15), but quantitative verification of this theory is not quite attainable since the viscoelastic constant of water is not known. Again, for lack of a better model, the jump model is used and, as in the previous case, practice has shown that this picture works reasonably well. The potential ζ of the slipping plane can be experimentally determined from, say, the electrophoretic mobility provided a number of conditions with respect to surface potential, double-layer thickness, and particle radius are met (16). At the same time it must be realized that any defect in the theory is reflected in the value obtained for ζ. For example, the theory (16) is based on homogeneous smooth surfaces and ignores surface conductance. If the surface were rough, a lower value for ζ would be found (17).

The crucial question for our present theme is: By how much do ψ_d and ζ differ? If ζ were (much) below ψ_d, this might point to an effect of the wall on the water molecules, reaching beyond the short range of specific interaction and rendering the fluid resilient against tangential disturbances. It would, at least in this sense, point to a long-range influence of the surface. On the other hand, if ψ_d and ζ could be proven to be identical, this would point to a surface effect persisting at most to a few molecular layers.

There are practically no systematic studies of the ψ_d–ζ problem in the literature of wide enough scope to convey some feeling of reliability. One of the main difficulties is that ψ_d is only a small fraction of the surface potential ψ_0. Consequently, the value calculated from an experimentally measured ψ_0 using an equation for the Stern layer is extremely sensitive to the choice of the parameters. Similar reasoning applies to an analysis in terms of charges. For any useful computation, it is at least mandatory to have an independent indication that the model chosen for the Stern layer is applicable. In turn, this would require experi-

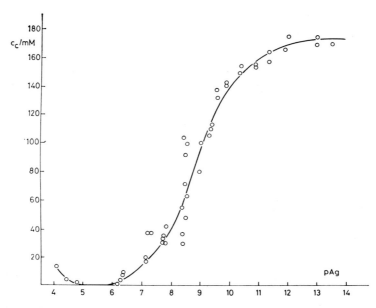

FIG. 1. Coagulation concentrations of AgI sols as a function of pAg, derived from log W − log c plots, temperature 25°C.

ments with a model system in which stability and electrokinetics can be systematically studied as a function of ψ_0 and/or the surface charge σ_0. This is a very strict criterion. It excludes systems where the value of ψ_0 is subject to argument (oxides, micelles), systems where the charge is due to adsorbed surfactants (if it is not known what happens to these charges upon particle encounter) and perhaps also latices: although their surface charge can be altered by choosing different conditions of polymerization, these different conditions might lead to differences in the surface structure. In the framework of this paper, it is not feasible to review adequately all the relevant literature. The AgI system, though far from perfect in all respects, fulfills at least some of the important requirements. It will be analyzed in the next section.

Before doing so, it may be good to point to one possible alternative, developed by Dukhin *et al.* (18, 19). The underlying idea is that ions in the stagnant layer retain their mobility, so

that in, say, low-frequency dielectric studies or studies involving surface conductance, the contributions of these ions must be taken into account. Dukhin counts all ions beyond the OHP as mobile. Consequently, liquid flow occurs only beyond the slipping plane, but ionic conduction occurs beyond the OHP. By measuring both simultaneously, the difference between ψ_d and ζ can in theory be established. A serious shortcoming of this theory is that it ignores Stern layers. Moreover, the experiments it requires are not easy. Except for a provisional attempt (20), this theory has not yet been applied to practical systems as far as this author is aware.

ANALYSIS OF THE AgI SYSTEM

As a model, silver iodide has several advantages. Stability and electrophoretic mobility have been measured as a function of ψ_0, whereas the relationship $\sigma_0(\psi_0)$ has also been established experimentally. Drawbacks are the polydispersity of the dispersion and the uncertainty whether the surface charge is homogeneously distributed, even after protracted aging.

Figure 1 gives critical coagulation concentrations as a function of pAg. The actual value obtained for c_0 depends somewhat on the nature of the sample and on the method of measurement. These results have been derived from log W — log c plots, where W is the so-called stability ratio. With respect to the technique, the classical method, using coagulation tubes (21), involves a mixture of orthokinetic and perikinetic coagulation. It produces results that are lower by ca. 25%. On the other hand, Tyndallometric measurements on *in statu nascendi* sols yield somewhat higher results (22). We decided to base our analyses on the data of Fig. 1, considering that with this technique a stability property is measured under pretty much the same conditions as where electrophoresis is done, i.e., a property of primary particles is investigated. The classical method is based on visual observation after a longer period over which already big clusters are formed. The *isn* technique must be rejected, since it involves growing sol particles, entirely different from the mature particles studied in electrophoresis. Notwithstanding these drawbacks, no serious errors would have been made if one of the alternative sets of data had been taken. The difference of 25% with the classical method is almost independent of ψ_0 and leads to an uncertainty in ψ_d of not more than 6%, which is hardly measurable.

Values of ψ_d were computed using the following set of equations:

$$W = \int_0^\infty \frac{\beta(u')}{(2+u')^2} \exp\left(\frac{V_t}{kT}\right) du' \Big/$$

$$\int_0^\infty \frac{\beta(u')}{(2+u')^2} \exp\left(\frac{V_A}{kT}\right) du', \quad [3]$$

$$\beta(u') = (6u'^2 + 13u' + 2)/(6u'^2 + 4u'), \quad [4]$$

$$u' = H'/(\Delta + a), \quad [5]$$

$$V_A = -\frac{A_{1(2)}}{6}\left(\frac{2}{s^2-4} + \frac{2}{s^2} + \ln\frac{s^2-4}{s^2}\right) \quad [6]$$

$$V_R = 64\pi\epsilon_0\epsilon\left(\frac{kT}{ze}\right)^2\left(\frac{a+\Delta}{s'}\right)\left[\tanh\left(\frac{ze\psi_d}{4kT}\right)\right]^2$$
$$\times e^{-\kappa(a+\Delta)(s'-2)}, \quad [7]$$

$$s = R/a, \quad [8]$$

$$s' = R/(a+\Delta), \quad [9]$$

$$H' = R - 2\Delta. \quad [10]$$

In [3], $\beta(u')$ is a factor accounting for the reduction in the rate of diffusion at close separation because of hydrodynamic interaction (23). $V_t = V_A + V_R$ is the total interaction energy. The new feature in these equations is the explicit introduction of Δ. The underlying idea is that V_A operates over a distance $H \cdot$ between the particle surfaces, whereas V_R acts only over $H' = H - 2\Delta$ since only the diffuse parts of the double layers are involved. Further symbols are the Hamaker constant $A_{1(2)}$, the particle radius a, the distance R between the particle centers, the relative dielectric constant ϵ of the medium, ϵ_0 being the permittivity of free space, and the reciprocal Debye length κ. Equation [6] is due to Hamaker (24), and [7] is our modification of an equation given by Verwey and Overbeek (25), probably the analytical equation with the largest range of validity (26).

The application of Eqs. [3]–[8] to compute ψ_d requires some comment. First, there is the problem that in our analysis *interaction* data are combined with *static* double-layer data. Arguments in support of this have been collected in Ref. (27).

The other point is that our analysis contains some circuity. The value to be derived for ψ_d depends somewhat on the value assigned to Δ. The consequence is that an assumption on the range of surface forces exerted on *ions* must be made before a conclusion can be arrived at on the range of surface forces on *solvent molecules*. This circuity is unavoidable and hence Δ must be taken as an adjustable parameter. In this work we have covered the range 0–0.54 nm (27, 28). Some circuity is encountered also in our using Eq. [6] with the bulk Hamaker constant. Formally, this could be read as if any influence of the surface on the adjacent water structure is ignored. However, in the computation a value for $A_{1(2)}$ was chosen which was derived from stability studies with the same sol under conditions where the other stability-determining factors are easily accessible (for instance, experiments at high temperature where the double layer is ideally diffuse (29)). In doing so, $A_{1(2)}$ becomes a composite parameter in which water structure contributions are already incorporated. The computations in this paper have been done with $A_{1(2)} = 3.4 \times 10^{-20}$ J. An uncertainty of ca. 20% either way would lead to a variability of ca. 10% in ψ_d which, in turn, would have no significant effect on our final conclusion concerning the ψ_d–ζ relationship. Finally, for a we took 50 nm. The sols are heterodisperse, but a variation of a between 10 and 90 nm affects ψ_d by not more than a few tenths of a millivolt (28).

Carrying out the computation yields a $\psi_d(pAg)$ graph, more or less parallel to the curve of Fig. 1. ψ_d decreases from zero at the pzc to about -23 mV at high pAg. This curve cannot yet be compared with experimental $\zeta(pAg)$ curves as the former data are taken at $c = c_c$, that is, at *variable* concentration, whereas the latter apply to *constant* concentration. Conversion into a $\psi_d(pAg)$ graph at constant c requires a relation between ψ_d and c or, for that matter, a Stern-layer adsorption isotherm. To that end we used

$$\frac{\theta}{1-\theta} = \frac{c}{55.5} \exp[(ze\psi_m/kT) - \phi], \quad [11]$$

with

$$\theta = |\sigma_m|/|\sigma_0| = (|\sigma_0| - |\sigma_d|)/|\sigma_0|, \quad [12]$$

$$\sigma_d = (8\epsilon_0\epsilon cRT/1000)^{\frac{1}{2}} \sinh (ze\psi_d/2kT), \quad [13]$$

$$\psi_m = \psi_0 - [(\Delta - \gamma)/\Delta](\psi_0 - \psi_d). \quad [14]$$

In these equations, ψ_m is the inner Helmholtz

plane (IHP) potential, ϕ the specific adsorption Gibbs energy, σ_m the charge density due to specifically adsorbed ions at the IHP, and γ the distance between the IHP and OHP, for which 0.095 nm was taken on the basis of model considerations (27, 28). The electrolyte concentration c in [13] is expressed in moles/liter. In the present analysis, the counterion valency was always $+1$. For any pAg and several values of c, values of σ_0 and ψ_0 are available from potentiometric titration (31).

Equation [11] is a modification of the original Stern equation (30). Essentially, it is the Langmuir equation, corrected for lateral electric interaction in the Bragg–Williams approximation. Using this equation implies that specific adsorption is localized and never superequivalent. For the AgI system, this is the most probable case. Simple cations, such as K^+, do not adsorb at uncharged AgI surfaces; they adsorb only if there are I^- sites on the surface. In passing, the author wants to stress that measurement of the tangential diffusivity of Stern ions still needs to be done. The outcome is important for the interpretation of surface conductance and the mechanism of particle interaction (constant charge or constant potential?).

It transpired from the computation that [11] applies moderately well over the higher pH range with $\phi_{K^+} = -4kT$ and $\Delta = 0.54$ nm. Closer to the pzc the calculation becomes progressively less accurate because of the low values of θ, but at the same time the value obtained for ψ_d at the desired concentration becomes less sensitive to the actual choice of ϕ. For the sake of giving an impression of the sensitivity of the final result upon ϕ, some calculations have also been done with other numerical values. Electrophoresis measurements are, as a rule, carried out at concentrations far below c_c; this again means that relatively large errors made in the calculation of ψ_d at $c = c_c$ lead to relatively accurate values of ψ_d at the lower concentration at which ζ has been measured.

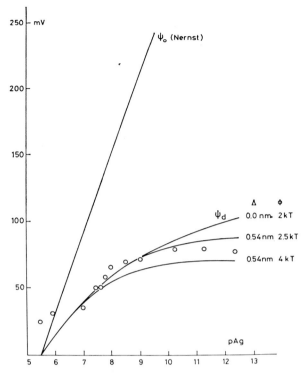

FIG. 2. Comparison of double-layer potentials for negatively charged AgI sols. ψ_0, surface potential according to Nernst; measuring points, electrokinetic potentials ζ; drawn curves, outer Helmholtz plane potentials ψ_d for different values of the Stern-layer thickness Δ and the Gibbs energy of specific adsorption ϕ.

For comparison purposes, ζ-potential measurements are needed on coarse mature AgI sols over the entire pAg range. The old results by Troelstra and Kruyt (32) suit this purpose. They have been confirmed within experimental error by Bijsterbosch (33).

Results are graphically collected in Fig. 2. This graph illustrates how sensitive $\psi_d(c)$ is to the choice of Δ and ϕ. In the $\Delta = 0$ case, the difference in the range of operation of V_A and V_R was ignored, i.e., in Eqs. [3]–[10], Δ was chosen zero, as is usually done, but in [11]–[14] $\Delta = 0.54$ nm was used.

The main conclusion is that within the range of reasonable values for Δ and ϕ, the OHP potential ψ_d and the electrokinetic potential are identical (the ζ-potential measure-

ments around the pzc are obviously incorrect (34)). The consequence is that with respect to the resistance against tangential motion there is no indication of any water structure formation beyond the Stern layer.

Vice versa, it follows from the comparison that a Stern-layer thickness of about 0.54 nm and a specific adsorption Gibbs energy somewhere between -2.5 and $-4\,kT$ are acceptable parameters for the double layer on AgI.

OTHER SYSTEMS

One other experiment allowing an accountable conclusion on the thickness of stagnant layers in tangential motion is the flow in thin liquid films. If a rigid film (i.e., a film with an inextendable surface) is pulled out of a surfactant solution, its thickness d depends as follows on the rate of pull-out v, the liquid viscosity η, the density ρ, and the surface tension γ:

$$d = 1.88(\eta^{\frac{2}{3}}v^{\frac{2}{3}}/\gamma^{1/6}\rho^{\frac{1}{2}}g^{\frac{1}{2}}). \qquad [15]$$

Here, g is the gravity constant. This equation has been derived by Frankel (35) and independently by Deryagin and Levi (36, 37). Experimental tests (38) have confirmed this law down to a few tens of a nanometer. Below these thicknesses, [15] becomes invalid because of double-layer repulsion and van der Waals compression (39). Extrapolation of the linear $d(v^{\frac{2}{3}})$ plots to $v = 0$ should give a zero intercept according to Eq. [15]. In practice, an intercept of 3.2 ± 1.6 nm was obtained for films of sodium dodecyl sulfate, rendered rigid by small admixtures of dodecyl alcohol (38). This finite intercept was attributed to the fact that the surfactant surface layers did not participate in the flow. Considering that 3.2 nm may well correspond to the thickness of two surfactant layers, inclusive of their Stern layers, it is again concluded that within a limit of 0.8 nm there is no indication of thick stagnant water layers.

CONSEQUENCES

The conclusion that, at least for two model systems, there are no thick stagnant water layers is gratifying. It simplifies a number of colloid-chemical analyses. For example, in theories involving surface conductance and/or low-frequency dispersion (18, 19) there is no need to consider the contribution to surface conductivity by the ions between the OHP and the slipping plane. Any surface conductivity in excess of the diffuse double-layer contribution (40) can be assigned to the Stern layer. Identifying ψ_d and ζ is common practice. This paper concludes that, within experimental error, this practice is viable.

At the same time, this paper concludes that accounting for different interaction distances of V_R and V_A is mandatory. This is not usually done. The importance of the Stern-layer properties is also underscored.

The final conclusion that there are no thick stagnant layers in tangential flow or, for that matter, in electrokinetics, is simple and rewarding but not necessarily appealing to everybody. For the time being it is based on only two model systems, one of which has been discussed extensively. In order to generalize this finding similar studies on other well-characterized model systems remain desirable.

ACKNOWLEDGMENTS

The author is indebted to Miss Jeanne Korteweg and Miss Olga van Hiele for performing the stability measurements and to Dr. J. N. de Wit for his assistance in the computer analysis of the double-layer data.

REFERENCES

1. Burgreen, D., and Nakache, F. R., *J. Phys. Chem.* **68**, 1084 (1964).
2. Churaev, N. V., and Deryagin, B. V., *Dokl. Akad. Nauk. SSSR* **169**, 396 (1966).
3. Hildreth, D., *J. Phys. Chem.* **74**, 2006 (1970).
4. Irmer, G., *Z. Phys. Chem. (Leipzig)* **254**, 137 (1973).
5. Levine, S., Marriott, J. R., and Robinson, K., *J. Chem. Soc. Faraday Trans. II* **71**, 1 (1975).
6. Levine, S., Marriott, J. R., Neale, G., and Epstein, N., *J. Colloid Interface Sci.* **52**, 136 (1975).
7. Oldham, I. B., Young, F. J., and Osterle, J. F., *J. Colloid Sci.* **18**, 328 (1963).
8. Rice, C. L., and Whitehead, R., *J. Phys. Chem.* **69**, 4017 (1965).
9. Churayev, N. V., Sobolev, V. D., and Zorin, Z. M., *Spec. Discuss. Faraday Soc.* **1**, 213 (1970), and ensuing discussion.
10. Derjaguin, B. V., and Churaev, N. V., *J. Colloid Interface Sci.* **49**, 249 (1974).
11. Henniker, J. C., *Rev. Modern Phys.* **21**, 322 (1949).
12. Deryagin, B. V., and Landau, L., *Acta Physicochim. URSS* **14**, 633 (1941).
13. Verwey, E. J. W., and Overbeek, J. Th. G., "Theory of the Stability of Lyophobic Colloids." Elsevier, Amsterdam/Houston/New York, 1948.
14. Vincent, B., Bijsterbosch, B. H., and Lyklema, J., *J. Colloid Interface Sci.* **37**, 171 (1971).
15. Lyklema, J., and Overbeek, J. Th. G., *J. Colloid Sci.* **16**, 501 (1961).
16. Wiersema, P. H., Loeb, A., and Overbeek, J. Th. G., *J. Colloid Interface Sci.* **22**, 78 (1966).
17. Dukhin, S. S., and Derjaguin, B. V., Electrokinetic phenomena, *in* "Surface and Colloid Science" (E. Matijević, Ed.), Vol. 7, p. 1. 1974.

John Wiley & Sons, New York/London/Sydney/ Toronto.

18. DUKHIN, S. S., AND SEMENIKHIN, N. M., *Kolloidn. Zh.* **31**, 36 (1970).

19. DUKHIN, S. S., *Spec. Discuss. Faraday Soc.* **1**, 158 (1970).

20. PILGRIMM, H., SONNTAG, H., AND DUKHIN, S. S., *Colloid Polym. Sci.* **253**, 750 (1975).

21. KRUYT, H. R., AND KLOMPÉ, M. A. M., *Kolloid Beih.* **54**, 484 (1942).

22. TEŽAK, B., MATIJEVIĆ, E., AND SCHULZ, K., *J. Phys. Chem.* **55**, 1558 (1951).

23. HONIG, E. P., ROEBERSEN, G. J., AND WIERSEMA, P. H., *J. Colloid Interface Sci.* **36**, 97 (1971).

24. HAMAKER, H. C., *Physica* **4**, 1058 (1937).

25. Ref. (13, p. 140).

26. BELL, G. M., LEVINE, S., AND MCCARTNEY, L. N., *J. Colloid Interface Sci.* **33**, 335 (1970).

27. LYKLEMA, J., AND DE WIT, J. N., *J. Electroanal. Chem.* **65**, 443 (1975).

28. DE WIT, J. N., Thesis, Agricultural University of Wageningen, 1975; *Meded. Landbouwhogesch. Wageningen* **75**, 14 (1975).

29. LYKLEMA, J., *Discuss. Faraday Soc.* **42**, 81 (1966).

30. STERN, O., *Z. Elektrochem.* **30**, 508 (1924).

31. LYKLEMA, J., *Trans. Faraday Soc.* **59**, 418 (1963).

32. TROELSTRA, S. R., AND KRUYT, H. R., *Kolloid Z.* **101**, 182 (1942).

33. BIJSTERBOSCH, B. H., Thesis, State University of Utrecht, 1965; *Meded.* Landbouwhogesch. **65**, 4 (1965).

34. SMITH, A. L., *in* "Dispersion of Powders in Liquids" (G. D. Parfitt, Ed.), (1969) Elsevier, Amsterdam/London/New York.

35. FRANKEL, S. P., *in* "Soap Films, Studies of Their Thinning, and a Bibliography" (K. J. Mysels, K. Shinoda, and S. P. Frankel, Eds.), p. 55. Pergamon, New York, 1959.

36. DERYAGIN, B. V., AND LEVI, S. M., "Fiziko-Khimiya Naneseniya Tonkikh Sloev na Dvizhuchshuyusya Podlozhki," Akad. Sci. U.S.S.R., Moscow, 1959.

37. DERYAGIN, B. V., *Zh. Eksp. Teor. Fiz.* **15**, 9 (1945).

38. LYKLEMA, J., SCHOLTEN, P. C., AND MYSELS, K. J., *J. Phys. Chem.* **69**, 116 (1965).

39. OVERBEEK, J. TH. G., *J. Phys. Chem.* **64**, 1178 (1960).

40. BIKERMAN, J. J., *Kolloid Z.* **72**, 100 (1935).

Effects of Vicinal Water on Colloidal Stability and Sedimentation Processes[1]

W. DROST–HANSEN

Laboratory For Water Research, Department of Chemistry, University of Miami, Coral Gables, Florida 33124

Received September 14, 1976; accepted October 13, 1976

Some of the evidence for the existence of vicinal water is reviewed. Such interfacial, structured water appears to be present at many (or most) solid/water interfaces. The structural effects appear to be manifested over distances of the order of several hundred molecular diameters (about 0.1 μm). Furthermore, the properties of the vicinal water are frequently observed to change rather abruptly at several temperatures (T_k) and the temperatures of these transitions appear, to a first approximation, to be independent of both the detailed chemical nature of the solid interface and the nature and concentration of solutes in the aqueous phase. Finally, some examples of the application of the idea of vicinal water structure to the stability of dispersed systems are discussed. Several examples from the literature suggest abrupt changes in colloidal properties at the temperatures where vicinal water has been shown to undergo structural transitions. The effects of vicinal water appear to be superimposed on such "classical" effects as London dispersion forces and electrical double layers. To achieve a comprehensive description of colloidal stability will require the inclusion of vicinal water, for instance, in any expression for the disjoining pressure.

I. INTRODUCTION

"Classical" theories of aqueous colloidal stability and sedimentation processes have for the most part ignored detailed considerations of the role of water structure and particularly the role of vicinal (interfacial) water structures, i.e., those structures, differing from the bulk, which may be induced by proximity to the solid/water interface. The purpose of this paper is to review briefly the evidence for the existence of structurally modified water at many water/solid interfaces, and, in turn, to discuss the likely role which such vicinal water may play in colloidal stability and sedimentation. The preliminary results of experiments from the author's laboratory on the effects of temperature on sedimentation are described together with a few of the relatively sparse observations of temperature effects on colloidal properties which have been reported by previous investigators.

[1] Contribution # 21, Lab. Water Res.

II. ASPECTS OF DISJOINING PRESSURE

There is little doubt that structurally modified water exists at many (or most) water/solid interfaces (1–4). However, very little agreement exists as to the nature of the change in structure, its extent, its relation to the chemical nature of the solid, or the influence on such structures of electrolytes, temperature, etc. Indeed, it is only recently that the notion of structural modification of interfacial water has gained some acceptance and respectability. One of the concepts which has proven most useful for a description of the interactions in thin aqueous films is the disjoining pressure, originally introduced by Deryaguin [(5); see also (1, 4)]. Clifford has pointed out that the disjoining pressure "does not necessarily imply modification of the liquid structure by the surface" (1, p. 79), and while this may be generally true, it does appear that in those cases where the liquid phase is aqueous, long-range structural effects do indeed occur.

Clifford separates the disjoining pressure into additive terms:

$$\pi = \pi_{LL} + \pi_{SL} + \pi_{el} + \pi_i, \qquad [1]$$

where π_{LL} and π_{SL} are due to van der Waals interactions (LL indicating a disjoining pressure of a thin liquid film in the absence of a solid, while SL indicates the disjoining pressure due to the influence of the solid on the liquid), π_{el} is that part of the disjoining pressure which is due to electrostatic effects from the double layers, and π_i is the part due to free-energy changes as the result of structural modifications in the liquid, induced by the solid. Thus, π_i includes, but is not restricted to, direct hydration of ionic and polar sites on the solid; dipole–dipole interactions, etc. It is important to note that there exists no reason to expect, on classical grounds, any abrupt changes in π_{LL}, π_{SL}, or π_{el} as a function of temperature. However, as shown by Peschel and Adlfinger (3), the disjoining pressure of water between quartz plates is not a monotonically changing function of temperature. Instead, minima and sharp maxima are observed in the temperature range from 0 to 60°C. As ion–dipole and dipole–dipole interactions must be expected to vary monotonically with temperature, the results by Peschel and Adlfinger can only imply that abrupt structural transitions occur in vicinal water and that *several* different structures must exist separately over the temperature interval from 0 to 60°C. This finding is consistent with the idea of vicinal water structures advocated for many years by the present author (6, 7). (Some of the evidence for interfacial water being structured differently from bulk water is described in Section III.)

In a recent paper, Deryaguin and Churaev (8) consider the structural component of the disjoining pressure. The authors first note that the structural component of the disjoining pressure has been studied in far less detail than the ionic-electrostatic part or the molecular part, because of our current lack of general structural theories of associated liquids. It is also necessary to take into account the effects

of overlapping double layers, such as will occur in more concentrated suspensions. Omitting for the present discussion the electrical double-layer effects, the disjoining pressure is taken to be $\pi = -(\partial F/\partial h)_T$; i.e., the variation with thickness of the free energy. As $dF = dU - TdS$ (where U is internal energy and S is the entropy), Deryaguin and Churaev call attention to the point that structural changes imply variations in both intermolecular bond energies and in the spatial arrangement. However, it should be noted that it seems unlikely that bond energies should vary abruptly and strongly at several discrete temperatures (in the interval from 0 to 60°C) and thus it seems far more probable that the observations by Peschel and Adlfinger (3) of peaks in the disjoining pressure are due primarily to variations in the entropy, i.e., ordering of the molecules. As discussed by the present author (2), it appears that rather different structures of water are stabilized below and above any one of the (four) thermal transition temperatures and that while the O–H–O bond strength remains essentially unchanged, the molecular ordering must change significantly on going from one temperature range to another. This being the case, many other properties of vicinal water are no doubt affected, such as density, dielectric properties, excess ultrasonic absorption, etc. Hence, the "steric factors" of colloidal stability must be expected to be highly temperature-sensitive in a rather "nonclassical" fashion. This aspect is discussed in Section V, based on the very limited data currently available in the literature.

III. EVIDENCE FOR VICINAL WATER STRUCTURING

In addition to measuring disjoining pressures, Peschel and Adlfinger have also measured the viscosity of thin films of water between two quartz plates (9, 10). From these measurements, it was possible to demonstrate that the viscosity of interfacial water was notably greater than the viscosity of bulk water and that the thickness of the modified layers was considerable: hundreds of molecular diameters. In addition, the viscosity varied in a highly unusual manner with temperature, exhibiting several maxima and minima in a "standard" Arrhenius plot.

Forslind (11) and Kerr and Drost-Hansen (12) demonstrated unusual temperature dependencies of the damping of an oscillating quartz capillary filled with water. The results were interpreted as evidence for anomalous, vicinal water structures at the quartz/water interface (13). Furthermore, the results could be correlated with unusual apparent entropies of surface formation, derived from measurements of surface tension in narrow capillaries, as a function of temperature (13, 14).

Recently, Wiggins (15) reported the results of careful measurements of ion distribution between a bulk phase and the water in narrow pores of a silica gel. The gel was equilibrated with equimolar amounts of sodium and potassium ions (present as either chlorides, iodides, or sulfates) at different temperatures. Apparent partition coefficients, λ, were defined as

$$\lambda_K = [K^+]_i/[K^+]_o \qquad [2]$$

and

$$\lambda_{Na} = [Na^+]_i/[Na^+]_o, \qquad [3]$$

where [ ] denote concentrations, and the subscripts i and o refer, respectively, to inside the pore or in the outer (bulk) solution. From these partition coefficients, a selectivity coefficient, K, was calculated:

$$K = K_{Na}^K = \lambda_K/\lambda_{Na} = f(T). \qquad [4]$$

The values obtained for K are shown in Fig. 1. Note that the values for K are invariably larger than 1 (in fact, generally in the range $1.2 < K < 1.8$), an observation which is likely of crucial importance to cell physiology (15, 16). Note also the sharp peaks, occurring at or very near those temperatures where Peschel and Adlfinger observed peaks in the disjoining pressure, and indeed at the same temperature where the present author has demonstrated that vicinal water undergoes anomalous (thermal) changes (2, 7, 17).

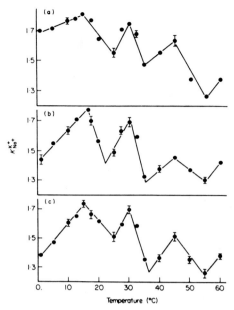

FIG. 1. Selectivity coefficient, K, of potassium/sodium ions between the pores of a silica gel and bulk solution as a function of temperature. Data by Wiggins (15).

Schufle et al. (18) have reported on the temperature dependence of surface conductivity (measured in Pyrex capillaries). The results suggested that a layer of water exists at the solid/water interface which notably affects the ionic conduction processes, and that the thickness of this layer appears to be of the order of 0.1 μm. Increased apparent values for the energy of activation of surface conductance near a variety of metal oxide/solution interfaces have also been reported by Morimoto and Kittaka (19).

Various unusual results have been obtained from dielectric studies on dispersed aqueous systems; see, for instance, Metzik et al. (20, 21). More recently, Ballario et al. (22, 23) have reported on anomalies in dielectric properties of aqueous suspensions of polystyrene spheres and also of bacterial spores. In both cases, the data show anomalous thermal behavior and these data were indeed interpreted in terms of vicinal water structuring. Berteaud *et al.* (24, 25) have reported on the dielectric properties of water and BSA solutions as a function of temperature. Both sets of data suggest thermal anomalies (i.e., more or less abrupt changes in slope of ϵ' as a function of temperature). While the authors did not elect to analyze the data in this manner, it appears certain, for instance, that the data for BSA show anomalous changes in the temperature dependence near 15 and 30°C. This finding is consistent with the ideas discussed in the present paper and in earlier papers by this author (2): at the interface between some macromolecules and water, vicinal water structuring may occur (as it does at many (or most) other "solid"/water interfaces). It is more difficult to understand the apparent anomalous thermal behavior which is indicated in the results reported for pure water. However, it is suggested here that the anomalies are again manifestations of interfacial structuring as the dielectric cells used for these measurements were remarkably small; for a 5 μl cell, the diameter is 156 μm. While it is obviously not suggested that water structuring occurs over distances of the order of 75 μm, a sufficiently large fraction (0.2%) of the total sample is still within, say, 0.1 μm from the confining walls of the sample tubes and this may be sufficient to affect the total values obtained, especially in view of the high sensitivity of the methods used.

Some time ago it was concluded from a study of ultrasonic velocity that structuring of water likely occurred near fully hydroxylated quartz particles ("Ludox," DuPont) with an average particle diameter of 22 nm (26); it was impossible to account for the ultrasonic velocity by classical theories without postulating the existence of (an extensive) "hydration hull." This study is remarkable as it constitutes an example of a very large surface-to-volume ratio system which did not in any obvious way exhibit thermal anomalies. At the time of publication, no explanation could be offered for this observation; however, recently Brunn, Sörensen, and the present author (1976, to be

published) have measured both ultrasonic velocities and excess ultrasonic absorption of aqueous suspensions of polystyrene spheres (91 nm average particle diameter). These data do show distinct anomalies in the velocities near 15 and 33°C. Returning to the ultrasonic velocity study on Ludox, it is possible that the data were simply not analyzed in sufficient detail to reveal the thermal anomalies which, in this case, may not have been particularly pronounced.

For many additional examples of evidence for interfacially modified water, see in particular the writings of Deryaguin (27–30). In the present context, however, only one of Deryaguin's ideas will be discussed; namely, the notion of spatial anisotropy of interfacial water. Thus, Deryaguin (28) notes the results by Deryaguin and Green-Kelly (29), who used optical birefringence to demonstrate structural anisotropy of montmorillonite swelling and deduced the existence of structural anisotropy on the basis of these data. Additional confirmation for the existence of structural anisotropy was derived from dielectric measurements (Kurbatov (31, 32); see (28)). However, perhaps the most convincing evidence for vicinal anisotropy comes from the study by Vignes and Dijkema (33), who determined the movement of water between an ice plug and a confining "solid." This effect is proposed as an explanation for the so-called frost-heaving of soils. The results were interpreted as due to an anisotropy of the pressure tensor, imposed by the anomalous properties of the interfacial water. The present author has offered (34) a conceptual visualization of this effect as shown in Fig. 2. As proposed by Vignes and Dijkema, "...anisotropy of pressure is developed where the tangential pressure is lowered and water is attracted but the normal pressure increased, whereby the ice is moved away from the wall." It would seem intuitively that the distance over which this effect operates must be considerable (i.e., perhaps 100 to 1000 molecular diameters rather than merely in a mono- or pauci-molecular layer)

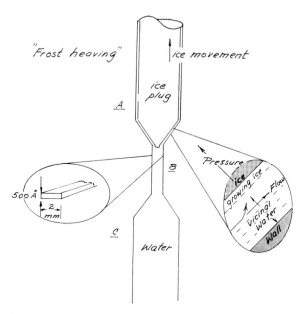

FIG. 2. Schematic representation of the ice/water/solid interface (under conditions of "freezing") implying anisotropy of vicinal water [see text and (33, 34)].

in order to account for the observed total net rates of flow (unless exhorbitant pressures were involved).

Other evidence for extensive ordering of interfacial water has been discussed by Low (35, 36), who attributes such structuring in clay/water systems to an expitaxial effect of the clay mineral surface, resulting in thick "ice-like" water structures.

Of the studies by the Russian school, attention is again directed to the extensive writings of Deryaguin and of Zorin, Churaev, and co-workers. Thus, Ershova, Zorin, and Churaev (37) have studied the thickness and nature of polymolecular adsorbed films on quartz. These authors observed that film thicknesses decrease from about 100 Å at 10°C to a monomolecular film above 65°C. The decrease in thickness is not linear; instead, a rapid decrease is observed from 10°C to about 25 to 30°C, with a fairly constant plateau (corresponding to about 35 Å thickness) up to 45 to 50°C. Unfortunately, the calculated values for

the film thickness may be somewhat in doubt as the assumption was made that the refractive index of these thin films does not differ from the value for bulk water. This is indeed a dangerous assumption; first, if structurally different films are stabilized by proximity to the solid, then the properties of such films must depend on the nature and extent of the structural changes. Furthermore, direct measurements of the index of refraction of water by Hawkes and Astheimer (38) have suggested that anomalous changes occur above 30 to 35°C. Frontas'ev (39) has also suggested anomalous temperature dependencies for the index of refraction of water and the present author (40) has interpreted this as yet another manifestation of a vicinal water effect. Similarly, it is proposed that the anomalies reported by Tilton and Taylor (41) for their high-precision refractive index data are also a manifestation of anomalous changes in the index of refraction of the water adjacent to

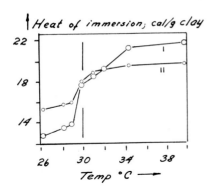

FIG. 3. Heat of immersion of Ca-bentonite as a function of temperature. Data by Slabaugh (50).

the solid/water interfaces of the measuring cell [see (42)].

IV. PARADOXICAL EFFECT

One remarkable feature of vicinal water is the fact that, to a first approximation, the temperatures at which the thermal anomalies occur are independent of the detailed chemical nature of the solid, and independent of both the nature and concentration of solutes (at least below, say, 0.5 M). It is not possible at present to offer a definite explanation for this phenomenon [for some speculations, see (43, 44)]. However, somewhat along the same lines, Clifford and Pethica (45) note that "the effect of varying porosity [i.e., of size of water domain] is seen to be much larger than the effect of varying the nature of the surface." The article by Clifford and Pethica presents a uniquely interesting overview of water in a great variety of dispersed systems. (The present author, however, takes issue with one of the notions of these authors; namely, that in some systems, such as micelles, "all effects on water structure can be adequately described in terms of one or two layers of 'bound' water near the micelle, ... Here also the existence of solid "ice" layers and of long-range effects can be ruled out.")

It is of interest also to quote Kay and Low (46) in this connection. From a study of the

heat of compression of clay–water mixtures, these authors concluded that "the properties of water in clay–water systems, even at high water contents, differ significantly from those of bulk water. Further, it is shown that only a small part of the difference can be attributed to the effect of exchangeable cations. Most of it must be attributed to the effects of particle surfaces." In a sense, the paradoxical effect offers a "unifying aspect" to the domain of colloidal stability, and indeed also allows the application of the notion of vicinal water to be carried into the field of cell biology.

V. COLLOIDAL STABILITY AND SEDIMENTATION PROCESSES

It is clear from the preceding sections that extensive structuring may occur in water near interfaces and it is, therefore, reasonable to inquire into the role of such vicinal water for colloidal stability and sedimentation processes. Unfortunately, only a very small number of studies have been reported in the literature of temperature effects on colloidal stability, and even fewer studies of temperature on sedimentation rates and volumes.

Before proceeding, it is important to note the general areas where vicinal water structuring may be expected to play a role. As discussed elsewhere (2, 43, 47), one apparent characteristic of vicinal water is the small energy differences between notably different structures. Recall, for instance, that the lattice energy differences between Ice Ih and Ice II is only 19 cal/g mole [see (48) or (49)]. For this reason, vicinal water structuring is not expected to play a significant role in "high-energy processes" where "steric" effects may be unimportant. In a sense, then, this constitutes an "exception" to the paradoxical effect. Thus, the stability of colloidal suspensions of materials with large surface potentials may not be measureably affected by vicinal water structures, while those suspensions with minimal energy barriers may be greatly influenced (similar types of considerations must apply to cellular systems).

An abrupt thermal transition has been reported by Slabaugh (50), who measured the heats of immersion in water of calcium bentonites. The observed heats depend on the pretreatment and the initial measurements revealed very sharply increasing values below 100°C. To characterize this sharp increase further, measurements were then made at closely spaced temperature intervals between 26 and 40°C. The results are shown in Fig. 3. Note the abrupt increase at 29 to 30°C. As the clay mineral itself is not known to undergo any type of transition at this temperature, it is reasonable to associate the observed effect with the structural change in vicinal water at 30°C.

The simultaneous effects of different salt concentrations and temperatures on interfacial systems have only rarely been investigated experimentally although such studies would have considerable practical applications. Low (51) has made some interesting observations on zeta potentials (ζ) for clay suspensions and corresponding "freezing concentration," c_f (i.e., the concentration of electrolyte at which freezing occurred at $-50°C$ in 12 hr). Sharp peaks in c_f and ζ occurred at the same concentrations for an Na-clay; namely, at 10^{-3} and 5.10^{-3} M (in other words, two peaks). This behavior was related by Low to the influence on the possible degree of supercooling, affected by particle–particle interaction. This idea is in excellent agreement with the much more recent observations by Peschel (52) that the disjoining pressure between two adjacent surfaces may exhibit relatively sharp, marked maxima and minima as a function of salt concentration (at 25°C). (Recall also the study by Peschel and Adlfinger on the disjoining pressure as a function of temperature.) In view of the peaks in zeta potential, freezing concentration, and disjoining pressure, all appearing at roughly the same concentration ranges, it is proposed that the former two phenomena reflect a change in interparticle distance between the clay platelets as the result of the changes in the (repulsive) disjoining pressure. Finally, in

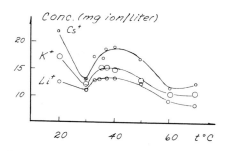

FIG. 4. Coagulation temperatures of $Fe(OH)_3$ sols for various cations (Li, K, Cs). Data by Baran et al. (59).

considering particle stability in salt solutions, one should also consider the possibility of transitions in ionic double layers, occurring rather abruptly, which have been proposed by Plesner (53) at certain solute concentrations.

Specific concentration-dependent anomalies have also been reported by Friend and Hunter (54) for vermiculite suspensions. These authors measured electrophoretic mobilities and converted the data to (calculated) zeta potentials. Ion-adsorption measurements were also made; for both parameters strongly concentration-dependent anomalies were observed. In both cases, the anomalies occurred at 10^{-2} M (bulk) concentration of electrolyte. These results may, as those obtained by Low on bentonites, reflect the type of "critical transitions" in the double layer proposed by Plesner. On the other hand, because of the greatly enhanced ion concentration at the mineral/solution interface, Friend and Hunter point to the possibility that the effect is induced by a more or less abrupt ion–water interaction, in a sense reflecting an unexpected transition due to a large-scale ion hydration phenomenon, such as discussed by Vaslow [(55); see also Drost-Hansen, (56)]. In this case, then, the effect involves direct water structuring at the interface, and it is for this reason one may expect the phenomenon to be strongly temperature sensitive. Thus, it is predicted that the anomalous electrophoretic mobilities may exhibit abrupt changes at the temperatures where vicinal water undergoes structural transitions.

As stressed earlier, the temperatures (T_k) at which the anomalies occur in the properties of vicinal water appear, to a first approximation, to be independent of both (A) the chemical nature of the substrate and (B) the nature and concentration of solutes. The latter observation immediately suggests that electrical double layers are *not* responsible for the thermal anomalies (and hence, the establishment of vicinal water structures). This point is extremely important as Aronson and Princen (57) have suggested that thick stable films of water on silica may be "substantially accounted for with electrical double-layer theory." See also the discussion of vicinal water thickness and electrical double layers by Schufle et al. (18). The substrate independence is more difficult to explain. The present author (2) originally suggested an "energy delocalization process" as a possible explanation, based on the notion that cooperativity among H bonds in (vicinal) "clusters" were somehow protected from disruptive thermal fluctuations by proximity to a surface (essentially *any* surface). More recently, the present author (44) has built on a suggestion by Clifford and by B. Drost-Hansen; namely, that geometric constraints might prevent the formation of (extensive) bulk water structures, and that these constraints, rather than the specific chemical nature of the solid, determines the "inducement" of vicinal structures.

Johnson et al. (58) have reported data for the rate of flocculation of polyvinyl toluene particles as a function of temperature. In an Arrhenius graph of these data, a distinct change in slope occurs near 43°C; i.e., one of the temperatures at which vicinal water undergoes a structural change. It is again tempting to suggest that this anomaly is due to "steric effects" on particle dynamics, related to the vicinal water structures.

Among the few papers which have dealt with the problem of colloidal stability and temperature, investigated at relatively closely spaced temperatures, is a recent one by Baran et al. (59). These authors measured the coagulation concentrations (using several alkali salts) on AgI, Sb_2S_3, and $Fe(OH)_3$ sols. The results are interesting although difficult to interpret theoretically. For the two negatively charged sols (AgI and Sb_2S_3), the coagulation concentrations respectively decreased rapidly (and monotonically) with increasing temperature (for AgI), while being essentially temperature independent for Sb_2S_3! On the other hand, for the positively charged $Fe(OH)_3$, the coagulation concentrations (due to LiCl, KCl, or CsCl) exhibit a notable minimum near 30°C, as shown in Fig. 4 (and perhaps a weak minimum around 60°C). The authors combine the notions of Deryaguin and those of the present author to ascribe the effect observed for the $Fe(OH)_3$ sol to a substantial role of the structural component of the disjoining pressure as affected by differing hydration structures. The aspect of this study which is difficult to explain is the temperature independence of the data obtained on the Sb_2S_3 sol, as well as the absence of overt thermal anomalies for the AgI sol. However, as discussed earlier in this section, vicinal water effects are not expected for those systems where strong interactions are dominant (i.e., in "high-energy" reactions).

In another study, Baran and Solomentseva (60) investigated the effects of temperature on the stability of AgI, Sb_2S_3, and Au sols to which were added various types of water-soluble polymers. Remarkably abrupt changes in the critical counterion concentrations are observed for both AgI and Sb_2S_3 sols as a function of temperature in the presence of sodium carboxymethylcellulose, and for Sb_2S_3 sols for polyethylene oxide and polyvinyl alcohol as well.

Somewhat similar results are reported for AgI sols in the presence of gelatin. Finally, red-gold sols in dilute polyethylene oxide solutions exhibited a sharp minimum in the critical counterion concentrations (Li^+, K^+, and Ca^+) at 30°C. The authors present an interesting discussion of these highly complex systems, considering both an important role for vicinal water structures and the effects of electrical double layers.

The effects of methylcellulose on another dispersed system has been reported by Lapan et al. (61). These authors determined the viscosity of a bentonite suspension as a function of temperature in the presence and absence of methylcellulose. The addition of the polymer leads to increased viscosities at lower temperatures, but larger temperature coefficients. Unfortunately, the measurements were apparently carried out using a simple Oswald-type viscometer and the study thus ignores any possible anomalous effects due to the likely non-Newtonian character of these suspensions. Lapan et al. also determined the yield value of such suspensions. These results are remarkable in showing sharp peaks at 60°C. The magnitude of the peak values varied with time and concentration of both polymer and clay, but in all four cases studied, the peak occurred at 60°C. This again suggests a critical role of vicinal water at this transition temperature.

Khentov et al. (62) have investigated the effects of temperature on structural changes in the dense part of the diffusion layer. The study involved the analysis of ion composition of drops formed by the breakup of gas bubbles. Unfortunately, such measurements are critically sensitive to the "purity" of the system; most any spurious nonelectrolytes present would tend to be concentrated at the air/liquid interface, and this may significantly alter the structure of the interface from a pure water (or pure ionic solution)/gas interface. In any event, the main result obtained by Khentov et al. is that the concentration of weakly hydrated anions in the "dense" part of the diffusion layer decreases near 30°C. The authors note the maximum in the disjoining pressure of water reported by Peschel and Adlfinger (3).

Recently, Ballario et al. (63) reported surface charge densities for polystyrene suspensions as a function of temperature. Although not stressed by these authors, their data suggest (in spite of relatively large experimental

TABLE I

Particle Sizes for Sedimentation Studies

Material	Size (μm)	Average (μm)
SiC 500	7.9– 24.3	13
Al$_2$O$_3$ 225	12.2–250	22.5
SiC 320	19.8– 40.0	29

errors) a minimum in the surface charge density near 15°C for a suspension of particles with a diameter of 0.091 μm. More recently, Bruun, Sörensen, and Drost-Hansen (1976, to be published), using the same type of suspension, observed a notable minimum in the ultrasonic excess absorption at the same temperature, and both observations must reflect the relatively abrupt changes in vicinal water structure at this temperature.

Barfod (64) has reported settling rates (as a function of concentration) in aqueous suspensions of Al$_2$O$_3$ and SiC particles (at 25°C). The particle sizes employed are listed in Table I. Barfod devised a model to predict the sedimentation rate. Characteristically, however, for SiC 320 the model breaks down in two concentration ranges; namely, below 0.01 Vol% and above 0.1 Vol%. It would appear that the former effect is likely due to the presence of vicinal water structures ("hydration hulls"), whereas the latter is due to the general effects of particle–particle interactions (hydrodynamic effects—possibly influenced by overlapping vicinal water structures). In either case, one would expect lower settling rates as are indeed observed. Barfod proposed a breakdown of the model because of the "co-movement" of water in the more concentrated suspensions, but he could not suggest an explanation in the case of the anomaly at the lower concentration range. It is regrettable that these measurements were not performed at several closely spaced temperature intervals, as such a study might have revealed a possible role of the viscosity and disjoining pressure anomalies discussed by Peschel and Adlfinger (3, 9).

Sedimentation volumes of glass beads in various water-containing or water-saturated

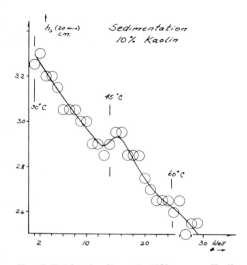

h_s (20 min.) cm.

Sedimentation
10% Kaolin

3.2

3.0

2.8

2.6

30°C

45°C

60°C

2   10   20   30 Well

FIG. 5. Height of sediment in 10% aqueous Kaolin suspension as function of temperature after 20 min of sedimentation (see text for details).

organic liquids (measured at two temperatures, 25 and 50°C) have been reported by Nakajima (65). In the case of 1% water dissolved in some lower alcohols, in an ester and in a ketone, the difference in sedimentation volume between 25 and 60°C is very small ($\sim 2\%$), the volume at 50°C always being smaller than the volume at 25°C. However, in water-saturated liquids of non-oxygen-containing polar and nonpolar organic solvents, the effect of temperature is remarkably large, with the decreases in volumes at 50°C ranging from 10 to (more often) near 50%. The author proposes that the decrease reflects the decrease in the amount of water absorbed at the higher temperatures. Nakajima also obtained values for the sedimentation volumes of the glass beads in pure water and water-saturated benzene. In this case, a large drop in sediment volume occurs in the water-saturated organic phase with increasing temperature and the volume approaches that for pure water somewhere between 40 to 60°C. Even if the explanation offered by Nakajima is correct—namely, that the "agglomeration" of glass beads in water-containing, "nonpolar" liquids is "merely" caused by the surface ten-

sion of the water layers adsorbed by the glass beads—recall the notable anomalies in apparent surface entropies observed at various temperatures, particularly at 30°C (14).

Pinvinskii and Kruglitskii (66) have reported on measurements of the effects of temperature on the rheological properties of rather concentrated, aqueous non-Newtonian suspensions. The relative viscosity of dilatant suspensions of quartz-glass as a function of temperature showed relatively broad maxima around 15 to 40°C, where the location and width of the maxima depended on the shear stresses. At higher temperatures a minimum was observed, again dependent on the shear stress, but generally centered around 60 to 70°C. For higher temperatures, the relative viscosity again increased, particularly rapidly for large shear stresses. Even more interesting is an Arrhenius graph of viscosity versus reciprocal absolute temperature for dilatant quartz suspensions in the article by Pinvinskii and Kruglitskii. This graph shows a distinct, abrupt change in slope between two straight line segments at approximately 30°C. The authors comment on this observation noting that "such a change in viscosity indicates a possible change in the structural mechanism of flow with increasing temperature . . . ." In view of the dimensional stability of the quartz particles themselves, it appears inescapable that the cause of the change is in the aqueous phase; more specifically, associated with the change in vicinal water structure at 30°C.

We have recently initiated a study of the effects of temperature on sedimentation rates and sedimentation volumes. Only a small part of the study has been completed at this time. However, data have been obtained for the apparent sedimentation volume in 10% aqueous suspensions of kaolin. The experiments are performed in a "polythermostat," a temperature gradient "incubator" originally developed by Drost-Hansen and Oppenheimer (67). The device consists of a thermally well-insulated aluminum bar (12 × 20 × 150 cm) with two sets of 30 "wells" (designed to accommodate standard size test tubes or culture tubes). The bar is heated in one end, cooled at the other end, each end maintained at constant (but different) temperature. The temperature gradient in the bar is nearly linear and calibrated before (and during) each run. Figure 5 shows the height of the sediment after 20 minutes of sedimentation as a function of temperature. (The abscissa is approximately linear in temperature. Three specific temperatures are indicated on the graph.) It is apparent that an anomaly exists near 45°C, but at present we are unable to separate possible contributions to this effect from anomalous viscosities of vicinal water around each clay platelet (as it settles in the column) and from anomalous disjoining pressures (see the discussion of the results obtained by Peschel and Adlfinger). Work on this problem continues in our laboratory.

ACKNOWLEDGMENTS

The author wishes to thank Ms. Ileana Tandron and Mr. Marcos Puga for their assistance in the experimental part of this study. The support by the U.S. Environmental Protection Agency through Grant R 80-38-26-01 is also gratefully acknowledged.

REFERENCES

1. CLIFFORD, J., in "Water—A Comprehensive Treatise" (F. Franks, Ed.), Vol. 5, p. 75. Plenum Press, New York, 1975.
2. DROST-HANSEN, W., in "Chemistry of the Cell Interface" (H. D. Brown, Ed.), p. 1. Academic Press, New York, 1971.
3. PESCHEL, G., AND ADLFINGER, K. H., Z. Naturforsch. 26a, 707 (1971).
4. PESCHEL, G., AND BELOUSCHEK, P., in "Supplement of Colloid and Polymer Science" (Paper presented at the 27th Meeting of the German Colloid Society, 1975.) In press.
5. DERJAGUIN, B. V., J. Phys. Chem. 3, 29 (1932).
6. DROST-HANSEN, W., Chem. Phys. Lett. 2, 647 (1969).
7. DROST-HANSEN, W., Ind. Eng. Chem. 61 (11), 10 (1969).
8. DERYAGUIN, B. V., AND CHURAEV, N. V., J. Colloid Interface Sci. 49, 249 (1974).
9. PESCHEL, G., AND ADLFINGER, K. H., J. Colloid Interface Sci. 34, 505 (1970).
10. PESCHEL, G., AND ADLFINGER, K. H., Naturwissenschaften 56, 558 (1969).
11. FORSLIND, E., Sv. Naturvidcnsk. 2, 9 (1966).
12. KERR, J. E. D., AND DROST-HANSEN, W., Dissertation (Kerr), University of Miami (see Ref. (13)), 1970.
13. DROST-HANSEN, W., J. Geophys. Res. 77, 5132 (1972).
14. DROST-HANSEN, W., Ind. Eng. Chem. 57 (April issue), 18 (1965).
15. WIGGINS, P. M., Clin. Exp. Pharmacol. Physiol. 2, 171 (1975).
16. WIGGINS, P. M., J. Theor. Biol. 52, 99 (1975).
17. DROST-HANSEN, W., in "L'eau et les systemes biologiques" (A. Alfsen and A.-J. Berteaud, Eds.), p. 177. Colloques Internationaux du C. N. R. S., No. 246, 1976.
18. SCHUFLE, J. A., HUANG, C.-T., AND DROST-HANSEN, W., J. Colloid Interface Sci. 54, 184 (1976).
19. MORIMOTO, T., AND KITTAKA, S., J. Colloid Interface Sci. 44, 289 (1973).
20. AFANAS'EV, N. V., METSIK, M. S., AND POPOVA, V. N., in "Research in Surface Forces" (B. V. Deryaguin, Ed.), Vol. 2, pp. 176, 181. Engl. Transl., Consultants Bureau, New York, 1966.
21. METSIK, M. S., PEREVERTAEV, V. D., LIOPO, V. A., TIMOSHTCHENKO, G. T., AND KISELEV, A. B., J. Colloid Interface Sci. 43, 662 (1973).
22. BALLARIO, C., BONINCONTRO, A., CAMETTI, C., AND D'AGOSTINO, S., Nuovo Cimento 6, 611 (1973).
23. BALLARIO, C., BONINCONTRO, A., AND CAMETTI, C., J. Colloid Interface Sci., (1975).
24. BERTEAUD, A.-J., HOFFMAN, F., AND MAYAULT, J.-F., J. Microwave Power 10, 309 (1975).
25. DAVID, R., BERTEAUD, A.-J., AND PERRON, R., in "L'eau et les systemes biologiques" (A. Alfsen and A.-J. Berteaud, Eds.), Colloques Internationaux du C. N. R. S., No. 246, Paris, p. 97. 1976.
26. YOUNGER, P. R., ZIMMERMAN, G. O., CHASE, C. E., AND DROST-HANSEN, W., J. Chem. Phys. 58, 2675 (1973).
27. DERYAGUIN, B. V., Discuss. Faraday Soc. 42, 109 (1966).
28. DERYAGUIN, B. V., Pure Appl. Chem. 10, 375 (1965).
29. DERYAGUIN, B. V., AND GREEN-KELLY, R., in "Research in Surface Forces" (B. V. Deryaguin, Ed.), Vol. 2, p. 119. Engl. Transl., Consultants Bureau, New York, 1966. (Note: Item 3 in "Conclusions" (p. 125) appears to be in error, suggesting isotropic rather than anisotropic structure.)
30. DERYAGUIN, B. V., See "Research in Surface Forces" B. V. Deryaguin, Ed), Vols. 3 and 4. Engl. Transl., Consultants Bureau, New York, 1971 and 1975, respectively.

31. KURBATOV, L. P., *Zh. Fiz. Khim.* **24**, 899 (1950).

32. KURBATOV, L. P., *Zh. Fiz. Khim.* **28**, 287 (1954).

33. VIGNES, M., AND DIJKEMA, K. M., *J. Colloid Interface Sci.* **49**, 165 (1974).

34. DROST-HANSEN, W., *Cryobiology* **12**, 552 (1975) (Abstract only).

35. LOW, P. F., *Advan. Agron.* **13**, 269 (1961).

36. RAVINA, T., AND LOW, P. F., *Clays Clay Miner.* **20**, 109 (1972).

37. ERSHOVA, G. F., ZORIN, Z. M., AND CHURAEV, N. V., *Colloid J. USSR* **37**, 190 (1975).

38. HAWKES, J. B., AND ASTHEIMER, R. W., *J. Opt. Soc. Amer.* **64**, 105 (1974).

39. FRONTAS'EV, V. P., AND SCHRAIBER, L. S., *J. Struct. Chem.* **6**, 493 (1965). (Engl. transl.).

40. DROST-HANSEN, W., *in* "Equilibrium Concepts in Natural Water Systems" (W. Stumm, Ed.; *Advan. Chem. Ser.* **67**, 70 (1967).

41. TILTON, L. W., AND TAYLOR, J. K., *J. Res. Nat. Bur. Stand.*, **20**, 419 (1938).

42. DROST-HANSEN, W., *in* "Proceedings of the First International Symposium on Water Desalination", Vol. 1, p. 382. U. S. Government Printing Office, Washington, D.C., 1967.

43. DROST-HANSEN, W., Paper presented at the XVth Solvay Conference on Electrostatic Interactions and the Structure of Water, June 1972. *Phys. and Chem. of Liquids*, 1977.

44. DROST-HANSEN, W., *Symp. Preprints, Div. Petrol. Chem., Amer. Chem. Soc.* **21**, 278 (1976).

45. CLIFFORD, J., AND PETHICA, B. A., "Hydrogen-Bonded Solvent Systems" (A. K. Covington and P. Jones, Eds.), p. 169. Taylor and Francis London, 1968.

46. KAY, B. D., AND LOW, P. F., *Clays and Clay Miner.* **26**, 266 (1975).

47. DROST-HANSEN, W., *in* "The Effects of Pressure on Organisms," Symposia of the Society for Experimental Biology, No. XXVI, p. 61 Cambridge Univ. Press, New York, 1972.

48. KAMB, B., *in* "Structural Chemistry and Molecular Biology" (A. Rich and N. Davidson, Eds.) p. 507. Freeman, San Francisco, 1968.

49. EISENBERG, D., AND KAUZMANN, W., "The Structure and Properties of Water," Oxford Univ. Press, New York, 1969.

50. SLABAUGH, W. H., *J. Phys. Chem.* **63**, 1333 (1959).

51. LOW, P. F., *in* "Water and Its Conduction in Soils," Highway Research Board, Special Report 40, p. 55. National Academy of Sciences/ National Research Council, Publ. 629, Washington, D.C., 1958.

52. PESCHEL, G., ADLFINGER, K. H., AND SCHWARZ, G., *Naturwissenschaften* **61**, 215 (1974).

53. PLESNER, I. W., AND MICHAELI, I., *J. Chem. Phys.* **60**, 3016 (1974).

54. FRIEND, J. P., AND HUNTER, R. J., *Clays and Clay Miner.* **18**, 275 (1970).

55. VASLOW, F., *J. Phys. Chem.* **67**, 2773 (1961).

56. DROST-HANSEN, W., See comments in Ref. (40, pp. 95–98).

57. ARONSON, M. P., AND PRINCEN, H. M., *J. Colloid Interface Sci.* **52**, 345 (1975).

58. JOHNSON, G. A., LECCHINI, S. M. A., SMITH, E. G., CLIFFORD, J., AND PETHICA, B. A., *Discuss. Faraday Soc.* **42**, 120 (1966). .

59. BARAN, A. A., SOLOMENTSEVA, I. M., AND KURILENKO, *Colloid J. USSR* **37**, 197 (1975). (Engl. Transl.)

60. BARAN, A. A., AND SOLOMENTSEVA, I. M., *Colloid J. USSR* **36**, 945 (1974).

61. LAPAN, B. T., KUCHER, R. V., ENAL'EV, V. D., AND GURENKO, L. V., *Colloid J. USSR* **31**, 200 (1969). (Engl. transl.).

62. KHENTOV, V. YA., GUNBIN, YU. V., AND FUKS, G. I., *Colloid J. USSR* **37**, 182 (1975). (Engl. transl.)

63. BALLARIO, C., BONINCONTRO, A., AND CAMETTI, C., *J. Colloid Interface Sci.* **54**, 415 (1976).

64. BARFOD, N., *Powder Technol.* **6**, 39 (1972).

65. NAKAJIMA, E., *Kolloid Z.* **175**, 60 (1961).

66. PIVINSKII, YU. E., AND KRUGLITSKII, N. N., *Colloid J. USSR* **37**, 899 (1975). (Engl. transl.)

67. DROST-HANSEN, W., AND OPPENHEIMER, C. H., *J. Bacteriol.* **80**, 21 (1960).

Water on Organosilane-Treated Silica Surfaces

A. C. ZETTLEMOYER AND H. H. HSING[1]

Center for Surface and Coatings Research, Lehigh University, Bethlehem, Pennsylvania 18015

Received September 30, 1976; accepted September 30, 1976

Both a fully hydroxylated silica (HiSil 233) and a partially hydroxylated silica (Cab-o-Sil M-5) were treated with organosilanes and a silazane. Water adsorption isotherms were then monitored on the resulting products. Reflectance IR spectroscopic measurements were used to follow the reactions of the surface hydroxyls with the silanes and to examine the subsequent water adsorption. The main objective was to ascertain the water susceptibilities of the different functional groups and the availability to water of any residual hydroxyls.

The silazane and silanes employed were: hexamethyldisilazane, vinyl-trichlorosilane, and the alkoxysilanes: γ-glycidoxypropyltrimethoxysilane, 3-mercaptopropyltrimethyoxysilane, and aminopropyltriethoxysilane. The latter three were reacted from the liquid phase, first hydrolyzing to the silanetriols, with additional complications from condensation reactions. The vinyl compound and the disilazane were reacted from the vapor phase.

Sharp differences between the results for the fully hydroxylated and the partially (25%) hydroxylated silica were found. The silanetriols upon heat treatment formed a multilayer polymer network on the latter and only a two-dimensional network on the former apparently due to hydrogen bonding to residual hydroxyls. The reacted silanes formed umbrellas over reacted hydroxyls on the former which were nevertheless available to water molecules. With the latter, however, a much more hydrophobic surface was produced. A number of details were also provided by the reflectance IR measurements.

INTRODUCTION

Water at oxide surfaces is influenced by the very character of the surface. On a fully hydroxylated surface such as HiSil 233, essentially each molecule of the first layer of water adsorbed is doubly hydrogen bonded to two surface hydroxyls (1). On sparsely hydroxylated surfaces, such as Cab-o-Sil M-5 (or on partially dehydroxylated HiSil), water molecules cluster about first adsorbed water molecules even before each surface hydroxyl takes up a water molecule (1).

NIR reflectance spectroscopy has been found to be a powerful tool to augment direct adsorption measurements in ascertaining the nature of the adsorbed water. It also has been found to be helpful in characterizing ligands such as

silanes reacted with the hydroxyls and in learning the effect of such reacted ligands on altering the milieu of any adsorbed water. Such studies as reported here could influence our understanding of ligand–water interactions (they are fixed in place on the surface as compared to their situation in membranes or in the solution phase) and our understanding of the influence of water on inorganic–organic composites. The last four compounds listed in Table II are used in the surface treatment of glass fibers for use in composites.

The NIR overtone region, we will see, is rich in interesting bands while lacking in background noise from the silica. At about 7300 cm^{-1}, the SiOH stretch overtone band (2ν) is diminished upon addition of water molecules while the neighboring broad H-bonded OH band at 7150 cm^{-1} increases. A band at about

[1] Present address: Ashland Chemical Company, P.O. Box 2219, L-142, Columbus, Ohio 43216.

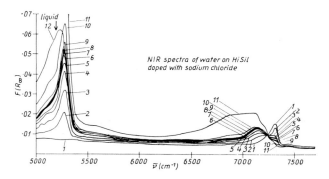

FIG. 1. The spectra obtained as water is adsorbed on partially dehydrated HiSil. The No. 1 spectrum shows the absence of absorption at 5300 cm^{-1} because this case represents the absence of physically adsorbed water. This band develops as water is adsorbed, while the broad band at about 7150 cm^{-1} grows at the expense of the sharp peak at 7300 cm^{-1}.

5300 cm^{-1} grows as water molecules are added to the surface because this combination band $(\nu + \delta)$ can only be due to molecular water and not to separate OH groups.

The exact location of this combination water band varies somewhat with the nature of the ligand on the surface but it has not been found to correspond to that for bulk water itself even when the surface is almost saturated. We examine this aspect toward the end of the paper. Many of the results were given at the Hungary meeting, September 1975, and published thereafter (2).

No previous workers have examined water interaction with silane-treated silica. The interaction of HMDS was reported in 1968 by Stark *et al.* (3), and the kinetics of the reaction with a poorly defined silica by Hertl and Hair (4) in 1971. Kiselev and his co-workers (5) in 1961 and 1964 examined the reaction of trimethylchlorosilane with silica, as did Snyder and Ward (6) in 1966, and Hair and Hertl (7) for all the methyl chlorosilanes. Stark *et al.* (3) in 1968 also examined the reaction of trimethylhydroxysilane, trimethylchlorosilane, and HMDS with silica. Lee (8) in 1968 examined the product of hydroxysilane treatment of glass slides by contact-angle measurements in an attempt to learn the configurations of the ligands on the surface. In 1970, Kharilo-

nov *et al.* (9) also reported on their studies of hydroxysilane reactions with silicas.

EXPERIMENTAL

We have pioneered the use of the NIR technique (1a) to study water adsorption on silicas. Since a loose bed of the powder is used, the problem of adherent particles and distorted surfaces as in pressed discs is minimized. Further, the temperature control is far better (within 0.2°C) than in transmission infrared where the temperature rise can be over 50°C. In addition, we scan with monochromatic radiation to reduce the heating effect. The addition of water vapor is made through an added port provided for the purpose. Figure 1 shows the NIR spectra obtained for one of the thermally treated HiSils. The addition of water diminishes the 7300 cm^{-1} band, as men-

TABLE I

Silica Samples

HiSil 233	123 m^2/g
Wet precipitated	Ar : 16.0 A^2
	1 OH / 7 A^2
Cab - o - Sil M - 5	178 m^2/g
Aerosil; SiCl$_4$	1 OH / 100 A^2

280

tioned previously, while increasing the neighboring H-bonded OH band at 7150 cm⁻¹. The band at 5300 cm⁻¹ develops as the water molecules are added to the surface. The data are taken at 1 nm intervals, then processed by a digital computer.

The silicas employed in this work are described in Table I.

The organosilanes employed in this study are depicted in Table II, along with the shorthand notation used hereafter. It was fortunate that both the HiSil and the Cab-o-Sil were employed; the two treated surfaces in each case behave quite differently toward water and the alkoxylated surfaces are quite different.

The Si(CH₃)₃ groups from the HMDS occupy about 40 Å², so there is just about one ligand for each OH group on Cab-o-Sil if the OH's are equispaced. Besides HMDS, only the unsaturated C=C compound was reacted from the vapor phase albeit at elevated temperatures.

The other three compounds were reacted from solution. The three silane triols were reacted with the silicas from water solutions. The starting compounds were the alkoxy compounds (Table II) which were hydrolyzed to the triols by the addition of several drops of acetic acid. After 3 hr of reaction with the silicas at room temperature, the solids were centrifuged and then pumped off at 50°C in a vacuum oven.

RESULTS AND DISCUSSION

First, the results for HMDS-treated silicas will be discussed. Then, the C=C-treated silicas

TABLE II

Organo Silicon Ligands

(CH₃)₃ Si N Si (CH₃)₃ H	HMDS
CH₂ = CH – Si Cl₃	C = C–
——— ‖ ———	
HS (CH₂)₃ Si (OCH₃)₃	HS –
CH₂–CH CH₂O (CH₂)₃ Si (OCH₃)₃ O	CH₂–CH – O
NH₂ (CH₂)₃ Si (OC₂H₅)₃	NH₂–

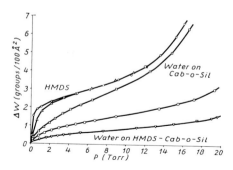

FIG. 2. The upper two curves are the HMDS isotherms for HiSil 233 and for Cab-o-Sil M-5 (△) at 25°C physical adsorption follow chemical adsorption. The lower three curves (discussed later) are water vapor adsorption isotherms for bare Cab-o-Sil (□), and HMDS-treated Cab-o-Sil: 0.92 ligands/100 Å (◇), and 1.75 ligands/100 Å (▽), all at 25°C. The latter coverage produces considerable hydrophobicity; by pumping the coverage is reduced from about 2.5 ligands/100 Å².

will be examined. Finally, results for the silanetriol-treated silicas will be given.

HMDS-treated silicas. An advantage of HMDS is that it reacts with the Sₛ–OH's (Sₛ is a surface silicon atom) from the vapor state even at room temperature. The surface OH's are titrated by the HMDS according to:

$$HMDS + 2 \geqslant S_s\text{–OH} \rightarrow$$
$$2 \geqslant Si_s\text{–O–Si(CH}_3)_3 + NH_3. \quad [1]$$

The question as to whether the NH₃ also reacts with the surface hydroxyls had not been discussed before our work. The cross-sectional area of the trimethyl ligand is about 40 Å²; therefore on Cab-o-Sil, if the sparse OH's are equispaced, each will react with HMDS (if little NH₃ is taken up). On the other hand, the surface density of OH's on HiSil is so high that many OH's remain after ligand coverage; the question of accessibility of the remaining OH's to water arises. These residual OH's might be regarded as being covered with an umbrella formed by the –Si(CH₃)₃ ligands.

Adsorption isotherms (the top two curves) for HMDS on both the HiSil and the Cab-o-Sil surfaces at 25°C in units of –Si(CH₃)₃ ligand groups/100 Å² are plotted in Fig. 2. From both

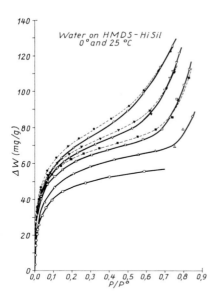

FIG. 3. Water vapor isotherms on HMDS-treated HiSil at four coverages: 0% HMDS (○), 29.6% (▽), 52.2% (□), and 100% (◇) at 25°C; solid points are for 0°C.

B points, the Si_s–O–$Si(CH_3)_3$ cross-sectional area is 40 Å², in accord with the estimate from the atomic model. It is interesting that the B point for the initially fully hydroxylated HiSil is reached at a lower HMDS pressure than it is for the Cab-o-Sil, suggesting that the HiSil possesses the more reactive OH's. The fact that the B point for the initially 1 OH/40 Å²-populated Cab-o-Sil is the same as that for the HiSil indicates that the surface OH's on Cab-o-Sil must be almost equidistantly spaced.

Water vapor isotherms at 0 and 25°C for HiSil are plotted in Fig. 3 for different amounts of chemisorbed HMDS from 0 to 100%. Controlled amounts of HMDS were introduced and then the amounts retained were ascertained gravimetrically after exhaustive pumping at room temperature. The HMDS percentages are based on the coverage of one ligand per 40 Å² as being 100%. The water isotherms are strongly Type II and retain a sharp knee even after high HMDS coverage. The organic ligand

does not inhibit the first-layer adsorption of water.

It is interesting to merge the HMDS and water vapor isotherms for HiSil as presented in Fig. 4 and Table III. They coincide in the monolayer region but diverge at higher relative pressures as might be expected. The chemisorbed ligands restrict the growth of the second and subsequent layers, the more so at higher HMDS coverages. (We shall learn below that the NIR results make this BET model doubtful, but comparison between isotherms in this manner is still useful.) However, the molecular area for the water molecules (Table I) is only about 7 Å², considerably smaller than that reported from this Laboratory heretofore (1b). This apparently small value for the co-area of water supports the contention of Sing (10) that HiSil contains a minor amount of micropores. While water would enter the micropores, of course the HMDS ligands would not. Heat treatment at 650°C or above removes these micropores.

Isosteric heats for water adsorption at three different HMDS coverages on HiSil are presented in Fig. 5. These curves overlap and only approach the heat of liquefaction at $\theta = 2$. The interaction of water with the underlying OH's is independent of the HMDS coverages. These values are higher than have been reported at 0.5 to 1.0 θ for partially dehydrated

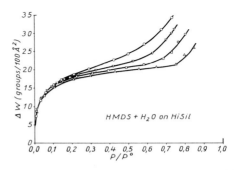

FIG. 4. Composite isotherms for water plus HMDS ligands on HiSil. The similar B points and repressed multilayer adsorption, especially at the higher ligand coverage, are expected. Composite monolayer values are also shown to agree in Table III.

HiSil (1c, d). Micropores may explain the high heats, and as mentioned heretofore, these are eliminated by the heat treatments used to produce the partially dehydrated samples.

Water adsorption isotherms for HMDS-treated Cab-o-Sil are compared with that for the bare Cab-o-Sil in Fig. 2. At the highest HMDS coverage after pumping to 10^{-6} Torr at 25°C, there are 1.75 ligands/100 $Å^2$ and the remaining OH's yield a water monolayer value of 0.58 by the BET model (probably some of which is adsorbed on top of adsorbed NH_3; see below). The latter value has been corrected to about 0.7 from the NIR studies reported below. The total of 2.45/100 $Å^2$ gives about 40 $Å^2$ per adsorbate unit as expected. This total is somewhat smaller, however, than the NIR monolayer value of 2.9 for water for the bare Cab-o-Sil reported below.

Examples of the NIR spectra obtained for HiSil at several increasing HMDS pressures are presented in Fig. 6. The peak intensities of the several bands are plotted against HMDS pressure in Fig. 7. The C–H stretch frequencies at 5730 and 5880 cm^{-1} produce curves similar to the adsorption isotherms as might be expected. The Si_sOH band (2ν) is shifted slightly from 7289 to 7241 cm^{-1}, apparently due to perturbation by the adsorbed ligands; this band diminishes with pressure and then levels off above 2 Torr in accord with the HMDS isotherm.

The band at 6530 cm^{-1} is assigned to the N–H stretch frequency. The development of this band parallels the diminution in the Si_sOH band. The evidence for the ammonia adsorption is, we believe, the first that has been reported. The shoulder which develops at 6610

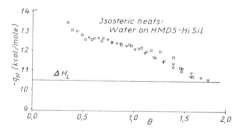

FIG. 5. Isosteric heats of adsorption on HMDS-treated HiSil: 0% (O), 29.6% (∇), 52.2% (□), showing almost no effect of the presence of ligands on the energetics of water adsorption. Band assignments are indicated here and in Table IV.

cm^{-1} is ascribed to N–H stretch in the buildup of the second layer of physically adsorbed HMDS. Thus, the complete picture of chemisorbed and physisorbed HMDS on HiSil is estsblished by this spectral analysis.

The HMDS-treated HiSil was further studied by obtaining spectra after evacuating to 10^{-6} Torr at room temperature and then at a sequence of elevated temperatures. Evacuation at room temperatures reduces all the bands except, of course, for a slight increase in the Si_sOH (2ν) band. The N–H(NH_s) band at 6530 cm^{-1} decreases only slightly indicating the strong adsorption of NH_3 to the surface OH's. Apparently, the main event upon evacuation at room temperature is the loss of physically adsorbed HMDS. Furthermore, little change in band intensities occurs up to 50°C. Large changes then occur up to 140°C which plateau at 170°C. The complete elimination of the N–H(NH_3) band at 170°C is of interest because the same temperature is required at 10^{-6} Torr to remove physically adsorbed water from HiSil. The shift of some 80 cm (6610 to 6530 cm^{-1}) suggests a strong perturbation in the decrease of the N–H vibration frequency from HMDS to NH_3 adsorption.

The NIR reflectance technique was also applied to the bare and HMDS-treated Cab-o-Sil surfaces. The spectra are displayed in Fig. 8. The assignments made for the bands are tabulated in Table IV along with those for the HiSil. The most striking discovery is that two

TABLE III

HMDS + H_2O on HMDS—HiSil

HMDS Groups/100A^2	HMDS %	H_2O's/100A^2	HMDS + H_2O/100A^2
0	0	15.8	15.8
0.73	29.6	14.8	15.6
1.28	52.2	14.1	15.4
2.45	100	13.0	15.5

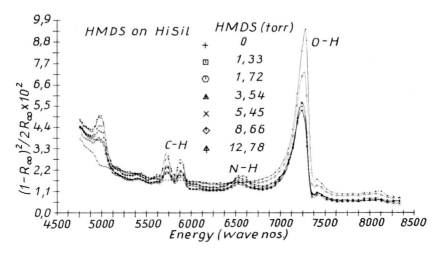

FIG. 6. Typical NIR reflectance spectra at a variety of pressures of HMDS adsorbed on HiSil at 25°C.

distinct bands due to Si–OH are found at 7246 and 7331 cm⁻¹. In the transmission mode, only a mixed (3665 and 3749 cm⁻¹) band has been reported (11) with a wider shoulder on the low-frequency side attributed to hydrogen bonded OH's or bulk OH groups. The results presented here indicate that hydrogen-bonded OH's either do not react with HMDS as suggested previously (4), or, as seems more likely to us, that bulk OH's are responsible for the low-frequency peak.

Additionally, the N–H(NH₃) band is both weaker on the Cab-o-Sil surface and is totally removed by evacuation at room temperature; its appearance at a higher frequency (6580 cm⁻¹) also suggests weaker interaction. Furthermore, the Si_sOH band (7331 cm⁻¹) is almost entirely quenched again indicating the rather equal spacing of the OH groups on the Cab-o-Sil. The band intensities versus the pressure are plotted in Fig. 9. In accord with the isotherms, these curves are not as sharp as for HiSil. The dashed lines refer to the changes on evacuation at room temperature. Both the C–H bands and the N–H band diminish on evacuation, indicating weaker interaction with the Cab-o-Sil surface.

NIR reflectance spectra were also measured to follow water adsorption on the bare and HMDS-treated silica surfaces. Since similar spectra were reported on the bare surfaces

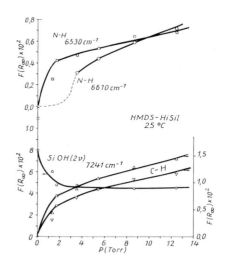

FIG. 7. Intensity of bands (Schuster–Kubelka–Munk function) versus HMDS pressure at 25°C from Fig. 6: ○, $Si_sOH(2\nu)$ band at 7289 cm⁻¹; ▽ and △, C–H bands at 5730 and 5880 cm⁻¹; □, N–H(NH₃) at 6530 cm⁻¹; and ◇, N–H (HMDS) at 6610 cm⁻¹.

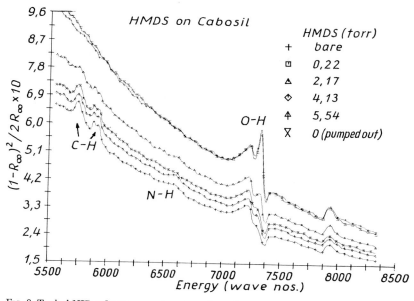

FIG. 8. Typical NIR reflectance spectra at a variety of HMDS pressures on Cab-o-Sil at 25°C.

heretofore (1a), these spectra are not displayed here. As noted before, water adsorption on the HMDS-treated HiSil is little affected by the HMDS treatment except for a reduction in the multilayer region.

The intensity curve (SKM function) for the 100% HMDS-treated HiSil is given in Fig. 10. The tailing off suggests clustering of the water

TABLE IV

NIR Bands: HMDS–Silicas

Freq. cm⁻¹	Surface	Group Assignments – Overtones
4975	Hisil & Cabosil	Adsorbed NH_3 – bending + assymetric stretch
5730 & 5880	Hisil & Cabosil	C–H stretch
6530	Hisil	N–H stretch – chemi. NH_3
6580	Cabosil	N–H stretch – chemi. NH_3
6610	Hisil	N–H stretch – physi. NH_3
7246	Cabosil	O–H stretch – hydrogen bonded
7289	Hisil	O–H stretch (2ν) – free OH's
7331	Cabosil	O–H stretch - free OH's

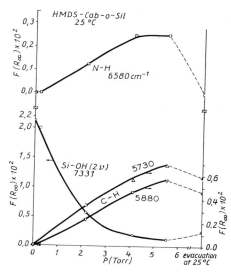

FIG. 9. Band intensities for HMDS adsorption on Cab-o-Sil from Fig. 8: ○, free OH band at 7331 cm⁻¹; △ and ▽, C–H bands at 5730 and 5880 cm⁻¹; □, N–H(NH_3) band at 6580 cm⁻¹. Dashed lines show changes caused by evacuation at room temperature to 10⁻⁶ Torr.

285

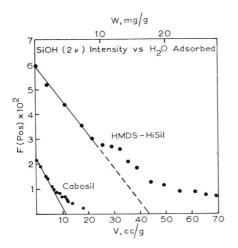

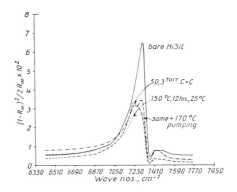

FIG. 11. Spectra showing both the perturbation of the OH-absorption peak (2ν) by physisorbed and by chemisorbed vinylsilane.

FIG. 10. Band intensities (S–K–M function) for water on 100% HMDS-treated HiSil and on bare Cab-o-Sil. The tailing-off is due to the clustering of water molecules around those first adsorbed.

molecules around molecules adsorbed at lower pressures (1a). And the extrapolated value is believed to be a more appropriate measure of the residual surface OH concentration than is given by the BET model or B point value from the isotherm. Of course, the diminution of the Si_sOH (2ν) peak indicates that adsorption occurs on the OH's below the silazane ligands; apparently also about 50% of the OH's are titrated before clustering becomes important as indicated by deviation from the

linear plot. Even at high water vapor relative pressures, the spectra indicate considerable difference of the adsorbed water from bulk water.

Water adsorption is too low on 100% HMDS-treated Cab-o-Sil to produce very useful results from the spectra. The intensity results on bare Cab-o-Sil are also plotted in Fig. 10 for comparison. The extrapolated monolayer value is 15.9 mg/g (2.9 molecules/100 Å²). This 18% difference from the B-point value is not too surprising; the clustering effect takes place here as well.

C=C-treated silicas. The vinylsilane was added to the silicas after pumping at 170°C for HiSil and at 100°C for Cab-o-Sil to remove physi-

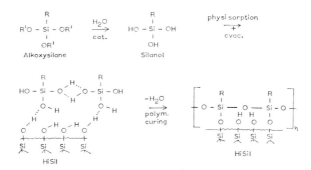

FIG. 12. Schematic diagram for the reaction of alkoxysilanes with a highly hydroxylated silica surface, HiSil. A two-dimensional polymer is eventually formed upon heat treatment.

286

cally adsorbed water. At room temperature no reaction occurred as indicated by the readily pumped-off C=C leaving the bare silica OH stretch peak.

However, the physically adsorbed C=C perturbed the OH stretch frequency on HiSil to 7230 cm^{-1} at about 50 Torr as seen in Fig. 11. After reaction at 170°C for 12 hr, the residual OH frequency was at 7252 cm^{-1} as perturbed by the reacted ligand (12). The fact that much OH still persists after reaction suggests that not all the Cl's are reacted, as might be expected. The reaction of the C=C compound with the surface OH's is much weaker than that of HMDS.

Cab-o-Sil reacted at 110 or 170°C for 12 hr also showed a small amount of residual OH.

For both silicas, the frequencies at 5900 and 6125 cm^{-1} are assigned to the C–H vibrations. These values are higher than reported for HMDS in accord with the fact that unsaturated groups yield higher frequency C–H vibrations (13).

Water adsorption will be discussed later.

Silanetriol-treated silicas. The reaction with HiSil is depicted in Fig. 12. Here the scheme suggests a one-layer addition in which the silanol groups condense to form what might

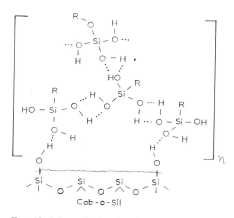

FIG. 13. Schematic showing three-dimensional polymer network that develops when extra OH's are not available, i.e., for Cab-o-Sil, to hold the silanetriol to the surface.

be regarded as a two-dimensional polymer. Hydrogen bonding with the unreacted surface OH's is believed to be responsible for holding the silanetriol to the surface to develop the two-dimensional effect. On the contrary, a three-dimensional network develops on Cab-o-Sil as shown in Fig. 13 when there are not too few residual OH's to confine the triol to one layer.

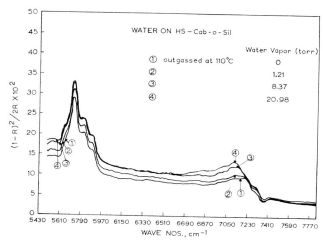

FIG. 14. NIR spectrum for HS-treated Cab-o-Sil showing the effect of increased water adsorption.

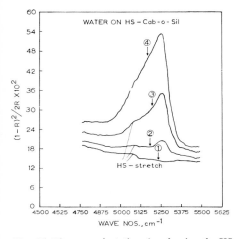

FIG. 15. The water $(\nu + \delta)$ region showing the HS-vibration shoulder which is finally swamped by the combination band of the water.

There are several pieces of evidence supporting these conclusions. First, the bands such as those for the C–H were about the same intensity for the silanetriol-treated HiSil as for HMDS-treated HiSil. A greater intensity occurred for the silanetriol-treated Cab-o-

Sil. An interesting support for this contention is also shown for the HS compound in Fig. 14, the left-hand portion magnified in Fig. 15, both for the Cab-o-Sil-treated surface. This band is very weak even for mercaptan compounds themselves and is finally overwhelmed here by the $(\nu + \delta)$ band for water at high coverages. This HS-stretch frequency cannot be detected in the case of the HS-treated HiSil. Thus, the contention that a thicker layer, perhaps seven or eight molecules thick as Lee has suggested (8), formed from the triols on Cab-o-Sil, whereas only about one layer formed on HiSil.

Second, this remarkable difference produced on the two substrates was also confirmed by the greater apparent agglomeration of the Cab-o-Sil when treated with the silanetriols. The cross linking of particles would reduce specific surface area. Measurements showed a reduction of 30–45% for the treated Cab-o-Sils, and only 10–15% for the treated HiSils.

The conclusion seems inescapable that there is a large difference between the behavior of

TABLE V

NIR Bands: Silane-treated Silicas

Freq. cm⁻¹	Ligand Surface	Assignments – Overtones
4948	NH₂ – HiSil NH₂ – Cab-o-Sil	NaH₂ – bending plus stretch combination band
5079	HS – Cab-o-Sil	H–S stretch
5670 5820	NH₂ – HiSil NH₂ – Cab-o-Sil	C–H stretch
5734 5821, 5886	HS – HiSil HS – Cab-o-Sil	C–H stretch
5737, 5858	CH₂–CH – HiSil \O	C–H stretch
6086	CH₂–CH – Cab-o-Sil \O	
5900, 5959 6125	C=C – HiSil C=C – Cab-o-Sil	C–H stretch
6540	NH₂ – HiSil NH₂ – Cab-o-Sil	N–H stretch
7225	NH₂ – Cab-o-Sil	O–H stretch –H bonded
7252	C=C – HiSil	O–H stretch – free OH's
7270	NH₂ – HiSil	same
7283	HS – HiSil	same
7300	C=C – HiSil	same
7331	C=C – Cab-o-Sil	same

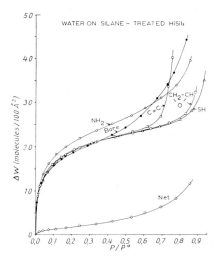

FIG. 16. Water adsorption isotherms for the silane-treated HiSils. Except for the higher adsorption on the NH₂-treated surface, the isotherms are similar in the B-point region. The "net" curve shows the difference in water adsorption between the polar NH₂-treated and the HS-treated surface. (The rise at 0.7 relative pressure for the C=C-treated surface is unexplained.)

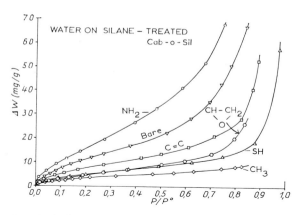

FIG. 17. Water adsorption isotherms for the silane-treated Cab-o-Sils. A distinct order to the polarity of these surfaces now emerges.

the sparsely OH-populated silica and the abundantly populated silica. It seems plausible that glass fibers might perform better in composites if first stripped of some of their surface OH's.

Table V presents the band assignments for the various silane-treated silicas other than those for HMDS (Table IV).

Water isotherms for the silane-treated HiSils are presented in Fig. 16. These have been corrected to areas determined for each. They show the same B point except for the NH₂-treated HiSil, which shows an enhanced specificity for water. The curve marked "net" is the difference from the HS-treated and the NH₂-treated HiSil curves. The R groups restricted the growth of the water multilayers except for the C=C case where the sharp rise at 0.7 relative pressure remains unexplained.

The silane-treated Cab-o-Sils yield unique water isotherms (Fig. 17) for each ligand. Except for the HMDS(CH₃)- and the C=C-treated Cab-o-Sil, the layers are thick so that there is no point in putting the results on a unit area basis. The NH₂-treated sample required three to five times the equilibrium time of the others for water adsorption indicating diffusion into the thick layer. The rating for the hydrophobicity of the ligands is distinct in the case of the Cab-o-Sil-treated surfaces be-

cause the surface hydroxyls have essentially disappeared. The presence of multilayers in the case of the silanetriols, however, makes a 1:1 analysis impossible. It is interesting that the C=C ligand showed considerably higher water attraction than the CH₃ ligand.

The peak frequencies for the water combination band for those silane-treated silicas for which we have clear values are listed in Table VI. The values for bulk water, ice, and critical-state water from Luck are also tabulated. Usually the peak values did not change ap-

TABLE VI
Peak Frequencies for Water $(\nu + \delta)$ Bands
at ca. 0.9 R.P. and 25°C

	Freq. cm⁻¹
Hi Sil	5236
HMDS - Hi Sil	5249
NH₂ - Hi Sil	5250
HS - Hi Sil	5352
C=C - Cab-o-Sil	5352
NH₂ - Cab-o-Sil	4888
CH₂ - CH - Cab-o-Sil (with O bridge)	5352
HS - Cab-o-Sil	5248
Bulk Water	5175
Ice	5150
Critical State	5291

preciably with increased water adsorption from 0.2 to 0.9 relative pressure, and they are nearer critical-state water than bulk water or ice. The NH_2 case is remarkable: its peak value is similar to the other cases for the HiSil where residual OH's apparently dominate the water adsorption, but the frequency is much lower for the Cab-o-Sil where the interaction is almost solely with the NH_2 groups.

REFERENCES

1. (a) KLIER, K., SHEN, J. H., AND ZETTLEMOYER, A. C., *J. Phys. Chem.* **77**, 1458 (1973).
   (b) BASSETT, D. R., BOUCHER, E. A., AND ZETTLE-MOYER, A. C., *J. Colloid Interface Sci.* **34**, 436 (1970).
   (c) BASSETT, D. R., BOUCHER, E. A., AND ZETTLE-MOYER, A. C., *J. Colloid Interface Sci.* **27**, 649 (1968).
   (d) ZETTLEMOYER, A. C., *J. Colloid Interface Sci.* **28**, 4 (1968).
2. HSING, H. H., AND ZETTLEMOYER, A. C., *Progr. Colloid Polym. Sci.* **61**, 54 (1976).
3. STARK, F. O., JOHANNSON, O. K., VOGEL, G. E., CHAFFEE, R. G., AND LACEFIELD, R. M., *J. Phys. Chem.* **72**, 2750 (1968).
4. HERTL, W., AND HAIR, M. L., *J. Phys. Chem.* **75**, 181 (1971).
5. (a) LYGIN, V. I., AND KISELEV, A. V., *Kolloid Zh.* **23**, 250 (1961).
   (b) DAVYDOV, V. Y., KISELEV, A. V., AND ZHURAVLEV, L. T., *Trans. Faraday Soc.* **60**, 2254 (1964).
6. SNYDER, L. R. AND WARD, J. W., *J. Phys. Chem.* **70**, 3941 (1966).
7. HAIR, M. L., AND HERTL, W., *J. Phys. Chem.* **73**, 2372 (1969).
8. LEE, L. H., *J. Colloid Interface Sci.* **27**, 751 (1968).
9. KHARILONOV, N. P., GLUSHKOVA, N. E., AND ZHUKOVA, A. S., *Izv. Akad. Nauk SSSR, Neorg. Mater.* **6**, 59 (1970).
10. BAKER, F. S., AND SING, K. S. W., *J. Colloid Interface Sci.* **55**, 3 (1976).
11. HAIR, M. L., "Infrared Spectroscopy in Surface Chemistry," Dekker, New York, 1967.
12. WALL, T. T., AND HORNIG, D. F., *J. Chem. Phys.* **43**, 2079 (1965).
13. BELLAMY, L. J., "Infrared Spectra of Complex Molecules." Methuen, London, 1958.

Rheology of Disperse
Systems

Orthokinetic Phenomena in Disperse Systems [1]

S. G. MASON

*Pulp and Paper Research Institute of Canada and Department of Chemistry,
McGill University, Montreal H3C 3G1, Canada*

Received July 30, 1976

Perikinetic phenomena, resulting from translational and rotational Brownian motion of the particles in fluid sols, have received much attention in the development of colloid science especially in connection with problems of stability. Of equal interest and importance are orthokinetic phenomena arising from particle motions when the fluid system undergoes flow such as shear.

This paper deals briefly with several of a wide range of orthokinetic effects which have been examined both theoretically and experimentally in pure (irrotational) and simple (rotational) shear flows. These include (1) two-body collisions of both neutral and interacting rigid spheres which can often result in capture, (2) the dispersion of particle aggregates, (3) configurational statistics of coiled flexible threads, and (4) oscillatory phenomena in nearly monodisperse dispersions of nonspherical particles.

The significance of the phenomena to a number of problems in colloid science is briefly discussed.

1. INTRODUCTION

Much of colloid science, and especially that dealing with the stability of sols, is based on perikinetic phenomena, i.e., those originating from Brownian translation and/or rotation of the dispersed particles*; since these motions are thermal in origin they follow random statistics. Another important set of phenomena is orthokinetic, i.e., resulting from particle movements caused by flow of the fluid medium in which they are dispersed; except when the motion is turbulent, the particle motions are highly ordered,* i.e., nonrandom and, as we will see, often possessing perfect memory.

For the past 25 years, we have been studying a variety of orthokinetic effects following in the footsteps of a few pioneers such as Einstein (1), v. Smoluchowski (2), Jeffery (3), Eirich (4), and Taylor (5), the last of whom continued his interest in the field for over 40 years until his recent death. In the past dozen or so years there has been a rapid growth in the subject to the point where there are now many workers in many countries.

In this account, however, I will dwell mainly on our own efforts and touch briefly on four of the many phenomena which we have examined in our theoretical and experimental studies (6),[2] some ideas of which are given and referenced in Table I. We have worked with a wide variety of particle types, sizes, and concentrations—ranging from wood-pulp fibers (which stimulated my interest in the work in the first place) to blood cells, to synthetic latex spheres, all of which can be seen and

[1] Plenary lecture, International Conference on Colloids and Surfaces, San Juan, Puerto Rico, June 21–25, 1976. The lecture was accompanied by a 16-mm cinefilm (silent), a copy of which (bearing the title of this paper and suitably subtitled) may be obtained on loan (for a nominal handling charge) from: The Librarian, Pulp and Paper Research Institute of Canada, 570 St. John's Boulevard, Pointe Claire, Quebec, Canada, H9R-3J9. Items in the text and Table I marked by an asterisk (*) are illustrated in the film.

[2] A detailed account of work up to 1966, including our own contributions, is given in Ref. (6). Our later work is reported in some of the remaining citations.

TABLE I

Scope of Work in Orthokinetics (with Partial List of Literature References)

A. Particles	B. Suspending fluids	D. Some phenomena
1. Types:	1. (Gases)	1. Translation, rotation, and
(a) Cellulose fibers (6)	2. Liquids:	deformation* (6)
(b) Flexible threads* (6, 7)	(a) Newtonian (6)	2. Collision kinetics* (6, 11, 22)
(c) Red blood cells (8) and	(b) Non-Newtonian* (19, 20)	3. Orientation kinetics* (15, 16)
platelets (9)	3. (Solids)	4. Stability: aggregation* (floc-
(d) Latex spheres* (<0.5 μm	C. Fields	culation) (6, 25), coales-
diam) (10, 11)	1. Shear:	cence (6)
(e) Fluid drops: gas and liquid	(a) Rotational* (6, 21, 22)	5. Dispersion of aggregates (6,
(6, 12); 2-phase drops	(b) Irrotational* (6, 21)	21)
(13)	2. Electric (6, 16)	6. Emulsification (21, 26)
(f) Model particles: spheres*	3. Shear and electric (6, 16)	7. Concentration changes at
(6, 14), rods (6, 15),	4. (Magnetic) (16)	boundaries (6, 27)
discs* (6, 15), spheroids	5. Gravitational (6) and	8. Memory effects* (28, 29)
(6, 16), etc.	centrifugal (23)	9. Optical properties (9, 30)
(g) (Micro- and macro-	6. Internal:	10. Electrical properties (31)
molecules)	(a) Coulombic (11, 25)	11. Macrorheology: viscosity
2. Translational and rotational	(b) van der Waals* (11, 25)	and normal stresses (6, 15,
Brownian motion:	(c) Entropic (24)	16)
(a) Negligible*: Br = ∞	(d) Molecular bridging* (24)	12. Oscillatory phenomena* (9,
(orthokinetic) (6, 15, 16)		15, 16, 31)
(b) Appreciable*: 0 < Br		E. Areas of application
< ∞ (ortho + peri-		1. Scientific: physical, biological,
kinetic) (10, 17)		earth sciences
3. Rigid and deformable* (6)		2. Technological: pulp and
4. Attraction/repulsion:		paper, synthetic polymers,
(a) Negligible* (6)		mineral processing, medical
(b) Appreciable* (6, 17)		and pharmaceutical, food-
5. Concentration:		stuffs, etc.
0 to 70% by volume (6, 18)		

photographed, often with the aid of a microscope. The bracketed items in the table were not actually studied but represent cases to which the principles of orthokinetics can be applied by extrapolation, e.g., to macromolecular solutions, aerosols, and solid composites. As indicated in Table I (Part E) the phenomena have proven of great interest to various sciences and technologies, and it might be argued are in the long run of much greater importance than those of purely perikinetic origin.

Before discussing specific examples, I would like to say a few words about simple flows in Newtonian liquids in which media most of our work has been done. Figure 1 shows schematically a generalized two-dimensional steady flow (32) in which the velocity components U_i along the X_i axes (in the absence of particles) are given by

$$U_1 = 0; \quad U_2 = \lambda G x_3; \quad U_3 = G x_2, \quad [1]$$

where $-1 \leq \lambda \leq +1$ is a dimensionless parameter. When $\lambda = -1$, the liquid undergoes pure rotation and no deformation; when $\lambda = +1$ the liquid undergoes pure deformation without rotation; when $\lambda = 0$ we have the familiar case of simple shear flow (a rotational flow of angular velocity $G/2$ (6) about the vorticity axis X_1) having a rate of shear G. Most of our work (both theoretical and experimental) has been conducted in flows corresponding to $\lambda = 0$, i.e., simple shear or Couette flow.

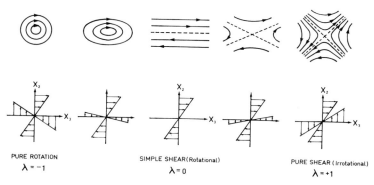

FIG. 1. Schematic of a family of two-dimensional steady flows of velocity components U_i given by [1] showing the streamline patterns at the top and the velocity components at the bottom. When $\lambda = -1$, the fluid undergoes pure rotation and no deformation; $\lambda = +1$ corresponds to pure shear: deformation without rotation; $\lambda = 0$ represents simple shear deformation with rotation $G/2$ about the X_1 axis (normal to the page). Most of our work has been done at $\lambda = 0$, with some at $\lambda = +1$ and a little over the whole range $-1 \leq \lambda \leq +1$. After (32).

When particles are present it is useful to define the dimensionless Brenner[3] number Br designating the ratio of ordered to disordered motion:

$$Br_t = Gb^2/D_t; \quad Br_r = G/2D_r, \quad [2]$$

where the subscripts t and r designate translation and rotation, b is a linear dimension (e.g., the radius of a sphere), and D the Fickian diffusion constant of the particles. Thus Br $= 0$, ∞ correspond respectively to perikinetic and orthokinetic conditions; $0 <$ Br $< \infty$ yields a combination of the two, of which the most familiar example is probably streaming double refraction (33). Most of what I will discuss is for Br $= \infty$.

Let me now mention briefly the behavior of rigid spheres in simple shear as illustrated in Fig. 2. It is worth mentioning that the foundations of modern macrorheological theory were laid by Einstein (1) when he calculated the additional dissipation of energy caused by the spheres and thus derived his classical viscosity equation. A single isolated sphere (shown in

[3] In previous papers we have designated this as the Peclet number (see e.g. (10)) but now believe that in the present context Br is more appropriate. See H. Brenner, J. Colloid Interface Sci. 34, 103 (1970).

the upper part of Fig. 2) translates at the same velocity U_3 as the suspending liquid at the X_2 coordinate of the sphere center and rotates about the X_1 axis with an angular velocity $\omega = G/2$ (6). The rotations are of great significance, especially when the particles are nonspherical when ω is variable, leading to preferred orientations in populations, and thus anisotropies, which can be important in a number of ways (3, 6, 15, 34).

The streamlines in the liquid in the neighborhood of the sphere are curvilinear, and are either open or closed, the two classes being separated by a limiting set of streamlines* which describe a pair of infinite three-dimensional surfaces symmetrical about the three principal planes; the equatorial limiting streamline is shown in Fig. 2 (upper). The predicted streamline pattern* has been confirmed in detail (36, 37).

Over the years we have developed a number of experimental devices for examining the behavior of particles in flows of the kind shown in Fig. 1, and particularly Couette flow, but these have been so widely described (6, 35) that I will mention only the recently developed microtube (10), which produces simple shear, and is of special value in studying particles

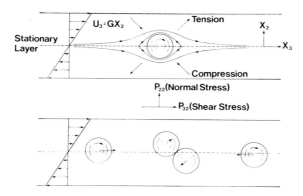

FIG. 2. Upper part: A rigid neutral sphere at Br = ∞ placed at the origin of a field of simple shear ($\lambda = 0$), where it remains and rotates* at $\omega = G/2$. Fluid streamlines near the sphere are symmetrical and are either open or closed, and are separated by a critical streamline which is shown for $X_1 = 0$. The surface experiences tensile and compressive stresses (as shown) and tangential stresses (not shown). To maintain the flow, macroscopic stresses must be applied at the boundaries, nine components in all, of which the most familiar is the stress P_{32}, which is related to the viscosity η by $P_{32} = \eta G$. Macrorheology is concerned with the prediction and measurement of the stresses P_{ij} (15). Lower part: When two spheres have different X_2's they translate relative to one another so that when the geometry is favorable (as in the case shown) they will approach one another and form a collision doublet (2, 6).

as small as 0.5 μm which can only be seen and photographed in a high-resolution microscope. In this apparatus (Fig. 3) we have a circular tube, which can be as small as 25 μm in diameter, arranged vertically with the suspension flowing through the tube to establish the well-known Poiseuille parabolic profile in which G varies linearly with radial distance from zero at the tube axis to reach extreme values at the wall of the tube. The particles are viewed through a fixed microscope and are rendered stationary in the field of view by moving the tube at a velocity equal and opposite to that of the particles which we wish to observe.*

2. TWO-SPHERE INTERACTIONS IN SIMPLE SHEAR ($\lambda = 0$)*

This is a subject which we have studied extensively, and represents the simplest of a series of orthokinetic rate processes which we have designated "the kinetics of flowing dispersions" (6, 15, 22) and which are often analogous to those occurring in molecular systems at Br = 0. An orthokinetic collision of

two neutral spheres is illustrated in the bottom half of Fig. 2 and assumes rectilinear approach of the pair until contact is made and a "collision doublet" is formed. Table II shows theoretical equations for such collisions compared with (1) bimolecular collisions in an ideal gas and (2) collisions in a colloidal sol due to translational Brownian motion, both perikinetic. It is seen that there is a close similarity between all three, with the collision frequency f per particle (molecule) being proportional to the number density N, the proportionality constant containing a measure of the relative speed of the colliding elements: the mean thermal speed of the molecules, a component of the translational diffusivity of the particles for the perikinetic collisions in the dispersion, and G for the orthokinetic collisions. The equation for orthokinetic frequency, due originally to v. Smoluchowski (2), was first confirmed experimentally about 25 years ago (6, 38) by direct observations along the X_2 axis (Fig. 2) even though it was realized that the approach trajectories are curvilinear and the spheres do not touch.

The complete details of the interaction of two neutral hard spheres at Br $= \infty$ have now been worked out theoretically and confirmed experimentally (14, 39), and are summarized schematically in Fig. 4 for equatorial interactions. The theory accounts for the existence of two kinds of doublet, one transient* and the other permanent or nonseparating.* The type of doublet formed depends upon the paths of approach, with a limiting trajectory which separates transient from permanent doublets. The two cases correspond more or less to the

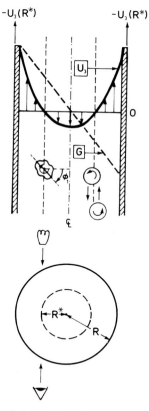

FIG. 3. Principle of the microtube apparatus (10) in which Poiseuille flow takes place through a cylindrical tube to produce a parabolic profile (solid line for U_3) with a known linear variation of G (dashed line). A region in the median plane at R* (usually containing a particle) is rendered stationary in the field of a fixed microscope by moving the tube at $-U_3(R^*)$.

TABLE II

Two-Body Collision Frequency of Hard Neutral Spheres (Rectilinear Approach)[a]

System	Br_t	f = Frequency per particle
Ideal gas (thermal collision)	0	$(16\pi)b^2 N[kT/\pi m]^{\frac{1}{2}}$
Fluid dispersion (Brownian collisions)	0	$8N[kT/3\eta]$
Fluid dispersion (shear collisions)	∞	$(32/3)b^3 NG$

[a] F = frequency/unit volume = $fN/2$, k = Boltzmann constant, N = particle (molecule) concentration, m = mass of molecule, T = temperature, η = viscosity of medium.

open and closed streamlines about a single sphere (Fig. 2, upper). The two-body encounters given by the Smoluchowski equation are those yielding transient doublets; recently we have reconciled the simple theory for f based on rectilinear approach with the exact theory based on curvilinear approach on open trajectories (22). Permanent doublets were first discovered experimentally (6) and have engaged our attention for many years; our interest in them has been rearoused by the new theory and improvements in experimental techniques (14).

An isolated doublet of neutral spheres will be either transient or permanent, and the transition from one to the other which would require the particle centers to cross the surface of limiting trajectories (Fig. 4) should theoretically be prohibited. However, we have observed such transitions experimentally and have shown that they can be brought to pass in a number of ways, including the following: (1) three-body collisions of initially separated spheres, with one of the spheres separating to leave a permanent doublet (40); (2) interparticle attraction from van der Waals forces (17),* or by induced Coulombic attraction in an externally applied electric field (41); (3) Brownian diffusion of a particle center across a limiting trajectory when $0 < Br_t < \infty$ (25); (4) surface asperities on spheres in close

297

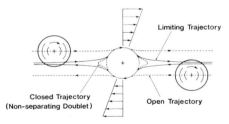

FIG. 4. Transient (open trajectory) and permanent (closed trajectory) doublets of spheres for the case shown in Fig. 2 (lower) using the exact theory of curvilinear approach (14, 39) shown for equatorial collisions. The central sphere is the exclusion sphere (14) which cannot be penetrated by either particle center. Shown also is the limiting trajectory which separates the two clases, and which yields a minimum gap width of $4.2 \times 10^{-5} \, b$. On reversing flow, the trajectories are retraced, a simple example of perfect memory exhibited in many ways when Br $= \infty$ (28, 29).

proximity, particularly during equatorial encounters where on a limiting trajectory the minimum separation is only $4 \times 10^{-5} \, b$ (14).*

These transitions are interesting to observe and are believed to be of considerable importance.

In our studies of transient doublets of neutral spheres at Br $= \infty$ we have calculated and confirmed experimentally equations for the distribution of doublet lives, the mean doublet life, the mean free path of single spheres, etc. from simple considerations of rectilinear approach (6); a number of similar equations have recently been derived from the exact curvilinear approach theory (41).

In sols stabilized by the dilute electrolytes to which the DLVO theory (42) applies approximately, there can exist an interaction potential (V_{int}) curve between two spheres of the type shown in Fig. 5 which produces attraction due to dispersion forces at very small and very large separations h when $(dV_{int}/dh) > 0$, repulsion from Coulombic forces at intermediate separations, and zero force when the repulsion and attraction balance one another at a value h_{eq} when $(dV_{int}/dh) = 0$. In simple

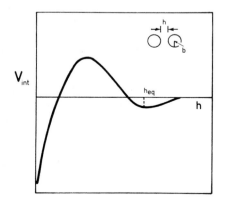

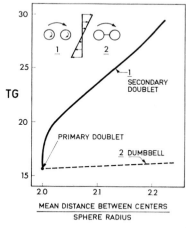

FIG. 5. Left: Interaction potential (schematic) from conventional DLVO theory when the interaction force is zero at the "secondary minimum" separation h_{eq} between a pair of spheres. After (42). Two kinds of permanent doublets can form for this case: primary (when $h = 0$) and secondary (when $h = h_{eq}$). Right: Calculated period of rotation of primary and secondary doublets, and of rigidly connected dumbells as functions of the distance of separation. The curve for the secondary doublet, in which the spheres can rotate relative to one another thus giving them a higher TG, is calculated for the case where β, a dimensionless parameter expressing the ratio of interaction/hydrodynamic forces between two spheres is large and TG for a given h_{eq} is independent of G, a condition which was satisfied in our experiments (11).

shear flow such a system will yield two kinds of permanent doublet which we have designated as primary, when $h = 0$ and the spheres touch and are thus rigidly connected, and secondary, when the mean distance of separation is h_{eq}. When such doublets, e.g., in a hydrosol of soap-free polystyrene latex spheres of radius $b = 1$ μm are viewed in the microtube (Fig. 3), they cannot be distinguished from one another* since $h_{eq} \simeq 100$ nm and both appear to be touching.* However, they can be identified by measuring the period of rotation T of the doublet about the vorticity axis (X_1 of Fig. 2) using the theoretical curves of Fig. 5, where it is seen that rigidly connected spheres, which cannot rotate relative to one another, have a lower period than those which can. At finite Br_r, rotary Brownian motion* will cause a dispersion to TG which can be allowed for by making a sufficiently large number of observations (11). Good agreement with theory was found for primary doublets, but discrepancies were found with secondary doublets from which we have tentatively concluded that the distribution of surface electric charges was not uniform (11). I believe that this technique, which is in its early stages of application, can be used to study various theoretical aspects of sol stability including entropic stabilization, and destabilization by macromolecular bridging (24, 25).

Figure 6 shows tracings of cinephotographs of two nearly linear aggregates undergoing orthokinetic capture as a result of van der Waals attraction.* By observing whether or not the aggregates, or segments of them, are rigid or deformable as they rotate in the shear field we can establish whether or not the adjacent spheres touch one another (25). The shapes of the rotating aggregates are also of considerable interest and significance (10).

3. DISPERSION OF PARTICLES*

This topic is of interest mainly to applied colloid scientists and engineers who are concerned with the hydrodynamic dispersion of

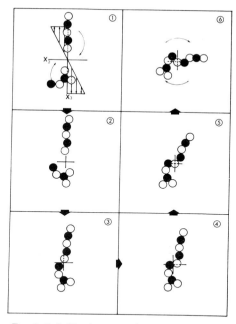

FIG. 6. Orthokinetic capture (frame 3) of two nearly linear aggregates each containing five polystyrene latex spheres of $b = 2$ μm. Some of the spheres touch neighbors ($h = 0$) and others do not ($h = h_{eq}$) as can be judged by whether or not segmental bending occurs during rotation in the shear flow. Traced from cinephotographs in the microtube apparatus.*

particles (e.g., wood pulp in papermaking processes, mixing, emulsification and homogenization, etc.) in cases where shearing motions play a central role. In standard treatments of the mechanisms involved (43) no distinction has been drawn between the efficacy of different kinds of shear.

To illustrate we consider the profound difference in dispersing efficiency between simple shear ($\lambda = 0$ in [1]) and pure shear ($\lambda = +1$), using for illustration an initially spherical aggregate of radius R_0 consisting of a concentrated suspension of plastic spheres in a viscous liquid, the spheres being sufficiently large and the suspension (which itself is fluid) sufficiently dilute that very few, if any, of the individual spheres are in contact with others. In simple shear, the aggregate rotates as

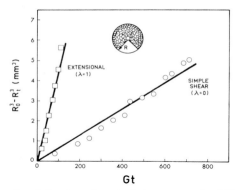

FIG. 7. Breakup of a spherical aggregate of a 60% suspension of $b = 50$ μm, plastics spheres in silicone oil, and suspended in the same liquid, R_0 (=1.8 mm) being the initial radius and R_t the radius at time t. Using the semiempirical relationship $(R_0{}^3 - R_t{}^3) = kGt$, values of k in these experiments were 7×10^{-6} and 44×10^{-6} cm³, respectively, for simple shear and extensional flows. After (21).

a coherent sphere of gradually diminishing radius, with single spheres gradually breaking away from the surface of the aggregate at angles ϕ_1 near 45° and 225° from the X_2 axis (Figs. 1 and 2) and then moving in closed orbits about the main aggregate in a manner similar to that in the closed streamlines around a single sphere (Fig. 2) until they continually cross the limiting streamlines to become completely dispersed.* In pure shear (or 2-D extensional flow) on the other hand, the aggregate does not rotate and the individual spheres are pulled off* at $\phi_1 = 45°$, 225° (Fig. 1) i.e., in the direction of principal extension and are immediately dispersed because there are no closed streamlines about a single sphere in such flow (14, 21, 32, 35). This leads to a much more rapid breakup of the aggregate as illustrated in Fig. 7, where the slope of the straight line which is a measure of the rate of decrease of aggregate volume at a given value of G is about six times greater for pure shear than for simple shear; the difference is attributable to the inhibiting effect of rotation in the latter field.

Another way of expressing the dispersing efficiency which will probably appeal more to

engineers is to use in place of G the power input W per unit volume required to overcome viscous flow of the medium (35) for flows following [1]:

$$W = \eta G^2(1 + \lambda)^2. \qquad [3]$$

For the system shown in Fig. 7, the rate of disintegration of the aggregate for a given W is $6/2 = 3x$ greater in extensional flow.

Similar but more complicated considerations may be expected to apply to the disintegration of aggregates whose constituent elements are held together by adhesive forces or by entanglement of elongated flexible particles such as fibers, or to the disruption of liquid droplets during emulsification. The point I wish to make is that the type of shear is of great importance, and should be specified when possible. The same is true of the fibrous crystallization of polymers from melts and supersaturated solutions in shear fields (44).

4. CONFIGURATION OF FLEXIBLE THREADS ($\lambda = 0$)*

Shear flow generates stresses on particles, as shown schematically in Fig. 2: when the particle is deformable its shape will change. To illustrate this effect, I have chosen long elastomer threads which in simple shear flow undergo continuous rotation (mainly about the X_1 axis) and form coils whose shape changes continuously. By taking cinephotographs si-

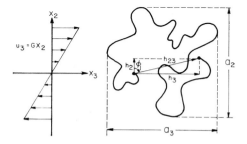

FIG. 8. Coiling of a rotating flexible thread in simple shear flow, and definition of end-to-end distances h_{ij} and maximum projected dimensions a_i. In the experiments the coil was simultaneously viewed and photographed by synchronized cinecameras directed along the X_1 and X_2 axes. After (7).

multaneously along the X_1 and X_2 axes (Fig. 2),* and then measuring the end-to-end distances h_{ij} and the maximum projected dimensions a_i (Fig. 8) frame by frame, we can determine the configurational statistics of the constantly changing coil dimensions: when the coils are sufficiently long, e.g., with an axis ratio in excess of 10^3, they follow Gaussian statistics to a surprisingly close degree (7) in the same way as dilute solutions of linear macromolecules, except that because of the action of the velocity gradient, the coil does not have spherical symmetry and its effective volume V is decreased with increasing G, as we would expect (Fig. 9).

We are currently extending these studies to entangled coils consisting of two or more filaments (at $\lambda = 0$, $+1$) in order to learn more about entanglement and disentanglement processes which, as implied earlier, are of interest in connection with papermaking.

5. OSCILLATORY PHENOMENA $(\lambda = 0)$*

If we take a single nonspherical particle, for example a disc, and place it in simple shear

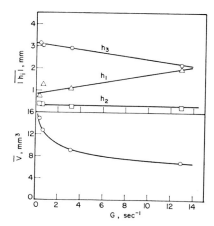

FIG. 9. Mean absolute values of end-to-end distances and the mean effective coil volume $\bar{V} = \pi a_1 a_2 a_3/4$ of a 40-μm-diam elastomer filament 2.5 cm long in 100 P. silicone oil in simple shear. Note the contraction in h_3, the expansion in h_1, and the decrease in $\bar{V}$ with G. After (7).

flow at $Br = \infty$ it will rotate in a fixed spherical orbit with a period of rotation TG about the vorticity axis X_1, as predicted in detail by

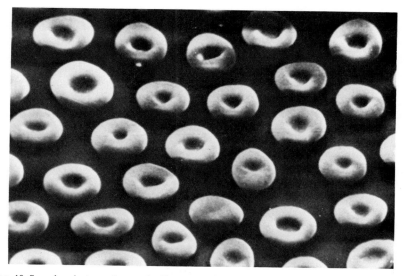

FIG. 10. Scanning electron micrograph of human erythrocytes showing their biconcave-disc shape and surprising monodispersity. The mean diameter is about 8.5 μm and maximum thickness is 2.4 μm. Courtesy of Professor S. Chien.

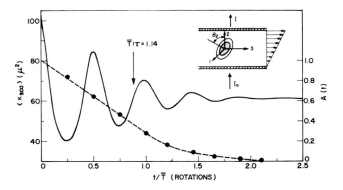

FIG. 11. Calculated oscillations (solid line) and amplitudes (broken line) of the optical scattering cross section of erythrocytes like those in Fig. 10 when the light beam is directed along the X_2 axis in simple shear (inset). The calculations were made for a mean equivalent spheroidal axis ratio $r_e = 0.4$ (SD = 0.07) and for an initially isotropic distribution of cell orientations. The period of oscillation and rate of damping calculated from τ_0 in [4] were in good agreement with experimental results in dilute systems at high Br when τ_1^{-1}, $\tau_2^{-1} = 0$ (9).

Jeffery (3) and confirmed experimentally (6).* This rotation leads to what at first sight is surprising—oscillatory behavior of any macroscopic property of a dilute suspension of reasonably monodisperse nonspherical particles which depends upon particle orientation(s); some examples are electrical (dielectric constant, conductivity), rheological, and optical properties (9, 15, 16, 28, 29, 31).

A good illustration, and the first to be established experimentally (9), is provided by the transmission coefficient of a dilute suspension of human erythrocytes, hardened to make them rigid and stable. The cells are biconcave discs and nearly monodisperse (Fig. 10). The optical transmission coefficient of a beam directed along the X_2 axis of simple shear flow (inset Fig. 11) undergoes a damped oscillation as a result of rotation of the individual cells whose optical scattering cross section, calculated from Rayleigh–Gans–Debye light scattering theory, is at a maximum when the beam "sees" the cells edge-on (which is contrary to what one would intuitively expect) and a minimum when face-on. The oscillations have a period $T/2$ [directly related to the mean ellipsoidal axis ratio $\bar{r}_e$ of the particles

(3, 6)] and are logarithmically damped with a relaxation time τ given by

$$\tau^{-1} = \tau_0^{-1} + \tau_1^{-1} + \tau_2^{-1}, \qquad [4]$$

where the components τ_i are due respectively to (1) the spread in r_e about $\bar{r}_e$, (2) n-body collisions, where $n = 2$ for sufficiently dilute suspensions, and (3) Br, of the axis of rotation of the particle which can be finite with erythrocytes. All three factors cause rotational-phase mixing (46) and thus damping of the oscillations. The experimental results (9) are in good agreement with theory (Fig. 11).

The oscillatory behavior has a number of interesting consequences. If Br $\to \infty$, i.e., rotary Brownian motion is negligible, the system has perfect memory: on reversing the flow after some time t, which theoretically could be anywhere in the range $0 < t < \infty$, the oscillations are fully restored. When $0 < \text{Br} < \infty$ the memory suffers some impairment which can also be caused by other factors (28). We are continuing these interesting studies of damped oscillations, memory impairment, and a number of other orthokinetic effects.

ACKNOWLEDGMENTS

This paper is based almost entirely on work done by graduate students (past and present) and other associates who have worked with me and whose contributions I acknowledge with pleasure. Too numerous to list by name, many can be identified as coauthors in the references cited below which bear my name.

REFERENCES

1. EINSTEIN, A., *Ann. Phys.* **19**, 289 (1906); **34**, 591 (1911).
2. VON SMOLUCHOWSKI, M., *Phys. Z.* **17**, 557, 583 (1916); *Z. Phys. Chem.* **92**, 129 (1917).
3. JEFFERY, G. B., *Proc. Roy. Soc. (London) Ser. A* **102**, 161 (1922).
4. EIRICH, F. R., BUNZL, M., AND MARGARETHA, H., *Kolloid Z.* **75**, 20 (1936).
5. TAYLOR, G. I., *Proc. Roy. Soc. (London) Ser. A* **138**, 41 (1932); **146**, 501 (1934).
6. GOLDSMITH, H. L., AND MASON, S. G., "Rheology: Theory and Applications" (F. R. Eirich, Ed.), Vol. 4, Chap. 2, pp. 85–250. Academic Press, New York, 1967.
7. OKAGAWA, A., AND MASON, S. G., *Canad. J. Chem.* **53**, 2689 (1975).
8. GOLDSMITH, H. J., AND MASON, S. G., CIBA Foundation Symposium on Circulatory and Respiratory Mass Transport (G. E. W. Wolstenholme and Julie Knight, Eds. p. 105.) Churchill, London, 1969.
9. FROJMOVIC, M. M., AND MASON, S. G., *Biochem. Biophys. Res. Commun.* **62**, 17 (1975).
10. VADAS, E. B., GOLDSMITH, H. L., AND MASON, S. G., *J. Colloid Interface Sci.* **43**, 630 (1973).
11. VAN DE VEN, T. G. M., AND MASON, S. G., *J. Colloid Interface Sci.*, **57**, 517 (1976).
12. TORZA, S., COX, R. G., AND MASON, S. G., *J. Colloid Interface, Sci.* **38**, 395 (1972).
13. TORZA, S., AND MASON, S. G., *J. Colloid Interface Sci.* **33**, 68 (1970).
14. ARP, P. A., AND MASON, S. G., *J. Colloid Interface Sci.* (in press).
15. OKAGAWA, A., COX, R. G., AND MASON, S. G., *J. Colloid Interface Sci.* **45**, 303 (1973).
16. OKAGAWA, A., COX, R. G., AND MASON, S. G., *J. Colloid Interface Sci.* **47**, 536 (1974).
17. VAN DE VEN, T. G. M., AND MASON, S. G., *Colloid Polym. Sci.*, In press.
18. VADAS, E. B., GOLDSMITH, H. L., AND MASON, S. G., *Trans. Soc. Rheol.*, **20**, 373 (1976).
19. GAUTHIER, F., GOLDSMITH, H. L., AND MASON, S. G., *Rheol. Acta* **10**, 344 (1971).
20. BARTRAM, E., GOLDSMITH, H. L., AND MASON, S. G., *Rheol. Acta* **14**, 776 (1975).
21. KAO, S. V., AND MASON, S. G., *Nature* **253**, 619 (1975).
22. ARP, P. A., AND MASON, S. G., *J. Colloid Interface Sci.* (in press).
23. ZIA, I. Y. Z., Ph.D. thesis, McGill University, Montreal, Canada, 1966.
24. TAKAMURA, K., GOLDSMITH, K. L., AND MASON, S. G., to appear.
25. VAN DE VEN, T. G. M., AND MASON, S. G., *J. Colloid Interface Sci.* **57**, 505 (1976).
26. MIKAMI, T., COX, R. G., AND MASON, S. G., *Int. J. Multiphase Flow*, **2**, 113 (1975).
27. KARNIS, A., AND MASON, S. G., *J. Colloid Interface Sci.* **23**, 120 (1967).
28. OKAGAWA, A., AND MASON, S. G., *Science* **181**, 159 (1973).
29. OKAGAWA, A., ENNIS, G., AND MASON, S. G., to appear.
30. SORRENTINO, M., AND MASON, S. G., *J. Colloid Interface Sci.* **41**, 178 (1972).
31. OKAGAWA, A., COX, R. G., AND MASON, S. G., to appear.
32. KAO, S. V., COX, R. G., AND MASON, S. G., to appear.
33. CERF, R., AND SCHERAGA, H. A., *Chem. Rev.* **51**, 185 (1952).
34. BURGERS, J. M., "Second Report on Viscosity and Plasticity," p. 113 ff. Academy of Sciences, Amsterdam, 1938.
35. KAO, S. V., AND MASON, S. G., to appear.
36. COX, R. G., ZIA, I. Y. A., AND MASON, S. G., *J. Colloid Interface Sci.* **27**, 7 (1968).
37. TORZA, S., HENRY, C. P., COX, R. G., AND MASON, S. G., *J. Colloid Interface Sci.* **35**, 529 (1971).
38. MANLEY, R. ST. J., AND MASON, S. G., *J. Colloid Sci.* **7**, 354 (1952).
39. BATCHELOR, G. K., AND GREEN, J. T., *J. Fluid Mech.* **56**, 375 (1972).
40. DARABANER, C. L., AND MASON, S. G., *Rheol. Acta* **6**, 273 (1967).
41. ARP, P. A., AND MASON, S. G., *Canad. J. Chem.*, in press.
42. VERWEY, E. J. W., AND OVERBEEK, J. TH. G., "Theory of Stability of Lyophobic Colloids," Elsevier, Amsterdam, 1948.
43. UHL, V. W., AND GRAY, J. B., "Mixing," Vols. I and II, Academic Press, New York, 1966.
44. MACKLEY, M. R., AND KELLER, A., *Polymer* **14**, 6, (1973).
45. GOLDSMITH, H. L., AND MARLOW, J., *Proc. Roy. Soc. (London) Ser. B* **182**, 351 (1972).
46. BATCHELOR, G. K., *Ann. Rev. Fluid Mech.* **6**, 227 (1974).

The Role of Inertia in Transport Properties of Suspensions

J. J. HERMANS

Chemistry Department, University of North Carolina, Chapel Hill, North Carolina 27514

Received May 3, 1976; accepted June 21, 1976

The role of inertia in the nonuniform motion of solids through liquids and its effect on the resistance to translatory or rotatory motion has been known for a long time, but the possible effect on solution properties has been considered only recently. Inertia plays a particularly important role when the velocity changes rapidly: Brownian motion. It is responsible for a slow decay in random force and velocity correlation functions. This decay has $t^{-\frac{3}{2}}$ character in the case of translation but $t^{-\frac{5}{2}}$ character in rotatory motion. Interestingly this is true also when the rotor is a rigid dumbbell; the hydrodynamic interaction of the two dumbbell ends completely eliminates the $\omega^{\frac{1}{2}}$ term in the expansion of the Fourier transform of the resistance factor in terms of frequency, ω. The contribution of suspended particles to the viscosity has been considered for linearly elastic dumbbells, and it was found that the relative effect of inertia is quite considerable in solvents of low viscosity, but this may be largely due to the fact that hydrodynamic interaction was neglected. The theory for rigid dumbbells is considered in some detail, but it will be left to a later publication to give a definitive answer as regards the effect of inertia.

INTRODUCTION

Theoretical and experimental work dealing with the hydrodynamic properties of solutions and suspensions has been concerned with the frictional resistance of particles against translational motion, the frictional torque in the case of rotation, hydrodynamic interaction between particles, orientation of elongated particles in an external field, viscosity of suspensions, flow birefringence, etc.

Even though phenomena which depend to a large extent on the occurrence of electrical charges (for example, electrophoresis, coagulation, or flocculation) will be omitted from consideration in this review, the literature is very extensive; only a few references are given.

Early work on the resistance experienced by a sphere when it moves through a viscous liquid at constant velocity, or when it rotates at constant angular velocity was extended to elongated particles by Overbeck (1), Jeffery (2), Gans (3), Haller (4), Kuhn (5), Oseen (6), and Burgers (7). More recently, Brenner

(8–13) has made many valuable contributions to the subject, while Mason and co-workers (14) have published extensively on theoretical and experimental aspects of elongated particles in shear flow and the hydrodynamic interaction of particles in such a flow. The special case of rigid dumbbells was considered in great detail by Bird and co-workers (15, 16).

As regards viscosity, Einstein's (17) well-known work concerned with a suspension of spheres was extended to rigid elongated particles by Eisenshitz (18), Boeder (19), Kuhn (5), Guth (20), Burgers (7), and Simha (21). A theoretical discussion of flow birefringence in a solution of elongated particles can be found in (22).

As is well known, there exists an enormous literature also on translation and rotation of polymer molecules of the coiling type, and on the viscosity of dilute solutions of these molecules. For the greater part, these theories are concerned with viscosity in simple shear, but lately other flow fields have been considered

J. J. HERMANS

as well, in particular the so-called elongational flow, which is of practical interest in fiber spinning. Some of the relevant references are (15, 16, 23–25).

In view of the fact that the literature mentioned in this brief summary is of necessity very incomplete, a few references to textbooks are added: Refs. (6, 7, 9, 14, 16, 26–29).

All the work referred to is based on the Navier–Stokes equations for imcompressible isotropic fluids:

$$\rho[\partial \mathbf{v}/\partial t + (\mathbf{v}\cdot\nabla)\mathbf{v}]$$
$$= -\nabla p + \eta\nabla^2\mathbf{v} + \rho\mathbf{F}, \quad [1]$$
$$\nabla\cdot\mathbf{v} = 0. \quad [2]$$

Here $\mathbf{v} = \mathbf{v}(\mathbf{r}, t)$ is the velocity of the liquid at position $\mathbf{r}$, p the pressure, $\mathbf{F}$ the external force per unit volume, ρ the density of the liquid, and η its viscosity. The present paper is restricted to phenomena in which the convective terms, which are quadratic in the velocity, may be ignored. This is permissible when Reynolds' number

$$\rho v R/\eta \quad [3]$$

is small compared to unity, R being a characteristic linear dimension. It *linearizes* Eq. [1], which becomes

$$\rho\partial \mathbf{v}/\partial t = -\nabla p + \eta\nabla^2\mathbf{v} + \rho\mathbf{F}. \quad [4]$$

At the end of the section on rotational Brownian motion, however, mention will be made of an interesting and somewhat disturbing problem in relation to the convective terms.

It is important to realize that almost all the work referred to thus far is based on the further assumption that in the linearized equation [4], the left-hand side may be set equal to zero. This is rigorous when the phenomenon studied is steady. For example, if a liquid flows past a stationary obstacle and has a time-independent velocity at large distances from this obstacle, a steady state is established in which the velocity at every point in space is independent of time. However, in theories of hydrodynamic interaction, shear viscosity, elongational viscosity, deformation

of particles in a flow field, etc., one is not concerned with steady-state situations. In these theories the omission of the term $\rho\partial\mathbf{v}/\partial t$ is motivated by the fact that accelerations of small particles in a viscous liquid are damped out very rapidly; the role played by the left-hand side of Eq. [4] is always restricted to very short times. A simple quantitative criterion for the validity of this statement is most easily formulated when the phenomenon studied is an oscillation with frequency ω, namely,

$$\rho\omega R^2/\eta \ll 1, \quad [5]$$

where R is a characteristic length. If the motion is not simply harmonic, we can always write the velocity of the liquid in the form of a Fourier transform with respect to time, and we can then say that the left-hand side in Eq. [4] may be omitted if, in the Fourier transform, the frequencies that contribute appreciably to the phenomenon satisfy the condition [5].

It is not true of course that the acceleration has been ignored in all earlier work on suspensons. Far from it. The resistance experienced by a sphere in oscillatory translational motion was calculated already by Stokes in 1851 (30). Its Fourier transform gives the resistance for arbitrary time-dependent velocity (31). The solution for the problem of the rotational oscillation of a sphere can be found in Ref. 32, p. 97. However, although such results were known, the possible consequences for other phenomena, such as viscosity, hydrodynamic interaction, etc., were not considered. Theories for such phenomena were always based on the hydrodynamic equations for steady states. This is true even for the theory of Brownian motion.

Brownian motion is mentioned specifically in this connection, because it is the phenomenon most strongly affected by the inertia of the liquid. This is because the role of inertia is greatest when the changes in velocity are rapid. In Brownian motion, a particle undergoes rapid erratic changes in its velocity and it may therefore be expected that the effect

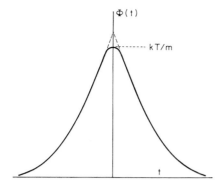

FIG. 1. Velocity autocorrelation function according to the "classical" theory.

a delta function

$$G(t) \equiv \langle X(t_1)X(t_1 + t) \rangle = c\,\delta(t); \qquad [7]$$

i.e., the Langevin equation describes a Markovian process, and it is found that the autocorrelation function for the velocity is exponential:

$$\phi(t) = \langle u(t_1)u(t_1+t) \rangle = A \exp(-\zeta_0|t|/m), \qquad [8]$$

where $A = \langle u^2 \rangle = kT/m$, in accordance with the equipartition principle.

The result in Eq. [8] is not meant to be valid at extremely short times. In fact $\phi(t)$ is by definition an even function and should therefore have zero slope at $t = 0$ (Fig. 1); there is a small time τ_0 below which the actual $\phi(t)$ deviates from the result given in Eq. [8]. The smaller the τ_0, the smaller the difference between the extrapolated value $\phi(0)$ and the actual value $\langle u^2 \rangle = kT/m$.

To make Eqs. [7] and [8] compatible with the Langevin equation we must take $c = 2kT\zeta_0$, and we observe that the Fourier time transform of $G(t)$ is

$$\hat{G}(\omega) = 2kT\tau_0, \qquad [9]$$

a result which is known as the fluctuation–dissipation theorem. The Fourier transform of a function $f(t)$ throughout this review is defined by

$$\hat{f}(\omega) = \int_{-\infty}^{\infty} dt\, f(t)e^{i\omega t}. \qquad [10]$$

There exists an abundant literature in connection with the Langevin equation; see, for example, Refs. (34–37). Equation [7] has often been formulated as an assumption, but the work of Kubo (37), Case (38), and others, makes it clear that the choice of $G(t)$ is not free once the frictional force in Eq. [6] has been specified. Eq. [7] is valid when the frictional force is proportional to the velocity. In the more general case in which the frictional factor is a linear operator, written as force $= \zeta_{op}\, u(t)$, the autocorrelation function for the random force obeys the following general-

of inertia will be particularly pronounced. Indeed, this was clearly recognized already by Lorentz (33), but the consequences were not worked out until much later.

TRANSLATIONAL BROWNIAN MOTION

To explain Lorentz's argument let us first consider the classical theory of Brownian movement is some detail. The simplest approach is the one that starts from the well-known Langevin equation

$$m\dot{u}(t) + \zeta_0 u(t) = X(t), \qquad [6]$$

where m is the mass of the Brownian particle, u its velocity, ζ_0 the frictional coefficient for a time-independent velocity, and X the random force. The dot denotes differentiation with respect to time. For Brownian motion in three dimensions, u and X are vectors, but this has no effect on the results because the components of the random force are not correlated.

In an ensemble of particles that start with velocity u_0 at time zero, the average of $X(t)$ is zero, and it follows that the average velocity in such an ensemble decays exponentially: $\langle u(t) \rangle = u_0 \exp(-\zeta_0 t/m)$. Here m/ζ_0 may be called relaxation time. The correlation between the random forces $X(t_1)$ and $X(t_1 + t)$ is of such short duration compared to the relaxation time that the autocorrelation function under stationary conditions may be treated as

ization of Eq. [9],

$$\hat{G}(\omega) = 2kT(\text{Re})\hat{\zeta}(\omega), \qquad [11]$$

where $\hat{\zeta}(\omega)$ is the Fourier transform of ζ_{op}; in other words, $\hat{\zeta}(\omega)\hat{u}(\omega)$ is the Fourier transform of the frictional force. The (generalized) Langevin equation for this case is

$$m\dot{u}(t) + \zeta_{op}u(t) = X(t), \qquad [12]$$

where the random force $X(t)$ now has the auto-correlation property represented by Eq. [11].

This formalism may be applied immediately to the hydrodynamic model, where the operator ζ_{op} (for rigid spheres that show no slip on the surface) is known from earlier work (30–32):

$$\zeta_{op} = \zeta_0 + \alpha\pi^{-\frac{1}{2}}\int_{-\infty}^{t} ds(t-s)^{-\frac{1}{2}}d/ds$$
$$+ \tfrac{1}{2}m_0 d/dt, \quad [13]$$

which has the Fourier transform

$$\hat{\zeta}(\omega) = \zeta_0 + \alpha(-i\omega)^{\frac{1}{2}} + \tfrac{1}{2}m_0(-i\omega). \quad [14]$$

Here ζ_0 is the friction coefficient for time-independent velocity ($6\pi\eta a$ for spheres of radius a), $m_0 = \tfrac{4}{3}\pi a^3\rho$ is the mass of liquid displaced by the particle, and

$$\alpha = \zeta_0(\rho a^2/\eta)^{\frac{1}{2}}. \qquad [15]$$

The argument given by Lorentz (33) to show that the second term in Eq. [13] plays an appreciable role can now be formulated very concisely as follows. The time in which the velocity of a Brownian particle changes appreciably is of order m/ζ_0, which means that the frequencies that are relevant in the Fourier transform are of order ζ_0/m. Thus the relative effect of the second term in Eq. [14] as compared to the first is determined by the ratio

$$(\alpha/\zeta_0)(\zeta_0/m)^{\frac{1}{2}} = (9m_0/2m)^{\frac{1}{2}},$$

which is of the order of unity if the densities of the Brownian particle and the medium are of the same order of magnitude.

The application of Eq. [11] shows that the fluctuation–dissipation theorem takes the form

$$\hat{G}(\omega) = 2kT[\zeta_0 + \alpha(\omega/2)^{\frac{1}{2}}], \qquad [16]$$

which means that at large times, $G(t)$ decays as $t^{-\frac{3}{2}}$. We note further that the force $\tfrac{1}{2}m_0 du/dt$ in Boussinesq's equation [13] plays no role in the result for $G(t)$; this is related to the fact that this force is not dissipative.

Equation [16] has also been derived (39, 40) by a method which bypasses the Langevin equation. To this end one adds fluctuating stresses to the linearized Navier–Stokes equations. The correlation properties of these fluctuating stresses were derived by Landau and Lifshitz (41) and Fox and Uhlenbeck (42) and these can be used to find the autocorrelation function for the random force.

It is not surprising that also the velocity autocorrelation function $\phi(t)$ shows a $t^{-\frac{3}{2}}$ tail at large times. Interestingly, this was first found in computer studies of molecular dynamics (43, 44) and has been discussed in a more general physical context by several authors (45, 46).

Another aspect of $\phi(t)$ that must be mentioned is the peculiar behavior at very short times. Indeed, the result derived by Hauge and Martin-Löf (40), Burgess (47), Zwanzig and Bixon (48), and others shows that $\phi(t)$ does not extrapolate to the equipartition value $m^{-1}kT$ but to $(m + \tfrac{1}{2}m_0)^{-1}kT$. There are several aspects to this problem that are worthwhile mentioning.

Taking into account that $G(t)$ is an even function, we may write Eq. [11] in the form

$$(\text{Re})\int_0^{\infty} dt G(t)e^{i\omega t} = kT(\text{Re})\hat{\zeta}(\omega). \quad [17]$$

This says nothing about the imaginary part. Molecular theories, however, suggest a relation between $\phi(t)$ and $G(t)$, namely,

$$mkT\phi(t) = -\int_0^{t} ds G(t-s)\phi(s), \quad [18]$$

and this is compatible (48, 49) with the equation

$$\int_0^{\infty} dt G(t)e^{i\omega t} = kT\hat{\zeta}(\omega). \qquad [19]$$

In other words, Eq. [17] should apply also to

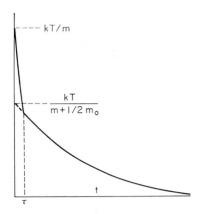

FIG. 2. Velocity autocorrelation function when inertia is taken into account.

the imaginary part. Clearly this cannot be true when Eqs. [14] and [16] are valid, because Eq. [16] shows that $G(t)$ does not contain m_0, whereas the right-hand side of Eq. [19] does contain such a term. One way to resolve the discrepancy is to assume that the term $\frac{1}{2}m_0\,du/dt$ in the resistance of the particle should not be included in the Langevin equation (49). Then Eq. [18] remains valid but Eq. [19] is not. Most authors, however, (for example, (40, 47, 48, 50)), take a different point of view, which is based on the behavior of the frictional resistance in compressible fluids.

When allowance is made for compressibility, it is found that the resistance for an oscillating particle is not much different from that in an imcompressible fluid (Eq. [14]) as long as the frequency of the oscillation is small compared to c/a, where c is the velocity of sound in the liquid and a the radius of the particle. However, when ω becomes of order c/a or larger, the resistance becomes much smaller and, for very large frequencies, goes to zero as ω^{-1}. We can also say that whenever the particle is accelerated, the resistance given by Eq. [14] does not fully develop until after a time of order $\tau = a/c$ (the time it takes a sound wave to travel over a distance of the order of the particle diameter).

The discrepancy found for incompressible liquids is thus considered to be an artifact. In real liquids, Eqs. [18] and [19] are believed to be valid, and the velocity autocorrelation function $\phi(t)$ undergoes an abrupt drop from the value kT/m to the value $kT/(m + \frac{1}{2} m_0)$ beween $t = 0$ and the very short time τ of order a/c. This τ would be zero if there were no compressibility at all, and in this limit Eq. [19] ceases to be valid.

It must be realized, however, that for very short times the hydrodynamic model may become inadequate. In this model the molecular interaction between the Brownian particle and the solvent molecules is replaced by macroscopic boundary conditions that are relevant to a macroscopic body in a continuous medium. For very short times this phenomenological description breaks down. In fact, the hydrodynamic model of Brownian movement cannot be taken seriously when the relaxation time m/ζ_0 is less than or of the same order as a/c. This sets a lower limit to the particle size for which the hydrodynamic model is applicable (49). In liquids with a viscosity of order 10^{-2} P, this limit lies near 10^{-6} cm.

ROTATORY BROWNIAN MOVEMENT

The relations discussed in the previous section are immediately applicable to rotatory motion if the mass m of the particle is replaced by its moment of inertia, its velocity $u(t)$ by the angular velocity $\Omega(t)$, and the random force $X(t)$ by a random torque. The autocorrelation function $Q(t)$ for this random torque statisfies the equation, analogous to [11],

$$\hat{Q}(\omega) = 2kT(\mathrm{Re})\hat{B}(\omega), \qquad [20]$$

where $\hat{B}(\omega)$ is the Fourier transform of the frictional coefficient for rotation, i.e., $\hat{B}(\omega)\hat{\Omega}(\omega)$ is the Fourier time transform of the frictional torque. Here again, B_{op} is supposed to be a linear operator. In the hydrodynamic model, therefore, it is to be derived from linearized hydrodynamics.

For particles of arbitrary shape there is coupling between translational and rotational

Brownian motion (51, 52), but for our purpose it is sufficient to consider the type of particle symmetry for which translation and rotation are independent stochastic processes, i.e., we assume that the dyadic of "coupling diffusivity" (51) is zero.

The function $\hat{B}(\omega)$ for a solid sphere, showing no slip, in an incompressible viscous liquid can be found in Ref. (32, p. 98). The expansion in powers of $\omega^{\frac{1}{2}}$ has the form

$$\hat{B}(\omega) = 8\pi\eta a^3 \left[1 - \frac{2}{3} i \frac{\rho a^2}{2\eta} \omega + \frac{2}{3}(1+i) \right.$$

$$\left. \times \left(\frac{\rho a^2}{2\eta}\right)^{\frac{3}{2}} \omega^{\frac{3}{2}} - \frac{4}{3}\left(\frac{\rho a^2}{2\eta}\right)^2 \omega^2 \cdots \right], \quad [21]$$

from which it follows that

$$\frac{\hat{Q}(\omega)}{2kT} = 8\pi\eta a^3 \left[1 + \frac{2}{3}\left(\frac{\rho a^2}{2\eta}\right)^{\frac{3}{2}} \omega^{\frac{3}{2}} \right.$$

$$\left. - \frac{4}{3}\left(\frac{\rho a^2}{2\eta}\right)^2 \omega^2 \cdots \right]. \quad [22]$$

As pointed out by Berne (53), this means that for large times $Q(t)$ decays as $t^{-\frac{5}{2}}$. A similar tail is found in the autocorrelation function for the angular velocity.

This result raises the interesting question whether the long-time tail for a rigid rod or rigid dumbbell decays at $t^{-\frac{5}{2}}$ or as $t^{-\frac{3}{2}}$. The motion of the individual beads at the end of the dumbbell has the character of translation, but the dumbbell as a whole is a rotating body. To be sure, if we assumed a frictional force $\zeta_{op}u(t) = L\zeta_{op}\Omega(t)$ for each of the beads separately, where ζ_{op} is given by Eq. [13], $\Omega(t)$ is the angular velocity and L is half the distance between the two beads, we would find immediately that the frictional torque is $L^2\zeta_{op}\Omega(t)$, i.e.,

$$B_{op} = L^2\zeta_{op}. \quad [23]$$

The Fourier time transform would then contain a term proportional to $\omega^{\frac{1}{2}}$ and thus the long time tail in $Q(t)$ would be proportional to $t^{-\frac{3}{2}}$. However, the derivation of Eq. [23] neglects all interaction between the beads. In reality there is hydrodynamic interaction, which has an important effect on the frictional torque. In the simplest approach to this problem the forces exerted on the liquid by the beads are treated as point forces and their effect on the flow of liquid is derived from Oseen's equations for time-dependent forces (6, 7, 54).

Specifically, let us consider a rigid dumbbell of length $2L$ which has its beads at the points $(x = L, y = 0, z = 0)$ and $(x = -L, y = 0, z = 0)$. Bead 1 exerts a force $-Y_1(t)$ on the liquid and this gives rise to a liquid velocity (in the y-direction) at the position of bead 2; we write this velocity $v_2(t) = -S_{op}Y_1(t)$, where S_{op} was derived already by Burgers (7). Likewise, the force $-Y_2(t)$ which bead 2 exerts on the liquid gives rise to a velocity $v_1(t) = -S_{op}Y_2(t)$ at the position of bead 1. The Fourier transform of S_{op} is

$$\hat{S}(\omega) = (32\pi i\rho L^3)^{-1}$$
$$\times \{1 - (1 + 2\kappa L + 4\kappa^2 L^2) \exp(-2\kappa L)\}, \quad [24]$$

where

$$\kappa = (-i\omega\rho/\eta).^{\frac{1}{2}} \quad [25]$$

If the dumbbell has a rotational velocity $\Omega(t)$, bead 1 has a velocity $L\Omega(t)$ in the y-direction and the force $-Y_1(t)$ is given by

$$-Y_1(t) = \zeta_{op}L\Omega(t) - \bar{\zeta}_{op}v_1(t). \quad [26]$$

Here ζ_{op} is the Boussinesq operator shown in Eq. [13]. As shown in recent work by Bedeaux and Mazur (50), the frictional coefficient in front of $v_1(t)$ differs from ζ_{op} by a term m_0d/dt:

$$\bar{\zeta}_{op} = \zeta_{op} + m_0d/dt. \quad [27]$$

In other words, when a particle moves in a liquid which has a time-dependent velocity $v(t)$, the force on the particle is not determined by the relative velocity; it contains an additional term $-m_0dv/dt$, which finds its origin in the fact that the mass m_0 of liquid which is displaced by the particle and which would have to be accelerated in the absence of the particle does not need to be accelerated when the particle is present. For details see Ref. (50).

Taking the Fourier transform of Eq. [26] and eliminating v by means of Eq. [24], the result for the torque $LY(t)$ is

$$L\hat{Y}(\omega) = -\hat{B}(\omega)\hat{\Omega}(\omega), \qquad [28]$$

where

$$\hat{B}(\omega) = L^2\hat{\zeta}(\omega)/(1 - \hat{S}(\omega)\hat{\zeta}\,\omega). \qquad [29]$$

Now this operator $\hat{B}(\omega)$ has the somewhat unexpected property that the expansion in powers of $\omega^{\frac{1}{2}}$ contains no first power. In other words, this term is completely eliminated by the hydrodynamic interaction of the beads. This restores the asympotic $t^{-\frac{5}{2}}$ behavior for rotating particles, but it immediately raises another question: The result suggests strongly that the asymptotic behavior of the correlation functions at large times is related to the fluid flow at relatively large distances from the particle where the liquid no longer sees any difference between a rotating sphere and a rotating dumbbell. However, it is well known that the flow at large distances is not described adequately by the linearized Navier–Stokes equations but requires the inclusion of the nonlinear convective terms. It is not expected that these will introduce a term proportional to $\omega^{\frac{1}{2}}$ in $\hat{B}(\omega)$, but the coefficient of $t^{-\frac{5}{2}}$ in the asymptotic behavior is probably not reliable and it is even quite conceivable that there is a small t^{-2} term.

INTRINSIC VISCOSITY OF FLEXIBLE MOLECULES

The transport property of suspensions which has recieved by far the greatest attention in the past is the contribution of the suspended particles to the viscosity. It is clearly of considerable fundamental interest to investigate the effect of the inertia of the liquid on this important property.

As a first approach to this question, Szu and Hermans (55), considered the linear elastic dumbbell, which has often been used as a simple model for flexible polymer molecules. The elastic force between the beads in this model is proportional to the distance between them,

with force constant

$$\gamma = 3kT/\langle L^2\rangle,$$

where $\langle L^2\rangle$ is the average square of the distance between the two beads. If hydrodynamic interaction is neglected, the generalized Langevin equations for the two beads will be

$$m\ddot{\mathbf{r}}_1 + \zeta_{op}\dot{\mathbf{r}}_1 - \bar{\zeta}_{op}\mathbf{v}_{01} + \gamma(\mathbf{r}_1 - \mathbf{r}_2) = \mathbf{F}_1, \quad [30]$$

$$m\ddot{\mathbf{r}}_2 + \zeta_{op}\dot{\mathbf{r}}_2 - \bar{\zeta}_{op}\mathbf{v}_{02} + \gamma(\mathbf{r}_2 - \mathbf{r}_1) = \mathbf{F}_2, \quad [31]$$

where $\mathbf{F}_1$ and $\mathbf{F}_2$ are the random forces, while $\mathbf{v}_{01}$ and $\mathbf{v}_{02}$ are the unperturbed velocities of the liquid at the positions of the beads. These velocities are determined by the flow field imposed on the liquid. For example, in simple shear with flow in the x-direction and velocity gradient q in the y-direction: $\mathbf{v}_{0i} = qy_i\mathbf{e}_x$ where $\mathbf{e}_x$ is the unit vector in the x-direction.

Note that the Eqs. [30], [31] differ from those given in Ref. (55) because they take into account what was explained in connection with Eq. [26]: ζ_{op} and $\bar{\zeta}_{op}$ are slightly different; according to a calculation carried out by Foister (56) this has only minor effects on the results obtained.

The ensemble average contribution of the elastic dumbbells to the viscosity of the suspension may be obtained either by calculating the energy dissipation or by the method advocated by Burgers (7). The result is as follows.

If the inertia of the liquid is ignored, i.e., if $\zeta_{op} = \bar{\zeta}_{op} = \zeta_0$, it is found that the viscosity is independent of the mass of the beads and is in agreement with earlier theories based on the diffusion (Fokker–Planck) equation. The inertia of the liquid has an effect which may amount to as much as 20% in liquids of low viscosity (55); this effect is due almost entirely to the second term on the right-hand side of Eq. [13] which depends on the history of the acceleration and corresponds to the term proportional to $\omega^{\frac{1}{2}}$ in Eq. [14]. The effect of the term $\frac{1}{2}m_0(-i\omega)$ in Eq. [14] is of the order of only 1%.

This relatively large influence of inertia on intrinsic viscosity is of fundamental interest, but it is necessary to make two remarks. In the first place, the relative role of inertia in solu-

311

tions of real polymer molecules is of much smaller magnitude. The extension of the theory from an elastic dumbbell to a pearl necklace of many beads connected by linear springs (Rouse–Zimm model (57, 58)) shows that the relative effect of inertia for the pearl necklace is much smaller than for the dumbbell. The physical reason for this lies in the fact that the deformation of a polymer molecule in the Rouse–Zimm model is described in terms of normal modes. The contribution of the molecule to the viscosity is determined to a large extent by the slow modes, and the inertia of the liquid plays of course a smaller role in the slow modes than in the fast ones. Indeed, it was found (59) that in the absence of hydrodynamic interaction, where the intrinsic viscosity is approximately proportional to N^2 (N being the number of beads in the molecule), the term which must be added to account for inertia is roughly proportional to the first power of N. Although this conclusion does not apply quantitatively when hydrodynamic interaction (with preaveraged Oseen tensor, see below) is included, it remains true qualitatively.

In the second place, the treatment in Ref. (55) neglected the hydrodynamic interaction of the two beads of the dumbbell. This may have a large and perhaps even an overriding effect. As explained in the section Rotatory Brownian Motion, the hydrodynamic interaction in the case of a rigid dumbbell eliminates completely the $\omega^{\frac{1}{2}}$ term in the rotational friction coefficient $\hat{B}(\omega)$, and it is to be remembered that it is this term which determined the large effect of inertia on intrinsic viscosity in the calculation for the elastic dumbbell.

Hydrodynamic interaction will introduce terms in Eqs. [30], [31] that are proportional to $|\mathbf{r}_1 - \mathbf{r}_2|^{-1}$, thus making the equations nonlinear. In the usual treatment of hydrodynamic interaction in polymer molecules, this difficulty is obviated by preaveraging the Oseen tensor (see, however, (60, 61)). If this procedure is followed for the elastic dumbbell, it is found (59) that the hydrodynamic interaction does indeed diminish the relative role of inertia in

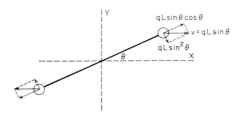

FIG. 3. Rigid dumbbell in a velocity gradient; two-dimensional model.

the intrinsic viscosity, but the result is not greatly different from that derived without hydrodynamic interaction. However, the theory for rigid dumbbells suggests that this may be due to the preaveraging.

INTRINSIC VISCOSITY OF RIGID RODS

For rigid rods the problem is nonlinear even when hydrodynamic interaction is ignored. To simplify the discussion, restriction is made to rigid dumbbells; the extension to rods is straightforward if one is willing to represent a rod by a series of beads. Furthermore, consideration is given to motion in two rather than three dimensions because this brings out all the essential points.

Imagine (Fig. 2) a rigid dumbbell in the xy-plane when the undisturbed velocity of the liquid is in the direction of the x-axis and is proportional to y:

$$v = qy.$$

Taking the origin at the center of the dumbbell, the radial and tangential components of v at the position of bead 1 are

$$v_r = qL \sin\theta \cos\theta; \qquad v_t = -qL \sin^2\theta. \qquad [32]$$

The velocity of the bead itself is zero in radial direction and $L\dot\theta$ in tangential direction. Thus, if hydrodynamic interaction between the beads is ignored, the liquid velocity relative to the bead has components v_r and $v_t - L\dot\theta$, and this gives rise to a rate of energy dissipation per dumbbell:

$$\dot\varepsilon = \zeta_0 q^2 L^2 \sin^2\theta \cos^2\theta$$
$$+ \zeta_0 L^2 (\dot\theta + q \sin^2\theta)^2. \qquad [33]$$

This is true if the role of inertia is ignored, so that the force is proportional to the instantaneous relative velocity.

To find the contribution to the viscosity, determine the ensemble average of $\dot{\varepsilon}$. The customary approach is to find the distribution function $f(\theta)$ from the diffusion equation

$$D df/d\theta + fq \sin^2 \theta = \text{constant}, \quad [34]$$

which expresses that the total flux in the direction of decreasing angle θ is independent of θ when a steady state has been reached; D is the rotational diffusivity, $-(D/f)df/d\theta$ is the average angular velocity due to diffusion and $-q \sin^2 \theta$ is the angular component for the fluid (v_t in Eq. [32]). Considering the limit of small gradient q, where the rotatory diffusion predominates, $f(\theta)$ may be expanded in powers of q/D,

$$f(\theta) = (2\pi)^{-1}[1 + (q/D)f_1(\theta) \\ + (q/D)^2 f_2(\theta) + \cdots],$$

and the constant on the right-hand side is written $a_1 q + a_2 q^2 + \cdots$. When one solves the equations for $f_1(\theta), f_2(\theta)$, etc., it must be taken into account that each of these functions is periodic and that the integral over θ is zero. This determines the constants $a_1, a_2, \ldots$ and leads to the result that the average angular velocity due to diffusion becomes $q(\sin^2 \theta - \frac{1}{2}) + O(q^2)$. When we add the angular velocity due to the flow of the liquid, we find that the average angular velocity for any given orientation is $-\frac{1}{2}q + O(q^2)$, where the second term will in general be a function of θ but the first one is not. It follows that the ensemble average angular velocity is likewise equal to $-\frac{1}{2}q + O(q^2)$.

So much for the "classical" treatment. To generalize the theory, two new aspects must be considered. One is the inclusion of hydrodynamic interaction between the beads, the other the inclusion of inertia effects. However, when inertia is taken into account, it is no longer permissible to use the diffusion equation [34] because this equation is valid only for times that are large compared to the relaxation time for rotatory Brownian motion. We will

then resort to the Langevin equation for the angular motion. It is easy to show that this leads to the same answers as before, when inertia is negligible, because in that case the Langevin equation becomes

$$I\ddot{\theta} + B(\dot{\theta} + q \sin^2 \theta) = M, \quad [35]$$

where I is the moment of inertia, B the frictional constant for rotation, and M the random torque. Now, if we take the ensemble average of Eq. [35] we find immediately that

$$\langle \dot{\theta} \rangle = - q \langle \sin^2 \theta \rangle = - \tfrac{1}{2}q + O(q^2), \quad [36]$$

because to lowest order in q the distribution is uniform, i.e., $\langle \sin^2 \theta \rangle = \frac{1}{2}$. The result [36] is exactly what was found on the basis of the diffusion equation.

Matters become more complicated if inertia is taken into account. The frictional coefficient B must now be replaced by an operator of the type discussed in the section on rotatory Brownian movement; compare Eq. [29]. This will not be discussed here in any detail; the complete treatment is reserved for a later publication (56).

REFERENCES

1. OVERBECK, A., *Crelle's J.* **81**, 62 (1876).
2. JEFFERY, G. B., *Proc. Roy. Soc. London Ser. A* **102**, 171 (1922).
3. GANS, R., *Ann. Phys.* (4) **86**, 654 (1928).
4. HALLER, W., *Kolloid Z.* **61**, 30 (1932).
5. KUHN, W., *Z. Phys. Chem. A* **161**, 427 (1932).
6. OSEEN, C. W., "Hydrodynamik," Akad. Verlagsgesellsch., Leipzig, 1927.
7. BURGERS, J. M., "Second Report on Viscosity and Plasticity," *Verh. Kon. Ned. Akad. Wetensch. Amsterdam*, Vol. XVI, No. 4, Chap. III, North Holland, Amsterdam, 1938.
8. BRENNER, H., *Chem. Eng. Sci.* **18**, 1 (1963); **19**, 599 (1964).
9. HAPPEL, J. AND BRENNER, H., "Low Reynolds Number Hydrodynamics," Prentice-Hall, Englewood Cliffs, N.J., 1965.
10. BRENNER, H., *J. Colloid Interface Sci.* **34**, 103 (1970).
11. BRENNER, H. AND CONDIFF, D. W., *J. Colloid Interface Sci.* **41**, 228 (1972); **47**, 199 (1974).
12. BRENNER, H., *Progr. Heat and Mass Transfer* **6**, 509 (1972).
13. BRENNER, H., *Int. J. Multiphase Flow* **1**, 195 (1974).

14. MASON, S. G. AND GOLDSMITH, H. L., *in* "Rheology," [F. R. Eirich, Ed.), Vol. 4, Academic Press, New York, 1967.

15. BIRD, R. B. AND WARNER, H. R., *Trans. Soc. Rheol.* **15**, 741 (1971).

16. BIRD, R. B., WARNER, H. R., AND EVANS, D. C., *Advan. Polym. Sci.* **8**, 1 (1971).

17. EINSTEIN, A., *Ann. Phys.* (4) **19**, 289 (1906); **34**, 591 (1911).

18. EISENSCHITZ, R., *Z. Phys. Chem. A* **158**, 85 (1932).

19. BOEDER, P. *Z. Phys.* **75**, 258 (1932).

20. GUTH, E. *Kolloid Z.* **74**, 147 (1936); **75**, 15 (1936).

21. SIMHA, R., *J. Phys. Chem.* **44**, 25 (1940).

22. PETERLIN, A. AND STUART, H. A., "Hand-und Jahrbuch der Physik," Vol. VIII, Abschn. I. B., Leipzig, 1943.

23. TAKSERMAN-KROZER, R., *J. Polym. Sci. C (Poly. Symp.)* **16**, 2845, 2855 (1967).

24. PETERLIN, A., *Pure Appl. Chem.* **12**, 563 (1966).

25. NICODEMO, L., MARRUCCI, G., AND HERMANS, J. J., *J. Polym. Sci. A-2* **10**, 1351 (1972).

26. STUART, H. A., "Die Physik der Hochpolymeren," Springer, Berlin, 1952.

27. HERMANS, J. J. (Ed.), "Flow Properties of Disperse Systems." North Holland, Amsterdam, 1953.

28. MILL, C. C. (Ed.), "Rheology of Disperse Systems." Pergamon, New York, 1959.

29. EIRICH, F. R. (Ed.), "Rheology: Theory and Applications", Vols. I, III, IV. Academic Press, New York, 1956.

30. BASSET, A. B., "Hydrodynamics," Vol. II, p. 260. Cambridge, 1888.

31. BOUSSINESQ, J., "Theorie Analytique de la Chaleur," Vol. II, p. 225 Gauthier-Villars, Paris, 1903.

32. LANDAU, L. D. AND LIFSHITZ, E. M., "Fluid Mechanics," p. 97 Pergamon, New York, 1959.

33. LORENTZ, H. A., Lessen over Theoretische Natuurkunde, V, Kinetische Problemen, Leiden, 1921.

34. WANG, M. C. AND UHLENBECK, G. E., *Rev. Mod. Phys.* **17**, 323 (1945).

35. CHANDRASEKHAR, S., *Rev. Mod. Phys.* **15**, 1 (1943).

36. WAX, N. (Ed.), "Noise and Stochastic Processes," Dover, New York, 1954.

37. KUBO, R., *Rep. Progr. Phys.* **29**, 255 (1966).

38. CASE, K. M., *Phys. Fluids* **14**, 2091 (1971).

39. CHOW, T. S. AND HERMANS, J. J., *J. Chem. Phys.* **56**, 3150 (1972).

40. HAUGE, E. H. AND MARTIN-LÖF, A., *J. Statist. Phys.* **7**, 259 (1973).

41. LANDAU, L. D. AND LIFSHITZ, E. M., "Statistical Physics," p. 523. Addison–Wesley, Reading, Mass., 1959.

42. FOX, R. F. AND UHLENBECK, G. E., *Phys. Fluids* **13**, 1893 (1970).

43. ALDER, B. J. AND WAINWRIGHT, T. E., *Phys. Rev. A* **1**, 18 (1970).

44. SUBRANNIAN, G., LEVIT, D., AND DAVIS, H. T., *J. Chem. Phys.* **60**, 591 (1974).

45. ERNST, M. H., HAUGE, E. H., AND VANLEEUWEN, J. M. J., *Phys. Rev. Lett.* **25**, 1254 (1970).

46. DORFMAN, J. R. AND COHEN, E. G. D., *Phys. Rev. Lett.* **25**, 1257 (1970).

47. BURGESS, R. E., *Phys. Lett. A* **42**, 935 (1973).

48. ZWANZIG, R. AND BIXON, M., *J. Fluid Mech.* **69**, 21 (1975).

49. CHOW, T. S. AND HERMANS, J. J., *Proc. Kon. Ned. Akad. Wetensch. Amsterdam B* **77**, 18 (1974).

50. BEDEAUX, D. AND MAZUR. P., *Physica* **78**, 505 (1974).

51. BRENNER, H., *J. Colloid Interface Sci.* **23**, 407 (1967).

52. CHOW, T. S., *Phys. Fluids* **16**, 31 (1973).

53. BERNE, B. J., *J. Chem. Phys.* **56**, 2164 (1972).

54. SZU, S. C. AND HERMANS, J. J., *J. Polym. Sci., Polym. Phys. Ed.* **11**, 1941 (1973); *J. Fluid Mech.* **66**, 385 (1974).

55. SZU, S. C. AND HERMANS, J. J., *J. Polym. Sci. Polym. Phys. Ed.* **12**, 1743 (1974).

56. FOISTER, R. T., Thesis, University of North Carolina, to appear.

57. ROUSE, P. E., *J. Chem. Phys.* **21**, 1272 (1953).

58. ZIMM, B. H., *J. Chem. Phys.* **24**, 269 (1956).

59. SZU, S. C., unpublished calculations.

60. PYUN, C. W. AND FIXMAN, M., *J. Chem. Phys.* **42**, 3838 (1965).

61. YAMAKAWA, H., *J. Chem. Phys.* **53**, 436 (1970). Most recent work in this connection: AKCASU, Z., AND GUROL, H., *J. Polym. Sci., Polym. Phys. Ed.* **14**, 1 (1976).

Macroscopic Transport Properties of a Sheared Suspension

L. G. LEAL

Chemical Engineering, California Institute of Technology, Pasadena, California

Received June 1, 1976; accepted October 13, 1976

The present paper is concerned with recent theoretical developments and a state-of-the-art summary of the framework for prediction of macroscopic transport properties of a sheared suspension. Included is (a) a derivation of the fundamental equation relating bulk conductive heat flux to microscale thermal properties of the suspension; (b) a discussion of the method of evaluation of this expression; (c) a theoretical framework based on the concept of hydrodynamic fluctuations for the inclusion of Brownian motion effects; and finally (d) specific evaluation of the effective thermal conductivity for a dilute suspension of drops in the limit of small local Peclet number, for a dilute suspension of rigid spheres in the large Peclet number limit, and for a dilute suspension of rigid spheroids in the limit of small Peclet number but including Brownian motion.

I. INTRODUCTION

The attempt to understand and predict the behavior of rheologically complex fluids is an old problem which has been the subject of an enormous amount of theoretical and experimental research. Since all complex fluids exhibit structure at some microscale, a logical initial objective in such studies would seem to be the development of relations between this structure and the bulk or macroscopic properties of the material. Here we consider one class of rheologically complex fluids where such a development is possible, namely suspensions or dispersions of small particles which are embedded in a second, "continuous" material. Suspensions are not only a technologically important class of fluids in their own right, but can also provide simple, idealized models for a much wider class of fluids including many macromolecular solutions (1).

Inherent in the description of a suspension as a fluid which consists of particles in a second continuous material is the important assumption that the characteristic dimension of the particles is large compared with molecular dimensions, yet still extremely small relative to the overall dimensions of the sample. In these circumstances, the macroscopic behavior of the whole suspension may be modeled in terms of an equivalent homogeneous continuum, and the microstructure consists of the relative positions, orientations, shapes, and sizes of the dispersed particles, as well as the parameters describing the material properties of the particles and suspending fluid. From an engineering point of view, modeling the suspension as a homogeneous continuum is advantageous since it allows use of the classical field equations at the bulk or macroscopic scale, for conservable quantities such as momentum, thermal energy, or molecular species. However, to obtain useful predictions at this level, it is imperative to obtain constitutive equations for the apparent diffusive flux of the conservable quantity in the presence of bulk or macroscopic gradients of its intensity. The attempt to derive such constitutive relationships for the mechanical or rheological properties of suspension-like materials has been the prime focus for much recent research. In many chemical engineering applications, however, the constitutive behavior of the material for heat or material transport is at least of equal importance, and it is to this topic that the majority of the present paper is devoted.

Two fundamentally different approaches exist for the development of constitutive relations. The most widely known is the phenomenological approach of classical, continuum mechanics, in which the detailed constitutive model is "derived" from an initial guess or hypothesis of the appropriate functional form between dependent and independent variables (e.g., stress and strain rate or their derivatives in the rheological problem). The considerable generality possible, which may at first appear to be a virtue, is in fact the major weakness of the phenomenological theories. Having postulated a constitutive form without reference to any specific material or class of materials, the theory has no way of predicting any of the material parameters of the model. Furthermore, it is impossible to determine whether the resulting model describes *any* real material without reference to an experimental program of great complexity. The second approach, which we shall term microstructural, begins with a physical description of a specific material, and then attempts to deduce its macroscopic properties by first analyzing its behavior in detail at the microscale, and then passing to the macroscale by an appropriate averaging process. For suspensions, this averaging process is, in general, a statistical one owing to the random distributions of particle motion, position, orientation, shape, and size which characterize the microstructural state. However, given the (instantaneous) microstructural state, the description of microscale behavior is itself a problem in classical continuum mechanics which can be analyzed (in principle) in a completely deductive fashion. The chief drawback is that the microstructure must be reasonably simple if the microscale problem is to be tractable for actual solution. On the positive side, however, the resulting theory not only provides a definite prediction of *form*

for the constitutive relation, but also allows all of the material parameters of the model to be explicitly evaluated. The relevant comparison between the phenomenological and microstructural approaches is that of specific predictive theories for a class of simple, but physically realizable materials on the one hand, and more general but nonpredictive models of hypothetical materials on the other. We believe that it is only from the existence of predictive relationships between microstructure and bulk or macroscopic properties for specific materials, that the necessary physical guidance will be forthcoming for further progress with the more general phenomenological theories.

The present paper is concerned with the microstructural derivation of constitutive models for the macroscopic, diffusive flux of heat (or molecular species) in a sheared suspension. Except for recent studies which I shall discuss in this paper, no prior work seems to have been done on this problem except for the limit of completely stationary dispersions, where a correct prediction of the effective thermal conductivity for a dilute dispersion of solid spheres was obtained over 100 years ago by Maxwell (2). Following Maxwell, fairly extensive work was also done for stationary dilute suspensions with nonspherical (mainly ellipsoidal) particles, and much later, for more concentrated suspensions of spheres (3), and for particles of arbitrary shape (4). A number of general results for the stationary case may be extracted from these various analyses. First, the microscopic or bulk conductive heat flux is simply the ensemble average of the microscale conduction flux, and the *effective* conductivity thus differs from that of the suspending fluid only when the conductivities of the fluid and particles have different values. Furthermore, the existence of an effective thermal conductivity tensor relating bulk heat flux and bulk temperature gradient can be rigorously proven. Finally, and of perhaps the greatest fundamental significance, the heat transfer properties of the suspension depend critically on its geometric microstructure. This means, presumably, that the heat transfer properties of the material will depend critically on the process by which the suspension is produced. One cannot discuss the thermal properties of a stationary suspension or dispersion, without first specifying the details of its manufacture: a rather unfortunate prospect.

A suspension which undergoes some given bulk flow will generally exhibit much more complex constitutive behavior for heat transfer. First, of all, the imposed bulk flow can cause considerable changes in the microstructure, which will depend critically on the specific flow type and rate of deformation. This dependence of microstructure on the type and strength of the flow will be reflected directly in the macroscopic thermal properties of the suspension. Furthermore, it may be anticipated that *convective* transport of heat on the microscale will provide an additional mechanism for enhanced heat flux in a suspension. Both the microscale temperature fields and the microstructure of the suspension will only respond to variations in the imposed velocity or temperature fields on a finite time scale and this introduces the further possibility for history-like effects in the thermal constitutive behavior. Finally, an additional fundamental difference from the stationary case is that the microstate of any real suspension will exhibit a dependence on earlier states only for a *finite* period of time. Thus, the microstructural state of most flowing suspensions will exhibit an intrinsically preferred configuration for each given type and strength of the undisturbed flow, which is *independent* of its initial manufactured state.

II. THE RELATION BETWEEN MACROSCALE HEAT FLUX AND THE MICROSTRUCTURAL STATE OF A SUSPENSION

1. The General Relation

We consider a suspension of neutrally buoyant particles in the presence of a general bulk shear flow and a bulk temperature field.

Our objective is the development of a constitutive equation which describes the effective thermal diffusivity of the suspension considered as an equivalent homogeneous material. The point of view adopted is the conventional one in the field of suspension rheology. We assume that the minimum dimension l of the particles is large compared to the intermolecular length scale α of the suspending medium. The latter may then be treated as a continuum and is modeled for present purposes as an incompressible Newtonian fluid in which a simple scalar Fourier heat conduction law is applicable.

At any arbitrary point in the suspension, when viewed on a length scale of order l, the local variables such as velocity, temperature, enthalpy, or conductive heat flux are random functions of time whose values at any instant depend upon the proximity and motion of suspended particles. The description of bulk or macroscopic quantities for the suspension thus becomes a problem of statistics. At the fundamental level, the most appropriate definition of bulk variables is as an ensemble average of the corresponding microscale quantities for a large number of realizations of the system. Instantaneous local values of the velocity, temperature, enthalpy, and conductive heat flux may then be expressed as a sum of the ensemble-averaged quantity, and an additional microscale or fluctuating component, i.e.,

$$u_i = \langle u_i \rangle + u_i',$$
$$T = \langle T \rangle + T',$$
$$h = \langle h \rangle + h' \qquad [1]$$
$$q_i = \langle q_i \rangle + q_i'.$$

By definition the averages of the fluctuating components are zero,

$$\langle u' \rangle = \langle T' \rangle = \langle h' \rangle = \langle q_i' \rangle = 0. \qquad [2]$$

As suggested in the Introduction, we wish to obtain an operational definition for the bulk conductive heat flux, Q_i, which is consistent with the thermal energy balance for the suspension, viewed as an equivalent homogeneous medium,

$$\frac{\partial\langle h\rangle}{\partial t} + \langle u_i\rangle\frac{\partial\langle h\rangle}{\partial x_i} + \frac{\partial Q_i}{\partial x_i} = 0. \qquad [3]$$

A convenient method of determining the proper definition of Q_i for this purpose, is simply to apply the same ensemble averaging used in [1] to the exact, instantaneous thermal energy balance which is applicable for each realization of the system,

$$\frac{\partial h}{\partial t} + u_i\frac{\partial h}{\partial x_i} + \frac{\partial q_i}{\partial x_i} = 0. \qquad [4]$$

Taking account of [1] and [2], as well as the continuity relation for an incompressible fluid

$$\partial u_i/\partial x_i = 0, \qquad [5]$$

the result is

$$\frac{\partial\langle h\rangle}{\partial t} + \langle u_i\rangle\frac{\partial\langle h\rangle}{\partial x_i} + \frac{\partial}{\partial x_i}\langle q_i + u_i'h'\rangle = 0. \qquad [6]$$

Comparing [6] and [3], it follows that

$$Q_i \equiv \langle q_i\rangle + \langle u_i'h'\rangle. \qquad [7]$$

With the suspension viewed as an equivalent homogeneous fluid, the bulk "conductive" heat flux thus consists of an ensemble average of the local, instantaneous *conductive* heat flux, plus an additional contribution due to the local, instantaneous *convective* heat flux as measured in a frame of reference which is translating with the ensemble average velocity, $\langle\vec{u}\rangle$.

To proceed further it is desirable to replace the ensemble averages in [7] with more easily calculable spatial (volume) averages. For this purpose, we assume that the suspension is statistically homogeneous on a scale of $O(V^{\frac{1}{3}})$, which is much larger than the microscale, l, yet still much smaller than the macroscale, L, over which significant variations occur in the bulk velocity or temperature gradients, or in the concentration of particles. In these circumstances, the ergodic hypothesis may be invoked to replace the ensemble

averages of [7] with volume averages, e.g.,

$$\langle q_i\rangle = \frac{1}{V}\int_V q_i dV.$$

These volume-averaged quantities vary only on a scale of $O(L)$, and are thus *point* quantities with respect to a macroscopic description of the material.

Starting with volume averages in [7], the volume integrals may be split into two parts, $\sum V_0$ and $V - \sum V_0$, representing the sum of the volumes V_0 of the individual particles in V and the volume of the surrounding fluid, respectively, and reexpressed using the Fourier heat conduction law in each phase to give

$$Q_i = -k\frac{\partial\theta}{\partial x_i} + \frac{k(1-m)}{V}\int_{\Sigma V_0}(\nabla T)_i dV$$

$$+ \frac{\rho C_p(\tau-1)}{V}\int_{\Sigma V_0}u_i'T'dV$$

$$+ \frac{\rho C_p}{V}\int_V u_i'T'dV. \qquad [8]$$

Here, $\partial\theta/\partial x_i$ is the macroscale temperature gradient, k and mk are the thermal conductivities of the fluid and particles, C_p and τC_p are the respective heat capacities, and ρ is the density (assumed equal for fluid and particles). Temperature variations of the physical properties have been neglected in writing Eq. [8].

At this stage of our development, it is useful to examine the general form of [8] in some detail, since certain qualitative features of the macroscopic behavior can be deduced without need for explicit evaluation of the various integral terms. We may note, as a preliminary observation, that Q_i consists of two parts; first, the heat flux which would exist in the suspending fluid alone if the same average temperature gradient were maintained, and second, various additional contributions to the bulk heat flux which are due to the presence of the particles. When the particle and fluid have identical thermal properties (i.e., $\tau = 1$ and $m = 1$) two of these particle contributions vanish identically. However, the third does not vanish, in general, as long as the suspension is undergoing some motion. It may be anticipated than that the bulk conductive heat flux of a flowing suspension will differ from that of a single homogeneous fluid even when the microscale thermal properties of the two phases are identical. It is only when the suspension is completely motionless, so that the last two terms in [8] vanish identically, that the particle contribution will vanish with equal properties (i.e., $m = 1$).

Further examination of [8] shows that two distinct, though related, effects of macroscopic bulk flow may be isolated. The first is the flow-induced change in the local temperature distributions for a given *microstructural configuration* by the action of local convective heat transfer. As the local temperature distribution is altered, so too are the contributions from each of the three volume integrals of [8]. At steady state, the bulk heat flux, Q_i, will clearly depend on both the type and strength of the bulk flow, since the local (microscale) temperature distribution is directly influenced by these parameters. Furthermore, since the local temperature fields will only respond to variations in the flow or bulk temperature profile on a finite time scale which depends on the termal properties of the two phases, the bulk heat flux will exhibit a *relaxation* time of the same order for transients.

The second effect of the flow is due to its influence on the microstructure of the suspension. In general, the particle shapes, relative positions, and orientations will all be changed from their equilibrium values to a degree which again depends critically on the type and strength of the bulk motion. For example, rigid nonspherical particles undergo flow-induced rotations, causing changes in the orientation distribution from that of a stationary suspension; deformable particles may change shape due to hydrodynamic interaction with the surrounding fluid; and the distribution of relative positions will change form under the action of hydrodynamically induced particle translation. All of these hydrodynamic-ally induced changes in microstructure are deterministic (initial value) processes, thus apparently causing the bulk heat flux to depend upon *all* past microstructural states of the material (i.e., a perfect "memory" in the usual rheological sense). Fortunately, in any real suspension, these flow-induced changes in microstructure are resisted by one or more "restoring" mechanisms, such as translational and rotational Brownian motion or particle elasticity, which return the system to its equilibrium or rest configuration in the absence of flow. The existence of these restoring mechanisms, each characterized by a finite time scale, *ensures* that the dependence of Q_i on previous microstructural states (including the initial state of "manufacture") is "fading" with time. Thus, changes in the bulk flow will produce transients in Q_i on one or more additional time scales, which are distinct from the thermal time scale discussed earlier. At steady state, the microstructural configuration represents a dynamic balance between the hydrodynamic forces and the restoring mechanisms. As a consequence, the corresponding bulk heat flux Q_i must again depend on the type of flow, and its strength, as well as those geometrical or mechanical properties of the particles or suspending fluid which influence the "strength" of the restoring mechanism.

The constitutive complexities for thermal energy transfer in a flowing suspension are thus formidable. Even at steady state, the thermal transport properties will no longer be a unique property of the material, but will depend critically on the nature of the flow. Furthermore, we can expect the bulk heat flux to be influenced by mechanical and thermal properties of the two constituent phases. In what follows we shall confine our attention to the steady-state case where some definite results have been obtained.

2. *Evaluation of the Bulk Heat Flux, Q_i, for Dilute Suspensions*

Expression [8] for the bulk conductive heat flux, Q_i, is not restricted to any particlar range of particle concentrations. However, its evaluation in the general case would require the existence of a solution to the many-body microscale problem for all of the n particles in a typical averaging volume V. Fortunately, as the limit of small volume fraction, $\phi \ll 1$, is approached, it can be shown (5) that the general expression for Q_i reduces to a power series in ϕ, with the term of $O(\phi^m)$ calculable from the solution of a microscale heat transfer problem for m particles, rather than all n. In particular, the first $O(\phi)$ correction for a dilute suspension is determined by considering the microscale problem for a single, isolated particle, subject to the macroscopic or bulk average temperature and velocity fields as boundary conditions at infinity. At $O(\phi^2)$, the microscale problem would involve two particles subject to the bulk temperature and velocity fields at infinity, but no calculations have been carried out beyond $O(\phi)$ for a sheared suspension. For present purposes, we therefore restrict our attention to a determination of the $O(\phi)$ contributions to the bulk conductive heat flux for a *dilute* suspension.

The small scale of the averaging volume V relative to spatial variations in the macroscale velocity or temperature gradients suggests that it is appropriate to consider locally linear temperature and velocity fields, e.g., $T_\infty = \alpha_i x_i$, as boundary values for the single-particle microscale calculations. Since hydrodynamic interactions are completely neglected at this $O(\phi)$ level of approximation, and the flow is creeping, one simplification which arises from the assumption of a linear velocity field is that the center of mass of a typical, transversely isotropic particle will translate with the local undisturbed mean velocity $\mathbf{u}_\infty$. In view of the assumed homogeneity of the suspension on the scale $O(V^{\frac{1}{3}})$, it is thus convenient to formulate the single particle problem in a frame of reference fixed to the particle, i.e., also translating with velocity $\langle u_i \rangle$. This shift of reference frame requires no modification of [8] since the simple translational motion of the particles makes no contribution to the bulk conductive heat flux.

It is important to note, however, that some

further restriction on the possible values α_j is required in order for this far-field behavior for the microscale temperature field to be compatible with the general advection–diffusion equation. In particular, as $|x| \to \infty$, it is required that

$$(\mathbf{u}_\infty)_i \alpha_i \equiv 0, \qquad [9]$$

where $(\mathbf{u}_\infty)_i$ represents the macroscale velocity vector. In the case of simple shear flow, which we will consider later in detail, we are thus restricted to temperature gradients which are orthogonal to the undisturbed flow. To calculate the bulk heat flux due to temperature gradients in the flow direction, it would be necessary to consider the *exact* macroscale velocity and temperature distributions locally, not just their *linear* approximations. This fact apparently precludes the existence of an effective conductivity tensor in the presence of a bulk flow, since the resulting bulk heat flux could not possibly be linearly related to ∇T. We shall be concerned here with the bulk heat flux due to gradients of the bulk temperature field which satisfy [9].

Two cases can be distinguished in the evaluation of Q_i for a dilute suspension depending on the presence of absence of randomness at the microscale. Let us assume, in what follows, that the particles of the suspension are identical in size and shape. If the motion and orientation of the particles should then also happen to be the same, the average contribution to Q_i per particle, i.e.,

$$\bar{Q}_i{}^* \equiv \frac{1}{n}\left(Q_i + k\,\frac{\partial \theta}{\partial x_i}\right),$$

could be calculated directly from Eq. [8] using the deterministic microscale fields for any single particle in V. The total particle contribution to Q_i in this completely deterministic case, is then simply $\bar{Q}_i{}^*$ multiplied by the number of particles, n, in the volume V. On the other hand, if the motions and orientations of the particles exhibit a random nature, as would most often be true, the (instantaneous) individual particle contributions to Q_i will be different for each particle and the average contribution per particle can then only be determined in a statistical sense. From the dilute suspension point of view of single particle calculations, the contribution to Q_i is calculated for a given arbitrary state (i.e. particle orientation and/or shape)—and then averaged over the distribution of possible states using the appropriate probability density function as a weighting factor. We shall denote this statistical averaging process by the use of an overbar. With this understanding the *particle* contribution to the bulk heat flux may be expressed in the form

$$Q_i^* = n\bar{Q}_i^* = \frac{k(1-m)\phi}{V_0} \int_{V_0} \overline{(\nabla T)}_i dV$$

$$+ \frac{\rho C_p(\tau - 1)\phi}{V_0} \int_{V_0} \overline{u_i'T'} dV$$

$$+ \frac{\rho C_p \phi}{V_0} \int_{V} \overline{u_i'T'}. \quad [10]$$

Here, V_0 is the volume of a particle (assumed identical for all particles), while V is the *fluid* volume which is assumed to extend to infinity in the single-particle microscale problem.

3. The Effects of Brownian Motion

One source of randomness at the microscale, for sufficiently small particles, is Brownian motion. Both the translation and rotation of a Brownian particle will exhibit a random component, and the effects of these random motions on the bulk heat flux must be accounted for in some rational manner.

Brownian motion is a reflection of the molecular nature of the suspending fluid at the scale of the small particles. Nevertheless, it is not necessary to surrender completely the benefits of a continuum approximation for the suspending fluid in order to model the Brownian motions of such a particle. A clear example which illustrates this point is the usual treatment of Brownian motion by the addition

of fluctuating forces to the systematic equations of motion for a particle in which classical hydrodynamics is used to model the mean mechanical interactions between fluid and particle. In the present context, however, our concern is not merely the random *particle* motions, but also the resulting motions and temperature variations in the suspending fluid. An alternative approach to the theory of Brownian motion which satisfies these objectives is the theory of hydrodynamic fluctuations, described by Landau and Lifschitz (6) and recently extended to include Brownian motions of a particle by several investigators (7, 8, 9, 10).

Let us first consider the theory of fluctuating hydrodynamics as applied to the suspending fluid, without particles. The main assumption of the theory is that the governing continuum equations of motion and thermal energy are valid for the *complete problem*, including fluctuations in all of the independent variables. Fluctuations in the stress tensor and heat flux vector occur not only due to fluctuations in the velocity and temperature gradients, but also as the result of *spontaneous* local stresses and heat fluxes, denoted by **s** and **q**, which are *independent* of each other, *and*, of the local velocity and temperature gradients (10). The quantities $\nabla \cdot \mathbf{s}$ and $\nabla \cdot \mathbf{q}$ drive fluctuations in the other hydrodynamical and thermal variables, and these fluctuations in the fluid, away from any macroscopic boundaries, are responsible for its dissipative *transport* properties. Relations between the statistical properties of **s** and **q** and the continuum transport parameters such as viscosity or thermal conductivity are contained in so-called fluctuation–dissipation theorems which may be derived from certain postulates of nonequilibrium thermodynamics (10).

The introduction of colloidal particles into the fluctuating fluid introduces a possible source of random motons on a *much longer* time scale than that characterizing $\nabla \cdot \mathbf{s}$. The key property of particles (whether solid, elastic, drops, etc.) which is responsible for these longer time-scale random motions is that they resist deformation of shape so that at least part of the fluctuating stress in the suspending fluid can be sustained with no local deformation. In these circumstances, the fluctuating stresses can contribute to random translational and rotational motions of the particles, but only on a relatively slow continuum time scale. For translational motion this time scale is a measure of the period required for a particle to slow down following an initial impulse, and is given by m/γ where m is the mass of the particle, and γ the fluid mechanical drag which acts on it. These random motions of the particle will also induce random motions in the suspending fluid with the same relatively slow time scale, which are superposed both on the random fluctuations, and on any deterministic local motions which may be present due to the bulk flow of the suspension. We shall refer to the random motions of particles and suspending fluid on the time scale m/γ (or its equivalent for rotational motion) as "Brownian," in order to distinguish them from the much more rapid fluctuations which occur at the molecular scale.

It is then convenient, following previous authors, to split the *fluid* velocity and stress fields for the single-particle microscale problem into three parts: deterministic contributions $\mathbf{U}$ and Σ associated with the motion of the suspension as a whole; Brownian contributions, $\mathbf{v}$ and $\tilde{\sigma}$, associated with the random continuum level motions of particles and fluid; and fluctuating terms $\mathbf{u}^{\ddagger}$ and $\mathbf{s}$ due directly to molecular fluctuations, i.e.,

$$\mathbf{u} = \mathbf{U} + \mathbf{v} + \mathbf{u}^{\ddagger},$$
$$\sigma = \Sigma + \tilde{\sigma} + \mathbf{s}. \qquad [11]$$

Governing equations for the individual fields, $\mathbf{U}$, $\mathbf{v}$, and $\mathbf{u}^{\ddagger}$ may easily be deduced from the equations of motion and boundary conditions for the system as a whole, once it is assumed that the Reynolds numbers for the motions $\mathbf{U}$, $\mathbf{v}$ and the magnitude of $\mathbf{u}^{\ddagger}$ are all small. Space restrictions prevent a comprehensive descrip-

tion here, but the interested reader may refer to (9, 10), or the forthcoming thesis by McMillen (11) where the theory is summarized in detail. For the present, we simply note that $\mathbf{U}$ is the deterministic, Stokes velocity field associated with a particle which is rotating under the action of the bulk flow.

To complete the formulation of the microdynamics problem, we need only specify the governing equations for Brownian translation and rotation of the *particle* (we shall assume that it is *rigid* so that its shape is fixed and known)

$$m\frac{\partial \tilde{\mathbf{u}}_p}{\partial t} = \int_{S_0} \mathbf{n} \cdot (\tilde{\sigma} + \mathbf{s})ds, \qquad [12a]$$

$$\mathbf{J}_p \cdot \frac{d\omega_p}{\partial t} = \int_{S_0} \mathbf{x} \wedge \mathbf{n} \cdot (\tilde{\sigma} + \mathbf{s})ds, \qquad [12b]$$

and the connection between this motion and that of the suspending fluid through the no-slip boundary condition $\mathbf{v} = \tilde{\mathbf{u}}_p + \omega_p \wedge \mathbf{x}$ at the particle surface. For purposes of subsequent discussion, we shall denote the random motions of the particle as $\mathbf{v}_p(\equiv \mathbf{u}_p + \mathbf{x} \wedge \omega_p)$ and its deterministic motions as $\mathbf{U}_p$, so that

$$\mathbf{u}_p = \mathbf{U}_p + \mathbf{v}_p, \qquad [13]$$

which is similar to [11] for the fluid. It may be noted that the force and torque contributions associated with the fluctuations of stress at the particle surface appear as white noise on the Brownian time scales characteristic of $\tilde{\mathbf{u}}_p$ and ω_p, and Eqs. [12a] and [12b] can thus be reduced to the traditional Einstein–Smoluchowski theory of Brownian motion. In this case, the only parameter describing the evolution of the particle's position or orientation distributions is the diffusion tensor, $\mathbf{D}_{Br}$, defined over the position and orientation space.

Finally, it is useful to consider the statistical averaging process, denoted in [9] by the overbar, applied to the velocity vectors $\mathbf{u}$ and $\mathbf{u}_p$. Since the average of the fluctuations $\mathbf{u}^{\ddagger}$ is zero even on the shorter Brownian time scale, the result is

$$\bar{\mathbf{u}} = \bar{\mathbf{U}} + \bar{\mathbf{v}}; \quad \bar{\mathbf{u}}_p = \bar{\mathbf{U}}_p + \bar{\mathbf{v}}_p. \quad [14]$$

One important implication of these relationships is that the random Brownian velocities $\mathbf{v}$ and $\mathbf{v}_p$ will have a *nonzero mean* whenever gradients exist in the statistical distribution of particle orientations.[1] As an example, we may consider axisymmetric particles which are characterized by a rotational diffusion coefficient, $\mathbf{D}_R$. In this case, the mean velocity fields $\bar{\mathbf{v}}$ and $\bar{\mathbf{v}}_p$ are associated with rotation of the axis of revolution of the particle at a rate

$$\mathbf{\Omega}_{p_{Br}} \equiv -\mathbf{D}_R \cdot \nabla_\phi \ln P(\boldsymbol{\phi}) \quad [15]$$

in a quiescent fluid. Here, P is the probability density function for particle orientation, which may be obtained by solution of a Fokker–Plank equation, representing the competition between purely hydrodynamic and Brownian rotations (for example, see (12) for the case of rigid spheroidal particles in a simple shear flow). A second implication of [14], which may appear surprising at first, is that the average $\bar{\mathbf{U}}$ is *not* simply $\mathbf{U}$. Instead, in the presence of Brownian rotation, the velocity field $\bar{\mathbf{U}}$ represents the motion associated with the deterministic rotation of a particle, calculated as a function of particle orientation and then averaged with the probability density function for orientation as a weighting factor. Again, for convenience, it will be useful to have a symbol for the difference between the orientation weighted mean velocity and the instantaneous value $\mathbf{U}$ for a particle with some specific orientation, i.e.,

$$\mathbf{U} = \bar{\mathbf{U}} + \mathbf{U}^*; \quad \mathbf{U}_p = \bar{\mathbf{U}}_p + \mathbf{U}_p^*.$$

We may note that *both* $\mathbf{U}^*$, $\mathbf{U}_p^*$ and $\mathbf{v}^*$, $^*\mathbf{v}_p$ are *random* variables with respect to the orientation average values, although they have quite different physical origins.

Let us now consider the effects of fluctuations and Brownian motions on the temperature distributions in the suspension, and on the

[1] Spatial variations of particle concentration, which could also lead to nonzero values for $\bar{\mathbf{v}}$, have been neglected by the assumption of local homogeneity in the present work.

bulk heat flux Q_i. We begin with the governing equations for the temperature distributions in the suspending fluid and particle. For simplicity, we again restrict our attentions to rigid, no-slip particles. The basic equations are (6)

$$\rho_T C_p \left\{ \frac{\partial T_p}{\partial t} + \mathbf{u}_p \cdot \nabla T_p \right\} = mk\nabla^2 T_p - \nabla \cdot \mathbf{q}_p,$$

$$\mathbf{x} \in V_0, \quad [16a]$$

$$\rho C_p \left\{ \frac{\partial T}{\partial t} + \mathbf{u} \cdot \nabla T \right\} = k\nabla^2 T - \nabla \cdot \mathbf{q},$$

$$\mathbf{x} \in (V - V_0). \quad [16b]$$

It is most convenient to consider the temperature distributions T and T_p, expanded in the manner of [11] and [13], to the forms

$$T = T + \tilde{T} + T\ddagger,$$
$$T_p = T_p + \tilde{T}_p + T_p\ddagger. \quad [17]$$

The last terms in these expressions are the temperature fluctuations arising from $\mathbf{q}$ and $\mathbf{q}_p$, which vanish identically when averaged over a Brownian time scale. The physical significance of T, T_p, $\tilde{T}$, and $\tilde{T}_p$ will be considered shortly.

To proceed further, we examine the equations and boundary conditions which govern the individual temperature distributions. Again, space limitations preclude a complete discussion (see (11)). However, it is important to note that, in contrast with the "systematic" Brownian *motions* which arise from the fluctuating force, there is no systematic heating or cooling of the particle on a time scale m/γ as a direct result of the heat flux fluctuations. The temperature distribution inside the particle, unlike the corresponding particle velocity, *can* react on the much shorter time scale of the fluctuations. As a result the fluctuations in $\mathbf{q}$ and $\mathbf{q}_p$ merely drive fluctuations in temperature ($T\ddagger$ and $T_p\ddagger$) which are coupled at the interface, but which produce no measurable effect on the bulk heat flux Q_i, except indirectly through the thermal conductivities to which they are related. It thus follows that the random parts of $T + \tilde{T}$ and $T_p + \tilde{T}_p$ exist solely as a consequence of the Brownian motions of the particle in a nonuniform ambient temperature distribution. Governing equations for these composite temperature fields can be derived from the full convection/diffusion equations by first linearizing with respect to the small velocity and temperature fluctuations, and then averaging over the Brownian time scale. The result is

$$\rho C_p \left(\frac{\partial (T + \tilde{T})}{\partial t} + (\mathbf{U} + \mathbf{v}) \cdot \nabla (T + \tilde{T}) \right)$$

$$= k \nabla^2 (T + \tilde{T}).$$

It is now convenient to depart somewhat from the notational precedence of Eqs. [11]–[15] and identify T and T_p as the complete orientation averaged temperature fields (i.e., including contributions from both the deterministic and mean Brownian motions of the particle) so that

$$\tilde{T} = \tilde{T}_p = 0. \qquad [18]$$

To obtain an equation for the mean field T, we average the preceding equation for $T + \tilde{T}$ over the orientation distribution as signified by the overbar. Taking account of [14] and [18], this gives

$$\rho C_p (\partial T / \partial t) + \overline{(\mathbf{U} + \mathbf{v})} \cdot \nabla T$$

$$= k \nabla^2 T - \rho C_p \nabla \cdot \overline{(\mathbf{U}^* + \mathbf{v}^*) \tilde{T}}. \qquad [19a]$$

Inside the particle, we may similarly derive the governing equation for T_p:

$$\tau \rho C_p \left(\frac{\partial T_p}{\partial t} + (\overline{\mathbf{U}_p + \mathbf{v}_p}) \cdot \nabla T_p \right)$$

$$= mk \nabla^2 T_p - \tau \rho C_p \nabla \cdot \overline{(\mathbf{U}_p^* + \mathbf{v}_p^*) \tilde{T}_p}. \qquad [19b]$$

Finally, in order to determine the circumstances in which the various terms of [19] are significant, we nondimensionalize the equations. For this purpose, we use the dimension of a particle, l, as a characteristic length scale. There are two relevant characteristic velocity scales. The first is the scale characteristic of the deterministic motions which are imposed

by the bulk motion of the suspension, V_{bulk}. If the bulk flow were a simple shear flow, this scale would be γl, where γ is the bulk shear rate. The scale V_{bulk} is relevant both to the deterministic fields, $\mathbf{U}$ and $\mathbf{U}_p$, and to the effective Brownian velocities, $\bar{\mathbf{v}}$ and $\bar{\mathbf{v}}_p$. The exact magnitudes of these effective velocities will, of course, be determined by [15] and the probability density function $P(\phi)$ resulting from the Fokker–Plank equation which is also nondimensionalized with the bulk velocity scale (12). The purely random velocity fields $\mathbf{v}^*$ and $\mathbf{v}_p^*$ are characterized by the second velocity scale, $V_{Br} \equiv (KT/m)^{\frac{1}{2}}$ of the Brownian motion.[2] We shall denote the ratio V_{Br}/V_{bulk} as ϵ. With these characteristic scales, Eqs. [19a] and [19b] become

$$\mathrm{Pe} \left(\frac{\partial T}{\partial t} + \overline{\mathbf{U}} \cdot \nabla T + \overline{\mathbf{v}} \cdot \nabla T \right)$$

$$= \nabla^2 T - \mathrm{Pe} \nabla \cdot \overline{[(\mathbf{U}^* + \epsilon \mathbf{v}^*) \tilde{T}]} \qquad [20a]$$

and

$$\mathrm{Pe}_p \left(\frac{\partial T_p}{\partial t} + \overline{\mathbf{U}_p} \cdot \nabla T_p + \overline{\mathbf{v}_p} \cdot \nabla T_p \right)$$

$$= \nabla^2 T_p - \mathrm{Pe}_p \nabla \cdot \overline{[(\mathbf{U}_p^* + \epsilon \mathbf{v}_p^*) \tilde{T}_p]}, \qquad [20b]$$

where

$$\mathrm{Pe} = \frac{V_{bulk} l}{(k/\rho C_p)}; \qquad \mathrm{Pe}_p = \frac{V_{bulk} l}{(mk/\rho \tau C_p)}.$$

Before discussing these equations, it is convenient to briefly reconsider Eq. [10] for the particle contribution to the bulk heat flux. In present terms

$$\mathbf{u}' = (\mathbf{u}_p - \mathbf{U}_\infty); \quad T' = (\mathbf{T}_p - T_\infty), \quad \mathbf{x} \in V_0,$$

and

$$\mathbf{u}' = (\mathbf{u} - \mathbf{U}_\infty); \quad T' = (\mathbf{T} - T_\infty), \quad \mathbf{x} \in (V - V_0)$$

where $\mathbf{U}_\infty$ and T_∞ are the undisturbed bulk fields and $\mathbf{u}, \mathbf{u}_p, \mathbf{T}, \mathbf{T}_p$ are given in Eqs. [11],

[2] Note that K is Boltzmann's constant.

[13], and [17]. Thus, in dimensionless terms,

$$Q_j^* = \alpha_j k \phi \left[(1 - m) \int_{V_0} \nabla T_p dV \right.$$

$$+ m\mathrm{Pe}_p \int_{V_0} [\overline{(\mathbf{U}_p + \mathbf{v}_p} - \mathbf{U}_\infty)(\mathrm{T}_p - T_\infty)$$

$$+ \overline{(\mathbf{U}_p^* + \mathbf{v}_p^*\epsilon)\tilde{T}_p}]dV$$

$$+ \mathrm{Pe} \int_{V-V_0} [\overline{(\mathbf{U} + \mathbf{v}} - \mathbf{U}_\infty)(\mathrm{T} - T_\infty)$$

$$\left. + \overline{(\mathbf{U}^* + \epsilon \mathbf{v}^*)\tilde{T}}]dV \right]. \quad [21]$$

Examination of [20] and [21] shows that Brownian motion has the potential for three distinct types of effects. First are those which we may term "direct" corresponding to an additional convective flux at the microscale due to correlation between the portions of the velocity and temperature distribution which appear random with respect to the orientation-weighted averages. These are represented by the last terms of [20a] and [20b], and of the second and third integrals of [21]. Second, the deterministic motion, $\mathbf{U}$ and $\mathbf{U}_p$, in the mean convection terms of these equations is averaged with respect to the orientation distribution for particle orientation, and the latter is affected by the Brownian rotations. Third, there is an additional effective convection velocity, $\overline{\mathbf{v}}$ and $\overline{\mathbf{v}}_p$, associated with the mean particle rotation, [15], which arises when Brownian rotation is superposed on a nonuniform orientation distribution. The latter two effects are, of course, not present if the particles are spherical, and in addition $\mathbf{U}^*$ and $\mathbf{U}_p^*$ will also be zero in that case. On the other hand, Brownian motion will still have an influence on Q_i through the correlations between the random translational velocities and temperatures, $\mathbf{v}^*$ and $\mathbf{v}_p^*$, and $\tilde{T}$ and $\tilde{T}_p$. This later effect will persist even if the suspension is stationary at the macroscale, provided only that the particles are free to move about in the suspending fluid. In general, then, Brownian motion will influence Q_i both in-

directly through changes in the mean temperatures T and T_p, and directly by means of the last terms of the second and third integrals in [21].

Whether these Brownian motion contributions are significant in a flowing suspension will clearly depend on the magnitudes of Pe, Pe_p, and ϵ. In particular, when compared to the magnitude of the Brownian motion effects associated with the deterministic and effective particle rotations, the direct convective flux associated with $\mathbf{v}^*$ and $\mathbf{v}_p^*$ may be neglected when $\epsilon \ll 1$. The condition $\epsilon \ll 1$ requires that particle be large enough, or dense enough, so that $(KT/m)^{\frac{1}{2}}$ is very small compared to the bulk imposed velocity scale, i.e.,

$$\rho l^3 \gg KT/V_{\mathrm{bulk}}. \quad [22]$$

We may note that this condition does not imply that Brownian motion effects will necessarily always be important when V_{bulk} is very small as it might first appear. Condition [22] is the condition for neglect of the direct Brownian heat flux compared to convective effects of the bulk motion. However, in a stationary, or near stationary suspension, the magnitude of Brownian motion effects must be compared to the pure *conduction* terms, i.e., for neglect of Brownian effects in this case

$$\mathrm{Pe}_{\mathrm{Br}}, \mathrm{Pe} \ll 1.$$

In terms comparable to [22] we thus require

$$\rho l^3 \gg KTl^2/(k/\rho C_p)^2. \quad [23]$$

For typical situations, this condition is satisfied, and the prior calculations for stationary suspensions which neglected the effect of translational Brownian motion (2, 3, 4), would seem to provide adequate first approximations.

III. CALCULATIONS OF THE BULK CONDUCTIVE HEAT FLUX FOR A FLOWING SUSPENSION

In order to provide detailed predictions for the bulk heat flux, it is necessary to evaluate [10] for specific imposed bulk flows and specific types of particles. Unlike the theories for completely stationary suspensions (composites) which we briefly reviewed in the Introduction, relatively few such calculations have been completed. The main difficulty as we can now see is that one must determine the detailed local temperature distribution at the scale of the suspended particles. We confine our attention to the dilute limit discussed in the previous section, where the problem reduces to that of a single particle. Furthermore, we begin by focusing our discussion on large particles where Brownian effects are small. Even with these simplifications, one must still obtain solutions of the full microscale thermal energy and momentum balances. In cases of practical significance, the microscale disturbance flow is creeping (i.e., the Reynolds number, $lV\rho/\mu$, is close to zero) and we shall assume that the local *velocity* field near a particle can be determined exactly. The problem thus reduces to the solution of [20], with all of the Brownian terms deleted, and $\mathbf{U}$, $\mathbf{U}_p$ specified as the deterministic local velocity fields. Two cases may be distinguished where a solution is possible, namely, the asymptotic limits Pe, $Pe_p \ll 1$ and Pe, $Pe_p \gg 1$. The case Pe, $Pe \ll 1$ is the limit of weak local convection relative to conduction, where the bulk heat flux is dominated by the stationary suspension results. In spite of the relatively weak effects of motion on the bulk heat flux, however, this limit is of considerable qualitative interest since calculations can actually be completed for a variety of particle shapes and flow types including time-dependent problems. The other limit Pe, $Pe_p \gg 1$ is of greater direct practical interest since the effect of convection is dominant at the microscale and the magnitude of the flow effects on the bulk heat flux therefore potentially *large*. The disadvantage for analytical investigation, however, is that very few problems of interest appear to be tractable. For example, in shear flow, even the problem of steady flow with spherical particles has not yet been completely resolved. In the remainder of this section, we outline the existing theories.

(a) *Spherical and nearly spherical liquid drops in linear shear flow with no Brownian*

motion and low local Peclet number. One problem which has been investigated quite completely is that of a dilute suspension of spherical or nearly spherical liquid drops which is undergoing a simple shearing motion in the presence of a linear temperature field with $(u_\infty)_i\alpha_i = 0$. The fluids in both phases are Newtonian, and equal in density. Furthermore, the Peclet number is assumed to be small, and Brownian motion effects negligible. The analysis of the microscale temperature distribution in this case then involves a straightforward application of the methods of matched asymptotic expansions, using the velocity fields and particles shapes calculated by Taylor (13) for the creeping motion of a drop in a simple shear flow. Details of the calculations of the bulk heat flux in this case have been reported elsewhere (14, 15) and will not be repeated here.

When the integrals in [10] are evaluated using the calculated microscale temperature distributions, an explicit expression is obtained for the bulk conductive heat flux. Here, we examine the component of the bulk heat flux in the same direction as the temperature gradient, first for the case of perfectly spherical particles where

$$Q_y{}^* = -k\alpha\Phi\left(\frac{3(m-1)}{m+2} + \left\{1.176\frac{(m-1)^2}{(m+2)^2}\right.\right.$$
$$+ \frac{(5\lambda+2)}{(\lambda+1)}\left[0.12\frac{5\lambda+2}{\lambda+1}\right.$$
$$\left.\left.\left. -.028\left(\frac{m-1}{m+2}\right)\right]\right\}Pe^{\frac{3}{2}} + O(Pe^2)\right). \quad [24]$$

The viscosity ratio μ_p/μ has been denoted λ.

The first term in [24] is the pure conduction result (i.e., for a stationary suspension) which was first obtained by Maxwell (2). The second term, of $O(Pe^{\frac{3}{2}})$, is the first *flow-induced* contribution to the bulk heat flux. Three points of special significance may be noted with regard to this term. First, as anticipated in Section IIa, we find that the presence of flow results in a nonzero particle contribution to the bulk heat flux even when the thermal conductivities of the two phases are identical. Second, it may be seen that the bulk conductive heat flux is enhanced by the existence of the shear flow for all k and k_p. In particular, the degree of flow enhancement depends not only on the material properties of the two fluids as in the case of a stationary suspension, but also on the shear rate (i.e., Pe). Thus, the corresponding effective conductivity $k^* \equiv -Q_y^*/\alpha$ is clearly *not* a *material* property for the suspension. Third, and finally, the critical importance of particle geometry is demonstrated by the fact that the bulk heat flux is first effected by flow at $O(\text{Pe}^{\frac{3}{2}})$, while the microscale temperature field is already modified significantly at $O(\text{Pe})$. This last anomaly is due to the special symmetry of an exactly spherical drop. Indeed, when small deformations of the drop are allowed, due to the presence of flow, the magnitude of the first flow-induced change in the thermal conductivity can occur at $O(\text{Pe})$ rather than $O(\text{Pe}^{\frac{3}{2}})$ (15). Two cases can be distinguished of small deformation, $O(\epsilon)$; one in which interfacial tension forces are dominant ($\epsilon \sim a\gamma\mu/\sigma \ll 1$), and the second in which the deformation process is restricted by a large internal (drop) viscosity ($\epsilon \sim 1/\lambda \ll 1$). Here, σ is the interfacial tension and a the undeformed drop radius. For the case of internal viscous forces controlling deformation, the result for the effective conductivity is found to consist simply of an $O(\epsilon)$ correction to the Maxwell result, and an $O(\epsilon\text{Pe}^{\frac{3}{2}})$ correction to the $O(\text{Pe}^{\frac{3}{2}})$ term in [24]. Thus, the droplet deformation only induces small corrections to the basic results [24] for a perfect sphere. On the other hand, for the case of deformation controlled by surface tension, the first deformation-induced contribution occurs at $O(\epsilon\text{Pe})$ (but is identically zero when $m = 1$). Thus, the $O(\text{Pe}^{\frac{3}{2}})$ result of [24] will *only* provide a uniformly valid first approximation to k^* when $\text{Pe}^{\frac{3}{2}} \gg \epsilon\text{Pe}$ (or $m \equiv 1$). In particular, the lack of any $O(\text{Pe})$ contribution to k^* in the expression [24] is now clearly seen to be an accident of the perfect sphere geometry which was assumed. Indeed, when $m \ll 1$, the deforma-

tion-induced term of $O(\epsilon\text{Pe})$ is the *dominant* flow contribution to the bulk conductive heat flux. A more detailed discussion of these results may be found in the references cited earlier, as well as in a recent review article by Leal and McMillen (16).

(b) *Spherical rigid particles in linear shear flow with no Brownian motions and large local Peclet number.* A second problem of considerable potential significance is the case of rigid spherical particles in linear shear flow, with Brownian motion still neglected, but the local Peclet number assumed to be large. In a recent paper, Nir and Acrivos (17) have studied this problem in detail, and we base the present discussion on their analysis.

The basic approach is, of course, unchanged but in this large Pe limit the calculation of a microscale temperature distribution is extremely complicated, even for spherical particles. Nir and Acrivos have used the general methods of matched asymptotic analysis for large Pe, to identify three distinct regions in the temperature field: one far from the particle where the streamsurfaces of the creeping flow solution are open; one generally nearer the particle (but extending to $\pm \infty$ along the axis which bisects the sphere and is parallel to the undisturbed flow) in which the streamsurfaces are closed; and one intermediate between the first two. For large Pe, a first approximation to the temperature distributions in the first two regions may be obtained from the condition that T is constant on streamsurfaces. In the outer, open streamsurface region, the variation of T in moving from one streamsurface to another is identical to that in the undisturbed field at infinity. Thus, a knowledge of the velocity field, and hence of the spatial variations of streamsurface variable E, is equivalent to a knowledge of the temperature distribution. In the inner region of closed streamsurfaces, the compatibility condition that the integrated heat flux across any closed surface be equal to zero at steady state, requires that the temperature distribution be just $T = $ constant. As Nir and Acrivos point out, this last fact has the rather striking consequence that the *bulk* conductive heat flux will be independent of the thermal properties of the particles, i.e., a suspension of perfectly conducting particles will have exactly the same bulk heat flux as a suspension of insulating particles provided that the other particle properties are the same. In fact, it may be shown that the leading contribution to the bulk heat flux in the direction of the bulk temperature gradient comes from the third region, which is a boundary layer lying between the regions of open and closed streamsurfaces. The governing equation for T, in this domain, can be derived, but it is complicated and has not yet been solved. Fortunately, with certain assumptions about the integrability of the solution, one can still provide an estimate of the bulk heat flux

$$Q_y{}^* = -k\alpha\Phi\{A_1\mathrm{Pe}^{1/11} + O(1)\}, \quad [25]$$

where A_1 is a constant still to be determined from the solution for T. Thus, in spite of the fact that $Q_y{}^*$ can, in principle, become very large for $\mathrm{Pe} \gg 1$ as we suggested in the introduction to this section, the rate of variation with Pe is extremely weak, and $Q_y{}^*$ is unlikely (for realistic values of Pe) to become much different from the $O(1)$ value found in the last section for pure conduction in a stationary composite. One *difference*, however, is that the contribution [25] is independent of the particle conductivity.

(c) *Rigid prolate spheroidal particles in simple shear flow with rotational Brownian motion and small local Paclet number.* Finally, we consider briefly a case recently completed by McMillen (11), of rigid spheroidal particles in simple shear flow with a significant degree of rotational Brownian motion.

In the absence of Brownian motion, the deterministic rotational motion of the particles is periodic, and described by the familiar orbit equations of Jeffrey (18). As a result, the corresponding microscale heat transfer problem is also time dependent, and a general solution impossible. Here, as in Section IIIa, we restrict our attention to the limiting case

of small Pe which *can* be solved by a straightforward but tedious perturbation scheme.

The governing microscale equations are [20], with the temperature fields represented as an asymptotic expansion in Pe (or $\mathrm{Pe_p}$). At first order, the problem is pure conduction. The first convection effects are included at the second order. This uncoupling of the order-one temperature fields from the deterministic or mean Brownian velocity fields simplifies the evaluation of T, T_p and the bulk heat flux, Q_i in two ways. First, the *direct* heat flux contributions to [19], [20], and [21] are all identically zero at $O(\mathrm{Pe})$ since $\tilde{T}$ and $\tilde{T}_\mathrm{p}$ are uncorrelated with (i.e., independent of) $\mathbf{U}^*$, $\mathbf{U_p}^*$, $\mathbf{v}^*$, and $\mathbf{v_p}^*$ at this first level of approximation. Second, since the governing equations (and boundary conditions) are all linear in T, it follows that the deterministic and mean Brownian convection contributions at $O(\mathrm{Pe})$ may be calculated independent of one another. McMillen's (11) approach for the deterministic convection contribution was to calculate the microscale temperature fields associated with the deterministic velocity field and a specific (but arbitrary) particle orientation; from this solution to determine the instantaneous and orientation-dependent contributions to the integrals in [10]; and then finally, to average these orientation-dependent quantities by integrating over the complete orientation space with the probability density function for orientation as a weighting factor. The additional convection contribution due to the orientation weighted Brownian rotation, was similarly determined by first considering the temperature field associated with steady rotation of the particle in a quiescent fluid; following this with a calculation of the instantaneous, orientation-dependent contributions to [10]; and finally again averaging the results using the orientation distribution function.

The probability distribution function for particle orientation is governed by a modified Fokker–Plank equation, and can be solved analytically for either strong or weak Brownian motion (12, 19). The calculations of McMillen (11) are restricted to the limit of strong

Brownian motion, $D/\gamma \gg 1$, where D is the rotational Stokes–Einstein diffusion coefficient for the particles.

Space restrictions do not permit a complete discussion of the results which were obtained by McMillen (11). Instead, we shall very briefly describe the behavior of the bulk conductive heat flux in the direction of the temperature gradient; i.e., across the flow in the direction of the velocity gradient. The general form for Q_y in this case, is

$$\frac{Q_y}{\alpha k} = 1 + \Phi \left\{ \left\langle \frac{Q_y}{\alpha k} \right\rangle_0^0 + \left[\frac{\gamma}{D}\right]^2 \left\langle \frac{Q_y}{\alpha k} \right\rangle_0^2 \right.$$

$$\left. + \text{Pe}\, \frac{\gamma}{D} \left\langle \frac{Q_y}{\alpha k} \right\rangle_1^1 + \cdots \right\}. \quad [26]$$

The first term in brackets represents the pure conduction contribution averaged against a completely random orientation distribution (i.e., dominant Brownian motion). The general behavior of this term, which is identically zero for $m = 1$, is in the direction of an enhancement of the bulk heat flux with increasing conductivity and prolateness of the particles. For $m < 1$, small to moderate negative contributions are found, but these are relatively insensitive to the precise particle geometry. For $m > 1$, on the other hand, the contribution is strongly positive and in fact increases without bound in the (unphysical) limit of infinitely long, perfectly conducting particles. The next term in [26], $O(\gamma/D)^2$, represents a modification of the pure conduction contribution caused by the preferential alignment of particles which occurs in the presence of a bulk shear flow. Since the action of shear is to align the particle, statistically, in the direction of flow, the *effect* of this modification is always to *decrease* the bulk heat flux across the flow, Q_y, by an amount which increases with prolateness and particle conductivity. The final term in [26] is the *combined* convective contribution corresponding to the deterministic *and* mean Brownian velocity fields, and is *zero* if $m = 1$. It may be noted, that there is *no* convective contribution at $O(\text{Pe})$ for a purely

random distribution of orientations. Furthermore the $O[(\gamma/D)\text{Pe}]$ contribution vanishes for spherical particles and remains very small for particle axis ratios up to approximately 10. For more elongated particles, however, the $O[(\gamma/D)\text{Pe}]$ contribution is mainly positive, becoming more so as m is increased, and, particularly for $m \gg 1$, is strongly increasing with increasing particle axis ratio. Detailed results for this and other components of the bulk heat flux are available elsewhere (11).

CONCLUSION

The results obtained here clearly indicate the complexity of the thermal constitutive behavior of a flowing suspension. The bulk conductive heat flux is seen to depend not only on the thermal properties of the two individual phases, but also on the strength and type of the bulk flow, and any other micromechanical properties of the particles which are involved in the microstructural configuration of the suspension. Although the mathematical difficulties of the microthermal energy problem have restricted the detailed analysis to the case of small concentration, with particles of simple shape in a simple linear shear flow, the complexity of the results in even limited situations should serve as an indication of the wide variety of thermal constitutive response that might be expected of any real fluid which exhibits a suspension-like structure. Included in such materials is the important class of dilute or moderately concentrated macromolecular solutions.

ACKNOWLEDGMENTS

The preparation of this paper was supported, in part, by the Research Corporation, and, in part, by the Petroleum Research Fund of the American Chemical Society. The author wishes particularly to acknowledge the contributions of T. J. McMillen. Discussions with H. Brenner were also useful in clarifying some of the issues discussed herein.

REFERENCES

1. Hinch, E. J. and Leal, L. G., *J. Fluid Mech.* **71**, 481 (1971).

2. MAXWELL, J. C., "Electricity and Magnetism," 1st ed., Oxford Univ. Press, London, 1873.

3. JEFFREY, D. J., *Proc. Roy. Soc. (London) Ser. A* **335**, 355 (1973).

4. ROCHA, A. AND ACRIVOS, A., *Quart. J. Mech. Appl. Math.* **26**, 217 (1973).

5. JEFFREY, D. J., *Proc. Roy. Soc. (London) Ser A* **338**, 503 (1974).

6. LANDAU, L. D. AND LIFSCHITZ, E. M., "Fluid Mechanics," pp. 523–529. Pergamon Press, London, 1959.

7. FOX, R. F. AND UHLENBECK, G. E., Phys. Fluids **13**, 1893 (1970).

8. CHOW, T. S. AND HERMANS, J. J., *J. Chem. Phys.* **56**, 3150 (1972).

9. HAUGE, E. H. AND MARTIN-LOF, A., *J. Statist. Phys.* **7**, 259 (1972).

10. HI: CH, E. J., *J. Fluid Mech.* **72**, 499 (1975).

11. McMILLEN, T. J., Ph.D. thesis, California Institute of Tech., Pasadena, 1976.

12. HINCH, E. J. AND LEAL, L. G., *J. Fluid Mech.* **52**, 683 (1972).

13. TAYLOR, G. I., *Proc. Roy. Soc. (London) Ser. A* **138**, 41 (1932).

14. LEAL, L. G., *Chem. Eng. Commun.* **1**, 21 (1973).

15. McMILLEN, T. J. AND LEAL, L. G., *Int. J. Multiphase Fow* **2**, 105 (1975).

16. LEAL, L. G. AND McMILLEN, T. J., *Archives of Mechanics/Archiwum Mechaniki Stosowanej*, **28**, 483 (1976).

17. NIR, A. AND ACRIVOS, A., *J. Fluid Mech.*, in press.

18. JEFFREY, G. B., *Proc. Roy. Soc. (London) Ser. A* **102**, 161 (1922).

19. LEAL, L. G. AND HINCH, G. J., *J. Fluid Mech.* **46**, 685 (1971).

The Constrained Brownian Movement of Spherical Particles in Cylindrical Pores of Comparable Radius

Models of the Diffusive and Convective Transport of Solute Molecules in Membranes and Porous Media

HOWARD BRENNER AND LAWRENCE J. GAYDOS

Department of Chemical Engineering, Carnegie-Mellon University, Pittsburgh, Pennsylvania 15213

Received July 26, 1976

Rigorous calculations are presented for the effect of a constraining circular cylindrical boundary upon the translational Brownian motion of an isolated spherical particle suspended in a Poiseuille flow. These results also apply to dilute, multiparticle systems. Results are presented in the format of modifications of the Taylor–Aris theory of dispersion, arising from nonzero values of the dimensionless parameter λ = sphere radius/tube radius. Differences between present, continuum-mechanical, results and those derived from global forms of the thermodynamics of irreversible processes are reconciled.

An outline of a more general theoretical framework is presented, with a view toward eventual applications to nonspherical particles and noncylindrical boundaries, such as occur, for example, in problems of aerosol deposition.

Contents. I. Introduction. II. Transport of Homogeneous Spherical Particles in Circular Cylinders (Circular cylinder). III. Moments of the Distribution Function (Zeroth-order moment; First-order moment; Second-order moment). IV. Phenomenological Coefficients (Convective and diffusive fluxes; Nondimensional forms; The case of no external potential). V. Hydrodynamic Calculations for Relatively Small Spheres ($\lambda \ll 1$) (Calculation of $\bar{d}_{11}$; Calculation of $\bar{u}^*$; Calculation of $\bar{d}_v$). VI. Nonspherical Particles and More General Boundary Shapes. VII. Discussion (Comparison with the results of Anderson and Quinn; Comparison with the results of DiMarzio and Guttman; General discussion; Qualitative comparison with the experiments of Small). Appendix A (Uniqueness). Appendix B (Direct Derivation of $\bar{D}_{11}$ and $\bar{U}^*$). Appendix C (Critique of the Anderson–Quinn Definition of Diffusivity). Acknowledgments. References.

I. INTRODUCTION

This paper is addressed to the combined diffusive and convective transport of an isolated, rigid, spherical, neutrally buoyant, Brownian particle suspended in a Poiseuille flow. In contrast to conventional convective–diffusion analyses of such problems, the novel feature of the analysis resides in the fact that the radius a of the Brownian solute molecule is comparable in size to the tube radius r_o; that is,

$$\lambda \overset{\text{def}}{=} a/r_o = O(1). \qquad [1.1]$$

As a result of this circumstance, hydrodynamic particle/wall interactions play a dominant role in modifying the results of conventional analyses, e.g., the Taylor (56, 57)/Aris (3) theory of solute dispersion in capillaries. Moreover, incorporation into the analysis of nonhydrodynamic particle/wall interactions, arising from wall potentials (e.g., London, van

331

der Waals, Debye double layer, etc.), provides modifications of existing theory.

The single-particle analysis applies without change to dilute systems of noninteracting solute particles. As such, the results to be deduced possess potential applications in modeling biological and physiological membrane transport processes (1, 2, 4, 27, 43, 44, 46–49, 54), as well as in furnishing a rationale for the observed outcome of gel permeation– and hydrodynamic–chromatographic separation experiments (5, 23–25, 35, 55).

The class of problems to be undertaken here has been attempted before (2, 23–25, 35, 46–48). Anderson and Quinn (2) consider, in essence, the *steady-state* transport of Brownian solute molecules between two reservoirs joined by a circular capillary tube of length L, and maintained at different uniform concentrations. Their analysis appears unduly restrictive, limiting itself exclusively to macroscopically steady-state situations. As a result, without a formal proof of the type furnished in the present paper, it is not immediately apparent that their expression for the flux may be applied to macroscopically unsteady situations. Moreover, their theory, by departing from a strictly continuum-mechanical view, and adopting a stance peculiar to irreversible thermodynamics, introduces a somewhat ad hoc "steric partitioning factor" to take account of the fact that the solute concentrations in the two reservoirs are not the same as those existing within the entrance and exit regions, interior to the capillary. As will be demonstrated, the introduction of such artificial and extraneous factors will prove superfluous in our mode of analysis. Indeed, if one insists upon the use of such factors in the alternate mode of analysis employed by Anderson and Quinn, one may regard our treatment as furnishing the correct *definition* of such steric factors. This comment is especially pertinent in the case of nonspherical particles, or where wall potentials exist, for the use of such steric factors in these cases appears to suffer from a fundamental ambiguity,[1] especially in non-equilibrium circumstances.

In addition to these deficiencies, their hydrodynamic analysis admittedly fails to take account of the translational–rotational "coupling" which occurs for particles in the immediate proximity of the tube wall. Finally, their treatment of the convective transport term arising from the Poiseuille flow is incomplete. In particular, it fails to take proper account of the Taylor–Aris contribution to the solute "diffusion" (i.e., dispersion) term in their expression for the total flux, consisting of separate diffusive and convective contributions.

This same problem has also been studied by DiMarzio and Guttman (23–25, 35), although in a different context. However, their analysis fails to account properly for hydrodynamic wall effects, in both the quiescent and convective cases. Quantitative differences between present results and those of Anderson and Quinn (2) and DiMarzio and Guttman (23–25, 35) will be discussed in Section VII. Even apart from these quantitative differences,

[1] The ambiguity introduced by the use of *global* irreversible-thermodynamic (2, 4, 27, 43, 44, 46–49) concepts may be thought of as arising in the following manner: From a continuum-mechanical view, one is strictly concerned, not with the translational diffusion and convection of *particles*, but rather with the transport of "*locator points*" (8, 9, 14, 15, 20) locked into the particles. These points serve to represent the "position" of the particle in space. Necessarily, the physics of the problem must ultimately show itself to be invariant to the arbitrary choice made for the location of such body-fixed origins. (Even for spherical particles one is not obliged to choose the sphere center as its locator point. Geometrically, this choice merely represents the simplest one from the point of view of the requisite algebraic manipulations.) The framework of *global* irreversible thermodynamics, which stresses reservoirs situated a finite distance apart, rather than continuous, pointwise variations, does not appear to possess the versatility necessary to distinguish between the actual particles and their "locator" representations. Therein lies the source of the ambiguity. (On the other hand, *local* irreversible thermodynamics (20), which deals with continuously varying fields, does appear to possess sufficient generality of structure to encompass locator-point-dependent quantities.)

however, our analysis is philosophically and conceptually simpler, since it focuses attention on the statistical "trajectory" of a single Brownian particle. As a result, the conceptual and computational scheme to be developed has potential applications in a variety of related areas, including aerosol deposition, surface diffusion, and the like.

An outline of the general structure of the present paper now follows. Rather than considering a steady-state situation, involving the entrance and exit of Brownian particles from reservoirs situated a finite distance apart, we consider instead the movement of a single Brownian particle, already contained within an infinitely long capillary tube. Elimination of these reservoirs obviates the need for introduction of "steric factors." Beginning at zero time, $t = 0$, with the sphere center located at some arbitrary initial longitudinal position z', lateral position r', and angular position ϕ' [with (r, ϕ, z) a system of circular cylindrical coordinates originating along the tube axis], in Section II we derive a convective–diffusion equation governing the temporal evolution of the conditional probability density function $P(r, \phi, z, t | r', \phi', z')$ for finding the sphere center at position (r, ϕ, z) at time t. In turn, this is replaced by a simpler axisymmetric, probability density function $f(r, z, t | r', z')$, which, by itself, proves to be sufficiently detailed for the purposes of carrying out the proposed program.

Owing to hydrodynamic wall effects, the particle diffusivity, governing the intensity of the translational Brownian movement of the sphere center in a quiescent fluid, is no longer a scalar, as in unbounded fluids; rather, it becomes a (symmetric) second-rank tensor or dyadic. This tensorial property follows directly from the Stokes–Einstein equations as a consequence of the fact that the hydrodynamic resistance of the particle to axial motion (of the sphere center), parallel to the tube axis, is different than that for transverse motion, normal to the tube axis. Moreover, the axial and transverse diffusivities are not constants, but are, rather, functions of the radial position

r of the sphere center, this being a consequence of the variation of particle resistance with distance from the boundary walls. In addition to this wall effect, which arises even in quiescent fluids, the sphere center does not translate with the fluid in its proximity in the Poiseuille flow case. Rather, it lags the fluid by an amount which depends, inter alia, upon its transverse position r. Again, this radial-position dependence of the translational slip velocity is a consequence of the variation of particle resistance with distance from bounding walls.

Analysis of *unsteady* particle movements, described by the time-dependent solution of the partial differential equation governing f, eliminates the apparent restriction of the Anderson–Quinn (2) analysis to steady-state transport processes. For a collection of non-interacting particles, the *concentration* distribution $c(r, z; t)$, corresponding to any arbitrary axixymmetric initial solute distribution $c(r', z'; 0)$ (or arbitrary "boundary conditions"), can be obtained by superposing solutions of fundamental, instantaneous, point-source-type initial distributions,

$$c(r', z'; 0) \equiv (2\pi r')^{-1}\delta(r - r')\delta(z - z')\delta(t),$$

of the type under consideration, with δ the Dirac delta function. It therefore follows that the expressions ultimately obtained for the mean axial particle velocity $\bar{U}^*$ and mean axial particle dispersivity $\bar{D}^*$ are properties only of the solute–solvent system (as well as of the steady mean fluid velocity $\bar{V}_m$ and tube radius r_o)—as modified by the presence of the bounding walls—and are independent of whether or not a steady state prevails.

Fortunately, in order to derive expressions relating $\bar{U}^*$ and $\bar{D}^*$ to the system parameters, it is unnecessary to actually solve the differential equation for f. Rather, as shown in Section III, it suffices only to derive the first few lower-order axial moments $\overline{(z - z')^m}$ $(m = 0, 1, 2)$ of this distribution function, and then only in the "long-time" limit, $t \to \infty$. More precisely, the time interval from the start of the experiment must be of sufficient duration to permit the (center of the) Brownian particle to have sampled all accessible radial positions, $0 \leq r \leq r_o - a$, a sufficient number of times to render the distribution function f sensibly independent of the initial transverse position r' of the particle. An appropriate criterion of long times is developed. In anthropomorphic terms, this time interval must be such that the particle effectively loses all "memory" of its initial position r'. In such circumstances, the axial moments $\overline{[z(t) - z']^m}$ are effectively independent of r', and are readily determined by appropriate modifications of the Taylor–Aris (3, 56, 57) theory of dispersion in capillary tubes, as generalized by Horn (40). In turn, knowledge of these asymptotic moments permits the derivation of explicit expressions for $\bar{U}^*$ and $\bar{D}^*$ in terms of the fundamental parameters characterizing the system. Additionally, it is shown that these quantities, derived from a single-particle analysis, also represent the phenomenological coefficients appearing in the respective expressions for the dispersive and convective fluxes in dilute multiparticle systems composed of noninteracting solute particles.

Results expressing $\bar{U}^*$ and $\bar{D}^*$ in terms of the system parameters are summarized in Section IV, and applied specifically to the case where no wall potential exists. This is done in a convenient nondimensional form.

Explicit numerical results are computed in Section V for the case $\lambda \ll 1$, where the spheres are relatively small compared with the tube radius. This represents the only case where sufficiently complete hydrodynamic data are available to bring the requisite numerics to fruition. Methods of singular perturbation theory and matched asymptotic expansions are required here to distinguish the central core region of the tube, where the "method of reflections" furnishes a satisfactory asymptotic result for $\lambda \ll 1$, from the region in the immediate proximity of the tube wall, where the hydrodynamic solution of the sphere/plane–wall problem yields a different asymptotic functional dependence for $\lambda \ll 1$.

The comparable theory for the transport of nonspherical particles in tubes, and near boundaries in general, is developed in Section VI. Unfortunately, at present there are insufficient hydrodynamic data to bring the numerics to fruition in any complete sense. It is possible, however, to perform the requisite calculations correctly to terms of lowest order in λ for axisymmetric particles (such as spheroids) without requiring these additional data. These results will be reported elsewhere.

The paper concludes in Section VII with a comparison of the present results with those of prior investigators. Applications to membrane transport processes and chromatographic separation processes are discussed.

II. TRANSPORT OF HOMOGENEOUS SPHERICAL PARTICLES IN CIRCULAR CYLINDERS

As in Fig. 1, consider a single spherical Brownian particle of radius a suspended in an infinitely long circular tube of radius r_o, through which a Poiseuille flow is occurring at mean velocity $\bar{V}_m$. With $\mathbf{R} \equiv (r, \phi, z)$ a system of circular cylindrical coordinates possessing an origin along the cylinder axis, let $P \equiv P(\mathbf{R}, t | \mathbf{R}')$ denote the conditional probability density that at time t the sphere center lies at the position $\mathbf{R}$, given that at time $t = 0$ it was located at the position $\mathbf{R}' \equiv (r', \phi', z')$. The quantity $P d^3\mathbf{R}$, with volume element $d^3\mathbf{R} = r\,dr\,d\phi\,dz$, then denotes the probability of finding the sphere center

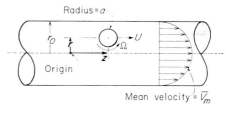

$$\lambda = a/r_0, \quad \beta = r/r_0$$

FIG. 1. Definition sketch. Force- and torque-free spherical particle suspended in a Poiseuille flow within a circular tube.

(at time t) somewhere within the volume $d^3\mathbf{R}$ centered about $\mathbf{R}$; that is, with its coordinates lying simultaneously in the ranges between r and $r + dr$, ϕ and $\phi + d\phi$, and z and $z + dz$.

Since the sphere center necessarily lies somewhere within the tube, P must satisfy the normalization condition

$$\int_{\mathbf{R}} P d^3\mathbf{R} = 1 \quad \text{for all } t > 0, \quad [2.1]$$

where the integration domain $\mathbf{R}$ corresponds to all accessible positions,

$$0 \leq r \leq r_o - a, \quad -\pi \leq \phi \leq \pi, \quad -\infty < z < \infty. \quad [2.2]$$

Explicitly,

$$\int_{z=-\infty}^{\infty} \int_{\phi=-\pi}^{\pi} \int_{r=0}^{r_o-a} P r \, dr \, d\phi \, dz = 1$$
$$\text{for all } t > 0. \quad [2.3]$$

Conservation and continuity of probability density requires that P satisfy the continuity equation (8, 9, 14, 15, 20)

$$(\partial P/\partial t) + \nabla \cdot \mathbf{J} = \delta(\mathbf{R} - \mathbf{R}')\delta(t)$$
$$\text{for } t > 0, \quad [2.4]$$

and

$$P = 0 \quad \text{for } t < 0, \quad [2.5]$$

with $\nabla \equiv \partial/\partial\mathbf{R}$; δ is the Dirac delta function, and $\mathbf{J} \equiv \mathbf{J}(\mathbf{R}, t \,|\, \mathbf{R}')$ the vector flux of probability density. The instantaneous source term appearing on the right-hand side of [2.4] expresses the fact that the particle was introduced into the fluid at time $t = 0$ at position $\mathbf{R}'$. That the coefficient of this source term is unity is demanded by the normalization condition [2.1].

The flux density vector is given by the constitutive relation (9, 14, 15)

$$\mathbf{J} = P\mathbf{U} - \mathbf{D} \cdot \nabla P + P\mathbf{M} \cdot \mathbf{F}. \quad [2.6]$$

The three terms appearing on the right respectively represent contributions to the flux arising from: (i) the convective particle flux, stemming from particle transport solely by the flowing fluid; (ii) the Brownian flux, arising from nonuniformities in the spatial distribution of sphere centers; (iii) the flux due to any external forces $\mathbf{F}$ exerted on the solute particles, e.g., by a wall potential.

It is important to recognize that Eqs. [2.4] and [2.5] focus attention upon the sphere center alone, and not, as such, upon the sphere as a particle. That is, these equations describe the motion of a single point. From a conceptual point of view, the situation can be visualized by imagining that the sphere is made of glass, possessing the same refractive index as the solvent, and that the sphere center is somehow distinguished by a visibly marked point. As the "transparent" sphere moves about, an experimentalist would then observe only the motion of this point. The sphere as a *finite* entity would be invisible to him.

The vector $\mathbf{U} \equiv \mathbf{U}(\mathbf{R})$ represents the hydrodynamic velocity that the neutrally buoyant sphere (center) possesses by virtue of the fluid motion alone, i.e., in the absence of Brownian motion and external forces. In general, because of wall effects and pressure gradients, this velocity will depart from the Poiseuille fluid velocity $\mathbf{V}(\mathbf{R})$ which would exist at the point $\mathbf{R}$ occupied by the sphere center if the sphere were not present in the flowing fluid (i.e., the "approach" velocity of the fluid).

Because of wall effects, the translational diffusivity dyadic $\mathbf{D} \equiv \mathbf{D}(\mathbf{R})$ of the sphere center will depend upon the radial position r of the latter. Indeed, owing to the infinite hydrodynamic resistance to translational motion of a sphere in contact with a wall—both for motion parallel (17, 18, 33) and perpendicular (7, 21) to the wall—this diffusivity necessarily goes to zero at $r = r_o - a$. From the Stokes–Einstein equation (8, 9), this translational diffusivity dyadic is related to the intrinsic, low Reynolds number, hydrodynamic resistance tensors for the sphere by the relation

$$\mathbf{D} = (kT/\mu)[{}^{t}\mathbf{K} - {}^{c}\mathbf{K}^{\dagger} \cdot {}^{r}\mathbf{K}^{-1} \cdot {}^{c}\mathbf{K}]^{-1}, \quad [2.7]$$

where k is the Boltzmann constant, T the absolute temperature, μ the solvent viscosity, and ${}^{t}\mathbf{K}$, ${}^{r}\mathbf{K}$, and ${}^{c}\mathbf{K}$, respectively, the translational, rotational, and coupling dyadics for motion of the sphere center. As a result of wall effects, these quantities vary with the transverse position r of the sphere center. Indeed, whereas the coupling dyadic $^c\mathbf{K}$ vanishes in an unbounded fluid, wherein wall effects are absent, this is no longer true in the presence of the cylindrical boundary. By the Nernst–Planck theorem, the mobility dyadic $\mathbf{M}$ is related to the diffusion dyadic by the expression (14, 15)

$$\mathbf{M} = \mathbf{D}/kT. \qquad [2.8]$$

It will be assumed in the sequel that any external forces exerted on the solute particle derive from a potential energy function $E \equiv E(\mathbf{R})$, such that the external force is given by

$$\mathbf{F}/kT = -\nabla E. \qquad [2.9]$$

The extraneous factor of kT conveniently serves to render E dimensionless. In these circumstances, the flux $\mathbf{J}$ can be written as

$$\mathbf{J} = \mathbf{U}P - \mathbf{D} \cdot (\nabla P + P\nabla E), \qquad [2.10]$$

or, what is equivalent,

$$\mathbf{J} = \mathbf{U}P - e^{-E}\mathbf{D} \cdot \nabla(Pe^E). \qquad [2.11]$$

The latter equation combines with [2.4] to yield

$$\partial P/\partial t + \nabla \cdot (P\mathbf{U}) = \nabla \cdot [e^{-E}\mathbf{D} \cdot \nabla(Pe^E)] \\ + \delta(\mathbf{R} - \mathbf{R}')\delta(t). \qquad [2.12]$$

This equation applies to homogeneous spherical particles near any bounding wall, and is not limited to transport in circular tubes. "Homogeneous" means that the external force on the sphere is independent of the sphere orientation, such as could otherwise arise, for example, in the case of dipolar (11, 16) spherical particles containing an embedded multipole.

Circular cylinder. As in Figs. 2a and 2b, the geometric symmetry of the sphere/cylinder combination is such that the joint configuration possesses two mutually perpendicular symmetry planes. From the comparable symmetry investigation of Happel and Brenner (37; cf. Eqs. [5.5]–[5.15] of that book), for a configuration possessing this symmetry (with Ox_1–Ox_2 and Ox_3–Ox_1 as the planes of reflection symmetry), the various resistance dyadics—which depend only upon the geometry of the

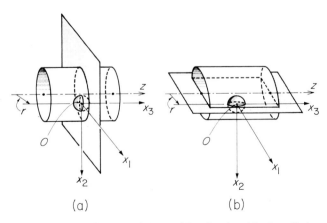

(a)   (b)

FIG. 2. Symmetry planes for the eccentrically positioned, sphere/circular-cylinder configuration. This configuration possesses, inter alia, two mutually perpendicular planes of reflection symmetry, each containing the center O of the sphere. Figure 2a shows the reflection plane, $x_3 = $ const (i.e., the Ox_1–Ox_2 plane), normal to the longitudinal cylinder axis. Figure 2b shows the meridonal reflection plane, $x_2 = $ const (i.e., the Ox_3–Ox_1 plane), containing the longitudinal axis of the cylinder.

system[2]—possess the representations

$$^t\mathbf{K} = \mathbf{i}_1\mathbf{i}_1{}^tK_{11} + \mathbf{i}_2\mathbf{i}_2{}^tK_{22} + \mathbf{i}_3\mathbf{i}_3{}^tK_{33}, \quad [2.13a]$$

$$^r\mathbf{K} = \mathbf{i}_1\mathbf{i}_1{}^rK_{11} + \mathbf{i}_2\mathbf{i}_2{}^rK_{22} + \mathbf{i}_3\mathbf{i}_3{}^rK_{33}, \quad [2.13b]$$

$$^c\mathbf{K} = \mathbf{i}_2\mathbf{i}_3{}^cK_{23} + \mathbf{i}_3\mathbf{i}_2{}^cK_{32}. \quad [2.13c]$$

It therefore follows from [2.7] that

$$\mathbf{D} = \mathbf{i}_1\mathbf{i}_1 D_{11} + \mathbf{i}_2\mathbf{i}_2 D_{22} + \mathbf{i}_3\mathbf{i}_3 D_{33}, \quad [2.14]$$

in which

$$D_{11} = (kT/\mu)^tK_{11}{}^{-1}, \quad [2.15a]$$

$$D_{22} = \frac{kT}{\mu}\left(^tK_{22} - \frac{^cK_{32}{}^2}{^rK_{33}}\right)^{-1}, \quad [2.15b]$$

$$D_{33} = \frac{kT}{\mu}\left(^tK_{33} - \frac{^cK_{23}{}^2}{^rK_{22}}\right)^{-1}. \quad [2.15c]$$

The Cartesian coordinates (x_1, x_2, x_3) represent a *local* coordinate system. These coordinates can be given global stature in terms of the circular cylindrical coordinate system (r, ϕ, z) by choosing

$$x_1 \equiv r, \quad x_2 \equiv \phi, \quad x_3 \equiv z. \quad [2.16]$$

With this choice, [2.14] can be written as

$$\mathbf{D} = \mathbf{i}_r\mathbf{i}_r D_\perp(r) + \mathbf{i}_\phi\mathbf{i}_\phi D_\phi(r) + \mathbf{i}_z\mathbf{i}_z D_{\shortparallel}(r), \quad [2.17]$$

in which

$$D_\perp \equiv D_{11}, \quad D_\phi \equiv D_{22}, \quad D_{\shortparallel} \equiv D_{33}. \quad [2.18]$$

Here, $(\mathbf{i}_r, \mathbf{i}_\phi, \mathbf{i}_z)$ are unit vectors in (r, ϕ, z) coordinates. The argument r appended to the coefficients D_i in [2.17] serves to indicate explicitly that these scalar translational diffusivities are, inter alia, functions of the distance r of the sphere center from the tube axis. More explicitly,

$$D_i/D_\infty = \text{function}(\beta, \lambda), \quad [2.19]$$

where

$$\beta = r/r_o, \quad [2.20]$$

[2] The sphere/cylinder combination is also invariant under rotation through 180° about the Ox_1 axis, the latter being the line of intersection of the two symmetry planes. However, from (37, p. 190), upon setting $\theta = 180°$, no additional restrictions arise among the coefficients $^{(\,)}K_{ij}$. In particular, no relation subsists between $^cK_{23}$ and $^cK_{32}$.

with λ as defined in [1.1], and

$$D_\infty = kT/6\pi\mu a, \quad [2.21]$$

the diffusivity of the sphere center in an unbounded fluid.

The undisturbed Poiseuille velocity field is given by

$$\mathbf{V} = \mathbf{i}_z V(r), \quad [2.22]$$

with

$$V = 2\bar{V}_m[1 - (r/r_o)^2]. \quad [2.23]$$

Since the center of a neutrally buoyant sphere suspended in a Poiseuille flow translates along a streamline of the undisturbed flow (10, 12, 17, 18, 37), parallel to the tube axis, its vector velocity necessarily takes the form

$$\mathbf{U} = \mathbf{i}_z U(r), \quad [2.24]$$

where

$$U/\bar{V}_m = \text{function}(\beta, \lambda). \quad [2.25]$$

Furthermore, attention will be confined to the case where the potential energy function E varies only with radial position, as would arise, for example, if the wall exerted an attractive or repulsive force on the sphere. Thus, $E \equiv E(r)$.

With these simplifications, [2.12] reduces to

$$\frac{\partial P}{\partial t} + U(r)\frac{\partial P}{\partial z}$$

$$= \frac{1}{r}\frac{\partial}{\partial r}\left[re^{-E(r)}D_\perp(r)\frac{\partial}{\partial r}(e^{E(r)}P)\right]$$

$$+ \frac{1}{r^2}D_\phi(r)\frac{\partial^2 P}{\partial\phi^2} + D_{\shortparallel}(r)\frac{\partial^2 P}{\partial z^2}$$

$$+ r^{-1}\delta(r - r')\delta(\phi - \phi')\delta(z - z')\delta(t),$$

$$(t > 0). \quad [2.26]$$

The above expression for the Dirac delta function in cylindrical coordinates is given by Friedman (29) and Morse and Feshbach (50).

In addition to the conditions already satisfied, $P \equiv P(r, z - z', \phi - \phi', t|r')$ is also required to satisfy the additional physical restrictions

$$P \to 0 \quad \text{as } |z - z'| \to \infty, \quad [2.27]$$

and

$$J_r = 0 \quad \text{at} \quad r = r_o - a, \qquad [2.28]$$

where

$$J_r = \mathbf{i}_r \cdot \mathbf{J} \qquad [2.29]$$

is the flux component normal to the wall. The latter condition expresses the impenetrability of the tube wall to passage by the sphere. From [2.10] and [2.24] the latter is equivalent to

$$D_\perp(\partial P/\partial r + PdE/dr) = 0$$
$$\text{at} \quad r = r_o - a. \qquad [2.30]$$

As will appear in Section V (cf. the footnote immediately preceding Eq. [5.94]), $D_\perp$ itself vanishes at $r = r_o - a$. Thus, the above flux condition will *automatically* be satisfied, provided that the term in parentheses is either finite at $r = r_o - a$, or —if it does become infinite at this radius—that it does so at a slower rate with r than $D_\perp$ goes to zero. Depending upon the explicit dependence of E on r, this may or may not impose restrictions upon the asymptotic behavior of P near the wall.[3]

As might have been anticipated by the obvious physical irrelevance of the angle ϕ, subsequent calculations do not require knowledge of P itself, but rather only of the conditional (axisymmetric) probability density function

$$f \overset{\text{def}}{=} (2\pi)^{-1} \int_{-\pi}^{\pi} P d\phi. \qquad [2.31]$$

Inasmuch as P is necessarily single-valued with respect to the angle ϕ, it readily follows upon integration of [2.26] and [2.5] that $f \equiv f(r, z, t \mid r', z') \equiv f(r, z - z', t \mid r')$ satisfies

[3] Suppose, for example, that near to the wall, E is dominated by a repulsive potential of the form $E = (\text{const})\epsilon^{-n}$ $(n > 0)$, where $\epsilon = [(r_o - a) - r]/a \ll 1$ is the dimensionless gap width. Since, from [5.94] and [5.95], $D_\perp$ goes to zero linearly with ϵ, Eq. [2.30] requires that, as $\epsilon \to 0$, P go to zero more rapidly than ϵ^n.

the differential equation

$$\frac{\partial f}{\partial t} + U(r) \frac{\partial f}{\partial z}$$

$$= \frac{1}{r} \frac{\partial}{\partial r} \left[r e^{-E(r)} D_\perp(r) \frac{\partial}{\partial r} (e^{E(r)} f) \right]$$

$$+ D_{11}(r) \frac{\partial^2 f}{\partial z^2} + (2\pi r)^{-1} \delta(r - r') \delta(z - z') \delta(t)$$

$$\text{for } t > 0, \qquad [2.32]$$

and

$$f = 0 \quad \text{for } t < 0. \qquad [2.33]$$

Furthermore, from [2.3], f satisfies the normalization condition

$$2\pi \int_{z=-\infty}^{\infty} \int_{r=0}^{r_o-a} f r dr dz = 1 \quad (t > 0). \qquad [2.34]$$

Additionally, from [2.27] and [2.28], it is also required that

$$f \to 0 \quad \text{as} \quad |z - z'| \to \infty, \qquad [2.35]$$

and

$$j_r = 0 \quad \text{at } r = r_o - a, \qquad [2.36]$$

in which $j_r = \mathbf{i}_r \cdot \mathbf{j}$, with

$$\mathbf{j} \overset{\text{def}}{=} (2\pi)^{-1} \int_{-\pi}^{\pi} \mathbf{J} d\phi, \qquad [2.37]$$

the flux density vector of the probability distribution f. Explicitly (cf. [2.11]),

$$\mathbf{j} = U f - e^{-E} \mathbf{D} \cdot \nabla(f e^E), \qquad [2.38]$$

with $\mathbf{D}$ given by [2.17] with the $\mathbf{D}_\phi$ term suppressed. Thus, the requirement [2.36] is equivalent to

$$D_\perp(r)(\partial f/\partial r + f dE/dr) = 0$$
$$\text{at } r = r_o - a. \qquad [2.39]$$

In regard to this boundary condition, the discussion following [2.30] is equally applicable in present circumstances, with f appearing in place of P.

As discussed in Appendix A, the system of Eqs. [2.32] to [2.36] possesses a unique solution. The normalization condition [2.34] is not strictly required for uniqueness, since it is already implicit in the delta function source term appearing in [2.32], when considered in conjunction with the "no-flux" conditions [2.35] and [2.36].

It readily follows from the sequences of equations defining f that the latter depends not on z and z' separately, but only upon the linear combination $z - z'$; that is,

$$f \equiv f(z - z', r, t | r'). \qquad [2.40]$$

Because of this fact, no loss in generality arises from choosing the origin of the coordinate system (r, ϕ, z) to lie along the axis $r = 0$, and to there coincide with the initial axial position of the sphere center at zero time. This is tantamount to choosing

$$z' = 0, \qquad [2.41]$$

whence f may now be regarded as being of the functional form

$$f \equiv f(z, r, t | r'). \qquad [2.42]$$

III. MOMENTS OF THE DISTRIBUTION FUNCTION

In principle, the system of differential equations and boundary conditions governing P can be solved. Of special interest are the lower-order moments of the axial position z of the sphere center, namely,

$$\bar{z} \stackrel{\text{def}}{=} \int_{z=-\infty}^{\infty} \int_{\phi=-\pi}^{\pi} \int_{r=0}^{r_o-a} z P r \, dr \, d\phi \, dz \Big/$$

$$\int_{z=-\infty}^{\infty} \int_{\phi=-\pi}^{\pi} \int_{r=0}^{r_o-a} P r \, dr \, d\phi \, dz, \qquad [3.1]$$

and

$$\overline{(z - \bar{z})^2}$$

$$\stackrel{\text{def}}{=} \int_{z=-\infty}^{\infty} \int_{\phi=-\pi}^{\pi} \int_{r=0}^{r_o-a} (z - \bar{z})^2 P r \, dr \, d\phi \, dz \Big/$$

$$\int_{z=-\infty}^{\infty} \int_{\phi=-\pi}^{\pi} \int_{r=0}^{r_o-a} P r \, dr \, d\phi \, dz. \qquad [3.2]$$

In view of the normalization condition [2.3] and the definition [2.31], these may be written alternatively in terms of f as

$$\bar{z} = 2\pi \int_{z=-\infty}^{\infty} \int_{r=0}^{r_o-a} z f r \, dr \, dz, \qquad [3.3]$$

and

$$\overline{(z - \bar{z})^2} = 2\pi \int_{z=-\infty}^{\infty} \int_{r=0}^{r_o-a} (z - \bar{z})^2 f r \, dr \, dz. \qquad [3.4]$$

As a consequence of the functional dependence of f upon the independent variables, expressed by [2.42], each of these moments depends upon the time t and the initial radial position r' of the sphere center. On physical grounds it is natural to expect that, after the elapse of sufficient time from the commencement of the experiment, these axial moments will become independent of r'. That is, for t sufficiently large, such that the inequality

$$D_o t / r_o^2 \gg 1 \qquad [3.5]$$

is satisfied, the Brownian particle will have sampled all radial positions r many times during the course of its concomitant axial travel, and hence have lost "memory" of its initial transverse position r'. Consequently, it is appropriate to anticipate that, at long times, $\bar{z}$ will possess the (r'-independent) asymptotic form

$$\bar{z} \sim \bar{U}^* t, \qquad [3.6]$$

with $\bar{U}^*$ a constant. The latter is to be identified as the mean axial velocity of the Brownian particle. Thus, we have that

$$\bar{U}^* = \lim_{t \to \infty} d\bar{z}/dt, \qquad [3.7]$$

wherein the limiting process $t \to \infty$ is meant to be understood in the sense of [3.5].

Similarly, from Einstein's relation for the mean-square displacement of a Brownian particle occurring during the time interval t, it may be anticipated that

$$\overline{(z - \bar{z})^2} \sim 2\bar{D}^* t, \qquad [3.8]$$

where the constant coefficient $\bar{D}^*$ is to be identified as the mean axial dispersivity of the suspended particle.[4] Thus,

$$\bar{D}^* = \tfrac{1}{2} \lim_{t \to \infty} d\overline{(z - \bar{z})^2}/dt. \qquad [3.9]$$

The asymptotic forms [3.6] and [3.8] will now be shown to be correct, in the sense that $\bar{z}$ and $\overline{(z - \bar{z})^2}$ will each be shown to vary linearly with t for sufficiently long times. Simultaneously, these demonstrations furnish explicit formulas for calculating $\bar{U}^*$ and $\bar{D}^*$ from the fundamental parameters characterizing the system. Such calculations constitute appropriate extensions of the Taylor–Aris theory to the case where wall effects and wall potentials are sensible.

Define the axial moments,

$$\mu_m(r, t \mid r')$$

$$= \int_{-\infty}^{\infty} z^m f dz \quad (m = 0, 1, 2, \ldots), \quad [3.10]$$

as well as the total moments,

$$M_m(t \mid r')$$

$$= 2\pi \int_0^{r_0 - a} \mu_m r dr \quad (m = 0, 1, 2, \ldots). \quad [3.11]$$

It will be supposed that

$$z^m f \to 0 \quad \text{as} \quad |z| \to \infty \qquad [3.12]$$

to secure convergence of the former integrals.[5] In view of [3.10], Eq. [3.11] is equivalent to

$$M_m = 2\pi \int_{z=-\infty}^{\infty} \int_{r=0}^{r_0 - a} z^m f r dr dz, \qquad [3.13]$$

[4] In the case of a quiescent fluid, wherein $\bar{U}^* = 0$ and $\bar{z} = 0$, $\bar{D}^*$ simply represents the mean axial molecular diffusivity, $\bar{D}_{11}$ (cf. [4.33]).

[5] Subsequent calculations show that this condition is met, since, at least for long times, it can be demonstrated that

$$f(z, r, t \mid r') \sim e^{-E(r)} (4\pi \bar{D}^* t)^{-\frac{1}{2}}$$
$$\times \exp[-(z - \bar{U}^* t)^2 / 4\bar{D}^* t],$$

independently of r', wherein $\bar{D}^*$ and $\bar{U}^*$ are constants.

upon interchanging the order of the integrations. In particular, for $m = 0$ it follows from [2.34] that

$$M_0 = 1 \quad \text{for all } t > 0, \qquad [3.14]$$

independently of the value of r'.

The differential equation satisfied by the moments μ_m may be derived by noting from [3.10] that

$$\partial \mu_m / \partial t = \int_{-\infty}^{\infty} z^m (\partial f / \partial t) dz.$$

With use of [2.32] this becomes, for $t > 0$,

$$\partial \mu_m / \partial t = -U(r) \int_{-\infty}^{\infty} z^m (\partial f / \partial z) dz$$

$$+ L_r \int_{-\infty}^{\infty} z^m f dz + D_{11}(r) \int_{-\infty}^{\infty} z^m (\partial^2 f / \partial z^2) dz$$

$$+ (2\pi r)^{-1} \delta_{m0} \delta(r - r') \delta(t),$$

in which

$$\delta_{m0} = 1 \quad \text{for } m = 0,$$
$$= 0 \quad \text{for } m \neq 0,$$

is the Kronecker delta, and L_r is the linear partial differential operator

$$L_r \psi = -\frac{1}{r} \frac{\partial}{\partial r} \left[r e^{-E(r)} D_\perp(r) \frac{\partial}{\partial r} (e^{E(r)} \psi) \right], \qquad [3.15]$$

for any function ψ. Integration by parts gives

$$\int_{-\infty}^{\infty} z^m (\partial f / \partial z) dz = z^m f \Big]_{z=-\infty}^{\infty} - m \int_{-\infty}^{\infty} z^{m-1} f dz.$$

Equation [3.12] shows that the first term on the right vanishes at both limits. Hence, with use of [3.10], it follows that

$$\int_{-\infty}^{\infty} z^m (\partial f / \partial z) dz = -m \mu_{m-1}. \qquad [3.16]$$

Similarly,

$$\int_{-\infty}^{\infty} z^m (\partial^2 f / \partial z^2) dz = m(m-1) \mu_{m-2} \qquad [3.17]$$

In combination, the above results show that μ_m satisfies the partial differential equation (for $t > 0$)

$$\partial \mu_m / \partial t - L_r \mu_m$$
$$= mU(r)\mu_{m-1} + m(m-1)D_{11}(r)\mu_{m-2}$$
$$+ (2\pi r)^{-1}\delta_{m0}\delta(r-r')\delta(t)$$
$$(m = 0, 1, 2, \ldots). \quad [3.18]$$

In particular, the first few moments satisfy

$$\partial \mu_0 / \partial t - L_r \mu_0 = \delta(r-r')\delta(t)/2\pi r, \quad [3.19]$$

$$\partial \mu_1 / \partial t - L_r \mu_1 = U(r)\mu_0, \quad [3.20]$$

$$\partial \mu_2 / \partial t - L_r \mu_2 = 2U(r)\mu_1 + 2D_{11}(r)\mu_0. \quad [3.21]$$

In addition, from [2.36] and [2.38], each μ_m is required to satisfy the "no-flux" boundary condition,

$$j_m = 0 \quad \text{at } r = r_o - a$$
$$(m = 0, 1, 2, \ldots), \quad [3.22]$$

where

$$j_m \overset{\text{def}}{=} -e^{-E(r)}D_\perp(r)(\partial/\partial r)(e^{E(r)}\mu_m) \quad [3.23]$$

represents the radial "flux" of the mth moment "substance." Moreover, from [2.33] and [3.10] there follows the requirement

$$\mu_m = 0 \quad \text{for } t < 0. \quad [3.24]$$

For $t > 0$, the above sequence of equations can, in principle, be solved recursively for the axial moments $\mu_m(r, t|r')$ within the circular region

$$0 \leq r \leq r_o - a. \quad [3.25]$$

The general methods of Appendix A may be employed to show that this system of equations possesses a unique solution for each m.

Zeroth-order moment. Upon setting $m = 0$, the first of these equations is seen to be isomorphic with the problem of nonconvective diffusion within the region [3.25], in the presence of the potential energy function $E(r)$. In terms of the flux j_m defined by [3.23], Eq. [3.19] adopts the form

$$\frac{\partial \mu_0}{\partial t} + \frac{1}{r}\frac{\partial}{\partial r}(rj_0) = \frac{\delta(r-r')\delta(t)}{2\pi r}.$$

Because of the no-flux condition, $j_0 = 0$ at $r = r_o - a$, a steady state is ultimately obtained, in which the zeroth moment "substance," initially introduced locally into the system at $r = r'$ by the source term in the above equation, is exponentially distributed over the circular region [3.25]. This steady-state solution corresponds to $j_0 = 0$ everywhere. From [3.23] this requires that the steady-state axial moment be of the form

$$\mu_0(r) = (K/2\pi)e^{-E(r)}, \quad [3.26]$$

with K a constant to be determined. To calculate the latter, introduce [3.26] into [3.11] with $m = 0$, and compare the result with [3.14]. This yields

$$1/K = \int_0^{r_o-a} e^{-E(r)}r\,dr. \quad [3.27]$$

It will now be demonstrated that the rate of approach to this steady state is exponentially rapid with time; that is, the solution of the *unsteady*-state equation [3.19] is asymptotically of the form

$$\mu_0 \sim (K/2\pi)e^{-E(r)} + \text{exp}, \quad [3.28]$$

where "exp" denotes terms that go to zero exponentially rapidly as $t \to \infty$. To prove this, we will write

$$\mu_0(r, t|r')$$
$$= e^{-E(r)}\Big[(K/2\pi) + \sum_{n=1}^{\infty} y_n(r|r')e^{\lambda_n t}\Big] \quad [3.29]$$

for the exact solution of [3.19] when $t > 0$, and demonstrate that the eigenvalues satisfy the inequality $\lambda_n < 0$. Substitute the above into [3.19]. For $t \neq 0$, the eigenvalues λ_n and eigenfunctions $y_n(r)$ are then determined by the solution of the total differential equation (for a fixed value of r'),

$$\lambda_n e^{-E} y_n - \frac{1}{r}\frac{d}{dr}\Big(re^{-E}D_\perp \frac{dy_n}{dr}\Big) = 0. \quad [3.30]$$

Moreover, from [3.22], [3.23], and [3.29], y_n has to satisfy the boundary condition

$$e^{-E}D_\perp dy_n/dr = 0 \quad \text{at } r = r_o - a. \quad [3.31]$$

Multiply the former by $ry_n dr$ and integrate over the region [3.25] to obtain

$$\lambda_n \int_0^{r_o-a} re^{-E} y_n^2 dr + \int_0^{r_o-a} re^{-E} D_\perp (dy_n/dr)^2 dr$$
$$- ry_n e^{-E} D_\perp (dy_n/dr) \Big]_{r=0}^{r_o-a} = 0,$$

upon integration by parts. The last term clearly vanishes at the lower limit, $r = 0$. Furthermore, [3.31] shows that it vanishes at the upper limit too. Consequently,

$$\lambda_n = - \int_0^{r_o-a} re^{-E} D_\perp (dy_n/dr)^2 dr \Big/$$
$$\int_0^{r_o-a} re^{-E} y_n^2 dr. \quad [3.32]$$

Each of the above integrands is nonnegative. This leads to the conclusion that $\lambda_n \leq 0$, in which the equality sign holds only for the case where $y_n = \text{const} = K/2\pi$, say. But this special case corresponds to the steady-state solution [3.26]. Accordingly, it follows that $\lambda_n < 0$, thereby demonstrating the contention [3.28].

First-order moment. Substitution of [3.28] into [3.20] yields the differential equation

$$\partial\mu_1/\partial t - L_r \mu_1$$
$$= (K/2\pi)e^{-E(r)} U(r) + \exp, \quad [3.33]$$

governing transport of the first-moment "substance." From [3.22] and [3.23], with $m = 1$, μ_1 is required to satisfy the no-flux boundary condition,

$$e^{-E(r)} D_\perp(r) (\partial/\partial r)(e^{E(r)} \mu_1) = 0$$
$$\text{at } r = r_o - a. \quad [3.34]$$

Formally, these equations correspond to a nonconvective, radial diffusion problem in the circular region [3.25], with a position-dependent source term, $(K/2\pi)e^{-E} U$, and a no-flux boundary condition on the cylinder "wall." Consequently, μ_1 must increase with time. It will therefore be assumed, subject to a posteriori verification, that μ_1 is asymp-

totically of the form

$$\mu_1 \sim (K/2\pi)e^{-E(r)} [\bar{U}t + B(r)] + \exp, \quad [3.35]$$

in which $\bar{U}$ and $B(r)$ are, respectively, a constant and function to be determined.

Introduction of the latter equation into [3.33] yields

$$\bar{U}e^{-E(r)} - \frac{1}{r} \frac{d}{dr} \left[re^{-E(r)} D_\perp(r) \frac{dB(r)}{dr} \right]$$
$$= U(r)e^{-E(r)}. \quad [3.36]$$

From [3.34] and [3.35], $B(r)$ has to satisfy the boundary condition

$$e^{-E(r)} D_\perp(r) dB/dR = 0 \quad \text{at } r = r_o - a. \quad [3.37]$$

In order to determine the value of the constant $\bar{U}$, multiply both sides of [3.36] by rdr and integrate over the region [3.25]. Since the term in square brackets in [3.36] vanishes at both limits of integration, it thereby follows that

$$\bar{U} = \int_0^{r_o-a} U(r)e^{-E(r)} rdr \Big/$$
$$\int_0^{r_o-a} e^{-E(r)} rdr. \quad [3.38]$$

The first integral of [3.36] for the function $B(r)$ is

$$re^{-E(r)} D_\perp(r)(dB/dr)$$
$$= \int_{r'=0}^{r} [\bar{U} - U(r')] e^{-E(r')} r' dr', \quad [3.39]$$

with r' a dummy variable of integration, not to be confused with the initial radial position of the source. A second integration then gives

$$B(r) = \int_{r''=0}^{r} \frac{dr''}{r'' e^{-E(r'')} D_\perp(r'')}$$
$$\times \int_{r'=0}^{r''} [\bar{U} - U(r')] e^{-E(r')} r' dr'$$
$$+ \text{const.} \quad [3.40]$$

The numerical value of the constant of integration will prove irrelevant in the subsequent theory.

That it has proved possible to obtain the constant $\bar{U}$ and function $B(r)$ provides a posteriori confirmation of the validity of the a priori assumption [3.35].

Introduction of [3.35] into [3.11], followed by use of [3.27], yields

$$M_1(t) \sim \bar{U}t + C + \exp, \qquad [3.41]$$

where C is the time-independent constant

$$C = K \int_0^{r_0-a} e^{-E(r)} B(r) r dr. \qquad [3.42]$$

Second-order moment. The axial moment μ_2 can be obtained by a procedure analogous to that employed above. However, as will now be demonstrated, the total moment M_2 can be calculated directly, solely from knowledge of μ_0 and μ_1. And it is only M_2 which is required in the subsequent theory, not μ_2 itself.

Differentiate [3.11] with respect to time and employ [3.18] to obtain

$$(2\pi)^{-1} dM_m(t)/dt = \int_0^{r_0-a} L_r \mu_m r dr$$

$$+ m \int_0^{r_0-a} U(r)\mu_{m-1} r dr$$

$$+ m(m-1) \int_0^{r_0-a} D_{11}(r)\mu_{m-2} r dr$$

$$+ \delta_{m0}\delta(t)/2\pi, \qquad [3.43]$$

for a fixed value of the initial position r'. However, from [3.15], there follows upon integration,

$$\int_0^{r_0-a} L_r \mu_m r dr$$

$$= \left[re^{-E(r)} D_\perp(r)(\partial/\partial r)(e^{E(r)}\mu_m) \right]_{r=0}^{r_0-a}$$

$$= 0. \qquad [3.44]$$

Vanishing of the bracketed term at the upper limit of integration is a consequence of the boundary condition [3.22]–[3.23]. Consequently,

$$(2\pi)^{-1} dM_m/dt = m \int_0^{r_0-a} U(r)\mu_{m-1} r dr$$

$$+ m(m-1) \int_0^{r_0-a} D_{11}(r)\mu_{m-2} r dr$$

$$+ \delta_{m0}\delta(t)/2\pi. \qquad [3.45]$$

This reveals, in general, that M_m can be computed solely from knowledge of the lower-order axial moments μ_{m-1} and μ_{m-2}. In particular, explicit knowledge of μ_m is not required for its determination.

For $m = 0$ this makes

$$dM_0/dt = \delta(t)$$

Integration of this gives

$$M_0 = 1 \quad \text{for all} \quad t > 0,$$

in agreement with [3.14].

For $m = 1$ we obtain

$$dM_1/dt = 2\pi \int_0^{r_0-a} U(r)\mu_0 r dr.$$

With use of [3.28] and [3.27] this becomes

$$dM_1/dt = \bar{U} + \exp, \qquad [3.46]$$

with $\bar{U}$ given by [3.38]. Integration yields

$$M_1 = \bar{U}t + \text{const} + \exp,$$

in accord with [3.41].

When $m = 2$, the result is

$$(2\pi)^{-1} dM_2/dt = 2 \int_0^{r_0-a} U(r)\mu_1 r dr$$

$$+ 2 \int_0^{r_0-a} D_{11}(r)\mu_0 r dr.$$

With the aid of [3.28] and [3.35], this may be written explicitly as

$$dM_2/dt = 2K\bar{U}t \int_0^{r_0-a} e^{-E(r)} U(r) r dr$$

$$+ 2K \int_0^{r_0-a} e^{-E(r)} U(r) B(r) r dr$$

$$+ 2K \int_0^{r_0-a} e^{-E(r)} D_{11}(r) r dr + \text{exp.}$$

Use of [3.38] and [3.27] thereby yields

$$dM_2/dt = 2\bar{U}^2t + 2K \int_0^{r_0-a} e^{-E(r)} U(r) B(r) r dr$$

$$+ 2\bar{D}_{11} + \text{exp,} \quad [3.47]$$

wherein

$$\bar{D}_{11} \stackrel{\text{def}}{=} \int_0^{r_0-a} e^{-E(r)} D_{11}(r) r dr \Big/$$

$$\int_0^{r_0-a} e^{-E(r)} r dr. \quad [3.48]$$

In order to put the remaining integral in [3.47] into a form which will prove useful in subsequent calculations, multiply [3.36] by $B(r) r dr$, integrate over the region [3.25], and integrate the resulting expression by parts to obtain

$$\int_0^{r_0-a} e^{-E(r)} U(r) B(r) r dr$$

$$= \bar{U} \int_0^{r_0-a} e^{-E(r)} B(r) r dr$$

$$+ \int_0^{r_0-a} r e^{-E(r)} D_\perp(r) (dB/dr)^2 dr$$

$$- \left[r e^{-E(r)} D_\perp(r) B(r) dB/dr \right]_{r=0}^{r_0-a}.$$

The bracketed term clearly vanishes at the lower limit. It vanishes too at the upper limit in consequence of the boundary condition [3.37]. Hence, with use of [3.42] in the first term on the right-hand side of the preceding expression, Eq. [3.47] adopts the form

$$dM_2/dt = 2\bar{U}^2t + 2\bar{U}C$$

$$+ 2(\bar{D}_{11} + \bar{D}_v) + \text{exp,} \quad [3.49]$$

in which

$$\bar{D}_v = \int_0^{r_0-a} e^{-E(r)} D_\perp(r) (dB/dr)^2 r dr \Big/$$

$$\int_0^{r_0-a} e^{-E(r)} r dr.$$

Upon using [3.39], the latter expression can be written more explicitly as

$$\bar{D}_v = \int_0^{r_0-a} \frac{H^2(r) dr}{r e^{-E(r)} D_\perp(r)} \Big/ \int_0^{r_0-a} e^{-E(r)} r dr, \quad [3.50a]$$

in which

$$H(r) = \int_{r'=0}^{r} [\bar{U} - U(r')] e^{-E(r')} r' dr'. \quad [3.50b]$$

Integration of [3.49] gives

$$M_2 = \bar{U}^2 t^2 + 2\bar{U}Ct + 2(\bar{D}_{11} + \bar{D}_v)t$$

$$+ \text{const} + \text{exp.} \quad [3.51]$$

The numerical value of the constant term in the above expression is not required for our purposes.

IV. PHENOMENOLOGICAL COEFFICIENTS

With the moments M_0, M_1, and M_2 now determined, the quantities $\bar{U}^*$ and $\bar{D}^*$ can be calculated. Comparison of [3.3] with [3.13] shows that

$$\bar{z} = M_1. \quad [4.1]$$

Similarly, beginning with [3.4], it readily follows that

$$\overline{(z - \bar{z})^2} = M_2 - 2M_1\bar{z} + \bar{z}^2$$

$$= M_2 - M_1^2, \quad [4.2]$$

upon employing [4.1]. Accordingly, from [3.7] and [3.9],

$$\bar{U}^* = \lim_{t \to \infty} dM_1/dt, \quad [4.3a]$$

344

and

$$\bar{D}^* = \frac{1}{2} \lim_{t \to \infty} \left(\frac{dM_2}{dt} - 2M_1 \frac{dM_1}{dt} \right). \quad [4.3b]$$

From the former, in conjunction with [3.41] or [3.46], it follows that

$$\bar{U}^* = \bar{U}, \quad [4.4]$$

with $\bar{U}$ given by [3.38]. Consequently, the mean axial velocity of the Brownian particle is

$$\bar{U}^* = \int_0^{r_o-a} e^{-E(r)} U(r) r \, dr \Big/ \int_0^{r_o-a} e^{-E(r)} r \, dr. \quad [4.5]$$

Likewise, from [4.4], [3.49], and [3.41], it follows that

$$\bar{D}^* = \bar{D}_{11} + \bar{D}_v, \quad [4.6]$$

with $\bar{D}_{11}$ and $\bar{D}_v$ given, respectively, by [3.48] and [3.50].

The quantity $\bar{D}_{11}$ represents the *direct*, molecular diffusion contribution to the axial dispersivity $\bar{D}^*$. In the absence of flow,[6] where $\bar{D}_v = 0$, this represents the sole contribution to $\bar{D}^*$.

Equations [4.5] and [4.6] constitute the generalization of the Taylor–Aris (3) results to include both wall effects and a wall potential. Indeed, subsequent calculations show that in the limit, where $\lambda = 0$, and when $E = 0$,

$$\bar{U}^* = \bar{V}_m, \quad \bar{D}_{11} = D_\infty,$$

and

$$\bar{D}_v = r_o^2 \bar{V}_m^2 / 48 D_\infty,$$

in complete accord with the results of Taylor and Aris (3).

Convective and diffusive fluxes. As will now be demonstrated, it is possible to place an alternative interpretation on the quantities $\bar{U}^*$ and $\bar{D}^*$ as being phenomenological coefficients in expressions for the area-averaged mean fluxes, as opposed to their present,

[6] In the case of no convective transport, we have, of course, that $\bar{U} = 0$, and $\bar{D}^* = \bar{D}_{11}$. This result can also be derived directly from [3.2], without going through the Taylor–Aris moment analysis. A technique for accomplishing this is described in Appendix B. (See also Appendix C for a *steady-state* derivation of this quantity.)

Brownian motion, status. Inasmuch as fluxes are closer to those quantities which are directly susceptible to experimental observation, such a digression is not inappropriate. [This is not to imply, however, that the quantities $\bar{U}^*$ and $\bar{D}^*$, pertaining to the motion of a *single* Brownian particle, are not themselves susceptible to direct experimental measurement (58, 59) by a modernization of the classical Brownian motion experiments of Perrin.]

In dilute systems of noninteracting Brownian particles, knowledge of the fundamental solution P of Eq. [2.26] permits one to calculate the solute concentration distribution, $c(r, \phi, z; t)$, from knowledge of the initial concentration distribution, $c(r, \phi, z; 0)$, via the superposition theorem (19, 29),

$$c(r, \phi, z; t)$$
$$= \int_{z'=-\infty}^\infty \int_{\phi'=-\pi}^\pi \int_{r'=0}^{r_o-a} c(r', \phi', z'; 0)$$
$$\times P(r, \phi, z, t | r', \phi', z') r' \, dr' \, d\phi' \, dr', \quad [4.7]$$

valid for linear systems of the type under consideration. Here, c denotes the number of sphere centers per unit volume of suspension. Similarly, in the axisymmetric case,

$$c(r, z; t) = 2\pi \int_{z'=-\infty}^\infty \int_{r'=0}^{r_o-a} c(r', z'; 0)$$
$$\times f(r, z, t | r', z') r' \, dr' \, dz', \quad [4.8]$$

where f represents the fundamental solution of [2.32]. Because of these quantitative connections, existing between concentration in dilute multiparticle systems and probability distributions in single particle systems, one may use the language and terminology appropriate to either Brownian motion theory or transport processes, whichever is more convenient. In particular, it is possible to speak synonomously of f as if it were either a particle "concentration" c or a probability density. Concomitantly, ∇f may be termed a "concentration" gradient. In other words, we may confound the symbols f and c.

The basic convective–diffusion equation for f (cf. [2.32]) can be written in the form

$$\frac{\partial f}{\partial t} + \frac{\partial j_z}{\partial z} + \frac{1}{r}\frac{\partial}{\partial r}(rj_r)$$

$$= \frac{\delta(r - r')\delta(z - z')\delta(t)}{2\pi r}, \quad [4.9]$$

with

$$j_z = U(r)f - D_{\shortparallel}(r)\partial f/\partial z, \quad [4.10]$$

and

$$j_r = -D_{\perp}(r)\left(\frac{\partial f}{\partial r} + f\frac{dE}{dr}\right)$$

$$= -e^{-E(r)}D_{\perp}(r)\frac{\partial}{\partial r}(e^{E(r)}f), \quad [4.11]$$

the axial and radial fluxes, respectively. In "concentration" terminology, the right-hand side of [4.9] constitutes an instantaneous point source of solute.

Multiply [4.9] by $2\pi r dr$ and integrate over the circular region [3.25] to obtain

$$\frac{\partial \bar{f}}{\partial t} + \frac{\partial \bar{j}_z}{\partial z} = \frac{\delta(z - z')\delta(t)}{\pi r_o^2}, \quad [4.12]$$

since rj_r vanishes at both the lower and upper limits of integration (cf. [2.36] and [2.39]). Here,

$$\bar{f}(z;t) \overset{\text{def}}{=} (1/\pi r_o^2)\int_0^{r_o-a} f 2\pi r dr, \quad [4.13]$$

and

$$\bar{j}_z(z;t) \overset{\text{def}}{=} (1/\pi r_o^2)\int_0^{r_o-a} j_z 2\pi r dr. \quad [4.14]$$

Multiply the numerator and denominator of [4.13] by dz to obtain

$$\bar{f} = (1/\Delta V)\int_{\Delta V} f dV, \quad [4.15]$$

with $dV = 2\pi r dr dz$ and $\Delta V = \pi r_o^2 dz$. Thus, $\bar{f}$ represents the number density of solute particles (i.e., sphere centers per unit of superficial volume) at the position z. In addition, since $dA = 2\pi r dr$ is a unit of cross-

sectional area of the cylinder, the quantity

$$N_z(z;t) \overset{\text{def}}{=} \int_0^{r_o-a} j_z 2\pi r dr$$

represents the total number of sphere centers crossing the plane $z = $ constant per unit time. Equation [4.14] then shows that $\bar{j}_z$ is the axial "macroscopic" solute flux (per unit of cross-sectional area) at the position z.

Equations [3.13] and [4.13] combine to give

$$M_m(t) = \int_{\bar{V}_\infty} z^m \bar{f}(z;t)d\bar{V}, \quad [4.16]$$

where

$$d\bar{V} = \pi r_o^2 dz. \quad [4.17]$$

The integration domain $\bar{V}_\infty$ corresponds to the entire volume of space interior to the infinitely long tube; that is,

$$\bar{V}_\infty \equiv \{-\infty < z < \infty\}$$
$$\oplus \{0 \leq r \leq r_o\}. \quad [4.18]$$

Since

$$\bar{f}(z;t)d\bar{V} \equiv dz\int_0^{r_o-a} f 2\pi r dr = d\bar{n}, \quad [4.19]$$

say, represents the number of sphere centers contained within the volume element $d\bar{V}$, then [4.16] shows that the quantity

$$M_m(t) = \int_{z=-\infty}^{\infty} z^m d\bar{n}(z;t) \quad [4.20]$$

represents the mth moment of the number distribution of solute molecules.

From a macroscopic point of view, the net solute transport process may be regarded as being one-dimensional, taking place entirely in the z direction; that is, we envision a convective–diffusive transport process occurring within the capillary tube, in which all fluxes, number densities, velocities, phenomenological coefficients, and the like are independent of radial position r within the tube, depending only upon z (and t). Equation [4.12] may thus be interpreted as constituting the differential equation describing an inhomogeneous, one-dimensional, *plug flow*–

diffusive transport process. The source term appearing therein corresponds to a "unit number" of sphere centers (per unit of cross-sectional area πr_o^2), initially distributed uniformly over the tube cross section at the position $z = z'$. (This delta function source term is precisely that used by others (32, 45, 53) in describing a comparable one-dimensional process.)

From this macroscopic, one-dimensional point of view, the macroscopic solute flux density $\bar{J}_z$ appearing in [4.12] may be expressed in the usual way, as the sum of convective and diffusive contributions:

$$\bar{J}_z = \bar{U}^* \bar{f} - \bar{D}^* \partial \bar{f}/\partial z, \qquad [4.21]$$

in which $\bar{U}^*$ and D^* are constants to be determined. Introduction of this flux expression into [4.12] yields

$$\frac{\partial \bar{f}}{\partial t} + \bar{U}^* \frac{\partial \bar{f}}{\partial z} = \bar{D}^* \frac{\partial^2 \bar{f}}{\partial z^2} + \frac{\delta(z)\delta(t)}{\pi r_o^2}, \qquad [4.22]$$

in which the arbitrary choice $z' = 0$ has once again been made for convenience. This differential equation describes the one-dimensional transport process in the region $-\infty < z < \infty$. It is to be solved for $\bar{f}(z; t)$ subject only to the condition that

$$\bar{f} \to 0 \quad \text{as } |z| \to \infty \qquad [4.23]$$

sufficiently rapidly with $|z|$. The concentration $\bar{c}(z; t)$ for any arbitrarily prescribed initial distribution $\bar{c}(z; 0)$ may then be derived, by superposition, from this fundamental solution as (cf. [4.8])

$$\bar{c}(z; t) = \int_{z'=-\infty}^{\infty} \bar{c}(z'; 0) \bar{f}(z - z'; t) dz', \qquad [4.24]$$

with z' a dummy variable of integration.

The moments $\bar{M}_m(t)$ of the solute number distribution $\bar{f}$, corresponding to the solution of [4.22] and [4.23], are

$$\bar{M}_m(t) \stackrel{\text{def}}{=} \int_{\bar{V}_\infty} z^m \bar{f}(z; t) d\bar{V}$$

$$(m = 0, 1, 2, \ldots), \qquad [4.25]$$

with $d\bar{V}$ and $\bar{V}_\infty$ given, as before, by [4.17] and [4.18], respectively. This expression may be compared with the exact moments, $M_m(t)$, of the distribution of sphere centers, given by [4.16]. Provided that sufficient time is allowed for the solute particles to sample all accessible radial positions many times, as per criterion [3.5], these different moments may be expected to be asymptotically equal, i.e.,

$$\bar{M}_m(t) \sim M_m(t) \quad \text{as } D_\infty t/r_o^2 \to \infty. \qquad [4.26]$$

This asymptotic relation furnishes the *ansatz* by means of which the coefficient $\bar{U}^*$ and $\bar{D}^*$, defined by [4.21], can be expressed in terms of the "microscopic" parameters [$U(r)$, $E(r)$, $D_\perp(r)$, $D_{\parallel}(r)$, r_o, and a] appearing in Eqs. [2.32] *et seq.*, which describe the local, or microscopic, distribution function f.

Time differentiation of [4.25] followed by use of [4.22] gives

$$dM_m/dt = -\pi r_o^2 \bar{U}^* \int_{-\infty}^{\infty} z^m (\partial \bar{f}/\partial z) dz$$

$$+ \pi r_o^2 \bar{D}^* \int_{-\infty}^{\infty} z^m (\partial^2 \bar{f}/\partial z^2) dz$$

$$+ \delta_{m0}\delta(t). \qquad [4.27]$$

Integration by parts in the manner of [3.16] and [3.17] eventually yields

$$d\bar{M}_m/dt = m\bar{U}^* \bar{M}_{m-1} + m(m-1)\bar{D}^* \bar{M}_{m-2} + \delta_{m0}\delta(t). \qquad [4.28]$$

For the case $m = 0$ the preceding reduces to $d\bar{M}_0/dt = \delta(t)$, integration of which gives

$$\bar{M}_0 = 1 \quad \text{for all } t > 0. \qquad [4.29]$$

Similarly, for $m = 1$,

$$\bar{M}_1 = \bar{U}^* t + C', \qquad [4.30]$$

is obtained, with C' an integration constant. For $m = 2$ the relation

$$\bar{M}_2 = (\bar{U}^*)^2 t^2 + 2\bar{U}^* C' t$$

$$+ 2\bar{D}^* t + \text{const} \qquad [4.31]$$

ultimately arises.

These results for $\bar{M}_0$, $\bar{M}_1$, and $\bar{M}_2$ may be compared with [3.14], [3.41], and [3.51], respectively. In particular, the asymptotic condition [4.26] will clearly be satisfied by choosing

$$\bar{U}^* = \bar{U} \qquad [4.32]$$

and

$$\bar{D}^* = \bar{D}_{11} + \bar{D}_v \qquad [4.33]$$

(as well as $C' = C$). Although the macroscopic and microscopic moments have only been matched for the cases $m = 0$, 1, and 2, it can be shown, by the use of techniques similar to those of Fried and Combarnous (28), that the above choice for $\bar{U}^*$ and $\bar{D}^*$ results in asymptotic matching of the moments to *all* orders in m.

Equations [4.32] and [4.33] are precisely the same as [4.4] and [4.6], respectively. Thus, this procedure demonstrates that the same expressions obtain for the mean solute velocity $\bar{U}^*$ and dispersivity $\bar{D}^*$, irrespective of whether one chooses to define these quantities by focusing attention on the Brownian movement of a single solute molecule, or by defining these quantities as the phenomenological coefficients appearing in the expression [4.21] for the flux in a dilute multiparticle system. That identical results are obtained from these two conceptually different lines of reasoning is, of course, gratifying, since each of the two points of view is independently susceptible to experimental confirmation.

Even further, if one defines the moments of the *macroscopic* distribution (cf. [3.3] and [3.4]),

$$\bar{z} = \int_{\bar{V}_\infty} z \bar{f} d\bar{V} \Big/ \int_{\bar{V}_\infty} \bar{f} d\bar{V}, \qquad [4.34]$$

and

$$\overline{(z - \bar{z})^2} = \int_{\bar{V}_\infty} (z - \bar{z})^2 \bar{f} d\bar{V} \Big/ \int_{\bar{V}_\infty} \bar{f} d\bar{V}, \qquad [4.35]$$

with $\bar{f}$ given by the solution of [4.22] and [4.23], it follows that

$$\bar{z} = \bar{U}^* t \qquad [4.36a]$$

and

$$\overline{(z - \bar{z})^2} = 2\bar{D}^* t. \qquad [4.36b]$$

With $\bar{U}^*$ and $\bar{D}^*$ given by [4.32] and [4.33],

these are asymptotically equivalent to [3.6] and [3.8]. This provides yet further confidence in the credibility of the calculations.

Nondimensional forms. The three principal parameters of interest are $\bar{U}^*$, $\bar{D}_{11}$, and $\bar{D}_v$. Expressions for these quantities are most conveniently given in dimensionless form by introducing the nondimensional *local* variables,

$$u = U/\bar{V}_m, \qquad [4.37a]$$

$$d_{11} = D_{11}/D_\infty, \qquad [4.37b]$$

$$d_\perp = D_\perp/D_\infty, \qquad [4.37c]$$

each of which is a function of λ and of β; these parameters are defined in [1.1] and [2.20], respectively; $\bar{V}_m$ is the mean Poiseuille velocity, and D_∞ is defined by [2.21]. Define the mean dimensionless parameters

$$\bar{u}^* = \bar{U}^*/\bar{V}_m, \qquad [4.38a]$$

$$\bar{d}_{11} = \bar{D}_{11}/D_\infty, \qquad [4.38b]$$

$$\bar{d}_v = 48\,\mathrm{Pe}^{-2}(\bar{D}_v/D_\infty), \qquad [4.38c]$$

where

$$\mathrm{Pe} = r_o \bar{V}_m/D_\infty \qquad [4.39]$$

is a Peclet number. The latter constitutes a global measure of convective to diffusive particle transport. The extraneous factor of 48 is introduced as a normalization factor such that, on the basis of the Taylor–Aris (3, 56, 57) analysis, $\bar{d}_v = 1$ at $\lambda = 0$. In terms of the above variables,

$$\bar{d}^* \overset{\mathrm{def}}{=} \bar{D}^*/D_\infty = \bar{d}_{11} + (\mathrm{Pe}^2/48)\bar{d}_v. \qquad [4.40]$$

Following from these definitions, in conjunction with [3.48], [3.50], and [4.5], are the expressions

$$\bar{u}^*(\lambda) = \int_0^{1-\lambda} e^{-E(\beta,\lambda)} u(\beta, \lambda)\beta d\beta \Big/ \int_0^{1-\lambda} e^{-E(\beta,\lambda)}\beta d\beta, \qquad [4.41]$$

$$\bar{d}_{11}(\lambda) = \int_0^{1-\lambda} e^{-E(\beta,\lambda)} d_{11}(\beta, \lambda)\beta d\beta \Big/ \int_0^{1-\lambda} e^{-E(\beta,\lambda)}\beta d\beta, \qquad [4.42]$$

and

$$\bar{d}_v(\lambda) = 48 \int_0^{1-\lambda} \frac{h^2(\beta, \lambda)d\beta}{\beta e^{-E(\beta,\lambda)}d_\perp(\beta, \lambda)} \bigg/$$

$$\int_0^{1-\lambda} e^{-E(\beta,\lambda)}\beta d\beta, \quad [4.43a]$$

with

$$h(\beta, \lambda) = \int_{\beta'=0}^{\beta} e^{-E(\beta',\lambda)}[\bar{u}^*(\lambda)$$
$$- u(\beta', \lambda)]\beta'd\beta'. \quad [4.43b]$$

In addition to depending upon λ, the mean quantities $\bar{u}$, $\bar{d}_{\shortparallel}$ and $\bar{d}_v$ also depend upon the dimensionless parameters entering into the potential energy function E. For example, when E derives from attractive and repulsive forces exerted by the wall upon the Brownian particle, E will depend, inter alia, upon the characteristic length scales L_a and L_r defining the respective ranges of the attractive and repulsive potentials. In this case, the functional dependence will be of the form

$$\bar{u}^*, \bar{d}_{\shortparallel}, \bar{d}_v = \text{functions } (\lambda, l_a, l_r), \quad [4.44]$$

with

$$l_a = L_a/a, \quad [4.45a]$$

$$l_r = L_r/a. \quad [4.45b]$$

The case of no external potential. Attention in this paper will be addressed only to the case where $E = 0$. Circumstances in which $E \neq 0$ will be taken up in subsequent papers in this series. With $E = 0$, the preceding equations reduce to the forms

$$\bar{u}^*(\lambda) = 2(1 - \lambda)^{-2} \int_0^{1-\lambda} u(\beta, \lambda)\beta d\beta, \quad [4.46]$$

$$\bar{d}_{\shortparallel}(\lambda) = 2(1 - \lambda)^{-2} \int_0^{1-\lambda} d_{\shortparallel}(\beta, \lambda)\beta d\beta, \quad [4.47]$$

and

$$\bar{d}_v(\lambda) = 96(1 - \lambda)^{-2} \int_0^{1-\lambda}$$

$$\times [h^2(\beta, \lambda)d\beta/\beta d_\perp(\beta, \lambda)], \quad [4.48a]$$

with

$$h(\beta, \lambda) = \int_{\beta'=0}^{\beta} [\bar{u}^*(\lambda)$$
$$- u(\beta', \lambda)]\beta'd\beta'. \quad [4.48b]$$

Evaluation of the preceding integrals requires knowledge of various low-Reynolds-number hydrodynamic resistance coefficients for an eccentrically positioned spherical particle in a circular tube. In dimensionless form, these coefficients are functions of β and λ. Requisite formulas for these hydrodynamic coefficients as a function of β are available only for the case of relatively small spheres, $\lambda \ll 1$ (i.e., $0 < \lambda \lesssim 0.15$) (17, 38, 39), and for relatively large spheres which almost fill the tube, $\lambda \approx 1$ (i.e., $\lambda \gtrsim 0.95$) (18). Theoretical results are not yet available for the intermediate range, $0.15 \lesssim \lambda \lesssim 0.95$, except for the case of a concentrically positioned sphere, $\beta = 0$ (6, 18, 30, 36, 41, 52). However, such isolated center-line results are of little use in evaluating the preceding integrals.

Numerical computations for the limiting case $\lambda \ll 1$ are presented in the next section. Analogous results for the closely fitting sphere case, $\lambda \approx 1$ (18), will be presented elsewhere.

V. HYDRODYNAMIC CALCULATIONS FOR RELATIVELY SMALL SPHERES ($\lambda \ll 1$)

For a specified value of λ satisfying the inequality

$$\lambda \ll 1, \quad [5.1]$$

no *uniformly valid* solutions, spanning the complete range (cf. [2.20]),

$$0 \leq \beta \leq 1, \quad [5.2]$$

exist for low Reynolds number hydrodynamic problems involving the motion of a spherical particle within an infinitely long circular cylinder. There do exist, however, two *locally-valid* solutions, which may be rigorously pieced together via the method of matched asymptotic expansions (60) to furnish a uniformly valid composite solution. These locally valid solutions encompass the respective ranges

$$1 - \beta \gg \lambda \quad [5.3]$$

and

$$1 - \beta = O(\lambda). \quad [5.4]$$

The first of these corresponds to the class of problems treated by Hirschfeld and Brenner

(38, 39) via the "method of reflections." The second corresponds to the class of problems treated by Bungay and Brenner (17), as a perturbation on the motion of a sphere moving in proximity to a single plane wall (33, 34), taking account of first-order wall-curvature effects (17).

The two classes of problems defined by the eccentricity parameter β ranges [5.3] and [5.4] may be clearly distinguished from each other by introducing the alternative, dimensionless, eccentricity parameter

$$\delta = l/a, \qquad [5.5]$$

with

$$l = r_o - r. \qquad [5.6]$$

The latter represents the distance of the sphere center from the cylinder wall. Where appropriate, this parameter, rather than β, may be utilized to specify the location of the sphere center. It is related to β by the expression

$$1 - \beta = \delta\lambda. \qquad [5.7]$$

By definition, δ therefore lies in the region

$$1 \leq \delta \leq \lambda^{-1}. \qquad [5.8]$$

The limiting value $\delta = 1$ corresponds to the situation in which the sphere touches the wall. The other limiting value, $\delta = \lambda^{-1}$, corresponds to the opposite case, where the sphere is located along the tube axis, $\beta = 0$. As $\lambda \to 0$, δ tends to infinity.

"Reflection" solutions of the type [5.3] are valid in the "central core" region (hereafter designated by the affix "c"), corresponding to

$$\delta \gg 1, \quad \text{with } \delta\lambda = O(1). \qquad [5.9]$$

In circumstances where this inequality (and [5.1]) apply, the sphere is relatively far from the tube wall compared with its radius. In contrast, the class of "sphere/plane–wall" solutions of the type [5.4] is valid in the "wall" region (hereafter designated by the affix "w")

$$\delta = O(1), \qquad [5.10]$$

wherein the sphere is relatively close to the tube wall in comparison to its radius.

The two classes of solutions encompassing the respective ranges [5.9] and [5.10], may, in the terminology of singular perturbation theory, be described as "inner" and "outer" expansions, respectively. These distinct expansions may be regarded as being asymptotically matched in their common domain of validity, intermediate between the two extreme ranges [5.9] and [5.10]. The latter, respectively, represent the regions of validity of the inner and outer expansions. Formally, the matching process may be accomplished by introducing a λ-dependent, intermediate variable, $\delta_* \equiv \delta_*(\lambda)$, defined such that, for a fixed value of λ satisfying the inequality [5.1], δ_* lies *simultaneously* in the regions of validity of both of these expansions; that is, in the overlap region. This can be done, for example, by choosing

$$1 \ll \delta_* \ll \lambda^{-1}, \qquad [5.11a]$$

or, more specifically, by defining (21)

$$\delta_* = \lambda^{-q}, \qquad [5.11b]$$

where q is a λ-independent constant, lying in the range

$$0 < q < 1, \qquad [5.12]$$

which does not include the end points, $q = 0$ and $q = 1$. The limiting value $q = 1$ corresponds to the core region [5.9], whereas the other limiting value, $q = 0$, corresponds to the wall region [5.10]. Consequently, for any value of q lying within the above range, δ_* straddles the intermediate region between the core and wall regions. Observe that

$$\delta_*\lambda \to 0 \quad \text{as } \lambda \to 0. \qquad [5.13]$$

Moreover, for a fixed value of λ, the wall and core regions are defined, respectively, by the δ ranges

$$1 \leq \delta < \delta_* \qquad [5.14]$$

and

$$\delta_* < \delta < \lambda^{-1}. \qquad [5.15]$$

Comparable to [5.11b], the intermediate, matching region may equally well be expressed in terms of an eccentricity variable $\beta_* \equiv \beta_*(\lambda)$, defined by the relation (cf. [5.7])

$$1 - \beta_* = \delta_* \lambda, \qquad [5.16]$$

with δ_* given by [5.11b]. In terms of this intermediate variable, the central core, or inner, region corresponds to the circular domain

$$0 \leq \beta < \beta_*, \qquad [5.17]$$

while the wall, or outer, region corresponds to the annular domain

$$\beta_* < \beta \leq 1. \qquad [5.18]$$

From [5.13] and [5.16] it follows that

$$\beta_* \to 1 \quad \text{as } \lambda \to 0. \qquad [5.19]$$

Calculation of $\bar{d}_{11}$. To terms of the first order in λ, Hirschfeld and Brenner (38, 39)—using the method of reflections—have solved, inter alia, the hydrodynamic Stokes flow problem of a freely rotating, eccentrically positioned sphere, translating through an otherwise quiescent fluid, parallel to the axis of a circular cylindrical tube. The force F required to cause the sphere center to translate with steady velocity U is given by the expression

$$F/F_\infty = [1 - \lambda f_{11}(\beta)$$
$$+ O\{\lambda^3/(1 - \beta)^3\}]^{-1}, \qquad [5.20]$$

where

$$F_\infty = 6\pi\mu a U \qquad [5.21]$$

is the Stokes force on the sphere in an unbounded fluid; $f_{11}(\beta)$ is a dimensionless function of β, available in tabular form (38, 39) over the range $\beta = 0(0.01)0.99$ to eight significant figures. [The error estimate in [5.20] derives from the wall-effect analysis of Cox and Brenner (22), as is discussed in detail by Gaydos (31).] For $\beta = 0$, $f_{11}(0) = 2.1044$. On the other hand, as $\beta \to 1$, the tabulated results, up to $\beta = 0.99$, extrapolate smoothly to show numerically that

$$f_{11}(\beta) \sim (9/16)(1 - \beta)^{-1} \quad \text{as } \beta \to 1. \qquad [5.22]$$

The latter algebraic form derives from the asymptotic, "method of reflections," expression for the force on a freely rotating sphere, translating through a quiescent fluid parallel

to a plane wall (33, 34), for the case where $l/a \gg 1$, with l the distance of the sphere center from the wall, as in [5.6]. In view of the asymptotic behavior indicated by [5.22], it is convenient to define a function $\alpha_{11}(\beta)$ by the relation

$$f_{11}(\beta) = (9/16)(1 - \beta)^{-1} + \alpha_{11}(\beta). \qquad [5.23]$$

The function $\alpha_{11}(\beta)$ is therefore of $O(1)$ with respect to β over the entire range, $\beta = 0$ to 1. It may be obtained numerically in the range $\beta = 0(0.01)0.99$ from the Hirschfeld–Brenner (38, 39) tabulation. The limiting value, obtained by a Lagrangian extrapolation technique, using the last five tabulated points $\beta = 0.95(0.01)0.99$, is $\alpha_{11}(1) = 0.2110$.

According to the Stokes–Einstein equation (8, 9), the particle diffusivity D is given by the expression

$$D/D_\infty = (F/F_\infty)^{-1}, \qquad [5.24]$$

where F refers to the force [5.20] on the torque-free sphere. In this manner it follows from [4.38b], [2.21], and [5.20] that the nondimensional diffusion coefficient for (freely rotating) translation of the sphere center parallel to the tube axis is

$$d_{11}(\beta, \lambda) = 1 - \lambda[(9/16)(1 - \beta)^{-1} + \alpha_{11}(\beta)]$$
$$+ O[(\lambda^3/(1 - \beta)^3], \quad (0 \leq \beta < \beta_*). \qquad [5.25]$$

Having been obtained by the method of reflections, this formula applies only within the core region, as indicated.

In the wall region, Bungay and Brenner (17) derive the analog of [5.20] for the force on the freely rotating sphere, translating in a quiescent fluid parallel to the tube axis, as

$$F/F_\infty = (6\pi a L_0^r)^{-1}(K_0^t L_0^r - K_0^r L_0^t) + O(\lambda), \qquad [5.26]$$

with $K_0^r = L_0^t$. The translational and rotational resistance coefficients, K_0 and L_0, utilized by Bungay and Brenner (17), which derive from the work of Goldman *et al.* (33), are functions only of the sphere radius a and of the dimensionless gap width δ. In terms of these coefficients, the dimensionless diffusivity is given by

$$d_{11}(\delta, \lambda) = w_{11}(\delta) + O(\lambda) \quad (\delta_* > \delta \geq 1), \quad [5.27]$$

wherein

$$w_{11}(\delta) = 6\pi a L_0{}^r (K_0{}^t L_0{}^r - K_0{}^r L_0{}^t)^{-1}. \quad [5.28]$$

This quantity is a function only of δ, and possesses the value unity in an unbounded fluid. Numerical values of the function $w_{11}(\delta)$ in the range $\delta = 1$ to ∞ are tabulated by Goldman et al. (33) for a freely rotating sphere, translating parallel to a single plane wall in an otherwise unbounded, quiescent fluid. Asymptotic values derived from this tabulation are[7]

$$w_{11} \sim 1 - (9/16)\delta^{-1} + (1/8)\delta^{-3}$$
$$\text{as } \delta \to \infty, \quad [5.29]$$

$$\sim 2/\ln (\delta - 1)^{-1} \quad \text{as } \delta \to 1. \quad [5.30]$$

The latter result shows explicitly that

$$d_{11} \to 0 \quad \text{as } \delta \to 1, \quad [5.31]$$

demonstrating that the axial diffusivity goes to zero as the sphere approaches contact with the wall.

In equivalent notation, [5.22] and [5.29] accord with each another; that is, both the Hirschfeld–Brenner (38, 39) and Bungay–Brenner (17) force expressions yield the same asymptotic diffusion result, namely,

$$d_{11} \sim 1 - (9/16)(1 - \beta)^{-1}\lambda, \quad [5.32]$$

in their common domain of validity. In this sense, the two solutions are asymptotically matched in the intermediate region, where the core and wall regions overlap.

Equation [4.47] may be written as

$$\bar{d}_{11} = 2(1 - \lambda)^{-2} I(\lambda), \quad [5.33]$$

in which

$$I(\lambda) = I_c(\beta_*, \lambda) + I_w(\beta_*, \lambda), \quad [5.34]$$

where

$$I_c(\beta_*, \lambda) = \int_{\beta=0}^{\beta_*} d_{11}(\beta, \lambda)\beta d\beta, \quad [5.35a]$$

and

$$I_w(\beta_*, \lambda) = \int_{\beta=\beta_*}^{1-\lambda} d_{11}(\beta, \lambda)\beta d\beta, \quad [5.35b]$$

represent the respective contributions to $I(\lambda)$ of the core and wall regions. Though each of the latter two integrals separately depends upon the value assigned to the parameter β_* (because of its appearance in the limits of integration), their sum cannot, owing to the somewhat arbitrary manner in which β_* is to be selected. Accordingly, the β_* (or δ_*) contributions appearing separately in I_c and I_w must cancel when the latter are summed to obtain $I(\lambda)$.

Introduction of [5.25] into [5.35a] yields, upon performing the indicated integrations,

$$I_c(\beta_*, \lambda) = (\tfrac{1}{2})\beta_*{}^2 + \lambda\left[(\tfrac{9}{16})\beta_* \right.$$

$$\left. + (\tfrac{9}{16}) \ln (1 - \beta_*) - \int_0^{\beta_*} \alpha_{11}(\beta)\beta d\beta \right]$$

$$+ E_1 + E_2 + E_3, \quad [5.36]$$

where

$$E_1 = O[\lambda^3/(1 - \beta_*)],$$
$$E_2 = O[\lambda^3/(1 - \beta_*)^2], \quad [5.37]$$
$$E_3 = O(\lambda^3),$$

are error terms arising from integration of the error term of $O[\lambda^3(1 - \beta)^{-3}]$ appearing in [5.25]. By means of [5.16], in the above, β_* may be replaced by $1 - \delta_*\lambda$. The error terms thus adopt the forms

$$E_1 = O(\lambda^2\delta_*{}^{-1}), \quad E_2 = O(\lambda\delta_*{}^{-2}), \quad [5.38]$$
$$E_3 = O(\lambda^3).$$

These may be rank ordered with respect to their relative magnitudes by use of [5.11b]. That is,

$$E_1 = O(\lambda^{2+q}), \quad E_2 = O(\lambda^{1+2q}), \quad E_3 = O(\lambda^3).$$

This makes

$$E_1/E_2 = O(\lambda^k), \quad E_3/E_2 = O(\lambda^{2k}),$$

where $k = 1 - q$. It follows from [5.12] that

$k > 0$, whence, in view of the inequality [5.1],

$$E_1 \ll E_2 \quad \text{and} \quad E_3 \ll E_2. \qquad [5.39]$$

Consequently, the error terms E_1 and E_3 are both negligible compared with E_2, and each may be neglected in comparison with the latter.

In addition, the remaining integral in [5.36] may be written as

$$\int_0^{\beta*} \alpha_{11}(\beta)\beta d\beta = C_1 - F_c(\beta*), \qquad [5.40]$$

where C_1 is the numerical constant (31),

$$C_1 = \int_0^1 \alpha_{11}(\beta)\beta d\beta = 0.3290, \qquad [5.41]$$

and

$$F_c(\beta*) = \int_{\beta=\beta*}^1 \alpha_{11}(\beta)\beta d\beta. \qquad [5.42]$$

However, by Taylor series expansion,

$$F_c(\beta*) = F_c(1) + (1 - \beta*)F_c'(1) + O[(1 - \beta*)^2]. \qquad [5.43]$$

But $F_c(1) = 0$ by definition. In addition, by Leibnitz' rule for differentiating an integral, $F_c'(\beta*) = -\alpha_{11}(\beta*)\beta*$. Consequently, $F_c'(1) = -\alpha_{11}(1) = -0.2110$. In view of [5.16], this shows that

$$F_c(\beta*) = O(\lambda\delta*). \qquad [5.44]$$

Hence,

$$\lambda \int_0^{\beta*} \alpha_{11}(\beta)\beta d\beta = \lambda C_1 + O(\lambda^2\delta*). \qquad [5.45]$$

Collecting these results together, and replacing $\beta*$ by $1 - \delta*\lambda$, gives

$$\begin{aligned} I_c(\delta*, \lambda) = {} & (1/2)(1 - \delta*\lambda)^2 \\ & + (9/16)\lambda \ln (\delta*\lambda) - C_1\lambda \\ & + (9/16)\lambda(1 - \delta*\lambda) \\ & + O(\lambda\delta*^{-2}) + O(\lambda^2\delta*). \qquad [5.46] \end{aligned}$$

Evaluation of I_w is best accomplished by changing from core variables to wall variables via the substitutions [5.7] and [5.16]. This makes

$$I_w(\delta*, \lambda) = \lambda \int_1^{\delta*} (1 - \delta\lambda)d_{11}d\delta, \qquad [5.47]$$

with d_{11} given to the first order in λ by [5.27]. In view of the asymptotic forms [5.29] and [5.30], it is convenient to rewrite w_{11} in the form

$$w_{11} = 1 - (9/16)\delta^{-1} - \Delta(\delta), \qquad [5.48]$$

with Δ a "slack" variable, possessing the asymptotic forms

$$\begin{aligned} \Delta(\delta) &\sim 7/16 \quad \text{as} \quad \delta \to 1, \qquad [5.49] \\ &\sim O(\delta^{-3}) \quad \text{as} \quad \delta \to \infty. \end{aligned}$$

Integration thereby gives

$$I_w(\delta*, \lambda)$$

$$= \lambda\left[(\delta* - 1) - (\tfrac{9}{16}) \ln \delta* - \int_1^{\delta*} \Delta(\delta)d\delta \right]$$

$$- \lambda^2\left[(\tfrac{1}{2})(\delta*^2 - 1) - (\tfrac{9}{16})(\delta* - 1) \right.$$

$$\left. - \int_1^{\delta*} \delta\Delta(\delta)d\delta \right] + o(\lambda). \qquad [5.50]$$

One may write

$$\int_{\delta=1}^{\delta*} \Delta(\delta)d\delta = C_2 - F_w(\delta*), \qquad [5.51]$$

where C_2 is the numerical constant (31),

$$C_2 = \int_{\delta=1}^\infty \Delta(\delta)d\delta = 0.0030, \qquad [5.52]$$

and

$$F_w(\delta*) = \int_{\delta*}^\infty \Delta(\delta)d\delta. \qquad [5.53]$$

From [5.511b], $\delta* \to \infty$ as $\lambda \to 0$. Hence, as regards dominant terms, only the larger values of $\delta*$ are of interest in calculating F_w. It thus follows from the second of equations [5.49] that

$$F_w = O(\delta*^{-2}). \qquad [5.54]$$

Similar arguments show that

$$\int_1^{\delta*} \delta\Delta(\delta)d\delta = O(1) \qquad [5.55]$$

353

with respect to λ. Equation [5.50] thereby becomes

$$I_w(\delta_*, \lambda) = \lambda[(\delta_* - 1) - (9/16)\ln \delta_* - C_2]$$
$$- \lambda^2[(1/2)(\delta_*{}^2 - 1) - (9/16)(\delta_* - 1)]$$
$$+ O(\lambda\delta_*{}^{-2}) + o(\lambda), \quad [5.56]$$

as regards dominant terms in λ.

Addition of the latter to [5.46] yields

$$I_e + I_w = (1/2) + (9/16)\lambda \ln \lambda$$
$$- [(7/16) + C_1 + C_2]\lambda + O(\lambda\delta_*{}^{-2})$$
$$+ O(\lambda^2\delta_*) + o(\lambda), \quad [5.57]$$

to terms of dominant order in λ. As has been pointed out in connection with [5.34], the preceding sum must show itself to be independent of δ_*. According to the theory of matched asymptotic expansions [see, for example, Ref. (21)], this necessarily comes about by the appearance of equal and opposite terms of orders $\lambda\delta_*{}^{-2}$ and $\lambda^2\delta_*$, which arise when higher-order terms in λ are included in the asymptotic expansions [5.25] and [5.27]. Their explicit presence here merely furnishes a definitive estimate of the accuracy of the expression for $I_e + I_w$ as regards orders of λ. In this context, Eq. [5.11a] shows that δ_* lies in the *open* interval $\lambda^{-1} > \delta_* > 1$, which does not include the end points. Consequently, the extreme ranges of λ, into which the two δ_* error terms in [5.57] fall, are bounded by the regions

$$O(\lambda) > O(\lambda\delta_*{}^{-2}) > O(\lambda^3)$$

and $\quad\quad\quad\quad\quad\quad\quad\quad\quad\quad\quad [5.58]$

$$O(\lambda) > O(\lambda^2\delta_*) > O(\lambda^2).$$

These terms are of smaller order in λ than those explicitly retained in [5.57]. Thus, it may be concluded from [5.34] that

$$I(\lambda) = (1/2) + (9/16)\lambda \ln \lambda$$
$$- [(7/16) + C_1 + C_2]\lambda + o(\lambda). \quad [5.59]$$

Substitution into [5.33] thereby yields

$$\bar{d}_{11} = (1 - \lambda)^{-2}[1 - (9/8)\lambda \ln \lambda^{-1}$$
$$- \{(7/8) + 2C_1 + 2C_2\}\lambda + o(\lambda)]. \quad [5.60]$$

Numerically,

$$\bar{d}_{11} = (1 - \lambda)^{-2}[1 - (9/8)\lambda \ln \lambda^{-1}$$
$$- 1.539\lambda + o(\lambda)], \quad [5.61a]$$

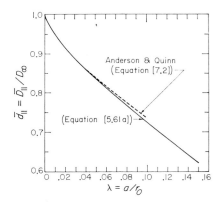

FIG. 3. Dimensionless mean axial molecular diffusion coefficient vs sphere/cylinder radius ratio.

or, less accurately,

$$\bar{d}_{11} = 1 - (9/8)\lambda \ln \lambda^{-1} + 0.4610\lambda$$
$$- (9/4)\lambda^2 \ln \lambda^{-1} + O(\lambda^2). \quad [5.61b]$$

A plot of Eq. [5.61a] is given in Fig. 3, where it is compared with a comparable result of Anderson and Quinn (2), as per the discussion of Section VII. The large effect of the tube wall in diminishing the molecular diffusivity below its value in an unbounded fluid is quite pronounced. Thus, for $\lambda = 0.01$, corresponding to a sphere/cylinder cross-sectional area ratio of only one part in 10 000 (10^{-4}), the diffusivity reduction is 5%!

Calculation of $\bar{u}^*$. Procedures similar to those employed above may be utilized to calculate the mean particle velocity from [4.46]. The calculations are most conveniently performed in terms of the (dimensionless) slip velocity,

$$u_s = 2(1 - \beta^2) - u. \quad [5.62]$$

In general, $u_s \geq 0$ for all β owing to the fact that the center of the neutrally buoyant, freely rotating, sphere lags the local Poiseuille flow in which it is suspended.

In the core region, this slip velocity is given by the method of reflections as (8, 10, 31, 37–39)

$$u_s = \tfrac{4}{3}\lambda^2 + O[\lambda^3/(1 - \beta)^2]$$
$$(0 \leq \beta < \beta_*). \quad [5.63]$$

On the other hand, in the wall region (17, 31),

$$u_s = 4\lambda A_{11}(\delta) + O(\lambda^2)$$
$$(\beta_* < \beta \leq 1 - \lambda), \quad [5.64]$$

with

$$A_{11} = \delta[1 - \chi(\delta)], \quad [5.65]$$

wherein χ is a function of δ, tabulated by Goldman *et al.* (34). The function A_{11} possesses the asymptotic properties[8]

$$A_{11} \sim (5/16)\delta^{-2} \quad \text{as } \delta \to \infty,$$
$$\sim 1 - O[1/\ln{(\delta-1)^{-1}}] \quad \text{as } \delta \to 1. \quad [5.66]$$

From [5.62] it follows, on integration, that

$$\int_0^{1-\lambda} u\beta d\beta = \tfrac{1}{2}(1 - \lambda)^2(1 + 2\lambda - \lambda^2)$$
$$- \int_0^{1-\lambda} u_s\beta d\beta. \quad [5.67]$$

Define

$$\int_0^{1-\lambda} u_s\beta d\beta = J_c + J_w, \quad [5.68]$$

wherein

$$J_c(\beta_*, \lambda) = \int_0^{\beta_*} u_s\beta d\beta, \quad [5.69a]$$

$$J_w(\beta_*, \lambda) = \int_{\beta_*}^{1-\lambda} u_s\beta d\beta. \quad [5.69b]$$

It follows readily from [5.63] that

$$J_c = \tfrac{2}{3}\beta_*^2\lambda^2 + O(\lambda^2\delta_*^{-1}). \quad [5.70]$$

In terms of δ, the wall integral is

$$J_w = \lambda \int_1^{\delta_*} [4\lambda A_{11}(\delta) + O(\lambda^2)](1 - \lambda\delta)d\delta$$

$$= 4\lambda^2 \int_1^{\delta_*} A_{11}(\delta)d\delta - 4\lambda^3 \int_1^{\delta_*} \delta A_{11}(\delta)d\delta$$

$$+ O(\lambda^3\delta_*), \quad [5.71]$$

in which error terms of orders $O(\lambda^3)$ and

[8] The second of Eqs. [5.66], in conjunction with [5.62] and [5.64], shows that the velocity u of the sphere center goes to zero as the sphere makes contact with the wall. This follows as a consequence of the fact that $1 - \beta^2 \sim 2\lambda + O(\lambda^2)$ as $\delta \to 1$.

$O(\lambda^4\delta_*^2)$ have been suppressed compared with the term of order $O(\lambda^3\delta_*)$ retained. Write

$$\int_1^{\delta_*} A_{11}(\delta)d\delta = C_3 - G_w(\delta_*), \quad [5.72]$$

with C_3 the numerical constant of $O(1)$ (31),

$$C_3 = \int_1^{\infty} A_{11}(\delta)d\delta = 0.32 \quad [5.73]$$

and

$$G_w(\delta_*) = \int_{\delta_*}^{\infty} A_{11}(\delta)d\delta. \quad [5.74]$$

As in the evaluation of the similar integral [5.53], it suffices to use the asymptotic form of A_{11} for large δ, given in [5.66]. This yields $G_w(\delta_*) = O(\delta_*^{-1})$, whereupon

$$\int_1^{\delta_*} A_{11}(\delta)d\delta = C_3 + O(\delta_*^{-1}). \quad [5.75]$$

Similarly,

$$\int_1^{\delta_*} \delta A_{11}(\delta)d\delta = O(1) + O(\ln \delta_*). \quad [5.76]$$

Upon collecting these results together,

$$J_w = 4C_3\lambda^2 + O(\lambda^2\delta_*^{-1}) + O(\lambda^3\delta_*), \quad [5.77]$$

follows, from which negligible error terms of orders λ^3, $\lambda^3 \ln \delta_*$, $\lambda^4\delta_*^2$, and λ^4 have been suppressed, compared with those error terms explicitly retained.

Addition of [5.70] and [5.77] gives, with use of [5.16], and elimination of negligible error terms,

$$J_c + J_w = (\tfrac{2}{3} + 4C_3)\lambda^2 + O(\lambda^2\delta_*^{-1}) + O(\lambda^3\delta_*). \quad [5.78]$$

Analogous to the discussion following [5.57], this sum must be independent of δ_*. Use of the same arguments employed there eventually shows that [5.68] possesses the form

$$\int_0^{1-\lambda} u_s\beta d\beta = (\tfrac{2}{3} + 4C_3)\lambda^2 + o(\lambda^2). \quad [5.79]$$

In combination with [5.67] and [4.4,6]

this makes

$$\bar{u}^* = 1 + 2\lambda - \lambda^2$$
$$- (\tfrac{4}{3} + 8C_3)(1 - \lambda)^{-2}\lambda^2 + o(\lambda^2), \quad [5.80]$$

or, less accurately,

$$\bar{u}^* = 1 + 2\lambda$$
$$- [(7/3) + 8C_3]\lambda^2 + o(\lambda^2). \quad [5.81]$$

Numerically,

$$\bar{u}^* = 1 + 2\lambda - 4.9\lambda^2 + o(\lambda^2). \quad [5.82]$$

A plot of [5.80] is given in Fig. 4, and discussed in Section VII. Note that the mean axial particle velocity $\bar{U}^*$ exceeds the mean velocity $\bar{V}_m$ of the fluid in which it is suspended. Furthermore, all other things being equal, $\bar{U}^*$ goes through a maximum as the particle size is increased.

Calculation of $\bar{d}_v$. The convective contribution to the total dispersivity $\bar{d}^*$ can be calculated by integration of Eqs. [4.48]. Write

$$\bar{d}_v(\lambda) = (1 - \lambda)^{-2}M(\lambda), \quad [5.83]$$

where

$$M(\lambda) = M_c(\beta_*, \lambda) + M_w(\beta_*, \lambda), \quad [5.84]$$

in which

$$M_c = 96 \int_0^{\beta_*} \frac{h^2}{\beta d_\perp} d\beta, \quad [5.85a]$$

$$M_w = 96 \int_{\beta_*}^{1-\lambda} \frac{h^2}{\beta d_\perp} d\beta. \quad [5.85b]$$

Analogous to [5.20], Hirschfeld and Brenner (38, 39) derive the following expression for the force on a sphere translating within the core region, perpendicular to the cylinder axis, through an otherwise quiescent fluid:

$$F/F_\infty = [1 - \lambda f_\perp(\beta) + O\{\lambda^3/(1 - \beta)^3\}]^{-1}$$
$$(0 \le \beta < \beta_*), \quad [5.86]$$

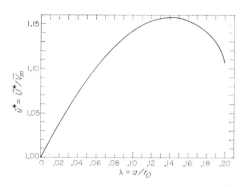

Fig. 4. Dimensionless mean axial particle velocity vs sphere/cylinder radius ratio. This graph represents a plot of Eq. [5.80].

with $f_\perp$ a tabulated function (38, 39), comparable to $f_\parallel$. For $\beta = 0$, $f_\perp(0) = 1.804$; at the opposite end of the range, where $\beta \to 1$,

$$f_\perp(\beta) = (9/8)(1 - \beta)^{-1} + \alpha_\perp(\beta), \quad [5.87]$$

with $\alpha_\perp(\beta)$ of $O(1)$ with respect to β over the entire β range. This limiting form corresponds to the "method of reflections" expression for the force on a sphere moving normal to a plane wall (7, 21). Thus, one can derive an expression for $d_\perp$ comparable to [5.25] by replacing the subscript $\parallel$ in the latter by $\perp$. Inverse expansion of the resulting expression thus yields

$$1/d_\perp = 1 + \lambda f_\perp(\beta) + \lambda^2 f_\perp^2(\beta) + o(\lambda^2), \quad [5.88]$$

valid within the core region. Because of algebraic complexities, we will not attempt here, or in subsequent calculations, to carry along the error estimate with the same precision as in prior subsections.

For the purposes of calculating M_c, the function h need only be known in the core region. From [4.48b], [5.62], [5.63], and [5.81], straightforward integration yields

$$h = -\tfrac{1}{2}\beta^2(1 - \beta^2) + \beta^2\lambda$$
$$- (\tfrac{3}{2} + 4C_3)\beta^2\lambda^2 + o(\lambda^2). \quad [5.89]$$

Introduction of this and [5.88] into [5.85a]

ultimately gives

$$M_c = 1 - 4(1 - \beta_*{}^2)^3 + 3(1 - \beta_*{}^2)^4$$

$$- 8\left[1 - 3\int_0^{\beta_*} (1 - \beta^2)^2\beta^3 f_\perp(\beta)d\beta\right.$$

$$- 3(1 - \beta_*{}^2)^2 + 2(1 - \beta_*{}^2)^3\bigg]\lambda$$

$$+ 8\bigg[-12\int_0^{\beta_*} (1 - \beta^2)\beta^3 f_\perp(\beta)d\beta$$

$$+ 3\int_0^{\beta_*} (1 - \beta^2)^2\beta^3 f_\perp{}^2(\beta)d\beta + 3\beta_*{}^4$$

$$+ (\tfrac{3}{2} + 4C_3)(3\beta_*{}^4 - 2\beta_*{}^6)\bigg]\lambda^2$$

$$+ o(\lambda^2). \quad [5.90]$$

In the usual manner, use of the substitution [5.16], followed by an error analysis, eventually yields,

$$M_c = 1 - 8(1 - 3C_4)\lambda$$
$$+ 4(7 + 8C_3 - 24C_5)\lambda^2 + o(\lambda^2), \quad [5.91]$$

wherein (31),

$$C_4 = \int_0^1 (1 - \beta^2)^2\beta^3 f_\perp(\beta)d\beta = 0.1724, \quad [5.92]$$

$$C_5 = \int_0^1 (1 - \beta^2)\beta^3$$

$$\times [1 - \tfrac{1}{4}(1 - \beta^2)f_\perp(\beta)]f_\perp(\beta)d\beta$$

$$= 0.2483. \quad [5.93]$$

Evaluation of M_w requires knowledge of $d_\perp$ and h only in the wall region. As regards $d_\perp$, the general methods of Bungay and Brenner (17) applied to the sphere/plane–wall analyses of Brenner (7) and Cox and Brenner (21) show that[9] (31)

$$d_\perp(\delta, \lambda) = w_\perp(\delta) + O(\lambda), \quad [5.94]$$

[9] $w_\perp$ is equal to $(F/F_\infty)^{-1}$, where F is the force experienced by the sphere in the neighborhood of the wall. Thus, in the notation of Brenner (7), $w_\perp = \lambda^{-1}$. Alternatively, in the notation of Cox and Brenner (21), $w_\perp = F^{-1}$. Note that the second of Eqs. [5.95] combines with [5.94] to show that the diffusivity $d_\perp$ approaches zero (linearly with the gap width) as the sphere approaches contact with the wall.

analogous to [5.27]. Asymptotic values are (7, 21)

$$w_\perp \sim 1 - (9/8)\delta^{-1} + O(\delta^{-3}) \quad \text{as } \delta \to \infty,$$
$$\sim (\delta - 1)[1 + O(\delta - 1)\ln(\delta - 1)^{-1}] \quad [5.95]$$
$$\text{as } \delta \to 1.$$

Tabulated values of $w_\perp$ are available (7, 21) over the entire range, $1 \leq \delta < \infty$.

To obtain an expression for h which is valid in the wall region, observe from [4.48a] that

$$\int_0^{1-\lambda} [\bar{u}^*(\lambda) - u(\beta', \lambda)]\beta'd\beta' = 0. \quad [5.96]$$

Consequently, [4.48b] may be written alternatively as

$$h(\beta, \lambda) = -\int_{\beta'=\beta}^{1-\lambda} [\bar{u}^*(\lambda) - u(\beta', \lambda)]\beta'd\beta'. \quad [5.97]$$

With the use of [5.62] this yields

$$h = \tfrac{1}{2}[(1 - \beta^2) - (2\lambda - \lambda^2)]$$
$$\times [(1 - \beta^2) + (2\lambda - \lambda^2) - \bar{u}^*(\lambda)]$$
$$- J_w(\beta, \lambda), \quad [5.98]$$

with (cf. [5.69b])

$$J_w(\beta, \lambda) = \int_{\beta'=\beta}^{1-\lambda} u_s(\beta', \lambda)\beta'd\beta'. \quad [5.99]$$

In the integrand of M_w, β varies only from β_* to $1 - \lambda$. Since $\beta' \geq \beta$ in the above integrand, it follows that in [5.99] we may properly employ the expression [5.64] for u_s, whose validity is limited to the wall region. As in the evaluation of the comparable integral [5.69b], this makes (cf. [5.77])

$$J_w(\beta, \lambda) = 4C_3\lambda^2 + o(\lambda^2). \quad [5.100]$$

Introduction of this, as well as [5.81], into [5.98] yields, upon transforming to wall variables,

$$h = -(\delta - 1)\lambda + o(\lambda), \quad [5.101]$$

valid in the wall region.

In addition, from [5.7], in the wall region,

$$1/\beta = 1 + \delta\lambda + O(\lambda^2). \quad [5.102]$$

357

Written in terms of wall coordinates, [5.85b] becomes, with use of the last two equations,

$$M_w = 96\lambda^3 \int_1^{\delta*} \frac{(\delta - 1)^2 d\delta}{w_\perp(\delta)} + o(\lambda^3). \quad [5.103]$$

The second of Eqs. [5.95] shows that $w_\perp$ tends to zero like $(\delta - 1)$ as $\delta \to 1$. Hence, the integrand of [5.103] approaches zero as $\delta \to 1$ like $(\delta - 1)$. The fact that the denominator of [5.103] vanishes at the lower limit of integration does not therefore give rise to any difficulty in performing the integration. The function $w_\perp$ increases monotonically from 0 to 1 as δ increases from 1 to ∞. Since $\delta_* \gg 1$, it follows that the main contribution to the integral [5.103] arises from the larger values of δ. Since $w_\perp \sim 1 + O(\delta^{-1})$ and $(\delta - 1)^2 \sim \delta^2$ at these large values of δ, a simple integration yields

$$M_w = 32\lambda^3\delta_*^3 + \cdots. \quad [5.104]$$

Upon adding this to [5.90], the term dis, played explicitly in [5.104] must, of necessity, be canceled by a comparable term in [5.90]. In particular, it is canceled by one of the terms arising when the second term of M_c, i.e., $-4(1 - \beta_*^2)^3$, is expressed in terms of $\delta_*\lambda$ via [5.16]. Omitting details (31), the final result obtained for [5.84] is

$$M(\lambda) = M_c + o(\lambda^2), \quad [5.105]$$

where M_c is the quantity given explicitly in [5.91]. From [5.83] it now follows that

$$\bar{d}_v = (1-\lambda)^{-2}[1-3.862\lambda+14.40\lambda^2+o(\lambda^2)], \quad [5.106]$$

or, less accurately,

$$\bar{d}_v = 1 - 1.862\lambda + 9.68\lambda^2 + o(\lambda^2), \quad [5.107]$$

for the convective dispersion parameter.

A plot of [5.106] is presented in Fig. 5. To the order in λ indicated, this indicates the existence of a minimum, occurring at a value of $\lambda \approx 0.075$. Whether this is a true minimum, or merely an artifice of the approximate nature of the calculation, awaits the comparable numerical calculation for the "closely fitting" sphere case (18).

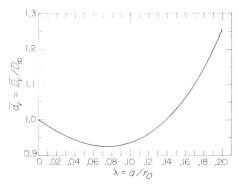

FIG. 5. Dimensionless convective contribution to the axial dispersivity vs sphere/cylinder radius ratio. This graph represents a plot of Eq. [5.106].

VI. NONSPHERICAL PARTICLES AND MORE GENERAL BOUNDARY SHAPES

The preceding calculations have dealt explicitly with spherical particles moving within circular cylinderical tubes. Clearly, it is desirable to extend these results to more general situations, involving nonspherical particles and/or boundaries of other shapes. Accordingly, a complete theoretical framework for achieving these aims is erected in this section. Only minimal additional conceptual apparatus is required beyond that needed to treat the more elementary sphere/cylinder problem already analyzed, although the numerical effort required to bring to fruition these more difficult problems can prove considerable. This more general theory transcends the specialized applications discussed in prior sections, and possesses ramifications with respect to a host of other related physical phenomena, including surface diffusion, aerosol deposition, and macromolecular separation schemes. Specifically, the range of phenomena falling within the purview of the subsequent theory encompasses those fields arising from the interaction between low Reynolds number flows, translational and rotational Brownian motions, external forces and couples exerted upon the Brownian particles, and hydrodynamic wall effects experienced by the latter.

The subsequent analysis may be regarded as an extension of the prior work of Brenner

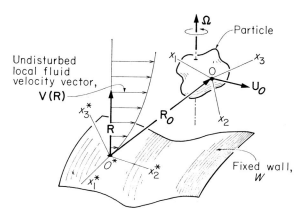

FIG. 6. Definition sketch. A particle of arbitrary shape, suspended in an undisturbed flow field $\mathbf{V}(\mathbf{R})$, translating and rotating in proximity to a fixed boundary W.

and Condiff (14, 15, 20), concerned with non-spherical Brownian particles in *unbounded* fluids, to include effects arising from the presence of bounding walls. The presence of such boundaries influences the probability distribution function σ of Brenner and Condiff by each of the following mechanisms: (i) It affects the hydrodynamic resistance of the particle, and hence the translational and rotational velocities $\mathbf{U}_o$ and $\mathbf{\Omega}$ with which the particle moves as it is transported by the flowing fluid; (ii) By virtue of the Stokes–Einstein equations (8, 9), the various diffusion tensors for the particle are now dependent upon both the position $\mathbf{R}_o$ and orientation ϕ of the particle relative to the boundaries; (iii) The wall itself may exert attractive or repulsive physicochemical forces upon a particle in its proximity.

Consider a rigid Brownian particle of arbitrary shape in proximity to a boundary W, as in Fig. 6. The instantaneous position of the particle relative to this boundary may be specified in terms of an origin O, rigidly affixed to the particle, which translates and rotates with it. This is termed the particle "locator point." Choose an arbitrary origin O^* fixed in the bounding wall. The position vector $\mathbf{R}_o = O^* \rightarrow O$ then serves to specify the instantaneous location of the particle in space. As

in earlier work (8, 9, 14, 15), the instantaneous orientation of the particle may be represented symbolically by the orientation angle "vector" ϕ, this being, for example, a triplet of Eulerian angles describing the orientation of a body-fixed Cartesian system of axes, Ox_j ($j = 1, 2, 3$), locked into the particle, relative to a space-fixed Cartesian system, Ox_j^*, fixed in the wall.[10]

The fundamental equations governing the diffusive and convective transport of the Brownian particle are the same as those given by Brenner and Condiff (14. 15), the only difference being that the various particle hydrodynamic resistance and diffusion tensors, which formerly depended only upon the particle orientation ϕ, now depend too upon the position $\mathbf{R}_o$ of the particle relative to the bounding wall (22, 31). That the basic *form* of the Brenner–Condiff equations remains unaltered is a direct consequence of the no-slip

[10] For some classes of problems, e.g., an axisymmetric particle in proximity to a plane wall or within a circular cylinder, the particle orientation can be specified more simply in terms of the orientation of a vector **e** (14, 15, 20), lying along the axis of revolution of the particle, relative to the normal to the plane wall or to the longitudinal axis of the circular cylinder, respectively. In such cases the subsequent equations simplify along the lines outlined for *unbounded* fluids in (14, 15, 20).

boundary condition satisfied by the un-perturbed local fluid velocity vector $\mathbf{V}(\mathbf{R})$ on the wall W. Here, $\mathbf{R}$ denotes the position vector of a fluid point relative to O^*.

Let

$$\sigma \equiv \sigma(\mathbf{R}_o, \boldsymbol{\phi}, t \mid \mathbf{R}_o', \boldsymbol{\phi}') \qquad [6.1]$$

denote the conditional probability density for simultaneously finding the particle (locator point) at position $\mathbf{R}_o$ and possessing an orien-tation $\boldsymbol{\phi}$ at time t, given that, initially (at time $t = 0$), the particle was located at $\mathbf{R}_o'$ and possessed an orientation $\boldsymbol{\phi}'$. Thus, with $d^3\mathbf{R}_o$ and $d^3\boldsymbol{\phi}$, respectively, denoting physical- and orientation-space volume elements, the quantity $\sigma d^3\mathbf{R}_o d^3\boldsymbol{\phi}$ denotes the probability of finding the particle locator point within the volume element $d^3\mathbf{R}_o$ centered at $\mathbf{R}_o$ and, simultaneously, possessing an orientation in the "region" between $\boldsymbol{\phi}$ and $\boldsymbol{\phi} + d^3\boldsymbol{\phi}$. Since the particle must possess *some* position and orientation, it follows that

$$\int_{\mathbf{R}_o} \int_{\phi} \sigma d^3\boldsymbol{\phi} d^3\mathbf{R}_o = 1 \qquad (t > 0). \quad [6.2]$$

The domain of integration, $\mathbf{R}_o \oplus \boldsymbol{\phi}$, is taken over all physically accessible orientations and positions, which—for most particle and wall geometries—will be "coupled."

Conservation and continuity of probability density leads to the fact that σ satisfies the differential equation (8, 9, 14, 15)

$$\frac{\partial \sigma}{\partial t} + \frac{\partial}{\partial \mathbf{R}} \cdot {}^t\mathbf{J}_o + \frac{\partial}{\partial \boldsymbol{\phi}_o} \cdot {}^r\mathbf{J}$$

$$= \delta(\mathbf{R}_o - \mathbf{R}_o')\delta(\boldsymbol{\phi} - \boldsymbol{\phi}')\delta(t) \quad (t > 0), \quad [6.3]$$

in addition to

$$\sigma = 0 \quad (t < 0). \qquad [6.4]$$

In the above, $\partial/\partial\mathbf{R} \equiv \partial/\partial\mathbf{R})_\phi$ denotes the physical-space gradient operator for a fixed particle orientation $\boldsymbol{\phi}$, whereas $\partial/\partial\boldsymbol{\phi}_o \equiv \partial/\partial\boldsymbol{\phi})_{\mathbf{R}_o}$ represents the pseudovector orien-tational gradient operator with respect to rotation about point O, each of these being with respect to a space-fixed reference frame.

The vector ${}^t\mathbf{J}_o$ and pseudovector ${}^r\mathbf{J}$ are, respectively, the translational flux of O and the rotational flux.

These fluxes are given by the constitutive relations (8, 9, 14, 15)

$$\begin{pmatrix} {}^t\mathbf{J}_o \\ {}^r\mathbf{J} \end{pmatrix} = \sigma \begin{pmatrix} \mathbf{U}_o \\ \boldsymbol{\Omega} \end{pmatrix} + \begin{pmatrix} {}^t\mathbf{D} & {}^c\mathbf{D}_o\dagger \\ {}^c\mathbf{D}_o & {}^r\mathbf{D}_o \end{pmatrix}$$

$$\cdot \left[\frac{\sigma}{kT} \begin{pmatrix} \mathbf{F} \\ \mathbf{T}_o \end{pmatrix} - \begin{pmatrix} \partial/\partial\mathbf{R} \\ \partial/\partial\boldsymbol{\phi}_o \end{pmatrix} \sigma \right], \quad [6.5]$$

in which

$$\mathbf{U}_o = d\mathbf{R}_o/dt, \quad \boldsymbol{\Omega} = d\boldsymbol{\phi}/dt, \qquad [6.6]$$

are a vector and pseudovector, respectively, representing the quasi-steady translational and angular velocities that a neutrally buoyant particle would possess by virtue of the fluid motion $\mathbf{V}(\mathbf{R})$ alone; that is, if no Brownian motion forces and torques, nor external forces and torques, acted upon the particle. In general,

$$\mathbf{U}_o \equiv \mathbf{U}_o(\mathbf{R}_o, \boldsymbol{\phi}), \quad \boldsymbol{\Omega} \equiv \boldsymbol{\Omega}(\mathbf{R}_o, \boldsymbol{\phi}). \quad [6.7]$$

These velocities are also an implicit function of time due to the dependence of $\mathbf{R}_o$ and $\boldsymbol{\phi}$ on t.

The 6×6 matrix of diffusion tensors is related to the matrix of particle hydrodynamic resistance tensors via the generalized Stokes–Einstein equation (8, 9)

$$\begin{pmatrix} {}^t\mathbf{D} & {}^c\mathbf{D}_o\dagger \\ {}^c\mathbf{D}_o & {}^r\mathbf{D}_o \end{pmatrix} = \frac{kT}{\mu} \begin{pmatrix} {}^t\mathbf{K}_o & {}^c\mathbf{K}_o\dagger \\ {}^c\mathbf{K}_o & {}^r\mathbf{K} \end{pmatrix}^{-1}. \quad [6.8]$$

In general, relative to a space-fixed observer, the resistance tensors, ${}^{(\ )}\mathbf{K}$, and, hence, the diffusion tensors, ${}^{(\ )}\mathbf{D}$, are explicit functions of $\mathbf{R}_o$ and $\boldsymbol{\phi}$, and, therefore, implicit functions of t. In contrast, in an unbounded fluid, these phenomenological tensors are constant relative to axes fixed in the particle and, hence, explicit functions only of $\boldsymbol{\phi}$; that is, they are in-dependent of $\mathbf{R}_o$. Moreover, whereas in unbounded fluids these tensors depend only upon the size and shape of the particle, in bounded fluids they also depend, inter alia, upon the size and shape of the boundary W (22, 31). Hence, though "coupling" terms, such as $^c\mathbf{K}_o$ and $^c\mathbf{D}_o$, vanish in unbounded fluids for centrally symmetric particles (provided that O is chosen to lie at the center of symmetry), this is no longer true when a bounding wall is present. Such terms are then nonzero, even for spherical particles (cf. [2.13c]).

The external vector force $\mathbf{F}$ and pseudo-vector torque $\mathbf{T}_o$ (about O) exerted on the Brownian particle will be assumed to derive from a (nondimensional) potential energy function $E(\mathbf{R}_o, \boldsymbol{\phi})$:

$$\begin{pmatrix} \mathbf{F} \\ \mathbf{T}_o \end{pmatrix} = -kT \begin{pmatrix} \partial/\partial\mathbf{R} \\ \partial/\partial\boldsymbol{\phi}_o \end{pmatrix} E. \quad [6.9]$$

In combination, the preceding relations yield the convective–diffusion equation

$$\frac{\partial\sigma}{\partial t} + \frac{\partial}{\partial\mathbf{R}} \cdot (\sigma\mathbf{U}_o) + \frac{\partial}{\partial\boldsymbol{\phi}_o} \cdot (\sigma\boldsymbol{\Omega}) = (\partial/\partial\mathbf{R}, \partial/\partial\boldsymbol{\phi}_o)$$

$$\cdot e^{-E} \begin{pmatrix} ^t\mathbf{D} & ^c\mathbf{D}_o\dagger \\ ^c\mathbf{D}_o & ^r\mathbf{D}_o \end{pmatrix} \cdot \begin{pmatrix} \partial/\partial\mathbf{R} \\ \partial/\partial\boldsymbol{\phi}_o \end{pmatrix} (\sigma e^E)$$

$$+ \delta(\mathbf{R}_o - \mathbf{R}_o')\delta(\boldsymbol{\phi} - \boldsymbol{\phi}')\delta(t)$$

$$(t > 0), \quad [6.10]$$

in the six-dimensional $\mathbf{R}_o \oplus \boldsymbol{\phi}$ hyperspace. This equation is to be solved subject to the no-flux condition,

$$\mathbf{v}\cdot {}^t\mathbf{J}_o = 0 \quad \text{on } \partial V, \quad [6.11]$$

across the physical-space boundary ∂V which bounds the physical volume V; $\mathbf{v}$ is a unit vector normal to W. In addition, when a portion of the physical boundary extends to infinity, it is also required that

$$\sigma \to 0 \quad \text{as } |\mathbf{R} - \mathbf{R}'| \to 0. \quad [6.12]$$

Equation [6.4] is also pertinent.

Equation [6.10] constitutes the counterpart of [2.12], to which it reduces for spherical particles, in which case questions of particle orientation no longer arise. This can be demonstrated formally by defining the marginal density, P, of σ, as

$$P(\mathbf{R}_o, t) = \oint_\phi \sigma(\mathbf{R}_o, \boldsymbol{\phi}, t)d^3\boldsymbol{\phi}, \quad [6.13]$$

as well as the total translation flux, irrespective of orientation,

$$\mathbf{J}_o(\mathbf{R}_o, t) = \oint_\phi {}^t\mathbf{J}_o(\mathbf{R}_o, \boldsymbol{\phi}, t)d^3\boldsymbol{\phi}. \quad [6.14]$$

Here, O refers to the sphere center. All orientations are then accessible, and the indicated integrations in the above are taken over all of the orientation space. As a consequence of the closure of this space, the following theorem obtains (14):

$$\oint_\phi (\partial\boldsymbol{\psi}/\partial\boldsymbol{\phi})d^3\boldsymbol{\phi} = 0, \quad [6.15]$$

for $\boldsymbol{\psi}$ any single-valued, orientation-dependent, tensor field of arbitrary rank. Application of this theorem to [6.3] and [6.5] gives

$$\frac{\partial P}{\partial t} + \frac{\partial}{\partial\mathbf{R}} \cdot \mathbf{J}_o$$

and

$$= \delta(\mathbf{R}_o - \mathbf{R}_o')\delta(t) \quad (t > 0), \quad [6.16]$$

$$\mathbf{J}_o = P\mathbf{U}_o - {}^t\mathbf{D} \cdot \left(\frac{\partial P}{\partial\mathbf{R}} - \frac{P}{kT}\mathbf{F}\right), \quad [6.17]$$

provided that the sphere is "homogeneous," so that $\mathbf{F}$ is independent of the sphere orientation; i.e., E is independent of $\boldsymbol{\phi}$, with $\mathbf{F} = -\partial E/\partial\mathbf{R}$. Equation [6.17] follows as a consequence of the facts that $\mathbf{U}_o$, $^t\mathbf{D}$, and $^c\mathbf{D}_o$ are independent of $\boldsymbol{\phi}$ for spherical particles.

The preceding equations are identical to [2.4] and [2.6] upon suppressing the subscript o and employing the more conventional notation $\partial/\partial\mathbf{R} \equiv \nabla$. Furthermore, [6.11] and [6.14] combine to show that $\mathbf{v}\cdot\mathbf{J}_o = 0$ on ∂V, in accord with [2.28]. These equations demonstrate that the distribution of sphere orientations is physically irrelevant with respect to the distribution of sphere centers in physical space—a result which, of course, was to be anticipated.

Solution of [6.10] requires, among other things, knowledge of $\mathbf{U}_o$ and $\mathbf{\Omega}$. These quantities are not generally the same as the undisturbed fluid velocity vector $\mathbf{V}(\mathbf{R}_o)$ and the undisturbed fluid vorticity pseudovector $\boldsymbol{\omega}(\mathbf{R}_o)$ $\equiv \frac{1}{2}\nabla \times \mathbf{V}|_{\mathbf{R}_o}$, respectively, owing to both wall effects (22, 31) and other hydrodynamic effects[11] (15). General procedures for determining the required translational and angular "slip" velocities, $\mathbf{V}(\mathbf{R}_o) - \mathbf{U}_o$ and $\boldsymbol{\omega}(\mathbf{R}_o) - \mathbf{\Omega}$, respectively, are discussed by Gaydos (31).

Equations [6.10] and [6.4], subject to boundary conditions [6.11] and [6.12], constitute the starting point for all future developments in this series. Their moments, e.g., for cylindrical ducts,

$$\bar{z} = \int_{\mathbf{R}_o} \int_\phi z\sigma d^3\boldsymbol{\phi} d^3\mathbf{R}_o \Big/$$

$$\int_{\mathbf{R}_o} \int_\phi \sigma d^3\boldsymbol{\phi} d^3\mathbf{R}_o \quad [6.18]$$

and

$$\overline{(z - \bar{z})^2} = \int_{\mathbf{R}_o} \int_\phi (z - \bar{z})^2\sigma d^3\boldsymbol{\phi} d^3\mathbf{R}_o \Big/$$

$$\int_{\mathbf{R}_o} \int_\phi \sigma d^3\boldsymbol{\phi} d^3\mathbf{R}_o, \quad [6.19]$$

permit the mean axial particle velocity $\bar{U}^*$ and dispersivity $\bar{D}^*$ to be computed from [3.7] and [3.9], respectively. Apart from purely algebraic complications, the main barrier to such calculations resides in the current lack of knowledge of the requisite hydrodynamic resistance tensors for nonspherical particles in proximity to boundaries. (See, however, Refs. (22, 31).) Future developments must therefore wait upon the computation of these tensors by the solution of appropriate, low-Reynolds-number flow problems. It is possible, however, to derive some limiting results to terms of lowest order

[11] Differences exist even in *unbounded* fluids, where no wall effects contravene. Thus, for example, $\Omega \neq \omega(\mathbf{R}_o)$ for a spheroid (42) or other body of revolution (12, 13) in a simple shear field. Likewise, $\mathbf{U}_o \neq \mathbf{V}(\mathbf{R}_o)$ in a simple shear field if the body of revolution lacks a center of symmetry (12, 26, 51).

in λ without requiring such solutions, at least for axisymmetric particles. These nonspherical particle results will be reported elsewhere for motion in circular cylindrical tubes.

For nonspherical particles, *physical* results such as $\bar{U}^*$ and $\bar{D}^*$ must prove to be independent of the choice of particle locator point O. This fact can provide an incisive prove in establishing the invariant structure of the underlying equations.

VII. DISCUSSION

The principal results of the present investigation are encompassed by Eqs. [5.61a], [5.82], and [5.106] for $\bar{d}_{11}$, $\bar{u}^*$ and $\bar{d}_v$, respectively.

Comparison with the results of Anderson and Quinn (2). Present results for the Brownian contribution $\bar{d}_{11}$ to the dispersivity $\bar{d}^*$ may be compared with those obtained by the above authors (2) via a steady-state analysis. Apart from strictly numerical questions pertaining to the efficacy of their integration scheme, the general formula utilized by them for computing this diffusivity differs from our Eq. [4.47] by their steric factor of $(1 - \lambda)^2$. In particular, the mean axial molecular diffusivity $\bar{D}$ is given by them as

$$\xi \overset{\text{def}}{=} \bar{D}/D_\infty = 2\int_0^{1-\lambda} d_{11}\beta d\beta, \quad [7.1]$$

in contrast to [4.47]. This dimensionless diffusivity ratio is related to $\bar{d}_{11}$ by the expression

$$\bar{d}_{11} = \xi/(1 - \lambda)^2. \quad [7.2]$$

In a semantic sense, it is tempting to attempt to rationalize this discrepancy in terms of the existence of two possible choices for the definition of the mean axial diffusivity $\bar{D}_{11}$ in Fick's law of diffusion, according as the particle number flux density is defined per unit of *total* pore cross-sectional area, πr_o^2, or per unit of *accessible* cross-sectional area, $\pi(r_o - a)^2 \equiv \pi r_o^2(1 - \lambda)^2$. This, however, begs the point, since the diffusivity may also be regarded as unambiguously defined by Einstein's relation (cf. [3.8] and [4.6]) for the mean-squared displacement of the sphere center, and this definition does not hinge upon such flux questions.[12] In particular, the upper limit of integration $r_o - a$ in [3.1] and [3.2] derives from the fact that $P = 0$ for $r > r_o - a$, and not from any areal questions. Thus, in our view, the definition of diffusivity adopted by Anderson and Quinn is fundamentally incorrect in a continuum-mechanical sense. This issue is further addressed in Appendix C, which reproduces their derivation in sufficient detail to make the source of the discrepancy transparent. The ultimate conclusion is that the Anderson–Quinn diffusivity definition is tenable only in a global, irreversible-thermodynamic context, but not in a local continuum-mechanical context. Interestingly, this disparity seems not to have surfaced before in discussions of membrane transport, perhaps because the duality disappears in the limit $\lambda = 0$.

Despite existing differences with respect to physical interpretation, it is nevertheless possible to make a numerical comparison. Values of ξ vs λ are presented in graphical form by Anderson and Quinn (2). In effect, by arbitrarily choosing $\delta_* = 2$, they have numerically integrated [7.1] by employing the results of Hirschfeld and Brenner (38, 39) in the core region and those of Goldman et al. (33) in the wall region. Having ignored "coupling" in the latter region (by assuming $^cK_{23} = 0$ in [2.15c]), their results cannot be strictly correct. However, since d_{11} is itself small in the immediate proximity of the wall, the resulting numerical error is not large. In general, the quantitative agreement, shown in Fig. 3, between [5.61a] and the numerical values of $\bar{d}_{11}$ derived from [7.2] is very good.

That the leading wall correction term in [5.61b] is logarithmic in λ, rather than linear, shows that the so-called centerline approxi-

[12] Furthermore, when a wall potential is included, the discrepancy between the two sets of results now involves a factor of $\Phi = (2/r_o^2) \int_0^{r_o-a} e^{-E} r\, dr$, in place of $(1 - \lambda)^2$. This is not purely an areal ratio, so that the disparity is necessarily more basic in nature.

mation (4, 46, 52),

$$\bar{d}_{11} = 1 - 2.104\lambda + O(\lambda^2), \qquad [7.3]$$

is fundamentally incorrect, at least in circumstances where no wall potential exists. The only situation in which such an approximation would be rigorously correct, arises in circumstances where a wall potential exists, possessing a sharp minimum in the neighborhood of the tube axis. If, for example, the potential energy function in [4.42] was of the Dirac delta function form,

$$E(\beta) = \delta(\beta), \qquad [7.4]$$

Eq. [7.3] would be unequivocally correct; for the sphere center would then possess zero probability of ever being located at any radial position other than $\beta = 0$; that is, [4.42] would reduce to

$$\bar{d}_{11} = d_{11}(\beta = 0). \qquad [7.5]$$

As regards the mean particle velocity $\bar{u}^*$, Anderson and Quinn (2)—once more employing a steady-state analysis—derive a general expression for this quantity, which again differs from ours by their steric factor. In particular, they take the mean axial particle velocity (made dimensionless with the mean Poiseuille velocity $\bar{V}_m$) to be given by the expression

$$\chi = 2 \int_0^{1-\lambda} u\beta\, d\beta, \qquad [7.6]$$

rather than [4.46]. These quantities are related by

$$\bar{u}^* = \chi/(1 - \lambda)^2. \qquad [7.7]$$

For χ, Anderson and Quinn (2) give

$$\chi \approx (1 - \lambda)^2 [2 - (1 - \lambda)^2]$$
$$\times [1 - \tfrac{2}{3}\lambda^2 - 0.163\lambda^3], \qquad [7.8]$$

an approximation which they recognize as being inadequate, owing to the fact that the last term in square brackets is based entirely upon a "centerline approximation" for the slip velocity u_s.

Again, we disagree with that χ is to be interpreted as the mean particle velocity, since the latter can be unambiguously defined by the change in mean particle position $\bar{z}$ with time (cf. [3.1] and [3.6]). However, as in the case of the axial diffusivity, their expression is, nevertheless, appropriate within the context of irreversible thermodynamics.

Anderson and Quinn (2) do not quantitatively consider the convective contribution $\bar{d}_v$ to the dispersivity $\bar{d}^*$, except to comment that in circumstances (Pe $\gg$ 1) where the "Taylor-diffusion" contribution, $\bar{d}_v$, is sensible compared with the "Aris molecular diffusion" contribution, $\bar{d}_{11}$, solute dispersion is unimportant anyway compared with solute convection. Being based solely on steady-state arguments, this conclusion lacks general validity.

Comparison with the results of DiMarzio and Guttman (23–25, 35). In a series of papers preceding the publication of Anderson and Quinn (2), the above authors address themselves to essentially the same class of problems, namely the convective–diffusive transport of particles of finite size through cylindrical pores. The context of the latter investigations lies, however, in the area of gel-permeation chromatographic separation processes occurring in packed beds, rather than in the area of membrane transport. Cylindrical pores were selected as furnishing capillary models of the interstitial space in such beds.

Though recognition is given to the pore volume exclusion, occasioned by the finite particle size, these authors do not strictly take account of hydrodynamic wall effects, except partially in regard to its effect upon the slip velocity in the core region. On the other hand, an attempt is made by them to model "particles" other than hard spheres—in particular, free-draining models of polymer molecules. Employing the same definition of $\bar{u}^*$ as in [4.46], but incorrectly using the expression [5.63] for the slip velocity over the *entire* cross section, they obtain the expression

$$\bar{u}^* = 1 + 2\lambda - (7/3)\lambda^3, \qquad [7.9]$$

for hard spheres. This contrasts with the correct result [5.81]. The disagreement in the

$O(\lambda^2)$ term stems, of course, from their failure to properly take account of the contribution to the hydrodynamic retardation resulting from the presence of the sphere in the immediate neighborhood of the wall. The $O(\lambda)$ term derives merely from a "cutoff" of the Poiseuille velocity distribution at $\delta = 1$, and hence is correctly given by their elementary analysis.

DiMarzio and Guttman also consider the Taylor–Aris convective contribution to the dispersion. For hard spheres they derive

$$\bar{d}_{11} = 1 \qquad [7.10]$$

and

$$\bar{d}_v = (1 - \lambda)^6 \equiv 1 - 6\lambda + 15\lambda^2 + O(\lambda^3), \qquad [7.11]$$

by considering only the volume exclusion, but not the hydrodynamic wall effects. The former of these may be compared with the correct expression [5.61b], and the latter with

$$\bar{d}_v = 1 - (6 - 24C_4)\lambda + (15 + 32C_3 + 48C_4 - 96C_5)\lambda^2 + o(\lambda^2), \qquad [7.12]$$

derived from [5.91], [5.105], and [5.83]. (Numerically, the latter is equivalent to [5.107].) Comparison of [7.12] with [7.11] clearly shows the errors resulting from failure to properly account for wall effects, such effects being embodied in the coefficients C_3, C_4, and C_5.

General discussion. Two of the more striking predictions of the preceding theory, at least for sufficiently small λ, are that: (i) On average, the Brownian particle moves faster than the fluid in which it is suspended; (ii) The larger the particle, the more rapidly it moves. Each of these conclusions follows as an immediate consequence of the first-order relation, $\bar{u}^* = 1 + 2\lambda$. Qualitative explanations of each of these phenomena are offered by DiMarzio and Guttman, namely: (i) Because of its finite size, the sphere (center) is not free to sample the slowest moving streamlines near to the wall, in contrast to a fluid particle; (ii) The larger the sphere, the larger the inaccessible portion of the cross-sectional area near to the walls.

The former of these accounts for the fact that $\bar{u}^* > 1$; the latter accounts for the fact that $d\bar{u}^*/d\lambda > 0$, i.e., for a fixed tube radius, $(\partial\bar{u}/\partial a)_{r_o} > 0$. The retarding effect of the wall upon $\bar{u}^*$ does not begin to manifest itself until terms of $O(\lambda^2)$ are considered in the analysis. This retardation accounts for the negative algebraic sign in the coefficient of the quadratic term in [5.82]. It will prove informative to establish whether or not these same conclusions persist for the case of large, closely fitting (18) spheres.

DiMarzio and Guttman first advanced these ideas as a qualitative explanation of the observed outcome of gel-permeation chromatographic separation experiments, wherein large solute molecules are generally observed to transverse the bed more rapidly than smaller ones. Whether or not this rationale is correct remains to be established by quantitative experiments. Alternative theories exist (5), purporting to explain this phenomenon on the basis of preferential adsorption and retention of the smaller molecules on bed grains, or the enhanced ability of smaller molecules to diffuse into narrowly constricted, dead-end pores.

Qualitative comparison with the experiments of Small (55). Qualitative support for some aspects of the prior theoretical analysis exists in the experimental observations of Small (55). His investigation pertains, inter alia, to the separation of colloidal particles of precisely defined sizes, suspended in ionic solvents, by passage through packed beds. Vis-à-vis these "hydrodynamic chromatography" (HDC) experiments, some of the alternative, nonhydrodynamic, gel-permeation chromatographic explanations of size fractionation, cited above, lose their credibility when going from macromolecular- to colloidal-sized particles.

Small's data convincingly accord with the theoretical conclusion that $\bar{u}^* > 1$; moreover, all other things being equal, the larger colloidal particles traverse the bed more rapidly than do smaller ones, as demanded by the theory. Quantitative comparisons are not possible in the absence of detailed knowledge of fluid velocity distributions in packed beds— particularly the extent to which circular cylindrical pores are likely to provide realistic models of the void-space distributions occurring within the interstices.

A decrease in the ionic strength of the solvent, corresponding to diminished electrolyte concentrations, manifests itself locally by an increased electrostatic wall repulsion. In turn, this biases the spatial probability distribution toward enhanced centerline particle densities, and hence toward an increased value of $\bar{u}^*$—this being a consequence of [4.41]. (For example, the extreme repulsive potential [7.4] gives $\bar{u}^* = u(\beta = 0)$, which, from [5.62] and [5.63], yields

$$\bar{u}^* = 2 - (\tfrac{4}{3})\lambda^2 + O(\lambda^3). \qquad [7.13]$$

This centerline velocity greatly exceeds the random distribution result, [5.82].) Small's (2) data support this inference, in that a diminution in ionic strength resulted in a monotonic increase in mean particle speed through the bed, all other things being equal.

At high ionic strengths, where little wall repulsion exists, Small (2) observes the existence of a maximum occurring in a plot of what is essentially $\bar{u}^*$ vs sphere radius a. While he suggests the possibility that this may be a consequence of attractive van der Waals forces coming to dominate over the now relatively weak, double-layer, repulsive forces, an equally credible, and simpler, explanation is possible on the basis of [5.82] or, more accurately, [5.80]. Differentiation of the latter predicts a maximum value of $\bar{u}^*$ at $\lambda \approx 0.14$, beyond which $\bar{u}^*$ *decreases* with increasing particle size. Of course this rough calculation neglects the higher-order λ terms in [5.82]. 'While, for the reasons cited earlier, this result cannot be quantitatively compared with Small's data, neither can it be dimissed a priori as the underlying cause of the observed maximum.

APPENDIX A: UNIQUENESS

It will be demonstrated here that the systems of equations

$$\partial f/\partial t + \nabla\cdot\mathbf{j} = \delta(\mathbf{R} - \mathbf{R}')\delta(t) \quad (t > 0), \quad [A.1]$$

and[13]

$$f = 0 \quad (t < 0), \quad [A.2]$$

with

$$\mathbf{j} = \mathbf{U}f - e^{-E}\mathbf{D}\cdot\nabla(e^{E}f), \quad [A.3]$$

possesses a unique solution, $f(\mathbf{R}, t|\mathbf{R}')$, provided that

$$\mathbf{v}\cdot\mathbf{j} = 0 \quad \text{on } \partial V, \quad [A.4]$$

and that

$$f \to 0 \quad \text{as } |z - z'| \to \infty \quad [A.5]$$

sufficiently rapidly with $|z - z'|$. Here,

$$\partial V = S_w \oplus S_{+\infty} \oplus S_{-\infty} \quad [A.6]$$

denotes the boundaries of the accessible fluid volume,

$$V = \{-\infty < z < \infty\}$$
$$\oplus \{0 \leq r \leq r_o - a\}, \quad [A.7]$$

interior to the infinitely long cylindrical tube; S_w, $S_{+\infty}$, and $S_{-\infty}$, respectively, represent the infinite cylindrical "tube wall" surface $\{-\infty < z < \infty; r = r_o - a\}$, the "exit" end of the tube $\{z = \infty; 0 \leq r < r_o - a\}$, and the "inlet" end of the tube $\{z = -\infty; 0 \leq r < r_o - a\}$. Furthermore, $\mathbf{v}$ denotes a unit normal vector on one of the above three surfaces. In the above,

$$\mathbf{D} = \mathbf{i}_r\mathbf{i}_r D_\perp(r) + \mathbf{i}_z\mathbf{i}_z D_{\shortparallel}(r), \quad [A.8]$$

$$\mathbf{U} = \mathbf{i}_z U(r), \quad [A.9]$$

and

$$E = E(r). \quad [A.10]$$

As discussed following [2.29], the boundary condition [A.4] may be automatically satisfied on the "wall" S_w, owing to the fact that $\mathbf{D}$

[13] Following Friedman (29) we can restrict ourselves exclusively to consideration of only the case $t > 0$ by defining $f(\mathbf{R}, t|\mathbf{R}') = f^+(\mathbf{R}, t|\mathbf{R}')H(t)$, with $H(t) = 1$ for $t > 0$, and $H(t) = 0$ for $t < 0$, the Heaviside unit step function, and f^+ the solution of the preceding differential equation for $t > 0$. However, we shall not pursue this alternative scheme.

and $\mathbf{U}$ vanish there. However, we shall not avail ourselves of this possibility.

The uniqueness proof starts with a demonstration that f automatically fulfills the normalization condition

$$\int_V fd^3\mathbf{R} = 1, \quad (t > 0), \quad [A.11]$$

by virtue of the presence of the source term in [A.1]. Integrate [A.1] over the volume V and employ the divergence theorem to obtain

$$\int_V (\partial f/\partial t)d^3\mathbf{R} + \oint_{\partial V} ds\mathbf{v}\cdot\mathbf{j} = \delta(t).$$

The surface integral vanishes in consequence of [A.4]. Hence,

$$(d/dt) \int_V f(\mathbf{R}, t)d^3\mathbf{R} = \delta(t)$$

results, upon suppressing the constant $\mathbf{R}'$ in the argument of f. Integration of this expression from $t = -|T|$ to any $t > 0$ yields

$$\int_V f(\mathbf{R}, t)d^3\mathbf{R} - \int_V f(\mathbf{R}, -|T|)d^3\mathbf{R} = 1$$
$$\text{for } t > 0, \quad [A.12]$$

since, by the properties of the delta function,

$$\int_{-|T|}^{t} \delta(t)dt = 1 \quad \text{for any } t > 0,$$

provided that $T \neq 0$. However, in view of [A.2],

$$\int_V f(\mathbf{R}, -|T|)d^3\mathbf{R} = 0,$$

for any $T \neq 0$. Introduction of this into [A.12] shows that the normalization condition [A.11] is indeed fulfilled for $t > 0$.

In order to prove uniqueness, let f_1 and f_2 each be solutions of [A.1] to [A.5], and define the function

$$f_0 = f_1 - f_2. \quad [A.13]$$

In view of the linearity of the system of

Eqs. [A.1] to [A.5], f_0 satisfies the homogeneous equation

$$\partial f_0/\partial t + \nabla \cdot \mathbf{j}_0 = 0 \quad (t > 0), \quad [A.14]$$

along with Eqs. [A.2] to [A.5], in which a subscript "0" is appended to f and $\mathbf{j}$. Here, $\mathbf{j}_0 = \mathbf{j}_1 - \mathbf{j}_2$. A uniqueness proof consists of demonstrating that this *homogeneous* system of equations possesses only the trivial solution $f_0 = 0$; for this makes $f_1 = f_2$, which, in turn, shows that all solutions of the system of f equations are necessarily identical.

(The uniqueness proof which follows applies equally well to the case of an *arbitrary* initial distribution, $f(\mathbf{R}, 0) = \eta(\mathbf{R})$, satisfying

$$\partial f/\partial t + \nabla \cdot \mathbf{j} = \eta(\mathbf{R})\delta(t), \quad [A.15]$$

in place of [A.1]. For if f_0 is defined as in [A.13], then f_0 once again satisfies [A.14].)

Multiply [A.14] by the function $e^E f_0 = \psi_0$, say, and apply an obvious vector identity to obtain

$$(\partial/\partial t)(\tfrac{1}{2}e^E f_0{}^2) + \nabla \cdot (\mathbf{j}_0 \psi_0) - \mathbf{j} \cdot \nabla \psi_0 = 0,$$

since E is independent of t. However, from [A.3],

$$\mathbf{j} = \mathbf{U} f_0 - e^{-E} \mathbf{D} \cdot \nabla \psi_0.$$

This makes

$$(\partial/\partial t)(\tfrac{1}{2}e^E f_0{}^2) + e^{-E}\nabla\psi_0 \cdot \mathbf{D} \cdot \nabla\psi_0$$
$$= f_0 \mathbf{U} \cdot \nabla\psi_0 - \nabla \cdot (\mathbf{j}_0 \psi_0). \quad [A.16]$$

By some obvious vector identities,

$$f_0 \mathbf{U} \cdot \nabla\psi_0 = \tfrac{1}{2}\mathbf{U} \cdot \nabla(e^E f_0{}^2) + \tfrac{1}{2}e^E f_0{}^2 \mathbf{U} \cdot \nabla E.$$

However, from [A.9] and [A.10], the vector fields $\mathbf{U}$ and ∇E are mutually perpendicular, whereupon $\mathbf{U} \cdot \nabla E = 0$. Moreover, it follows from [A.9] that $\nabla \cdot \mathbf{U} = 0$, whence by vector identity

$$\mathbf{U} \cdot \nabla(e^E f_0{}^2) = \nabla \cdot (\mathbf{U} e^E f_0{}^2).$$

In this manner [A.16] becomes

$$(\partial/\partial t)(\tfrac{1}{2}e^E f_0{}^2) + e^{-E}\boldsymbol{\chi} \cdot \mathbf{D} \cdot \boldsymbol{\chi}$$
$$= \nabla \cdot (\tfrac{1}{2}\mathbf{U}e^E f_0{}^2 - \mathbf{j}_0 e^E f_0),$$

with $\boldsymbol{\chi} = \nabla\psi_0$.

The preceding equation may be integrated over the volume V, and the divergence theorem

applied to the term on the right-hand side, to obtain

$$dJ/dt + \int_V e^{-E}\boldsymbol{\chi} \cdot \mathbf{D} \cdot \boldsymbol{\chi} d^3\mathbf{R} = \oint_{\partial V} ds\mathbf{v} \cdot \tfrac{1}{2}\mathbf{U}e^E f_0{}^2$$

$$- \oint_{\partial V} ds\mathbf{v} \cdot \mathbf{j}_0 e^E f_0, \quad [A.17]$$

in which

$$J(t) = \tfrac{1}{2}\int_V e^E f_0{}^2 d^3\mathbf{R}. \quad [A.18]$$

The last integral on the right-hand side of [A.17] vanishes by virtue of the boundary condition [A.4]. As regards the remaining integral on the right, [A.5] shows that $f_0 = 0$ on $S_{+\infty}$ and $S_{-\infty}$. In addition, on S_w, $\mathbf{v} = \mathbf{i}_r$, with $\mathbf{i}_r$ a unit radial vector in cylindrical coordinates. In view of [A.9], this makes $\mathbf{v} \cdot \mathbf{U} = 0$ on S_w in consequence of the orthogonality of the unit vectors $\mathbf{i}_r$ and $\mathbf{i}_z$. Thus, the first surface integral on the right-hand side of [A.17] vanishes too. Consequently,

$$dJ/dt = -\int_V e^{-E}\boldsymbol{\chi} \cdot \mathbf{D} \cdot \boldsymbol{\chi} d^3\mathbf{R}. \quad [A.19]$$

Since the diffusivity dyadic $\mathbf{D}$ is necessarily a symmetric positive-definite form, it follows that

$$\boldsymbol{\chi} \cdot \mathbf{D} \cdot \boldsymbol{\chi} \geq 0 \quad \text{for all} \quad \mathbf{R} \in V,$$

for *any* arbitrary vector field $\boldsymbol{\chi}$. Thus

$$dJ/dt \leq 0. \quad [A.20]$$

However, from its definition, J is itself nonnegative. Therefore, integration of the inequality [A.20] from $t = -|T|$ to any $t > 0$ yields

$$J(t) - J(-|T|) \leq 0. \quad [A.21]$$

However, from [A.2] and the definition [A.18], it follows that

$$J(-|T|) = 0 \quad \text{for any} \quad T \neq 0.$$

This makes[14]

$$J(t) \leq 0. \qquad [A.22]$$

On the other hand, the integrand of [A.18] is nonnegative, requiring that

$$J(t) \geq 0. \qquad [A.23]$$

The last two equations can be simultaneously satisfied if, and only if,

$$J(t) = 0. \qquad [A.24]$$

Since the integrand of [A.18] is nonnegative, this condition can be satisfied if, and only if,

$$f_0 = 0 \qquad [A.25]$$

at each point $\mathbf{R} \in V$, and for all t. Uniqueness is thereby demonstrated.

This uniqueness proof, while suggestive, is not wholly rigorous in the absence of a priori estimates assuring convergence of all the integrals arising in the proof. Indeed, in a related, heat-conduction, context, Carslaw and Jaeger (19) cite a counterexample which shows that a priori uniqueness proofs of this nature must be regarded with some degree of caution. A posteriori uniqueness proofs, which are free of this flaw, require, however, at least partial knowledge of a complete solution.

APPENDIX B: DIRECT DERIVATION OF $\bar{D}_{11}$ AND $\bar{U}^*$

The expression [3.48] for the mean molecular diffusivity $\bar{D}_{11}$ can be derived directly, without the necessity of a moment analysis of the Taylor–Aris type. The physical significance of $\bar{D}_{11}$ is such that an expression for it can be obtained by considering the case of no flow, $\bar{V}_m = 0$.

In the absence of convection, [3.2] reduces to

$$\bar{z^2} = \int_V z^2 P dV \bigg/ \int_V P dV, \qquad [B.1]$$

with V corresponding to the region [2.2], and $dV \equiv d^3\mathbf{R} = r dr d\phi dz$. From [2.1] the integral in the denominator is a constant, independent of time. Hence, time differentiation of the above yields

$$d\bar{z^2}/dt = \int_V z^2(\partial P/\partial t)dV \bigg/ \int_V P dV.$$

With use of [2.4] (choosing $z' = 0$, as in [2.41]) this becomes

$$d\bar{z^2}/dt = - \int_V z^2 \nabla \cdot \mathbf{J} dV \bigg/ \int_V P dV, \qquad [B.2]$$

where, with $\mathbf{U} = 0$ in [2.10],

$$\mathbf{J} = -\mathbf{D} \cdot (\nabla P + P\nabla E). \qquad [B.3]$$

By some elementary vector identities,

$$z^2 \nabla \cdot \mathbf{J} = \nabla \cdot (\mathbf{J}z^2) - 2zJ_z,$$

where $J_z = \mathbf{i}_z \cdot \mathbf{J}$. Application of the divergence theorem thereby yields

$$\int_V z^2 \nabla \cdot \mathbf{J} dV = \oint_{\partial V} d\mathbf{s}_V \cdot \mathbf{J}z^2$$
$$- 2 \int_V zJ_z dV. \qquad [B.4]$$

From the requirement that $\mathbf{v} \cdot \mathbf{J} = 0$ on the surface ∂V bounding V (cf. the comparable condition imposed upon f in [A.4]), the surface integral is zero. In addition, from [B.3], [2.17], and the fact that E is independent of z, it follows that

$$J_z = -D_{11}(r)\partial P/\partial z. \qquad [B.5]$$

Hence, we obtain,

$$d\bar{z^2}/dt = -2 \int_A \int_{z=-\infty}^{\infty} D_{11}(r)z(\partial P/\partial z)dzdA \bigg/ \int_V P dV,$$

---

[14] The procedure utilized in deriving [A.22] may appear contrived, since it depends upon using the definition of J in [A.18] for $t < 0$. A more satisfactory procedure is to note from [A.1] that the *initial* condition imposed upon f is equivalent to the requirement that $f(\mathbf{R}, 0) = \delta(\mathbf{R} - \mathbf{R}')$. Thus, $f_1 = f_2 = \delta(\mathbf{R} - \mathbf{R}')$ at $t = 0$. From [A.13] this requires that $f_0 = 0$ at $t = 0$. In turn, this makes $J(0) = 0$ in [A.18]. Thus, derivation of [A.22] from [A.20] now follows by integration of the latter from $t = 0$ to $t > 0$, to obtain $J(t) - J(0) \leq 0$. Since $J(0) = 0$, this furnishes an alternative proof of [A.22], which has only required use of the relation [A.18] for $t \geq 0$.

where A denotes the accessible cross-sectional area of the tube, and $dA = rdrd\phi$ is an element of cross-sectional area. However, upon integration by parts,

$$\int_{-\infty}^{\infty} z(\partial P/\partial z)dz = zP\Big]_{z=-\infty}^{\infty} - \int_{-\infty}^{\infty} Pdz. \quad [\text{B.6}]$$

In view of the condition [2.27], the leading term of the above vanishes at both limits. Consequently,[15]

$$d\bar{z^2}/dt = 2\int_V D_{11}(r)PdV \Big/ \int_V PdV. \quad [\text{B.7}]$$

This may be written alternatively in terms of the function f, defined in [2.31], as

$$d\bar{z^2}/dt = 2\int_{z=-\infty}^{\infty}\int_{r=0}^{r_o-a} D_{11}(r)frdrdz \Big/$$

$$\int_{z=-\infty}^{\infty}\int_{r=0}^{r_o-a} frdrdz. \quad [\text{B.8}]$$

To evaluate this integral, define the function (cf. [3.10])

$$\mu_0(r, t\,|\,r') = \int_{z=-\infty}^{\infty} f dz, \quad [\text{B.9}]$$

so that

$$d\bar{z^2}/dt = \int_0^{r_o-a} D_{11}(r)\mu_0 rdr \Big/ \int_0^{r_o-a} \mu_0 rdr. \quad [\text{B.10}]$$

As demonstrated in connection with [3.28],

$$\mu_0 \sim (K/2\pi)e^{-E(r)} + \exp \quad \text{as } t \to \infty,$$

where K is the constant [3.27]. Therefore,

$$d\bar{z^2}/dt \sim 2\bar{D}_{11} + \exp \quad \text{as } t \to \infty, \quad [\text{B.11}]$$

[15] In the limiting case where $\lambda = 0$, $D_{11} = D_\infty = \text{const}$, independent of r. In these circumstances, [B.7] reduces immediately to $d\bar{z^2}/dt = 2D_\infty$. Integration, subject to the initial condition $\bar{z^2} = 0$ at $t = 0$, then gives $\bar{z^2} = 2D_\infty t$, which is Einstein's result, valid when wall effects are absent.

with $\bar{D}_{11}$ the constant given by [3.48]. Integration gives

$$\bar{z^2} \sim 2\bar{D}_{11}t + \text{const} + \exp. \quad [\text{B.12}]$$

By the Einstein analysis, $\bar{D}_{11}$ is therefore to be interpreted as the mean axial diffusivity.

The result [B.11], with $\bar{D}_{11}$ given by [3.48], also follows directly from [B.7] by recognizing that for "long" times (cf. [3.5]), the probability density P necessarily adopts the Gaussian form,

$$P \sim e^{-E(r)}\bar{P} \quad (t > 0), \quad [\text{B.13}]$$

with

$$\bar{P}(z, t) = (4\pi\bar{D}_{11}t)^{-\frac{1}{2}}\exp(-z^2/4\bar{D}_{11}t), \quad [\text{B.14}]$$

in which $\bar{D}_{11}$ is some constant. Substitution of [B.13] into [B.7] then gives

$$d\bar{z^2}/dt = 2\int_{r=0}^{r_o-a} e^{-E(r)}D_{11}(r)rdr$$

$$\times \int_{-\infty}^{\infty}\int_{-\pi}^{\pi} \bar{P}(z, t)d\phi dz \Big/ \int_{r=0}^{r_o-a} e^{-E(r)}rdr$$

$$\times \int_{-\infty}^{\infty}\int_{-\pi}^{\pi} \bar{P}(z, t)d\phi dz. \quad [\text{B.15}]$$

This immediately yields [B.11], with $\bar{D}_{11}$ given by [3.48], upon canceling the comparable $\bar{P}$ integrals in the numerator and denominator.

This alternative scheme for determining the mean molecular diffusivity $\bar{D}_{11}$, which does not require use of dispersion theory, nor determination of the higher-order moments μ_1 and μ_2, should prove to be of considerable utility in determining the mean molecular diffusivity for situations involving the diffusion of *nonspherical* Brownian particles in tubes.

Moreover, even in cases where convective transport is sensible, this same technique may be employed to calculate the mean particle velocity $\bar{U}^*$, again without having to invoke the elaborate machinery of dispersion theory. Thus, beginning with [3.1], one may proceed successively to derive the relations

$$d\bar{z}/dt = \int U(r)P dV \bigg/ \int P dV$$

$$= \int U(r)f dV \bigg/ \int f dV$$

$$= \int U(r)\mu_0 r dr \bigg/ \int \mu_0 r dr \qquad [B.16]$$

$$\sim \int e^{-E(r)}U(r)r dr \bigg/ \int e^{-E(r)}r dr$$

$$= \bar{U}^*,$$

(cf. [4.5] and [3.7]), in agreement with prior results. In a sense, the ability to determine $\bar{U}^*$ in this manner is intimately related to the facts that, in dispersion theory: (i) the first total moment M_1 may be determined solely from knowledge of μ_0, as in [3.45], and (ii) that M_1 is related to $\bar{z}$ by [4.1].

Similarly to [B.13], the last of Eqs. [B.16] can be derived from the first by recognizing that for long times, [B.13] is once again applicable, but with

$$\bar{P}(z, t) = (4\pi\bar{D}^*t)^{-\frac{1}{2}}$$
$$\times \exp[-(z - \bar{U}^*t)^2/4\bar{D}^*t], \quad [B.17]$$

where $\bar{D}^*$ and $\bar{U}^*$ are constants. On substitution into the first of Eqs. [B.16] it now follows that

$$d\bar{z}/dt = \int_{r=0}^{r_o-a} e^{-E(r)}U(r)r dr$$

$$\times \int_{-\infty}^{\infty}\int_{-\pi}^{\pi} \bar{P}(z, t)d\phi dz \bigg/ \int_{r=0}^{r_o-a} e^{-E(r)}r dr$$

$$\times \int_{-\infty}^{\infty}\int_{-\pi}^{\pi} \bar{P}(z, t)d\phi dz. \quad [B.18]$$

This yields the penultimate member of Eqs. [B.16] upon cancellation of the comparable $\bar{P}$ integrals in the numerator and denominator of [B.18].

It is unfortunate that these simple, and straightforward, techniques do not appear to lend themselves equally well to the determination of $\bar{D}_v$ (the convective contribution to the dispersivity $\bar{D}^*$).

APPENDIX C: CRITIQUE OF THE ANDERSON–QUINN (2) DEFINITION OF DIFFUSIVITY

The central issue in the Anderson–Quinn (2) definition of mean axial diffusivity may be put into perspective by elaborating on their steady-state derivation of this quantity.

Consider a steady-state diffusive transport process occurring between two infinitely large reservoirs maintained at different uniform concentrations, spaced a distance of L units apart, and connected by a capillary tube of radius r_o, as in Fig. 7. Solute concentrations in the reservoirs are denoted by C_0 and C_L, respectively, with $C_0 > C_L$.

In the absence of convective transport, the equations governing the number density $c = c(r, z)$ $(0 \leq r \leq r_o, 0 < z < L)$ of sphere centers at any point within the tube are taken to be (cf. [2.38])

$$\nabla \cdot \mathbf{j} = 0, \qquad [C.1]$$

$$\mathbf{j} = -e^{-E(r)}\mathbf{D} \cdot \nabla(ce^{E(r)}), \qquad [C.2]$$

satisfying the boundary condition

$$j_r = 0 \quad \text{at } r = r_o - a. \qquad [C.3]$$

Here, $\mathbf{D}$ is given by [A.8]. Though primary interest centers on the case $E = 0$, it is enlightening to carry along this factor, since its presence clearly reveals the confusion which appears to exist in the Anderson–Quinn (2) analysis between the local concentration c and the mean concentration $\bar{c}$.

The above system of equations possesses the solution

$$c(r, z) = e^{-E(r)}(K - Gz),$$
$$\text{for } 0 \leq r \leq r_o - a, \quad [C.4a]$$

$$= 0 \quad \text{for } r_o - a < r \leq r_o, \quad [C.4b]$$

with

$$\mathbf{j} = \mathbf{i}_z j_z(r), \qquad [C.5]$$

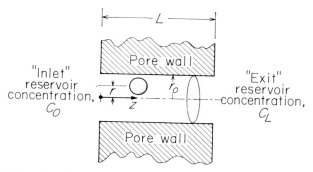

FIG. 7. A circular cylindrical pore connecting two, "well-stirred," effectively infinite reservoirs. The distance z is measured from the pore entrance.

wherein

$$J_z(r) = e^{-E(r)}D_{11}(r)G,$$

$$\text{for } 0 \leq r \leq r_o - a, \quad [C.6]$$

$$= 0, \qquad \text{for } r_o - a < r \leq r_o.$$

The quantities K and G are constants.

The number N_z of spheres per unit time crossing any plane $z = \text{const}$ is, by definition,

$$N_z = \int_0^{r_o} j_z(r)2\pi r\,dr$$

$$= 2\pi G \int_0^{r_o-a} e^{-E(r)}D_{11}(r)r\,dr. \quad [C.7]$$

This quantity is a constant, independent of z. Thus, the number of solute particles being transported per unit time per unit of superficial tube cross-sectional area is (cf. [4.14])

$$\bar{J}_z \overset{\text{def}}{=} (N_z/\pi r_o^2)$$

$$= (2G/r_o^2) \int_0^{r_o-a} e^{-E(r)}D_{11}(r)r\,dr. \quad [C.8]$$

Derivation of an appropriate expression for the macroscopic axial diffusivity requires that the constant G in the above be expressed in terms of the macroscopic concentration gradient, $d\bar{c}/dz$. Consider an elementary slice dz of tube length. As in [4.19], the volume-average concentration $\bar{c}$ is defined as

$$\bar{c}(z)\pi r_o^2 dz = dz \int_{r=0}^{r_o} c(r,z)2\pi r\,dr, \quad [C.9]$$

whence $\bar{c}$ is also the area-averaged concentration. In view of [C.4a] and [C.4b], this yields

$$\bar{c}(z) = (K - Gz)(1/\pi r_o^2) \int_0^{r_o-a} e^{-E(r)}2\pi r\,dr. \quad [C.10]$$

Differentiation gives

$$G = -\frac{d\bar{c}}{dz} \Big/ \frac{1}{\pi r_o^2} \int_0^{r_o-a} e^{-E(r)}2\pi r\,dr. \quad [C.11]$$

Introduction of this into [C.8] makes

$$\bar{J}_z = -\bar{D}_{11}d\bar{c}/dz, \quad [C.12]$$

wherein

$$\bar{D}_{11} \overset{\text{def}}{=} \int_0^{r_o-a} e^{-E(r)}D_{11}(r)r\,dr \Big/ \int_0^{r_o-a} e^{-E(r)}r\,dr. \quad [C.13]$$

In dimensionless form, the latter is equivalent to [4.42]. Equation [C.12] is regarded as being the appropriate macroscopic form of Fick's law, with $\bar{D}_{11}$ the macroscopic diffusivity.

In the Anderson–Quinn analysis, which pertains to the case $E = 0$, [C.4a] and [C.4b] are replaced by

$$c(r,z) = K - Gz,$$

$$\text{for } 0 \leq r \leq r_o - a, \quad [C.14a]$$

$$= 0, \qquad \text{for } r_o - a < r \leq r_0. \quad [C.14b]$$

so that

$$G = -dc/dz, \qquad [C.15]$$

at least in the accessible region, $0 \leq r \leq r_o - a$. On substitution into [C.8] this makes, for $E = 0$,

$$\bar{J}_z = -\bar{D}dc/dz, \qquad [C.16]$$

with

$$\bar{D} = (2/r_o^2) \int_0^{r_o-a} D_{11}(r)r\,dr. \qquad [C.17]$$

The latter is Anderson and Quinn's expression for the diffusivity. In contrast, for the case $E = 0$, [C.13] is larger than this quantity by an amount $1/(1 - \lambda)^2$.

The difficulty with the Anderson–Quinn interpretation lies in the mistaken assumption that the local concentration c in [C.14] is independent of r, and hence a function only of z. As revealed jointly by [C.14a] and [C.14b], this is not the case. Rather, c is a function of r. Accordingly, the quantity dc/dz appearing in [C.16] is ambiguous, since

$$(\partial c/\partial z)_r = -G, \quad \text{for } 0 \leq r \leq r_o - a, \\ = 0, \quad \text{for } r_o - a < r \leq r_o. \qquad [C.18]$$

It is apparent that the problem of interpretation arises from confusing the discontinuous, *local* concentration gradient, dc/dz, with the continuous macroscopic gradient, $d\bar{c}/dz$. The point is not moot, for in the case where $E \neq 0$, Eq. [C.4a] clearly shows that c is a function of r, even in the region $0 \leq r \leq r_o - a$. In such circumstances, Eq. [C.16] would make no sense when attempting to write down a macroscopic form of Fick's law. In contrast, [C.12] continues to maintain its unambiguous status.

Despite our criticism of the Anderson–Quinn definition of diffusivity in the context of a pointwise, continuum-mechanical approach to the diffusive transport process, their definition is consistent with the more conventional, global approach to membrane transport processes, as embodied in the thermodynamics of irreversible processes (2, 27, 43, 44, 46–49, 54). Here, the interreservoir "concentration gradient,"

$$\Delta C/L \stackrel{\text{def}}{=} (C_0 - C_L)/L, \qquad [C.19]$$

serves as the "driving force" in defining the diffusivity. That is, in such global theories the macroscopic "diffusivity" $\bar{D}$ is defined by the relation

$$\bar{J}_z = \bar{D}\Delta C/L. \qquad [C.20]$$

The uniform solute concentrations, C_0 and C_L, existing in the two reservoirs are not, in general, identical to the respective *local* solute concentrations c prevailing at the cross sections $z = 0+$ and $z = L-$, situated just inside of the tube entrance and exit. Indeed, in the case where a wall potential exists, Eq. [C.4a] shows unequivocally that the homogeneous reservoir concentrations cannot possibly be equal to the position-dependent (i.e., inhomogeneous) local entrance and exit concentrations, $c(r, 0)$ and $c(r, L)$, respectively. (The same conclusion obtains in the absence of a wall potential, since Eqs. [C.14a] and [C.14b], en toto, show c to be dependent on radial position r.) Rather, as demonstrated at the conclusion of this Appendix, the reservoir concentrations are related to the *mean* solute concentrations prevailing at the capillary inlet and exit via the relations

$$\bar{c}(0) = \Phi C_0 \quad \text{and} \quad \bar{c}(L) = \Phi C_L, \qquad [C.21]$$

where Φ is the (dimensionless) steric partitioning factor,

$$\Phi = (2/r_o^2) \int_0^{r_o-a} e^{-E(r)}r\,dr, \qquad [C.22]$$

and $\bar{c}(z)$ is defined by [C.9]. In the absence of a wall potential, this steric factor reduces to

$$\Phi = (1 - \lambda)^2. \qquad [C.23]$$

Physically, the latter represents the fraction, $\pi(r_o - a)^2\Delta z/\pi r_o^2\Delta z$, of the total volume which is accessible to occupancy by a sphere center within the elementary "slice" Δz.

In consequence of [C.19],

$$\Delta C/L = [\bar{c}(0) - \bar{c}(L)]/\Phi L, \qquad [C.24]$$

whereupon [C.20] adopts the form

$$\bar{J}_z = \bar{D}[\bar{c}(0) - \bar{c}(L)]/\Phi L. \qquad [C.25]$$

It is a consequence of [C.11] that $d\bar{c}/dz$ = const, independent of z. Hence,

$$(d\bar{c}/dz) = -[\bar{c}(0) - \bar{c}(L)]/L, \qquad [C.26]$$

whence [C.25] becomes

$$\bar{J}_z = -(\bar{D}/\Phi)(d\bar{c}/dz). \qquad [C.27]$$

Comparison with [C.12] shows that

$$\bar{D} = \Phi \bar{D}_{11}. \qquad [C.28]$$

Since $\bar{D}_{11}$ is given by [C.13], it follows that $\bar{D}$ is given by [C.17] (for $E = 0$). Thus, the irreversible-thermodynamic definition of the diffusivity accords with that of Anderson and Quinn (2) and their antecedents. While this definition clearly has much to commend it in an experimental context, it is at odds with the continuum-mechanical definition of diffusivity, especially with the Einstein relation [3.8]. Indeed, the latter relation is wholly devoid of physical meaning within an exclusively irreversible-thermodynamic framework.

Equation [C.21] is most simply derived (31a) by considering a single reservoir at thermodynamic equilibrium with the contents of a pore of radius r_o, as in Fig. 8. In the absence of convection, an equation comparable to [2.11] for the local concentration c of sphere centers may be invoked to show that

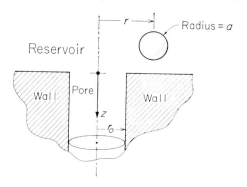

the equilibrium, no-flux condition, $\mathbf{j} = 0$, can be sustained by a Boltzmann distribution,

$$c(r, z) = Ce^{-E(r,z)}, \qquad [C.29]$$

with C a constant. This gives the local solute concentration c at any point in the reservoir or within the pore. The potential energy function E is to be selected in the following manner:

(i) $E = \infty$ for all sphere-center positions which would require that the sphere surface penetrate the wall. Most, but not all, of these positions are described by the domains $(-\infty < z + a < 0, r_o < r < \infty)$ and $(0 < z < \infty, r_o - a < r < \infty)$. This potential is a fictitious one, assuring the impermeability of the wall to penetration by the spherical solute molecules.

(ii) $E = 0$ for all "non-wall penetrating" sphere-center positions within the reservoir $(z < 0, 0 \le r < \infty)$. This choice has the effect of making $c = C = $ const at all such points in the reservoir. Thus, the constant C appearing in [C.29] corresponds to the homogeneous reservoir concentration.

(iii) $E = E(r)$ for all "accessible" sphere-center positions $(0 < z < \infty, 0 \le r \le r_o - a)$, within the pore. Here, $E(r)$ is the true pore-wall potential appearing in the previous theoretical development.

From the definition [C.9] of the mean solute concentration $\bar{c}(z)$ at any cross section $z > 0$ within the pore, it thus follows by integration of [C.29] that

$$\bar{c} = \Phi C = \text{const} \quad \text{for all} \quad z > 0, \qquad [C.30]$$

with Φ given by [C.22]. This equation relates the mean pore concentration $\bar{c}$ to the reservoir concentration C, and thereby furnishes a proof of [C.21].

Q.E.D.

ACKNOWLEDGMENTS

This research was supported by the National Science Foundation under Grant 34855. A portion of this work was performed while H.B. was on sabbatical leave in the Chemical Engineering Department at the California Institute of Technology as a Sherman Mills Fairchild Distinguished Scholar. Thanks are due them for providing a congenial and hospitable atmosphere.

REFERENCES

1. ANDERSON, J. L. AND MALONE, D. M., *Biophys. J.* **14**, 957 (1974).
2. ANDERSON, J. L. AND QUINN, J. A., *Biophys. J.* **14**, 130 (1974).
3. ARIS, R., *Proc. Roy. Soc. (London) Ser. A* **235**, 67 (1956).
4. BEAN, C. P., *in* "Membranes—A Series of Advances" (G. Eisenman, Ed.), Vol. 1, pp. 1–54. Dekker, New York, 1972.
5. BLY, D. D., *Science* **168**, 527 (1970).
6. BOHLIN, T., *Trans. Roy. Inst. Technol. (Stockholm)* **155**, 63 pp. (1960).
7. BRENNER, H., *Chem. Eng. Sci.* **16**, 242 (1961).
8. BRENNER, H., *in* "Advances in Chemical Engineering: Volume 6" (T. B. Drew, J. W. Hoopes, Jr., and T. Vermuelen, Eds.), pp. 287–438. Academic Press, New York, 1966.
9. BRENNER, H. *J. Colloid Interface Sci.* **23**, 407 (1967).
10. BRENNER, H., *J. Fluid Mech.* **43**, 641 (1970).
11. BRENNER, H., *J. Colloid Interface Sci.* **32**, 141 (1970).
12. BRENNER, H., *in* "Progress in Heat and Mass Transfer" (G. Hetsroni, S. Sideman, and J. P. Hartnett, Eds.), Vol. 6, pp. 509–574. Pergamon, New York, 1972.
13. BRENNER, H., *Int. J. Multiphase Flow* **1**, 195 (1974).
14. BRENNER, H. AND CONDIFF, D. W., *J. Colloid Interface Sci.* **41**, 228 (1972).
15. BRENNER, H. AND CONDIFF, D. W., *J. Colloid Interface Sci.* **47**, 199 (1974).
16. BRENNER, H. AND WEISSMAN, M. H., *J. Colloid Interface Sci.* **41**, 499 (1972).
17. BUNGAY, P. M. AND BRENNER, H., *J. Fluid Mech.* **60**, 81 (1973).
18. BUNGAY, P. M. AND BRENNER, H., *Int. J. Multiphase Flow* **1**, 25 (1973).
19. CARSLAW, H. S. AND JAEGER, J. C., "Conduction of Heat in Solids," 2nd ed., pp. 35–38, 256. Oxford, London, 1959.
20. CONDIFF, D. W. AND BRENNER, H., *Phys. Fluids* **12**, 539 (1969).
21. COX, R. G. AND BRENNER, H., *Chem. Eng. Sci.* **22**, 1753 (1967).
22. COX, R. G. AND BRENNER, H., *J. Fluid Mech.* **28**, 391 (1967).
23. DiMARZIO, E. A. AND GUTTMAN, C. M., *Polym. Lett.* **7**, 131 (1970).
24. DiMARZIO, E. A. AND GUTTMAN, C. M., *Macromolecules* **3**, 131 (1970).
25. DiMARZIO, E. A. AND GUTTMAN, C. M., *J. Chromatogr.* **55**, 83 (1971).
26. DORREPAL, J. M., The Stokes resistance of a spherical cap to translational and rotational motions in a linear shear flow. *J. Fluid Mech.*, in press.

27. DUBOIS, R. AND STOUPEL, E., *Biophys. J.*, in press
28. FRIED, J. J. AND COMBARNOUS, M. A. *in* "Advances in Hydroscience" (V. T. Chow, Ed.), Vol. 7, pp. 169–282. Academic Press, New York, 1971.
29. FRIEDMAN, B., "Principles and Techniques of Applied Mathematics," pp. 143, 291, 296 Wiley, New York, 1956.
30. FROST, P. A., AND HARPER, E. Y., *SIAM Rev.* **18**, 62 (1976).
31. GAYDOS, L. J., Ph.D. Thesis, Carnegie–Mellon University. 1977.
31a. GIDDINGS, J. C., KUCERA, E., RUSSEL, C. P., AND MYERS, M. N., *J. Phys. Chem.* **72**, 4397 (1968).
32. GILL, W. N. AND SANKARASUBRAMANIAN, R., *Proc. Roy. Soc. (London) Ser. A* **327**, 191 (1972).
33. GOLDMAN, A. J., COX, R. G., AND BRENNER, H., *Chem. Eng. Sci.* **22**, 637 (1967).
34. GOLDMAN, A. J., COX, R. G., AND BRENNER, H., *Chem. Eng. Sci.* **22**, 653 (1967).
35. GUTTMAN, C. M. AND DiMARZIO, E. A., *Macromolecules* **3**, 681 (1970).
36. HABERMAN, W. L. AND SAYRE, R. M., "Motion of Rigid and Fluid Spheres in Stationary and Moving Liquids Inside Cylindrical Tubes." David W. Taylor Model Basin Rept. No. 1143, 67 pp., U.S. Navy Dept., Washington, D.C., 1958.
37. HAPPEL, J. AND BRENNER, H., "Low Reynolds Number Hydrodynamics," Nordhoff, The Netherlands, 1973.
38. HIRSCHFIELD, B. R., "A Theoretical Study of the Slow, Asymmetric Settling Motion of an Arbitrarily-Positioned Particle in a Circular Cylinder." Ph.D. Thesis, New York University, 1972.
39. HIRSCHFIELD, B. R. AND BRENNER, H., *Int. J. Multiphase Flow*, in press.
40. HORN, F. J. M., *AIChE J.* **17**, 613 (1971).
41. IWAOKA, M. AND TSUTOMU, I., *Chem. Eng. J.*, in press.
42. JEFFERY, G. B. *Proc. Roy. Soc. (London) Ser. A* **102**, 161 (1922).
43. KATCHALSKY, A. AND CURRAN, P. F., "Non-Equilibrium Thermodynamics in Biophysics." Harvard Univ. Press, Cambridge, Mass., 1965.
44. LAKSHMINARAYANAIAH, N., "Transport Phenomena in Membranes," pp. 325–331. Academic Press, New York, 1969.
45. LEVENSPIEL, O. AND BISCHOFF, K. B., *in* "Advances in Chemical Engineering" (T. B. Drew, J. W. Hoopes, Jr., and T. Vermuelen, Eds.), Vol. 4, pp. 95–198. Academic Press, New York, 1963.
46. LEVITT, D. G., *Biophys. J.* **15**, 533 (1975).
47. LEVITT, D. G., *Biophys. J.* **15**, 553 (1975).
48. LIGHTFOOT, E. N., BASSINGTHWAIGHTE, J. B., AND GRABOWSKI, E. F., *Biophys. J.*, in press.

49. MANNING, G. S., *Biophys. Chem.* **3**, 147 (1975).

50. MORSE, P. M. AND FESHBACH, H., "Methods of Theoretical Physics," pp. 825, 830. McGraw–Hill, New York, 1953.

51. NIR, A. AND ACRIVOS, A., *J. Fluid Mech.* **59**, 209 (1973).

52. PAINE, P. L. AND SCHERR, P., *Biophys. J.* **15**, 1087 (1975).

53. SANKARASUBRAMANIAN, R. AND GILL, W. N., *Proc. Roy. Soc. (London) Ser. A* **329**, 479 (1972).

54. SATTERFIELD, C. N., COLTON, C. K., AND PITCHER, W. H., JR., *AIChE J.* **19**, 628 (1973).

55. SMALL, H., *J. Colloid Interface Sci.* **48**, 147 (1974).

56. TAYLOR, G. I., *Proc. Roy. Soc. (London) Ser. A* **219**, 186 (1953).

57. TAYLOR, G. I., *Proc. Roy. Soc. (London) Ser. A* **225**, 473 (1954).

58. VADAS, E. G., "The Microrheology of Colloidal Dispersions," Ph.D. Thesis, McGill University, 1975; see also VADAS, E. G., COX, R. G., GOLDSMITH, H. L., AND MASON, S. G., "The Microrheology of Colloidal Dispersions. II. Brownian Diffusion of Doublets of Spheres," *J. Colloid Interface Sci.*, **57**, 308 (1976).

59. VADAS, E. G., GOLDSMITH, H. L., AND MASON, S. G., *J. Colloid Interface Sci.* **43**, 630 (1973).

60. VAN DYKE, M., "Perturbation Methods in Fluid Mechanics" (annotated edition). Parabolic, Stanford, Calif., 1975.

61. WAKIYA, S., DARABANER, C. L., AND MASON, S. G., *Rheol. Acta* **6**, 264 (1967).

Stability and Instability

Stability and Instability in Disperse Systems

R. H. OTTEWILL

School of Chemistry, University of Bristol, Bristol BS8 1TS, England

Received August 19, 1976; accepted October 14, 1976

The balance of the forces of interaction between colloidal particles determines, in general, whether a colloidal dispersion remains stable or undergoes coagulation or flocculation. A review of interaction forces is given and it is then shown how such forces can be measured experimentally using either macroscopic systems or particulate dispersions. Particular attention is given to the repulsive forces which arise from electrostatic charges and/or adsorbed layers. From an understanding of these forces some rationalization can be made of the many phenomena which constitute colloid and interface science.

INTRODUCTION

The colloidal state was first clearly recognized by Graham (1) in 1861. Since then a considerable amount of research has been devoted to understanding why, at low electrolyte concentrations, many colloidal particles remain in a dispersed state while at higher electrolyte concentrations, they clump together. The latter process, which is termed coagulation, demonstrates that the system has become unstable. The early work of Schulze (2) ca. 1882, Linder and Picton (3), and Hardy (4) demonstrated the sensitivity of colloidal dispersions to the addition of electrolytes. Hardy (5) in 1900 showed that the stability of sols was connected with the electrophoretic mobility of the particles and clearly demonstrated that the valency of the ion opposite in charge to that of the sol particles, determined the ability of an electrolyte to coagulate a sol and that the effectiveness increased rapidly with increase in valency. These observations formed the basis of the so-called Schulze–Hardy rule.

During the period which followed, many notable scientists including Zsigmondy, Freundlich, Ostwald, Tourila, and Kruyt attempted to understand the process of coagulation. In addition, Smoluchowski (6) in 1917 developed a theory for the kinetics of coagulation by treating it as a diffusion-controlled process. Later in 1934, Fuchs (7) developed the theory of diffusion in a force field which led to the idea of a stability ratio. Despite the progress up to this point, Ostwald (8) in 1938 wrote:

> I believe that a colloid chemist, if asked today to explain the coagulation of a lyophobic hydrosol by electrolytes will make a rather unhappy face. Most presumably, when explaining simple and well-investigated cases as, for example, the flocculation of arsenious trisulphide sols with neutral salts, a conscientious colloid chemist will even voice a warning to the effect that this matter is not as simple as it looks.

An accurate observation since, in fact coagulation may well be one of the most difficult of the many phenomena that we are aware of in colloid science to explain in detail since it can involve a number of kinetic processes, e.g., diffusion of the particles, viscous drainage on approach, and also changes of charge and/or potential. Nevertheless, any theories which we develop must lead ultimately to an understanding of coagulation and flocculation processes. The distinction between the terms coagulation and flocculation has been discussed

379

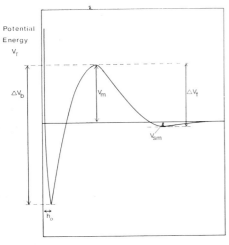

Potential Energy V_T

Distance, h

FIG. 1. Schematic form of the curve of total potential energy, V_T, against the distance of surface separation, h, for interaction between two particles. V_m = height of primary maximum; V_{sm} = depth of secondary minimum; ΔV_f = energy barrier to a forward transition; ΔV_b = energy barrier to a backward transition.

elsewhere (9) and will not be considered further here.

During the 1930's a clearer idea of the role of the electrical double layer in stabilizing colloidal particles began to emerge, particularly in the work of Verwey, Kruyt, and Derjaguin. In a classic paper (10) Langmuir in 1938 showed that when an overlap of double layers occurred, with two flat plates at the same electrostatic potential, then a repulsive pressure ensued between them.

Also in the 1930's London (11) indicated the quantum mechanical origin of dispersion forces between apolar molecules and the subsequent work of Bradley (12), de Boer (13), and Hamaker (14) extended these ideas to interaction between particles. Whereas the force between molecules varied inversely as the seventh power of the separation distance, that between thick flat plates was found to vary as the inverse third power, thus constituting a "long range" van der Waals attractive force.

A considerable advance was made in the

1940's when in the theories of Derjaguin and Landau (15) and Verwey and Overbeek (16, 17), now frequently termed the DLVO·theory, the simple pairwise addition of the potential energy of electrostatic repulsion, V_R, and van der Waals attraction, V_A, was assumed. The total potential energy of interaction, $V_T(=V_R + V_A)$ with the addition of a short-range Born repulsion energy gave a curve of potential energy against distance, h, which exhibits certain characteristic features (see Fig. 1). At short distances, a deep potential-energy minimum occurs, the position of which decides the distance of closest approach, h_0. At intermediate distances the electrostatic repulsion makes the largest contribution and hence a maximum occurs in the potential-energy curve, V_m. At larger distances, the exponential decay of the electrical double layer term causes it to fall off more rapidly than the power law of the attractive term and a second minimum appears in the curve, V_{sm}.

THE STABILITY–INSTABILITY APPROACH

The Critical Coagulation Concentration

The form of the potential-energy curve (Fig. 1) immediately gave an explanation for the *stability* behavior of a lyophobic dispersion. When the primary maximum had a large positive value, say substantially greater than 10 kT, then the system was kinetically stable owing to the large activation energy (ΔV_f) opposing a transition into the primary minimum. However, as V_T tended to zero or became negative the transition was facilitated and the system became unstable. Thus theoretically (17) convenient conditions for the onset of *instability* were defined by

$$V_T = 0 \quad \text{and} \quad \partial V_T/\partial h = 0. \quad [1]$$

This led immediately to the fact that the electrolyte concentration at which these conditions were fulfilled gave a theoretical critical coagulation concentration for platelike particles which was proportional to

$v^{-6} A^{-2}$ for high surface potentials,

$\psi_0^4 v^{-2} A^{-2}$ for small surface potentials,

where v = the valency of the counterion of the coagulating electrolyte, A = Hamaker constant, and ψ_e = the surface potential of importance in coagulation.

The fact that a critical coagulation concentration can be obtained in this way, showing a dependence on the counterion valency (cf. the Schulze–Hardy rule), and that a qualitative fit is obtained with experimental data is one of the successes of the DLVO theory.

The Stability Ratio W

In an elegant paper in 1917 Smoluchowski (6) analyzed the kinetics of the coagulation process. He found that the rate of disappearance of primary particles in the initial stages of coagulation could be written as

$$-dN/dt = kN_0^2, \quad [2]$$

where N_0 = the number of particles per unit volume initially present and k is a rate constant. For rapid coagulation, i.e., coagulation in the absence of an energy barrier, then, $k = k_0 = 8\pi DR$ where D = the diffusion coefficient of a single particle and R = the collision radius (18).

In a subsequent analysis, Fuchs (7) showed that if diffusion in the presence of an energy barrier is considered, i.e., slow coagulation, then k can be put equal to k_0/W and Eq. [2] rewritten as

$$-dN/dt = (k_0/W)N_0^2. \quad [3]$$

The rate of coagulation in the early stages can be determined experimentally using either direct particle counting or light scattering (19). Since the rate remains constant in the rapid coagulation region, making the assumption for this situation that $W = 1$ leads to a value of k_0, we can obtain values of k and W in the slow coagulation region. A typical example of the type of experimental data obtained (19) is shown in Fig. 2; the sharp change of gradient at a particular electrolyte concentration establishes this as the critical coagulation concentration.

It was pointed out by Verwey and Overbeek (17) that the factor W, normally called the

stability ratio, is related to V_T through the equation

$$W = 2a \int_0^\infty \exp(V_T/kT)dh/(h + 2a)^2. \quad [4]$$

As a further development Reerink and Overbeek (20) showed that the gradient of the log W against log C_e curve (C_e = electrolyte concentration) could be given as

$$(d \log W)/(d \log C_e)$$
$$= -2.06 \times 10^7(a\gamma^2/v^2), \quad [5]$$

where a = particle radius, v = the valency of the counter-ion and

$$\gamma = [\exp (ve\psi_e/2kT) - 1]/$$
$$(\exp (ve\psi_e/2kT) + 1].$$

Whether ψ_e should be taken as the Stern potential or the zeta potential is not yet clearly established, although the two quantities have similar magnitudes.

Equation [5] can be subjected to direct experimental test in that, for comparable values of ψ_e and hence γ and the same electrolyte, the gradient $(d \log W)/(d \log C_e)$ should be directly dependent on the radius of the particle. A detailed study using a series of

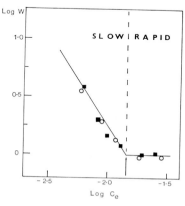

FIG. 2. Log W against log C_e for a polystyrene latex (diameter = 423 nm) in barium nitrate solutions (19). ■, results obtained by light scattering; ○, results obtained using a particle counter. ↑, critical coagulation concentration.

monodisperse polystyrene latices did not show the expected dependence on particle radius (19). The results, in fact, appeared to show that $(d \log W)/(d \log C_e)$ was independent of radius.

A more precise treatment of the kinetic process (21–23) has not solved the major discrepancy between experiment and theory. This problem is discussed in more detail by Overbeek (24).

A further essential point which arises from Fig. 1 concerns the reversibility of the coagulation process. It is clear that if the energy barrier to the backward reaction (repeptization), ΔV_b, is much larger than ΔV_f, then the coagulation stage will be essentially irreversible. Moreover, time is an important variable—prolonged contact leading to "welding" of the particles. On the other hand if $\Delta V_b \ll \Delta V_f$, repeptization can occur. This topic is also discussed by Overbeek (24).

INTERACTION FORCES

As an alternative approach to understanding the kinetics of coagulation or seeking to justify the validity of the Schulze–Hardy rule, different questions can be asked. For example, let us pose the following questions:

(1) What is the nature of the interparticle forces between particles and what is their origin?

(2) What is the distance dependence of the forces?

(3) What is the magnitude of the forces?

(4) How do they act in various types of dispersion?

(5) When are they all important and when are only some of them important?

(6) How can these forces be measured?

During the last 50 years, our knowledge of the basic forces involved in colloidal systems of all types has become substantially clearer and we can suggest now a list of possible forces:

(i) those arising from the surface potential, ψ_0, and the consequent counter-ion cloud, i.e., electrostatic forces

(ii) those arising from electromagnetic fluctuations, i.e., van der Waals or dispersion forces

(iii) "Born" forces—usually very short range and repulsive

(iv) "Steric" forces—usually dependent on the geometry and conformation of molecules (frequently adsorbed) in an interfacial region adjacent to the particle

(v) those forces arising from the removal, displacement, or rearrangement of solvent molecules in an interfacial region, i.e., the so-called solvation forces

(vi) magnetic forces.

As mentioned earlier the addition of (i) and (ii) provides a combination which forms the basis of the DLVO theory. With the addition of (iii) we obtain a close-range primary minimum (see Fig. 1). "Steric" effects probably occur much more widely than was recognized in early work, and their nature, which is becoming more deeply understood following the work of Mackor (25), Napper (26), Hesselink (27), Dolan and Edwards (28), and others, is discussed in detail by Napper (29). The question of solvation is only understood in a very rudimentary fashion. Its role, however, is of crucial importance in many systems and a deeper understanding of the kinetic and thermodynamic parameters involved in solvent effects is urgently required. Magnetic effects are usually very specific and only apply to a limited number of systems.

Measurement of Interaction Forces

The precise measurement of the force of interaction between particles in a dispersion requires that the internal force of interaction should be balanced by an external applied force so that a position of equilibrium is reached at a particular distance of separation which can be measured by a physical technique. This is illustrated in Fig. 3, where the condition of equilibrium per square centimeter

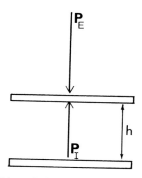

FIG. 3. Schematic diagram showing balance of pressure in a plate–plate interaction. P_E = external applied pressure; P_I = internal pressure; h = distance of surface separation.

of plate is given by

$$P_I + P_E = 0, \qquad [6]$$

with P_I and P_E the internal and external pressures, respectively; for the present purpose it will be assumed that there is no additional hydrostatic pressure. If the potential energy of interaction, V_T, is also expressed per square centimeter of surface, then

$$P_I(h) = -\partial V_T/\partial h. \qquad [7]$$

The pressure P_I can be subdivided, assuming additivity, into the various components discussed above, and hence,

$$P_I = P_{el} + P_A + P_B + P_{st} + P_{so}, \qquad [8]$$

where P_{el} = electrostatic contribution, P_A = van der Waals term, P_B = Born repulsion, P_{st} = steric effect, and P_{so} = solvation. A pressure due to mechanical or frictional forces would also be required for nonstable systems when the particles come into contact.

The schematic representation given in Fig. 3 can be translated into experiment provided that surfaces can be found which are sufficiently smooth so that the forces measured are not a function of surface roughness. This figure also demonstrates the point that colloidal phenomena are not confined to those which occur between particles, interactions between macroscopic surfaces are part of colloid science

when the separation distances are such that the forces discussed above become operative. In order to compare experiment and theory each force requires independent study. The elegant work of Israelachvili and Tabor has shown how van der Waals forces between macroscopic surfaces can be studied with air as the medium between the plates (30). The present contribution will deal with interactions between surfaces and particles in liquid media. Under these conditions the van der Waals forces are much weaker and the longer range repulsive forces, i.e., electrostatic and steric, can be studied under conditions where the attractive forces are negligible.

ELECTROSTATIC REPULSIVE FORCES

Macroscopic Surfaces

An ingenious suggestion was made by Roberts and Tabor (31) to the effect that if the surface of a transparent rubber sphere was used as a macroscopic surface, it would deform under pressure when brought into close proximity to a rigid glass surface and conform to the surface profile of the glass. This was suc-

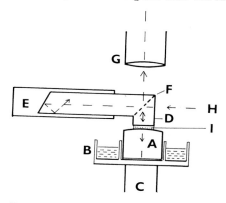

FIG. 4. Diagram of apparatus used for measuring interaction forces between macroscopic surfaces (32). (A) Transparent polyisoprene rubber with spherical cap; (B) support cup for rubber; (C) micrometer drive unit to control movement of rubber surface; (D) glass prism; (E) prism clamp and Rayleigh horn; (F) beam splitter; (G) microscope for visual observation and intensity measurement; (H) incident laser beam; (I) solution of surface active agent.

cessfully used by Roberts and Tabor (31) to study electrostatic interactions between glass and rubber surfaces through an aqueous medium with adsorbed layers of surface active agent at the solid–liquid interfaces. This work has been extended recently (32) using the apparatus shown in Fig. 4. The spherical cap of rubber, A, was pushed toward the rigid glass prism D with a solution of sodium dodecyl sulphate between the two surfaces. Distortion of the rubber was observed through the microscope, G, from changes in the Newton's rings formed by the glass–liquid–rubber interfaces. The thickness of the liquid film in the biplanar region was determined from reflectance measurements and the applied pressure was obtained from the biplanar area using an independent calibration (32).

The results obtained, in the form of pressure against the thickness of the aqueous layer, are shown in Fig. 5 for solutions of sodium dodecyl sulphate at two different ionic strengths. It can be seen that, as anticipated from theory, at the lower ionic strength 6×10^{-3}, repulsion starts to become a significant effect at a distance of 40 nm, whereas at 2×10^{-2} this occurs at ca. 26 nm. This demonstrates the long-range

nature of electrostatic repulsive forces and their dependence on ionic strength.

It was shown by Langmuir (10) that the electrostatic pressure between two flat plates, each having the same surface potential, was given by

$$P_{el} = 2n_0 kT (\cosh u - 1), \qquad [9]$$

where $n_0 = $ the number of ions per cubic centimeter of each type, $k = $ Boltzmann constant, $T = $ absolute temperature, and u is given for a 1:1 electrolyte by

$$u = e\psi_{h/2}/kT, \qquad [10]$$

where $e = $ fundamental unit of charge and $\psi_{h/2} = $ the electrostatic potential midway between the plates.

For weak interaction, $\kappa h > 2$, and constant surface potential ψ_0, eq. [9] reduces to

$$P_{el} = 64 n_0 kT \gamma_0^2 \exp(-\kappa h), \qquad [11]$$

with

$$\gamma_0 = [\exp(ve\psi_0/2kT) - 1)/ \\ (\exp(ve\psi_0/2kT) + 1]$$

and where $\kappa = $ Debye–Huckel reciprocal screening distance. Theoretical curves for $\psi_0 = 100$ and 250 mV are also given in Fig. 5. As can be seen there appears to be reasonable agreement between theory and experiment at low pressures but as the pressure increases the experimental pressure becomes less than the theoretical for $\psi_0 = 100$ mV. Although the general form of the electrostatic repulsion appears to be confirmed it appears likely that the condition of constant surface potential is not fulfilled. This could be a consequence of some desorption of surfactant from the surfaces on close approach and/or deionization of some surface ionic groups (32).

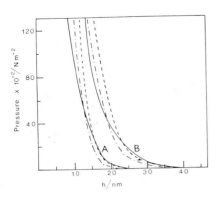

FIG. 5. Pressure against h curves. ———, experimental results using 6×10^{-3} mole dm^{-3} sodium dodecyl sulphate plus 1.4×10^{-2} mole dm^{-3} sodium chloride (A) and 6×10^{-3} mole dm^{-3} sodium dodecyl sulfate (B). Theoretical curves under the same electrolyte conditions for comparison (Eq. [11]); —·—·, $\psi_0 = 100$ mV; ---, $\psi_0 = 250$ mV.

Colloidal Dispersions of Platelike Particles

The concept of pressure measurement between flat plates can be directly extended to dispersions using arrays of aligned parallel plates. A suitable material for such studies has been the clay sodium montmorillonite, where

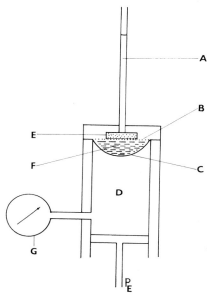

FIG. 6. Diagram of apparatus used for measuring interaction forces between colloidal particles (36). (A) Capillary for dispersion medium; (B) membrane permeable to dispersion medium; (C) elastic rubber membrane; (D) hydraulic fluid; (E) porous disc; (F) colloidal dispersion; (G) gauge for measurement of applied pressure, P_E.

the plates are sufficiently thin (ca. 10 Å) that not only are the van der Waals forces negligible but also surface roughness is not a problem.

A specially designed compression cell (Fig. 6) was designed to provide both pressure and distance measurements and the data obtained using this cell by Barclay (33, 36) and Callaghan (34) at different electrolyte concentrations are shown in Fig. 7. In these experiments much-closer-range interactions can be measured in that the thickness of the aqueous core has been reduced to ca. 2 nm, i.e., approximately 10 molecular layers of water, and pressures of the order of 100 atm have been generated by electrostatic repulsion. For strong interactions at constant potential Eq. [11] is not applicable and h is related to u by the

integral equation,

$$h = -\frac{2}{\kappa} \int_z^u \frac{dy}{[2(\cosh y - \cosh u)]^{\frac{1}{2}}}, \quad [12]$$

where $z = ve\psi_0/kT$. The potential at any distance other than the surface or the midplane is given by ψ, and hence the general variable y is given by $ve\psi/kT$. In the case of clay plates a constant-charge model is in closer conformity with the nature of the particles since the surface charge arises as a consequence of isomorphous substitution in the lattice (35).

The experimental data show the type of electrolyte dependence predicted by the DLVO theory but a more detailed comparison shows that quantitatively a reasonable fit is only obtained with both constant-potential and constant-charge models at long separation distances only (36). However, after the simple theory is corrected for the fact that the measurements were carried out in a system of finite volume with partition of electrolyte between the clay system and an external finite

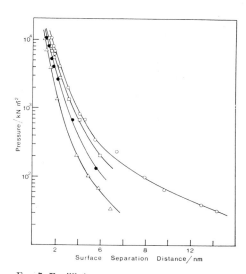

FIG. 7. Equilibrium pressure against the distance of surface separation between the plates, h, using sodium montmorillonite in sodium chloride solutions (33, 34) of different concentration. $\bigcirc$, 10^{-4} mole dm^{-3}; $\square$, 10^{-3} mole dm^{-3}; $\bullet$, 10^{-2} mole dm^{-3}; $\triangle$, 10^{-1} mole dm^{-3}.

reservoir, good agreement is obtained between experiment and theory (37). The implication appears to be that in this type of flat-plate constant-charge system a two-body interaction model provides an adequate explanation for electrostatic repulsion in a many-body system.

The basic ideas of the DLVO model also provide a good explanation for the stability of thin foam lamellae. A considerable body of work exists on this subject which has been reviewed elsewhere (38).

Colloidal Dispersions: Non-plate-like Particles

Development of model colloidal dispersions. Once studies are extended to non-plate-like particles the geometrical form of the particles becomes an important parameter. Moreover, with disperse systems it is frequently important that the particles, in addition to all having the same shape, should also all have the same size, i.e., the systems should be monodisperse. The recent excellent work of Matijević and his co-workers has provided methods for preparing many inorganic materials as sols containing monodisperse particles; these include alumina (39), chromia (40), copper hydroxide (41), and ferric hydroxide (42). Work of this type underlines the importance of chemical reactions in colloid science and the role of chemical complexing in the formation and stability of colloidal dispersions is emphasized by Professor Matijević (43).

Simultaneously, there have been rapid developments in the formation of organic systems, especially in the preparation of monodisperse polymer latices over a wide range of particle sizes (44, 45). These dispersions, now frequently referred to as polymer colloids (46), usually contain perfectly spherical particles in which the polymer is totally insoluble in the dispersion medium.

Interactions between spherical particles. Since they have spherical geometry and their surface charge can be controlled to a large extent by the method of preparation, polymer latices constitute excellent systems for the study of interaction forces between spherical particles (36). An interesting property of these latices is that under certain conditions of particle size, volume fraction, and electrolyte concentration the particles form an ordered array (36, 47, 48, 49). When the center-to-center distance between the particles is of the order of the wavelength of light, a beautiful irridescence is observed. This type of array resembles a crystal lattice with the latex particles as the structural units and the diffraction of visible light obeys the Bragg relationship

$$m\lambda_0 = 2d_{hkl} \sin \phi_{hkl}, \qquad [13]$$

where m = the order of the diffraction, λ_0 = the wavelength of the incident light *in vacuo*, d_{hkl} = the interplanar spacing in the lattice, and ϕ_{hkl} = the Bragg reflecting angle for a particular set of ordered planes. The intensity peaks observed experimentally are very sharp and the angle at the intensity maximum, ϕ_{hkl}, after due correction for refraction in the observation cell, can be obtained with a high degree of precision, particularly at the lower volume fractions; increased accuracy in obtaining the interlayer spacing, d_{hkl}, can be obtained by making observations at various wavelengths. Frequently, the experimental observations can be indexed on the basis of a face-centered or body-centered cubic lattice (50).

The study of interparticle forces in latex systems has been undertaken using an apparatus similar to that used for clay systems (Fig. 6) but modified to allow simultaneous diffraction measurements to be made. Thus curves of pressure were obtained as a function of the volume fraction of the latex or, after diffraction analysis, as a function of a d_{hkl} spacing. An example, obtained by Parentich (51), in 10^{-5} mole dm^{-3} sodium chloride, is shown in Fig. 8. At low volume fractions the pressure changes with volume fraction and then, over an intermediate region the curve runs almost parallel to the abscissa. The implication appears to be that the lattice, although ordered in this region, resembles a "liquid state." At the lower volume fractions, however, no diffraction effects are observed, i.e., the particles are diffusing randomly in Brownian motion in a state which might be termed a "colloidal vapor." At high volume fractions (see Fig. 8) the pressure rises very steeply implying that a nearly incompressible state has been reached, i.e., by analogy a "crystal" of colloidal particles (36, 50).

The ordering of latex particles can also occur in very dilute dispersions (52). Polystyrene latices containing particle swith a number average radius of 23.1 nm, in the volume fraction range 7.46×10^{-5} to 3.08×10^{-4}, after treatment with mixed-bed ion-exchange resin for two weeks, were found to exhibit a well-defined diffraction maximum in the angular light scattering intensity envelope indicative of spatial ordering of the particles (52). Under these conditions the time average intensity, $\langle I_\theta \rangle$, of the plane-polarized scattered beam at a scattering angle θ is given by

$$\langle I_\theta \rangle = \langle I_\theta' \rangle S(\theta), \qquad [14]$$

where $\langle I_\theta' \rangle$ = the intensity of the disordered dispersion, acting as an independent incoherent elastic scattering system, and $S(\theta)$ = the structure factor. Theoretically, the latter is given by

$$S(\theta) = 1 + (4\pi N/\mathbf{K}) \int_0^\infty [g(r) - 1] r$$
$$\times \sin \mathbf{K} r dr, \quad [15]$$

with N = the number of particles per unit volume, r = the center-to-center interparticle separation, and $g(r)$ is the particle–pair distribution function. $\mathbf{K}$ is the scattering vector and is given by

$$\mathbf{K} = (4\pi n_0/\lambda_0) \sin \theta/2, \qquad [16]$$

with n_0 = refractive index of the scattering medium and λ_0 = the wavelength of the incident radiation *in vacuo*. The Fourier inversion of Eq. [15] is straightforward, giving

$$N[g(r) - 1] = (1/2\pi^2 r) \int_0^\infty [S(\theta) - 1] \mathbf{K}$$
$$\times \sin \mathbf{K} r d\mathbf{K}. \quad [17]$$

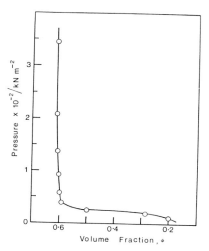

FIG. 8. Applied pressure against volume fraction for a polystyrene latex in 10^{-5} mole dm^{-3} sodium chloride (51). Particle diameter = 182 nm.

For noninteracting scattering particles $S(\theta) = 1$ and hence the scattering intensity is $\langle I(\theta)' \rangle$.

The curve of $S(\theta)$ against $\mathbf{K}$ for a latex at a volume fraction of 1.74×10^{-4} is shown in Fig. 9a. The distribution function obtained from these data, i.e., $N[g(r) - 1]$ is shown in Fig. 9b plotted against r. The latter curve shows a clear peak corresponding to a definite nearest-neighbor shell. The value of r at the maximum of this peak is 580 nm indicating a particle separation of 25 radii. The point where the probability of finding a second particle is zero, i.e., where $g(r) = 0$ and hence $N[g(r) - 1] = -N$, occurs at approximately 330 nm. At this position the repulsive potential energy is clearly rising very steeply and hence the form of the curve suggests that the particles are located in a potential-energy well with relatively steep sides. The only operative force under these conditions appears to be that of electrostatic repulsion.

In addition to the conventional time average light scattering measurements, photon correlation spectroscopy was used to examine these dilute latices (52). In dispersions in 8×10^{-2} moles dm^{-3} sodium chloride solutions where the latex particles showed no evidence of order,

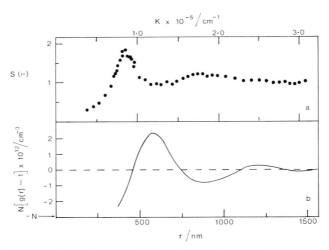

FIG. 9. (a) Experimental structure factor $S(\theta)$ against scattering vector **K**. (b) Data after Fourier inversion to give a radial distribution function, $N[g(r) - 1]$ against the center-to-center separation distance, r. Polystyrene latex particles (diameter = 23.1 nm) in 10^{-6} mole dm^{-3} electrolyte.

the translational diffusion coefficient of the particles was found to be $8.56 \pm 0.43 \times 10^{-8}$ cm^2 sec^{-1} corresponding to an average particle radius of 25 ± 1.3 nm; a value in good agreement with the number average value of 23.1 ± 2.5 nm obtained by electron microscopy on the dried system, when allowance is made for the different averages determined by the two methods. Measurements on the ion-exchanged ordered system gave evidence for at least two diffusional modes, one, observed at short times, being strongly dependent on the structure of the dispersion. A more detailed analysis of this mode has been given by Pusey (53), who assumed that the instantaneous velocity of a particle could be written as a sum of two components, a rapidly fluctuating Brownian component arising from particle–solvent molecule interactions and a slowly fluctuating "drift-velocity" component due to interparticle interactions.

The interactions in these dilute dispersions at very low electrolyte concentrations, i.e., in the region of $10^{-6} M$, and several orders of magnitude below the critical coagulation concentration, show an interesting feature in that some samples showed phase separation into an

ordered latex-rich phase and a latex-dilute phase, a behavior similar to that observed by Bernal and Fankuchen (54) with tobacco mosaic virus particles and that observed by several sets of workers with more concentrated latex systems (55, 56). It seems improbable in view of the long interparticle distances involved that the source of attraction could be the van der Waals force and an answer must be sought to explain why strongly interacting particles apparently undergo a spontaneous contraction to a phase of smaller volume. From the two-particle interaction approach of DLVO the forces between the particle would clearly be repulsive and the system would occupy the whole volume available, particularly as with such small particles Brownian motion would maintain the particles in the disperse state and gravitational effects would be small. The apparent phase change would imply that in the total particle ensemble, either a favorable entropy gain or a negative enthalpic contribution or a combination of both could occur on phase separation. It is not certain yet in these systems that conditions of constant potential or constant charge are maintained as a function of volume fraction;

FIG. 10. Scanning electron micrograph of a polystyrene latex sample after compression to a high volume fraction.

release of bound water molecules could occur with deionization of surface groupings on increase of volume fraction at very low ionic strengths.

Work of this type may provide an insight into the problem of dealing with concentrated systems and the changes in adsorption equilibrium which might occur as particles approach (57). At the present time there is not a completely definitive treatment for the thermodynamics of a total colloidal system which includes many-particle interactions and solvent–particle interactions. Such treatments are important if progress is to be made. Some interesting calculations, however, have recently been made by van Megan and Snook (58, 59), using a Monte Carlo approach, to determine the osmotic pressure of a concentrated latex dispersion.

In summary, it appears that with latex systems analogies to the various states of matter can be observed. In very dilute systems or in salt solutions where the electrostatic repulsive forces are weak and unrestricted Brownian motion is possible, the "vapor" state occurs, at intermediate volume fractions there is evidence from diffraction studies of "liquid-like" structures, and at high volume fractions, scanning electron microscope studies show the presence of a "crystalline" arrangement of particles (Fig. 10).

STERIC STABILIZATION

In the discussion so far, we have discussed essentially smooth surfaces with the charge residing on the surface, i.e., "bare particles." In many situations, however, particularly in practical dispersions the particles are coated with an adsorbed layer of a surface active agent or a polymeric material. In the former case the adsorbed layer may be reasonably homogeneous with an approximately equal number of adsorbed segments per unit volume, such that the concentration profile is a

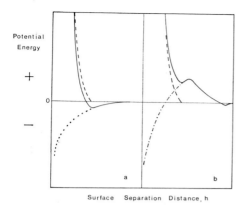

Surface   Separation   Distance, h

FIG. 11. Schematic potential-energy diagrams for sterically stabilized particles. (a) In the absence of electrostatic repulsion: V_S, – – –; V_A, $\cdots\cdots$; $V_T = V_S + V_A$, ——. (b) In the presence of electrostatic repulsion: V_S, – – –; $V_R + V_A$, —·—·—; $V_T = V_S + V_R + V_A$, ——.

step function. In the case of a polymeric material, either nonionic or ionic, this will not usually be so. The concentration profile will be distance dependent and the distribution of segments on the surface, e.g., in loops, trains, and tails, will not be homogeneous, i.e., the "bare surface" of the native particle has become "hairy." Thus adsorbed layers, either homogeneous or "hairy," modify the particle–particle interaction forces.

The effect is not new and it was well known in the early part of the century that the adsorption of macromolecules could confer stability on colloidal dispersions at electrolyte concentrations where in the absence of the macromolecule coagulation would have occurred. The use of gelatin as a "protective" colloid was well known to Zsigmondy (60) and Freundlich (61). The term *steric stabilization* for this effect appears to have been used initially by Heller and Pugh (62) in 1954 and at about the same time the possibility of steric hindrance between adsorbed chain molecules was recognized by Mackor (25), Mackor and van der Waals (63), and Koelmans and Overbeek (64).

The basis of steric stabilization is discussed in detail by Napper (29), but we should recog-

nize here that if we call V_S the interaction energy between particles due to steric hindrance, then the total potential energy of interaction, V_T, can be rewritten as

$$V_T = V_R + V_A + V_S. \qquad [18]$$

The effect of V_S is shown schematically in Fig. 11 for the interactions which occur in the presence and absence of electrostatic interactions. It is immediately clear from this qualitative picture that the entry of particles into a deep primary minimum is made virtually impossible by the presence of steric interactions.

Possibly the nearest experimental case so far considered for a homogeneous layer on a particle is that which occurs when a nonionic surface active agent, e.g., dodecylhexaoxyethylene glycol monoether, $C_{12}H_{25}(CH_2CH_2O)_6OH$, is adsorbed on to a spherical polystyrene latex particle (65, 66). Once a monolayer had adsorbed on to the latex particles, at temperatures well below the cloud point of the surface active agent, it was found with small particles (radius ~ 20 nm) that the dispersion was stable to the addition of electrolyte. An ultracentrifuge technique was developed to determine the thickness of the adsorbed layer and this gave

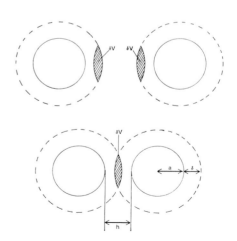

FIG. 12. Volume-overlap model. Schematic illustration of the overlap of adsorbed layers (thickness δ) on the approach of two spherical particles of radius a (65). Overlap volume $= \delta V = 2\pi(\delta - h/2)^2(3a + 2\delta + h/2)/3$.

a value slightly longer (50 ± 10 Å) than the extended length of the extended surface active agent molecule (ca. 40 Å), suggesting extensive solvation of the layer.

The data were interpreted in terms of volume overlap of the adsorbed layers (65), as shown in Fig. 12. Using the volume statistical model of Flory (67) and the overlap-volume idea of Fischer (68) it was shown that the free energy of interaction was given by

$$V_S = \Delta G_S = RT \frac{4\pi c^2}{3V_1 \rho_2^2} (\psi_1 - \chi_1) \left[\delta - \frac{h}{2}\right]^2$$
$$\times \left(3a + 2\delta + \frac{h}{2}\right), \quad [19]$$

and the force of interaction by

$$F_S = -\frac{d(\Delta G_S)}{dh} = RT \frac{4\pi c^2}{3V_1 \rho_2^2} (\psi_1 - \chi_1)$$
$$\times \left[\delta - \frac{h}{2}\right]\left(3a + \frac{3\delta}{2} + \frac{3h}{4}\right), \quad [20]$$

where c = concentration of material in the adsorbed layer expressed as g cm^{-3}, ρ_2 = density of the adsorbed material, V_1 = molar volume of the solvent molecules, δ = thickness of the adsorbed layer (Fig. 12), ψ_1 = an entropy parameter which ideally can be taken as 0.5, and χ_1 = a quantity characterizing the interaction of the adsorbed material with solvent molecules.

Although this simple model is open to criticism it does show directly that the free energy and force of interaction can be written essentially as a product of two terms, i.e.,

a thermodynamic term $\times$ a geometric term

The thermodynamic term involving ψ_1 and χ_1 shows that the enthalpic and entropic interactions of the solvent with the groupings in the adsorbed molecules and the changes in enthalpy and entropy involved on overlap must be considered. The geometric terms show clearly that the thickness of the adsorbed layer, δ, is a very important parameter. Further developments have been made by Napper (26,

29), Doroszkowski and Lambourne (69, 70), and Bagchi (71).

Possibly the simplest systems for studying steric hindrance effects between adsorbed layers are those carried out in hydrocarbon media of low dielectric constant, e.g., dodecane. Under these conditions the number of ions present per unit volume is so low that electrical double layers are essentially nonexistent, and the predominant effects are steric repulsion and van der Waals attraction. In this environment polymer latices again provide suitable model systems for experimental studies. Poly(12-hydroxy stearic acid) forms a suitable steric stabilizing group and methods have been developed whereby molecules of this type can be initially adsorbed, and then chemically condensed, onto spherical polymethylmethacrylate latex particles (70, 74). Dodecane is a good solvent for the poly(12-hydroxy stearate) chains and the latter therefore project outwards from the surface in a solvated state.

Doroszkowski and Lambourne (69, 70) first used systems of this type to study steric stabilization in a 2-dimensional array. A monolayer of latex was spread at an oil–water interface on a surface balance and then compressed. A steep increase in surface pressure occurred on compression and from the area occupied per particle on the surface the thickness of the stabilizing layer was found to be 13 nm; the latter value was found to be independent of the core particle size. Independent hydrodynamic measurements gave values for the stabilizer-layer thickness of 7–9 nm (72, 73). The surface-balance studies were found to be in agreement with the simple model (see above) of interpenetration of adsorbed layers after some allowance had been made for the redistribution of material in the adsorbed layer (69, 70).

The first three-dimensional experimental examination was made by Cairns et al. (74) using a latex with a particle core of poly(methyl methacrylate) and a stabilizer shell of poly-(12-hydroxy stearic acid). The latex was dispersed in dodecane. A cell similar in design to

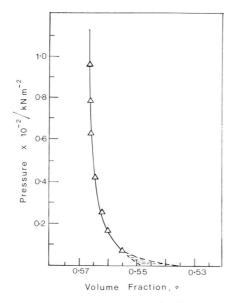

FIG. 13. Pressure against volume fraction for compression of a sterically stabilized poly(methyl methacrylate) latex. Core diameter of the particles = 155 nm.

that shown in Fig. 6 was used to measure the pressure developed in the system as a function of the volume fraction of the core particles. The experimental results are shown in Fig. 13. At low volume fractions the resistance to compression was negligible. At a volume fraction ϕ of about 0.55, however, the resistance to compression increased considerably with small changes of ϕ and at $\phi = 0.566$ very considerable resistance to compression was experienced. At this point $\partial P / \partial \phi \to \infty$ and the system became virtually incompressible. At $\phi = 0.566$ the total distance of separation between the surface of the core particles was found to be 14.5 nm, corresponding to a contribution of 7.25 nm from each shell. The extended length of a poly(12-hydroxy stearic acid) chain is about 9 nm and using this value it can be estimated that the shell surfaces would just touch at a core volume fraction of 0.53. These data suggest that if the chains are extended in the shell layer, then either some interpenetration of the layer must occur under pressure,

or alternatively that the shell layers themselves are compressed, i.e., the "denting" model of Bagchi applies (71). It is clear, however, that it is the interaction of the poly(12-hydroxy stearic acid) chains in the outer region of the shell layer which causes the strong repulsion between the particles. The experimental results confirm that the forces which arise as a consequence of steric stabilizing layers are qualitatively of the form predicted by Eq. [20] and that they are of short range unlike electrostatic repulsive forces in dilute aqueous electrolyte solutions. In the latter context it is interesting to compare the results given in Fig. 13 with those of Figs. 7 and 8.

Equations [19] and [20] underline the importance of obtaining the thickness of the adsorbed layer. However, in the case of inhomogeneous layers, such as those formed by adsorbed polymers, information is also needed about the spatial distribution of segments between the surface and the outer boundary of the adsorbed layer with the solution phase. The conformation of polymers at interfaces and methods for examining adsorbed macromolecules is discussed by Eirich (75). Such studies together with thermodynamic studies on polymer–solvent interactions and their dependence on composition and temperature are important for a more detailed understanding of steric stabilization and flocculation by polymers.

One of the classical protective colloids, as mentioned earlier (60, 61), is gelatin. Despite its widespread use little information has been obtained on the thickness of the adsorbed layers formed by this material. Recently, however, studies have been made of the optical thickness of adsorbed layers of alkali-processed gelatin on to the (111) face of a single crystal of silver bromide (76). The measurements were made using ellipsometry, as a function of pH, in 10^{-2} mole dm^{-3} potassium nitrate solution at pBr 3 and 40°C. Figure 14 shows both the adsorption excess and the optical thickness. It is clear that both the amount adsorbed and the thickness reach a maximum value just below the isoelectric point of 4.8. On the acid side the thickness decreases from about 40 nm at pH 4.5 to 15 nm at pH 3.5 while between pH 5.0 and 9.0 the thickness remains essentially constant at ca. 30 nm. At 40°C and at its isoelectric point gelatin molecules exist in solution in the random coil form. From these studies it appears that the molecules retain a conformation close to this in the adsorbed state and that the thickness of the layers formed by adsorbed gelatin molecules clearly accounts, at least in part, for the effectiveness of gelatin as a protective solid. Away from the isoelectric point the spatial distribution of charges in the adsorbed layers will provide electrostatic repulsion between the particles in addition to steric repulsion.

INSTABILITY WITH POLYMER MOLECULES

In the last section the use of polymeric molecules as a means of providing steric stabilization was discussed and as indicated for such stabilization to occur "complete" coverage of the particle surface by the polymer is required in order to provide the steric barrier. Conversely, however, polymer molecules can also be very effective in inducing instability in disperse systems. This effect was also well known to early workers in the field, who discovered that the addition of a very low concentration of a hydrophilic colloid, e.g., gelatin, to an aqueous dispersion frequently caused flocculation. The effect was termed sensitization and was discussed in some detail by Brossa and Freundlich (77). It clearly occurs under conditions where the particle surfaces are incompletely covered by polymeric molecules. Moreover, if the root mean square end-to-end distance of the polymeric molecule is sufficiently long it can span the distance across the energy barrier (see Fig. 1) and attach its ends to the surfaces of two particles despite the electrostatic repulsion. This is now usually termed bridging flocculation (78) and can be obtained using either polyelectrolytes or nonionic polymers (79, 80). A number of factors are important in this phenomenon, e.g., the

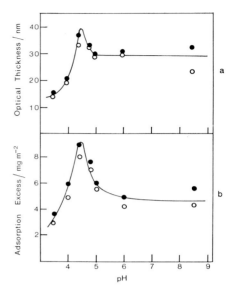

FIG. 14. The adsorption of gelatin on to a (111) silver bromide surface as a function of pH from 10^{-2} mole dm^{-3} potassium nitrate at pBr 3. (a) Optical thickness against pH; (b) adsorption excess against pH; ○, value after $2h$; ●, value after $12h$.

size of the polymer, the size of the particle, the ionic environment, etc., but at the present time no complete theory exists to describe the effect quantitatively and no precise measurements have been made of the energies and forces involved.

CONCLUDING REMARKS

In this paper I have attempted to give some insight into the nature of the forces which act between colloidal particles and to show how they can be measured. I have also attempted to show how these forces control the "stability or instability" of a dispersion. Although substantial progress has been made in the last 50 years in understanding the physical and chemical reasons for many colloid phenomena, much still remains to be learned before we can always control the balance of forces to produce a desirable dispersion or abolish an undesirable one.

ACKNOWLEDGMENTS

The work described in this paper has primarily been carried out at the University of Bristol during the period 1966–1976. It is therefore a personal view. It is a pleasure to acknowledge most gratefully my indebtedness to my collaborators during this period for their enthusiastic support and discussion, in particular, Drs. L. Barclay, R. J. R. Cairns, I. C. Callaghan, J. W. Goodwin, A. H. Harrington, T. Maternaghan, A. Parentich, P. N. Pusey, J. N. Shaw, and G. Wiese. I should also like to thank the Science Research Council and I. C. I. Limited for the financial support of a number of the projects described.

REFERENCES

1. GRAHAM, T., *Phil. Trans.* **151**, 183 (1861).
2. SCHULZE, H., *J. Prakt. Chem.* **25**, 431 (1882); **27**, 320 (1883).
3. LINDER, S. E., AND PICTON, H., *J. Chem. Soc.* **61**, 148 (1892); **71**, 568 (1897).
4. HARDY, W. B., *Proc. Roy. Soc.* **66**, 110 (1900).
5. HARDY, W. B., *Z. Phys. Chem.* **33**, 385 (1900).
6. SMOLUCHOWSKI, M. VON, *Z. Phys. Chem.* **92**, 129 (1917).
7. FUCHS, N., *Z. Phys.* **89**, 736 (1934).
8. OSTWALD, W., *J. Phys. Chem.* **42**, 981 (1938).
9. OTTEWILL, R. H., *in* "Colloid Science" (Specialist Periodical Report), Vol. 1, p. 173. The Chemical Society, London 1973.
10. LANGMUIR, I., *J. Chem. Phys.* **6**, 873 (1938).
11. LONDON, F., *Z. Phys.* **63**, 245 (1930).
12. BRADLEY, R. S., *Phil. Mag.* **13**, 853 (1932).
13. DE BOER, J. H., *Trans. Faraday Soc.* **32**, 10 (1936).
14. HAMAKER, H. C., *Physica* **4**, 1058 (1937).
15. DERJAGUIN, B. V., AND LANDAU, L., *Acta Physicochim. URSS* **14**, 633 (1941).
16. VERWEY, E. J. W., *Chem. Weekbl.* **39**, 563 (1942).
17. VERWEY, E. J. W., AND OVERBEEK, J. TH. G., "Theory of the Stability of Lyophobic Colloids," Elsevier, Amsterdam, 1948.
18. KRUYT, H. R., "Colloid Science," Vol. 1. Elsevier, Amsterdam, 1952.
19. OTTEWILL, R. H., AND SHAW, J. N., *Discuss. Faraday Soc.* **42**, 154 (1966).
20. REERINK, H., AND OVERBEEK, J. TH. G., *Discuss. Faraday Soc.* **18**, 74 (1954).
21. DERJAGUIN, B. V., AND MULLER, V. M., *Dokl. Phys. Chem.* **176**, 738 (1967).
22. SPIELMAN, L. A., *J. Colloid Interface Sci.* **33**, 562 (1970).
23. HONIG, E. P., ROEBERSEN, G. J., AND WIERSEMA, P. H., *J. Colloid Interface Sci.* **36**, 97 (1971).
24. OVERBEEK, J. TH. G., *J. Colloid Interface Sci.* **58**, 408 (1977). Invited lecture, Puerto Rico Symposium.
25. MACKOR, E. L., *J. Colloid Sci.* **6**, 492 (1951).
26. NAPPER, D. H., *Ind. Eng. Chem., Prod. Res. Dev.* **9**, 467 (1970); *M. T. P. Int. Rev. Sci., Ser. II 1*, Chap. 3 (1975).
27. HESSELINK, F. TH., *J. Phys. Chem.* **75**, 65, 2094 (1971).
28. DOLAN, A. K., AND EDWARDS, S. F., *Proc. Roy. Soc. Ser. A.* **337**, 509 (1974); **343**, 427 (1975).
29. NAPPER, D. H., *J. Colloid Interface Sci.* **58**, 390 (1977). Invited lecture, Puerto Rico Symposium.
30. ISRAELACHVILI, J. N., AND TABOR, D., *Progr. Surf. Membr. Sci.* **7**, 1 (1973).
31. ROBERTS, A. D., AND TABOR, D., *Proc. Roy. Soc. Ser. A* **325**, 323 (1971).
32. HOUGH, D. B., AND OTTEWILL, R. H., *J. Colloid Interface Sci.* **58**, in press, (1977). Puerto Rico Symposium, 1976.
33. BARCLAY, L. M., AND OTTEWILL, R. H., *Spec. Discuss. Faraday Soc.* **1**, 138 (1970).
34. CALLAGHAN, I. C., AND OTTEWILL, R. H., *Faraday Discuss. Chem. Soc.* **57**, 110 (1975).
35. VAN OLPHEN, H., "An Introduction to Clay Colloid Chemistry," Wiley, London, 1963.
36. BARCLAY, L., HARRINGTON, A. H., AND OTTEWILL, R. H., *Kolloid Z. Z. Polym.* **250**, 655 (1972).
37. CALLAGHAN, I. C., Ph.D. Thesis, University of Bristol, England 1975.
38. BUSCALL, R., AND OTTEWILL, R. H., *in* "Colloid Science" (Specialist Periodical Report), Vol. 2, p. 191. The Chemical Society, London, 1975.
39. BRACE, R., AND MATIJEVIĆ, E., *J. Inorg. Nucl. Chem.* **35**, 3691 (1973).
40. MATIJEVIĆ, E., LINDSAY, A. D., KRATOHVIL, S., JONES, M. E., LARSON, R. I., AND CAYEY, N. W., *J. Colloid Interface Sci.* **36**, 273 (1971).
41. McFADYEN, P., AND MATIJEVIĆ, E., *J. Inorg. Nucl. Chem.* **35**, 1883 (1973).
42. MATIJEVIĆ, E., SAPIESZKO, R. S., AND MELVILLE, J. B., *J. Colloid Interface Sci.* **50**, 567 (1975).
43. MATIJEVIĆ, E., *J. Colloid Interface Sci.* **58**, 374 (1977). Invited lecture, Puerto Rico Symposium.
44. KOTERA, A., FURUSAWA, K., AND TAKEDA, Y., *Kolloid Z. Z. Polym.* **239**, 677 (1970).
45. GOODWIN, J. W., HEARN, J., HO, C. C., AND OTTEWILL, R. H., *Colloid Polym. Sci.* **252**, 464 (1974).
46. FITCH, R. M., "Polymer Colloids." Plenum, New York, 1971.
47. LUCK, W., KLIER, M., AND WESSLAU, H., *Ber. Bunsenges. Phys. Chem.* **67**, 75, 84 (1963).
48. LUCK, W., KLIER, M., AND WESSLAU, H., *Naturwissenschaften* **14**, 485 (1963).
49. HILTNER, P. A., AND KRIEGER, I. M., *J. Phys. Chem.* **73**, 2386 (1969).
50. OTTEWILL, R. H., *Progr. Colloid Polym. Sci.* **59**, 14 (1976).
51. PARENTICH, A., to be published.
52. BROWN, J. C., PUSEY, P. N., GOODWIN, J. W., AND

OTTEWILL, R. H., *J. Phys. A: Gen. Phys.* **8**, 664 (1975).

53. PUSEY, P. N., *J. Phys. A: Gen. Phys.* **8**, 1433 (1975).

54. BERNAL, J. D., AND FANKUCHEN, I., *J. Gen. Physiol.* **25**, 111, 147 (1941).

55. KRIEGER, I. M., AND HILTNER, P. A., *in* "Polymer Colloids," p. 63. Plenum, New York, 1972.

56. HACHISU, S., KOBAYASHI, Y., AND KOSE, A., *J. Colloid Interface Sci.* **42**, 342 (1973).

57. ASH, S. G., EVERETT, D. H., AND RADKE, C., *J. Chem. Soc. Faraday Trans. II* **69**, 1256 (1973).

58. VAN MEGAN, W., AND SNOOK, I., *J. Colloid Interface Sci.* **53**, 172 (1975).

59. VAN MEGAN, W., AND SNOOK, I., *J. Chem. Soc. Faraday Trans. II* **72**, 216 (1976).

60. ZSIGMONDY, R., *Z. Anal. Chem.* **40**, 697 (1901).

61. FREUNDLICH, H., "Colloid and Capillary Chemistry." Methuen, London, 1926.

62. HELLER, W., AND PUGH, T. L., *J. Chem. Phys.* **22**, 1778 (1954).

63. MACKOR, E. L., AND VAN DER WAALS, J. H., *J. Colloid Sci.* **7**, 535 (1952).

64. KOELMANS, H., AND OVERBEEK, J. TH. G., *Discuss. Faraday Soc.* **18**, 52 (1954).

65. OTTEWILL, R. H., AND WALKER, T., *Kolloid Z. Z. Polym.* **227**, 108 (1968).

66. OTTEWILL, R. H., AND WALKER, T., *J. Chem. Soc. Faraday Trans. I.* **70**, 917 (1974).

67. FLORY, P. J., *J. Chem. Phys.* **10**, 51 (1952).

68. FISCHER, E. W., *Kolloid Z.* **160**, 120 (1958).

69. DOROSZKOWSKI, A., AND LAMBOURNE, R., *J. Polym. Sci.* **C34**, 253 (1971).

70. BARRETT, K. E. J., "Dispersion Polymerization in Organic Media." Wiley, London, 1975.

71. BAGCHI, P., *J. Colloid Interface Sci.* **47**, 86, 100 (1974); **50**, 115 (1975).

72. DOROSZKOWSKI, A., AND LAMBOURNE, R., *J. Colloid Interface Sci.* **26**, 214 (1968).

73. BARSTED, S. J., NOWAKOWSKA, J., WAGSTAFF, I., AND WALBRIDGE, D. J., *Trans. Faraday Soc.* **67**, 3598 (1971).

74. CAIRNS, R. J. R., OTTEWILL, R. H., OSMOND, D. W. J., AND WAGSTAFF, I., *J. Colloid Interface Sci.* **54**, 45 (1976).

75. EIRICH, F. R., *J. Colloid Interface Sci.* **58**, 423 (1977). Invited lecture, Puerto Rico Symposium.

76. MATERNAGHAN, T. J., AND OTTEWILL, R. H., *J. Photogr. Sci.* **22**, 279 (1974).

77. BROSSA, A., AND FREUNDLICH, H., *Z. Phys. Chem.* **89**, 306 (1915).

78. LA MER, V. K., *Discuss. Faraday Soc.* **42**, 248 (1966).

79. KITCHENER, J. A., *Br. Polym. J.* **4**, 217 (1972).

80. FLEER, G. J., KOOPAL, L. K., AND LYKLEMA, J., *Kolloid Z. Z. Polym.* **250**, 689 (1972).

The Role of Chemical Complexing in the Formation and Stability of Colloidal Dispersions [1]

EGON MATIJEVIĆ

Institute of Colloid and Surface Science and Department of Chemistry, Clarkson College of Technology, Potsdam, New York 13676

Received June 8, 1976; accepted September 28, 1976

The role of chemical complexing in the preparation of monodispersed metal hydrous oxides of iron, aluminum, and titanium is discussed. Specifically, the effect of anions on the size, shape, and composition of the colloidal particles is emphasized. A reaction mechanism is offered which explains the nucleation and growth process involved in the formation of uniform spherical particles of titanium dioxide in acidic solutions containing sulfate ions. To illustrate the influence of chemical complexing in the solution and at the interface on the stability of colloidal dispersions, the interactions of hydrolyzed metal ions and of chelates with latexes and silver halide sols are described. Finally, new data on the adhesion of monodispersed chromium hydroxide particles on glass are given.

INTRODUCTION

The existence of complex species as intermediates in the formation of colloidal particles in electrolyte environments was recognized a long time ago. Indeed, embryos and nuclei are assumed to be generated through specific aggregations of solutes.

The role of chemical complexing in solutions and at solid/liquid interfaces in the interpretation of dispersion stability has been less frequently emphasized. Yet the significance of such reactions in a number of colloid phenomena has become increasingly obvious, even in seemingly simple systems consisting of "inert" particles in solutions of "neutral" electrolytes. The vast majority of the sols, particularly those in most applications, are composed of solids with surfaces of rather diverse chemical composition, which are suspended in media containing a mixture of solutes of varying degrees of complexity. Interactions of these species at the solid/solution interfaces are

frequently responsible for the stability of dispersions, so that the properties of a sol can only be explained if the chemical processes in solution and at particle surface are fully elucidated. Conversely, one can exploit complexing phenomena either to prepare colloidal systems of certain desired characteristics or to stabilize or coagulate a sol with a minimum amount of additives. The success of such endeavors will greatly depend upon the understanding of the role of specific chemical reactions involved in the particle formation and/or stability. This requires a quantitative knowledge of the composition of the electrolyte environments and of the solid/solution interfaces, a task which is not always easily accomplished. In recent years considerable progress has been made in explaining various chemical aspects of interfacial phenomena. An earlier paper (1) contains a comprehensive review of the subject, whereas this paper describes some new developments in the author's laboratory which shed further light on the role of chemical complexing in colloid formation and stability.

[1] Supported in part by the National Science Foundation, Grant CHE 74-09402 A02.

MONODISPERSED METAL HYDROUS OXIDES

a. Effect of Anions

Over the past few years procedures have been developed for the preparation of colloidal dispersions of metal hydrous oxides consisting of particles of considerable uniformity in size and shape. Such sols have been obtained with chromium (2, 3), aluminum (4), iron (5), and copper (6), by keeping salt solutions of the respective metals at elevated temperatures for various periods of time. The particle shape and composition depend most strongly on the pH and on the nature of the *anions* contained in the aging systems. Significantly, the ions responsible for the characteristics of the particles frequently were not found in the solid phase, indicating that solute complexes involving such anions must act as precursors to the nucleation stage. For example, the presence of sulfate ions is essential in order to produce uniform colloidal spheres of chromium hydroxide, yet sulfate is not a constituent species in the final particles (7).

Since each metal ion in aqueous solution forms different hydrolysis products and complexes with anions other than hydroxyl, it is obvious that the chemical mechanism of precipitation of metal hydroxides will vary from case to case. The availability of reproducible "monodispersed" metal hydrous oxide sols makes it possible to study in detail the relationship between the complex formation in electrolyte solutions and the composition and morphology of particles nucleated and grown in such environments.

Several series of electron micrographs are offered to illustrate the effects of various experimental parameters, and particularly of the nature of the anions, on the shape and composition of the metal hydroxide particles generated in different salt solutions.

Figure 1 represents two ferric hydrous oxide systems obtained under identical conditions except that in (a) $Fe(NO_3)_3$ and in (b) $Fe(ClO_4)_3$ was used. Figure 2 exemplifies the effect of altering experimental conditions on the particles formed in the $FeCl_3+HCl$ solutions. Note that the sol illustrated in Fig. 2a was obtained at the same concentrations as those in Fig. 1, thus, offering another example of the influence of anions on the particle morphology. It is also of interest that scanning electron micrographs of the system 2b show the particles to be spherical. The precipitated solids differ not only in shape, but also in

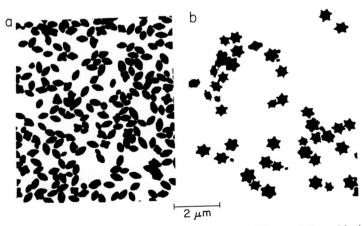

$2 \mu m$

FIG. 1. Electron micrographs of particles obtained by aging two different solutions of ferric salts at 100°C for 24 hr; in both cases the original pH was 1.3–1.4 and the final pH was 1.1–1.2. Concentrations: (a) System $Fe(NO_3)_3 + HNO_3$: $[Fe^{3+}]$, 0.018 M; $[NO_3^-]$, 0.104 M. (b) System $Fe(ClO_4)_3 + HClO_4$: $[Fe^{3+}]$, 0.018 M; $[ClO_4^-]$, 0.104 M.

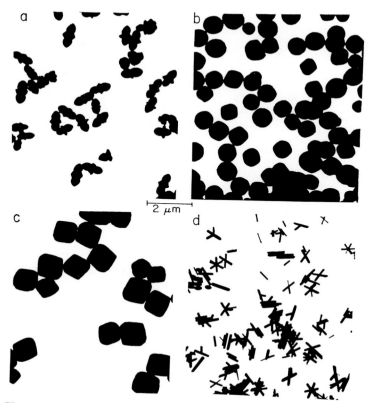

FIG. 2. Electron micrographs of particles obtained in solutions of FeCl$_3$ + HCl under the following conditions:

	[Fe^{3+}] (M)	[Cl$^-$] (M)	pH (initial)	pH (final)	Temp. of aging (°C)	Time of aging
(a)	0.018	0.104	1.3	1.1	100	24 hr
(b)	0.315	0.995	2.0	1.0	100	9 days
(c)	0.09	0.28	1.65	0.88	100	24 hr
(d)	0.09	0.28	1.65	0.70	150	6 hr

chemical composition. X-ray analysis indicates all the systems in Figs. 1 and 2 to be consistent with α-Fe$_2$O$_3$ except for the rods in Fig. 2d, which correspond to β-FeOOH.

Similar anionic effects on colloid particle formation have been observed when aluminum salt solutions were aged at elevated temperatures. The electron micrograph shown in Fig. 3 gives the perfectly spherical and amorphous particles obtained in an aluminum sulfate solution. Scanning electron micrographs (Fig. 4) of colloidal dispersions generated in aluminum chloride and aluminum perchlorate solutions, respectively, show that the particles are of unusual structure complexity, yet are still rather uniform in size. Electron diffraction patterns of the sols presented in Fig. 4 are consistent with that of γ-AlOOH (diaspore).

In a previous study (4) it was established that the spherical particles of aluminum hydroxide illustrated in Fig. 3 contain considerable amounts of sulfate ions. However, these are readily exchangeable for hydroxyl groups if the pH is raised to the neutral or a slightly alkaline range. Obviously, a different chemical mechanism is responsible for the formation of aluminum hydroxide sols than that elucidated for chromium hydroxide sols prepared under similar conditions; yet in both cases amorphous spherical particles are produced.

b. Basic Ferric Sulfate Sols

The third case in which sulfate ions play a crucial role refers to the formation of ferric hydrous oxide sols. Indeed, the best-defined systems so far, chemically, structurally, and morphologically, have been obtained by heating ferric salt solutions which contained sulfate ions. Figure 5 gives a transmission and a scanning electron micrograph of two such dispersions. In both examples, the solids have been identified as basic ferric sulfates corresponding to the alunite-type minerals (5). The sol shown in Fig. 5a contains particles of hexagonal symmetry which have a chemical composition $Fe_3(SO_4)_2(OH)_5 \cdot 2H_2O$. The scanning electron micrograph (Fig. 5b) clearly indicates two types of crystals, one hexagonal and the other monoclinic. Two different structures of the cited symmetries were also identified by X-ray crystallography and chemically they are consistent with formulations $Fe_3(SO_4)_2(OH)_5 \cdot 2H_2O$ and $Fe_4(SO_4)-(OH)_{10}$, respectively.

In view of the reproducibility of preparation and exceedingly good characterization of the basic ferric sulfate sols, they represent ideal model systems for the study of the chemical mechanism of their formation. For this purpose, it is necessary to know the composition and the concentration of all species that may be involved in the making of the particle. As unexpected as it may sound, few data are available on the hydrolysis of ferric ions as a function of temperature and these are for rather restricted conditions. As a matter of fact, only one study could be found for

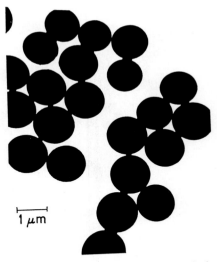

Fig. 3. Electron micrograph of aluminum hydrous oxide particles obtained by aging for 48 hr at 97°C a 2×10^{-3} M solution of $Al_2(SO_4)_3$. Initial pH, 4.0; final pH, 2.5.

temperatures as high as 51°C (8), which is 20 years old. Equally scarce is the information on sulfate complexes with ferric ions at different temperatures.

We have undertaken a comprehesive program of determining the composition of the complex species and their formation constants in aqueous solutions containing ferric ions in the absence and in the presence of sulfuric acid over the temperature range of 25 to 80°C at controlled ionic strengths (9). This investigation also involved the measurements of the formation constants of $NaSO_4^-$ and HSO_4^- as a function of temperature and ionic strength. Spectrophotometric, conductometric, and potentiometric techniques have been used.

Our findings are that solutions of ferric salts at low pH contain only relatively simple solute complexes, i.e., $FeOH^{2+}$, $Fe(OH)_2^+$, and $Fe_2(OH)_2^{4+}$. In addition, in the presence of SO_4^{2-} ions, species $FeSO_4^+$ and $FeHSO_4^{2+}$ have been identified. If higher polynuclear ferric solutes are present, their concentration must be too small to be detected by the techniques employed.

Fig. 4. Scanning electron micrographs of aluminum hydrous oxide particles obtained by aging (a) for 24 hr at 150°C a 5×10^{-3} M solution of $AlCl_3$ (initial pH 3.7, final pH 2.1); (b) for 25 hr at 125°C a 5×10^{-3} M solution of $Al(ClO_4)_3$ (initial pH 3.8, final pH 2.1).

Fig. 5. (a) Electron micrograph of basic ferric sulfate sol particles prepared by aging a solution 0.18 M in $Fe(NO_3)_3$ and 0.53 M in Na_2SO_4 in an oil bath at 80°C for 1.5 hr (pH: 1.7); (b) scanning electron micrograph of a basic ferric sulfate sol prepared by aging a solution 0.18 M in $Fe(NO_3)_3$ and 0.27 M in $NaSO_4$ in an oil bath at 80°C for 2 hr (pH: 1.5).

401

Table I gives the formation constant for $FeOH^{2+}$ as a function of temperature and of the ionic strength.

The formation constants for $FeSO_4^+$ and $FeHSO_4^{2+}$ have been determined so far at one ionic strength (2.67 M), which corresponds to the conditions giving the monodispersed basic ferric sulfate particles. The values at three different temperatures are listed in Table II.

Considering only the most abundant species in acidic solutions containing ferric and sulfate ions at elevated temperatures, which we have now established, one can tentatively suggest the following overall chemical processes in the formation of monodispersed basic ferric sulfates (9):

$$FeOH^{2+}+2FeSO_4^++6H_2O \rightarrow$$
$$Fe_3(OH)_5(SO_4)_2 \cdot 2H_2O+4H^+ \quad [1]$$

for hexagonal particles, and

$$3Fe(OH)_2^++FeSO_4^++4H_2O \rightarrow$$
$$Fe_4(SO_4)(OH)_{10}+4H^+ \quad [2]$$

for monoclinic particles.

We have work in progress which may help the elucidation of the various steps in the formation of the above colloidal solids.

c. Titanium Dioxide Sols

Recently we succeeded in the preparation of titanium dioxide sols consisting of spherical particles of a rather narrow size distribution. To achieve this, titanium salt solutions had to be aged at elevated temperatures for ex-

tensive periods of time (weeks). Also, there was a considerable induction period, often a week or longer, before the first solid phase could be detected. Importantly, such systems were produced only in the presence of sulfate ions, yet these could not be found in the particles. Figure 6 illustrates a sol, obtained on aging at 98°C for 37 days, a solution of 0.1 M $TiCl_4$ in 5.67 M HCl which contained sulfate ions to give a ratio $[SO_4^{2-}]:[Ti^{4+}] = 1.9$. An X-ray analysis gave a particle structure consistent with that of rutile. A series of experiments carried out under identical conditions, except that the sulfate-to-titanium ratio was altered, showed that the particle size became larger with increasing concentration of sulfate ions.

TABLE II

Formation Constants, K_1^*, for $FeSO_4^+$ and $FeHSO_4^{2+}$ Complexes ($\mu = 2.67\ M$)

t (°C)	$K^*_{FeSO_4^+}$	$K^*_{FeHSO_4^{2+}}$
25	84	8
55	240	5
80	1020	2

TABLE I

Values of the Formation Constant K_1^* for $Fe^{3+} \rightleftharpoons FeOH^{2+} + H^+$ ($K_1^* \times 10^3$)

t (°C)	μ (M)					
	2.67	2.25	1.89	1.00	0.50	0.10
25	1.16	1.23	1.24	1.47	1.75	2.46
33	1.54	1.63	1.68	1.93	2.38	3.46
40	2.15	2.21	2.26	2.66	3.33	4.90
55	3.59	3.85	4.08	4.67	6.33	8.89
70	6.19	6.74	6.98	8.63	11.1	15.3
80	8.82	9.52	10.5	11.9	16.1	23.2

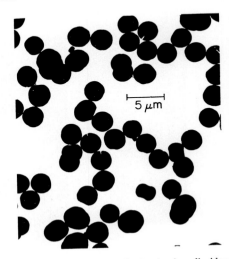

5 μm

FIG. 6. Electron micrograph of a titanium dioxide sol obtained by aging for 37 days at 98°C a solution of 0.1 M $TiCl_4$ in 5.76 M HCl which contained sulfate ions to give a ratio $[SO_4^{2-}]:[Ti^{4+}]=1.9$.

The results of this study, to be given in detail elsewhere (10), were used to derive a mechanism of titanium dioxide formation. It is suggested that the aging of titanium salt solutions causes the hydrolysis and polymerization of titanium(IV) ions which finally result in solid phase formation in the following series of processes:

$$Ti^{4+} + p\ OH^- \rightleftharpoons Ti(OH)_p^{(4-p)+}, \qquad [3]$$

$$2Ti(OH)_p^{(4-p)+} \rightleftharpoons (2p-2)OH^-$$

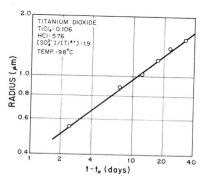

$$E \rightleftharpoons H_2O + \underset{\diagup}{\overset{\diagdown}{-}}Ti-O-Ti\underset{\diagdown}{\overset{\diagup}{-}}(\equiv F), \qquad [5]$$

$$E + F \rightarrow TiO_2\ crystals. \qquad [6]$$

It is further assumed that Eq. [4] represents the rate-determining step and that the species F corresponds to the critical nucleus. On the basis of these assumptions,

$$[F] = k[Ti^{4+}]^2\ [OH^-]^2, \qquad [7]$$

and at constant pH

$$N = k'[Ti^{4+}]^2, \qquad [8]$$

where N is the number of nuclei.

The role of sulfate ions is to be found in their strong binding with titanium(IV) ions.

The resulting complex solutes slowly decompose on heating to yield the titanium(IV) ions which are used up in the particle growth. Thus, one can describe an equilibrium

$$Ti_a(SO_4)_b^{(4-2b)+} \rightleftharpoons aTi^{4+} + bSO_4^{2-}, \qquad [9]$$

for which

$$K = [Ti^{4+}]^a[SO_4^{2-}]^b / [Ti_a(SO_4)_b^{(4-2b)+}]. \qquad [10]$$

If S and T represent the total analytical concentrations of sulfate and titanium(IV), respectively, and S is large so that the complex $Ti_a(SO_4)_b^{(4-2b)+}$ is the dominant Ti(IV) solute species,

$$[Ti_a(SO_4)_b^{(4-2b)+}] = T/a, \qquad [11]$$

$$[SO_4^{2-}] = S - bT/a. \qquad [12]$$

Hence, from Eq. [10]

$$[Ti^{4+}] = (KT/a)^{1/a}(S - bT/a)^{-b/a}. \qquad [13]$$

It was shown in this work (10) that S/T must be at least equal to 0.5 for the particles to form. This suggests that $b/a = 0.5$, and therefore

$$[Ti^{4+}] = (KT/a)^{1/a}(S - T/2)^{-\frac{1}{2}}. \qquad [14]$$

Combining this result with the equation

$$d(\tfrac{4}{3}\pi r^3 N)/dt = ck_5[Ti(OH)_p^{(4-p)+}]^2, \qquad [15]$$

which is the steady-state assumption for spherical particles (k_5 = rate constant for dimerization in reaction [4], and c = a constant of proportionality), yields eventually

$$r = c'(S - T/2)^{\frac{1}{3}}(t - t_0)^{\frac{1}{3}}, \qquad [16]$$

where c' is a constant, if pH and the concentration of titanium ions are kept constant.

Figure 7 gives a plot of the data which indeed gives a slope of $\frac{1}{3}$ as required by Eq. [16].

Thus, it would seem that the proposed mechanism in its simple form indeed describes well the finding with the monodispersed titanium dioxide sols.

TITANIUM DIOXIDE
TiCl₄ : 0.106
HCl : 5.76
[SO₄²⁻]/[Ti⁴⁺] : 1.9
TEMP : 98°C

RADIUS (μm)

t−t₀ (days)

FIG. 7. Plot of the modal radius of the spherical titanium dioxide particles against the time of aging according to Eq. [16]. $[SO_4^{2-}]/[Ti^{4+}]$: 1.9. Other conditions are the same as in Fig. 6.

d. Effect of Polymers

Despite the obvious importance of the effects which dissolved polymers may have on the formation and growth of particles in an aqueous environment, only a very few studies on the subject have been reported and the results are by no means conclusive (11).

A systematic investigation has been carried out in which the chromium hydroxide sols were generated in the presence of various dextrans. These polymers are particularly convenient, because they can be obtained in rather narrow molecular weight fractions in nonionic, anionic, and cationic forms. The chromium hydroxide system was chosen, since on aging of chrom alum solutions at moderate temperatures ($\sim$75°C) stable and chemically inert monodispersed sols of spherical particles are readily produced (2, 3).

Dextrans were added to chrom alum solutions and the characteristics of the chromium hydroxide sols formed on aging have then been investigated as a function of the molecular weight and the concentration of the polymers. Nonionic dextrans of molecular weights 4×10^4 and 5×10^5 in concentrations less than 0.1% by weight had little influence on the generation of chromium hydroxide particles; i.e., their

shape and size were the same as those obtained in pure chrom alum solutions. If concentrations of these dextrans exceeded 0.1% the modal particle diameter decreased markedly. Similar effects were observed with nonionic dextrans of the molecular weight of 2×10^6 except that the smaller particles resulted already at an addition of 0.01%. The phenomena described are attributed to an increase in viscosity of the media, in which particles were grown, on addition of polymers.

Considerably greater influence was found when dextran sulfate 2000 (molecular weight $\sim$4 $\times$ 10⁶) was used in chrom alum solutions. Figure 8 shows that only 0.001% of this polyelectrolyte causes a drastic reduction in the size of chromium hydroxide particles. The modal diameter became even smaller when the dextran sulfate 2000 content was increased to 0.01%. However, at still higher concentrations of the polyelectrolyte the sol changed its characteristics: the particles became rather irregular, larger, and of much broader distribution. Such dramatic effects indicated strong interactions between the anionic polyelectrolyte and the cations in solution. To investigate the reaction of dextran sulfate with chromium ions, dialysis equilibration studies were made

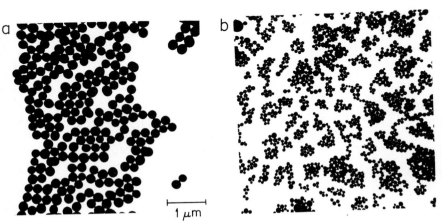

Fig. 8. Electron micrographs of chromium hydroxide sols prepared by aging for 26 hr at 75°C a solution of 4×10^{-4} M KCr(SO₄)₂ in the absence (a), and in the presence of 0.001% by weight of sodium dextran sulfate 2000 (b).

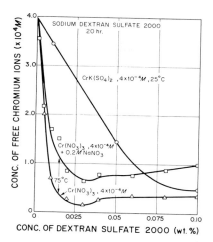

Fig. 9. The concentration of "free" chromium ions after dialysis against different concentrations of sodium dextran sulfate for 20 hr solutions of (a) 4×10^{-4} M $KCr(SO_4)_2$ at 25°C (○); (b) 4×10^{-4} M $Cr(NO_3)_3$ at 75°C (△); and (c) 4×10^{-4} M $Cr(NO_3)_3 + 0.2$ M $NaNO_3$ at 75°C (□).

using a two-compartment cell separated by a regenerated cellulose dialysis membrane. One compartment contained the polyelectrolyte solution and the other the chromium salt solution. The dialysis was carried out at room temperature and at 75°C, the latter being the condition of sol formation.

Figure 9 shows that dextran sulfate strongly binds chromium ions, particularly at higher temperatures. At 25°C, chrom alum was the electrolyte in the dialysis experiments, whereas at 75°C only $Cr(NO_3)_3$ could be employed, because the colloidal particles would form in the presence of free sulfate ions over the extended periods of time needed for the equilibration. Two sets of data are given at 75°C, which differ inasmuch as 0.2 M $NaNO_3$ was present in one of the systems to eliminate a possible Donnan effect. The sharp decrease in the concentration of the soluble chromium species on addition of dextran sulfate explains the decrease of particles size formed on aging of chrom alum solutions with the polymer as compared to the particles obtained in the pure chromium salt solution. Earlier it was

shown that the chromium hydroxide particles became smaller as the concentration of the chrom alum solutions used in the aging experiments was lowered (2, 12). An attempt to quantitatively relate the particle size to the free chromium ion concentration in the presence and in the absence of dextran sulfate has not been successful. The particles obtained when the anionic polyelectrolyte was added are somewhat smaller than expected. However, this is not surprising as the comparison is not completely justified. For the reasons mentioned above, it is not feasible to obtain the data for binding of chromium ions by dextran sulfate in the presence of free sulfate ions at elevated temperatures. Yet, there is no doubt that elimination of chromium ions from solutes by the anionic polyelectrolyte is the dominating cause for the change in the particle size when such polymer is added to the electrolyte solution.

It is noteworthy that there is a minimum in the curves representing chromium ion binding as a function of the dextran sulfate concentration. This minimum becomes more pronounced at larger equilibration times and it coincides well with the change in the nature of the sol above a certain polymer concentration which was mentioned earlier.

THE ROLE OF COMPLEX CHEMISTRY IN COLLOID STABILITY

One of the essential factors affecting the properties of colloidal dispersions is the adsorption of various solutes, most importantly of counterions and of solvent molecules, at the particle/solution interface. The nature and the quantity of adsorbed species will determine the double layer composition, which is in turn responsible for sol stability. The adsorptivity of ions onto solid particles depends on many factors, some of which are still not understood. If a chemical bond can be formed between a solute and the solid, the amount taken up is, as a rule, significant. Numerous examples have been offered which show that the complex species adsorb more readily than simple ions or molecules. In view of this, considerable effort has been made to establish the quantitative relationships between the reactions of complexes at interfaces and colloid stability, in itself a major undertaking. To illustrate some of the progress made, new data will be described in the systems consisting of lyophobic particles suspended in the solutions containing hydrolyzed metal ions and metal chelates, respectively.

a. Effect of Hydrolyzed Metal Ions

Every polyvalent cation undergoes hydrolysis in aqueous solution and most metal ions of 3+ and 4+ charge are hydrolyzed already in rather acidic media. It has been established without exception, that hydroxylation of counterions causes a significant enhancement in their adsorptivity and, consequently, influences the stability of a colloid dispersion. Indeed, the adsorption may be sufficient to reverse the negative charge of particles to positive with a resulting sol restabilization. Since polyvalent metal ions, particularly aluminum and iron(III), are present in most natural environments and used in countless applications, the effects of their hydrolysis on adsorption and colloid stability has been extensively investigated. Due to the still inadequate information on various solution species involved, a quantitative interpretation of these phenomena is yet to be worked out.

We have now carried out a rather systematic study on adsorption of aluminum solutes on different lyophobic colloids (various latexes and silver halides) as a function of aluminum salt concentration, pH, and the nature of the anions. Furthermore, the rate of desorption of hydrolyzed aluminum ions was investigated as affected by the same parameters.

Figure 10 gives the change in zeta-potentials in dependence of pH of a PVC latex at three different concentrations of Al(NO₃)₃. In each case, a pronounced charge reversal is observed; the isoelectric point shifted toward lower pH values as the concentration of the aluminum salt increased. The zeta-potential of the latex itself was independent of pH. The adsorption

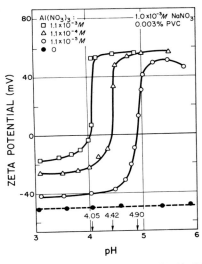

FIG. 10. Zeta-potential of a polyvinylchloride (PVC) latex (0.003% by weight) as a function of pH in the absence ($\bullet$), and in the presence of Al(NO₃)₃: 1.1×10^{-3} M ($\square$), 1.1×10^{-4} M ($\triangle$), and 1.1×10^{-5} M ($\bigcirc$). Arrows indicate the isoelectric points. All systems contained 1.0×10^{-3} M NaNO₃.

of aluminum ions on the same colloid under identical conditions was also measured colorimetrically using the eriochrome cyanine R dye (13). Taking the equilibrium constants of hydrolyzed aluminum species as recently summarized (14), it was possible to carry out quantitative evaluation of interaction of these counterions with the hydrophobic latex particles.

The total adsorption of aluminum solute species can be written in terms of known solute complexes as

$$\Gamma_{Al_{(total)}} = \Gamma_{Al^{3+}} + \Gamma_{Al(OH)^{2+}} + \Gamma_{Al_8(OH)_{20}^{4+}} + \Gamma_{Al(OH)_3} + \Gamma_{Al(OH)_4^-}. \quad [17]$$

Using the model of adsorption of hydrolyzed metal ions as proposed by James and Healy (15), the standard change in free energy of adsorption of species i

$$\Delta G^0_{ads_i} = \Delta G^0_{coul_i} + \Delta G^0_{solv_i} + \Delta G^0_{chem_i}, \quad [18]$$

where subscripts denote the coulombic, solvation, and chemical contributions.

TABLE III

Calculated Adsorption Model Parameters for Aluminum Solutes on PVC Latex

Species	$\Delta G^0_{coul_i}$ (kJ/mole)	$\Delta G^0_{solv_i}$ (kJ/mole)	$\Delta G^0_{ads_i}$ (kJ/mole)	K'_{ads_i}
Al^{3+}	-44	154	110	5×10^{-20}
$AlOH^{2+}$	-29	68	39	2×10^{-7}
$Al_s(OH)_{20}^{4+}$	-58	4	-54	3×10^9
$Al(OH)_3$	0	0	0	1

The individual terms were evaluated from the known equations, i.e.,

$$\Delta G^0_{coul_i} = z_i F \Delta \psi_x, \quad [19]$$

where

$$\Delta \psi = \frac{2RT}{zF} \ln$$

$$\times \left[\frac{(e^{zF\psi_0/2RT} + 1) + (e^{zF\psi_0/2RT} - 1)e^{-\kappa a}}{(e^{zF\psi_0/RT} + 1) - (e^{zF\psi_0/RT} - 1)e^{-\kappa a}} \right],$$

$$[20]$$

in which a is the radius of the hydrated counterion and all other terms have the usual meanings. Furthermore,

$$\Delta G^0_{solv_i} = \frac{z_i^2 e^2 N}{16\pi\epsilon_0} \left[\frac{1}{r_i + 2r_w} - \frac{r_i}{2(r_i + 2r_w)^2} \right]$$

$$\times \left(\frac{1}{\epsilon_{int}} - \frac{1}{\epsilon_b} \right) + \left(\frac{z_i^2 e^2 N}{32\pi\epsilon_0} \right) \left(\frac{1}{r_i + 2r_w} \right)$$

$$\times \left(\frac{1}{\epsilon_{solid}} - \frac{1}{\epsilon_{int}} \right), \quad [21]$$

where r_i and r_w are the radii of the counterion i and of water, and ϵ_{int}, ϵ_b,

and ϵ_{solid} are the dielectric constants of the interfacial, bulk solution, and solid phase, respectively. The calculated parameters using the above equations are summarized in Table III, in which the "prime" is used to indicate the neglect of the $\Delta G^0_{chem_i}$ term in the evaluation of $\Delta G^0_{ads_i}$. Obviously, the complex octamer should be by far the most strongly adsorbed species. It was not possible to evaluate $\Delta G^0_{chem_i}$ independently; therefore, this was disregarded in the calculations. The inclusion of the chemical energy term would have a favorable effect on the interactions of all species.

Table IV gives the adsorption densities and the average charge per aluminum atom at the isoelectric point, based on direct adsorption measurements and the knowledge that the charge density of PVC latex, as determined by potentiometric titrations, $\sigma_0 = 3.4 \times 10^{13}$ charges/cm². Also included in the table are the calculated values obtained from a simple adsorption isotherm

$$\theta = \Sigma_i K_i M_i / (1 + \Sigma_i K_i M_i) \quad [22]$$

and

$$\Gamma_{Al} = (\theta/N_A)S, \quad [23]$$

in which θ is the fraction of covered adsorption sites, M_i is the equilibrium molar concentration

TABLE IV

Experimental and Calculated Adsorption Densities of Aluminum Solute on PVC Latex

Initial conc. of $Al(NO_4)_3$ (M)	pH (IEP)	pH_p (Al(OH)₃)	Experimental		Calculated	
			Γ_{Al} (moles/cm²)	Average charge/Al atom	Γ_{Al} (moles/cm²)	Charge/Al atom
5.0×10^{-5}	4.60	5.00	1.2×10^{-9}	0.05	6.4×10^{-10}	0.08
1.1×10^{-5}	4.90	5.20	3.8×10^{-10}	0.15	3.9×10^{-10}	0.13

of species i and K_i is the corresponding equilibrium constant for the adsorption process; N_A is the Avogadro number and S is the number of surface sites per unit surface area of the adsorbent. Finally, the table contains the pH at which solid Al(OH)₃ would precipitate in the absence of the latex.

The agreement between the calculated and experimental adsorption densities is in general satisfactory, indicating that the model is at least a good first approximation to the actual physical process.

The average charges shown in the table are determined by dividing the residual latex charge by the product of the Avogadro number and the adsorption density. The fact that the average charge is so low should be interpreted to mean that the adsorbed polynuclear complexes bind hydroxyl ions from the interfacial region leading to further polymerization. The experimentally determined adsorption density in the systems containing 5.0×10^{-5} M Al(NO₃)₃ is higher than that calculated for a monolayer. This can be understood if one considers the structural characteristics of the adsorbed polynuclear complex. If the latter is three-dimensional with respect to the bound aluminum ions, the adsorbed sheet would consist of more than one layer of aluminum ions.

Figure 10 shows convincingly that, as the hydrolysis of aluminum ions progresses with increasing pH, the amount adsorbed increases. It was of interest to establish if the process can be reversed, i.e., could the ions desorb on acidification. Figure 11a gives the mobility as a function of pH for a silver bromide sol, the charge of which was reversed by aluminum nitrate. Several curves designate measurements taken at different times after acidification. It is obvious that the removal of aluminum species from the surface is a very slow process, which can be understood in terms of the discussion of the previous section. If the desorption was conditioned only on reversing the hydrolytic reaction by lowering the pH, the decomplexed ions would diffuse away rather fast. The fact

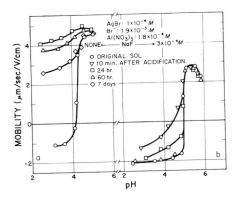

FIG. 11. Electrophoretic mobilities of a negatively charged silver bromide sol (AgBr: 1×10^{-4} M, excess Br⁻: 1.9×10^{-3} M) in the presence of 1.8×10^{-4} M Al(NO₃)₃ as a function of pH. (a) Original sol (○); other curves are for sols the charge of which was first reversed by Al(NO₃)₃ at the highest pH of the plot and subsequently acidified to various lower pH value. The measurements were carried out at different times of acidification. (b) The same systems as described above except that 3×10^{-4} M NaF was added to the silver bromide sol.

that the adsorbed aluminum complexes are further hydrolyzed on the surface to give essentially an aluminum hydroxide precipitate requires a dissolution of the latter which is normally a time consuming process.

It is to be expected that adding to the system an anion, which complexes aluminum to a nonadsorbing species, would affect the charge reversal and accelerate the desorption process. This was found when sodium fluoride was introduced into the system as described above (Fig. 11b). The isoelectric point is shifted toward a higher pH value and the removal of the adsorbed aluminum ions from the particle surface on acidification is quite rapid. A considerable effect of fluoride ions on the coagulation ability of aluminum ions was demonstrated earlier (16). The present data substantiate the previous findings.

b. Effect of Metal Chelates

The use of metal chelates in studies of colloid stability has a number of advantages: one can select very stable complexes of known charge, size, shape, and other properties eliminating, thus, many uncertainties encountered in the work with hydrolyzed counterions.

As an illustration, two cases will be discussed here.

It was found that the addition of a very small amount of 2,2'-dipyridyl (3.0×10^{-6} M) had a striking effect on the stability of a negatively charged silver bromide sol. Figure 12 gives two coagulation curves (high Rayleigh ratio, RR, = unstable sol) showing that this system can be destabilized by nickel ions, in concentrations orders of magnitude lower than is the critical coagulation concentration of Ni^{2+} for the same sol in the absence of 2,2'-dipyridyl. Also in both cases a pronounced maximum is observed, i.e., below and above certain concentrations of Ni^{2+}, the sols

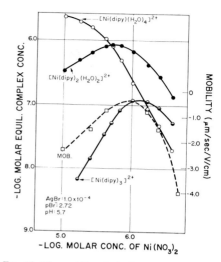

FIG. 13. The mobility ($\square$) of a silver bromide sol (AgBr, 1.0×10^{-4} M; pBr, 2.72; pH, 5.7) *in statu nascendi* in the presence of 2,2'-dipyridyl (3.0×10^{-6} M), as a function $Ni(NO_3)_2$ concentration. The equilibrium concentrations of the complex ions present are also shown. $[Ni(dipy)(H_2O)_4]^{2+}$, $\bigcirc$; $[Ni(dipy)_2(H_2O)_2]^{2+}$, $\bullet$; and $[Ni(dipy)_3]^{2+}$, $\bullet$.

remain stable. Such behavior is typical when charge reversal, due to the adsorption of counterions, takes place. However, electrokinetic mobility data show that the sols remain negatively charged over the entire Ni^{2+} concentration range, and that the minimum in mobility coincides with a maximum in RR.

The "sensitization" of the silver bromide sols toward nickel ions by the addition of dipyridyl can be explained if one considers the formation of complexes between the metal ion and the ligand molecules. The concentrations of these can be readily calculated from the literature data on the stability of the nickel chelates. The composition of the solution at pH 5.7 as a function of $Ni(NO_3)_2$ concentration is given in Fig. 13 along with the mobilities of silver bromide particles in the same environment. It can be seen that the concentration of the mono complex continually increases as the $[Ni^{2+}]$ rises. These results indicate that the species adsorbing on the sol are $[Ni(dipy)_2-(H_2O)_2]^{2+}$ and $[Ni(dipy)_3]^{2+}$. The complex

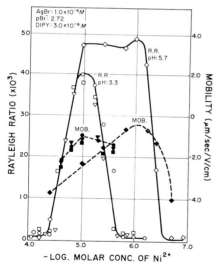

FIG. 12. Coagulation curves of a silver bromide sol (AgBr, 1.0×10^{-4} M; pBr, 2.72) as a function of $Ni(NO_3)_2$ in the presence of a constant amount of 2,2'-dipyridyl (3.0×10^{-6} M) at two different pH ($\bigcirc$, $\triangledown$, $\square$, 3.3; $\diamond$, 5.7). Dashed lines and blackened symbols give the corresponding mobilities. Circles and diamonds represent a sol *in statu nascendi*, whereas triangles and squares are for a sol aged in the presence of dipy for 30 min and 3 hr, respectively, before addition of $Ni(NO_3)_2$.

ion $[Ni(dipy)(H_2O)_4]^{2+}$ is apparently too heavily hydrated to interact with the hydrophobic silver bromide particles. Analogous results were obtained at pH 3.3, except at this lower pH the calculated concentration of the tris complex is extremely low and the entire stability situation is dominated by the bis chelate ion (17).

This example again illustrates the fact that one or two solute species out of a larger number in a mixture can dominate the stability of a sol.

Once the complex chemistry of a system is understood it is possible to apply the double layer theory to a colloidal dispersion. The case to be described consists of a silver iodide sol in the presence of tris(1,10-phenanthroline)-cobalt(III) ion, $[Co(phen)_3]^{3+}$. Figure 14 gives the zeta-potential of the sol particles as a function of the added complex ion.

An analysis showed (18) that these data are consistent with a model which assumes the chelate ions to be adsorbed in the outer Helmholtz plane (OHP) of the Stern layer. Furthermore, it was possible to calculate the specific adsorption potential of this complex on silver iodide surface assuming the Langmuir–Stern model. This was found to be equal to $\phi = -14.1$ kT.

To test the validity of the assumption that the Stern–Langmuir isotherm is applicable to the described system, the charge density per unit area at the OHP, σ_1, was calculated using the standard double layer expressions, i.e.

$$\sigma_1 = -\sigma_0 - \sigma_2, \qquad [24]$$

$$\sigma_0 = K(\psi_0 - \psi_\sigma), \qquad [25]$$

and

$$\sigma_2 = \frac{-\kappa\epsilon kT}{2\pi z e}\sinh\left(\frac{ze\psi_\sigma}{2kT}\right), \qquad [26]$$

where σ_0 is the charge per unit area at the solid surface, and σ_2 is the charge density of the diffuse part of the double layer; ψ_0 and ψ_σ are the potentials at the solid surface and in the Stern plane, respectively, ϵ is the permittivity of the bulk solution, and κ is the Debye–Hückel reciprocal length while the other symbols have their usual meaning.

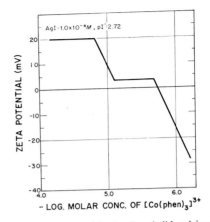

FIG. 14. Zeta-potentials of a silver iodide sol *in statu nascendi* (AgI, 1.0×10^{-4} M; pI⁻, 2.72) as a function of the concentration of added $[Co(phen)_3]^{3+}$.

This was compared with calculations using the Stern–Langmuir adsorption isotherm

$$\sigma_1 = \frac{zeN_s}{1 + (1/x)\exp[(ze\psi_\sigma/kT) + (\phi/kT)]}, \qquad [27]$$

where N_s is the number of available adsorption sites and z and x are the valency and mole fraction of the adsorbing species. ψ_σ is usually taken to be equal to the electrokinetic potential ζ, and N_s is determined by titrations.

Table V gives a comparison of the calculated values for σ, by the two methods described, which shows that the agreement is indeed excellent.

PARTICLE ADHESION

The problem of adhesion of colloidal particles on different substrates is of fundamental importance in many technological areas; for example, in corrosion of metals. The strength of adhesion depends on the nature of the particle and substrate surfaces as well as on the composition of the electrolyte environment. The work reported so far has been difficult to interpret because of the rather poorly defined systems used in such investigations. The

410

TABLE V

The Charge Densities Associated with the Stern Model for the Flat Electrical Double Layer[a]

ψ_δ (mV)	$[Co(phen)_3]^{3+}$ (moles/liter)	$\sigma_0{}^b$ ($\mu F/cm^2$)	σ_2 ($\mu F/cm^2$)	σ_1 ($\mu F/cm^2$)	$\sigma_1{}^{SLc}$ ($\mu F/cm^2$)
0.0	1.82×10^{-6}	-3.23	0	3.23	3.23
-1.0	1.74×10^{-6}	-3.23	0.01	3.22	3.20
-3.0	1.58×10^{-6}	-3.21	0.03	3.18	3.14
-5.0	1.45×10^{-6}	-3.20	0.05	3.15	3.11
-10.0	1.16×10^{-6}	-3.16	0.10	3.06	3.03
-15.0	9.33×10^{-7}	-3.13	0.15	2.98	2.97
-20.0	7.50×10^{-7}	-3.07	0.20	2.89	2.90

[a] The values were calculated for selected values of ψ_δ between 0 and -20 mV. The ψ_δ values were taken from the data presented in Fig. 14 and correspond to the concentrations of $[Co(phen)_3]^{3+}$ indicated above. $pI^- = 2.72$, $\psi_0 = -462$ mV, $\phi = -14.05$ kT.

[b] σ_0, σ_2, and σ_1 were calculated using Eqs. [24], [25], and [26], respectively.

[c] $\sigma_1{}^{SL}$ denotes the σ_1 values calculated according to the Stern–Langmuir adsorption isotherm, Eq. [27].

availability of well-characterized metal hydrous oxide dispersions has made it possible to study the problems of particle adhesion in a more quantitative way.

Using the packed column technique (19) we have investigated the adhesion of monodispersed spherical colloidal particles of chro-

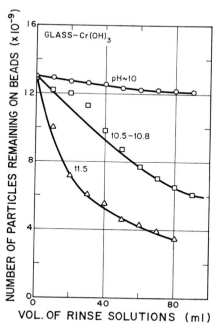

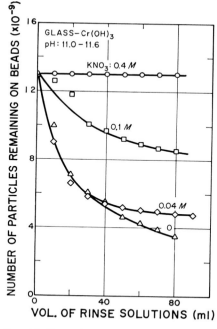

FIG. 15. Desorption of monodispersed chromium hydroxide particles (modal diameter 0.28 μm) from glass on repeated elution with 10-ml volumes of rinse solutions of a given pH. Each curve is for a different rinse pH.

FIG. 16. Desorption of monodispersed chromium hydroxide particles (modal diameter 0.28 μm) from glass on repeated elution with 10-ml volumes of rinse solutions containing different concentrations of KNO_3. The pH was in all cases kept between 11.0 and 11.6.

mium hydroxide on glass beads as a function of pH and the ionic strength. The procedure consisted of passing at low pH (3.1) a "monodispersed" chromium hydroxide sol (particle modal diameter 0.28 μm) over a bed of glass beads packed in a Chromaflex chromatographic column. Under these conditions, the particles, which are positively charged, quantitatively adhere to the substrate. The bed was then washed of free acid using a rinse solution adjusted to pH $\sim$7. No particles were removed in this process. The columns so prepared were treated with solutions of various pH values and ionic strengths in 10 ml rinse cycles at an average time of 8 min per cycle. The concentration of the described particles in the effluent was determined by light scattering.

Figure 15 is a plot of the number of particles remaining on the glass beads in the column as a function of the number of rinse cycles; each curve is for a different pH of the rinse solutions (adjusted by NaOH). As one would expect the adhesion is strongly pH dependent; detachment of particles increases as the pH rises, which is easily understood as at high pH both the particles and the glass beads are negatively charged. Interestingly, the zero point of charge of the chromium hydroxide sol used is $\sim$7.5, yet even at pH 10, the particles remain almost quantitatively bound to the surface. This may be either due to a different interfacial pH or to the fact that a certain amount of bonding took place by condensation of the surface silanolic groups of the glass with the =CrOH groups on the particle surface. The so-formed ≡Si–O–Cr= bonds may be difficult to break in moderately alkaline solutions. Figure 16 shows the effect of the addition of a "neutral" electrolyte, in this case, KNO$_3$. Most surprisingly, the detachment of particles can be completely suppressed, even at the very high pH, if a sufficient amount of salt is added. Since a rather high concentration of the salt is needed to prevent particle desorption this effect can be attributed to the reduction of repulsion forces between the particles and the substrate.

ACKNOWLEDGMENTS

This presentation has been based on the work of the following of my associates: Mr. M. Budnik, Dr. D. L. Catone, Mr. J. Kolakowski, Mr. M. Onofusa, Mr. R. S. Sapieszko, Mr. P. Scheiner, Mr. W. Scott, and Mr. D. W. White, to whom I am indebted for their efforts and enthusiasm. Fruitful discussions with Professors L. Meites and R. Patel on some aspects of this work are gratefully acknowledged.

REFERENCES

1. MATIJEVIĆ, E., *J. Colloid Interface Sci.* **43**, 217 (1973).
2. DEMCHAK, R., AND MATIJEVIĆ, E., *J. Colloid Interface Sci.* **31**, 257 (1969).
3. MATIJEVIĆ, E., LINDSAY, A. D., KRATOHVIL, J., JONES, M. E., LARSON, R. I., AND CAYEY, N. W., *J. Colloid Interface Sci.* **36**, 273 (1971).
4. BRACE, R., AND MATIJEVIĆ, E., *J. Inorg. Nucl. Chem.* **35**, 3691 (1973).
5. MATIJEVIĆ, E., SAPIESZKO, R. S., AND MELVILLE, J. B., *J. Colloid Interface Sci.* **50**, 567 (1975).
6. McFADYEN, P., AND MATIJEVIĆ, E., *J. Colloid Interface Sci.* **44**, 95 (1973).
7. BELL, A., AND MATIJEVIĆ, E., *J. Inorg. Nucl. Chem.* **37**, 907 (1975).
8. MULAY, L. N., AND SELWOOD, P. W., *J. Am. Chem. Soc.* **77**, 2693 (1955).
9. SAPIESZKO, R. S., PATEL, R. C., AND MATIJEVIĆ, E., *J. Phys. Chem.*, submitted.
10. MATIJEVIĆ, E., BUDNIK, M., AND MEITES, L., *J. Colloid Interface Sci.*, in press.
11. LINDSAY, A. D., MATIJEVIĆ, E., AND KRATOHVIL, J. P., *Colloid Polym. Sci.* **253**, 581 (1975).
12. ONOFUSA, M., M.S. thesis, Clarkson College, 1976.
13. SCHULL, K. E., AND GUTHAM, G. R., *J. Amer. Water Works Assoc.* **59**, 1456 (1967).
14. HAYDEN, P. L., AND RUBIN, A. J., *in* "Aqueous-Environmental Chemistry of Metals," Ann Arbor Science, Ann Arbor, Mich., 1974.
15. JAMES, R. O., AND HEALY, T. H., *J. Colloid Interface Sci.* **40**, 65 (1972).
16. MATIJEVIĆ, E., KRATOHVIL, S., AND STICKLES, J., *J. Phys. Chem.* **73**, 564 (1969).
17. CATONE, D. L., AND MATIJEVIĆ, E., *J. Colloid Interface Sci.*, **55**, 476 (1976).
18. CATONE, D. L., Ph.D. thesis, Clarkson College, 1975.
19. CLAYFIELD, E. J., AND LUMB, E. C., *Discuss. Faraday Soc.* **42**, 285 (1960).

Steric Stabilization

D. H. NAPPER

Department of Physical Chemistry, University of Sydney, N.S.W. 2006, Australia

Received April 12, 1976; accepted August 4, 1976

Steric stabilization is a generic term that embraces all aspects of the stabilization of colloidal particles by nonionic macromolecules. The technological exploitation of steric stabilization dates back at least 4000 years to the preparation in ancient Egypt of fresco paints and inks. Despite this long history, the mechanisms whereby polymers impart stability have only been recognized in the past two decades.

Three types of experiment have proved useful in discriminating between the various competing theories of steric stabilization. These are studies of incipient flocculation, measurements of the repulsive interactions between stable particles, and stability in polymer melts.

Experiments suggest that two separate regions of close approach must be distinguished: the interpenetration domain (characterized by an interparticle separation of between one and two contour lengths of the stabilizer chains) and the interpenetration plus compression domain, which is entered on even closer approach. Particle interactions in the interpenetrational region arise primarily from the mixing of segments and solvent; the elastic interactions are negligible. This mixing term is evaluated in the first generation theories via the Flory–Huggins theory. It may be positive or negative. As compression occurs on closer approach, the elastic contributions also become significant. These may be evaluated by statistical thermodynamics and are always positive.

The foregoing concepts correctly predict the observed practical limit to stability, which is the attainment of theta-conditions for the stabilizing chains. Except for very short chains, the van der Waals attraction is rarely important in determining stabilization by polymers. Stabilization in polymer melts, however, arises primarily from elastic effects so that for these systems van der Waals forces become the major source of attraction. Elastic contributions increase abruptly, even under incipient flocculation conditions; consequently, Brownian collisions sample primarily the interpenetrational domain. The distance dependence of the repulsion between stable particles can be predicted with a fair degree of success.

INTRODUCTION

Steric stabilization is a generic term that encompasses all aspects of the stabilization of colloidal particles by nonionic macromolecules. The phenomenon has had an especially long history of technological exploitation. It dates back some five millennia to the dynasties of such pharaohs as Cheops and Chephren, whose pyramids can still be visited today. It was during those alchemical times that inks (and papyrus) were first developed to solve an early problem in communication. Ancient Egyptian inks were prepared by dispersing carbon black (formed by combustion) in water using natural steric stabilizers, such as gum arabic, egg albumin, or casein (from milk) (1).

Traditionally colloid scientists have referred to stabilization by natural macromolecules as "protection." There is presumably an electrostatic component to stability imparted by natural macromolecules because they are usually charged. Heller and Pugh (2) appear to have been the first to use the term "steric stabilization" which may be differentiated from protective action by the absence of any electrostatic component. The word "steric" in this context is perhaps somewhat misleading because the phenomenon is not closely related to the more familiar steric effects between nonbonded atoms or groups that are beloved by organic chemists. The latter arise from electron–electron and electron–nuclei interactions and are usually repulsive. Clearly steric stabilization embraces a significantly wider range of fundamental thermodynamic causes. It seems conceivable that classical steric effects could impart colloid stability; indeed, they may well be operative in stabilizing some lipid bilayers that contain few or no solvent molecules.

Steric stabilization has been exploited through the ages in the preparation of both water-based and, since Roman times, oil-based paints and inks. It is especially useful in nonaqueous media where electrostatic stabilization is less successful. It can also be effective in media of high ionic strengths. Indeed, because steric stabilizers are often relatively insensitive to the presence of electrolytes, it is scarcely surprising that the phenomenon is important in biological systems (e.g., in the stabilization of blood, milk, etc.). Stabilization by nonionic macromolecules has also become of increasing technological significance (e.g., in pharmaceutical emulsions, in the dispersal of oil spills, in the domestic use of nonionic surfactants, in dry cleaning, etc). Steric stabilization is likely to be even more widely exploited because of its versatility in being equally effective in aqueous and nonaqueous dispersion media.

Despite the long history of technological exploitation of steric stabilization, it is only in the past two decades that some understanding of the mechanisms by which polymers can impart colloid stability has emerged.

THE PREPARATION OF STERICALLY STABILIZED DISPERSIONS

A comprehensive monograph (3) has appeared recently that describes in detail most of the various methods of preparation of sterically stabilized latices that have been developed to date. The patent literature is especially well covered. For our purposes all

that need be noted is that there are two general procedures presently available.

The first method (3, 4) generates the latex particles in the presence of a steric stabilizer. This may be considered to be an *in situ* method of particle formation. The stabilizer may also be generated *in situ* by the addition of a suitable precursor; alternatively, it may be fully preformed before addition to the reaction mixture.

The second general method (5) consists of adding the stabilizer to an already formed colloidal dispersion (or vice versa). In the latter case a significant electrostatic component to stability may be present because the first-formed particles are usually electrostatically stabilized.

A critically important question, both from the practical and theoretical viewpoints, is the types of polymers that constitute the best steric stabilizers. Considerable experience (3) shows that the stabilizers that provide best stability are amphipathic block or graft copolymers. The most effective stabilizers consist of both anchor groups and stabilizing moieties. The stabilizing moieties must be soluble in the dispersion medium to be effective whereas the anchor groups function most efficiently if nominally insoluble in the dispersion medium (3).

The purpose of the anchor polymer is to prevent the stabilizing moieties attached to one particle from moving away from the interaction zone on the approach of a second particle (3). The stabilizing moieties could in principle escape from the stress of the interaction zone by two different mechanisms (6): first, the stabilizer molecules could move laterally around the surface of the particle while remaining attached to the particle; or secondly, the stabilizers may in some circumstances be desorbed from the particle surface. Lateral movement can usually be prevented by ensuring that the surfaces are fully coated. Desorption can be eliminated by attaching the stabilizing moieties to suitably insoluble polymer chains that anchor, either chemically or physically, the moieties to the particles. Of course, the manufacturers of nonionic detergents have long recognized the need for anchor groups, without necessarily being aware of all the physicochemical principles involved so far as stability is concerned.

Stabilizing moieties that are unanchored or not strongly anchored can, in fact, impart some stability to colloidal dispersions (7). Such dispersions will, however, flocculate before the thermodynamic limit to stability is reached, as a result of stabilizer displacement or desorption. The pioneering studies of steric stabilization by Heller and co-workers (2, 8, 9) were performed on such unanchored macromolecules. Unfortunately, quantitative theoretical description of such complex behavior seems to be out of the question at present. For this reason, we shall confine our discussion to systems that are stabilized by strongly anchored moieties.

EXPERIMENTAL STUDIES OF INCIPIENT INSTABILITY

As with electrostatically stabilized dispersions, there are two general aspects of the stability of sterically stabilized dispersions that can usefully be investigated. These two domains cover long-term stability and incipient instability. Studies of the incipient instability of model dispersions appear to have outpaced

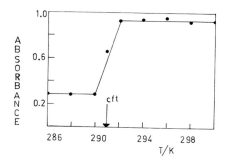

Fig. 1. Characteristic increase in turbidity of a poly(vinyl acetate) dispersion, stabilized by poly(ethylene oxide), on reaching the cft in an aqueous electrolyte solution.

studies on stable dispersions. These will therefore be discussed first.

The Critical Flocculation Point

Most of the studies on incipient instability have been performed on model polymer latices. These latices are reasonably monodisperse and can be prepared, even in aqueous media, in an apparently uncharged state (4). The particles can be regarded as being fully coated by stabilizing moieties that are strongly anchored to the particle surface. When these conditions are fulfilled, patterns of behavior emerge that are common to both aqueous and nonaqueous dispersion media (6).

The simplest way to induce instability in sterically stabilized dispersions is to reduce the solvency of the dispersion medium for the stabilizing moieties. This may be achieved through temperature changes (4, 5, 10–12), pressure changes (13), or a combination of both (14). Another method is to add nonsolvent for the stabilizing moieties to the dispersion medium (10). When the solvency is reduced, the dispersions often exhibit a very sharp transition from long-term stability to catastrophic instability (4, 10). An example of this is displayed in Fig. 1 (4). This shows that the transition from long-term stability to fast flocculation occurs over only a few degrees Celsius when poly(ethylene oxide) stabilized latices are heated in aqueous electrolyte. Note that such flocculation is commonly reversible (4, 10). If the solvency of the dispersion medium for the stabilizing chains is significantly improved (e.g., by cooling), spontaneous redispersion of the latex particles usually occurs.

The critical flocculation point (cfpt) at which flocculation first becomes observable is referred to as a critical flocculation temperature (cft) or a critical flocculation pressure (cfp), according to which intensive variable is used to induce flocculation. If flocculation results from the addition of nonsolvent, the corresponding terms are the critical flocculation volume (cfv) (for a liquid nonsolvent) and the critical flocculation concentration (cfc) (for a solid nonsolvent).

The strong temperature dependence of the stability of many sterically stabilized dispersions contrasts markedly with the relative insensitivity of electrostatically stabilized dispersions to changes in temperature. The temperature dependence of stability provides a clue to the thermodynamic processes that govern stabilization (15, 16). One very characteristic feature of sterically stabilized systems is the diversity of their responses to temperature changes. Some dispersions flocculate on heating (4, 16) whereas others flocculate on cooling (10, 16). Still others cannot, at least in principle, be flocculated at any accessible temperature (12). To date, only one system has been prepared (polyacrylonitrile latex particles stabilized by polystyrene in methyl acetate) that flocculates both on heating and cooling (17). Dispersions that flocculate either on heating or on cooling can be prepared in both aqueous and nonaqueous dispersion media. Ottewill (18) has rightly noted that entropic stabilization is more common in nonaqueous media whereas enthalpic stabilization is more frequently encountered in aqueous media.

The temperature dependence of the Gibbs free energy of close approach (ΔG_R) for two sterically stabilized particles is given by

$$\partial(\Delta G_R)/\partial T = -\Delta S_R, \qquad [1]$$

where ΔS_R is the corresponding entropy change. Since ΔG_R must change from being positive to being negative in passing from the stability to the instability domain, the sign of ΔS_R may be inferred by whether flocculation is induced by heating or cooling. The various combinations of signs for ΔS_R and ΔH_R (the corresponding enthalpy change) that lead to a positive value for ΔG_R, and thus to stability, are summarized in Table I.

The simplest way to discuss the different possibilities is in terms of the relationship $\Delta G_R = \Delta H_R - T\Delta S_R$. If both ΔH_R and ΔS_R are positive but $\Delta H_R > T\Delta S_R$, the enthalpy change on close approach opposes flocculation whereas the entropy change promotes it. Be-

TABLE I

Types of Steric Stabilization

ΔH_R	ΔS_R	$|\Delta H_R|/ T|\Delta S_R|$	ΔG_R	Type	Flocculation
+	+	>1	+	Enthalpic	On heating
−	−	<1	+	Entropic	On cooling
+	−	$\gtrless 1$	+	Combined enthalpic–entropic	Not accessible

cause the enthalpic contribution to the free energy of close approach is dominant, this is referred to as enthalpic stabilization. The dominance of the enthalpic contribution to the free-energy changes over that of the entropic contribution will clearly be reduced if the temperature is increased. Accordingly enthalpically stabilized dispersions are characterized by flocculation on heating. A contrasting situation to enthalpic stabilization occurs if both ΔH_R and ΔS_R are negative and $|\Delta H_R| < T|\Delta S_R|$. In this case, the entropy term opposes flocculation whereas the enthalpy term favors it. The dominance of the entropic contribution to the free energy of close approach suggests that this possibility be termed entropic stabilization. It is clearly characterized by flocculation on cooling. The third category is characterized by ΔH_R being positive and ΔS_R being negative so that both the enthalpy and entropy terms contribute to stability. Dispersions stabilized by combined enthalpic–entropic stabilization, as this case is called, cannot in principle be flocculated at any accessible temperature. Flocculation may, however, be induced on changing the temperature because a transition to enthalpic or entropic stabilization may conceivably occur.

Three important points must, however, be noted with regard to the foregoing discussion. First, the signs of ΔS_R and ΔH_R are determined unequivocally only in the vicinity of the cfpt because ΔG_R necessarily changes sign only on passing through the cfpt. How far from the cfpt these inferences remain valid is critically dependent upon the nature of the stabilizing moieties and the dispersion medium; for some systems the signs of ΔH_R and ΔS_R

are unchanged even several hundred degrees from the cfpt. The need for caution, however, is brought home forcibly by recalling the behavior of the latex cited above that flocculates both on heating and on cooling.

The second important point to note is that ΔG_R refers to the free-energy change associated with that part of the steric barrier that is sampled during Brownian collisions. As shall be seen, incipient flocculation appears to be associated primarily with the interpenetrational domain of close approach. The domain characterized by interpenetration plus compression is not ordinarily accessible to any significant extent in the course of a Brownian collision because of the rapid rise in the elastic repulsion on compression.

The final point to note is that the preceding thermodynamic discussion is model independent. Osmond et al. (19) have claimed that the classification of dispersions as outlined above depends upon the assumption of some specific model (in actual fact that due to Fischer (20)). This assertion is clearly incorrect. The only bases needed to specify the general thermodynamic origin of stability near to the cfpt is Eq. [1], plus some elementary classical thermodynamic notions. No model-dependent assumptions are required to establish the signs of ΔS_R and ΔH_R. Of course, any physical interpretation of the signs of ΔS_R and ΔH_R must necessarily be associated with model-dependent concepts.

Table II lists the sterically stabilized dispersions that have been prepared and whose classification near to the cfpt seems reasonably well assured (13). Note that both entropic and enthalpic stabilization have been observed in both aqueous and nonaqueous media. Several other points deserve comment. First, it is clear that a polymer may be an enthalpic stabilizer in one dispersion medium and an entropic stabilizer in another. For example, PEO is an enthalpic stabilizer in water (4) and an entropic stabilizer in methanol (5). Likewise polyisobutylene is an enthalpic stabilizer in isopentane (14) and an entropic stabilizer in n-pentane (21). In the latter case, the difference between the two dispersion media is the rather subtle chemical difference

TABLE II

Classification of Sterically Stabilized Dispersions Near to the Critical Flocculation Point

Stabilizer	Dispersion medium		Classification
	Type	Example	
Poly(laurylmethacrylate)	Nonaqueous	n-Heptane	Entropic
Poly(12-hydroxystearic acid)	Nonaqueous	n-Heptane	Entropic
Polystyrene	Nonaqueous	Toluene	Entropic
Polyisobutylene	Nonaqueous	n-Heptane	Entropic
Poly(ethylene oxide)	Nonaqueous	Methanol	Entropic
Polystyrene	Nonaqueous	Methyl acetate	Entropic
Polystyrene	Nonaqueous	Methyl acetate	Enthalpic
Polyisobutylene	Nonaqueous	2-Methylbutane	Enthalpic
Poly(ethylene oxide)	Aqueous	0.48 M MgSO$_4$	Enthalpic
Poly(vinyl alcohol)	Aqueous	2 M NaCl	Enthalpic
Poly(methacrylic acid)	Aqueous	0.02 M HCl	Enthalpic
Poly(acrylic acid)	Aqueous	0.2 M HCl	Entropic
Polyacrylamide	Aqueous	2.1 M (NH$_4$)$_2$SO$_4$	Entropic
Poly(vinyl alcohol)	Mixed	Dioxan/water	Combined
Poly(ethylene oxide)	Mixed	Methanol/water	Combined

TABLE III

Comparison of Theta-Temperature with Critical Flocculation Temperature

Stabilizer	Molecular weight	Dispersion medium	Cfpt/°K	θ/°K
Poly(ethylene oxide)	10,000	0.39 M MgSO$_4$	318 ± 2	315 ± 3
Poly(ethylene oxide)	96,000	0.39 M MgSO$_4$	316 ± 2	315 ± 3
Poly(ethylene oxide)	1,000,000	0.39 M MgSO$_4$	317 ± 2	315 ± 3
Poly(acrylic acid)	9,800	0.2 M HCl	287 ± 2	287 ± 5
Poly(acrylic acid)	51,900	0.2 M HCl	283 ± 2	287 ± 5
Poly(acrylic acid)	89,700	0.2 M HCl	281 ± 1	287 ± 5
Poly(vinyl alcohol)	26,000	2 M NaCl	302 ± 3	300 ± 3
Poly(vinyl alcohol)	57,000	2 M NaCl	301 ± 3	300 ± 3
Poly(vinyl alcohol)	270,000	2 M NaCl	312 ± 3	300 ± 3
Polyacrylamide	18,000	2.1 M (NH$_4$)$_2$SO$_4$	292 ± 3	—
Polyacrylamide	60,000	2.1 M (NH$_4$)$_2$SO$_4$	295 ± 5	—
Polyacrylamide	180,000	2.1 M (NH$_4$)$_2$SO$_4$	280 ± 7	—
Polyisobutylene	23,000	2-methylbutane	325 ± 1	325 ± 2
Polyisobutylene	150,000	2-methylbutane	325 ± 1	325 ± 2

between two positional isomers; yet it leads to a reversal of the roles of the enthalpy and entropy contributions to stabilization. Secondly, the evidence to support the classification of dispersions as combined enthalpic–entropic stabilization is necessarily indirect because it is not possible to approach the cfpt directly. Finally, note that although only one dispersion has been demonstrated to date to exhibit flocculation both on heating and on cooling, all dispersions in principle should display similar behavior, if only a sufficiently wide range of temperature and pressure could be scanned.

The next question that must be resolved is whether the critical flocculation point can be identified in some general way. The answer is found experimentally to be the same irrespective of whether we are considering aqueous or nonaqueous systems; or whether flocculation is induced by a change in temperature or by a change in pressure, or by the addition of nonsolvent. The critical flocculation point correlates strongly with the corresponding theta (θ)-point of the stabilizing moieties in free solution (4, 10–16). The θ-point is variously defined from both the conformational

and the thermodynamic viewpoint. The different definitions are not always equivalent. For our purposes the definition of a θ-point that is most germane is a thermodynamic one. A θ-point will be defined as one where the second viral coefficient of the polymer chains vanishes (22). Experimental methods for the absolute determination of θ-points include light scattering and osmometry (22). Quicker, but perhaps less soundly based methods have been developed by Elias (23, 24) and Cornet and van Ballegooijen (25). Nevertheless, these quicker methods give good agreement with the absolute methods (22) and they can be shown to have a thermodynamic basis (25).

Shown in Table III is a comparison of some θ-temperatures and critical flocculation temperatures. The θ-point is a measure of the magnitude of the segment–solvent interaction. As such the θ-point ought to be essentially independent of the molecular weight of the polymer. The PEO data provide good evidence to support this assertion. Changes in the molecular weight of the stabilizing moieties of three orders of magnitude scarcely altered the measured θ-temperature. Neither were the cfts of PEO stabilized latices significantly influenced by comparable changes in molecular weight. The strong correlation between the θ-temperatures and the cfts is apparent from Table III. Similar strong correlations have been observed between θ-pressures and cfps (13), and θ-volumes and cfvs (10). Note that the data are not sufficiently precise to allow any statement to be made with regard to stability under θ-conditions. All that can be concluded is that θ-conditions represent a practical thermodynamic limit to the stability of some model sterically stabilized dispersions.

From the experiments (4, 10) that have been conducted thus far, it would appear that the critical flocculation point is essentially independent of the size of the core particles, the nature of the disperse phase (if polymeric), and the nature of the anchor polymer (if insoluble in the dispersion medium). Such independence would be expected if the cfpt correlated with the θ-point. The cfpt, however, does depend upon the extent of the coverage of the surface. The correlation between the cfpt and the θ-point is only observed (4, 10) if the surface is fully covered (i.e., the plateau of the adsorption isotherm has been attained). At lower surface coverages, flocculation occurs in dispersion media that are better solvents than θ-solvents. This may be a consequence of lateral movement of the stabilizer, of desorption or even, perhaps, of bridging flocculation by the stabilizing chains.

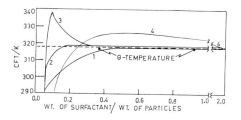

FIG. 2. Plots of the cft versus concentration of stabilizer for latex particles stabilized by poly(ethylene oxide) of molecular weight 10,000 at pH = 4.65. Curve 1, poly(vinyl acetate) lattices (no carboxylic acid groups); curves 2 and 3, low carboxyl and high carboxyl polystyrene lattices, respectively; curve 4, poly-(styrene-co-acrylic acid) latex particles.

Enhanced Steric Stabilization

We have noted in the foregoing description that the general pattern of incipient flocculation behavior is disrupted if the stabilizing moieties are only weakly attached to the surface. A different pattern of behavior may also result if the stabilizing chains interact too strongly with the surface (27). Strong interactions between the chains and the surface can produce anchoring at many points along the stabilizing chain. This is termed multipoint anchoring. One consequence of multipoint anchoring is that the free-solution thermodynamics may no longer be relevant. No correlation between the cfpt and the θ-point for the chains in free solution would then be expected.

So far only one system exhibiting this type of behavior has been investigated in any detail (27). The strong interactions in that case were those between stabilizing poly(ethylene oxide) chains and carboxylic acid groups. The carboxylic acid groups were at the surfaces of polystyrene latex particles. It was found that at suitably low pH values, such latices were stabilized by poly(ethylene oxide) in dispersion media that were significantly worse solvents than θ-solvents. The low pH values were apparently necessary to ensure that H-bonding was possible between the protonated carboxylic acid groups and the ether oxygens of the stabilizing moieties.

Figure 2 displays the cft of such latices as a function of the solution concentration of the stabilizer. The cft exceeded the θ-temperature significantly at solution concentrations corresponding to incomplete surface coverage. This was interpreted as resulting from multipoint anchoring. At higher solution concentrations, the adsorption of more stabilizer resulted in a reduction in multipoint anchoring. The frequently encountered correlation between the cfpt and the free solution θ-point reappeared at the higher solution concentrations.

Stability in worse than θ-solvents is termed enhanced steric stabilization. The magnitude of the enhancement (i.e., the penetration into the worse than θ-solvent domain) was found to depend upon such factors as the pH, the molecular weight of the stabilizing moieties, and the density of surface groupings. The enhancement decreased with increasing pH, increasing molecular weight, and fewer surface carboxylic acid groupings. These trends are just those expected for a reduction in multipoint anchoring.

Experiments on enhanced steric stabilization provide some clues as to the range of validity of the assumption that the thermodynamic properties of the stabilizing moieties in free solution are relevant to stabilizing chains attached to an interface.

STABILIZATION IN POLYMER MELTS

Recent experiments have shown that it is possible to prepare stable latices in polymer melts (28). The pattern of flocculation behavior is again different from that observed near to the θ-point. It was found that there was apparently an upper limit to the size of the latex particles that could be stabilized by a given molecular weight stabilizer. Some results are shown in Table IV for stabilization in poly(ethylene oxide) melts. The molecular weight of the stabilizer was the same as that of the melt polymer. The higher this molecular weight, the larger was the maximum particle size that could be stabilized. Thus, in the melt, incipient instability appears to depend critically upon the stabilizer molecular weight and the particle radius. As discussed above, neither of these factors was found to be important in studies on incipient instability near to the θ-point. Different patterns of behavior suggest fundamental differences in the primary origins of stability in the two types of dispersions.

FLOCCULATION BY HIGH CONCENTRATION OF POLYMER

In the foregoing presentation of the experimentally observed phenomenology associated with sterically stabilized dispersions, it was noted that latices stabilized by PEO could be prepared in pure water and also in pure

TABLE IV

Apparent Maximum Latex Particle Sizes in PEO Melts at 70°C

Molecular weight	Maximum particle size/nm		
	Experimental	Flory theory + Monte Carlo	Dolan and Edwards
600	30–60	50	92
1500	100–150	100	145
6000	300–500	500	290

PEO melt. It is surprising to find, therefore, that the addition of PEO to the aqueous dispersion medium of a PEO stabilized latex in water may induce flocculation. This was first demonstrated by Li-in-on et al. (29), who showed that the stability of such latices exhibited a minimum as the volume fraction of added PEO increased. The position of the minimum depended critically upon the molecular weight of the added PEO but typically occurred at volume fractions in the range 0.3–0.8. The lower molecular weight polymers were less effective in inducing flocculation. It should be noted that the instability domain was quite broad. Moreover, if water is added to polymer latices prepared in a PEO melt, flocculation is also observed (28). Whether such phenomena are exhibited by all sterically stabilized systems remains to be established. However, Li-in-on et al. (29) reported qualitatively similar results for polystyrene latices stabilized by polyisoprene.

THEORIES OF STERIC STABILIZATION

There does not appear, at least to date, to be significant disagreement about the experimental aspects of incipient instability. However, formulation of a physically realistic theoretical framework within which these data may be interpreted is decidedly controversial.

The Unimportance of the London–van der Waals Forces

In the DLVO theory for electrostatic stabilization, the London–van der Waals dispersion forces play a central role as the primary driving force towards coagulation. It is tempting to assume, by analogy, that dispersion forces are also very important in determining the incipient instability of sterically stabilized latices. This may well be true, e.g., for very low molecular weight chains; for very large particles; for incompletely covered surfaces; or for poorly anchored stabilizers. Moreover, it seems likely that the stability of latices prepared in polymer melts is probably determined primarily by the van der Waals interactions between the core particles (28).

It appears to be extremely unlikely, however, that the incipient instability behavior of model systems, which was discussed above is a consequence of dispersion forces. Any realistic calculation of the magnitude of the London forces between particles stabilized by chains of molecular weight greater than say, 10 000, strongly suggests that the dispersion interactions are negligibly small. This conclusion holds at the ordinary distances of close approach sampled in a Brownian collision but may break down if the particles are forced by compression to approach more closely. It should be noted that Meier (30), Hesselink et al. (31), and Edwards and co-workers (32, 33) ascribe central importance to the dispersion interactions between the core particles. In our view, their assumed values (of order 10^{-19}–10^{-20} J) for the Hamaker "constant" (which may, in electrolyte-free systems, be variable) are at least an order of magnitude too large for polymer latex particles (34).

If the London dispersion forces are not responsible for incipient instability then alternative attractive interactions must be invoked to account for flocculation. It is well known (35) that polymer segments become self-attracting in worse than θ-solvents. Thus segmental interactions appear to be responsible not only for stability but also for incipient flocculation near to the θ-point.

The Two Domains of Close Approach

In order to understand flocculation behavior, it seems to be imperative to distinguish between two domains of close approach of two sterically stabilized colloidal particles. Let L represent the contour length in solution of the highest molecular weight stabilizing moiety and d be the minimum distance of close approach of the surfaces of the core particles. If $d \geqslant 2L$, then no direct interaction is seemingly possible between the polymer chains attached to the different particles. If, however, $L \leqslant d < 2L$ then the polymer chains on the opposing surfaces may, in principle at least, undergo segmental interpenetration. Not until $d < L$ can there be compression of the chains attached to one surface by the impenetrable surface of the other particle; in principle, therefore, both interpenetration and compression may occur in this domain. It is convenient in what follows to refer to the $L \leqslant d < 2L$ region of close approach as the interpenetration domain and to the inner region, $d < L$, as the interpenetration plus compression domain. Note that L may often be conveniently regarded as the barrier layer thickness rather than the contour length, without altering the foregoing arguments.

Of course, whether interpenetration occurs exclusively (or even at all) in the interpenetration domain will depend critically upon both the kinetic and thermodynamic parameters that govern any specific act of close approach. The approaching macromolecules may have insufficient time to undergo interpenetrational relaxation and if so, some compression may occur. Interpenetration is clearly one extreme possibility in this domain; it seems likely to correspond to the lowest free-energy state because any compressional interactions result from the exclusion of the segments completely from some elemental volume rather than from part thereof.

THE GENERAL PHYSICAL BASES FOR STERIC STABILIZATION

For our purposes, the simplest way to view the behavior of polymers in solution is to regard them as occupying two different spaces. The first is a thermodynamic space in which the polymer segments occupy an excluded volume that may be positive, negative, or zero. In addition, the segments exist in real space in which their real volumes are necessarily always positive. The first space gives rise to what Flory (35) termed "mixing" effects, whereas the second space gives rise to "elastic" contributions. These two spaces are at least weakly coupled because the coordinates of the centers of mass of the segments in real space are determined by excluded volume effects. Flory (35) used these bases in his classic treatment of the excluded volume effect.

The mixing free energy on interpenetration or compression can be positive or negative according to the sign of the segmental excluded volume. On the other hand, the elastic free-energy term is always positive for movements away from the mean equilibrium conformation. The total free-energy change is obtained by simply summing the mixing and elastic terms.

The propriety of separating these two effects has been called into question (33). Nevertheless, the Flory theory for the intramolecular expansion factor α has proved (36) to be fairly accurate when compared with experiment; it has not greatly been improved upon over the years, despite the expenditure of significant effort (36). In our view, the application of this approach to steric stabilization ought to provide a realistic first-order theory. Note, however, that Osmond *et al.* (19) have put forward the contrary view without apparently providing compelling substantiating arguments.

The Interpenetration Domain

In what follows, we will assume that the stabilizing macromolecules have sufficient time to interpenetrate if the core particles approach one another. As the polymer chains interpenetrate, solvent molecules will be forced out of the interpenetration volume. There will clearly be a free energy associated with this solvent exclusion. Whether there is a significant elastic contribution in this domain is still an open question. Undoubtedly, given sufficient time, the interpenetrated chains will relax into new conformations different from those in the absence of interpenetrations. Consequently, it

seems indisputable that there will be an elastic contribution. But the volume fraction of real space occupied by the segments of a polymer molecule is quite small on average (usually $\leqslant 0.1$) and will be even smaller in the peripheral regions of the chains that are involved in interpenetration. It seems unlikely, therefore, that there is a significant elastic contribution to stabilization in the interpenetration domain. Osmond *et al.* have, however, contended in a recent review that there is a significant elastic contribution (19).

The problem of the interpenetration of two polymer molecules in solution has been treated by Flory and Krigbaum (37), using the Flory–Huggins theory for polymer solutions. Their theory assumes that the segment density distribution functions of the chains are the same before and after interpenetration. Strictly speaking, this cannot be correct, if the macromolecules have sufficient time to relax to their equilibrium conformations. Because the density of polymer segments is likely to be quite small and to be symmetrically displaced about the interpenetration zone, it seems probable that this effect is quite small. Recent calculations by Dolan and Edwards (33) support this viewpoint. They found relatively small perturbations of one chain by the presence of another. It is our opinion that this effect is likely to be insignificant compared with the gross inadequacies of the Flory–Huggins theory. It seems pointless to include these relaxation effects without introducing more sophisticated theories of polymer solutions, such as the Flory–Prigogine free-volume theory.

According to the Flory–Krigbaum theory, the free energy of interpenetration of two chains brought from infinite separation to a separation d may be written as

$$\Delta G = 2kT(V_{\mathrm{s}}^2/V_1)(\tfrac{1}{2} - \chi_1) \int \rho_{\mathrm{d}}\rho_{\mathrm{d}}' dV, \quad [2]$$

where V_{s} = volume of segment, V_1 = volume of a solvent molecule, χ_1 = interaction parameter, ρ_{d} and ρ_{d}' are the segment density distribution functions of the two chains, respectively. The integration is taken over the entire volume of interpenetration.

The adaptation of Eq. [2] to steric stabilization is almost, but not quite, trivial. Consider two parallel flat plates to which stabilizing chains are irreversibly attached. The problem is dramatically simplified if we assume that the segment density parallel to the surface is constant. This is an intuitively satisfying assumption but it is an approximation at best. The free energy of interpenetration per unit area (ΔG_{FP}^{I}) then becomes

$$\Delta G_{FP}^{I} = 2kT(V_{s}^{2}/V_{1})\nu^{2}i^{2}(\tfrac{1}{2} - \chi_{1})$$

$$\times \left(\int_{d-L}^{L} \hat{\rho}_{d}\hat{\rho}_{d}'dx \right), \quad [3]$$

where d is now the distance of separation of the flat plates. The segment density distribution functions have been normalized so that

$$\int_{0}^{L} \hat{\rho}_{d}dx = 1$$

and νi is the total number of segments in the volume normal to unit surface area of one flat plate at infinite separation. The x-direction is chosen normal to the surface.

The evaluation of the integral in Eq. [3] can be achieved, either numerically or analytically, once the nature of the segment density distribution functions has been specified. Both the constant segment density (38) and a Gaussian type of distribution (39) function have been evaluated analytically.

A flat-plate potential (FP) can be transformed into a sphere (S) potential for thin layers and large particles by using the Deryagin integration (40) procedure:

$$\Delta G_{S} \sim \pi a \int_{d_{0}}^{\infty} (\Delta G_{FP})dd, \quad [4]$$

where a = particle radius and d_{0} = minimum distance of separation between the surfaces of the particles. Thus, for the interpenetration of two constant segment densities, we find that

$$\Delta G_{S} \sim 2\pi N_{A}\omega^{2}(v_{2}^{2}/\bar{V}_{1})(\tfrac{1}{2} - \chi_{1})aSkT. \quad [5]$$

Here $S = 2(1 - d_{0}/2L)^{2}$, ω = weight of stabilizing moieties per unit surface area, and v_{2} = partial specific volume of the moieties (38).

One virtue of Eq. [5] is that it predicts the observed flocculation near to the θ-point. For solvents better than θ-solvents, $\chi_{1} < \tfrac{1}{2}$ and the free energy of close approach is positive. On the other hand, for worse than θ-solvents, $\chi_{1} > \tfrac{1}{2}$ and the free energy of close approach in the interpenetration domain becomes negative. If sufficiently large, this attraction between the attached chains can lead to flocculation. Thus the foregoing approach would predict flocculation in somewhat worse than θ-solvents. Numerical calculations suggest that in many systems flocculation should occur within a few degrees of the θ-temperature (39). This would explain the observed correlation between the cfpt and the θ-point.

The foregoing theory is also able to explain the observed temperature dependence of stability. The thermodynamic parameter ($\tfrac{1}{2} - \chi_{1}$) may be rewritten in terms of the classical entropy (ψ_{1}) and enthalpy (κ_{1}) parameters through the relationship:

$$\tfrac{1}{2} - \chi_{1} = \psi_{1} - \kappa_{1} = \psi_{1}(1 - \theta/T), \quad [6]$$

where θ = theta-temperature.

All the evidence currently available suggests that for model dispersions, the signs of ΔS_{R} and ΔH_{R} may be equated to those of $-\psi_{1}$ and $-\kappa_{1}$, respectively. Since the signs of both ψ_{1} and κ_{1} may assume both positive and negative values, the whole gamut of behavior summarized in Tables I and II becomes readily explicable. Of course, at this point, our discussion of ΔH_{R} and ΔS_{R} has become model-dependent. The values of ψ_{1} and κ_{1} that are relevant here are those determined experimentally. Any attempt to explain the signs and magnitude of ψ_{1} and κ_{1} will demand a more elaborate theory than the Flory–Huggins theory. It is stressed yet again that the signs of ΔH_{R} and ΔS_{R} are determined near to the cfpt in a rigorous thermodynamic and model-independent fashion by the temperature dependence of stability.

The strong correlations observed between the cfpt and the θ-point provides comforting reassurance of the insignificance of elastic effects in the interpenetration domain. Elastic interactions are independent of the solvency of the dispersion medium and if strong, would demand significantly worse solvency than θ-solvents to induce flocculation. This is not found experimentally.

Osmond et al. (19) have argued that the correlation between the cfpt and the θ-point is rather fortuitous, being an artefact of the method of determining θ. This claim overlooks the fact that the correlation is observed irrespective of the method used to determine θ. Perhaps the most precisely determined θ-temperature is that for polystyrene in cyclohexane. The value $\theta = 307 \pm 1°K$ is generally agreed upon. March (41) has shown that the reversible cft for water-in-cyclohexane emulsions sterically stabilized by well-anchored polystyrene is $308 \pm 3°K$. This type of experiment suggests that the correlation between the cfpt and θ is far from being a consequence of artefactual serendipity.

The Interpenetration Plus Compression Domain

If two spheres approach such that $d_0 < L$, then both compression and interpenetration of the stabilizing chains may occur. Intermolecular interpenetration gives rise to a mixing term as described above. Compression of the chains attached to one surface by the impenetrable opposing surface produces two effects: the first is an intramolecular self-interpenetration, which is accompanied by a corresponding mixing term; the second effect is the configurational compression, which results in an elastic term. Of course, the aforementioned intermolecular interpenetration term must be calculated for compressed chains.

The total free-energy change in this domain may, therefore, be split up into three terms:

$$\Delta G^{I+C} = \Delta G^{MIX}(\text{inter}) + \Delta G^{MIX}(\text{intra})$$

$$+ \Delta G^{ELASTIC} \quad [7]$$

The first two terms may be written down on sight from Eq. [2]. Thus, for flat plates,

$$\Delta G_{FP}{}^{MIX}(\text{inter}) = 2kT(V_s{}^2/V_1)(\tfrac{1}{2} - \chi_1)v^2 i^2$$

$$\times \left(\int_0^d \hat{\rho}_d \hat{\rho}_d{}' dx \right).$$

Moreover,

$$\Delta G_{FP}{}^{MIX}(\text{intra}) = 2kT(V_s{}^2/V_1)(\tfrac{1}{2} - \chi_1)v^2 i^2$$

$$\times \left(\int_0^d \hat{\rho}_d{}^2 - \int_0^d \hat{\rho}_\infty{}^2 dx \right),$$

where the factor of 2 now allows for the presence of two surfaces. Note that the intramolecular interpenetration term is derived from the intermolecular term simply by setting $\hat{\rho}_d = \hat{\rho}_d{}'$ since it is self-interpenetration. Of course, allowance must be made in this case for the mixing term due to self-interpenetration at infinite separation.

The importance of an elastic contribution to steric stabilization was probably first stressed by Jäckel (42), although he appears to have used the term in a somewhat different sense from that employed here. The simplest method by which to calculate the elastic contribution seems to be that developed by Flory (35) in the theory of rubber elasticity. This allows different types of segment density distribution functions to be studied. For a Gaussian distribution function, the Flory approach yields:

$$\Delta G_{FP}{}^{ELASTIC} = 2kTv[(\delta^2 - 1)/2 - \ln \delta], \quad [8]$$

where $\delta = $ compression ratio ($=d/L$ in this case).

The theory for spheres is once again developed through Deryagin integrations. A typical formula for spheres, if we assume a symmetrical Gaussian distribution function for polymer chains between two flat plates, is

$$\Delta G_s{}^{I+C} \sim 2\pi N_A \omega^2 (v_2{}^2/\bar{V}_1)(\tfrac{1}{2} - \chi_1) a S k T$$

$$+ 2\pi L v a k T (\tfrac{2}{3} - \delta_0/2 - \delta_0{}^3/6 + \delta_0 \ln \delta_0),$$

$$[9]$$

where

$$S = 0.5 + 2.3484 \ln (L/d_0)$$

$$- 0.498(L - d_0)/\langle r^2 \rangle^{\frac{1}{2}}.$$

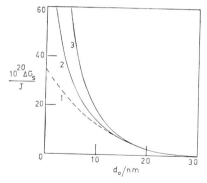

FIG. 3. The distance dependence of the repulsion between polyacrylonitrile latex particles stabilized by polystyrene in toluene. Curve 1, data of Doroszkowski and Lambourne; curves 2 and 3, theory of Smitham, Evans, and Napper with mixing term only (2) and with mixing plus elastic term (3).

Note that in Eq. [9], $\delta_0 = d_0/L$ and $\langle r^2 \rangle^{\frac{1}{2}} = $ rms end-to-end length in free solution.

A comparison of the predictions of Eqs. [5] and [9] with the experimental results of Doroszkowski and Lambourne (43) is presented in Fig. 3. These refer to polyacrylonitrile particles stabilized by polystyrene in toluene. There are no major adjustable parameters in this model. The agreement seems reasonable at larger distances of separation. On close approach, however, the theory predicts much stronger interactions than were observed experimentally.

Note that in polymer melts the large values for $\bar{V}_1$ render the mixing terms very small. Stability in those systems appears to derive primarily from elastic interactions (28). This accounts very well for the different phenomenology observed. The above mentioned theory predicts, that there is a maximum particle size that can be stabilized by a given low molecular weight polymer. Of course, it is necessary in these systems to include the London–van der Waals dispersion attraction between the core particles, for these interactions constitute the dominant constraint on stability. Provided that a realistic, Monte Carlo predicted value for the "elastic thickness" of the steric barrier is adopted, the theory predicts fairly accurately the maximum particle that can be stabilized in polymer melts. The Edwards theory (33), however, to be discussed below, will probably prove in the long run to be a more powerful and physically satisfying approach.

The elastic contribution to the free energy of close approach rises quite abruptly once the particles are sufficiently close for elastic effects to become significant (39). Three important consequences follow from this. First, the repulsive elastic term is dominant in very close approach, even in worse than θ-solvents. For this reason, most sterically stabilized particles appear unlikely to exhibit the primary minimum so important to the understanding of the stability of electrostatically stabilized dispersions. Accordingly, the ubiquitous statements in textbooks of colloid science that all dispersions of colloidal particles are thermodynamically unstable may require qualification. Sterically stabilized latices in dispersion media that are good solvents for the stabilizer may well be thermodynamically stable. If that is so, the classical textbook statement must be restricted in its application to electrostatically stabilized dispersions.

The second point to be made in this context follows from the first. Because the elastic contribution exceeds thermal energy so abruptly, the predominant region of close approach sampled during a Brownian collision will be the interpenetration domain, together with that part of the interpenetration plus compression domain that admits of negligible elastic contributions (39). This implies that only mixing terms will be important in determining incipient flocculation. The body of experimental data available to date supports this inference overwhelmingly. It is for this reason that elastic contributions to steric stabilization prove experimentally to be somewhat elusive. They are perhaps most easily perceived experimentally in melt systems or in the compression of flocculated latices.

Finally, it should be noted that incipient flocculation near to the θ-point appears to take place in a potential energy minimum that is analogous to the secondary minimum in electrostatically stabilized systems. It must be regarded as a pseudo-secondary minimum, however, because, as discussed above there is no apparent primary minimum (see Fig. 3). Therefore the pseudo-secondary minimum is determined primarily by the mixing parameters. The rheological data of Hunter et al. (44) on latices flocculated near to the θ-point provides strong support for this viewpoint.

OTHER THEORIES OF STERIC
STABILIZATION

Many of the theoretical concepts discussed above were developed by Evans, Smitham, and the author. This approach provides the only theory that has been compared with experiment in at least three different stability regions: these are incipient flocculation near to the θ-point; long-term (thermodynamic?) stability; and stability and incipient flocculation in polymer melts. The predictions of this theory of steric stabilization are in excellent qualitative agreement with the results of experiment in all three areas. The quantitative predictions of the theory are also in fair agreement with experiment.

Only one type of experiment is not easily encompassed by the foregoing approach. This is the flocculation that is induced by the addition of moderate to high concentrations of polymer to the dispersion medium. This phenomenon is, however, brought within the ambit of our approach if it results from the concentration dependence of the interaction parameter, which may well become greater than or equal to $\frac{1}{2}$. The phenomenon would then fall into the category of flocculation near to the θ-point. Certainly this interesting observation requires more detailed study.

Other theories of steric stabilization, which differ either in matters of detail or in general approach, have been proposed. They all appear, with perhaps one notable exception, to fail in some important respect when compared with experiment. We note that in the area of steric stabilization, there seems to have been some degree of reluctance to compare the predictions of theory with the results of experiment. The merit of any theory, however, must surely be directly related to its ability to comprehend the results of a wide range of experiments. Space permits but a brief review of these theories, all of which have contributed, to a greater or lesser extent, to the development of the subject.

Entropy Theories

Mackor (45, 46) was perhaps the first to attempt to calculate the repulsive energy barrier in steric stabilization. His primitive, yet germinal, theory estimated the loss of configurational entropy of a rigid rod, freely jointed to one flat plate, on the approach of a second parallel flat plate. The free-energy change was calculated directly from the decrease in entropy for it was assumed that enthalpy effects could be ignored. Of course, such calculations, which ignore the solvent completely, are unlikely to possess any quantitative predictive value. In addition, most steric stabilizers are synthetic macromolecules that exhibit a high degree of external flexibility. Note that Bagchi and Vold (47) have extended Mackor's theory to spheres.

Clayfield and Lumb (48–52) in a series of papers have elaborated upon the approach of Mackor. They have allowed for the flexibility of the stabilizing chains by computer simulation, using Monte Carlo methods. A cubic lattice was adopted and it was claimed that the excluded volume effect was incorporated into the models by not allowing more than one segment to occupy any given lattice site. Osmond et al. (19) have asserted that this approach is the best that has been developed thus far.

We would contest the assertion of Osmond et al. (19). First, it is by no means clear physically to what process the loss of entropy calculated by Clayfield and Lumb actually refers. The claim that the excluded volume effect was allowed for might suggest that it was part of the entropy of compression associated with the mixing term. Yet the solvent was virtually ignored. There is absolutely no doubt, therefore, that the approach fails to predict the observed incipient flocculation near to the θ-point. Further, the Monte Carlo method is a notoriously ineffectual procedure for treating problems in the thermodynamics of polymer solutions. In our conceptual framework, it seems that what Clayfield and Lumb actually calculated is an elastic contribution to the free energy of close approach.

Fischer's Solvency Theory

Fischer (20) was perhaps the first to stress the importance of the molecules of the dispersion medium in steric stabilization. He recognized that in the interaction zone, the molecules of the dispersion medium (hereafter termed the solvent) may have a chemical potential different from that in the external bulk phase. If this chemical potential is less than that in the bulk phase, solvent molecules from the bulk phase flow into the interaction zone and force the particles apart. The key role of the solvent in steric stabilization thus becomes apparent.

Fischer, however, only considered the interpenetration of two layers of constant segment density. His simple theory makes no allowance for compressional interactions and ignores elastic effects completely. But, because elastic effects are so repulsive, interpenetration is the main effect apparent in Brownian collisions. Accordingly, the Fischer theory predicts the observed incipient flocculation near to the θ-point. It does not, however, comprehend stability in polymer melts.

Entropy Plus Solvency Theories

Meier (30) proposed that the entropy and solvency concepts should be added to obtain the total interaction. He regarded the mixing contribution as an osmotic effect and the elastic term as one deriving from volume restriction. Hesselink et al. (31) subsequently adopted Meier's terminology. In our view, this terminology is confusing. There are contributions to the mixing term that result from the

restriction of the available volume by the approaching impenetrable surface. The Gibbs–Duhem relationship, moreover, assures that elastic contributions may give rise to differences in the chemical potential of the solvent in the interaction zone, i.e., to osmotic effects. The terms "mixing" and "elastic," as cited by Flory, seem to be more appropriate.

Meier (30) estimated the elastic term by application of the diffusion equation and the mixing term by using Flory–Krigbaum theory. His results, however, are in error because the polymer chains were, in fact, allowed to penetrate the impenetrable barriers. Subsequently, Hesselink et al. (31) corrected this mistake. Their calculations predicted (a) that the boundary between stability and instability should be very sensitive to the molecular weight of the stabilizing moieties, (b) that stability should be observed in significantly worse than θ-solvents, and (c) that no correlation should be evident between the cfpt and the θ-point. None of these predictions appears to be realized in incipient instability studies.

In attempting to trace the source of this discrepancy between theory and experiment, Evans and the author (53) tentatively attributed it to an erroneous adoption of standard states. Osmond et al. (19) have argued against this ascription of the discrepancy. We now accept that our tentative proposal is incorrect, although not for the reasons adduced by Osmond et al. As noted by the latter authors, both Meier (30) and Hesselink et al. (31) invoke London dispersion forces to explain incipient instability. They assume values for the Hamaker constant that are at least an order of magnitude too large for latex particles. If the value of the Hamaker constant is reduced to a more realistic figure, the predictions of Hesselink et al. become even more discordant when compared with experiment. We now believe that the discrepancies can be traced to the use of segment density distribution functions that extend to infinity, rather than being truncated at the contour length. As a result, any distinction between the interpenetration and the interpenetration plus compression domains becomes blurred. There results a significant elastic contribution at all separational distances of interest. Such an elastic term would destroy any correlation between the θ-point and the cfpt, and would be strongly molecular weight dependent, as is predicted by the calculations of Hesselink *et al*.

Bagchi (54–58) has extended the earlier ideas that he published with R. D. Vold (47, 59, 60) on what he has termed a "denting" model. This model assumes that when the stabilizing chains attached to one particle come in contact with those attached to another particle, the chains are compressed as if they had come in contact with an impenetrable surface. Such an assumption is intuitively disquieting in the light of the small segment densities involved, at least in the peripheral zones. The incipient flocculation results seem to be quite decisive in eliminating the denting models for that case at least. A strong elastic contribution would again be implicit in the denting model, obscuring any correlation between the cfpt and the θ-point. Note, too, that Bagchi applies the Flory–Huggins theory in a self-inconsistent fashion. He retains the logarithmic form for the entropy of mixing term; yet he uses experimental data derived on the assumption that the logarithmic term is truncated after the second power of the volume fraction. This is not a pedantic objection, for such an inconsistency demands that χ_1 be ca. 0.6 for incipient flocculation. As a result, any correlation between the cfpt and the θ-point is totally obscured.

Dolan and Edwards (32, 33) have recently provided an important signpost to how the second generation of entropy plus solvency theories will be developed. All the theories described in this section have demanded significant digital computation. Simple analytical formulas were missing. Of course, the theory developed in the present author's laboratory is essentially of an entropy plus solvency type; this study, moreover, has been directed toward the generation of analytical formulas. It is a

first-generation theory, however, for it uses the now obsolete Flory–Huggins theory (albeit in an updated, pragmatic fashion) and separately calculates the elastic and mixing terms.

Dolan and Edwards (32, 33) have shown how it is possible to solve the diffusion equation for a polymer attached at one end to an impenetrable flat surface by exploiting either the method of images or an eigenfunction expansion method. Approximate analytic expressions can then be generated for the loss of entropy on approach of a second flat surface. The original theory was developed in the absence of excluded volume effects but subsequently these were incorporated into the theory by a powerful self-consistent field device. Unfortunately, to date no comparison between theory and experiment has been published for the case where excluded volume effects are important. As analytical formulas are still lacking for that case, resort to digital computation is mandatory. The method does appear, however, to offer significant potential for calculating, without artificial separation, the mixing and elastic terms. In addition, the theory developed for the elastic contributions can be expressed approximately in closed form; when due allowance is made for the curvature of the particles, these formulas provide a realistic basis for estimating the maximum particle size that can be sterically stabilized in polymer melts (61).

CONFORMATION OF MACROMOLECULES AT AN INTERFACE

The conformation of macromolecules at an interface is important in the theory of steric stabilization because it determines the segment density distribution function. This in turn controls the distance dependence of the steric repulsion in stable systems. Note that incipient flocculation is apparently insensitive to the segment density distribution function.

The theories (62, 63) developed for the adsorption of a homopolymer at an interface could well lead to the inference that adsorption at various points along the chain would produce a flattened stabilizer conformation at the particle surface. This inference is almost surely incorrect. All the experimental evidence available thus far for sterically stabilized latices suggests that the macromolecules are in an extended conformation normal to the interface (64–68). This extension exceeds significantly that predicted for a 1-D random walk normal to an impenetrable barrier (69).

The foregoing experimental observation may be rationalized if we recall that block and graft copolymers are the most effective steric stabilizers. Hence the free energy of adsorption is made up of contributions from both the anchor polymer and the stabilizing moieties:

$$\Delta G^{ads} = \Delta G^{anchor} + \Delta G^{stab}.$$

Clearly, ΔG^{ads} may be negative, even if ΔG^{stab} is positive. It requires only that ΔG^{anchor} be sufficiently negative. That will be true if the chains are suitably well anchored, as is required for steric stabilization. This is not to argue that the stabilizing moieties adopt, in all cases, a thermodynamically unfavorable conformation. Rather it is a cautionary warning of the difficulty of trying to predict, by minimization of its free energy, the conformation of a stabilizing moiety. The presence of the anchor polymer may well vitiate this approach.

We originally ascribed the additional extension at an interface to the compression of the stabilizing moieties by close packed contiguous chains (38). Monte Carlo calculations on the problem of a polymer-in-a-box fail, however, to substantiate this hypothesis, at least at the normal packing densities observed experimentally (70). The extension normal to the interface appears to be a consequence of excluded volume effects. This conclusion is also supported by the self-consistent field approach of Dolan and Edwards (33). If this conclusion is correct, previous theories of macromolecular conformations at interfaces, which ignore excluded volume effects, are likely to be substantially in error (71, 72).

CONCLUSIONS

This review of the present status of steric stabilization is intended to convey some sense of the dramatic improvement in our understanding of the principles governing steric stabilization. Hopefully, such primitive notions as ascribing steric stabilization to "hydration effects" have been swept away and replaced by more precise physical concepts. That is not to say that much does not remain to be done. Incipient instability near to the θ-point is now fairly well characterized experimentally. The distance dependence of the repulsive barrier for stable dispersion needs further experimental characterization. This should allow the now emerging theories of steric stabilization to be tested in more detail. In this process, the first generation theories will undoubtedly be superseded by more complete descriptions of steric stabilization.

ACKNOWLEDGMENTS

It is a pleasure to express my gratitude to my co-workers, Messrs. R. Evans, J. B. Smitham, R. Feigin, and G. March, each of whom has contributed significantly to my understanding of the subject. I am grateful to the ARGC for financial support of our studies. I thank the Australian–American Educational Foundation for the award of a Fulbright–Hays Senior Scholar's Travel Grant.

REFERENCES

1. "Encyclopaedia Brittanica," Vol. 12, p. 257, 1968; Vol. 17, p. 38, 1968.
2. HELLER, W., AND PUGH, T. L., J. Chem. Phys. **22,** 1778 (1954).
3. BARRETT, K. E. J., "Dispersion Polymerization in Organic Media." Wiley, London, 1975.
4. NAPPER, D. H., J. Colloid Interface Sci. **32,** 106 (1970).
5. NAPPER, D. H., AND NETSCHEY, A., J. Colloid Interface Sci. **37,** 528 (1971).
6. NAPPER, D. H., AND HUNTER, R. J., MTP Int. Rev. Sci. Ser. 1 **7,** 280 (1972).
7. FLEER, G. J., AND LYKLEMA, J., J. Colloid Interface Sci. **46,** 1 (1974).
8. HELLER, W., AND PUGH, T. L., J. Chem. Phys. **24,** 1107 (1956).
9. HELLER, W., Pure Appl. Chem. **12,** 249 (1966).
10. NAPPER, D. H., Trans. Faraday Soc. **64,** 1701 (1968).

11. NAPPER, D. H., *J. Colloid Sci.* **29**, 168 (1969).

12. NAPPER, D. H., *Kolloid Z. Z. Polym.* **234**, 1149 (1969).

13. EVANS, R., NAPPER, D. H., AND EWALD, A. J., *J. Colloid Interface Sci.* **51**, 552 (1975).

14. EVANS, R., AND NAPPER, D. H., *J. Colloid Interface Sci.* **52**, 260 (1975).

15. NAPPER, D. H., *Ind. Eng. Chem. Prod. Res. Develop.* **9**, 467 (1970).

16. EVANS, R., DAVISON, J. B., AND NAPPER, D. H., *Polym. Lett.* **10**, 449 (1972).

17. EVANS, R., Ph.D. Dissertation, University of Sydney, Australia 1976.

18. OTTEWILL, R. H., *Ann. Rep. Progr. Chem. (Chem. Soc. London) A* **66**, 212 (1969).

19. OSMOND, D. W. J., VINCENT, B., AND WAITE, F. A., *Colloid Polym. Sci.* **253**, 676 (1975).

20. FISCHER, E. W., *Kolloid Z.* **160**, 120 (1968).

21. BUECHE, F., *J. Colloid Interface Sci.* **41**, 374 (1972).

22. ELIAS, H.-G., AND BÜHRER, *in* "Polymer Handbook" (J. Brandrup and E. H. Immergut, Eds.), 2nd ed., p. IV–157. Wiley, New York, 1975.

23. ELIAS, H.-G., *Makromol. Chem.* **33**, 140 (1959).

24. ELIAS, H.-G., *Makromol. Chem.* **50**, 1 (1960).

25. CORNET, C. F., AND VAN BALLEGOOIJEN, H., *Polymer* **7**, 293 (1966).

26. NAPPER, D. H., *Polymer* **10**, 181 (1969).

27. DOBBIE, J. W., EVANS, R., GIBSON, D. V., SMITHAM, J. B., AND NAPPER, D. H., *J. Colloid Interface Sci.* **45**, 557 (1973).

28. SMITHAM, J. B., AND NAPPER, D. H., *J. Colloid Interface Sci.* **54**, 467 (1976).

29. LI-IN-ON, F. K. R., VINCENT, B., AND WAITE, F. A., *Amer. Chem. Soc. Symp. Ser.* **9**, 165 (1975).

30. MEIER, D. J., *J. Phys. Chem.* **71**, 1861 (1967).

31. HESSELINK, F. TH., VRIJ, A., AND OVERBEEK, J. TH. G., *J. Phys. Chem.* **75**, 2094 (1971).

32. DOLAN, A. K., AND EDWARDS, S. F., *Proc. Roy. Soc. London Ser. A* **337**, 509 (1974).

33. DOLAN, A. K., AND EDWARDS, S. F., *Proc. Roy. Soc. London Ser. A* **343**, 427 (1975).

34. EVANS, R., AND NAPPER, D. H., *J. Colloid Interface Sci.* **45**, 138 (1973).

35. FLORY, P. J., "Principles of Polymer Chemistry," Cornell Univ. Press, Ithaca, N. Y., 1953.

36. YAMAKAWA, H., "Modern Theory of Polymer Solutions," Harper and Row, New York, 1971.

37. FLORY, P. J., AND KRIGBAUM, W. R., *J. Chem. Phys.* **18**, 1086 (1950).

38. SMITHAM, J. B., EVANS, R., AND NAPPER, D. H., *J. Chem. Soc. Faraday Trans. I.* **71**, 285 (1975).

39. EVANS, R., SMITHAM, J. B., AND NAPPER, D. H., *Kolloid Z. Z. Polym.*, to appear.

40. DERYAGIN, B. V., *Kolloid Z.* **69**, 166 (1934).

41. MARCH, G., M.Sc. Qualifying Thesis, University of Sydney, Australia, 1975.

42. JÄCKEL, K., *Kolloid Z. Z. Polym.* **197**, 143 (1964).

43. DOROSZKOWSKI, A., AND LAMBOURNE, R., *J. Colloid Interface Sci.* **43**, 97 (1973).

44. HUNTER, R. J., NEVILLE, P. C., AND FIRTH, B. A., *Amer. Chem. Soc. Symp. Ser.* **9**, 193 (1975).

45. MACKOR, E. L., *J. Colloid Sci.* **6**, 492 (1951).

46. MACKOR, E. L., AND VAN DER WAALS, J. H., *J. Colloid Sci.* **7**, 535 (1952).

47. BAGCHI, P., AND VOLD, R. D., *J. Colloid Interface Sci.* **33**, 405 (1970).

48. CLAYFIELD, E. J., AND LUMB, E. C., *J. Colloid Interface Sci.* **22**, 269 (1966).

49. CLAYFIELD, E. J., AND LUMB, E. C., *J. Colloid Interface Sci.* **22**, 285 (1966).

50. CLAYFIELD, E. J., AND LUMB, E. C., *J. Colloid Interface Sci.* **47**, 6 (1974).

51. CLAYFIELD, E. J., AND LUMB, E. C., *J. Colloid Interface Sci.* **47**, 16 (1974).

52. CLAYFIELD, E. J., AND LUMB, E. C., *J. Colloid Interface Sci.* **49**, 489 (1974).

53. EVANS, R., AND NAPPER, D. H., *Kolloid Z. Z. Polym.* **251**, 409 (1973).

54. BAGCHI, P., *J. Colloid Interface Sci.* **41**, 380 (1972).

55. BAGCHI, P., *J. Colloid Interface Sci.* **47**, 86 (1974).

56. BAGCHI, P., *J. Colloid Interface Sci.* **47**, 100 (1974).

57. BAGCHI, P., *J. Colloid Interface Sci.* **50**, 115 (1975).

58. BAGCHI, P., *Amer. Chem. Soc. Symp. Ser.* **9**, 145 (1975).

59. BAGCHI, P., AND VOLD, R. D., *J. Colloid Interface Sci.* **38**, 652 (1972).

60. BAGCHI, P., AND VOLD, R. D., *J. Colloid Interface Sci.* **41**, 397 (1972).

61. SMITHAM, J. B., AND NAPPER, D. H., *J. Chem. Soc. Faraday Trans. I.* **72**, 2425 (1976).

62. VINCENT, B., *Advan. Colloid Interface Sci.* **4**, 193 (1974).

63. LIPATOV, YU. S., AND SERGEEVA, L. M., "Adsorption of Polymers." Wiley, New York, 1974.

64. OSMOND, D. W. J., AND WALBRIDGE, D. J., *J. Polym. Sci. Polymer Symposia* **30**, 381 (1970).

65. DOROSZKOWSKI, A., AND LAMBOURNE, R., *J. Colloid Interface Sci.* **26**, 214 (1968).

66. DOROSZKOWSKI, A., AND LAMBOURNE, R., *J. Polym. Sci. Polymer Symposia* **34**, 253 (1971).

67. BARSTED, S. J., NOWAKOWSKA, L. J., WAGSTAFF, I., AND WALBRIDGE, D. J., *Trans. Faraday Soc.* **67**, 3598 (1971).

68. OTTEWILL, R. H., AND WALKER, T. W., *Kolloid Z. Z. Polym.* **227**, 108 (1968).

69. DI MARZIO, E. A., AND MCCRACKIN, F. L., *J. Chem. Phys.* **43**, 539 (1965).

70. SMITHAM, J. B., AND NAPPER, D. H., *J. Polym. Sci. Polymer Symposia*, submitted for publication (1976).

71. HESSELINK, F. TH., *J. Phys. Chem.* **73**, 3488 (1969).

72. HESSELINK, F. TH., *J. Phys. Chem.* **75**, 65 (1971).

Recent Developments in the Understanding of Colloid Stability

J. TH. G. OVERBEEK

Van't Hoff Laboratory, University of Utrecht, Transitorium III,
Padualaan 8, Utrecht, The Netherlands

Received June 10, 1975; accepted August 20, 1976

The stability of suspensions and emulsions against coagulation is governed by the forces between the particles. The main forces are: van der Waals attraction, electrostatic repulsion, and the repulsion due to the interaction of adsorbed large molecules. It is stressed that coagulation and redispersion are rate phenomena. Smoluchowski's classical theory on coagulation as modified by Fuchs to include interaction forces between particles has been the basis of theories of colloid stability. Corrections to these theories are needed to account for the hampering of Brownian motion when the particles are close together. Furthermore it is pointed out that the duration of a Brownian collision is long enough to allow redistribution of the double layer and of the conformation of adsorbed macromolecules, but too short for adaptation of the surface charge or of the adsorption of macromolecules. Repeptization requires the existence of a distance of closest approach between particles which is of the order of one or two solvent molecules.

I. INTRODUCTION

In suspensions and emulsions, in general in dispersions of fine particles in a liquid, frequent encounters between particles occur due to Brownian movement, to gravity (creaming, sedimentation), and to convections. Whether such encounters result in permanent contact or whether the particles rebound and remain free is determined by the forces between them. A dispersion is *stable in the colloid-chemical sense* when its particles remain permanently free. In dilute dispersions it is sufficient to consider only interactions between pairs of particles. In very concentrated systems it may be necessary to take also multiparticle interactions into account.

On the basis of pair interaction and considering only electrostatic repulsion and van der Waals attraction a statisfactory understanding of the stability of hydrophobic suspensions (also called *electrocratic* systems) was reached in the Forties (1, 2). In particular, a theoretical explanation of the extremely large effect of the valence of the counterions on the

stability (rule of Schulze (3) and Hardy (4)) could be given.

An important extension was reached when protective action and sensitization of dispersions were interpreted as due to the adsorption of long molecules: originally, proteins, gums, and other polysaccharides; later, nonionic detergents, synthetic polymers, and polyelectrolytes. After a rudimentary, but essentially correct, theory by Mackor and van der Waals (5) a fairly large number of authors have worked out the theory in much greater detail (6–11). Since this subject is treated in the paper of D. H. Napper on "Steric Stabilization" we shall limit our discussion here to a few brief remarks.

Another feature of the stability of colloidal dispersions is the role of chemical reactions, in particular chemical complexing studied extensively by Matijević (12).

Then we should mention briefly new tools that have been introduced into the study of colloid stability. The availability of isodisperse latices (13) of different particle sizes and chemical nature has made it easier to obtain quantitative comparisons between theory and experiments. Many new spectroscopic techniques such as NMR, ellipsometry, and laser-beat methods have been applied successfully to colloid and surface chemistry.

Although obviously coagulation is a rate process, nevertheless colloid theory has shown a tendency to concentrate its attention more on the thermodynamics (surface thermodynamics) behind the rate processes (adsorption equilibria, electrochemical equilibria) than on the rate processes themselves.

In this paper I want to discuss *coagulation as a rate process* and in particular I want to show that it is really a combination of several processes each occurring at its own rate and that these rates may vastly differ in some cases. I shall also spend some time discussing the inverse process, *redispersion* or *repeptization* and the rates involved in it.

II. FORCES BETWEEN TWO PARTICLES

As mentioned in the Introduction, encounters between particles are frequent and the result of such an encounter is determined by the forces between the particles. We distinguish three types of forces.

a. Van der Waals forces. Van der Waals forces are always attractive between particles of the same nature. Hamaker (14) derived equations for these forces on the basis of additivity of van der Waals energies between pairs of atoms or molecules, and assuming these energies to be proportional to the inverse sixth power of the distance. Casimir and Polder (15) introduced the influence of retardation, which changes the power law to an inverse seventh power at large separations. For the special case of metals they also could abandon the assumption of additivity by using continuum considerations. Lifshitz (16) generalized the continuum treatment to dielectrics. In his treatment the interaction energy between macroscopic bodies is the result of charge fluctuations in these bodies.

The law describing the change of the force (or energy) with the distance is accurately

known, except for the precise value of the proportionality constant (Hamaker constant, A, retarded constant, B) and the characteristic length connected with the transition from nonretarded forces (small distances) to retarded forces (large distances). In most practical examples of colloid stability only the nonretarded van der Waals forces are important.

In the special case of the attraction between two spherical particles of radius, a, at a distance, H, (distance between the centers = $2a$ + $H = R$) Hamaker, considering only nonretarded forces, found for the energy of attraction, V_{attr},

$$V_{attr} = -\frac{A}{6}\left(\frac{2a^2}{R^2 - 4a^2} + \frac{2a^2}{R^2} + \ln\frac{R^2 - 4a^2}{R^2}\right), \quad [1]$$

where A depends on the properties of the particles and of the dispersion medium.

This somewhat awkward equation can be approximated by using expansions in powers of H/a (small distances) or a/R (large distances) but these expansions converge rather slowly. Expansions which converge more rapidly are obtained by introducing for small distances

$$L = a + (3H/4) \quad [2]$$

and for large distances

$$Q^4 = R^2(R^2 - 4a^2). \quad [3]$$

In these new variables Eq. [1] can be approximated quite accurately by

$$V_{attr}\,(H/a < 1) = -\frac{AL}{12H}\left\{1 + \frac{2H}{L}\ln\frac{H}{L}\right.$$
$$\left. - \frac{15}{16}\left(\frac{H}{L}\right)^2 - \frac{3}{32}\left(\frac{H}{L}\right)^3\right\} \quad [4]$$

and

$$V_{attr}\,(a/R < 1) = -\frac{16Aa^6}{9Q^6}\left\{1 - \frac{6}{5}\left(\frac{a}{Q}\right)^4\right.$$
$$\left. + \frac{18}{7}\left(\frac{a}{Q}\right)^8\right\}. \quad [5]$$

At $H < 0.4a$ $(R < 2.4a)$, Eq. [4] is accurate to better than 0.1% and at $H > 0.4a$, Eq. [5] is accurate to better than 0.5%. At $H < 0.1a$ the first two terms of Eq. [4] are already accurate to better than 1.5% and at $H > 1.6a$ $(R > 3.6a)$ the first term of Eq. [5] alone is accurate to 1% or better.

The existence of the logarithmic term in Eq. [4] causes $V_{attr} = -AL/12H$ or $V_{attr} = -Aa/12H$ to be rather poor approximations unless H/a is very small.

  b. Electrostatic forces. Electrostatic forces, due to the interaction of the electrical double layers surrounding the particles, always lead to a repulsion between particles if they are of the same chemical nature and have surface charges and surface potentials of the same sign and magnitude.

This repulsion is due to the interaction of the ions adsorbed at the surfaces, mitigated by the space charges in the double layers. It also contains an entropic contribution, since the ionic concentrations in the double layers vary during interaction. Furthermore, insofar as the adsorption of ions at the surfaces changes, this contributes a chemical term to the interaction energy.

When the surface charge is generated by the adsorption of *potential determining ions*, the surface potential, ϕ_0, is determined by the activity of these ions and remains constant during interaction at least if complete adsorption equilibrium is maintained. In that case interaction occurs at *constant surface potential*. Lack of adsorption equilibrium, or other mechanisms to generate the surface charge or simply a shortage of potential determining ions (very low concentrations) may cause the interaction to take place at *constant surface charge* or at some situation intermediate between constant charge and constant potential.

Whereas Van der Waals forces fall off as an inverse power of the separation between the particles and have a range comparable to the particle size, the electrostatic repulsion falls off as an exponential function of the distance and has a range of the order of the thickness of the electrical double layer (equal to the

Debye–Hückel length, $1/\kappa$)

$$1/\kappa = (\epsilon\epsilon_0 RT/F^2 \sum c_i z_i^2)^{\frac{1}{2}}, \qquad [6]$$

where ϵ_0 is the permittivity of the vacuum, ϵ is the relative permittivity (=dielectric constant) of the dispersion medium, R is the gas constant, T is the absolute temperature, F is the Faraday constant, c_i and z_i are the concentration and the charge number of the ions of type i in the dispersion medium.

No simple expression can be given for the energy of repulsion but the following expressions are reasonable approximations (17).

For small surface potentials (ϕ_0) and small κa (extension of double layer larger than particle radius),

$$V_{rep} = 2\pi\epsilon\epsilon_0 a\phi_0^2 \frac{\exp(-\kappa H)}{1 + H/2a}. \qquad [7]$$

For large κa, symmetrical electrolytes (only one electrolyte with ions of charge number $+z$ and $-z$), and relatively large distances (κH large),

$$V_{rep} = 32\pi\epsilon\epsilon_0 a(RT\gamma/zF)^2$$
$$\times \ln[1 + \exp(-\kappa H)], \qquad [8]$$

with

$$\gamma = \tanh(zF\phi_0/4RT). \qquad [9]$$

For small values of ϕ_0 and small values of $\exp(-\kappa H)$ this expression simplifies to

$$V_{rep} = 2\pi\epsilon\epsilon_0 a\phi_0^2 \exp(-\kappa H). \qquad [10]$$

In the above expressions it is assumed that during the approach of two particles the surface potentials ϕ_0 remain constant. It is also possible to calculate the repulsion when instead of the surface potential ϕ_0 the surface charge, Q, remains constant. For large separations (small interaction) the interaction energy is independent of the choice, Q constant or ϕ_0 constant, but at small distances the repulsion is stronger at constant charge than at constant potential. Frens (18) has described a method of deriving the interaction energy at constant charge, when that at constant potential is given.

Just as in the case of the van der Waals forces, the dependence of the electrostatic repulsion on the distance is well known, but there is uncertainty about the absolute value, since the value of ϕ_0 in Eqs. [7]–[10] is often not known. In the first place, the equations as given above do not take account of the finite size of the ions and the potential to be used in calculating the repulsion is the potential, ϕ_δ, in the *Stern plane* (plane of closest approach of the ions to the surface) rather than the surface potential ϕ_0. In cases, such as AgI–aqueous solution, where the surface potential is governed by an equilibrium of potential determining ions, ϕ_0 may be rather accurately known. Even then ϕ_δ can only be estimated from ϕ_0, when the double-layer structure has been analyzed e.g., by capacity determinations. In some other cases the surface charge density, σ, may be well known and assuming a simple diffuse double-layer structure, without incomplete dissociation or specific adsorption of ions, ϕ_δ may be calculated. In most cases, however, the nearest practical approximation to ϕ_δ is the ζ-potential and this presupposes that the Stern plane and the slipping plane are (nearly) identical.

c. Forces connected with adsorbed macromolecules. The third group of forces between particles, which is relatively well understood now, is due to the interaction of long-chain molecules adsorbed on the particles. When two particles, each carrying an adsorbed layer of dangling chains (or loops), approach one another closely, two effects lead to repulsion. In the first place each chain—if its extended length is larger than the distance between the surfaces—loses some of its otherwise available conformations, and thus its contribution to the free energy of the system is increased, resulting in a repulsive force ("volume restriction" effect). Moreover, when the chains belonging to the two particles start to overlap, this amounts to a local increase of the concentration and again to an increase in the free energy ("osmotic" effect). As a rule the quantitative influence of the osmotic effect is the more important of the two. Many authors have contributed to the quantitative calculation of these effects (5–11). The repulsion increases

with the chain length, with the quality of the solvent for the chains (distance from the θ-temperature), and with the number of chains per unit area. The repulsion as a rule is quite steep, also compared with the slope of the van der Waals force.

A difficulty in the application of this effect to practical cases is that the protective chains must be well soluble in the solvent, but the molecules as a whole must contain parts that are adsorbed to the particles (anchor groups) and since adsorption from solution is always a competition between adsorbate and solvent it is often difficult to predict how strong the adsorption will be, quite apart from the technical difficulty of preparing molecules with two parts having partially conflicting properties.

From the above it will be clear that the same type of adsorbed molecules *may lead to attraction* rather than to repulsion either if they contain two or more anchor groups adsorbed on different particles or if the solvent is changed, so as to be below the θ-temperature.

d. Combination of forces. The combination of van der Waals attraction and electrostatic repulsion leads in general to interaction curves with a maximum and a minimum if the energy is plotted versus the distance. Both at small and at large distances the van der Waals energy, proportional to H^{-x} (where x varies from 1 to 7) surpasses the repulsion, which is proportional to $\exp(-\kappa H)$. But at intermediate distances the repulsion energy may be the larger of the two. Electrolyte concentrations and valence act via $\kappa (\sim (\sum c_i z_i^2)^{\frac{1}{2}})$ on the interaction curves. Figure 1 gives an example how the total interaction depends on the electrolyte concentration. If we assume that an energy barrier of 10 kT is enough to prevent two particles from colliding, we see that under the conditions of Fig. 1 c must be smaller than $0.001\ M$ ($\kappa < 10^6$ cm^{-1}) for stability.

The combination of van der Waals forces with "steric" repulsion (interaction of adsorbed chains) again leads to a minimum in the energy at large separations since the van der Waals energy decays more slowly than the steric energy. But, as a rule, if the repulsion

POTENTIAL ENERGY OF INTERACTION

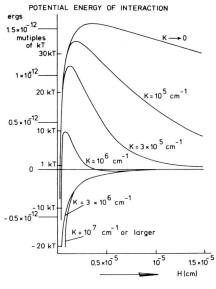

$a = 10^{-5}$ cm ; $A = 10^{-12}$ ergs ; $\psi_0 = \dfrac{RT}{F} \approx 25.6$ mV .

FIG. 1. Energy of interaction of two spherical particles as a function of the distance, H, between the surfaces. For monovalent ions c (in moles liter^{-1}) $= 10^{-15} \, \kappa^2$ (in cm^{-1}).

is strong enough to overcome the van der Waals attraction at some intermediate distance repulsion is found at all shorter distances.

Figure 2 taken from Evans and Napper's work (10) shows how sensitive the repulsion is to the quality of the solvent.

We shall not go into the effects of the simultaneous presence of all three types of forces.

III. RATE OF COAGULATION

III. 1. Classical Theory

Suspensions and emulsions are never stable in the thermodynamic sense. Their large interfacial area will always lead to coarsening by recrystallization or droplet growth, possibly slowly, but inexorably. Whether coarsening by coagulation occurs rapidly, slowly, or hardly at all will depend on the interaction forces discussed in the previous section. The fre-

quency of Brownian encounters determines the maximum rate of coagulation in the absence of forces. The maximum energy of repulsion acts as an activation energy decreasing the rate of coagulation.

Rate effects are therefore clearly important as a sensitive test of our understanding of the interaction of colloidal particles. They are also important in practical applications. A suspension coarsening with a half-life of 5 years is stable for most practical purposes. If its half-life is 1 week, it is stable enough for many laboratory uses, and if its half-life is 10 minutes it is a good object for investigating rates of coagulation.

Already in 1916 Smoluchowski (20) developed an admirable theory of the rate of coagulation. He described coagulation as a bimolecular reaction, obeying the equation

$$-dn/dt = k_r n^2, \qquad [11]$$

where n is the number of particles per unit volume, t the time, and k_r the rate constant.

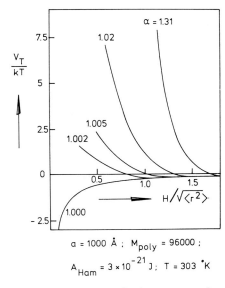

FIG. 2. Energy of interaction between two spheres. α is the parameter introduced by Flory (19), giving the expansion factor of a coiled chain and intimately connected with the quality of the solvent.

He calculated the rate constant for *rapid co-agulation* from the frequency of Brownian collisions assuming no interaction between the particles except a short-range force leading to permanent contact on the first encounter. An additional assumption that the particle dimension is large compared to the mean free path of the fluid molecules is always true in hydrosols but is not necessarily so in aerosols.

The collision frequency was derived by considering the mutual rate of diffusion of two spherical particles of equal size toward one another. The collision frequency appears to be independent of the size since the slower Brownian motion of the larger particles is exactly compensated by their larger collision cross section. Furthermore, when the particles are of different, but not too widely different, size the collision frequency is only slightly modified. Even the collision frequency between small molecules, as determined from the rates of diffusion-controlled reactions, is of the same order of magnitude as that between colloidal particles (21).

Smoluchowski expressed his results in terms of the half-life of the suspension, i.e., the time $T_{\frac{1}{2}}$, in which the number of particles was reduced to one-half the original value, n_0. He found

$$T_{\frac{1}{2}} = 1/k_r n_0 = 1/8\pi Da n_0, \qquad [12]$$

where D is the diffusion coefficient of the particles and a their radius. Using the expression

$$D = k_B T/6\pi\eta a, \qquad [13]$$

where k_B is the Boltzmann constant and η the viscosity of the medium we further have

$$T_{\frac{1}{2}} = \frac{3\eta}{4k_B T n_0} \approx \frac{2 \times 10^{11}}{n_0} \text{ cm}^{-3} \text{ sec}, \quad [14]$$

where the last equality holds for water (or any other liquid with the same viscosity) and ambient temperature.

For coagulation, slowed down by the presence of an energy barrier, Smoluchowski wrote simply

$$T_{\frac{1}{2}} \text{ (slow)} = T_{\frac{1}{2}} \text{ (rapid)}/\alpha, \qquad [15]$$

where α is the fraction of the number of collisions leading to permanent contact.

In 1934 Fuchs (22) showed how α could be related to the energy of interaction by considering the encounter of two particles as a diffusion in a field of force. He found

$$T_{\frac{1}{2}} \text{ (slow)} = \frac{1}{8\pi Da n_0}$$

$$\cdot 2a \int_{2a}^{\infty} \exp(V/k_B T) \frac{dr}{r^2}, \quad [16]$$

where V is the energy of interaction at a distance r between the centers of the two particles.

Since rates of coagulation can be measured rather accurately, especially if isodisperse latices are used and since a great deal is known or supposed to be known about the interaction energy, Eq. [16] and some of its consequences form a sensitive test of our understanding of interactions in suspensions, the more so since attraction and repulsion must be nearly balanced in order to obtain rates that are neither too slow, nor too fast.

The factor by which coagulation is slowed down is called W.

$$W = 1/\alpha = 2a \int_{2a}^{\infty} \exp(V/k_B T) dr/r^2. \quad [17]$$

Since V is approximately proportional to the particle radius, a, it is expected that, other things being equal (in particular the surface potential ϕ_0 or ϕ_δ and the electrolyte concentration), the coagulation of large particles is slowed down more strongly than that of small particles (23). This, however, has not been confirmed by experiments (23, 24). The experiments show more nearly that W is independent of particle size.

This discrepancy between theory and experiment has led to a reexamination of a number of aspects of the theory.

It may not be superfluous to point out here that the usual determination of W by light scattering or light absorption is based on the assumption that the optical properties of pairs and higher aggregates are the same in rapid and in slow coagulation. This assumption, although plausible, may not always be correct.

Theoretical calculations of W only apply to pairs formed from primary particles and therefore the comparison between theory and experiment has to be limited to the initial phase of the coagulation. Similarly, for the absolute value of the collision frequency it is essential to work with initially isodispersed sols, use only the very first part of the coagulation, and have an accurate theory of the optical properties of single particles and of doublets.

III. 2. Modified Collision Theory

Following a remark made by Deryagin in 1966 (25) a number of authors (Deryagin and Muller (26), Spielman (27), Honig, Roebersen, and Wiersema (28), and independently of these, Deutch and Felderhof (29)) have recalculated the collision frequency, taking into account that the last part of the approach of two particles is slowed down because it is difficult for the liquid to flow away from the narrow gap remaining between the particles. This effect is calculated quantitatively by estimating the mobility or diffusion coefficient along the line interconnecting the centers of the particles as a function of the distance, $H = R - 2a$, between the particles. The problem has been solved exactly by Brenner (30) but it leads to a fairly complicated expression. Honig et al. (28) give a very useful approximation in the form of a rational function.

$$\frac{D(H)}{D(H \to \infty)} = \frac{2H}{a} \left(\frac{1 + 3H/2a}{1 + 13H/2a + 3H^2/a^2} \right)$$
$$= \frac{1 + 2a/3H}{1 + 13a/6H + a^2/3H^2}. \quad [18]$$

Since D approaches zero as $2H/a$ when the particles are close together, actual Brownian encounters could never occur, unless there is some attraction at least proportional to a/H.

Van der Waals attraction satisfies this requirement and therefore collisions do occur, but their frequency depends on the value of the van der Waals forces.

The simplest check on this extension of the collision theory is the absolute measurement of the rate of rapid coagulation of a suspension of isodisperse spheres. This has been performed by Lichtenbelt (31), who determined the rate of coagulation of a series of isodisperse polystyrene latices, using a stopped-flow technique with extremely rapid mixing of latex and coagulating electrolyte (LaCl$_3$, 0.0058 mole liter^{-1}; BaCl$_2$ and MgCl$_2$, 0.05 mole liter^{-1}; NaCl, 0.4 mole liter^{-1}) and interpreting the increase in turbidity on the basis of a Rayleigh–Gans–Debye calculation of the light scattering by doublets.

The essence of his results can be expressed in the value of k_r, as introduced in Eq. [11], and comparing this value with the value k_r (Smol) it should have according to Smoluchowski's theory.

$$k_r \text{ (Smol)} = 4k_B T/3\eta. \quad [19]$$

Table I gives the comparison. Table I mentions two sets of values of k_r depending on whether the particle diameter derived by Dow from their electronmicroscopic measurements is accepted as correct, or assuming that the diameters are about 5% smaller as appears to follow from quite a number of remarks in the literature (32–37).

Whatever the correct choice may be, it is obvious that Lichtenbelt found k_r to be lower than the Smoluchowski value, independent of the particle size, and in agreement with the value based on Eq. [18] if Hamaker constants as mentioned in Table 1 are accepted as correct. The value $A = 7.0 \times 10^{-21}$ J is in agreement with the Lifshitz theory and of the same order as some other estimates for the van der Waals interaction of polystyrene particles in water.

We may thus consider the hydrodynamic correction to the collision theory as expressed in Eq. [18] to be confirmed by experiment.

Unfortunately this does not help us to solve the original discrepancy about the influence of the particle size on the factor W (Eq. [17])

TABLE I

Rate Constant, k_r, at 20°C of Rapid Coagulation Found with Experiments on Latices
and Compared to the Smoluchowski Value, k_r (Smol) = $4k_BT/3\eta$

(Dow) diameters of polystyrene spheres	$k_r \times 10^{12}$ cm^{-3} sec^{-r} Dow diameters assumed to be correct	$\dfrac{k_r}{k_r(\mathrm{Smol})}$	$k_r \times 10^{12}$ cm^{-3} sec^{-r} Dow diameters minus 5%	$\dfrac{k_r}{k_r(\mathrm{Smol})}$
$2a$ = 91 nm	3.2	0.60	2.6	0.49
$2a$ = 109 nm	2.95	0.55	2.35	0.44
$2a$ = 176 nm	3.4	0.63	2.8	0.52
$2a$ = 234 nm	2.65	0.50	2.25	0.42
$2a$ = 312 nm[a]	1.95[a]	0.37[a]	1.65[a]	0.32[a]
$2a$ = 357 nm	2.75	0.51	2.3	0.43
Average value[a]	3.0	0.56	2.45	0.46
Theoretical				
$A_{\mathrm{Ham}} = 7.0 \times 10^{-21}$ J	3.0	0.56		
$A_{\mathrm{Ham}} = 1.5 \times 10^{-21}$ J			2.45	0.46
Smoluchowski	5.4	1.00	5.4	1.00

[a] Results of the 312 nm latex are not included in the average.

since, if the hydrodynamic correction is now applied to the case of slow coagulation, it leaves the influence of a on W nearly unchanged.

This is dramatically illustrated in Fig. 3, where calculated values of log W are plotted against $\frac{1}{2}$ log c_{el} (c_{el} = electrolyte concentration) for a series of values of G (which is proportional to the particle radius a). The theoretical curves depend strongly on G and thus on a whereas the measurements by Ottewill

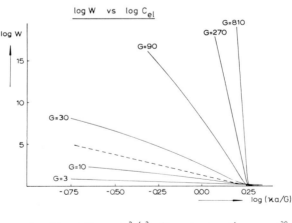

$$\log(\varkappa a/G) = -0.53 + \log(z^3/\gamma^2) + \tfrac{1}{2}\log C_{el} \cdot \mathrm{l\ mol}^{-1} \qquad A = 10^{-20} \mathrm{J}$$

$$G \simeq \frac{\varepsilon\varphi_\delta^2}{2k_BT}\cdot a \qquad \text{Ottewill and Shaw} \quad z = 2$$
$$a = 600\ \text{Å} - 4230\ \text{Å}. \quad \varphi_\delta = 2 - 20\ \text{mV}$$

FIG. 3. Double logarithmic plot of the delay factor, W, against $\varkappa a/G$ for various values of G. $G \simeq \varepsilon\phi_\delta^2 a/2k_BT$; $A = 10^{-20}$ J. The dotted curve corresponds to the average of the experimental values of Ottewill and Shaw (24).

and Shaw (24) on the average follow one line (dotted in Fig. 3) although they worked with particle diameters ranging from 120 to 846 nm and values of G (assuming $\phi_\delta = \zeta$) varying from about 2 to over 100. Further refinements by Roebersen and Wiersema (38) taking the initial non-steady state during coagulation into account did not help to explain the discrepancy.

The whole matter deserves further effort, both from the theoretical and from the experimental side. Experimentally measuring with higher and more varied surface potentials than those used by Ottewill and Shaw, varying the charge number of the coagulating electrolyte, and possibly using sols other than polystyrene latex (e.g., gold-sols) would be worthwhile. Theoretically the influence of variations of the surface potential and/or surface charge with electrolyte concentration, and slow rates of charge adjustment should be reconsidered.

III. 3. Rate Effects in a Single Collision

In calculating the interaction between colloidal particles it is usually assumed that complete equilibrium exists between the particle surface and the surrounding liquid (dispersion medium) at any separation between the particles. This, however, is not necessarily correct, since the adjustment of equilibrium takes time and this time may be longer than the time involved in a collision or even than the coagulation time.

Let us therefore examine the rates of the separate processes involved in coagulation and start with electrocratic systems.

The time needed for *adding a coagulant* and mixing depends on the technique used and may vary from a millisecond (turbulent mixing in a small chamber as used in the study of fast reactions) to a few seconds or more (conventional mixing techniques).

The time needed for adjusting the structure of the double layer (*relaxation time of the double layer*) is equal to the average time needed for displacement of ions across the double layer

$$t_{relax} \sim \frac{(1/\kappa)^2}{2D_i} \simeq \frac{5 \times 10^{-11} \text{ sec}}{c/\text{mole liter}^{-1}}$$

$$\sim 10^{-8} \text{ sec}, \quad [20]$$

where D_i is the ionic diffusion coefficient. The numerical values given apply to aqueous solutions at room temperature and a concentration of about 5 mM.

If the double-layer equilibrium involves adsorption of potential determining ions, the surface potential tends to stay constant and after a change in composition of the solution or in a Brownian collision the *surface charge has to be adjusted*.

Surface charge densities, σ, are of the order of 1 μC/cm^2 and the rate of charge adjustment is determined by the exchange current density, i_0.

For exchange current densities, 1 A cm^{-2} is a high value and 10^{-10} A cm^{-2} a low one. The time for charge adjustment may therefore vary from

$$t_{charge} \sim \frac{\sigma}{i_0} = \frac{10^{-6} \text{ C cm}^{-2}}{(10^{-10} - 1)\text{A cm}^{-2}}$$

$$= 10^{-6} \text{ to } 10^4 \text{ sec.}$$

Finally the *time of a Brownian collision* is the average time needed for a particle to travel through the thickness of the double layer. This time is given by

$$t_{Brown} = ((1/\kappa)^2/2D_p) = 3\pi\eta a/\kappa^2 k_B T, \quad [21]$$

where D_p is the diffusion coefficient of the particles.

For a particle with a radius of 1000 Å and $1/\kappa$ varying from 1 to 10 nm (i.e., electrolyte concentrations from 10^{-1} to 10^{-3} M) in water at room temperature

$$t_{Brown} \simeq 10^{-5}\text{–}10^{-7} \text{ sec.}$$

Coagulation times may vary widely depending on particle size and concentrations, but in order to be studied in the laboratory they have to be considerably longer than the mixing time, say 0.1 sec and up.

Comparing the different time scales we may conclude that double-layer adjustment is fast even when compared to the short time of the Brownian encounter, but the charge adjustment will often be too slow to follow the single encounter. It may be even too slow to adjust the charge within the coagulation time. Consequently, in calculating the interaction energy of two particles the condition of constant charge is probably a better approximation than that of constant potential and intermediate cases may be met in practice.

For the interaction of particles with *adsorbed long chains* similar considerations apply (11). Relaxation times involved in the adjustment of the conformation of an adsorbed chain or loop can be estimated from oscillatory flow measurements and have been found by Thurston *et al.* (39) to be 2×10^{-4} sec for polystyrene of $M = 50\,000$ dissolved in a liquid with a viscosity of 2 P. The time of adjustment of the loop size distribution (i.e., of the adsorption) is at least several minutes (40). The time involved in a Brownian encounter may now be considered as the time needed for a particle to travel over a distance equal to the thickness of the adsorbed layer. Taking the layer thickness as 6 nm, the particle radius as 100 nm, and the viscosity again as 2 P, we have

$$t_{\text{Brown}} = \frac{(6 \text{ nm})^2}{2D_p} = \frac{(6 \text{ nm})^2 \cdot 3\pi\eta a}{k_B T}$$

$$\approx 2 \times 10^{-3} \text{ sec.} \quad [22]$$

We conclude that the conformation adjustment can follow the collision process, but the rates of adsorption and loop size redistribution are too slow.

IV. REPEPTIZATION

IV. 1. Facts and Simple Theory

It is obviously both of practical and theoretical importance to know whether a flocculated system can be made to redisperse (repeptize) by changing the composition of the medium and with little or no stirring or grinding.

In the case of stabilization by adsorbed polymers flocculation can be caused by adding a nonsolvent for the polymers or by changing the temperature. Repeptization after return to the original conditions is the expected behavior unless desorption or recrystallization of the particles occur.

With electrocratic sols redispersion after flocculation is not expected. As a matter of fact other names for these sols are "irreversible" or "irresoluble." From the most simplistic point of view redispersion is impossible since at $H = 0$ (contact between particles) the van der Waals energy is minus infinity. But even if we realize that the van der Waals minimum is not infinitely deep, Fig. 4 shows a more essential reason for the lack of redispersion. If the potential energy diagram for the approach of two particles shows the usual maximum and two minima, then, after lowering the maximum by the addition of salt, flocculation occurs in the left-hand side, deep, so-called primary minimum. Then after eliminating the flocculating electrolyte the original energy diagram is restored, but for repeptization the particles still have to overcome the energy barrier, which, if high enough to prevent coagulation,

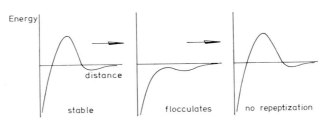

FIG. 4. Potential energy diagrams for a pair of particles at short distances, showing that after flocculation restoration of the original diagram does not lead to redispersion.

is certainly high enough to prevent passage from the other side.

Nevertheless repeptization of electrocratic systems is quite frequent and when it occurs it usually requires very little mechanical action, especially if only a short time is allowed to pass between coagulation and repeptization. We mention clays; ferric and many other oxides and hydroxides for which repeptization is often a step in the normal way of preparing a sol; HgS, which after precipitation and washing finally "runs through the filter." AgI can be repeptized and the Carey Lea silver sol is prepared and cleaned by repeptization. It looks as if repeptization is the rule rather than the exception, at least with monovalent coagulating electrolytes.

Repeptization can be demonstrated and measured quantitatively with a simple technique proposed by Frens (41). Coagulation is carried out in the sample cell of a spectrophotometer. After a preselected degree of coagulation (preselected extinction) is reached the sample is diluted, e.g., five times, and the resulting extinction is compared to one-fifth of the extinction of the original sol before and

TABLE II

Coagulation and Repeptization of a Negative AgI-sol with Bivalent Ions[a]

Coagulate with	Dilute with	Result
115 mM KNO$_3$	0.4 mM Ba(NO$_3$)$_2$	Repeptization
1.67 mM Ba(NO$_3$)$_2$	30 mM KNO$_3$	No repeptization

[a] Final concentrations in both experiments: 24 mM KNO$_3$ and 0.32 mM Ba(NO$_3$)$_2$.

after coagulation. If the extinction corresponds to that of the sol before coagulation repeptization is complete; if it corresponds to the extinction of the sol after coagulation repeptization is absent. Figure 5 illustrates this type of measurement.

It was found that after coagulating a negative AgI-sol with KNO$_3$ repeptization was nearly complete after dilution with about two-thirds (or less) of the coagulation concentration. Repeptization after coagulation with bivalent cations did not occur, unless the coagulating ions were first exchanged with a large excess of monovalent ones. Table II shows that a sol coagulated with monovalent ions repeptizes after dilution with a low concentration of bivalent ions, but if the bivalent ions are used for coagulation, dilution with a low concentration of monovalent ions does not result in repeptization, although the final concentrations are the same in the two experiments.

IV. 2. Theoretical Considerations

For understanding repeptization both thermodynamic and kinetic aspects are important. In the first place spontaneous repeptization can only occur if the free energy decreases in the process. Moreover the activation energy should be low or absent. Figure 6 shows this in energy vs distance diagrams. In the middle sketch curves for two situations are given for which repeptization does not occur, the lower curve because it would require an increase in free energy, the upper one, although it would lead to an energy decrease, because the activation energy is too high. Then the question presents itself: How is it possible to obtain an energy curve such as that in the r.h.s. diagram

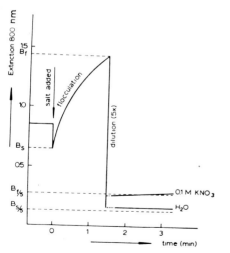

Fig. 5. Extinction, B, versus time for an aqueous AgI-sol, coagulated with KNO$_3$ and diluted with either 0.1 M KNO$_3$ (no repeptization) or with water (nearly complete repeptization).

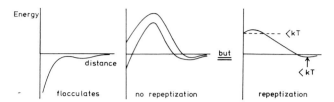

FIG. 6. Energy vs distance diagrams comparable to Fig. 4. Only the r.h.s. diagram corresponds to repeptization.

of Fig. 6? On the basis of van der Waals attraction and electrostatic repulsions this is only possible if the particles in the flocculated state remain separated by a finite distance that we will call 2δ, as if each of the particles is covered with a layer of solvent, that can not be pushed away, at least not in the time between flocculation and redispersion. Moreover, a high surface potential, ϕ_δ, and a low Hamaker constant, A favor redispersion. Here kinetic arguments mentioned in the previous section may be invoked. After adding the coagulating electrolyte, the double layer is compressed, the surface potential is lowered (at constant charge), and the surface charge will increase only slowly to restore the original surface potential, ϕ_0. Even after charge adjustment, the potential at the Stern-plane, ϕ_δ, will still remain below its value in the original solution before the addition of electrolyte.

After dilution of the electrolyte in the co-agulated sol, the double layer will expand and the surface potential and the Stern potential will shoot up, reaching values above the final equilibrium values which they will subsequently reach only slowly after a loss of surface charge.

It is therefore a consequence of the whole conception that flocculation occurs at low ϕ_δ and repeptization at high ones. Frens and Overbeek (18) made a number of calculations in this area and found that very modest values of δ (e.g., 0.2 or 0.3 nm), when combined with quite reasonable values of A and of ϕ_δ are sufficient to explain both coagulation and re-dispersion at realistic concentrations. Figures 7 and 8 show two energy diagrams, one for flocculation (depth of minimum about 80 kT) and one for repeptization (energy decrease about 135 kT, energy barrier about 1 or 2 kT).

The curious influence of the order of addition of monovalent and divalent counterions shown in Table II can be explained in the following way. The situation after flocculation either with KNO_3 or with $Ba(NO_3)_2$ corresponds to the l.h.s. diagram of Fig. 6. Immediately after diluting the KNO_3 flocculate with $Ba(NO_3)_2$ only a small fraction of the $Ba(NO_3)_2$ is close to the double layers and exchanges against KNO_3, but the double layer is still mainly composed of K^+ ions. The concentration is such that an energy diagram of the type

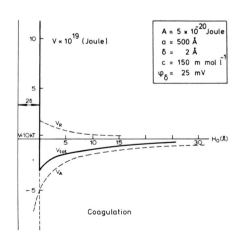

FIG. 7. An energy diagram leading to coagulation. V_A = Van der Waals attraction, Eq. [1] or [4]; V_R = electrostatic repulsion at constant charge; δ is the thickness of the Stern-layer; 1–1 electrolyte; $V_{tot} = V_A + V_R$.

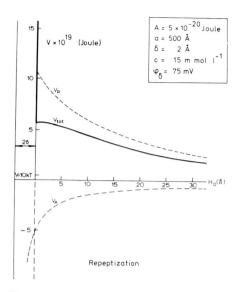

FIG. 8. An energy diagram for repeptization. Note the higher potential (75 mV instead of 25 mV) and lower electrolyte concentration (15 mM instead of 150 mM) as compared with Fig. 7.

in the r.h.s. of Fig. 6 results, repeptization follows and only after repeptization further exchange of Ba^{2+} against K^+ occurs which modifies the energy diagram in the direction of the middle curves of Fig. 6, allowing the sol to remain stable. If, however, the $Ba(NO_3)_2$ flocculate is diluted with KNO_3, Ba^{2+} in the double layer is only for a small part exchanged against K^+ and although the double layer expands and the surface potential goes up, the resulting energy diagram is of the middle type of Fig. 6. So there is no repeptization, although ion exchange of Ba^{2+} against K^+ continues and the final double-layer structure is the same for both experiments.

It should be obvious, and further calculations confirm this, that the shape of the energy curves is very sensitive to the values of δ, ϕ_δ, and A. It would therefore seem to be profitable to investigate repeptization in more detail, especially with the now available isodisperse systems, and determine exactly the conditions where repeptization will just take place. This should be a good approach to obtaining more quantitative information on the distance of closest approach 2δ, the Stern potential ϕ_δ, and the Hamaker constant, A.

V. EMULSIONS

With emulsions one has to distinguish coagulation, usually followed by rapid creaming, and coalescence. Emulsions that are only coagulated or that have creamed are easily redispersed with a little shear, because the individual droplets are fairly large (except in microemulsions). Since the droplets are large the whole energy diagram is on a large scale and the secondary minimum may be deep enough to be the cause of coagulation.

Coalescence, that is, the flowing together of drops to form one large coherent mass, is more difficult to understand, especially since we may have to accept the existence of a tightly adhering layer of the dispersion medium around the droplets, just as such a layer was essential for the understanding of repeptization in the case of solid particles.

The process of coalescence is formally similar to the breaking of foam lamellae for which Vrij (42) and Sheludko (43) have developed a theory, based on the spontaneous growth of long wave fluctuations in the surface. Also with foams, however, this mechanism may either lead to the breaking of the lamellae or to the formation of ultrathin black soap films. It is not yet clear whether this same approach will help in the understanding of emulsion coalescence, but it is worth a try. Coalescence is certainly one of the still open problems in colloid science.

VI. NEW APPLICATIONS OF COLLOID STABILITY AND PROBLEMS WAITING FOR A SOLUTION

The number of applications of colloid stability is constantly growing but on the whole this is more in the direction of extending fields of application already known than in completely new applications. It is true that a few fields, such as pharmaceutical and agricultural applications of suspensions and emulsions to the distribution and regulation of the rate of uptake of active substances, are growing rapidly. Applications to biological and to environmental problems, e.g., the dispersion of oil spills by emulsification, get more attention.

Nonaqueous dispersion media and microemulsions, which, by the way, do not fit well into the present theoretical conceptions, are attracting many investigators.

There are of course still a number of problems that seem to be more or less ready for a solution but have not yet been solved. I want to mention—

    a. understanding the influence or lack of influence of particle size on slow coagulation,

    b. making more use of the rate of coagulation and of the conditions for redispersion to obtain a better quantitative knowledge on van der Waals forces and double-layer structure,

    c. understanding the conditions for coalescence of emulsions,

    d. obtaining better quantitative understanding (and prediction!) of the anchoring of polymers at interfaces,

    e. obtaining better quantitative data on the influence of chainlength and solubility (α) of steric stabilizers,

    f. better understanding of the density of flocs and being able to control it, since easy filtration requires open flocs but for transport and storage a high density is preferable.

Summarizing, we may state that the main recent developments in the understanding of colloid stability are the theoretical explanation of protective action and sensitization (steric repulsion by long chains) and a much greater awareness of the rate effects occurring in stable, coagulating, and redispersing colloids.

REFERENCES

1. Deryagin, B. V., *Trans. Faraday Soc.* 36, 203, 730 (1940); Deryagin, B. V., and Landau, L., *Acta Physicochim. URSS* 14, 633 (1941); *J. Expt.*

Theor. Phys. (in Russian) 11, 802 (1941); 15, 662 (1945).
2. Verwey, E. J. W., *Chem. Weekbl.* 39, 563 (1942) (in Dutch); *Philips Res. Rep.* 1, 33 (1945); Verwey, E. J. W., and Overbeek, J. Th. G., *Trans. Faraday Soc.* 42B, 117 (1946); Verwey, E. J. W., and Overbeek, J. Th. G., "Theory of the Stability of Lyophobic Colloids," Elsevier, Amsterdam, 1948.
3. Schulze, H., *J. Prakt. Chem.* (2) 25, 431 (1882); 27, 320 (1883).
4. Hardy, W. B., *Proc. Roy. Soc. London* 66, 110 (1900); *Z. Phys. Chem.* 33, 385 (1900).
5. Mackor, E. L., *J. Colloid Sci.* 6, 492 (1951); Mackor, E. L., and van der Waals, J. H., *J. Colloid Sci.* 7, 535 (1952).
6. Fischer, E. W., *Kolloid Z.* 160, 120 (1958).
7. Clayfield, E. J., and Lumb, E. C., *J. Colloid Interface Sci.* 22, 285 (1966).
8. Meier, D. J., *J. Phys. Chem.* 71, 1861 (1967).
9. Ottewill, R. H., and Walker, T., *Kolloid Z. Z. Polym.* 227, 108 (1968).
10. Napper, D. H., *Trans. Faraday Soc.* 64, 1701 (1968); *J. Colloid Interface Sci.* 32, 106 (1970); Evans, R., and Napper, D. H., *Kolloid Z. Z. Polym.* 251, 329, 409 (1973).
11. Hesselink, F. Th., *J. Phys. Chem.* 73, 3488 (1969); 75, 65 (1971); Hesselink, F. Th., Vrij, A., and Overbeek, J. Th. G., *J. Phys. Chem.* 75, 2094 (1971).
12. Matijević, E., *in* "Twenty Years of Colloid and Surface Chemistry, The Kendall Award Adresses" (K. J. Mysels, C. M. Samour, and J. H. Hollister, Eds.), p. 283. American Chemical Society, Washington D. C., 1973.
13. Vanderhoff, J. W., van den Hul, H. J., Tausk, R. J. M., and Overbeek, J. Th. G., *in* "Clean Surfaces: Their Preparation and Characterization for Interfacial Studies" (G. Goldfinger, Ed.), p. 15. Dekker, New York, 1970.
14. Hamaker, H. C., *Rec. Trav. Chim.* 55, 1015 (1936); 56, 3, 727 (1937); *Physica* 4, 1058 (1937).
15. Casimir, H. B. G., and Polder, D., *Phys. Rev.* 73, 360 (1948); *Nature* 158, 787 (1946); Casimir, H. B. G., *Proc. Kon. Ned. Akad. Wetensch.* 51, 793 (1948).
16. Lifshitz, E. M., *Dokl. Akad. Nauk. SSSR* 97, 643 (1954); *Sov. Phys. JETP* 2, 73 (1956); Dzyaloshinskii, I. E., Lifshitz, E. M., and Pitaevskii, L. P., *Advan. Phys.* 10, 165 (1961).
17. Verwey, E. J. W., and Overbeek, J. Th. G., "Theory of the Stability of Lyophobic Colloids," p. 137 ff., p. 152. Elsevier, Amsterdam, 1948.
18. Frens, G., and Overbeek, J. Th. G., *J. Colloid Interface Sci.* 38, 376 (1972).
19. Flory, P. J., "Principles of Polymer Chemistry," p. 600. Cornell Univ. Press, Ithaca, N. Y., 1953.

20. SMOLUCHOWSKI, M., *Phys. Z.* **17**, 557, 585 (1916); *Z. Phys. Chem.* **92**, 129 (1917).

21. HALPERN, J., *J. Chem. Educ.* **45**, 373 (1968).

22. FUCHS, N., *Z. Phys.* **89**, 736 (1934).

23. REERINK, H., "De Uitvloksnelheid als Criterium voor de Stabiliteit van Zilverjodidesolen," Thesis, Utrecht, The Netherlands, 1952; REERINK, H., AND OVERBEEK, J. TH. G., *Discuss. Faraday Soc.* **18**, 74 (1954).

24. OTTEWILL, R. H., AND SHAW, J. N., *Discuss. Faraday Soc.* **42**, 154 (1966).

25. DERYAGIN, B. V., *Discuss. Faraday Soc.* **42**, 317 (1966).

26. DERYAGIN, B. V., AND MULLER, V. M., *Dokl. Akad. Nauk. SSSR* (Engl. transl.) **176**, 738 (1967).

27. SPIELMAN, L. A., *J. Colloid Interface Sci.* **33**, 562 (1970).

28. HONIG, E. P., ROEBERSEN, G. J., AND WIERSEMA, P. H., *J. Colloid Interface Sci.* **36**, 97 (1971).

29. DEUTCH, J. M., AND FELDERHOF, B. U., *J. Chem. Phys.* **59**, 1669 (1973).

30. BRENNER, H., *Chem. Eng. Sci.* **16**, 242 (1961).

31. LICHTENBELT, J. W. TH., RAS, H. J. M. C., AND WIERSEMA, P. H., *J. Colloid Interface Sci.* **46**, 522 (1974); LICHTENBELT, J. W. TH., PATHMA-MANOHARAN, C., AND WIERSEMA, P. H., *J. Colloid Interface Sci.* **49**, 281 (1974).

32. DEŽELIC, G., AND KRATOHVIL, J. P., *J. Colloid Sci.* **16**, 561 (1961).

33. PHILLIPS, D. T., WYATT, P. J., AND BERKMAN, R. M., *J. Colloid Interface Sci.* **34**, 159 (1970).

34. VAN DEN HUL, H. J., AND VANDERHOFF, J. W., *in* "Polymer Colloids" (R. M. Fitch, Ed.), p. 1. Plenum, New York, 1971.

35. DAVIDSON, J. A., AND COLLINS, E. A., *J. Colloid Interface Sci.* **40**, 437 (1972).

36. LEE, S. P., TSCHARNUTER, W., AND CHU, B., *J. Polym. Sci. (Phys.)* **10**, 2453 (1972).

37. COOKE, D. D., AND KERKER, M., *J. Colloid Interface Sci.* **42**, 150 (1973).

38. ROEBERSEN, G. J., AND WIERSEMA, P. H., *J. Colloid Interface Sci.* **49**, 98 (1974).

39. THURSTON, G. B., AND SCHRAG, J. L., *J. Polym. Sci., Part A-2* **6**, 1331 (1968); THURSTON, G. B., AND MORRISON, J. D., *Polymer* **10**, 421 (1969).

40. KIPLING, J. J., "Adsorption from Solutions of Non Electrolytes," pp. 135, 230. Academic Press, London, 1965; STROMBERG, R. R., *in* "Treatise on Adhesion and Adhesives" (R. L. Patrick, Ed.), Vol. I, p. 101. Dekker, New York, 1967.

41. FRENS, G., AND OVERBEEK, J. TH. G., *J. Colloid Interface Sci.* **36**, 286 (1971).

42. VRIJ, A., *Discuss. Faraday Soc.* **42**, 23 (1966); VRIJ, A., AND OVERBEEK, J. TH. G., *J. Amer. Chem. Soc.* **90**, 3074 (1968).

43. SCHELUDKO, A., *Proc. Kon. Ned. Akad. Wetensch. B* **65**, 87 (1962).

The Conformational States of Macromolecules Adsorbed at Solid–Liquid Interfaces

FREDERICK R. EIRICH

Department of Chemistry, Polytechnic Institute of New York, Brooklyn, New York 11201

Received October 25, 1976; accepted November 8, 1976

The adsorption isotherms and adsorbate layer thicknesses of about 20 different, dilute, polymeric systems were determined. With very few exceptions, all were found to form monolayers, which in the majority, consisted of individual macromolecules. This conclusion stems from the observation that the layer thicknesses vary with the square root of the molecular weight and, if the supernatant changes, change in the same way as the intrinsic viscosity. It follows further that the macromolecules are held, by relatively few attachments, as solvent-pervaded coils of dimensions similar to those of coils that are free in solution. This is true for organic and aqueous solutions, and polar and nonpolar adsorbants and adsorbates. Polyelectrolytes of high group density may be pulled entirely onto the surface by salt formations; stiff or oligomeric molecules with reactive end groups may become anchored at one end only, or lie flat on the surfaces. Peptides and proteins exhibit the same pattern of behavior.

INTRODUCTION AND METHODOLOGY

The adsorption of macromolecules plays a part in a wide variety of natural, inorganic and biological, and technological processes (see Table I), and is also of major theoretical interest. Limiting the enquiry at first to adsorption from dilute solutions, one will want to understand the manner, the energetics, and the kinetics of macromolecular attachment, the density of adsorption, the nature of any exchange reactions, and the dependence of these features on the structures of surface, solution, and macromolecules, and on such variables as concentration, temperature, pH, electrolytes, and cosolutes. The following is a brief review, mostly of work carried out at the Polytechnic Institute.

Among the many methodologies (1, 2) e.g.), to establish the amounts adsorbed per area as a function of concentration at constant temperature, i.e., the adsorption isotherms, the simplest one is the determination of the moles of solute (the adsorbate) missing from the solution, after mixing the latter with the solid (the adsorbent) and allowing adsorption equilibrium to be reached. The thicknesses, δ, of the adsorbed layers are measured either optically by ellipsometry, or hydrodynamically by finding the changes in particle or capillary dimensions after adsorption; or, under some conditions, spectroscopically which methods may also yield information on the chemistry of the site–adsorbate interaction.

Unfortunately, no experimental method as yet provides direct information on the segmental density distribution within the adsorbed layer, although variations of the isotherms and of the adsorbed layer thicknesses with conditions may give some clues (1, 2). These are best derived from observing characteristic features of the adsorption isotherms, e.g., the initial slope which reflects the distribution tendency of the solute between surface and solution and thus has the nature of an "affinity" (A); and the amounts held

447

TABLE I

Adsorption of (or onto) polymers plays a key role in:	
Adhesion	Corrosion
Coatings	Aging of composites
Paints	Crack resistance
Lamination	Drag reduction
Reinforcement	Textile finishing
Emulsions	Flocculants
Suspensions	Chromotography
Detergent action	Soil structure
Flotation	Films and membranes
Drilling and cutting	Biological agglutination
Solid lubricants	Immune reactions
Crystallization	Cell recognition
Precipitation	Drug direction
Agglomeration	Genetic reproduction

TABLE II

General Energetics of Adsorption
(Opposite Signs for Desorption) on Powders

Process	ΔH	Δs	Symbol[a]
Particle dispersion	$+$	$+$	S–S
Primary particle (surface) solvation (desolvation on adsorption)	$\mp$	$\mp$	S–L
Adsorption proper	$-$	$-$	S–Ad
Adsorbate desolvation	$\pm$	$+$	Ad–L
Interadsorbate contact (on interface)	$\pm$	$-$	Ad–Ad
Solvent cohesion	$-$	$+$	L–L
Extension or contraction of molecular chains	$\pm$	$\pm$	Conformational changes

[a] S, solid surface; L, (liquid) solvent; Ad, adsorbate.

per unit surface area extrapolated to high concentrations, the so-called capacity (C), which defines the density of the immediate surface layer and thus gives information on interadsorbate packing (repulsive), or condensation (attractive), phenomena (see Fig. 1, and Tables II and III). $A/C = k$ is the

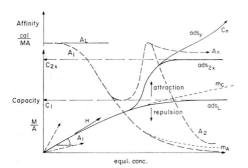

FIG. 1. Schematic representation of three isotherms (full lines): ads_L, Langmuir monolayer; ads_{2x}, Double layer; ads_F, Freundlich, BET, or multilayer adsorption. The plateau values for ads_L and ads_{2x}, i.e., the capacities C_1 and C_2, are shown as extrapolations to the ordinate. The slopes of the isotherms, the affinities, A, are shown as dashed curves: A_L for the Langmuir, $A_{1,2}$ real mono and double layer; A_n for a multilayer isotherm. The initial value for A_1, divided by C_1, gives the affinity constant, k, in Langmuir terms, the tangents at $c \to 0$ represent Henry's law; m_c and m_a are capacity and affinity for the continuing adsorption of a monolayer of adsorbate with compressible molecules.

"affinity constant" which can, at least in principle, be derived from thermodynamics. Together with the thicknesses, δ, at high surface coverage, the C's allow calculation of the density of the interphase. A, C, and δ are then studied as functions of molecular weight (MW), of solvent (pH, cosolutes), of temperature, and of the nature of the surface. The rate of shear over moderate ranges (10–500 sec⁻¹) does not seem to matter.

The differences in the rates of adsorption and desorption (hysteresis) sometimes make it difficult to establish true equilibrium, but solvent changes or displacement methods are usually capable of distinguishing between true physical adsorption, chemisorption, and vari-

TABLE III

Isotherms[a]

Freundlich:	$m/a = Ac^n$
Langmuir:	$\theta/(1 - \theta) = Ac$
	$m/a = Ac/(1 + kc)$
or	$c/(m/a) = c/C + 1/A; k = A/C$
Fowler:	$c/(m/a) = c/C + (1/A) \exp[(-\chi - \theta w)/RT]$
Frisch, Simha, Eirich:	
(FSE)	$\{[\theta e^{k'\theta}/(1 - \theta)][p(1 - \theta)/\theta e^{-1(1-\theta)}]^p\}^{\bar{\nu}} = Kc$
For $p \to 0$:	$[\theta e^{k'\theta}/(1 - \theta)]^{\bar{\nu}} = Kc$

[a] m/a, moles/surface area; c, equilibrium concentration; C, m/a for $c \to \infty$; k, affinity constant; χ, adsorbate-adsorbent energy; w, interadsorbate energy at surface; K, k', constants; l, degree of polymerization; p, probability factor; A, initial slope of isotherm; θ, fraction of surface area covered; $\bar{\nu} = pl$, average number of segments attached at one contact.

ous intermediate forms. A comparison of the temperature coefficients of (especially macromolecular) solubility and of adsorption allow some conclusion on the nature of the accompanying ad- or desorption processes of the solvent on surface and solute. Table II, and again Fig. 1, show the energetics to be expected from the mutual interaction of surface (S), solvent (L), and adsorbate (Ad). Thus, although rising temperatures reduce all individual adsorption steps, if solvent desorption with temperature from surface or from the solute happens to be very marked, the attachment (adsorption) of the solute may, in apparent anomaly, increase. This has frequently been observed (1, 3) to be the case for macromolecules.

FEATURES OF MACROMOLECULAR ADSORPTION

Early established features of macromolecular adsorption (1–7) to be accounted for in any attempts at theoretical understanding were: indications of rather weak dependencies on molecular weight on smooth surfaces, the existence of larger amounts of adsorption than flat layers could be expected to contain, very large affinities, extremely large adsorption hysteresis, and isotherms which rise to

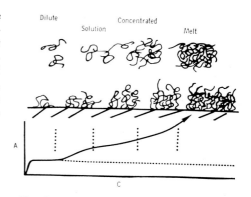

Fig. 3. Conformation and coil interpenetration of polymer chains at the interface as a function of polymer solution concentrations, or from melt (basic conformation is that of (d) of Fig. 2); corresponding adsorption isotherms are shown below as Adsorption vs. Concentration plot (14).

apparent plateaus by a single step. These have been interpreted to signify polymeric adsorption in the form of monolayers, sometimes lying flat on the surface, sometimes extending as loops into the solvent (3). No decision could be made in the absence of data on layer thicknesses, since a flat deposition, and sinusoidal, or random-coil, protrusion into the solvent may give plateau heights approximately independent of molecular weight. Figure 2 depicts some projected forms of chain deposition (3, 8–14), while Fig. 3 shows schematically for case (d), the increase in monolayer thickness and density, as well as the corresponding adsorption isotherms as a function of increasing coil compression and/or interpenetration at higher and higher polymer concentrations, and potential multilayer formation (3, 7, 14).

On the basis of these observations and conjectures and existing knowledge of macromolecular chain conformations, considering further their restrictions when held to solid surfaces, and the accompanying desorption of solvent, and applying the principles of statistical thermodynamics, the first comprehensive theory of adsorption of flexible chains from dilute solution (15, 16) not only covered the

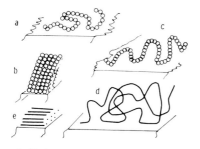

Fig. 2. Various models of macromolecular chain deposition at a solid–solvent interface: (a) Chain, lying totally on interface; (b) layer of chains (packed) standing on end; (c) partially adsorbed chain, extending about equally sized loops into the solution; (d) chain adsorbed by short sections, with loops of random length extending into the solution, loop length α MW$^{0.5}$; (e) straight and rigid chains (packed) lying flat on interface (14).

known observations, but predicted such later essential findings (12, 13, 17–19) as the dependence of the thickness of the adsorbed layers on the square root of the molecular weight, i.e., adsorption in the form of monolayers of only partly attached and moderately deformed coils, as shown in Fig. 2d. Over the years, this theory has been much debated, has been improved, alternative theories developed by many other authors (20–26), special features and limiting cases analyzed; much experimental work was stimulated (7–10, 27–29) which added a great deal of information but also introduced new complications for our understanding.

The driving factor of adsorption, as always, is the change in free energy of the system, made up of the changes in enthalpies and entropies of every one of the three components: surface, solvent, and adsorbate. During the deposition of the adsorbate, these six components together must add up to a negative value, otherwise adsorption will not occur. As the adsorption progresses, interadsorbate forces soon lead to either one, or the other, of two opposing courses: if the adsorbate particles on the surface attract one another, their condensation provides an additional negative free-energy term (mitigated by desolvation) and the adsorption process will lead by a two-dimensional phase transition to a dense surface layer (condensation), or to the formation of multilayers and eventually to the nucleation of epitactic growth and precipitation of the adsorbate. Alternatively, if the adsorbate molecules on the surface (in the interphase space of the adsorbed layer) are rigid and repel one another, then the declining negative free energy per increment of adsorption leads to saturated monolayers, seen as plateaus of the isotherms (Figs. 1 and 3). Freundlich and Langmuir's isotherms, respectively, are the idealized formulations of the former and the latter alternatives, with the extension of Henry's law (the situation for low surface coverage where rate and extent of adsorption are not yet affected by

adsorbate interference) as the dividing line (Fig. 1). In this context, the BET isotherm may be seen to be a more sophisticated Freundlich type, whereas Fowler's equation is a modification of Langmuir's assumptions, adding a term for intra-interphase attraction (30, 31) (Table III). Neither of these equations takes the role of the solvent into account explicitly, even though, certainly at higher fractions of surface coverage, desolvation of interphase and adsorbate may play a major part.

Turning to polymer coils, one has to account for substantial changes in conformational energy, since a transition from a three-dimensional to an at least partially two-dimensionally constrained state causes entropy losses besides the drop in adsorption enthalpy. Unless the conformational changes becomes severe, however, the will modify, but will not decide, the on–off character of adsorption. The theoreticians have become preoccupied with conformational changes because of the latters' relation to fold-coil–helix transitions, and because of the need to gain insight into the structure of the adsorbed layer, e.g., the ratio of segments held on the surface vs those in loops, and the lengths of the runs (15, 16, 20–26). Thus the role of the surface chain conformations occupies much of the theoretical literature on adsorption (1, 2). In any case, since decreasing entropy due to increasing conformational constraints on the molecular coils on adsorption will increasingly diminish the free energy of adsorption per segment, one can see that this circumstance alone will favor adsorption of individual coils by only a small number of segments.

All these complications make it obvious why it is a most formidable task to formulate a generally acceptable model of adsorption and to cast it into a mathematical form. Moreover, the general features of polymeric adsorption will be substantially modified for individual polymers by their specific properties based on monomer, or side-group size, polarity or charge, by tacticity, internal rota-

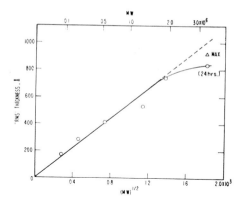

FIG. 4. Root-mean-square thickness in the plateau region of polystyrene, adsorbed on chrome ferrotype plate as a function of the square root of the molecular weight, in cyclohexane, 36.4°C (17).

tional potential, molecular weight distribution, etc., which determine their ability to place segments into various arrays on the surface. Major deviations from the idealized process of adsorption of individual macromolecules must be expected from the intermolecular interactions at high degrees of surface coverage, when the energies of polymer–polymer interaction become comparable to those of polymer–surface interaction, adsorbed molecules block the access of further molecules and, eventually, aggregates of polymer coils become attached to the surface. The theoretical predictions so far developed are quite precarious, not only for these more complicated, realistic, cases; they disagree even for the ideal case of adsorption from dilute solutions and low surface coverage (1, 2). Thus, while obviously leaning toward the use of FSE-type isotherms, I will refer our results not so much to individual models but will describe our experimental exploration of specific macromolecular systems, and then extrapolate toward some of the more general features of macromolecular adsorption, such as the nature of the isotherms, the factors which influence affinity, capacity, and layer thickness, the effects of solvents, cosolutes, of electrolytical charges, pH, the nature of the interface, and the relation between molecular conformation, interphase and density, and between δ and the radius of gyration, R_G, or the intrinsic viscosity, $[\eta]$.

RESULTS AND DISCUSSION

The principal findings relevant to our theoretical understanding, some already men-

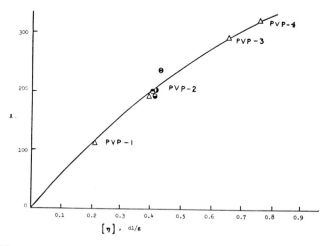

FIG. 5. Thicknesses of adsorbed layers of poly(N-vinyl pyrrolidone) vs intrinsic viscosity, in aqueous salt solutions (19).

451

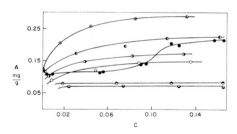

FIG. 6. Equilibrium adsorption of polyvinyl acetate on tin powder as a function of solvent and molecular weight. Milligrams of polymer adsorbed per gram of adsorbent (A) versus equilibrium concentration (C). Molecular weights ranged from the upper to the lowest isotherm, from 900 to 140×10^{-3}; solvents from good to poor, the latter giving higher isotherms; see (3).

tioned, are:

(1) The adsorption of polymer molecules from dilute organic or aqueous nonelectrolytic solutions proceeds most frequently to a plateau of saturation of the order of several condensed monolayers of monomer units, with a low dependence on the molecular weight (1, 3, 6, 11, 12). Also frequently, one

observes thicknesses of 100–1000 Å for the absorbed layer, δ, i.e., of about 2× the radius of gyration of the polymer coils in equilibrium solution (11–13, 17–19, 27), i.e., approximately proportional to $MW^{0.5}$ (Figs. 4, 5). In such cases, then, the polymer covers the surface in the form of, at most moderately distorted, random coils, eventually packed to complete monolayers. This form of adsorption, close to being of the Langmuir type (Fig. 6), indeed permits characterization of the isotherms by an initial slope A, and a plateau (capacity), C. The slopes are found to be much steeper than for small molecules, expressing a very high affinity and/or space requirement. Compared with this, the C values are somewhat depressed and, after a practical halt, may rise again slowly, up to much higher concentrations.

(2) The most telling independent supporting evidence for the formation of monolayers of coils extending into the solution is the parallelism, often even a proportionality with a coefficient near unity, between δ and the intrinsic viscosity $[\eta]$ of a polymer in

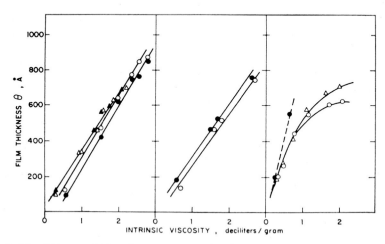

FIG. 7. Film thickness at high surface coverage as a function of intrinsic viscosity. Left: Poly(vinyl acetate) in benzene (circles) at 30°C (open) and 50°C (full); triangles (open and full) in 2-butanone at 30 and 50°C, respectively. Center: Poly(methyl methacrylate) in benzene, points same as on left. Right: Poly(styrene) in cyclohexane at 34.2°C (full points); in benzene at 30°C (open circles), 50°C (triangles).

given solvents, irrespective of whether $[\eta]$ changes with molecular weight, solvent, or temperature (Fig. 7) (11–13). This proves that in these systems the hydrodynamically effective dimensions of molecular coils do not change greatly on adsorption. For very high molecular weights, δ increases (except in theta solvents) more slowly than $[\eta]$, which is probably indicative of a flattening of very large and flexible coils on adsorption (1, 11–13, 17–19).

(3) The levels of the plateaus rise only very slowly with concentration, up to amounts of 5–10 monolayers of segments, beyond which point irregularities due to adsorption of macromolecular aggregates begin to show (2, 3). Over the same range, the adsorbed layer thickness increases only by fractions (1, 11, 14, 17). The most likely interpretation for both observations is that, up to the outer reaches of the loops, the segmental density does not fall off very rapidly with distance, δ, from the surface, that tangential flow through the loops is greatly retarded by viscous resistance, and that the individual adsorbed coils are being laterally compressed at higher concentrations, without greatly affecting the average loop length. A recent theoretical analysis by Silberberg (32) is not incompatible with this interpretation of our data.

(4) The parallelism between δ and $[\eta]$ does not hold when δ grows either much faster, or slower, than α MW$^{0.5}$. The former happens for rather extended, dangling chains of low affinity, or for polyelectrolytes of low charge attraction, the latter for strong affinities, e.g., due to molecular group reactions between adsorbate and adsorbant surface, such as polyethers on polycationic particles, polyacrylamide on strongly hydrogen bonding, or PVP on negative, surfaces and, in particular, for the adsorption of polyelectrolytes on surfaces of opposite net charge. Examples for the latter cases of "binding" by ion pairing are: Poly(sulfonates) and poly(acrylates) on calcium carbonate (12),

poly(acrylic acid) on anatase (33, 34), poly-(ethanolamine methacrylate) or acid gelatins on a negatively charged latex, etc. (35). In these cases, A is usually too high to be measured (see, however, recent work by Ottewill on polyelectrolyte copolymers (36)), while C is rather low because of the repulsion of parallel ion pairs, and δ is of the magnitude of the thickness of the polyelectrolyte chains themselves, i.e., of the order of 10 Å. As a rule, the adsorption falls off sharply when either surface or polyelectrolyte charges decrease by changes in pH or increasing salt concentrations, and also when a weak polyacid or base becomes too strongly charged and thus becomes too soluble for adsorption, or neutralizes the surface with too few molecules to be analytically measurable. Conversely, if a polyelectrolyte becomes deprived of its charges and therefore insoluble, the adsorption will increase sharply, prior to precipitation (37). Thus, A and C are not only large for a strong drop in free energy on combining adsorbent and adsorbate, but also when the adsorbent facilitates the separation of an energetically unfavorable solution.

(5) Interesting observations are made, if a polymeric adsorbate is capable of more than one mechanism of adsorption. If poly-(styrene sulfonate) (PSS) is adsorbed on purified carbon black which contains multivalent cations on its surface, a strong dependence of adsorption on pH, with a maximum near neutrality, is observed (38). We think that in acid dispersion the cations are displaced from the carbon surface by H$^+$ ions, allowing only a minor adsorption of PSS, whereas, when the cations are retained, strong adsorption of PSS ensues. At very alkaline pH's, the OH$^-$ ions compete with the sulfates, and the PSS adsorption drops again to the low values of hydrophobic bonding between the carbon itself and the PSS backbone (Fig. 8). The latter low values are also obtained if PSS is adsorbed nonionically on the same, now nonpolar, carbon after purification which removed oxygen

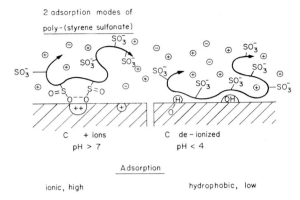

FIG. 8. Schematic representation of dual way of adsorption of poly (styrenesulfonate) on charcoal. Left: ionic bond formations on carbon, holding surface cations. Right: hydrophobic adsorption on acidified or purified carbon.

and/or the cations; moreover, this adsorption of PSS on purified carbons is now pH independent. In agreement with these observations, we find that it is easier to disperse pure than impure carbons by PSS in a wide range of aqueous solutions. Interestingly, we find (33, 34, 38) that synthetic polyanionic electrolytes adsorb extremely weakly on glass, presumably because of weak interactions between silica and vinyl backbones, and very wide spacings of adsorbed charges on a dielectric, even if hydrated, surface (38).

(6) As another, even more specific, instance of combined ionic and general physical adsorption, oligopeptides are bound on acidic clays, e.g., montmorillonite, in their ammonium form as a function of their base strengths by replacing exchangeable cations on the negative sites of the basal faces (39, 40). For monovalent cations, this is a 1:1 exchange for charge and site, i.e., one oligomer each per surface charge, anchored by its terminal NH_3^+ group, but for multivalent ions and their correspondingly fewer sites, the oligomer adsorption is at first $\frac{1}{2}$ for divalent, and $\frac{1}{3}$ for trivalent surface ions. It requires higher solution concentrations for these isotherms to reach, stepwise, the full displacement of the multivalent surface cations, equal to that of the monovalent ions.

We can surmise what the structure of these adsorbed layers might be, even though we have no method to observe directly their thicknesses which should be proportional to the oligopeptide length, were the chains to stand on end. The isotherm capacities can not be used either for estimating thicknesses, since the C values would increase with molecular weight in any case. Our guides are the affinity values: for end-on adsorption, A should vary to a minor degree, but for flat adsorption, i.e., interaction between the whole molecule and the surface, A should increase strongly with molecular weight. This is actually observed (see Fig. 9). Thus, again, there are two types of binding, one specifically ionic, the other of a more diffuse, van der Waals type. For cationic peptides on clays, the latter supplements the ionic process which by itself, by pairing sites, charges, and adsorbate molecules, would limit the adsorption. For increasingly larger polymer molecules, the ion pairing loses in importance vis à vis the ever increasing van der Waals forces per macromolecule, and thus the whole surface can be covered irrespective of the number of charged sites. As a result, the capacity for macromolecules lies by a factor of about 10 above that for mono- or dipeptides (see again Fig. 9) (40). A parallel in-

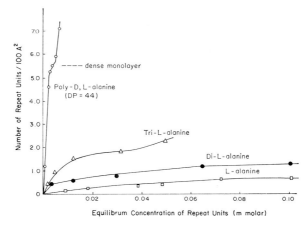

FIG. 9. Adsorption isotherms of polyalanine and oligomers on sodium–montmorillonite (25°C), pH ≈ 3.

crease in adsorption, but accompanied by a more fundamental change in its nature, i.e., a change with molecular weight from flat-lying chains to coils billowing into the solvent, has been observed for synthetic polymers in organic solvents (3, 11, 17).

(7) Changing the emphasis of the investigation from solute–surface interaction to that of solute–solvent effects, we measured the adsorption of some polypeptides, e.g., poly(β-benzyl-L-glutamate) (PBLG) of MW = 250 000, poly(proline) (PP) and poly(hydroxyproline (PHP) each of MW = 10 000, on glass as a function of solvents in which the degree of helicity would differ (41). Our method, measuring changes in capillary diameters, as a result of the adsorption, by determining the changes in flow times and viscosities, was here augmented by measuring optical rotation and circular dichroism to ascertain the conformational states of the peptides in the solutions from which the adsorption isotherms were determined on powder of the same glass as that of the capillaries.

The isotherms were again found to exhibit plateaus, but as a special feature, on occasion, additional steps or rises were observed (Fig. 10). A combination of capacities and our thickness measurements also frequently

reveals the presence of very thin, or of very thick, surface layers. In a number of cases the latter were quite condensed, as shown by the surface areas per peptide molecule, which were the same as for their monomeric units. In other words, when the peptides in their helical form acted as stiff rods, they were

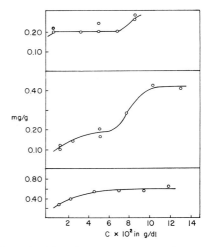

FIG. 10. Adsorption isotherm of poly(γ-benzyl-L-glutamate), and of poly(β-benzyl-L-aspartate) on glass powder at 30°C from helix-supporting solvents CHCl$_3$, CH$_2$Cl·CH$_2$·CH$_2$Cl). Amounts adsorbed vs. equilibrium concentration.

455

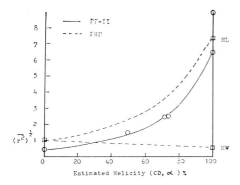

FIG. 11. Schematic trends of thicknesses, δ (in units normalized to $(\overline{r^2})^{\frac{1}{2}}$), and of helix length (HL) and width (HW), with percentage of helicity estimated from optical rotation and circular dichroism.

either lying flat on the surface or standing on end. This situation, which one would also intuitively expect, is in agreement with the theory of Hoeve (10). On the other hand, one finds that in strong H-bond breaking solvents which open the helices into kinked structures, and further to random coils, the adsorbed polypeptides form layers of intermediate thicknesses, commensurate with their dimensions as coils. Figures 10 and 11 give examples of isotherms and of changes in layer thicknesses (41).

In these cases, then, the glass acts as a relatively inert surface and the conformation of the adsorbates and the thickness of the adsorbed layers are dominated by the interaction between solute and solvent. It is interesting that, for our polypeptides here, and similarly in some instances of poly(vinyl-alcohol) adsorption before (42), bi- or multi-layer adsorption develops, apparently on top of a first adsorbed dense layer reinforced by extensive hydrogen bonding and/or by two-dimensional condensation.

(8) A combination of all three factors—surface effects, influence of solvent, and specific polymer structure—was seen when we studied the adsorption of two gelatins on glass as part of a broad investigation into bioadhesion and tissue mineralization (43–48).

One of the gelatins was a single-stranded alkali-treated calfskin gelatin, C-1, of MW = 90 000 and an isoelectric point (IEP) of 4.7. The other was an acid-treated pigskin "parent" gelatin, P-1, with an IEP of 8–9 and MW of 300 000, i.e., with the three strands of the original collagen still hanging together and, as shown by polarimetry, still possessing some residual internal order at the measuring temperature of 30°C (43).

The adsorption of both samples was studied over a pH range of 2.5–11. The isotherms exhibited regular initial rise and saturation ranges, fitting reciprocal Langmuir plots, with amounts adsorbed of the order of 1–3 complete monolayers of amino acid residues for the C-1, and 1–6 for P-1 (43). The thicknesses measured by the capillary method, and with very similar results (36) by ellipsometry, again compare well, over a part of the pH range, with the R_G values calculated from intrinsic viscosities. At their isoelectric points, e.g., the ratio of the thicknesses of the C-1 to the P-1 layer is again close to that of their $MW^{0.5}$. Thus, we see another case of monolayers of somewhat compressed macromolecular coils. The larger capacity values for P-1 are most likely due to its structure, i.e., that the molecules retain a number of multistranded or still helical sections within the coil, per number of surface contacts.

We are, however, also presented with some new features, in particular that δ and molecular dimensions derived from $[\eta]$ may run mutually inverse courses. Further, although gelatin, as an ampholite, has its smallest dimensions around IEP, we find the adsorbed layer thicknesses there to be low, but by no means at their lowest. The A and C values do have their maxima quite close to the IEP (see below). The maxima of δ lie at about one-half of the steep rise in $[\eta]$, where the molecular dimensions expand fastest due to the repulsion of the increasing number of intramolecular charges of equal sign (Figs. 12, 13) (43).

For an explanation, one would expect that the affinities and capacities for ampholite

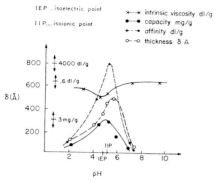

FIG. 12. Adsorption of alkaline-gelatin on glass powder as function of pH, at 30°C. δ at capacity plateaus.

adsorption on a weakly charged surface such as glass should be highest around the IEP, for once because all the various van der Waals, polarization, and electric charge forces are then most concentrated within the contracted molecular cross section, and so interact strongest with the surface. Additional reasons are that at the IEP the ampholytic solubility is lowest and the electrostatic interadsorbate repulsion is at its minimum; these are all factors which combine to cause maximal attraction and packing, and thus capacity. In reality, one finds the maxima of both A and C shifted somewhat from the IEP to the isoionic point (IIP), i.e., C-1 to the right and P-1 to the left, where they carry their greatest total number of charges

and thus presumably interact most strongly with a dielectric substrate.

Complementary to the reasons given above for the position of the maxima, the capacities fall off at either side of the IIP–IEP range. On the acid side this occurs, notwithstanding stronger electrostatic attraction to the glass still carrying negative charges because the electrical inter- and intraadsorbate repulsion demonstrated by the rapid rise in $[\eta]$ diminishes the number of adsorbed molecules per surface area. This is, of course, also true on the alkaline side, but the falling-off in A and C is even stronger, because of the increasingly negative net charge of the gelatine molecules opposite the negative glass.

The δ values reflect the changes in the number and dimensions of the adsorbate molecules, generally running parallel to the capacity curve and, to the right of IEP, also to the $[\eta]$ curve. Two features should be noted: for C-1, the thicknesses on the negative side rise at first while the capacities drop, reflecting the rapid expansion of the adsorbate with pH, and thus its greater extension away from the surface, overcompensating the decrease in density of the adsorbed layer. The same is not true for P-1, where over the same alkaline range the $[\eta]$ values decrease. Both gelatins show a rapid drop of δ on the acid side, parallel to the capacities, but counter to the courses of $[\eta]$. I assume this means that, although the coil dimensions still increase and are likely to reach further

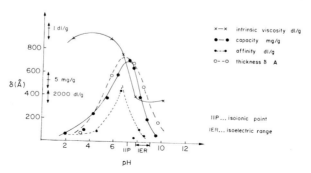

FIG. 13. Adsorption of acid-gelatin on glass powder as function of pH, at 30°C. δ at capacity plateaus.

out into the solution, the adsorbate population dwindles here rapidly so as to create sufficiently thinned-out layers to deepen the hydrodynamic penetration and lower the effective hydrodynamic thickness (43, 48–50).

(9) The measurements of gelatin adsorption were extended to different surfaces, and to the action of added specific ions and electrolytes, with results which will be published elsewhere (43, 48). They were found to be further complicated by the supplementary action of ions like Ca^{2+}, and Fe^{3+}, on the adsorption sites and on the gelatin itself. Altogether, in all the above accounts, I have limited the discussion to the simplest three-component systems, or to those in which additional components play known parts, or are assumed to be negligible. In the majority of practical cases this is no longer true, and inherent or adventitious impurities which frequently are very surface active may play major roles, which in every case demand detailed and specific study.

CONCLUSIONS

I have discussed the above findings, and particularly the last group of data, in some detail to illustrate the great extent of experimental information which we possess, in part well established, but also frequently conjectural. The picture which emerges is one of great complexity and it makes it clear why in most systems the specific features are too dominant to be fitted into the mold of existing general theories of physical adsorption of macromolecules, sophisticated as they are. Fortunately, it remains justifiable and sufficiently informative to analyze the data in terms of monolayer isotherms, i.e., of the parameters derivable from their course, and of independent measurement of the monolayer thicknesses. Combining these with the available information on the physical chemistry of adsorbates and adsorbent, a working understanding of the nature and the underlying mechanisms of the adsorption processes emerges which also furnishes clues to the many varieties of adsorption phenomena, and to the limitations of the existing theories.

The most important conclusion concerning chainlike macromolecules as adsorbates is that, unless they are anchored by only one end group, or are forced to lie with all their segments in the interface, as, e.g., in some cases of strong ion-pair formation of polyelectrolytes, or of stiff helices, the first adsorbed layer consists of molecules which are only partly attached to the interface, but otherwise extend into the solution with which the are in equilibrium. All observations corroborate this deduction, in particular the observation that changes in the supernatant have parallel consequences for free and adsorbed molecules alike. Thus, thanks to their length and flexibility, macromolecules may combine features of the adsorbed and solution states (3, 13, 17, 32, 41, 43, 50).

All theories, and FSE in particular, include Langmuir isotherms as special cases, e.g., for single-point adsorption of coils with no interaction other than short-range repulsion. The close resemblance, then, of so many macromolecular adsorption plots to Langmuir isotherms to the point where the reciprocal plots yield useful parameters in the form of "affinities" and "capacities" is surprising. We deduce from this behavior that during the first, very dilute, regime, the polymer coils attach themselves extensively in such a way that even for substantially incomplete surface coverage more than the number of polymer units necessary for a monolayer coverage is held to the surface in the form of partly flattened (random) coils. Further deposition causes not only more interpolymer contact by fitting in more chains against an already high segmental concentration within a coil diameter from the surface, but also a reduction in surface contacts per macromolecule, as well as an incipient coil compression. Thus, adsorption approaching a surface coverage of $\theta \to 1$ for undisturbed random coils requires disproportionately larger solution concentrations, i.e., gives rise to a very flat second regime of the isotherm, an apparent plateau. The compressibility of the coils then permits a gradual densification of the adsorbed layer, without destroying the monolayer character or increasing the layer thickness more than fractionally, until the coil compressibility drops to a value at which resistance to further compression cancels the remaining coil–surface attraction. At this point, a true plateau seems to exist for the adjoining concentration range, as judged by the fit of inverse Langmuir plots. Over this plateau the resistance against further compression, or to the addition or penetration by more polymer, is at a constant high; it is this range of surface density which offers the best steric stabilization of suspensions (12, 49).

Going beyond this, double-layer formation appears to be difficult because of the distance of a second layer from the original interface and because of the repulsion between coils. Still-higher solution concentrations, thus, are likely, next, to enhance coil interpenetrations, followed by the formation and adsorption of adsorbate aggregates, then by epitactic condensation, and, last, by nucleation of multilayers anchored to molecules entrapped in the lower layer.

If surfaces and molecules combine very different functionalites, specifics of surface–group interactions such as covalent bonding, ion or dipole pairing, chelation, ion exchange, or hydrophobic bonding become important. In a related manner, since every adsorption step involves two desorption steps of the solvent or cosolutes, the specific energetics of solvation of adsorbant and adsorbate will be a major influence.

ACKNOWLEDGMENT

The financial assistance of NIDR through a series of Research and Trainings Grants (DE-00099 and DE-01769) is gratefully acknowledged.

The financial support of the National Institutes of Dental Research and of NASA, Division Planetary Biology, of parts of this study is gratefully acknowledged.

REFERENCES

1. Stromberg, R. R., in "Treatise on Adhesion and Adhesives, I" (R. Patrick, Ed.). Dekker, New York, 1967.
2. Lipatov, Yu. S., and Sergeeva, L. M., "Adsorption of Polymers." Wiley, New York, 1974.
3. Koral, L., Ullman, R., and Eirich, F., J. Phys. Chem. 62, 541 (1958).
4. Jenkel, E., and Rumbach, B., Z. Elektrochem. 55, 612 (1951).
5. Fendler, H. G., Rohleder, H., and Stuart, H. A., Makromol. Chem. 18, 383 (1955).
6. Oehrn, O. E., J. Polym. Sci. 19, 199 (1956).
7. Peterson, C., and Kwei, T. K., J. Phys. Chem. 65, 1330 (1961).
8. Gottlieb, M. H., J. Phys. Chem. 64, 427 (1960).
9. Fontana, B. J., and Thomas, P. J., J. Phys. Chem. 65, 480 (1961).
10. Thies, C., Peyser, P., and Ullman, R., in "Proceeding of the 4th International Congress Surface Activity, 1964"; Perkel, R., and Ullman, R., J. Polymer. Sci. 54, 127 (1961); Ellerstein, S., and Ullman, R., J. Polym. Sci. 55, 123 (1961).
11. Rowland, F., Dissertation, Polytechnic Institute of Brooklyn, 1963.
12. Rowland, F., Bulas, R., Rothstein, E., and Eirich, F., Ind. Eng. Chem. 57, 49 (1965).
13. Rowland, F., and Eirich, F., J. Polym. Sci. A1 4, 2401 (1966).
14. Eirich, F. R., in "Factors in Interface Conversion for Polymer Coatings" (P. Weiss, Ed.), G. M. Symposium. American Elsevier, New York, 1968.
15. Simha, R., Frisch, H., and Eirich, F., J. Phys. Chem. 57, 584 (1953); J. Chem. Phys. 25, 365 (1953); J. Polym. Sci. 29, 3 (1958).
16. Frisch, H., and Simha, R., J. Phys. Chem. 58, 507 (1954); J. Chem. Phys. 27, 702 (1957).
17. Stromberg, R. R., Passaglia, E., and Tutas, D. J., J. Res. Nat. Bur. Stand. A 67, 431 (1963).
18. Stromberg, R. R., Tutas, D. J., and Passaglia, E., J. Phys. Chem. 69, 3955 (1965).
19. Wadman, J. W., Dissertation, Polytechnic Institute of Brooklyn, 1966.
20. Silberberg, A., J. Phys. Chem. 66, 1872 (1962); 66, 1884 (1962); J. Chem. Phys. 46, 1105 (1967); 48, 2835 (1968).
21. Hoeve, C., J. Chem. Phys. 43, 3007 (1965); 44, 1505 (1966).
22. Hoeve, C., DiMarcio, E., and Peyser, P., J. Chem. Phys. 42, 2558 (1965).
23. Hoeve, C., Preprints, International Symposium on Macromolecular Chemistry, Toronto, 1968.
24. Roe, R., J. Chem. Phys. 43, 1951 (1965); Proc. Nat. Acad. Sci. U.S.A. 53, 50 (1965).

25. DI MARCIO, E., AND RUBIN, R., *Amer. Chem. Soc. Polym. Prepr.* **11**, 1239 (1970).

26. DI MARCIO, E. A., *J. Chem. Phys.* **42**, 2101 (1965).

27. GARVEY, M. J., TADROS, TH. F., AND VINCENT, B., *J. Colloid Interface Sci.* **49**, 57 (1974).

28. HARA, K., AND TIMOTO, T., *Kolloid Z. Z. Polym.* **237**, 297 (1970).

29. PARFITT, R. L., AND GREENLAND, D., *Clay Miner.* **8**, 305 (1970).

30. EIRICH, F., "Chimica Macromol., Varenna 1961," Consiglio Naz. Ricerche, Roma, 1963.

31. FOWLER, R., AND GUGGENHEIM, E., *in* "Statistical Thermodynamics," Chap. 10, Cambridge Univ. Press, London/New York, 1939.

32. SILBERBERG, A., *Faraday Discuss. Chem. Soc.*, No. 59, p. 203 (1975).

33. LOPATIN, G., AND EIRICH, F. R., *in* "Proceedings of the 2nd International Congress on Surface Activity, B," p. 97.

34. LAURIA, R., Dissertation, Polytechnic Institute of Brooklyn, 1962.

35. CHOUGH, E., M.Sc. thesis, Polytechnic Institute of Brooklyn, 1968.

36. OTTEWILL, R. H., private communication, 1976.

37. SCHMIDT, W., AND EIRICH, F. R., *J. Phys. Chem.* **66**, 1907 (1962).

38. EARL, C., Dissertation, Polytechnic Institute of Brooklyn, 1971.

39. VAN OLPHEN, H., "Introduction into Clay Chemistry," Interscience, New York, 1963.

40. HSU, S. C., Dissertation, Polytechnic Institute of New York, 1977.

41. CHAO, H. Y., Dissertation, Polytechnic Institute of New York, 1974.

42. BOBA, R., M.Sc., thesis, Polytechnic Institute of Brooklyn, 1962.

43. KUDISH, A., Dissertation, Polytechnic Institute of Brooklyn, 1972.

44. EIRICH, F. R., *in* "Proceedings of the 2nd Workshop, National Institute of Dental Research, 1965."

45. EIRICH, F. R., *in* "Proceedings of the Brookdale Workshop on Dental Adhesives, New York University, 1975."

46. EIRICH, F. R., *in* "Biocompatibility of Implant Materials" (D. F. Williams, Ed.), Pitman (Med.), London, 1976.

47. KATCHER, J., M.Sc. thesis, Polytechnic Institute of New York, 1974.

48. KUDISH, A., AND EIRICH, F. R., In preparation.

49. MATHERNAGHAN, T. J., AND OTTEWILL, R. H., *J. Photogr. Sci.* **22**, 279 (1974).

50. KILLMAN, E., AND WELGAND, H., *Macromol. Chem.* **132**, 239 (1970).

Membranes

Deuterium NMR and Spin Label ESR as Probes of Membrane Organization

IAN C. P. SMITH, GERALD W. STOCKTON, ALEXANDER P. TULLOCH,[1] CARL F. POLNASZEK, AND KENNETH G. JOHNSON

Division of Biological Sciences, National Research Council of Canada,[2] *Ottawa, Canada K1A 0R6*

Received June 2, 1976; accepted August 26, 1976

A series of investigations of the mobility and organization of lipids in model and natural biological membranes using electron spin resonance (ESR) and nuclear magnetic resonance (NMR) of nitroxide- and deuterium-labeled lipids, respectively, was performed. In most cases, parallel NMR and ESR experiments were carried out in order to compare the advantages and disadvantages of each technique. Both methods report the same qualitative picture of the lipid bilayer: high order and low mobility for acyl chain segments near the carboxyl groups of the fatty acids and much less order and more rapid motion near the terminal methyl groups. However, the quantitative results differ markedly and it appears that ESR spin labels are unable to report the fine details of order and mobility gradients in lipid bilayers, possibly because of nitroxide-induced perturbations and the difficulty in resolving the spatial and temporal information from ESR spectra. The ESR method has the distinct advantages of superior sensitivity and speed where a qualitative answer is sufficient. In studies of deuterium- and nitroxide-labeled lipids in the plasma membrane of *Acholeplasma laidlawii*, both methods yield spectra containing characteristics of the gel-to-liquid crystal phase transition but only the NMR result is quantitatively consistent with the *broad* transition found from differential thermal analysis. The influence of cholesterol on acyl chain ordering in egg lecithin was studied as a function of the position of the deuterium or nitroxide label. Cholesterol increases the order parameter for all chain positions but the effect is greatest for positions 2–12. The ^2H NMR results are discussed in terms of geometrical changes in the bilayer due to cholesterol. No evidence for a specific complex between lecithin and cholesterol is apparent. Multilamellar dispersions and single bilayer vesicles of phospholipids are compared as models for membrane processes: they yield equivalent information about mobility and order in the bilayers when the effects of vesicle overall rotation are correctly taken into account.

INTRODUCTION

Biological membranes are a complex mixture of lipids, proteins, and enzymes whose architecture contributes greatly to their biological function. It appears that membranes differ widely in their organization and properties according to their function and the species in which they exist. Detailed models for the arrangement of the various components in particular membranes, such as that of the erythrocyte, have been proposed (1). Consistent features of most models include the phospholipids organized in fluid bilayers with protein (enzymes) located on the outer or inner surfaces of the bilayer, or spanning the bilayer and communicating between the inside and the outside of the membrane (2). Also available is considerable evidence that some phospholipid is not fluid, but exists as a rigid boundary layer about proteins and enzymes (3).

It is apparent that a structural approach to membrane function is difficult. A wide variety of techniques have been applied, but many of them suffer from the difficulty that

[1] Prairie Regional Laboratory, National Research Council, Saskatoon, Canada S7N 0W9.

[2] N.R.C.C. Publication Number 15810.

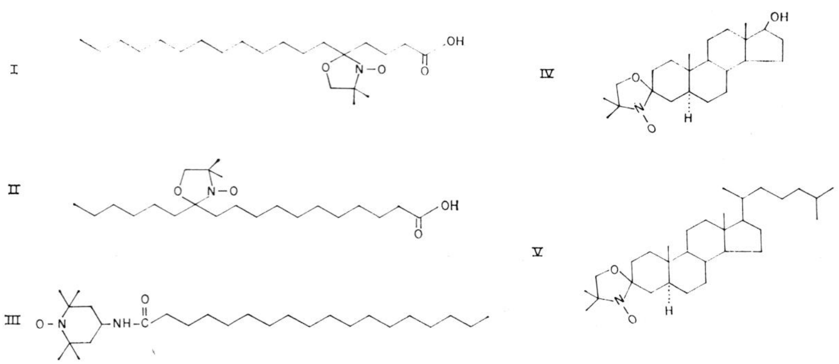

FIG. 1. Molecular formulas of some common ESR lipid spin probes: I, 5-doxyl-stearic acid; II, 12-doxyl-stearic acid; III, 4-stearamide-1-oxyl-2,2,6,6-tetramethyl piperidine; IV, 3-doxyl-androstane-17β-ol; V, 3-doxyl-cholestane.

they observe an average property of the overall membrane or a blurred response from many similar but not identical sites. It is evident that techniques must be developed which observe membrane components on the molecular level one at a time.

An extremely successful approach to this problem is the use of nitroxide-labeled membrane components. The ESR spectra of the spin-labeled compounds are very sensitive to rate of motion and degree of organization within a system (4), and are detectable at very low concentrations. The formulas of some typical spin-labeled lipids are shown in Fig. 1. Spin-labeled lipids suffer from one principal disadvantage—their spectra report the properties of the labeled compound and the properties of the unlabeled compound must be inferred by implication. The nitroxide-containing moiety constitutes a bulky, hydrophilic probe in an organized, hydrophobic environment and therefore probably perturbs its immediate environment to some extent. To calibrate the spin label technique we have undertaken ^{2}H NMR studies and parallel spin label studies on a variety of systems.

^{2}H NMR SPECTRA

Deuterium has a lower magnetic moment than the more familiar ^{1}H isotope and there-fore resonates at a lower frequency for a given applied magnetic field. Some properties of ^{2}H are given in Table I. Its detection sensitivity is far below that of ^{1}H, and approximately six orders of magnitude lower than that of ESR probes. The low natural abundance of ^{2}H is highly desirable; in experiments with specifically labeled compounds no background signals from unlabeled components are expected. The presence of a quadrupole moment in ^{2}H is a particular advantage since relaxation behavior and spectral shapes are dominated by a relatively simple mechanism.

In Fig. 2 the differences between a quadrupolar and nonquadrupolar nucleus are represented. For spin $\frac{1}{2}$ nuclei such as ^{1}H, ^{13}C, and ^{31}P the nuclear charge distribution is spherically symmetric. This is not true for nuclei with spin I greater than $\frac{1}{2}$: ^{2}H, $I = 1$; ^{23}Na, $I = \frac{3}{2}$. In the latter nuclei the distribution of nuclear charge departs from spherical symmetry and the nuclei possess an electric quadrupole moment, Q. This electric moment can interact (couple) with an asymmetric electronic charge distribution about the nucleus. The magnitude of the interaction depends on the size of the quadrupole moment Q and on the electric field gradient at the nucleus due to the asymmetric distribution of charge in the electronic environment. The latter is characterized by the three components of the second derivative of the electric potential V of the electronic environment. For carbon–deuterium sp^3-hybridized bonds the surrounding electric field is axially symmetric and one need only consider the z-component of the field gradient q_{zz}, henceforth represented as q. Thus, the magnitude of the interaction between the quadrupole moment Q and the electronic environment is given by e^2qQ/h, the quadrupole coupling constant, where e is the charge on the electron and h is Planck's constant. For a carbon–deuterium sp^3 bond the quadrupole coupling constant has a value of approximately 170 kHz (31).

Figure 3 demonstrates the effects of the quadrupole coupling constant on the energies of the various spin states of the ^{2}H nucleus. When the surrounding electric field is spherically symmetric $q = 0$ and there is no effect on the Zeeman energies of the two degenerate spin transitions. Thus one would obtain a single ^{2}H resonance for a C–^{2}H moiety. In the case of a nonzero field gradient, all energy levels are perturbed and the two allowed transitions have different energies. The difference between these energies depends upon the angle θ between the C–^{2}H vector and the applied magnetic field (as well as upon the magnitude of q), as shown in the expression for the quadrupole Hamiltonian at the top of Fig. 3. Thus, the quadrupole splittings obtained from the ^{2}H NMR spectrum vary with the orientation of the carbon–deuterium bond with respect to the applied magnetic field. For an organized system of molecules such as the lipid bilayer in a biological membrane, the quadrupole splittings provide a facile measure of the degree of orientational order in the system.

TABLE I

Properties of ^{2}H

Natural abundance	0.015%
Resonance frequency (23 kG)	15.4 MHz
Spin	1
Sensitivity relative to ^{1}H	0.965%
Quadrupole moment ($\times 10^{24}$, cm^2)	0.00273

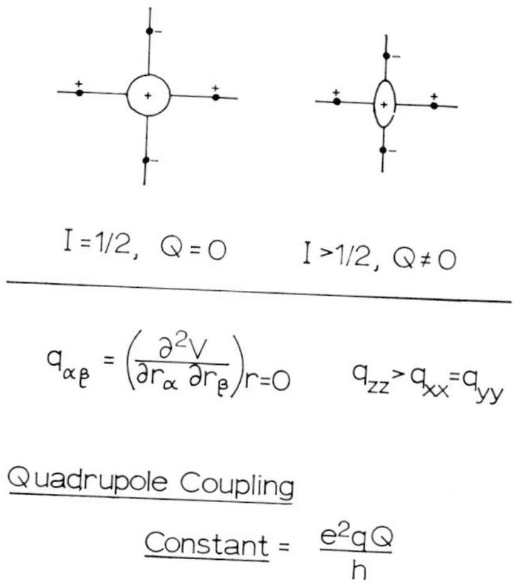

$$I = 1/2, \quad Q = 0 \qquad I > 1/2, \quad Q \neq 0$$

$$q_{\alpha\beta} = \left(\frac{\partial^2 V}{\partial r_\alpha\, \partial r_\beta}\right)_{r=0} \qquad q_{zz} > q_{xx} = q_{yy}$$

Quadrupole Coupling

$$\text{Constant} = \frac{e^2qQ}{h}$$

FIG. 2. Schematic of the properties of a nonquadrupolar and a quadrupolar nucleus. The charges about the nucleus represent the asymmetric charge distribution of the nuclear environment.

PROPERTIES OF PHOSPHOLIPID BILAYERS AND THEIR INFLUENCE ON ^{2}H NMR SPECTRA OF LABELED LIPIDS

A representation of a phospholipid bilayer is given in Fig. 4. The polar head groups are represented by circles and the long fatty acid chains by the wiggly lines. To determine the usefulness of ^{2}H NMR in biological systems we have studied a variety of model phospholipid bilayer systems. We have labeled fatty acid chains and inserted them into phospholipid systems as well as incorporating them covalently into phospholipid molecules (5–9). The former procedure has been used in order to have a direct comparison with the results of ESR spin label studies where fatty acids are the most popular probes. The latter yield a direct measure of the properties of the phospholipid molecules. Similar studies have been performed with great success by Seelig and co-workers (10–13).

At the outset it is important to distinguish between the spatial and temporal aspects of membrane organization. Both are described in Fig. 5. In a perfectly organized phospholipid bilayer all fatty acid chains would be

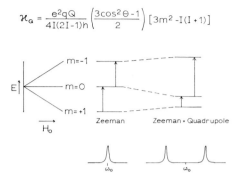

$$\mathcal{H}_Q = \frac{e^2 qQ}{4I(2I-1)h} \left(\frac{3\cos^2 \Theta - 1}{2} \right) [3m^2 - I(I+1)]$$

FIG. 3. The nuclear Zeeman energy levels of a quadrupolar nucleus ($I = 1$) for the cases where the electric field gradient at the nucleus (q) is zero (left) or finite (right). The Hamiltonian describing the interaction between the quadrupole moment Q and the electric field gradient q for a partially ordered system (S_{CD}) is shown at the top of the figure; m is the quantum number for the z-component of the spin angular momentum I.

completely extended and all C–^{2}H (henceforth described as C–D) bonds of the methylene groups would make equal angles β with the normal to the membrane surface. However, if *gauche* rotamers about C–C bonds are allowed, a wide variety of orientations of the C–D moieties is possible. A description of the distribution of angles is given by the order parameter S_{CD}, which is the average of the function $(3\cos^2 \beta - 1)/2$ over all allowed orientations. The bond order parameter may be converted to a molecular order parameter by multiplication with the angular transformation appropriate to the type of C–D bond (for the fatty acid chains $S_{mol} = -2S_{CD_2}$, $S_{mol} = -6S_{CD_3}$; for the choline N–CD$_3$, $S_{mol} = -18S_{CD_3}$). Thus for perfect order, $S_{CD_2} = -0.5$, $S_{mol} = 1$, and for perfect disorder, $S_{mol} = 0$.

The temporal aspect of the membrane organization involves the rate of interconversion between allowed conformations. If this is rapid on the ^{2}H NMR time scale (greater in frequency than the difference between the largest and smallest potential quadrupole splitting for different conforma-

tions), one observes a splitting that is the average over all possible conformations and is a direct measure of the order parameter. The ^{2}H relaxation times (spin–lattice, T_1; spin–spin, T_2) are determined by this rate of isomerization as well as by the magnitude of S_{CD} (8, 14). As this rate becomes slower the ^{2}H resonance lines become broader and may in some cases become difficult to observe with high-resolution NMR spectrometers. A variety of motions can contribute to the rate of positional interconversion: rotations about individual C–C bonds; motions of the entire fatty acid chains in directions parallel and perpendicular to the normal to the membrane surface; translational diffusion of entire phospholipid molecules; and in some cases the rate of overall rotation of the membrane structure itself. The first type of motion may be very different from one position on the fatty acid chain to another. Thus the values of T_1, T_2, and the resonance linewidths for C–D moieties can depend strongly on the position of deuteration in the phospholipid molecule (8).

THE ^{2}H NMR POWDER SPECTRUM

Biological membranes are very large structures on the scale of molecular dimensions, with overall diameters of the order of microns. A schematic diagram of a membrane is shown

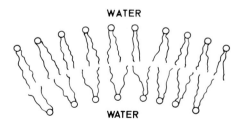

FIG. 4. A representation of a bilayer of phospholipids The polar head groups (glycerol, choline, ethanolamine, serine, etc.) are portrayed by the circles. The wiggly lines represent the fatty acyl chains with various degrees of order (populations of *gauche* and *trans* rotamers about C–C bonds) and packing.

in Fig. 6. For a given orientation of a C–D moiety with respect to the membrane normal a variety of orientations (θ) with respect to the applied magnetic field (H) are possible, each with its characteristic quadrupole splitting. If the membrane is rotating slowly on the ^{2}H NMR time scale (less than the frequency difference between the possible quadrupole splittings), all these splittings will be observed in the ^{2}H NMR spectrum. The resultant spectrum is thus a superposition of many component spectra. Nevertheless, the resultant powder-type spectrum is characterized by discontinuities at D_q and $2D_q$, representative of the quadrupole splittings for $\theta = 90$ and $0°$, respectively, and the bond order parameter can be obtained from D_q as

$$|S_{CD}| = 4D_q/3(e^2qQ/h). \qquad [1]$$

With increasing rate of overall membrane rotation, the shape of the powder spectrum changes as shown in Fig. 7 (8, 14). At intermediate rates (such as 10^4 sec^{-1}) the separation between the major peaks is less than D_q, and D_q must be estimated by spectral simu-

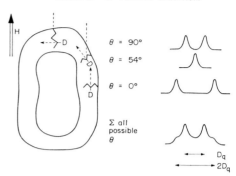

ORIGIN OF ^{2}H POWDER SPECTRUM

FIG. 6. Schematic of a large membrane showing the various possible angles (θ) between a carbon–deuterium bond and the applied magnetic field (H) for a given conformation of the C–D moiety. The ^{2}H NMR spectra for three possible orientations are shown. Note that for $\theta = 54°$ no quadrupole splitting is observed. The resultant spectrum for this system is the sum of spectra for all possible θ, as shown at the bottom right.

lation. At fast rates, the spectra collapse to single resonances for which no splitting is evident and only relaxation times T_1 and T_2 can be measured.

MODEL MEMBRANE SYSTEMS—SONICATED AND UNSONICATED PHOSPHOLIPID DISPERSIONS

Two types of systems popular in model membrane studies are represented in Fig. 8. The multilamellar system on the left is obtained by gentle agitation of phospholipid dispersions. It is an onion-like structure comprising many concentric bilayers separated by aqueous layers 20–30 Å thick. Its overall rate of rotation is very slow on the ^{2}H NMR time scale and powder-type spectra are always obtained. The liposome system on the right is prepared by ultrasonic agitation of the lamellar dispersion. It consists of single bilayer vesicles of small diameter whose rate of overall rotation is rapid on the ^{2}H NMR time scale and yields single-resonance ^{2}H NMR spectra (5, 6, 8). It has been found experimentally that there are twice as many phos-

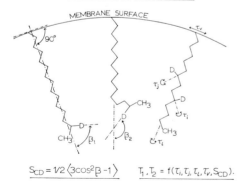

ORDER & MOBILITY IN MEMBRANES

$$S_{CD} = 1/2 \langle 3\cos^2\beta - 1 \rangle \qquad T_1, T_2 = f(\tau_i, \tau_j, \tau_k, \tau_v, S_{CD}).$$

FIG. 5. Possible conformations and degrees of motional freedom for a C–D bond in a membrane bilayer. The axis about which ordering takes place is shown as the normal to the membrane bilayer in accord with the angular dependence of the ^{2}H NMR spectra of oriented multibilayers (6). The correlation times τ are for the possible motions available to the C–D bond.

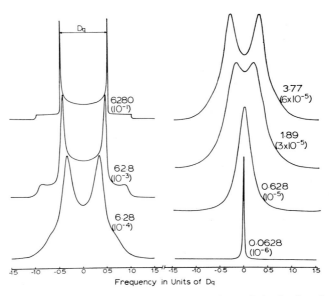

Fɪɢ. 7. The influence of the correlation time for vesicle tumbling τ_v (in brackets) on the shape of the ²H NMR spectrum of a system with quadrupole splitting $D_q = 10$ kHz, and intrinsic linewidth 10 Hz. The quantity $2\pi D_q \tau_v$ is shown above the value of τ_v (8).

pholipid molecules in the outer monolayer as there are in the inner monolayer due to the molecular packing requirements (the diameter of a vesicle is only three to four times greater than the thickness of the bilayer (15, 16)). It has been suggested that this leads to large differences in the degree of molecular order and in the rates of orientational interconversion within the bilayers of sonicated and unsonicated phospholipid systems (17). We shall consider mainly unsonicated lamellar dispersions as their spectra are similar to those expected from biological membranes, but we shall discuss the differences between the ²H NMR spectra of sonicated and unsonicated systems with respect to possible differences in spatial and motional characteristics of the phospholipid molecules.

THE CHOLINE METHYL GROUP IN EGG LECITHIN AND THE LIQUID CRYSTAL— GEL TRANSITION

In Fig. 9 we show the spectrum of the CD₃ groups of the choline head group in egg lecithin. Details of experimental methods are given in (8). From 55 to −5°C the spectra are powder type with measurable quadrupole splittings that indicate the expected increase in molecular order with decreasing temperature. Note also that the components become broader with decreasing temperature due to decreasing rates of motion within the bilayers. On passing from −15 to −25°C the resonances become too broad to observe on a high-resolution spectrometer. Over this range the phospholipid has passed from the liquid crystal to the gel state. In the latter state the rate of molecular motion is slow compared to that for the liquid crystalline phase. In contrast to the ESR spin probes, it is difficult to observe gel-state lipid by ²H NMR.

THE GRADIENT OF MOLECULAR ORDER WITHIN PHOSPHOLIPID BILAYERS

One usually regards the polar groups of phospholipids in a bilayer as being arranged in a *quasi*-ionic lattice which holds the phospholipids together and stabilizes the bilayer structure. Given this semirigid arrangement of the polar groups, what sort of variation in the population of *gauche* and *trans* conformers about C–C bonds would one expect on proceeding down the fatty acid chain towards the terminal methyl group? Intuitively one might expect the *gauche* populations to increase, but the nature of the dependence on *n*, the number of carbon atoms from the carbonyl carbon of the fatty acid moiety, is not clear. Figure 10 shows the dependence of the molecular order parameter on *n* for a series of specifically deuterated stearic acids intercalated in lamellar dispersions of egg lecithin at 30°C (circles) and 55°C (squares). The order parameter is constant and large, as is the population of *trans* conformers about the C–C bonds, up to carbon atom 10, and then decreases monotonically to a small value at carbon-18. This surprising behavior has been suggested by the theoretical calculations of Marčelja (18). He postulates that the disturbing effect of a *gauche⁺* rotamer can be essentially overcome if the following two segments have the *trans* and *gauche⁻* conformations. This feature known as a "kink" (18). To conserve the plateau of S_{mol} versus *n* the kink probability must be higher near the polar head-group

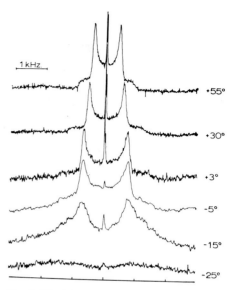

FIG. 9. The 15.4-MHz spectra of choline-d_9 lecithin derived from egg yolk phosphatidylethanolamine by addition of three CD_3 groups to the ethanolamine amino group. The lecithin was 32 mg/ml in distilled H_2O and 500–3000 free-induction decays were time-averaged before Fourier transforming on a modified Varian XL-100 spectrometer (8).

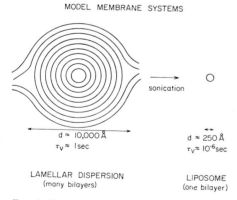

MODEL MEMBRANE SYSTEMS

sonication

d ≈ 10,000 Å
τ_V ≈ 1 sec

d ≈ 250 Å
τ_V ≈ 10^{-6} sec

LAMELLAR DISPERSION
(many bilayers)

LIPOSOME
(one bilayer)

FIG. 8. Representation of the relative sizes and structures of two common model membrane systems: left, lamellar multibilayer dispersions; right, single-bilayer liposomes.

region and decrease rapidly beyond carbon-10. Note that raising the temperature by 25° causes an overall decrease in order, with retention of the general profile of order versus *n*. Similar results have been obtained for dipalmitoyl lecithin in the liquid crystal state (11).

The triangles in Fig. 10 are the order parameters obtained from spin-labeled stearic acid in oriented multibilayers of egg lecithin (19). Although the absolute values of the order parameters derived from spin probes are lower than those determined by ²H NMR, as might be expected due to the bulky nature of the nitroxide-containing moiety, the decline in order parameter with increasing *n* is indicated. A similar study on dipalmitoyl lecithin demonstrated the same general conclusion, but the spin probe data failed to show the plateau in order versus *n* up to carbon-10 (11).

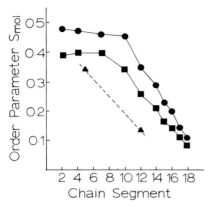

FIG. 10. Molecular order parameters as a function of the position from the carboxy-terminal of the fatty acid chain of specifically labeled stearic acid in egg lecithin lamellar dispersions: ●, ^{2}H NMR, 30°C; ■, ^{2}H NMR, 55°C; ▲, ESR spin probes, 30°. The ^{2}H NMR data are from (8); the ESR data are from (19).

Clearly the spin probe data give a reasonable qualitative estimate of the behavior, but are inaccurate in reporting fine details.

A further explanation for the differences between order parameters derived via spin probes and ^{2}H NMR could lie in the different time scales of the two types of experiment (20). If a further disordering existed which was symmetrical about the bilayer normal, and which had a lifetime of 10^{-6}–10^{-8} sec (short on the ^{2}H NMR time scale but long on the ESR time scale), a different transformation of S_{CD} to S_{mol} would lower the ^{2}H values. Studies are in progress using saturation transfer ESR experiments (which are sensitive to motions in this range) on spin-labeled lipids to determine whether or not such short-lived disorders exist.

THE INFLUENCE OF CHOLESTEROL ON ORDERING IN BILAYERS OF EGG LECITHIN

A perdeuterated fatty acid probe such as stearic acid-d_{35} yields ^{2}H NMR spectra in which separate quadrupole splittings can be resolved for each region with a distinct order parameter (9). This greatly reduces the num-

ber of experiments which must be done in a detailed study of the influence of cholesterol on the ordering profile. Figure 11 shows the molecular order parameters measured in egg lecithin dispersions as a function of cholesterol concentration (9). Positions 2–11, which lie on the plateau of the order profile, respond steeply and in equivalent fashion. From position 12 to position 18 the slope of the dependence decreases markedly. An explanation of this phenomenon lies in the rigidity of the ring structure and the flexibility of the hydrocarbon tail of cholesterol. If one assumes that the hydroxyl group of cholesterol is located in the bilayers in the region of the glycerol backbone of the phospholipids, the rigid ring structure of the cholesterol molecule extends as far as positions 9–11 (depending on the degree of extension of the fatty acyl chains). Thus, as the degree of order is constant in this region in the absence of cholesterol, and as the ring structure is rigid, one expects an effect of cholesterol which is similar for carbons 2–11. The flexibility of the hydrocarbon tail of cholesterol imposes a lesser constraint on positions 12–18.

The simple dependences shown in Fig. 11 are not consistent with the presence of a

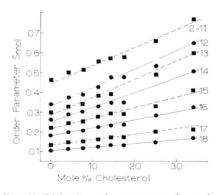

FIG. 11. Molecular order parameters for various positions on stearic acid in egg lecithin as a function of mole% of incorporated cholesterol (9). The points were measured using stearic-d_{35} acid and verified by comparison with some corresponding points from specifically deuterated stearic acids (8).

specific $1:2$ or $1:1$ complex of cholesterol with lecithin (21), if one can assume that the degree of acyl chain order would be different in the complex from that in the absence of cholesterol. For a complex which is short-lived on the ^{2}H NMR time scale a curvature in the plots of Fig. 11 is expected; in the case of a long-lived complex two quadrupole splittings for each position on the acyl chain would be present in the spectra.

From the individual order parameters one can calculate the populations of *gauche* and *trans* rotamers at each position and hence infer the effective length of the acyl chains. As the degree of order increases, so does the length of the acyl chain. The estimates of acyl chain length, and its increase with greater cholesterol concentration, determined by ^{2}H NMR (9) and X-ray diffraction (22) are in good agreement.

A comparison of the results from ^{2}H NMR and ESR spin probes in this system cannot be made at present. This is due to a lack of spin probe order parameters for positions 13–17.

MEMBRANES OF *ACHOLEPLASMA LAIDLAWII* ENRICHED BIOSYNTHETICALLY WITH ^{2}H-LABELED FATTY ACIDS

The microorganism *Acholeplasma laidlawii* provides an excellent biological membrane in which to incorporate ^{2}H-labeled fatty acids biosynthetically because it does very little to change the lengths of fatty acids supplied in the growth medium. Figure 12 shows the ^{2}H NMR spectra of the plasma membranes of *Acholeplasma laidlawii* B grown on a medium enriched with palmitic-16d_3 acid (7). A fatty acid analysis of the lipids reveals that 79% of the incorporated fatty acids are 16:0. Thus the order parameters derived from the spectra of Fig. 12 are those for position 16 in the membrane lipids. At temperatures greater than that of growth (37°C) the quadrupole splitting decreases until at 55°C the two components have coalesced. This indicates a decreasing order parameter

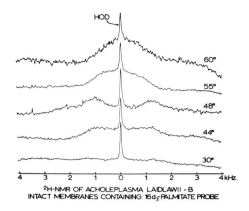

^{2}H-NMR OF ACHOLEPLASMA LAIDLAWII - B
INTACT MEMBRANES CONTAINING 16d_3-PALMITATE PROBE

Fig. 12. The 15.4-MHz ^{2}H NMR spectra of palmitic-16d_3 acid incorporated biosynthetically into the plasma membrane of *Acholeplasma laidlawii* (7). The membranes were 300 mg/ml aqueous β-buffer (0.5 M NaCl, 0.025 M Tris–HCl, pH 8.5). Up to 2.5×10^5 free-induction decays were time-averaged before Fourier transformation on a modified Varian XL-100 spectrometer.

and a concomitant decrease in membrane thickness. Over the range 44–30°C the resonances broaden considerably until the spectrum is almost too broad to detect. This is due to a liquid crystal–gel state transition of the membrane lipids which has been observed by differential thermal analysis (23). Expansion of the vertical scale of the 30°C spectrum reveals that at this temperature approximately 20% of the lipids are still in the liquid crystalline state.

Comparison of the variation of order parameter with temperature (30–60°C) for *Acholeplasma laidlawii* membranes and egg lecithin vesicles reveals a much steeper dependence of membrane thickness on temperature for the former. This suggests that the loss of viability for *Acholeplasma* at low and high temperatures originates from the high proportion of gel-state lipid (low temperatures) or extremely thin plasma membranes (high temperatures). The gel state of the plasma membrane has a very low passive permeability, whereas the thinner membranes have a very high permeability (24).

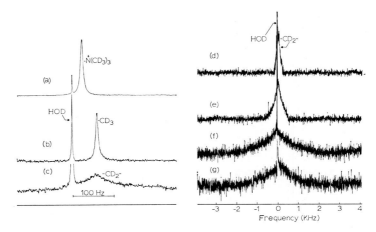

Fɪɢ. 13. The 15.4-MHz ²H NMR spectra of specifically deuterated lipids in single-bilayer vesicles of egg lecithin at 30°C: (a) choline-d_9-lecithin; (b) stearic-18-d_3 acid; (c) stearic-17d_2 acid; (d) stearic-16d_2 acid; (e) stearic-15d_2 acid; (f) stearic-10d_2 acid; (g) stearic-2d_2 acid. The spectra were transformed on a modified Varian XL-100 spectrometer from accumulations of 5×10^2 to 2.5×10^5 free-induction decays (8).

Parallel ESR spin label studies have been done on the *Acholeplasma* membranes using spin-labeled fatty acids (7). In contrast to ²H NMR, it is quite easy to obtain ESR spectra of gel-state lipids. The ESR order parameters decrease rapidly with temperature, and show a sharp discontinuity at 37°C due to the gel–liquid crystal phase transition. The transition indicated by the ESR data is much sharper than that found by either ²H NMR (7) or differential thermal analysis (23). This may be due to the difficulty of separating motional and ordering effects in the ESR spectra (25, 26), of distinguishing discrete spectra from spin labels in coexisting gel and liquid crystalline environments, and to the possible migration of the fatty acid spin label from gel to liquid crystalline regions (32).

A wide variety of ²H NMR experiments is now possible for the *Acholeplasma* membrane. Currently under investigation are the profile of order versus position in the fatty acyl chains, the ordering effect of cholesterol, the influence of membrane proteins, and the disruptive effect of antibiotics.

THE ORGANIZATION OF PHOSPHOLIPIDS IN LIPOSOMES

The sonicated phospholipid dispersion, represented on the right of Fig. 8, is important because it is a popular model membrane system. Its popularity with NMR spectroscopists is based on the considerably better resolution that is obtained in the spectra relative to that obtained from lamellar dispersions. The ²H NMR spectra of a series of specifically deuterated stearic acids intercalated in egg lecithin liposomes are presented in Fig. 13 (8). They comprise single resonances whose widths decrease rapidly with distance of the labeled position from the polar head group. The absence of a quadrupole splitting is due to the rapid rate of rotation of the vesicles with respect to the applied magnetic field (10^6 sec^{-1}) compared to potential quadrupole splittings of up to 30 kHz (see Fig. 7). For these resonances the spin–lattice relaxation times, T_1, and the spin–spin relaxation times, T_2^*, can be measured (5, 6, 8). The T_1 values are determined mainly by rapid motions within the membrane, but depend in part on the degree of order (8, 14). The

T_2^* values, which can be related to the resonance widths by $W = 1/\pi T_2^*$, depend strongly on the degree of order S_{CD} and on the liposome correlation time τ_V,

$$W = \frac{1}{\pi T_{1e}} + \frac{9\pi}{20}(e^2qQ/h)^2 S_{CD}{}^2 \tau_V, \quad [2]$$

where the T_{1e} are the measured values of the spin–lattice relaxation times (8). Thus, if the correlation time τ_V remains constant, the linewidths can be readily interpreted in terms of values of S_{CD} which decrease rapidly with distance of the labeled position from the polar head group. The linewidths in the sonicated system can then be compared directly with the order parameters determined for the unsonicated system, as shown in Fig. 14. Deviations from linearity in plots of W versus S_{CD}^2 would be indicative of differences in order between the phospholipids of the sonicated and unsonicated systems. The data are most accurate for positions 15–18 due to the relatively narrow resonances observed in either the sonicated or unsonicated systems for these positions. The largest measurable deviation from linearity occurs at position 17; this corresponds to a difference in order parameter of 30%. The T_1 values for position 18 are 0.32 and 0.3 sec for sonicated and unsonicated systems, respectively, indicating a negligible difference in the rates of rapid motion at this position. Differences in order between the two systems for the other positions must lie within the error bars indicated in Fig. 14. Thus, we conclude that there are small differences in the degree of order between the two systems, with the order parameters being lower in the liposomes (based on the data for positions 16 and 17). These minor differences are to be expected, especially in the interface of the terminal methyl groups, in view of the packing of two phospholipid molecules in the outer monolayer to every one in the inner monolayer. These conclusions are in accord with the results of recent laser Raman studies on the same system (27). Spin label studies on

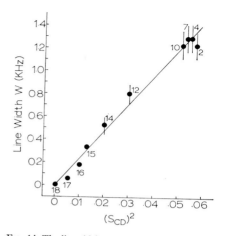

FIG. 14. The linewidths W for specifically deuterated stearic acid in single-bilayer vesicles of egg lecithin at 30°C versus S_{CD}^2 determined directly from unsonicated multilamellar dispersions of the same lipid (8). The numbers above the points indicate the position labeled; vertical bars represent the estimated experimental uncertainties except when these are less than the width of the circles identifying the data points.

sonicated and unsonicated egg lecithin containing 5-doxyl stearic acid (I in Fig. 1) have been reported (28). They indicated a maximum difference in order parameter of 7%, with that for the sonicated system being the lower.

The influence of cholesterol on the ^2H NMR spectra of sonicated and unsonicated egg lecithin has also been determined (8). It has been established previously that addition of cholesterol results in an increase in the diameter of single bilayer vesicles (29), and a correction term must be added to τ_V in Eq. [2] to compare the linewidths for the sonicated system with the order parameters for the unsonicated system. When this is done, a linear relationship between the linewidth and the square of the order parameter is obtained for the stearic acid-$18d_3$ probe. This confirms the conclusion above that the differences in molecular organization between sonicated and unsonicated egg lecithin are small and justifies the use of single-bilayer vesicles as model membranes.

CONCLUSIONS

Deuterium NMR of labeled lipids constitutes an accurate method to determine molecular order parameters for lipids in membranes. The synthetic routes to specifically labeled lipids are relatively simple and the cost of starting materials is low. No background signal from unlabeled components interferes with spectral interpretation. The spin–lattice relaxation times T_1 provide a good measure of the rates of rapid motion whereas the spin–spin relaxation times T_2 are sensitive to slow motions as well as to molecular order. The ^2H data provide a means to determine the reliability of ESR spin label results. In most comparisons made to date the ESR spin label conclusions appear to be qualitatively correct, although the absolute values of parameters measured usually differ significantly from those found by ^2H NMR. The relative advantage of the ESR method is its detection sensitivity, even for spin probes in gel-state lipid.

A disadvantage of ^2H NMR is its low detection sensitivity. This difficulty is diminishing continuously with improvement in spectrometer design, the use of larger sample tubes, and the introduction of modified techniques for data acquisition such as the solid echo (30). The disadvantages of ESR spin labels are the perturbations they produce in the system under investigation and the difficulty in separating the order and mobility contributions to the ESR spectra without resort to complex spectral simulations.

ACKNOWLEDGMENTS

We are deeply indebted to our colleagues who have been involved in the development of the ^2H NMR technique over the past four years: Dr. K. W. Butler, Mr. R. Cyr, Dr. F. Hasan, Dr. L. C. Leitch, Dr. A. R. Quirt, Dr. H. Saitô, Dr. S. Schreier, and Mr. L. Turner. We are also very grateful to Dr. K. Schaumburg for the initial inspiration.

REFERENCES

1. STECK, T. L., *J. Cell Biol.* **62**, 1 (1974).
2. SINGER, S. J., AND NICHOLSON, G. L., *Science* **175**, 720 (1972).
3. JOST, P., GRIFFITH, O. H., CAPALDI, R. A., AND VANDERKOOI, G., *Biochim. Biophys. Acta* **311**, 141 (1973).
4. "Spin Labeling—Theory and Applications" (L. J. Berliner, Ed.). Academic Press, New York, 1976.
5. SAITÔ, S., SCHREIER-MUCCILLO, S., AND SMITH, I. C. P., *FEBS Lett.* **33**, 281 (1973).
6. STOCKTON, G. W., POLNASZEK, C. F., LEITCH, L. C., TULLOCH, A. P., AND SMITH, I. C. P., *Biochem. Biophys. Res. Commun.* **60**, 844 (1974).
7. STOCKTON, G. W., JOHNSON, K. G., BUTLER. K. W., POLNASZEK, C. F., CYR, R., AND SMITH, I. C. P., *Biochim. Biophys. Acta* **401**, 535 (1975).
8. STOCKTON, G. W., POLNASZEK, C. F., TULLOCH, A. P., HASAN, F., AND SMITH, I. C. P., *Biochemistry* **15**, 954 (1976).
9. STOCKTON, G. W. AND SMITH, I. C. P., *Chem. Phys. Lipids* **17**, 251 (1976).
10. SEELIG, J. AND SEELIG, A., *Biochem. Biophys. Res. Commun.* **57**, 406 (1974).
11. SEELIG, A. AND SEELIG, J., *Biochemistry* **13**, 4839 (1974).
12. SCHINDLER, H. AND SEELIG, J., *Biochemistry* **14**, 2283 (1975).
13. SEELIG, A. AND SEELIG, J., *Biochem. Biophys. Acta* **406**, 1 (1975).
14. POLNASZEK, C. F., STOCKTON, G. W., AND SMITH, I. C. P., *J. Amer. Chem. Soc.* In preparation.
15. JOHNSON, S. M., BANGHAM, A. D., HILL, M. W., AND KORN, E. D., *Biochim. Biophys. Acta* **233**, 820 (1971).
16. BYSTROV, V. F., DUBROVINA, N. I., BARSUKOV, L. I., AND BERGELSON, L. D., *Chem. Phys. Lipids* **6**, 343 (1971).
17. SEITER, C. H. A. AND CHAN, S. I., *J. Amer. Chem. Soc.* **95**, 7541 (1973).
18. MARČELJA, S., *Biochim. Biophys. Acta* **367**, 165 (1974).
19. SCHREIER-MUCCILLO, S., MARSH, D., DUGAS, H., SCHNEIDER, H., AND SMITH, I. C. P., *Chem. Phys. Lipids* **10**, 11 (1973).
20. MCCONNELL, H. M., *in* "Spin Labeling—Theory and Applications" (L. J. Berliner, Ed.), Chap. 13, p. 525. Academic Press, New York, 1976.
21. ENGELMAN, D. M. AND ROTHMAN, J. E., *J. Biol. Chem.* **247**, 3694 (1972).
22. LECUYER, H. AND DERVICHIAN, D. G., *J. Mol. Biol.* **45**, 39 (1969).
23. STEIM, J. M., TOURTELLOTTE, M. E., REINERT, M. E., AND MCELHANEY, R. N., *Proc. Nat. Acad. Sci. U.S.A.* **63**, 104 (1969).
24. DE GIER, J., MANDERSLOOT, J. G., HUPKES, J. V., MCELHANEY, R. N., AND VAN BEEK, W. P., *Biochim. Biophys. Acta* **233**, 610 (1971).
25. CANNON, B., POLNASZEK, C. F., BUTLER, K. W., ERIKSSON, L. E. G., AND SMITH, I. C. P., *Arch. Biochem. Biophys.* **167**, 505 (1975).

26. POLNASZEK, C. F., *Quart. Rev. Biophys.* In press.
27. MENDELSOHN, R., SUNDER, S., AND BERNSTEIN, H. J., *Biochim. Biophys. Acta* **419**, 563 (1976).
28. MARSH, D., PHILLIPS, A. D., WATTS, A., AND KNOWLES, P. F., *Biochem. Biophys. Res. Commun.* **49**, 641 (1972).
29. GENT, M. P. N. AND PRESTGARD, J. H., *Biochemistry* **13**, 4027 (1974).
30. DAVIES J., H., JEFFEREY, K. R., BLOOM, M., VALIC, M. I., AND HIGGS, T. P., *Chem. Phys. Lett.* **42**, 390 (1976).
31. MANTSCH, H. H., SAITÔ, H., AND SMITH, I. C. P., *in* "Progress in the Nuclear Magnetic Resonance Spectroscopy" (J. W. Emsley, J. Feeney, and L. H. Sutcliffe, Eds.). Pergamon Press, London, in press.
32. BUTLER, K. W., TATTRIE, N. H., AND SMITH, I. C. P., *Biochim. Biophys. Acta* **363**, 351 (1974).

The Proteins of Membranes

S. J. SINGER

Department of Biology, University of California at San Diego, La Jolla, California 92093

Received June 8, 1976; accepted August 12, 1976

Proteins constitute the largest fraction by weight of most membranes, and their properties are critical to membrane functions. A brief discussion is given of some salient theoretical and experimental advances of the last decade in our understanding of the structures of proteins in membranes, which have led to the fluid mosaic model of membrane structure, and to new insights into the molecular mechanisms of membrane functions.

INTRODUCTION

The existence of a lipid bilayer in biological membranes, as championed by Danielli (1), has been amply confirmed by a large body of experimental evidence. This lipid bilayer forms the matrix that defines the thickness of the membrane and its properties as a permeability barrier to ionic and hydrophilic substances. In addition to the lipids of membranes, however, proteins are also present. In fact, proteins constitute the largest fraction by weight in many membranes (2). Furthermore, it is clear that it is the proteins that mediate most of the important biochemical functions that membranes perform, such as the specific transport of ions and other hydrophilic small molecules through membranes, all enzymatic activities including electron transport and oxidative phosphorylation, the specific binding of many hormones to cell surfaces and the transduction of this binding into chemical signals across the membrane, and many others. In other words, while the lipids primarily provide the structural framework of the membrane, it is the proteins which carry out the chemical processes that are specialized to membranes. An understanding of the structure and the special characteristics of membrane proteins is therefore essential to an understanding of how membranes work.

It is understandable that not much progress was made in this direction at a time when the problem of protein structure was generally poorly understood. In 1959, Kauzmann (3), in an article that is now a classic, made hydrophobic interactions a byword for protein chemists. The first X-ray diffraction analysis of the three-dimensional structure of a protein molecule came in 1960, and commercial circular dichroism equipment became available only a little over 10 years ago. In the early 1960's, therefore, the analysis of protein structure had become sufficiently advanced to apply these developments to the problem of proteins in membranes.

In the ensuing years, explosive progress has been made at both the conceptual and experimental levels. Among the important conceptual advances were: (a) The recognition that membrane proteins could be divided into two categories, peripheral and integral (4), or extrinsic and intrinsic (5). (In what follows, the terminology we introduced will be used.) (b) The hypothesis that integral proteins are of many different kinds, but that all consist of globular, amphipathic molecules (6, 7), either partially embedded in, or spanning, the lipid interior of the membrane. (c) The proposal that under physiological conditions, the membrane is in effect a two-dimensional solution of the integral proteins in the fluid lipid bilayer, as embodied in the fluid mosaic model of membrane structure (Fig. 1) (4, 8).

PERIPHERAL AND INTEGRAL PROTEINS

The criteria we have suggested (4, 9) to discriminate between these two categories of membrane proteins are given in Table I. The distinction is an important one, because the structures and properties of peripheral and integral proteins are predicted to be widely different. Peripheral proteins are expected to be essentially similar to cytoplasmic soluble proteins in amino acid composition, three-dimensional structure, and solubility. The association of a peripheral protein with a membrane is suggested (9) to be due to specific binding to an integral protein at a site where the latter molecule protrudes from the membrane into the aqueous phase. An example of a peripheral protein is cytochrome C of mitochondrial inner membranes. This protein is readily dissociated intact from its membrane by raising the ionic strength. Once dissociated, its characteristics are indistinguishable from many water-soluble proteins. There is good evidence (summarized in Ref. 9) that cytochrome C is attached to the mitochondrial inner membrane at sites on both cytochrome C reductase and cytochrome oxidase molecules, both of which are integral proteins that very likely protrude from one surface of the inner membrane.

On the other hand, integral proteins are proposed to be amphipathic molecules, with a hydrophilic segment containing all the ionic amino acid residues of the protein and a hydrophobic segment devoid of such ionic residues. Such an integral protein would be embedded in, and interact with, the lipid interior of the membrane by its hydrophobic segment, with its hydrophilic segment exposed to water. If the protein has two hydrophilic segments with an intervening hydrophobic one of appropriate size, it could span the thickness of the membrane and protrude from both membrane surfaces.

The localization of ionic residues to the hydrophilic segments of these molecules, and their

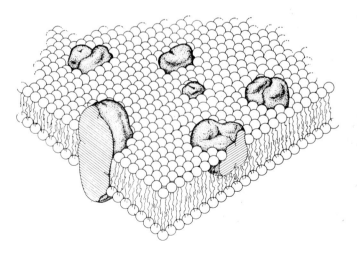

FIG. 1. The fluid mosaic model of membrane structure (8), in schematic three-dimensional and cross-sectional views. The solid bodies with stippled surfaces represent the molecules of amphipathic integral proteins, either partway embedded in the fluid lipid bilayer, or spanning the entire bilayer. The circles represent the ionic and polar head groups of the phospholipid molecules; the wavy lines represent the fatty acid chains.

THE PROTEINS OF MEMBRANES

TABLE I

Criteria for Distinguishing Peripheral and Integral Membrane Proteins (9)

Property	Peripheral protein	Integral protein
Requirements for dissociation from membrane	Mild treatments sufficient: high ionic strength, metal ion chelating agents	Hydrophobic bond-breaking agents required: detergents, organic solvents, chaotropic agents
Association with lipids when solubilized	Usually soluble free of lipids	Usually associated with lipids when solubilized
Solubility after dissociation from membrane	Soluble and molecularly dispersed in neutral aqueous buffers	Usually insoluble or aggregated in neutral aqueous buffers

absence from hydrophobic segments, would be features that distinguished a membrane integral protein from a water-soluble protein. Molecules of soluble proteins have a more-or-less uniform distribution of ionic amino acid residues over their surfaces, and therefore interact with water all over their surfaces (i.e., are soluble).

At the time this proposal was made (6, 7), no integral proteins had been sufficiently well studied to support or refute it. The proposal was based on thermodynamic considerations, and recognized the importance of hydrophobic and especially of hydrophilic interactions to membrane structure (4). By hydrophilic interactions is meant the strong thermodynamic preference of ionic and polar structures to be in an aqueous environment rather than a nonpolar one, if given a choice between the two.

Since the time this structural proposal was made, several integral proteins have been isolated from membranes in a pure state and have been subjected to structural analysis, with results that strongly support the proposal. Two examples will be discussed briefly. One is cytochrome b_5. This protein is a component of an electron transport chain found in endoplasmic reticulum membranes inside eukaryotic cells. Some 15 years of study were required to elucidate the basic structural features of this protein molecule, but here we discuss only the end result (10). The molecule is a single polypeptide chain of 152 amino acid residues. The 104 amino acids start-

ing from the NH_2-terminal end of the chain can be cleaved from cytochrome b_5 in the intact membrane by the action of any of a wide variety of proteases. The cleaved-off portion, containing the heme group of the protein, now behaves just like a water-soluble globular protein. It has been crystallized and its three-dimensional structure has been determined by X-ray crystallography (11). However, a 48 amino acid COOH-terminal fragment is left behind in the membrane upon such proteolytic cleavage. The entire cytochrome b_5 molecule can be extracted from the membrane by suitable detergent treatment, and upon removal of the detergent, it aggregates in the aqueous solution. Amino acid analysis shows that the COOH-terminal fragment is depleted of ionic amino acid residues and is enriched in hydrophobic ones. The detergent-extracted intact cytochrome b_5 molecule can spontaneously rebind to membranes, but the 104 amino acid residue fragment cannot. All of this information is entirely consistent with the proposal that cytochrome b_5 is an amphipathic molecule, with its NH_2-terminal 104 amino acid hydrophilic fragment protruding from the membrane and the COOH-terminal 48 amino acid hydrophobic fragment largely embedded in the membrane. Even more interesting, the three-dimensionally amphipathic structure is the result of a *linear* amphipathy; i.e., the linear amino acid sequence exhibits a hydrophilic end and a hydrophobic end. Apparently

479

the two ends fold up independently of one another and are connected by an intervening stretch of chain which is particularly susceptible to proteolytic attack (Fig. 2a).

Another example of an integral membrane protein that has been extensively studied is the protein glycophorin of the erythrocyte membrane (12). This protein is a glycoprotein, containing 60% carbohydrate and 40% polypeptide. The polypeptide is a single chain of 131 amino acid residues. All of the carbohydrate is in the form of short oligosaccharide chains which are covalently attached to several amino acid residues which are *all* within the first 80 residues from the amino-terminus of the polypeptide chain. This highly hydrophilic segment is known to protrude from the exterior-facing surface of the erythrocyte membrane. This hydrophilic segment is followed by a linear stretch of about 22 amino acid residues none of which is ionic, and which is thought to span the thickness of the membrane as a single α-helical segment. This relatively hydrophobic short segment is in turn followed by a hydrophilic

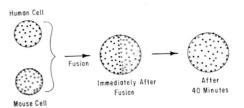

FIG. 3. A representation of the Frye–Edidin experiment (16), showing the rapid intermixing in the plane of the membrane of the surface molecules of two different cells that have been made to fuse together.

sequence to the COOH-terminus of the chain, most of which appears to protrude from the interior-facing surface of the membrane. Glycophorin is therefore a membrane-spanning integral protein, with two hydrophilic segments exposed on opposite surfaces of the membrane, separated by a hydrophobic segment embedded in, and spanning, the membrane.

Integral proteins of membranes are now being isolated in a pure state in appreciable numbers and we may expect a great surge of data on the amino acid sequences and membrane structures of their molecules in the coming decade.

A structure which we have predicted (4, 9) exists for some integral proteins in membranes is depicted schematically in Fig. 2b. It is a specific aggregate of a small number (2, 3, or 4) of identical or similar polypeptide chains which spans the thickness of the membrane. Such a subunit aggregate might have a narrow water-filled channel running down the central axis of the aggregate, as do many water-soluble subunit-aggregate enzymes, e.g., cytoplasmic malate dehydrogenase (13). Such protein-lined channels, or pores, may play an important role in membrane transport (4, 9), as is discussed below. At this time, the existence of integral proteins that have structures such as is depicted in Fig. 2b has not been firmly established, but the chemical data for the Na^+, K^+-dependent adenosinetriphosphatase (the Na pump) (14) are at least consistent with this structure.

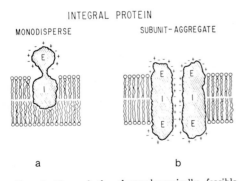

FIG. 2. Two of the thermodynamically feasible structures of integral proteins, depicted schematically in cross section (9). (a) A single-chain monodisperse protein, with its hydrophilic segment (E) protruding from the membrane and its hydrophobic segment (I) embedded in it, and a region of the chain linking the two segments (resembling the cytochrome b_5 structure, see text). (b) A subunit aggregate structure, shown here as a dimer of two similar but not identical chains, with a continuous water-filled channel running down the axis of the aggregate.

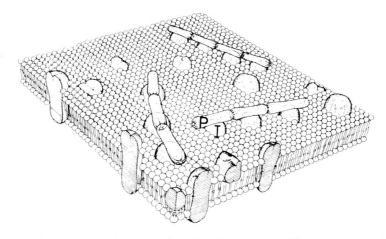

FIG. 4. A schematic view of the cytoplasmic surface of a membrane, with molecules of a peripheral protein (P) attached to molecules of an integral protein (I). The self-aggregation of P molecules, while attached to I molecules, could provide a mechanism to restrict the lateral mobility of I molecules in the plane of the membrane (17).

THE FLUID MOSAIC MODEL OF MEMBRANE STRUCTURE

The intercalation of globular amphipathic protein molecules (Fig. 2) into a lipid bilayer generates a mosaic structure (15) in the plane of the membrane. Furthermore, if the lipid bilayer is in a fluid, or liquid-crystalline, state under physiological conditions, as has been demonstrated to be true for a variety of membranes by several physical methods, this mosaic is a *fluid mosaic* (Fig. 1). The significance of this occurred to us at about the same time that a remarkable paper by Frye and Edidin (16) appeared. In this work, the authors demonstrated that when two different cells are fused, and the surface membranes of the two cells are thereby made contiguous with one another (Fig. 3), the components of the two membranes completely intermix in the plane of the membrane in a matter of minutes at 37°C. These and many other studies have confirmed the proposal that the basic organization of most membranes is a solution of integral proteins in a two-dimensional bilayer lipid solvent. Where specialized structures appear in membranes, which indicate that random mixing of these membrane components does not occur, some mechanism must be operating to prevent mixing. One such mechanism was suggested in our original paper on the fluid mosaic model (4), and involves the self-association of a *peripheral* protein, each molecule of which is attached to an *integral* protein molecule in the membrane (Fig. 4). Such an arrangement could serve to restrict the lateral mobility of the integral protein in the membrane (17).

The concept of membranes as two-dimensional solutions has already had a considerable impact on our ideas of how membranes work, and by what mechanisms individual biochemical functions of membranes are carried out (18). As we learn more about the structure of membranes and their protein components, the fluidity of membranes is certain to play an increasingly important role in our understanding of membrane function and cellular dynamics.

THE ASYMMETRY OF MEMBRANES AND MEMBRANE TRANSPORT

It is a remarkable fact that despite the rapid lateral mobility of protein and phospholipid components in the plane of most

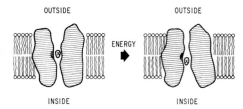

OUTSIDE   OUTSIDE

ENERGY

INSIDE   INSIDE

FIG. 5. A schematic representation of the "aggregate-rearrangement" mechanism of the translocation event in active transport through a membrane (4, 9, 24). It is proposed that a specific site for a small hydrophilic ligand X exists on the surface of the water-filled channel formed by an integral protein that is present as a subunit-aggregate in the membrane (Fig. 2b). Some energy-yielding process then drives the quaternary rearrangement of the subunits, which translocates the binding site and X from one side of the membrane to the other.

membranes, there is essentially no rotation of these components across the membrane from one surface to the other. For example, a recent estimate of the *minimum* half-time for the rate of transmembrane rotation of a phospholipid molecule in a synthetic phospholipid bilayer vesicle is 80 days (19). The absence of such rotations is, in our opinion (4, 8, 9), responsible for the *maintenance* of the molecular asymmetry that has been found to exist across membranes. The integral proteins of membranes appear to have a specific orientation in the direction perpendicular to the membrane, and the individual classes of phospholipids also appear to be asymmetrically distributed in the two half layers of membrane bilayers. We have argued that the free energies of activation required to move ionic groups of the amphipathic proteins or the phospholipids through the hydrocarbon interior of the membrane to reach the opposite surface are large enough to account for these very slow transmembrane rotations. In this view, the asymmetry of membranes is a result of the initial asymmetric insertion of integral proteins and asymmetric synthesis of phospholipids in the membrane bilayer, and the subsequent retention of the asymmetry.

The molecular asymmetry of membranes is clearly a very important feature of membrane function. As one example, the asymmetry is central to Mitchell's chemiosmotic theory of oxidative phosphorylation in mitochondria and photophosphorylation in chloroplasts (20). In this theory, the enzymes of the electron transport chain are presumed to be asymmetrically oriented in the membrane, in such a manner that the individual electron transfer reactions produce a proton electrochemical gradient across the membrane, and it is this gradient that provides the free energy to drive the phosphorylation of ADP to ATP.

The asymmetry of membranes and its maintenance also bear on the problem of the specific transport of small hydrophilic molecules and ions (ligands) through membranes. One molecular mechanism for this process that has been widely entertained in recent years is the so-called rotating carrier mechanism. In this mechanism, the transport protein binds the ligand on one side of the membrane, then rotates or diffuses across the membrane to release the ligand on the other side. For the reasons just given, this mechanism seems highly unlikely, and there is now direct evidence that contradicts it (21–23). As an alternative to the rotating carrier mechanism, we have proposed the "aggregate-rearrangement" mechanism (4, 9, 23), unaware that in essence it had earlier been suggested by Jardetzky (24). In this mechanism (Fig. 5), the transport protein is a subunit aggregate spanning the thickness of the membrane (Fig. 2b), with a continuous water-filled channel running down the axis of the aggregate. Such an aggregate can exist in at least two quaternary arrangements in one of which the ligand binding site is accessible from only one side of the membrane, and in the other arrangement, from the other side. In the case of active transport against a gradient of the ligand, some source of energy other than that of ligand binding is required to effect this rearrangement.

There is as yet little evidence to support

this mechanism of transport, but it is at least consistent with all of our present information.

In this brief discussion of the proteins of membranes, their structures, dispositions, and dynamic arrangements in the membrane, we have presented a progress report of some of the work of the past decade. Membrane structure is now one of the most active areas in molecular biology, and the next decade should enormously extend our understanding of this important subject.

ACKNOWLEDGMENT

Original studies that are discussed in this paper were supported by U.S. Public Health Service Grants AI-06659 and GM-15971.

REFERENCES

1. DAVSON, H. AND DANIELLI, J. F., "The Permeability of Natural Membranes," 2nd ed., Cambridge Univ. Press, London, 1952.
2. KORN, E. D., *Annu. Rev. Biochem.* **38**, 263 (1969).
3. KAUZMANN, W., *Advan. Protein Chem.* **14**, 1 (1959).
4. SINGER, S. J., *in* "Structure and Function of Biological Membranes" (L. I. Rothman, Ed.), p. 145. Academic Press, New York, 1971.
5. VANDERKOOI, G., *Ann. N. Y. Acad. Sci.* **195**, 6 (1972).
6. LENARD, J. AND SINGER, S. J., *Proc. Nat. Acad. Sci. U.S.A.* **56**, 1828 (1966).
7. WALLACH, D. F. H. AND ZAHLER, P. H., *Proc. Nat. Acad. Sci. U.S.A.* **56**, 1552 (1966).
8. SINGER, S. J. AND NICOLSON, G. L., *Science* **175**, 720 (1972).
9. SINGER, S. J., *Annu. Rev. Biochem.* **43**, 805 (1974).
10. STRITTMATTER, P., ROGERS, M. J., AND SPATZ, L., *J. Biol. Chem.* **247**, 7188 (1972).
11. MATTHEWS, F. S., ARGOS, P., AND LEVINE, M., *Cold Spring Harbor Symp. Quant. Biol.* **36**, 387 (1972).
12. TOMITA, M. AND MARCHESI, V. T., *Proc. Nat. Acad. Sci. U.S.A.* **72**, 2964 (1975).
13. HILL, E., TSERNOGLOU, D., WEBB, L., AND BANASZAK, L. J., *J. Mol. Biol.* **72**, 577 (1972).
14. KYTE, J., *J. Biol. Chem.* **250**, 7443 (1975).
15. GLASER, M., SIMPKINS, H., SINGER, S. J., SHEETZ, M., AND CHAN, S. I., *Proc. Nat. Acad. Sci. U.S.A.* **65**, 721 (1970).
16. FRYE, L. D. AND EDIDIN, M., *J. Cell Sci.* **7**, 313 (1970).
17. PAINTER, R. G., SHEETZ, M., AND SINGER, S. J., *Proc. Nat. Acad. Sci. U.S.A.* **72**, 1359 (1975).
18. SINGER, S. J., *Advan. Immunol.* **19**, 1 (1974).
19. ROSEMAN, M., LITMAN, B. J., AND THOMPSON, T. E., *Biochemistry* **14**, 4826 (1975).
20. MITCHELL, P., *Biol. Rev.* **41**, 445 (1966).
21. KYTE, J., *J. Biol. Chem.* **249**, 3652 (1974).
22. MARTINOSI, A. AND FORTIER, F., *Biochem. Biophys. Res. Commun.* **60**, 382 (1974).
23. DUTTON, A., REES, E. D., AND SINGER, S. J., *Proc. Nat. Acad. Sci. U.S.A.* **73**, 1532 (1976).
24. JARDETZKY, O., *Nature (London)* **211**, 969 (1966).

Effects of Bivalent Cations and Proteins on Thermotropic Properties of Phospholipid Membranes

Implications for the Molecular Mechanism of Fusion and Endocytosis

DEMETRIOS PAPAHADJOPOULOS

Experimental Pathology, Roswell Park Memorial Institute, Buffalo, New York 14263

Received September 27, 1976; accepted October 28, 1976

Phospholipid membranes undergo thermotropic transitions from the solid to the liquid crystalline state, which represent the melting of the acyl chains. The temperature (T_c) and enthalpy (ΔH) of the transition are influenced by the acyl chain length, double bonds, and head-group, and are characteristic for each phospholipid. The effects of different metal ions and proteins on both T_c and ΔH of several phospholipid membranes have been studied by differential scanning calorimetry. Calcium and magnesium both induce an increase in T_c of acidic phospholipids by approximately 10°C. In addition, calcium but not magnesium induces the appearance of a new phase transition at much higher temperatures, and an isothermic phase separation from a mixed membrane of neutral and acidic phospholipids. Hydrophobic proteins that tend to penetrate into the lipid bilayer have no effect on the T_c, but produce a decrease in ΔH proportional to their concentration in the membrane. On the contrary, proteins that bind to the interface by ionic bonds and do not penetrate into the bilayer have either no effect or increase the T_c, and also produce a large increase in ΔH. A third category of proteins that binds primarily by ionic bonds, but also expands phospholipid monolayers, produces a substantial decrease of both T_c and ΔH indicating a marked fluidization effect. It is clear from these results that different proteins and metal ions can affect profoundly the properties of lipid bilayers in a highly specific manner related to the type of interaction involved in each case. On the basis of the above results it is proposed that phase separations and density fluctuations induced isothermally by divalent metal ions and proteins are intimately involved in such phenomena as membrane fusion and the initial shape changes observed during endocytosis.

INTRODUCTION

It is now well established with both natural (1–4) and artificial membrane systems (5–7) that lipid phase transitions can have a large effect on the activity of various membrane-bound enzymes. The usual case is that the enzyme activity is diminished considerably at temperatures below the transition of the lipid bilayer from fluid to solid (T_c). The question that has not been answered as yet is whether lipid phase transitions play any important role during the normal functioning of membranes of living cells. More specifically it is not known yet whether changes in membrane fluidity and phase transitions play a regulatory role in living cells by controlling membrane enzyme activity or whether changes

in membrane fluidity are responsible for functional changes during pathological conditions. Such a role has been proposed for the case of atherosclerosis (8) and neoplastic transformation (9) although definitive data relating to the role of lipid phase transitions in mammalian cells are still missing.

The lack of evidence on the role of lipid phase transitions is due to several reasons: Phospholipids extracted from natural membranes exhibit the solid-to-fluid transition at temperatures usually below the ambient temperature. This is particularly true for membranes of higher organisms, while the T_c of lipids from certain bacteria could be only a few degrees below ambient temperature. The presence of cholesterol complicates the situ-

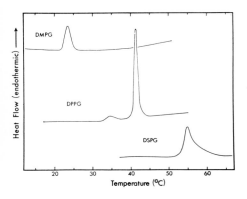

FIG. 1. Differential scanning calorimetry (DSC) thermograms of various phosphatidylglycerols dispersed in NaCl (100 mM) buffer, pH 7.4.

ation by producing a broadening of the temperature region for the transition and by eliminating the transition altogether at molar ratios of 0.5 to 1.0 (cholesterol to phospholipid). Phospholipids extracted from mammalian membranes exhibit single but very broad transitions (10) indicating partial mixing of the different components. It is possible, however, that the same lipids before extraction were segregated into discrete domains on the plane of the membrane with their own characteristic transition temperatures. It is also possible that the transition temperature of individual lipids could be modified by its interaction with other membrane components such as metal ions and proteins. Such interactions could modify sufficiently the properties of specific regions of the lipid bilayer in membranes, so that the phase changes occurring in that region would be at different temperatures from the average lipid in the rest of the membrane.

Because of all the above considerations it is very difficult to obtain direct information from biological membranes, relating specific functions to lipid phase transitions occurring at temperatures above the T_c of the average lipid components. Through the use of model systems, however, it has become possible to investigate the possible interactions that

might be expected between specific lipids, proteins, and metal ions. Several important findings established during the last few years enhance considerably our understandings of such interactions: the phase transition temperatures are defined primarily by the chain length, unsaturation, branching, etc. (11) but the head-group is also very important (12). Divalent metal ions can increase the T_c of acidic lipids by 10°C or more (7, 12, 13), inducing isothermal phase transitions. Calcium can induce isothermal phase separations from mixture of phospholipids (12, 14). Isothermal phase separations of acidic phospholipids can also be induced by positively charged proteins (15) and polypeptides (16). Interacting proteins can decrease the T_c of phospholipid bilayers (17, 18) which is a fluidizing effect, but can also increase the T_c or have no effect on it, depending on the type of interaction involved (18). Certain membrane proteins can interact with lipid molecules within the lipid bilayer producing an "annulus" or "halo" of relatively strongly bound lipid surrounding the protein (19). This boundary lipid could be different from the average lipid composition of the membrane.

In this paper we will review the recent evidence on the interactions of lipid bilayers with metal ions and proteins and we will then utilize the accumulated data as the basis of an explanation of certain events occurring during the phenomena of membrane fusion and phagocytosis.

MATERIALS AND METHODS

All the lipids discussed in this paper were synthesized in this laboratory by methods described previously (7, 12) except for phosphatidylethanolamine, which was obtained from Fluka AG (Switzerland). Proteins were either purchased or purified, as discussed previously (18). Differential scanning calorimetry was carried out with a Perkin–Elmer DSC-2 as before (12, 18).

RESULTS AND DISCUSSION

It is well established from studies with various lecithins that the aliphatic chain length plays a very important role in defining the T_c of the solid-to-liquid crystalline phase transition (11). Data on the thermotropic transitions of a series of phosphatidylglycerols (PG) are presented in Fig. 1.[1] It can be seen that the T_c increases by 14–17°C for each two-carbon unit added to each chain. Both the main transition and the premelt are evident with all three lipids. In all these respects the PG lipids behave remarkably similar to the phosphatidylcholines (PC). As will be discussed later, however, the PG are very sensitive to their ionic environment, while the PC are not.

The effect of various head-groups on the thermotropic properties of phospholipids with the same acyl chains is shown in Fig. 2. Several interesting differences and similarities become apparent. The PC and PG both have much lower T_c than the other analogs. The similarity in the T_c of PC and PG is intriguing since one is zwitterion (PC) and the other is carrying a fully ionized group under these conditions (PG). As a first approximation, the presence of a nonneutralized charged group would be expected to decrease the T_c of PG compared to PC due to electrostatic repulsions at the plane of the membrane (13). The problem was recently discussed at length (12) and it was proposed that lateral attractive interactions such as hydrogen bonding between neighboring molecules in PE and phosphatidic acid (PA) might account for the higher T_c. However, this would still not explain the high T_c of PA at pH 9.0 when it is carrying two charges per molecule, or the behavior of the diglyceride analog, which also has a high T_c. Data

[1] Abbreviations: DMPC, dimyristoylphosphatidylcholine; DMPG, dimyristolyphosphatidylglycerol; DMPA, dimyristoylphosphatidic acid; DMPE, dimyristoylphosphatidylethanolamine; DPPG, dipalmitoylphosphatidylglycerol; DSPC, distearoylphosphatidylcholine; DSPG, distearoylphosphatidylglycerol; PS, phosphatidylserine (bovine brain).

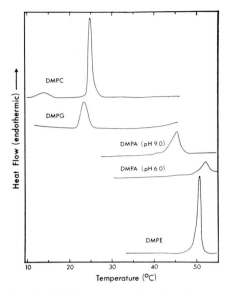

FIG. 2. DSC thermograms of various dimyristoyl phospholipids suspended in NaCl (100 mM) buffer at pH 7.4 except for DMPA which was dispersed at pH either 9.0 or 6.0 as indicated.

from phosphatidylserine (20) indicate that it also has a higher T_c than the homologous PC. In any case, the shift of the T_c of the PA following changes in pH (12, 13) indicates that it is possible to induce phase transitions isothermally by changes in the pH of the aqueous environment.

The effect of changes in the ionic strength and metal ion content on the thermotropic properties of an acidic phospholipid (PG) is shown in Fig. 3. Changing the ionic strength from 0.01–0.1 M (NaCl) has only a small effect, mainly shifting the main transition up by approximately 1°C (12). However, the addition of relatively low concentrations of divalent metals has a very large effect (12, 13, 21). Low concentrations of Mg^{2+} (1 mM) shift the T_c by approximately 10°C (from 41 to 50°C, Curves 1, 2). Higher concentrations produce only a small further shift to 52°C at 5 mM (Curve 3) and 53.5°C at 20 mM (not shown). Calcium has a much more pronounced effect, shifting the T_c to 56.5°C

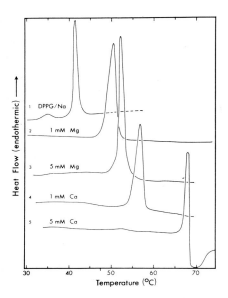

Fig. 3. DSC thermograms of dipalmitoylphosphatidylglycerol in the presence of magnesium and calcium. DPPG was dispersed in NaCl (100 mM) buffer, pH 7.4, and then transferred into a larger aliquot of the same buffer containing in addition: 1 mM MgCl$_2$ (Curve 2); 5 mM MgCl$_2$ (Curve 3); 1 mM CaCl$_2$ (Curve 4); 5 mM CaCl$_2$ (Curve 5).

It appears that the effect of divalent metals at low concentrations which increase the T_c by approximately 10°C is in reasonably good agreement with the expected effect of neutralization of the negative charges (13). The effects at higher concentrations however, which lead to the formation of cochleate cylinders and show significant qualitative differences between Ca^{2+} and Mg^{2+} (12, 22) must be related to specific interactions, which are still not well understood. In any case these effects appear to be related to the calcium-induced fusion phenomena between phospholipid membranes described recently (24).

When two phospholipids of different T_c are mixed in chloroform prior to their dispersion in an aqueous phase, the resultant membranes exhibit either one broad transition intermediate in temperature between that of the pure components, or two independent transitions. In the former case, it is considered that the two components are mixed, while in the latter case the two components separate out in different domains. Such phase separation has been shown to occur in mixed lipid systems with widely different T_c such as dioleoylphosphatidylcholine and distearoylphosphatidylcholine (DSPC) (25, 26). When two phosphatidylcholines vary only by two carbons in length they give only one intermediate T_c and this characteristic has been used recently to study the problem of fusion between separate phospholipid vesicles (14, 24). PS will also mix with DSPC as shown in Curves 1 and 3 of Fig. 4 when codispersed in NaCl buffer. The same mixture, however, shows a discrete phase separation when dispersed in NaCl buffer containing CaCl$_2$ or when CaCl$_2$ is added to preformed PS/DSPC vesicles (12, 14). Curve 4 of Fig. 4 shows the appearance of an endothermic peak at the temperature of the pure DSPC indicating the separation of this component from the original mixture. As was shown earlier (12), Ca^{2+} at that concentration (10 mM) will bind to pure PS and form crystalline membranes at that temperature and give no endothermic peak at the temperature regions

at relatively low concentrations (1 mM, Curve 4). At higher concentrations (5 mM, Curve 5) Ca^{2+} induces a qualitatively different effect, with a transition at much higher temperatures (68°C) and an exothermic component. Reheating of the DPPG sample in the presence of 1 mM Ca^{2+} gives a broad endothermic peak at 54°C. Other endothermic peaks (at higher temperatures) are also obtained in the presence of high concentrations of both Mg^{2+} (>5 mM) and Ca^{2+} (>1 mM) which are not present during reheating of the samples. Such metastability of PG membranes in the presence of Ca^{2+} and Mg^{2+} (at much higher concentrations) was reported earlier (22) and seems to be related to the formation of cylindrical structures (21–23) which were identified in the case of PS and Ca^{2+} as composed of folded lamellae forming cochleate cylinders (23).

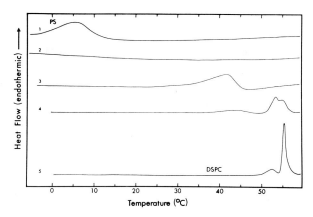

FIG. 4. DSC thermograms indicating a phase separation induced by Ca^{2+} in mixed membranes. Curve 1: phosphatidylserine (PS) dispersed in NaCl (100 mM) buffer, pH 7.4. Curve 2: Same as in 1 except buffer contained $CaCl_2$ (1 mM). Curve 3: A mixture of PS with DSPC (2:1 molar ratio) dispersed in NaCl buffer. Curve 4: Same PS/DSPC mixture dispersed in same buffer by sonication (1 hr at 42°C) then incubated for 1 hr in the same buffer containing in addition $CaCl_2$ (10 mM) at 42°C. If EDTA is added following this treatment (11 mM, pH 7.4) the thermogram obtained is similar to that shown in Curve 3. Curve 5: DSPC dispersed in NaCl (100 mM) buffer, pH 7.4.

tested. Phase separation of PS from PC by Ca^{2+} has also been reported on the basis of ESR data (27).

Calcium-induced phase separations have also been obtained with phosphatidic acid–PC mixtures (12, 16) but only at high pH (8.0) and not at pH 6.5 (12). Mixtures of PG with PC under similar conditions do not yield phase separations (unpublished observations) and even the PS/PC mixtures will give separations only when the Ca^{2+} concentration exceeds a threshold value which depends on the percentage of PS in PC (12, 14).

Membranes composed of PS and PC usually show smooth surfaces in freeze fracture electron micrographs when dispersed in NaCl. However, following the addition of Ca^{2+} the same membranes appear in freeze fractures as "patched" with stepped structures of irregular size and shape (Vail and Papahadjopoulos, unpublished observations). This indicates that the separated PC domains coexist laterally with the PS–Ca domains in the same membrane, and the domain size is in the range of 500–2000 Å. The coexistence of PS and PC domains is also indicated by the

reversibility of the calcium-induced phase separation, following chelation of the calcium by EDTA (14). If the separated PC component had formed completely separate phases in separate vesicles it would not be mixed again with PS following the addition of EDTA, since PC is not reactive to either Ca or EDTA at these concentrations. As reported earlier (12) the addition of Mg^{2+} or a decrease of the pH of the aqueous phase does not produce phase separation between PS and PC, indicating that the Ca effect discussed above is not simply due to charge neutralization. Phase separation appears to be related to vesicle fusion, as elaborated elsewhere (14).

The calorimetric studies of Steim and colleagues (28) have established that there is a general similarity between the thermotropic transition of intact membranes from mycoplasma, mitochondria, and microsomes with those of the extracted lipids when suspended in aqueous salt solutions. The interpretation of these and other (29, 30) studies, however, is complicated by the lack of detailed information concerning the effects of different

489

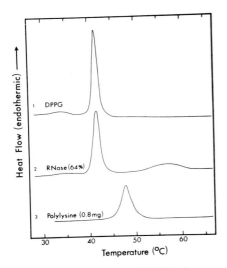

FIG. 5. DSC thermograms of DPPG in the presence of various proteins: (1) DPPG alone (3 μmole in 1.5 ml total volume) dispersed in NaCl (10 mM) buffer, pH 7.4 (2). Same as above, in the presence of RNase (64% of total dry weight after centrifugation. (3) Same as above, except in the presence of 0.8 mg of poly-L-lysine.

properties of DPPG are shown in Fig. 5. In this, as in the following figure, the proteins or polypeptide was added to the aqueous salt solution and was present initially during the suspension of the dry lipid. The amount of protein bound to the lipid membranes was determined as described (18) and is given in the figures. It appears from Fig. 5, Curve 2, that the binding of the (basic) protein, ribonuclease, to the acidic phospholipid membrane PG, has no appreciable effect on the midpoint T_c even at high percentages. However, the enthalpy of transition (ΔH) was found to be increased considerably in the presence of the protein (from 7.9 to 9.5 kcal/mole). Curve 3 in Fig. 5 indicates that the presence of polylysine produced a considerable increase in the T_c, concomitant to an increase of ΔH from 7.9 to 11 kcal/mole.

The above effects of these two peptides were interpreted as stabilization of the lipid bilayer through simple electrostatic binding

types of lipid–protein interactions on the thermotropic properties of the lipids. Several earlier studies (17, 21, 30–32) which have reported various effects of proteins on the molecular motion and other properties of lipid bilayers have not related the observed effects to the specific type of interaction involved in each case. More recently the interaction of phospholipid membranes with several proteins and peptides was examined by several techniques involving differential scanning calorimetry, vesicle permeability, and monolayer penetration (18). By a correlation of the results obtained it was possible to describe three different types of interactions involving simple surface adsorption, or surface adsorption with partial penetration, or complete penetration of the phospholipid bilayer.

The characteristics of simple surface adsorption were satisfied by the interactions of ribonuclease and poly-L-lysine. The effects of these two polypeptides on the thermotropic

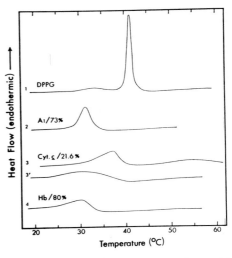

FIG. 6. DSC thermograms of DPPG in the presence of various proteins. Details as in Fig. 5. (1) DPPG alone; (2) in the presence of basic myelin protein (A₁, 73% of total dry weight after centrifugation); (3) in the presence of cytochrome c (21.6% by weight); (3′) same as in (3), except the sample heated (to 60°C) and cooled (to 0°C) two times before run; (4) in the presence of hemoglobin (80% by weight).

490

at the interphase (18). The conclusion was supported by the inability of both ribonuclease and polylysine to increase the permeability of negatively charged lipid vesicles (33) and their inability to expand monolayers of acidic phospholipids at the air–water interface at a pressure >25 dyn/cm (34).

The effects of three other water-soluble proteins on phospholipid phase transition is shown in Fig. 6. Here, in all three cases the proteins induce a considerable decrease in the midpoint T_c, broadening of the endothermic peak, and considerable decrease in the ΔH (18). The same proteins have been shown to be able to expand acidic phospholipid monolayers at high film pressure (34) and also induce a very large increase in the permeability of acidic phospholipid vesicles to small molecules (18, 33–35). These effects have been interpreted as due to "partial penetration" of the lipid bilayer by the protein and consequent "deformation" of the packing of the phospholipid acyl chains (18, 39). It has been proposed that this "fluidizing" effect is induced by the formation of nonpolar contacts between the protein and the lipid at the lipid–water interface. Earlier X-ray diffraction data with similar systems had indicated that the protein produces an increase in the overall width of the membrane (36–39) with a concomitant decrease in the width of the lipid region (37, 38), although the latter point assumes no interdigitation of the lipid and protein layers (38). The "fluidization" of the bilayer seen here, is in agreement with the "thinner" lipid bilayer postulated earlier (37–39).

Incorporation of a "hydrophobic" membrane protein such as the major apoprotein from myelin proteolipid (40, 41) produces relatively small effects on the thermotropic properties of lipid membranes. As shown in Fig. 7, this protein (lipophilin, (42)) even when it is incorporated to a high percentage has no effect on the midpoint T_c of either PC or PG. The ΔH of the transition is decreased, however, in linear relationship to the percentage of protein per total weight of membrane (41).

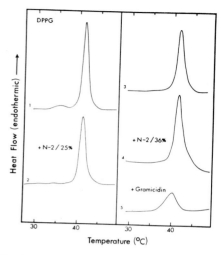

FIG. 7. DSC thermograms of DPPG in the presence of a hydrophobic protein (lipophilin): (1) DPPG alone in NaCl (100 mM) buffer, pH 7.4; (2) DPPG (3 μmole) solubilized in chloroform–methanol–water (10/5/1, v/v/v) in the presence of 0.72 mg of lipophilin (N-2), 25% protein by dry weight in buffer after centrifugation. (3) DPPG alone in NaCl (10 mM) buffer, pH 7.4. (4) DPPG (3 μmole) dispersed in 3 ml of NaCl (10 mM) buffer, pH 6.5 containing 0.5 mg N-2 (36% protein by dry weight after centrifugation). (5) DPPG (3 μmole) mixed in chloroform with 0.5 μmole Gramicidin A, and dispersed in 3 ml of NaCl (100 mM) buffer, pH 7.4. All the above samples were incubated at 42°C for 1 hr before brief contrifugation at room temperature (22–25°C) and calorimetry.

Similar results were obtained irrespective of whether the initial mixing of the lipophilin with the lipid was accomplished in chloroform/methanol, water, or buffer. Lipid membranes incorporating lipophilin have been shown to contain "intramembrane particles" in freeze-fracture electron microscopy (40, 41), an appearance which is taken to indicate the location of membrane-embedded proteins.

The decrease in ΔH induced by the lipophilic protein without concomitant effect on the midpoint T_c was interpreted in terms of complete embedding of this protein into the lipid bilayer and relatively strong association with the surrounding phospholipid molecules

DEMETRIOS PAPAHADJOPOULOS

which are thus not allowed to participate in the cooperative motion of the free lipids in the rest of the bilayer (18). This interpretation was strengthened by more recent ESR data indicating two lipid domains within the bilayers containing the lipophilin (42). Similar proposals were made earlier concerning the existence of a relatively strongly bound "halo" or "annulus" of lipid surrounding the embedded protein molecules of cytochrome oxidase (19) and calcium ATPase (43). It is possible that the annular lipid could have different chemical composition from the average mixed lipid bilayer, although no evidence of this effect has been obtained with the above proteins. The ESR evidence however, indicates that the lipid bound to the protein is relatively immobilized compared to the lipid molecules in the rest of the bilayer (19, 42) and could thus be considered to have different packing density and "fluidity" compared to the rest of the bilayer.

Implications for Membrane Fusion and Phagocytosis

All the data discussed above relating to the effects of metal ions and proteins on the thermotropic properties of lipid bilayers are relatively recent findings and their relationship to the function of biological membranes is still not well established. However, certain recent information of the architecture of cell membranes, and cell shape changes in response to membrane active molecules make it possible to relate the newly understood properties of lipid bilayers to specific cell membrane functions. It is now well accepted that the majority of the lipid in biological membranes behaves as in pure bilayers (44). Moreover, recent evidence from several laboratories indicates that cell plasma membrane is asymmetric, both in terms of lipids and proteins (45). Of the phospholipids, it appears that the phosphatidylserine, which is the main acidic species in mammalian cells, is located predominantly inside (in the cytoplasmic side), while the neutral phospholipid

PC and sphingomyelin are located outside (46, 47). Other evidence shows that PS molecules are located close enough to react with bifunctional reagents, which indicates the possibility for formation of clusters or separate domains (48). A large percentage of the membrane proteins (at least in the erythrocyte) are located on the inside (49, 50). It has also been shown that other proteins penetrate through the lipid bilayer (51, 52) with the amino terminal end of the peptide carrying the carbohydrate and sialic acid residues exclusively on the outside of the plasma membrane (52). Finally, it appears that both proteins and lipids are moving rapidly on the plane of the membrane, and that the lipid bilayer is accessible for interaction with large macromolecules added to the bulk phase, both from outside and inside (45).

The existence of lipid asymmetry in the erythrocyte membrane has been employed recently to explain the cell shape changes produced by membrane-active drugs (53). The observations were that local anesthetics, which are positively charged amphipathic molecules, produce "cupping" of erythrocytes, while amphipathic negatively charged drugs produce "crenated" erythrocytes. Sheetz and Singer explained the difference by a preferential localization of the drugs on the inside (positively charged local anesthetics) or the outside (negatively charged drugs) monolayer of the erythrocyte lipid bilayer (53). This preferential localization of the drug molecules would be expected from the known charge asymmetry of the phospholipids. Once established however, such preferential localization would induce expansion of one of the two monolayers assuming an isopieric system. This differential expansion would be expected to lead to either an invagination ("cupping"), if the drug is localized in the internal monolayer or a protrusion ("crenation"), if the drug is localized on the external monolayer. This explanation was called the bilayer-couple hypothesis (53).

The behavior of the two monolayers of the lipid bilayer of cell membranes as two separate systems with characteristic phase transitions and phase separations has been already suggested by Fox and colleagues as an explanation for the complex temperature dependency of certain membrane enzymes (54). Although the independence of the thermotropic properties of the two monolayers in a bilayer has not been studied or shown experimentally, it appears reasonable to expect that an asymmetric bilayer will exhibit a complex thermotropism to reflect the phase transitions of each of the monolayers. It is also reasonable to expect that isothermic phase separations induced by metal ions and proteins can occur independently in the internal or external monolayer of the lipid bilayer of the cell plasma membrane. We will now examine in detail the expected changes in membrane shape and ability to fuse as a result of lateral phase separations and clustering, which might occur isothermally in one or the other of the two monolayers.

Membrane fusion is a key event involved in numerous exocytotic phenomena which are part of the secretory mechanism of eucariotic cells (55). It is also generally accepted that Ca^{2+} (but not Mg^{2+}) is intimately involved in the process of membrane fusion (55, 56). Mg^{2+} is usually found in high concentrations inside the cell (cytoplasm) while Ca^{2+} is low inside, high outside the cell. Moreover, an influx of Ca^{2+} is usually associated with exocytotic events such as the acetylcholine release in presynaptic nerve endings (57). Although numerous intracellular events can be triggered by the transient increase in intracellular Ca^{2+} it is possible that the fusion of the secretory vesicles with the plasma membrane is a direct consequence on the phase separation that Ca^{2+} could be inducing at the "internal" monolayer of the plasma membrane. This suggestion is based on recent evidence on the fusion of lipid vesicles composed of PS and PS/PC (14, 23). This evidence shows that fusion occurs only in the presence of Ca^{2+} above a threshold concentration, and only under conditions where a phase separation is occurring. Mg^{2+} which

can induce vesicle aggregation (58) does not induce fusion or phase separation, except under special conditions (24, 58). It is therefore reasonable to suggest that the influx of Ca^{2+} (produced by an initial membrane depolarization or the action of a hormone on its receptor) induces a phase separation of PS which is normally localized at the internal monolayer. Collision of secretory vesicles in regions of domain boundaries would be followed by fusion, because of the unstable nature of such boundaries in phase separated systems (58). Phase separation could also be occurring on the "external" monolayer of the secretory vesicles as they approach close to the periphery of the cell and the plasma membrane due to the gradient of Ca^{2+} concentration. Such vesicles would then be further susceptible to fusion as they collide with the plasma membrane. This proposal is partly based on recent data indicating the Mg^{2+}, which does not support fusion by itself, will enhance the ability of Ca^{2+} to induce fusion and phase separation (Papahadjopoulos, Vail, Newton, unpublished observations).

Phagocytosis and the membrane invagination involved in the early stages of endocytosis could also be induced by lipid phase separations and clustering in one of the two monolayers. In the following paragraphs we will examine three hypothetical situations in which the binding of a particle at the cell periphery can induce condensation or expansion of specific lipid domains in either the external or internal monolayer. The basic principle is similar to the "bilayer-couple" hypothesis proposed earlier as an explanation for the membrane shape changes induced in erythrocytes by surface-active drugs (53). In the present case, however, the changes in membrane shape are not induced by insertion of extraneous molecules into the lipid bilayer, but by isothermal lateral phase separations, clustering, or local perturbations of lipid packing density induced by divalent cations or proteins in response to the binding of the particle.

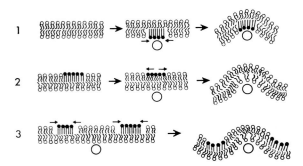

FIG. 8. Diagramatic representation of the effects of isothermic lipid phase separations on membrane shape: (1) Membrane initially fluid, with a "solid" domain induced on external monolayer at point of particle attachment. (2) Mixed membrane containing a solid domain on the internal monolayer initially. Particle attachment induces expansion and fluidization of the domain. (3) Fluid membrane initially, with "solid" domains forming on the internal monolayer at the periphery of the area where the particle is attached.

Figure 8 is a diagrammatic representation of some of the changes that might be anticipated as a particle collides with the external monolayer of the cellular plasma membrane. In situation 1 the adsorption of the particle on the surface induces a condensation of the neighboring domain of the lipid monolayer or a clustering of lipid molecules that are in a more condensed state than the rest. This could be achieved either by charge neutralization or by specific binding to lipid molecules that tend to form more condensed monolayers, or by binding specifically to proteins that aggregate in the region along with their motion-inhibited "annular" lipids. In either case, the result of the creation of a more condensed domain at the external monolayer would be an invagination and the beginning of the engulfing of the particle.

In situation 2, Fig. 8, the binding of the particle induces the expansion of a preexisting condensed cluster at the "internal" monolayer. Invagination would then follow as a result of the differential expansion of the two monolayers at that point.

Finally, as shown in situation 3, Fig. 8, the binding of the particle could induce the formation of condensed clusters in the perimetry of the area where the particle comes in contact but the clusters are formed only in the internal monolayer. This differential condensation would tend to form protrusions in the cell membrane region around the particle, which is eventually engulfed.

Although the above suggestions are hypothetical, the conceptual framework is based on recent evidence which shows that Ca^{2+} and to a lesser extent Mg^{2+} and other changes in the ionic environment of the aqueous phase can induce isothermic phase transitions and phase separations in phospholipid bilayers (12–16, 27). Changes in erythrocyte shape induced by an increase in intracellular Ca^{2+} concentration (59) can similarly be explained by phase changes induced within the inner lipid monolayer of the erythrocyte membrane, and are of obvious relevance to the above suggestions concerning membrane shape changes during endocytosis.

ACKNOWLEDGMENTS

I wish to thank Mr. Tom Isac for his expert technical assistance on the synthesis of phospholipids and calorimetry. I also thank Drs. A. Sanfeld for helpful discussions. Supported by the National Institutes of Health, Grant 5 Ro1 GM 18921.

REFERENCES

1. OVERATH, P., SCHAIRER, H. V., AND STOFFEL, W., *Proc. Nat. Acad. Sci.* **67**, 606 (1970).

2. WILSON, G. AND FOX, C. F., *J. Mol. Biol.* **55**, 49 (1971).

3. ESFAHANI, M., LIMBRICK, A. R., KNUTTON, S., OKA, T., AND WAKIL, S., *Proc. Nat. Acad. Sci. U.S.A.* **68**, 3180 (1971).

4. RAISON, J. K., *Bioenergetics* **4**, 285 (1973).

5. KIMELBERG, H. K. AND PAPAHADJOPOULOS, D., *Biochim. Biophys. Acta* **282**, 277 (1972).

6. PAPAHADJOPOULOS, D. AND KIMELBERG, H. K., in "Progress in Surface Sciences" (S. G. Davison, Ed.), Vol. 4, Part 2, p. 141. Pergamon, Oxford, 1973.

7. KIMELBERG, H. K. AND PAPAHADJOPOULOS, D., *J. Biol. Chem.* **249**, 1071 (1974).

8. PAPAHADJOPOULOS, D., *J. Theor. Biol.* **43**, 329 (1974).

9. INBAR, M. AND SHINITZKY, M., *Proc. Nat. Acad. Sci. U.S.A.* **71**, 4229 (1974).

10. BLAZYK, J. F. AND STEIM, J. M., *Biochim. Biophys. Acta* **266**, 737 (1972).

11. LADBROOKE, B. D. AND CHAPMAN, D., *Chem. Phys. Lipids* **3**, 304 (1969).

12. JACOBSON, K. AND PAPAHADJOPOULOS, D., *Biochemistry* **14**, 152 (1975).

13. TRAUBLE, H. AND EIBL, H., *Proc. Nat. Acad. Sci. U.S.A.* **71**, 214 (1974).

14. PAPAHADJOPOULOS, D., POSTE, G., SCHAEFFER, B. E., AND VAIL, W. J., *Biochim. Biophys. Acta* **352**, 10 (1974).

15. BIRRELL, G. B. AND GRIFFITH, O. H., *Biochemistry* **15**, 2925 (1976).

16. GALLA, H.-J. AND SACKMANN, E., *Biochim. Biophys. Acta* **401**, 509 (1975).

17. CHAPMAN, D., URBINA, J., AND KEOUGH, K. M., *Biochim. Biophys. Acta* **249**, 2512 (1974).

18. PAPAHADJOPOULOS, D., MOSCARELLO, M., EYLAR, E. H., AND ISAC, T., *Biochim. Biophys. Acta* **401**, 317 (1975).

19. JOST, P., GRIFFITH, O. H., CAPALDI, R. A., AND VANDERKOOI, G., *Biochim. Biophys. Acta* **311**, 141 (1973).

20. MACDONALD, R. C., SIMON, S. A., AND BAER, E., *Biochemistry* **15**, 885 (1976).

21. VERKLEIJ, A. J., DEKRUYFF, B., VERVERGAERT, P. H. J. TH., TOCANNE, J. F., AND VANDEENEN, L. L. M., *Biochim. Biophys. Acta* **339**, 432 (1974).

22. VERVERGAERT, P. H. J. TH., DEKRUYFF, B., VERKLEIJ, A. J., TOCANNE, J. F., AND VAN-DEENEN, L. L. M., *Chem. Phys. Lipids* **14**, 97 (1975).

23. PAPAHADJOPOULOS, D., VAIL, W. J., JACOBSON, K., AND POSTE, G., *Biochim. Biophys. Acta* **394**, 483 (1975).

24. PAPAHADJOPOULOS, D., VAIL, W. J., PANGBORN, W. A., AND POSTE, G., *Biochim. Biophys. Acta.* **448**, 265 (1976).

25. PHILLIPS, M. C., HAUSER, H., AND PALTAUF, F., *Chem. Phys. Lipids* **8**, 127 (1972).

26. SHIMSCHICK, E. AND MCCONNELL, H. M., *Biochemistry* **12**, 2351 (1973).

27. OHNISHI, S.-I. AND ITO, T., *Biochemistry* **13**, 881 (1974).

28. STEIM, J. M., TOURTELLOTTE, M. E., REINERT, J. C., MCELHANEY, R. N., AND RADER, R. L., *Proc. Nat. Acad. Sci. U.S.A.* **63**, 104 (1969).

29. OVERATH, P. AND TRAUBLE, H., *Biochemistry* **12**, 2625 (1973).

30. CHAPMAN, D. AND URBINA, J., *FEBS Lett.* **12**, 169 (1971).

31. BUTLER, K. W., HANSON, A. W., SMITH, I. C., AND SCHNEIDER, H., *Canad. J. Biochem.* **51**, 980 (1973).

32. HAMMES, G. G. AND SCHULLERY, S. E., *Biochemistry* **9**, 2555 (1970).

33. KIMELBERG, H. AND PAPAHADJOPOULOS, D., *J. Biol. Chem.* **246**, 1142 (1971a).

34. KIMELBERG, H. K. AND PAPAHADJOPOULOS, D., *Biochim. Biophys. Acta* **233**, 805 (1971b).

35. CALISSANO, P., ALEMA, S., AND RUSCA, G., *Biochim. Biophys. Acta* **255**, 1009 (1972).

36. PAPAHADJOPOULOS, D. AND MILLER, N., *Biochim. Biophys. Acta* **135**, 624 (1967).

37. GULIK-KRZYWICKI, T., SHECHTER, E., LUZZATI, V., AND FAURE, M., *Nature* **223**, 1116 (1969).

38. RAND, R. P., *Biochim. Biophys. Acta* **241**, 823 (1971).

39. BLAUROCK, A. E., *Biophys. J.* **13**, 290 (1973).

40. VAIL, W. J., PAPAHADJOPOULOS, D., AND MOSCA-RELLO, M. A., *Biochim. Biophys. Acta* **345**, 463 (1974).

41. PAPAHADJOPOULOS, D., VAIL, W. J., AND MOSCA-RELLO, M. A., *J. Membr. Biol.* **22**, 143 (1975).

42. BOGGS, J. M., VAIL, W. J., AND MOSCARELLO, M. A., *Biochim. Biophys. Acta.* In press.

43. WARREN, G. B., HOUSLAY, M. D., METCALFE, J. C., AND BIRDSALL, N. J. M., *Nature (London)* **255**, 684 (1975).

44. SINGER, S. J. AND NICOLSON, G. L., *Science* **175**, 720 (1972).

45. SINGER, S. J., *Ann. Rev. Biochem.* **43**, 805 (1974).

46. ZWAAL, R. F. A., ROELOFSEN, B., AND COLLEY, C. M., *Biochim. Biophys. Acta* **300**, 159 (1973).

47. BRETSCHER, M. S., *J. Mol. Biol.* **71**, 523 (1972).

48. MARINETTI, G. V. AND LOVE, R., *Biochem. Biophys. Res. Commun.* **61**, 30 (1974).

49. GUIDOTTI, G., *Ann. Rev. Biochem.* **41**, 731 (1972).

50. JULIANO, R. L., *Biochim. Biophys. Acta* **300**, 341 (1973).

51. BRETSCHER, M., *J. Mol. Biol.* **59**, 351 (1971).

52. MARCHESI, V. T., TILLACK, T. W., JACKSON, R. L., SEGREST, J. P., AND SCOTT, R. E., *Proc. Nat. Acad. Sci. U.S.A.* **69**, 1445 (1972).

53. SHEETZ, M. P. AND SINGER, S. J., *Proc. Nat. Acad. Sci. U.S.A.* **71**, 4457 (1974).

54. WISNIEWSKI, B. J., PARKES, J. G., HUANG, Y. O., AND FOX, C. F., *Proc. Nat. Acad. Sci. U.S.A.* **71**, 4381 (1974).

55. POSTE, G. AND ALLISON, A. C., *Biochim. Biophys. Acta* **300**, 421 (1973).

56. RUBIN, P., "Calcium and the Secretory Process," Plenum, New York, 1975.

57. QUASTEL, D. M. J., *in* "Synaptic Transmission and Neuronal Interaction," pp. 23–43. Raven, New York, 1974.

58. PAPAHADJOPOULOS, D., VAIL, W. J., NEWTON, C., NIR, S., JACOBSON, K., POSTE, G., AND LAZO, R., *Biochim. Biophys. Acta.* In press.

59. DUNN, M. J., *Biochim. Biophys. Acta* **352**, 97 (1974).

Photoelectric and Magneto-Orientation Effects in Pigmented Biological Membranes [1,2]

FELIX T. HONG

The Rockefeller University, New York, New York 10021

Received August 26, 1976; accepted October 29, 1976

Photoelectric and magneto-orientation effects in pigmented biological membranes are analyzed and discussed with special reference to three different levels of order in the membrane structures. The kinetic results of photoelectric responses from an artificial pigmented bilayer lipid membrane/aqueous redox system, measured by a tunable voltage clamp method are analyzed in terms of an equivalent circuit model which contains a novel chemical capacitance. The molecular basis of the equivalent circuit and the concept of chemical capacitance is established by a combined kinetic and electrostatic calculation, using Gouy–Chapman diffuse double layer theory. The generality of the concept of chemical capacitance is demonstrated by the ability of the model to explain previously reported data on a variety of pigmented membranes. The theory is further applied to photoelectric effects in three important photobiological systems: photosynthetic membranes of chloroplasts, visual disc membranes of rod outer segments, and purple membranes of *Halobacterium halobium*. Structure–function correlation is also stressed. Magneto-orientation effects in isolated rod outer segments and in *Chlorella* cells are discussed on the basis of summed diamagnetic anisotropy. The relation of this effect to the mechanism of bird orientation and navigation in the terrestrial magnetic field is speculated upon.

I. INTRODUCTION

Ample evidence has accumulated that biological membranes exist as highly ordered structures surrounding every cell and lining various intracellular organelles. The biological membrane was initially postulated as an invisible boundary, which separates the contents of a living cell from its environment, and which possesses differential permeability to various solutes (1, 2). This functional definition of biological membranes has been gradually enriched with ever-increasing structural details. The joint efforts of membranologists from diverse backgrounds and disciplines, using various biochemical and biophysical techniques

(3), culminated in the fluid mosaic model of biological membranes (4, 5). This model consists of a basic framework of lipid bilayer into which various integral proteins and other membrane-bound components are inserted. Evidence from X-ray diffraction (6, 7), differential scanning calorimetry (8), electron spin resonance (9), and nuclear magnetic resonance (10–12) strongly indicate that the majority of membrane lipids are organized into a fluid bilayer configuration at physiological temperatures, the polar head groups being exposed to the aqueous phases on both sides of the membrane with the hydrocarbon tails buried inside the membrane. Partial or full penetration by integral proteins makes the lipid bilayer discontinuous. Usually, these integral proteins are oriented with respect to the bilayer and are asymmetrically arranged so that the membrane has distinct sidedness (13). Likewise, various kinds of phospholipids are distributed asymmetrically between the

---

[1] Based on the 1976 Victor K. LaMer Award Lecture delivered on June 24, 1976, at the International Conference on Colloids and Surfaces, San Juan, Puerto Rico.

[2] Dedicated to the memory of the late President Emeritus Detlev W. Bronk of The Rockefeller University.

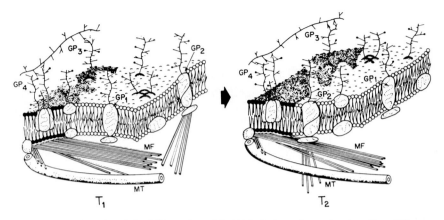

FIG. 1. Modified version of the fluid mosaic model of biological membrane structure. T_1 and T_2 represent different points in time ($T_1 < T_2$). The phospholipids are represented by circles (polar head groups) with two tails (hydrocarbon chains); certain hypothetical integral membrane protein components partially penetrate the membrane formed by a fluid lipid bilayer matrix, while others span the entire bilayer. Glycoproteins GP1 are "unanchored" and free to diffuse laterally in the membrane. Glycoproteins GP2 may be "anchored" or relatively impeded by a microfilament–microtubule cytoskeletal assemblage (MF–MT) but, under certain conditions, can be displaced by these cytoskeletal elements in an energy-dependent process. Motility of glycoprotein GP3 and GP4 is controlled by outer surface peripheral components and also transmembrane linkage to the cytoskeletal elements. In addition, GP3 and GP4 could be sequestered into a specific lipid domain indicated by the shaded area. (Reproduced and modified from Nicolson (16)).

two constituent monolayers (14). This ordered membrane structure exists in a dynamic state (15, 16). Some integral proteins and most phospholipids are capable of rotational and lateral diffusion. Some proteins can move only with restraints. Still others can be actively moved by specific intracellular cytoskeletal systems. A current version of fluid mosaic membrane model is shown in Fig. 1. Excellent reviews on membrane structures are available (13, 14, 16).

Throughout the current conception of the biological membranes is the principle of molecular order. In the present paper, we illustrate three levels of order in membrane structures with photoelectric and magneto-orientation effects in photobiological systems. The first level of order is the organization into a membrane phase of 40–85 Å in thickness and two well-defined membrane–water interfaces. This condition imparts the biological membrane with both high electrical resistance and

capacitance. These properties have been proposed as evidence in support of the existence of the cell membrane as a distinct structural entity (17), and are indeed the physical basis of all bioelectric phenomena. Besides the well-known excitability properties of nerve and muscle membranes, there is another class of bioelectric phenomenon, known collectively as the photoelectric effects (18).

We shall demonstrate that the first level of order at membrane interfaces is sufficient to give rise to photoelectric effects in an artificial bilayer lipid membrane (BLM) (19), which contains lipid-soluble pigment and separates two chemically asymmetric aqueous phases. Because of this segregation of reactants into a membrane phase and two aqueous phases, photochemical reactions can proceed aniso-tropically and vectorially in space and thus give rise to macroscopic electric effects. The latter are impossible to observe in the corresponding isotropic photoreactions in solution.

We believe that similar photoelectric events underlie processes occurring in natural photosystems such as visual photoreceptor membranes of rods and cones (20), photosynthetic membranes of cholorplasts (21), and purple membranes of *Halobacterium halobium* (22). However, the photoelectric effects in these systems might involve a higher level of order, namely, orientation and asymmetry of pigments within the membranes. This new level of order allows photochemistry to proceed vectorially even in the absence of any asymmetry in the aqueous phases.

When membranes containing oriented integral components are further organized into parallel-stacked multimembrane structures, a third level of order is established. Again, this order gives rise to striking macroscopic effects unobservable in disordered assemblies of the same components. Anisotropy of individual oriented molecules becomes macroscopically manifest via additive summation of individual small effects. This is the basis of the magneto-orientation effect in aqueous suspensions of isolated rod outer segments (23–25) and of unicelluar algae *Chlorella* (26, 27). In our subsequent presentation, we shall analyze photoelectric and magneto-orientation effects with special reference to the three levels of order in membrane structures mentioned above. Their biological significance will also be discussed.

II. PHOTOELECTRIC EFFECTS IN ARTIFICIAL PIGMENTED BILAYER LIPID MEMBRANES (BLM)

In 1968, Tien (28, 29) discovered a photovoltage from a BLM, which was formed from a chloroplast extract. It soon became apparent that a large photovoltage could be elicited in an asymmetrical pigmented BLM system, i.e., different concentrations of electron donors and/or acceptors are present in the two aqueous solutions (30). Similar photoelectric effects are observed in pigmented BLM of different compositions (31–53). Both steady-state and transient photoelectric responses have been studied. The light sources range from

tungsten lamps which provide a continuous light output to pulsed laser lights which give light pulses less than 1 μsec in duration. Light pulses with duration of intermediate ranges have also been used to analyze photolelectric transients from pigmented BLM.

Initial studies on photoelectric effects of pigmented BLM were carried out by a direct transplantation of electrophysiological techniques (54) which were responsible for triumphs in neurophysiology in the past quarter century. These methods fall into two categories. The first category is the open circuit measurement, in which the measuring instrument has sufficiently high input impedance to draw negligible current from the pigmented BLM. Included in this category is the current clamp method, which has various other names such as galvanostatic or constant current method. The second category is the short-circuit measurement, in which the instrument has negligible input impedance compared with that of the membrane. Included in this category is the voltage clamp method, which is also called the potentiostatic or constant voltage method. Hong and Mauzerall (44, 46, 55–57) have pointed out the inadequacy of these conventional methods, and have proposed a null current method for steady-state measurements (44, 55, 56), and a tunable voltage clamp (TVC) method for transient measurements (46, 55–57). The TVC method is a nonideal voltage clamp measurement, with the nonideality of a nonzero but small access impedance of the measuring device taken into account in a complete circuit analysis (including the measuring instrument). A TVC measuremeut can be carried out anywhere between a zero access impedance (short circuit) and an infinite access impedance (open circuit). In practice, it is closer to a short-circuit condition, and the access impedance can be adjusted (or "tuned") to match the source impedance of the pigmented BLM so as to optimize the measurement. Hence, the name tunable voltage clamp method. Detailed discussions on the TVC method appeared elsewhere (55–57). The advantage of

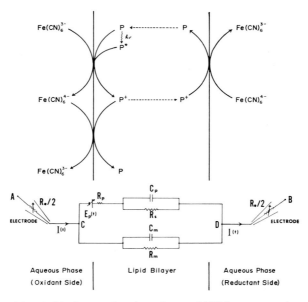

FIG. 2. Coupled interfacial photoreactions in a pigmented BLM—aqueous redox system and the equivalent circuit. The uncharged pigment in its ground state P, or excited state P* and the pigment monocation P⁺ are lipid-soluble and are confined to the membrane, while the ferricyanide ion, $Fe(CN)_6^{3-}$, and the ferrocyanide ion, $Fe(CN)_6^{4-}$, are located exclusively in the aqueous phases. The pigments, both charged and uncharged, are mobile inside the fluid bilayer, as indicated by dotted arrows, and can react with the aqueous redox reagents at either interface. The oxidant side contains predominantly the oxidant ferricyanide, and the reductant side predominantly the reductant ferrocyanide. This pigmented BLM— aqueous redox system is represented by the equivalent circuit (bottom half) in its photoelectric response. The pair of electrodes (plus intervening electrolyte) have a total effective access resistance R_e. The rest of symbols are explained in the text. (Reproduced and modified from Hong and Mauzerall (46)).

such a measurement will become apparent later.

The data of photoelectric responses of pigmented BLM will be analyzed in terms of the concept of chemical capacitance (46, 55, 56). Alternative theoretical models have been proposed (32, 37, 39, 58–60). The subject of photoelectric BLM has been reviewed (56, 61–68).

A. A Simple Pigmented BLM System

A typical pigmented BLM system in which the membrane-bound pigment reacts with water-soluble components in the two aqueous phases is depicted in Fig. 2. Those BLM systems in which the photosensitive elements are water-soluble are not included in this presentation (61, 64). The membrane contains lipid-soluble magnesium porphyrins, such as magnesium octaethylporphyrin and long-chain aliphatic esters of magnesium mesoporphyrin IX. The two aqueous phases contain the oxidant potassium ferricyanide and/or the reductant potassium ferrocyanide in addition to KCl or NaCl solutions. Usually, a redox gradient is imposed across the BLM so that the oxidant (acceptor) side contains more oxidant (electron acceptor) than reductant (electron donor), and the reductant (donor) side contains more reductant than oxidant. The ferricyanide concentration in the oxidant side is always at a concentration of 10 mM or more; further increase in its concentration has no effect on the photoresponses. This system proves to be simple and reproducible and, therefore, is highly suitable for quantitative studies. In fact, it is perhaps the simplest of its kind. The method of forming the BLM is standard (19). The method of preparing long-chain esters of magnesium mesoprophyrins has been published elsewhere (65).

The chemistry of both the ground state and the excited state of magnesium porphyrins has been well characterized (69–71). Side reactions which lead to destruction of the pigment are negligible. Therefore, the chemical reactions shown in Fig. 2 are quantitatively cyclic. In the present system (Fig. 2), the segregation of the pigment from the electron donor and acceptor is fairly complete because of their extremely low solubility, respectively, in water and in membrane. Since these electron transfer reactions are limited in spatial range (71), they must be essentially interfacial in nature. Since the BLM is extremely thin, and since the pigment molecules are mobile inside the liquid-like interior, the interfacial redox reactions at the two interfaces are coupled by diffusion of the membrane-bound pigment cation (P^+) and neutral pigment (P). However, this transmembrane diffusion coupling is considerably slower than the interfacial electron transfer reactions and thus is the rate-limiting step for a net electron transfer from the reductant to the oxidant side, both in the dark and in the light. A steady-state dark current and, under continuous illumination, a steady-state photocurrent flowing from the reductant side to the oxidant side can be observed under short-circuit condition. We have shown that this steady-state photocurrent is carried exclusively by pigment cation (P^+) rather than by other nonspecific ions such as K^+, Cl^- (44, 55, 56). Since the present system is linear with respect to the light stimulus, we shall henceforth ignore the dark reactions, which merely contribute a time-independent dark current across the BLM.

Both steady-state continuous light responses and transient pulsed light responses can be simulated by the equivalent circuit shown at the bottom of Fig. 2 (46, 55, 56). A plain membrane can be represented by a simple RC circuit: the membrane resistance (R_m), and the membrane capacitance (C_m). Photoreactions between the membrane-bound pigment and the aqueous electron acceptor and donor generate a parallel electric channel with a seat of photoemf (E_p) and its internal resistance (R_p). Since high concentration of ferrocyanide enhances the rate of the reverse electron transfer ($P^+ + e^- \rightarrow P$), the concentration of photogenerated P^+ is significant only at the oxidant interface and is negligible at the reductant interface (45, 55, 56). Hence, both E_p and R_p are mainly localized at the oxidant interface. Photogenerated P^+ at the oxidant interface has two different options of being reduced back to the ground state P. In one of the options, the photogenerated P^+ diffuses across the BLM to the reductant interface, becomes reduced to P, and then diffuses back to the oxidant interface. One cycle of these reactions causes a net transfer of one electron from the reductant side to the oxidant side. This process is represented by the photo-current flowing through the transmembrane pigment resistance (R_s). Since the transmembrane diffusion is rate-limiting, the resistance R_s contributes to the overwhelmingly major portion of the steady-state pigment resistance as measured by the null current method. In the other option, the photogenerated P^+ gets reduced at the same (oxidant) interface where it was formed. The reaction cycle is completed with no net electron transfer across the membrane. As we shall see later, this process is tantamount to charging and discharging of a capacitance, the chemical capacitance (C_p). The fraction of photocurrent flowing through C_p will be called the ac photocurrent and the fraction flowing through R_s will be called the dc photocurrent. The ac photocurrent can be resolved if a light pulse or light step with sufficiently fast risetime is used, and if the recording instrument has the required time resolution. By using a dye laser pulse of 300 nsec and a tunable voltage clamp circuit with an instrumental time constant of 1.5 μsec,

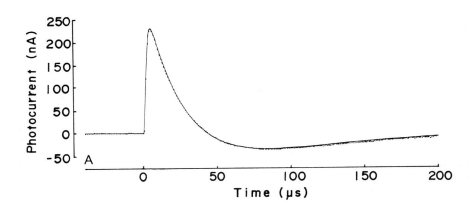

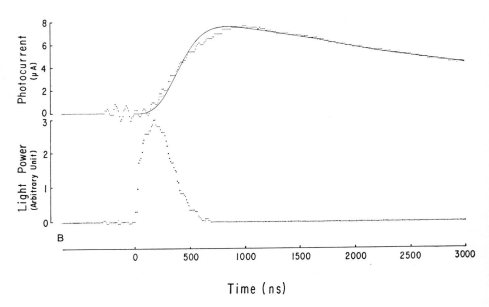

FIG. 3. Measured (dotted curve) and computed (smooth curve) photoelectric responses of a pigmented BLM to a dye laser pulse (0.3 μsec) by means of the TVC method. (A) The BLM is made from an egg lecithin–cholesterol–decane mixture which also contains a 5 mM magnesium mesoporphyrin IX di-n-amyl esters. The BLM separates two aqueous phases with 20 mM potassium ferricyanide and 0.5 mM potassium ferrocyanide in the oxidant side and 20 mM potassium ferrocyanide in the reductant side. Both aqueous phases also contain 1 M NaCl and 10 mM phosphate buffer at pH 7.2. The exciting laser beam is focused to illuminate 19% of the thin bilayer of area 1.7 mm². The instrumental time constant is 1.5 $k\Omega$. The access impedance is 5.1 $k\Omega$. Temperature is 26°C. The input parameters for computation are: $R_m = 1.5 \times 10^9$ Ω, $C_m = 8.2$ nF, $R_p = 34$ $k\Omega$, $C_p = 1.3$ nF, $R_s = 10^9$ Ω. (B) The BLM and the aqueous phases have the same composition as A except the pigment is the diethyl ester and that KCl replaces NaCl in the aqueous solutions. The membrane area is 2.5 mm². The instrumental time constant is 150 nsec. The access impedance is 380 Ω. The dye laser pulse (measured by a photodiode) is shown at the lower half of the figure. (Reproduced from Hong and Mauzerall (46, 57)).

Hong and Mauzerall (46, 55, 56) recorded a photocurrent from a pigmented BLM as shown in Fig. 3A. The photocurrent decays to a level below the (dark) baseline and returns at a slower rate to the baseline eventually. Furthermore, the time-integrated photocurrent responses decay nearly to the baseline, i.e., the area bound by the curve of photocurrent response above the baseline is nearly equal to that below the baseline. That is, the transient photocurrent in Fig. 3A is mainly a displacement current, namely, the ac photocurrent. In fact, it is the manifestation of the interfacial event at a *single* interface, the oxidant inter-

face (45, 55, 56). The dc photocurrent is not absent, but is relatively minuscule in its amplitude, and lasts only the duration of the light pulse.

The equivalent circuit shows that the photocurrent is in parallel with the ionic current which flows in the membrane resistance R_m. This parallelism is established by a shunting experiment (44, 55, 56). We shall see that this parallelism also exists in visual photoreceptor membranes.

We shall demonstrate that the equivalent circuit in Fig. 2 predicts the correct time course of the photocurrent measured by the TVC

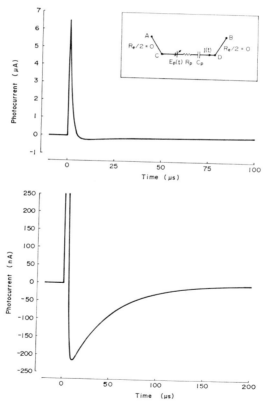

FIG. 4. Computed photoelectric current response as measured by ideal voltage clamp method. Input parameters for computation are from the data of Fig. 3A. Instrumental time constant is 1.5 μsec. The relaxation has a positive component with a time constant of 1.5 μsec and a negative component with a time constant of 44 μsec (τ_p). (Reproduced from Hong and Mauzerall (57)). An equivalent circuit for the ideal voltage clamp measurement at high frequency is included in the inset. See text for detail.

method, using experimentally determined circuit parameters. In order to appreciate the TVC method, let us first consider a hypothetical case of an ideal voltage clamp measurement, i.e., with zero access impedance. The photocurrent response of the equivalent circuit of Fig. 2 to an extremely short light pulse will be examined. During the light-induced transient, the voltage across the membrane (points C and D in the equivalent circuit) is kept equipotential. Therefore, no currents flow through R_m and C_m. Furthermore, the photocurrent flows predominantly through C_p instead of R_s during the brief transient. Thus, we need only consider an equivalent circuit shown in the inset of Fig. 4. Circuit analysis gives the photocurrent response

$$I(t) = \frac{E_p(t)}{R_p}$$
$$- \frac{1}{\tau_p} \int_0^t \frac{E_p(u)}{R_p} \exp\left(\frac{u-t}{\tau_p}\right) du, \quad [1]$$

where

$$\tau_p = R_p C_p. \quad [2]$$

If the light pulse is so brief relative to τ_p that E_p approximates a delta function, the photocurrent will appear as a sharp positive spike followed by a negative exponential component which relaxes with a time constant τ_p (Fig. 4). Notice that we obtain only one relaxation time constant τ_p.

Now, if a small electrode access resistance R_e is introduced, the voltage clamp circuit will maintain constant zero potential across the electrode input (points A and B in Fig. 2). The membrane voltage (across points C and D in Fig. 2) will deviate from zero during the transient. Thus, we can no longer ignore the elements R_m and C_m. Interaction of the photochemical channel and the membrane RC results in a distortion of the measured photocurrent, which is now given instead by (46,

55–57)

$$I(t) = \frac{1}{R_e C_m \left(\dfrac{1}{\tau_s} - \dfrac{1}{\tau_l}\right)} \left[\left(\frac{1}{\tau_s} - \frac{1}{R_s C_p}\right)\right.$$

$$\times \int_0^t \frac{E_p(u)}{R_p} \exp\left(\frac{u-t}{\tau_s}\right) du$$

$$-\left(\frac{1}{\tau_l} - \frac{1}{R_s C_p}\right)$$

$$\left. \times \int_0^t \frac{E_p(u)}{R_p} \exp\left(\frac{u-t}{\tau_l}\right) du\right], \quad [3]$$

where

$$\frac{1}{\tau_s} = \frac{1}{2}\left\{\left(\frac{1}{R_p C_m} + \frac{1}{\tau_p} + \frac{1}{\tau_m}\right)\right.$$

$$+ \left[\left(\frac{1}{R_p C_m} + \frac{1}{\tau_p} + \frac{1}{\tau_m}\right)^2\right.$$

$$\left.\left. - 4\left(\frac{1}{R_p C_m R_s C_p} + \frac{1}{\tau_p \tau_m}\right)\right]^{\frac{1}{2}}\right\}, \quad [4]$$

$$\frac{1}{\tau_l} = \frac{1}{2}\left\{\left(\frac{1}{R_p C_m} + \frac{1}{\tau_p} + \frac{1}{\tau_m}\right)\right.$$

$$- \left[\left(\frac{1}{R_p C_m} + \frac{1}{\tau_p} + \frac{1}{\tau_m}\right)^2\right.$$

$$\left.\left. - 4\left(\frac{1}{R_p C_m R_s C_p} + \frac{1}{\tau_p \tau_m}\right)\right]^{\frac{1}{2}}\right\}, \quad [5]$$

$$\frac{1}{\tau_p} = \frac{1}{R_p C_p} + \frac{1}{R_s C_p}, \quad [6]$$

$$\frac{1}{\tau_m} = \frac{1}{R_e C_m} + \frac{1}{R_m C_m}. \quad [7]$$

Here, R_s is included for completeness. In actual calculation, all terms which contain the factor $1/R_s$ can be replaced by zero without affecting the accuracy. Note that the photocurrent response to a delta function photoemf consists of two exponential components. The two time constants will be referred to as the short (τ_s) and the long (τ_l) time constants. They are neither τ_p, the intrinsic relaxation time constant, not τ_m, the charging time constant, but are complex mixtures of τ_p and τ_m and other combinations of various RC parameters. This is the consequence of interaction between the photochemical event and the membrane RC. It can be shown that R_p and C_p can be calculated uniquely by solving the simultaneous equations [4] and [5], after inserting measured values of τ_s, τ_l, R_e, R_m, R_s, and C_m (55–57). The system is thus completely characterized by experiments and no indeterminancy remains. We not only obtain the intrinsic relaxation time (τ_p), but also decompose it into R_p and C_p. The TVC measurement thus provides more information than the *ideal* voltage clamp measurement. Yet, the nonideality due to the small access impedance causes no serious error (55–57).

The data shown in Figs. 3A and B were measured by the TVC method with effective access resistances of 5 kΩ and 380 Ω, respectively, and with overall instrumental time constants of 1.5 μsec and 150 nsec, respectively. The computed photoresponses based on the equivalent circuit in Fig. 2 with experimental determined parameters are shown superimposed on measured photoresponses in Figs. 3A and B. In these computations, the photoemf is assumed to follow the same time course as that of the stimulating light pulse.

Figure 3B merits a few comments. The sigmoid-shaped rise phase of the photocurrent is accounted for by the low-pass filter effect of the instrumental time constant of 150 nsec and the finite width of the photoemf. This result indicates that the electron transfer from the excited state of the pigment occurs in less than 100 nsec. In the electrophysiologists' jargon, the latency of the photocurrent, if any, must be considerably shorter than 100 nsec. Also, notice that an access resistance of 380 Ω cannot be considered negligible in this measurement. Otherwise, the photocurrent should fall below the baseline no later than 150 nsec after the cessation of the light pulse. Although dc resistance of the pigmented membrane can

be as high as 10^9 to 10^{10} Ω, its impedance at megahertz range drops well below 10^6 Ω. Attempted measurement of short-circuit currents with instruments of input impedance as high as 10^6 or 10^7 Ω give results actually closer to open-circuit than short-circuit conditions (39, 47).

Equations [3]–[7] predict that, as the effective access resistance is increased, the two relaxation time constants τ_s and τ_l are prolonged, and the amplitude of the photocurrent is diminished. These are actually observed, and the time courses of photocurrent vary with the effective acesss impedance as predicted by the equivalent circuit of Fig. 2, using the identical set of membrane and pigment RC parameters (56, 57).

As R_e approaches a value which is large compared with those of R_m and R_s, the photocurrent approaches that of an open circuit measurement except, of course, for a factor of the access resistance R_e. In this case, the short time constant will be within a factor of 2 of the intrinsic relaxation time constant τ_p, and the long-time constant will be approximately equal to the membrane RC constant (R_mC_m) (56, 57). This represents a case of overinteraction between the photoevent and the membrane RC. Open circuit measurement with insufficient instrumental resolution could end up with measuring just the membrane RC (see Section VG in Ref. (56)). The condition of optimal TVC measurements is when the access impedance matches the source impedance of the membrane, i.e., $\tau_m = \tau_p$, or $R_e \cong R_p$.

B. The Molecular Basis of Chemical Capacitance and the Equivalent Circuit Model

The agreement of the equivalent circuit with experimental data only suggests the molecular interpretation presented in the previous section, but does not prove the existence of a chemical capacitance beyond doubt. To fill this gap, Hong (56, 72) attempted to derive the equivalent circuit from accepted physical principles. It is essentially a combined kinetic

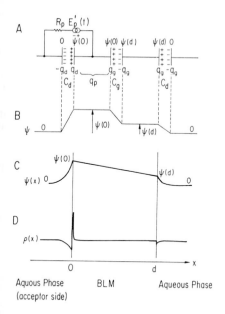

A

$R_p \; E_p'(t)$

$0 \quad \psi(0) \quad | \quad \psi(0) \; \psi(d) \quad | \quad \psi(d) \; 0$

$-q_d \; q_d \quad q_g \; -q_g \quad q_g \; -q_g$

$C_d \quad q_p \quad C_g \quad C_d$

B

ψ

$\psi(0)$

$\psi(d)$

C

$\psi(x)$

$\psi(0)$

$\psi(d)$

D

$\rho(x)$

$0 \qquad\qquad d \qquad\quad x$

Aquous Phase   BLM   Aqueous Phase

(acceptor side)

FIG. 5. Schematic diagrams explaining the result of combined kinetic and electrostatic calculations. (A) An equivalent circuit obtained from interpreting the result of the calculation. The charge q_p of the photogenerated P^+ is conveniently divided into two fractions. Notice that the polarity of charges on the geometric capacitance C_g and on the two double layer capacitances C_d dictates the way the photoemf $E_p(t)$ and its internal resistance R_p are connected with these capacitances. (B) Schematic potential profiles across the capacitances in A. (C) Potential profiles across the pigmented BLM. (D) Charge density distribution across the pigmented BLM. The coordinate across the membrane is shown at the bottom. The thickness of the membrane is d. (Reproduced from Hong (56).)

and electrostatic calculation based on the interfacial photoreaction scheme shown in Fig. 2, and the Gouy–Chapman diffuse double layer theory (73).

The transmembrane motion of pigment is much slower than the interfacial charge transfer reaction. Therefore, we can regard the photogenerated pigment cations as a sheet of fixed surface charge at the oxidant interface during the brief transient of photoinduced interfacial charge transfer. Furthermore, both ionic mobility and ionic concentration are

much higher in the aqueous phases than in the membrane phase. Hence, the interfacial charge transfer relaxation is much slower than the ionic cloud relaxation in the aqueous phases (<1 nsec), but is much faster than the ionic cloud relaxation in the membrane phase (>1 msec). As a consequence, the ionic distribution in the aqueous phases is always in quasi-equilibrium during the transient, while there is no charge screening in the membrane phase. The kinetic calculation and the electrostatic calculation can, therefore, be carried out separately.

The electrostatic calculation provides the instantaneous potential profile and space charge distribution profile across the BLM and the adjacent aqueous phases under short-circuit (ideal voltage clamp) condition (Fig. 5). In contrast to an externally applied voltage, the photoreaction injects photocurrent right at the (oxidant) interface (Fig. 5A). Therefore, the ac photocurrent can be further divided into two fractions. Only the fraction that flows through C_g and the opposite C_d (at the reductant side) is detected externally. The other fraction that flows through the adjacent C_d (at the oxidant side) is not detected externally. As seen by the photoemf $E_p'(t)$, the adjacent double layer capacitance C_d is in parallel, but the opposite double layer capacitance C_d and the geometric capacitance C_g are in series. As far as the externally measured photocurrent is concerned, the circuit shown in Fig. 5A is equivalent to the simpler one in Fig. 4. The latter is the *irreducible* equivalent circuit (56). The connection between the two equivalent circuits is provided by the equations

$$C_p = C_d + \left(\frac{1}{C_g} + \frac{1}{C_d}\right)^{-1}, \quad [8]$$

and

$$E_p(t) = \left(\frac{C_p - C_d}{C_p}\right) E_p'(t), \quad [9]$$

where

$$C_g = \epsilon_m/4\pi d, \quad [10]$$

$$C_d = \epsilon/4\pi L. \quad [11]$$

In Eqs. [10] and [11], ϵ_m and ϵ are the dielectric constants of the BLM and water, respectively, d is the BLM thickness, and L is the Debye length of the aqueous phases. Equation [8] indicates that the chemical capacitance C_p is physically distinct from the membrane capacitance C_m. The latter is a series combination of two double layer capacitances and one geometric capacitance (74). Equation [9] indicates that the photoemf $E_p(t)$ is attenuated by a factor $(C_p - C_d)/C_p$ because only the fraction of ac photocurrent that flows across the BLM is detected. Thus, the ac photocurrent is not observed in sufficiently thick membranes, because C_g is so small that $C_p = C_d$. As the membrane is thinning, the photocurrent grows in amplitude (Hong, unpublished observation).

Interfacial chemical kinetics can be solved for the case of linear light response (low light intensity) and pseudo-first-order relaxation (sufficient ferrocyanide concentration at the oxidant side). The result indicates that the photoemf is proportional to the light intensity, if the interfacial resistance R_p is regarded as time independent. This justifies our assumption that the photoemf follows the same time course as that of the light pulse. The result of kinetic calculation also indicates that the intrinsic rate constant $1/\tau_p$ can be interpreted as the pseudo-first-order rate constant of the reverse electron transfer reaction:

$$P^+_{(BLM)} + Fe(CN)_6{}^{4-}{}_{(aq)} \rightarrow$$

$$P_{(BLM)} + Fe(CN)_6{}^{3-}{}_{(aq)}. \quad [12]$$

Photoresponses of pigmented BLM with increasing ferrocyanide concentration in the oxidant side is shown in Fig. 6A. The rate constants $1/\tau_p$ calculated from these data are plotted against the corresponding ferrocyanide concentration in Fig. 6B. As expected from pseudo-first-order regimes, the plot gives an straight line through the origin. The slope $(4.3 \pm 0.5) \times 10^7 \ M^{-1} \ sec^{-1}$ is the apparent second-order rate constant of the reverse electron transfer reaction (55, 56).

Thus, the TVC method provides a simple, rapid, and sensitive electrical measurement of

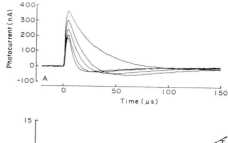

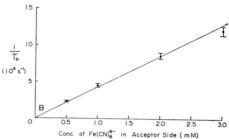

FIG. 6. Dependence of photoelectric relaxation on the ferrocyanide concentration in the oxidant aqueous phase. (A) Photoelectric current responses measured by the TVC method with an access impedance of 4.5 kΩ and an instrumental time constant of 1.5 μsec. The BLM contains diethyl esters of Mg mesoporphyrin. Both aqueous phases contain 0.1 M KCl. The reductant side contains no redox reagents. The oxidant side contains 10 mM potassium ferricyanide and various amounts of potassium ferrocyanide: in the order of diminishing amplitude, 0, 0.5, 1, 2, 3 mM. The light source is the same as in Fig. 3. Temperature is 26°C. (B) Plot of intrinsic relaxation rate constant vs the ferrocyanide concentration of the oxidant side. Data are from Fig. 6A. The slope of the linear fit gives the apparent second-order rate constant: $(4.3 \pm 0.5) \times 10^7 \ M^{-1} \ sec^{-1}$. (Reproduced from Hong (56).)

interfacial chemical kinetics of charge transfer reactions.

C. Generality of the Concept of Chemical Capacitance

The generality of the concept of chemical capacitance is best demonstrated by examining transient photoelectric responses of BLM containing other pigments. Since the conditions that we invoked in the derivation of chemical capacitance are valid in most of these pigmented BLM systems, a chemical capacitance ought to be observable under suitable

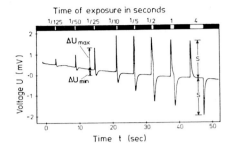

FIG. 7. Open-circuit photovoltage response of a BLM adsorbed on one side with a cyanine dye. The BLM is made from egg lecithin and *n*-octane and is bathed in 0.01 *M* NaCl. A cyanine dye (37) is added to one side of the aqueous solutions. The reference electrode is on the side with the dye. Open-circuit voltage is measured by a Keithley electrometer. The light source is a 50-W halogen–tungsten lamp equipped with a shutter to give rectangular pulses from 1/125 to 4 sec. $S =$ maximum value of ΔU at long duration of exposure. (Reproduced from Ullrich and Kuhn (37).)

conditions. This conclusion is applicable not only to redox electrode-type pigmented BLM (in which electrons are transferred), but also to pigmented BLM in which a proton is transferred. Also, the asymmetry that gives rise to a unidirectional net charge transfer need not be an aqueous chemical (redox or pH) gradient. It can be due to a voltage gradient (33, 34, 48). It can also be a structural asymmetry such as is believed to exist in visual membranes (75) and in the purple membranes of *Halobacterium halobium* (22).

Ullrich and Kuhn (36, 37) studied a BLM adsorbed with a cyanine dye (A) at one interface. Illumination with flashes of varying duration generates the (open-circuit) photovoltage shown in Fig. 7. These authors proposed a model in which electron transfer from an electron donor E to the adsorbed dye A results in the formation of (oriented) dipole E^+A^-. The light-induced oriented dipoles polarize the electrolyte in the aqueous phases and this gives rise to the characteristic time course of the photovoltage in Fig. 7. This model, commonly referred to as the *dipole model*, has also been applied by Trissl and Läuger (39) to a chlorophyll-containing BLM

coupled to aqueous oxidized cytochrome *c*, and by Froehlich and Diehn (48) to riboflavin-containing BLM. There is a subtle difference between the dipole model and our kinetic molecular model described in the previous section. In our model of interfacial charge transfer, the aqueous component ferrocyanide is not "attached" to or "paired up" with the membrane-bound pigment cation (P^+), but is free to diffuse in the adjacent aqueous phase. Therefore, the aqueous component (ferrocyanide) does not form an array of oriented dipoles with the membrane-bound pigment cation (P^+). Our view is supported by the observed deviation of photocurrent relaxation from pseudo-first-order kinetics at zero ferrocyanide concentration in the oxidant aqueous phase (55). Although we cannot exclude the possibility of dipole formation in systems other than ours, the data presented by the above-mentioned authors can be satisfactorily interpreted on the basis of our model without invoking the "attachment" of the aqueous component to the membrane-bound pigment. In fact, the data shown in Fig. 7 are qualitatively expected from our equivalent circuit with R_s set to infinity, since the cyanine dye does not cross the BLM. Another example of the presence of a chemical capacitance is provided by a report of Schadt on a retinal-BLM, and a vitamin A acid-BLM (47). The qualitative difference of photoresponses from these two systems can be explained by different relative magnitudes of the membrane and pigment parameters (56). Again, there is no need to propose two separate models for the two systems. The chemical capacitance may thus serve as a unifying concept in interpreting photoelectric responses from different pigmented BLM systems. The apparent relaxation time observed under most circumstances is the result of interaction between the chemical capacitance and the ordinary membrane capacitance. It is only through analysis of the entire system in terms of chemical capacitance that meaningful relaxation time can be obtained. For example, open circuit measurement, in principle, provides two apparent relaxation time constants. However, because of limited time resolution of the light pulse, the measurement provides only the slow time constant alone, which is nothing but the membrane RC time constant (56, 57). Such measured data, if taken at face value, would be extremely misleading.

III. PHOTOELECTRIC EFFECTS IN NATURAL PIGMENTED MEMBRANES

Photoelectric effects have been found in most pigment-containing membranes in animals and plants. However, we shall restrict our discussion to three important photobiological systems: the photosynthetic membranes of chloroplasts in higher plants, the visual photoreceptor membranes of rod outer segments in eyes, and the purple membranes of *Halobacterium halobium*.

A. Photosynthetic Membranes of Chloroplasts

The photosynthetic apparatus for the conversion of solar to chemical energy is located in the membranes of the chloroplasts: flattened sac-like vesicles, which Menke (76) has called thylakoids. The complex molecular organization of these membranes has only begun to be unraveled (77, 78). The major membrane-bound pigments are various forms of chlorophylls, which are magnesium porphyrins (79, 80). Some photopigments are organized into pigment–protein complexes, which are called reaction centers. Still others are organized as antennas, which can transfer absorbed light energy to the reaction centers. Components that constitute the electron transport carriers and other accessory pigments are also present. In higher plants these components are further organized into a photosystem I and a photosystem II. Evidence indicates that these components are oriented and distributed asymmetrically across the photosynthetic membranes (77, 78). Thus, in photosystem I, and, possibly also in photosystem II, the electron acceptors are located close to the outer membrane interface, and the electron donors close to the inner interface. This oriented and asymmetric internal membrane structure provides the sidedness that is required for externally observable photoelectric effects in these membranes.

The photoelectric effects in chloroplast membranes have been studied by indirect methods such as electrochromic absorption changes (21, 81). Direct intracellular recording of photoelectric responses was made possible only recently (82–86). Extracellular recording from nonoriented (87) and oriented (88) chloroplast suspensions has also been made.

Mitchell's (89, 90) chemiosmotic theory has played a central role in the studies of photophosphorylation in chloroplast thylakoids. The theory involves two hypotheses. The first is that anisotropic and asymmetric arrangement of reaction centers and electron transport carriers results in an outward electron transfer and a stoichiometrically related net translocation of protons from outer aqueous phase to inner aqueous phase of the thylakoids. The second is that this proton gradient is utilized to drive another anisotropic and asymmetric system of ATPase to form ATP from ADP and inorganic phosphates.

Jagendorf and Hind (91) first demonstrated such a proton uptake in thylakoids. Jagendorf and Uribe (92) also demonstrated formation of ATP in response to an artificially created proton gradient across thylakoid membranes. Based on interpretation of light-induced electrochromic absorbance change, Witt and co-workers (21, 81) proposed in 1968 that the two photosystems are anisotropically posed in the membrane so that light-induced charge transfer across the membrane results in a membrane potential which is inside positive. Furthermore, the decay of this potential is closely correlated with the decay of transmembrane proton gradient and the formation of ATP.

A straightforward comparison of the photoelectric events in photosynthetic membranes with those in our simple model pigmented BLM is difficult, because of the complexity of photochemistry and structure in photosynthetic membranes. There are, nevertheless, certain basic features in common. The following discussion with specific reference to results in pigmented BLM must be regarded as speculative rather than conclusive.

The molecular events in photosynthetic membranes will be grouped into two categories: one is primary photoelectric events, the other is the secondary "ionic" events.[3] The photo-induced charge separation and inward proton translocation due to photosystems I and II belong to the primary photoelectric events. Although the two photosystems are arranged *in parallel* spatially, they are connected *in series* electrically (77, 78), being two seats of photoemf with the same polarity in series. The secondary "ionic" events include (1) the field-driven outward flux of protons that is coupled to ATP formation, (2) ionic fluxes through spontaneous leakage, and (3) ionic fluxes due to action of uncouplers or ionophores such as gramicidin D and valinomycin, or through leakage after osmotic shock. All these secondary events are themselves electrically in parallel with one another, but can often be conveniently lumped together into a single ionic membrane resistance (R_m) and be characterized by at least one relaxation time constant ($R_m C_m$). One or more intrinsic relaxation time constants (similar to τ_p) represent the reversal of the primary photoelectric events of photosystems I and II. This view is in accordance with the original scheme of Mitchell (89). It is also consistent with several lines of experimental evidence. It has been found that phosphorylation and action of ionophores accelerate the decay of the proton gradient and the decay of the membrane potential (21, 81, 93–96). It has also been observed that pulsed-light-induced membrane potential decays in two time constants (81, 96, 97). Although a different interpretation has been proposed (96, 97), we suspect that this biphasic relaxation might be, at least partially, explicable by the presence of a chemical capacitance: under

open-circuit conditions, the fast component represents the relaxation of the photoelectric events (reverse reactions of the two photosystems), and the slow component is the relaxation of the secondary "ionic" event (cf. Section IIC). This view is supported by the observation that phosphorylation and shunting action of ionophores accelerate the slow component but apparently do not affect the fast component (21, 81). The presence of a chemical capacitance is further suggested by the appearance of an "undershoot" (decay below the baseline) in the intracellular recording of photovoltage responses (83, see also Fig. 1 on Ref. 86). The apparent lack of latency (risetime < 20 nsec) of the fast component again suggests that the primary photoevent is ac-coupled across the membrane (21). We expect that an intracellular recording with the TVC method at reduced access impedance might shed some light on this problem.

B. Visual Photoreceptor Membranes

The photoreceptor cells of vertebrates, rods (dim-light vision) and cones (color vision), transduce light energy into neural impulses. The visual pigments, rhodopsin (rods) and iodopsins (cones), are various membrane-bound proteins with covalently attached retinals as chromophores (98). These pigments are localized in the parallel-stacked multiple membranes of outer segments of rods and cones. In rods, these multiple membranes exist as free-floating discs enclosed by the cell membrane (99). In cones, these multiple membranes remain continuous with one another and with the cell membrane (99). From studies of birefringence and linear dichroism in intact rod outer segments, it was found that rhodopsin is oriented in the disc membranes with the transition moment of its retinal chromophore lying in the plane of the membranes (100–103). The rhodopsin is free to rotate around an axis perpendicular to the plane of the membrane (104, 105) and free to diffuse laterally in the plane of the membrane (106, 107). Early

[3] This classification is not to be confused with the conventional terminology of primary and secondary electron transfers.

X-ray diffraction data indicate that rhodopsin is on the intradisc side in the membrane and is globular in shape (75, 108). However, evidence from antibody (109), chemical labeling (110–112), and neutron diffraction studies (113–116) as well as X-ray diffraction performed on magnetically oriented rod outer segments (117) indicate that rhodopsin is located closer to the extradisc phase than the intradisc phase. Energy transfer measurements (118) as well as X-ray diffraction on oriented rod outer segments (117) indicate that rhodopsin has an elongated shape. Although a definite answer to the problem of rhodopsin localization may not be available, the asymmetry of the disc membrane is well established. Models with rhodopsin molecules distributed as two populations on both sides of the membrane can be rejected with certainty. Again, the anisotropic and asymmetric arrangement of rhodopsin in membranes ensures unidirectionality of photoelectric effects in visual membranes. The molecular aspect of visual pigments and membranes has been reviewed (119).

The electroretinogram (ERG) of the eyes has been known for over a century (20). However, the primary photoelectric event in visual photoreceptors was discovered only 12 years ago by Brown and Murakami (120). It is commonly referred to as the early receptor potential (ERP) or fast photovoltage because of its apparent lack of latency following the stimulation of a light flash (Fig. 8). The ERP has a cornea-positive R_1 component and a cornea-negative R_2 component (121, 122). The ERP is extremely resistant to all kinds of rough treatments including drastic change of ionic environment (120, 123–125) and seems to depend only on the integrity of the membranes with oriented visual pigments (126). In contrast, the late receptor potential (ERG a wave) is sensitive to metabolic inhibitors and to change of ionic environment (120, 123–125). The LRP represents widespread response of the entire rod to light stimulation and is related to the nonlinear ionic conduction mechanism (127–131). These features strongly suggest that the ERP and the LRP are connected

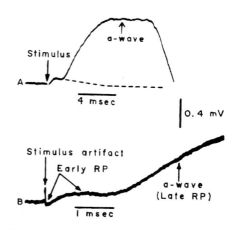

FIG. 8. The early and the late receptor potentials of the cynomolgus monkey. Responses recorded by a tungsten microelectrode at about the level of the inner segments in the retina, with the reference electrode in the vitreous humor of the eye. The stimulus was an intense 20-μsec flash. In record A, the ERP may be seen to intervene between the stimulus flash and the rising phase of the a-wave. The time course of a decay of the ERP, when isolated by cessation of artificial respiration, is shown by the dashed line. In record B, the stimulus artifact consists only of the positive tip which is labeled. Thus the ERP is biphasic, and is followed by the LRP, which gives the rising phase of the a-wave. (Reproduced from Brown (20).)

electrically in parallel, and the LRP is thus a secondary ionic event (cf Section IIIA). The major characteristics of the ERP are listed in Table I. It must be pointed out that all these features are shared by the transient (ac) photoelectric response of our model pigmented BLM (56). The conventionally accepted mechanism of the ERP is the oriented dipole mechanism: light-induced formation of oriented electric dipole due to intramolecular charge transfer accompanying conformational change during bleaching of rhodopsin (132–134). The oriented dipole mechanism is consistent with our present knowledge of visual pigment and visual membranes, but it is not the only possibility. The striking similarity between the ERP and the ac photocurrent in our model pigmented membrane suggests that an interfacial charge transfer mechanism

TABLE I

Characteristics of the ERP (from Hong (56))

1. It is of membrane origin.
2. Its action spectrum is that of the absorption spectrum of the membrane-bound pigment.
3. Its amplitude is linear with the intensity of stimulating light up to saturation.
4. There is no detectable latency.
5. It is independent of the ionic conduction mechanism.
6. It is a displacement current.

instead could generate an ERP-like photosignal just as well (56).[4] Since the two mechanisms are not mutually exclusive, it is conceivable that both mechanisms might exist simultaneously, or that one mechanism works for the R_1 component and the other mechanism works for the R_2 component. Clearly, further quantitative studies are required to resolve this problem. However, we shall look for clues already present in the literature. The only known charge transfer reaction in rhodopsin bleaching processes is the metarhodopsin I to metarhodopsin II transition, which involves a net uptake of one proton (135, 136). It is, therefore, possible that the R_2 component is generated by an interfacial proton transfer mechanism. In fact, the R_2 component has been correlated in time with the metarhodopsin I to metarhodopsin II transition (132). There is no known charge transfer reaction occurring at the same stage as the R_1 component, but one cannot resist asking whether there might be a charge (electron or proton) transfer reaction taking place so fast that the R_1 component is the only clue of its existence.

The central problem in visual photoreceptor physiology is the coupling mechanism that links the event of photon absorption by a visual pigment molecule and the event of electrical response of the visual membrane (137). Between these two events lies a huge factor of amplification that is responsible for

[4] Preliminary discussion on an alternative mechanism of the ERP has been presented by Hong in the Peter A. Leermakers Symposium on the Chemistry of Vision, held on May 5–6, 1976 at the Wesleyan University, Middletown, Connecticut.

the exquisite sensitivity of a visual photoreceptor. The fact that, in sharp contrast to the LRP, the ERP has no latency and is a linear function of light (121, 138) suggests that ERP might be associated with processes prior to the amplification mechanism.

As for the role of the ERP in visual transduction, two possibilities exist. First, the ERP is not on the main path of visual transduction and is merely a parallel phenomenon (epiphenomenon) accompanying the visual transduction process. Second, the action of the ERP directly leads to the amplification process. Visual physiologists seem to favor overwhelmingly the first possibility (139). However, our study of model systems indicates that the local electric field at the membrane interface or inside the visual membranes due to the ERP might not be so small as we were led to believe (56). Therefore, it remains possible that the *local* electrical effect might be sufficiently intense to operate a gating mechanism, which controls the release of a mediator such as Ca^{2+} (131, 140–142). On the other hand, even if ERP turns out to be just an epiphenomenon, it could be a gating current[5] accompanying a gating action. That is, it is an electrical manifestation of changing electric dipole moment of an opening and closing gate.

C. Purple Membranes of Halobacterium Halobium

A fascinating photobiological system which shares partially characteristics of chloroplasts and visual membranes is provided by the purple membrane of *Halobacterium halobium*.[6] This membrane contains only one species of protein, bacteriorhodopsin (22, 144), a chromoprotein with covalently attached retinal as the chromophore, quite similar to rhodopsin (144). In contrast to the visual pigment systems, the purple membrane functions as a photon energy

[5] The gating current found in squid axon is also a displacement current (143).

[6] The rest of the membranes in *H. halobium* are called the red membranes, which contain a red pigment bacterioruberin, cytochromes, flavoproteins, ATPase, and other components of the respiratory chain (22).

converter rather than as a photon energy sensor (22, 145). X-ray diffraction (146) and electron microscopy (147) indicate that the purple membrane has a hexagonal lattice structure approaching that of protein crystals in regularity. The bacteriorhodopsin is oriented with the transition moment of its chromophore lying in the plane of the purple membrane (146). The pigment is arranged asymmetrically with respect to the plane of the membranes, conferring upon the latter a distinct sidedness (22).

Under anaerobic conditions the bacteria acidify the medium when exposed to light, suggesting that the bacteriorhodopsin acts as a light-driven proton pump, which translocates protons out of the cell (148). In addition, exposure to light restores the ATP content of the bacteria, which has been depleted anaerobically in the dark (149). Functionally, the bacteria carry out photophosphorylation in a way similar to chloroplasts in plants (see Section IIIA). The mechanism of proton translocation per se differs drastically from that in chloroplasts however; conspicuously absent from the purple membrane are the elaborate electron transport chains (150–153). The proton translocation is accomplished solely by the cyclic photoreactions of the bacteriorhodopsin. Racker and Stoeckenius (154) have reconstituted phospholipid vesicles from purple membranes and phospholipid, and observed light-induced proton uptake. When oligomycin-sensitive ATPase from beef heart mitochondria is also incorporated in the vesicles, ATP can be formed in the light (154). Mitchell was remarkably correct in postulating that a proton gradient is the sole intermediate between the primary photoevent and the ATP formation (89). The opposite polarity of light-induced proton translocation in reconstituted vesicles is explained by the fact that the purple membrane has been turned inside out. This latter fact, indicating a definite sidedness of the purple membrane, has been established by freeze-fracture electron micrographs (22, 150).

As for the detailed mechanism of proton translocation, Lewis and, co-workers (155) have shown cyclic deprotonation and protonation of the Schiff base linkage in bacteriorhodopsin, and have suggested that these processes might be directly involved in the light-induced proton translocation, possibly through transfer to and from other groups located nearer the surfaces. Further experiments are required to elucidate the mechanism of the vectorial translocation of protons across the purple membranes. An interesting question to ask is whether the cyclic protonation–deprotonation processes can be reversed so that the proton is translocated in the reverse direction (from extracellular space to cytoplasm), albeit with a lower reaction rate. If this is the case, a chemical capacitance should be observable under suitable conditions. In fact, electrical measurements in reconstituted planar lipid bilayer containing oriented bacteriorhodopsin have been reported (49, 50). Open-circuit measurement of such membranes in response to rectangular light pulse indicates that the characteristic "overshoot" of the "on-response" and "undershoot" of the "off-response" appear only when uncoupler CCCP or an external shunt is added. This result strongly indicates the presence of a chemical capacitance, which is made observable by a reduction of the access impedance by CCCP or an external shunt. It also suggests that at least part of the cyclic reaction is reversible so that a proton taken up (or released) at one interface can be released (or taken up) at the same interface (ac proton current) or at the opposite interface (dc proton current). However, this ac proton current must be small compared with the dc proton current, otherwise the characteristic "on-overshoot" and "off-undershoot" should have appeared in the membranes that are not treated with CCCP or shunted (see Section VG of Ref. (56)).

D. Comments on Photoelectric Effects in Natural Pigmented Membranes

The photoelectric effects in pigmented membranes are conveniently classified as ac photoresponses and dc photoresponses. It is the dc component that effects a net transfer of protons across the purple membranes and the photosynthetic membranes. In contrast, the ERP in visual membranes is an ac photoresponse. Since transmembrane protonation–deprotonation of the Schiff base linkage in bacteriorhodopsin may be responsible for the dc photoresponse in purple membranes, it is possible that interfacial movement of the same Schiff base proton in rhodopsin might be the physicochemical basis of the R_2 component of the ERP.[7]

From our previous discussion, it is clear that the ultrathinness of the membrane is responsible not only for efficient transmembrane chemical coupling (dc photoresponse) but also for efficient transmembrane electrical coupling (ac photoresponse). The charge movement may be confined to but one interface; nevertheless, it also polarizes the opposite interface. Thus, the chemical capacitance is a direct consequence of the order at two membrane–water interfaces and the juxtaposition of these two interfaces.

We have analyzed presently available data of photoelectric effects in three important photobiological membranes and demonstrated the usefulness of chemical capacitance as a unifying concept in interpreting these data. We have repeatedly emphasized the sensitivity and superiority of a short-circuit measurement in studying photoelectric effects in pigmented membranes (55–57). Not only is the signal-to-noise ratio improved, but also the data are more easily analyzed. Thus, the TVC method remains the method of choice in such studies.

IV. MAGNETO–ORIENTATION EFFECT

A. Magneto-Orientation Effect in Isolated Rod Outer Segments

Chalazonitis and co-workers (23) reported in 1970 that isolated frog rod outer segments in aqueous suspension can be oriented with a

[7] There may be more than one proton movement in each direction during the metarhodopsin I to metarhodopsin II transition (135).

homogeneous magnetic field of 10 kG. The equilibrium orientation is in parallel with the applied field. Furthermore, the two ends of the rod appear completely equivalent in the magnetic field. This latter observation immediately excludes mechanisms based on ferromagnetism, ferrimagnetism, antiferromagnetism, and magnetic dipole moment generated by membrane currents (24, 55). Thus, the magnetism in the rod outer segment must be either diamagnetic or paramagnetic. In either case, the magneto-orientation effect can be due to (1) magnetic anisotropy, (2) "form" anisotropy, or (3) inhomogeneity of the applied magnetic field (55).

Our conservative estimation indicates that, unless the inhomogeneity of the magnetic field is unreasonably large, the magnetic orientation energy of a rod due to this influence falls well below the kT energy (4.1×10^{-14} ergs at 25°C). No orientation effect could be observed because of thermal fluctuations of the rod (24, 55).

As for the second possibility, orientation of an isotropic object can sometimes occur in a homogeneous magnetic field because of the geometry of the object. Magnetization of an isotropic rod in a homogeneous field leads to creation of a local magnetic field inside the rod, which is the sum of the applied field and the induced field, and is not homogeneous in general. This phenomenon is called "form" anisotropy. Again, our estimation of the orientation energy of a rod indicates that the orientation effect is unobservable because of thermal fluctuations (55). We are thus left with the mechanism of magnetic anisotropy.

A rod outer segment of frog contains a regular parallel stacking of approximately 1900 discs (156). From our previous discussion, we see that the majority of phospholipid and rhodopsin molecules are oriented with respect to the axial direction of the rod. If any or both of these two molecules are magnetically anisotropic, the individual small anisotropy will be summed up *additively*, and the whole rod may have quite a respectable magnetic anisotropy, with an overall cylindrical symmetry. Thus, the magnetic susceptibility tensor will have one principal axis coincident with the rod axis. The other two axes will be in the plane of disc membranes and equal (degenerate), because of the cylindrical symmetry. Thus, the magnetic potential energy of a rod in a constant and homogeneous magnetic field of magnitude H can be readily shown to be

$$U = -\tfrac{1}{2}V(\chi_a - \chi_r)H^2 \cos^2\theta, \quad [13]$$

where χ_a and χ_r are the *axial* and the *radial* principal volume susceptibilities, respectively, of the *oriented* anisotropic molecules in a rod, V is the *effective* volume of these molecules and θ is the angle between the direction of the magnetic field and the axis of the rod. The magnetic potential energy at $\theta = 90°$ has been conveniently assigned as the zero reference. Contributions from the isotropic or the randomly oriented anisotropic constituents of the rod are thus excluded from Eq. [13]. A stable equilibrium orientation occurs when the magnetic potential energy is at a minimum. Thus, the rod will orient parallel with the applied field if $\chi_a - \chi_r > 0$, but will orient perpendicular to the applied field if $\chi_a - \chi_r < 0$. Since there is no evidence of paramagnetism in rhodopsin and in phospholipid, and since the observed stable orientation is parallel to the magnetic field, we shall consider diamagnetic anisotropy. In this case, Eq. [13] predicts that $\chi_r < \chi_a$ (<0).

Further tests of this mechanism must also include estimation of orientation time and thermal fluctuation, as Chalazonitis and co-workers reported that the orientation is completed within 2 min and their accuracy of observation is within $\pm 1°$ (23, 25). The complete equation of rotational motion can be written and solved numerically (24, 55). However, we have found that an excellent approximation can be obtained by dropping the inertial term. We thus have

$$\frac{d\theta}{dt} = -\frac{V(\chi_a - \chi_r)H^2}{2\zeta}\sin 2\theta, \quad [14]$$

where ζ is the rotational frictional coefficient of the rod in water. Equation [14] states that the

maximal angular velocity occurs at $\theta = 45°$. The time course of orientation is given by

$$\ln\tan\theta = \ln\tan\theta_0 - \frac{V(\chi_a - \chi_r)H^2}{\zeta}t, \quad [15]$$

where θ_0 is the orientation at $t = 0$. Equation [15] states that the orientation time is portional to ζ but is inversely proportional to the factor $V(\chi_a - \chi_r)H^2$. The fluctuation of the orientation of a rod observed in a plane containing the magnetic field vector is given by

$$\Delta\theta_{rms} = \left[\frac{kT}{V(\chi_a - \chi_r)H^2}\right]^{\frac{1}{2}}, \quad [16]$$

provided that $V(\chi_a - \chi_r)H^2 \gg kT$.

Note the factor $V(\chi_a - \chi_r)$, which we shall call summed anisotropy,[8] appears in each of Eqs. [13]–[16]. Instead of being expressed per unit volume, $(\chi_a - \chi_r)$ may also be expressed per unit molecule if V is regarded as the total number of anisotropic oriented molecules. In this case, V will be 3×10^9 for rhodopsin or 10^{13} for phospholipid. A simple calculation quickly convinces us that it is impossible to orient a single molecule of rhodopsin or phospholipid since thermal fluctuation will be much too large and the orientation time much too long. However, when a large number of such molecules are oriented in an ordered structure such as disc membranes in a rod, the orientation time will be shortened by at least 10^9 times and thermal fluctuation reduced by at least 10^4 times. Our numerical estimation indicates that a mechanism due to diamagnetic anisotropy of rhodopsin or phospholipid is entirely compatible with an observable magneto-orientation effect. We have attributed the major contribution of this summed anisotropy to rhodopsin rather than phospholipid. Our view is supported by Becker (88).

[8] When more than one species of oriented molecules contributes to anisotropy, the *summed anisotropy* must be expressed as a linear sum $\sum_i V_i(\chi_{ai} - \chi_{ri})$, where the subscript i refers to each species of oriented anisotropic molecules.

Chagneux and Chalazonitis (25) have subsequently carried out more experiments to test our proposed mechanism of diamagnetic anisotropy (see Fig. 9A). Figure 9A shows maximal angular velocity (slope) at 45° orientation as predicted by Eq. [14]. If the data of Fig. 9A are replotted on a scale of ln tan θ vs time, Fig. 9B is obtained. Within the accuracy of the data, the orientation time course is in agreement with the prediction of Eq. [15]. The slope of the fitted straight line gives an anisotropy of 1.2% if all the anisotropy is attributed to rhodopsin. Recent experiments of Chalazonitis and Chagneux (personal communication, 1976) show that the magnetic anisotropy decreases by 20 to 25% upon complete bleaching of a dark-shaped rod outer segment.

B. Magneto-Orientation Effect in Chlorella Cells

Another example of magneto-orientation effect due to diamagnetic anisotropy is given by a unicellular green alga *Chlorella*. A *Chlorella* cell is spherical with only one single chloroplast attached to it (e.g., Fig. 5 of Ref. (157)). Geacintov and co-workers (26) studied the chlorophyll *a* fluorescence of an aqueous suspension of *Chlorella pyrenoidosa* in a homogeneous magnetic field of 10 kG or more. They found that the fluorescence quantum yield is enhanced by 4–9% if the exciting light beam is parallel to the applied magnetic field, but is decreased by 4–9% if the light beam is perpendicular to the applied field. These authors initially interpreted this effect as a consequence of reorientation of pigment molecules in the magnetic field (26). Our previous discussion indicates that the orientation of individual molecules required a much more intense field because of thermal fluctuation. Because of the regular parallel stacking of thylakoid membranes in the chloroplast, we suggested that the effect might be due to orientation of the entire chloroplast, that is, the whole cell, since the chloroplast is fixed within the *Chlorella* cell (24). This interpretation was confirmed by subsequent experiments of Gea-

cintov and co-workers (27, 88) in which they studied the effect of viscosity of the suspending medium on fluorescence depolarization of chlorophyll *a* (Fig. 10). As expected, the relaxation times of fluorescence polarization slow down with increasing viscosity of the medium. Furthermore, the relaxation data predict an average cell size that agrees with actual measurement under a light microscope (27, 88). These authors have concluded that the magneto-orientation effect in *Chlorella* cells is due to diamagnetic anisotropy of oriented components in chloroplast membranes (27, 88, 158). Alternative interpretation in terms of para-, ferro-, antiferro-, and ferrimagnetism as well as reorientation of individual molecules has been excluded (88). Using *Chlorella* samples stained with rhodamine B, they further determined that the chloroplast membrane planes are oriented perpendicular to the applied magnetic field (88, 158). By studyiug both the linear dichroism and the fluorescence depolarization of magneto-oriented *Chlorella*, they found that the "red" Q_y transition moment (660 $\sim$ 680 nm) of chlorophyll lies in the lamellar plane of membranes, that the porphyrin ring planes tilt away from the lamellar plane by an angle greater than 45° and that carotenoid molecules are oriented with their long axis in the lamellar plane (158). The short-wavelength spectroscopic forms of chlorophyll are not found to be oriented (158). Similar results in spinach chloroplast were reported by Breton and co-workers (159), who, in addition, also found that the tryptophan residues[9] of integral proteins are oriented with their aromatic rings perpendicular to the lamellar plane.

Previous reports on chlorophyll orientation *in vivo* and on the degree of fluorescence polarization have been ambiguous because of extensive intermolecular energy transfer among chlorophyll molecules. Geacintov and co-workers (160–162) pointed out that the

---

[9] Similar orientation of the tryptophan residues of rhodopsin in disc membranes of bovine rod outer segments has been found (88).

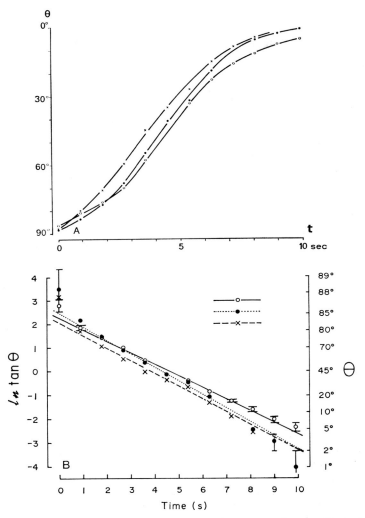

FIG. 9. The angular position of three isolated frog rod outer segments of comparable size in Ringer solution as a function of time under a homogeneous magnetic field of 10 kG. In A, the maximal slope occurs at 45°. (Reproduced from Chagneux and Chalazonitis (25); the ordinate has been relabeled to conform to our present definition of the angle θ, which is complementary to the angle defined in Ref. (25).) In B, the data in A is replotted as ln tan θ vs time. The angles θ are indicated on the right ordinate. The straight lines are least-square fit on the original scale of A rather than on the transformed scale to avoid bias due to points at two extremes, where error is aggravated. The error bars correspond to an error of $\pm 1°$ in reading the angular position of the rods. Given $H = 10^4$ G and $\zeta = 3.1 \times 10^{-10}$ g cm^{-2} sec^{-1} (25), the slopes of the straight lines give values of summed anisotropy of 1.53×10^{-18} (O), 1.78×10^{-18} (●), and 1.67×10^{-18} (✗), respectively (average, 1.66×10^{-18}). Assuming an effective volume as 1.4×10^{-10} cm^3 for rhodopsin and an averaged value of anisotropy as 10^{-6}, the diamagnetic anisotropy amount to 1.2%. If the anisotropy were attributed to phospholid (effective volume 12.67×10^{-10} cm^3), the anisotropy would be 0.13% (cf. Ref. (25)).

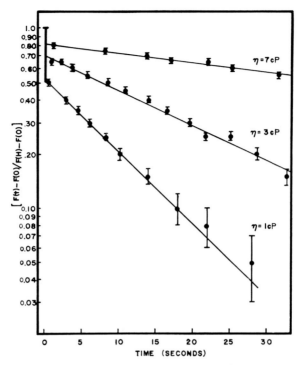

FIG. 10. Viscosity dependence of the relaxation of the magnetic-field-induced changes in the fluorescences in a suspension of *Chlorella* cells. $F(H)$, fluorescence at $H = 10.5$ kG; $F(0)$, fluorescence in zero field. The analyzer was oriented perpendicular to H. $F(t)$, decay of fluorescence from $F(H) \to F(0)$ after field had been removed. Drop in fluorescence during the 4 sec of magnetic field shutoff time is indicated by vertical bar at $t = 0$. The viscosity of the suspending aqueous medium is varied by adding a high molecular weight polymer (Ficoll). (Reproduced from Geacintov *et al.* (27).)

intrinsic polarization, which reflects the energy transfer between different pigment molecules and their degree of mutual orientation can be measured unambiguously only if (1) oriented membranes are used and the fluorescence is viewed along a direction perpendicular to the lamellar plane of the membranes, and (2) the excitation is confined to the Q_y absorption band of chlorophyll *in vivo*. This new approach has revealed a higher degree of orientation of chlorophyll *in vivo* than was previously thought (158, 159).

C. Comments on Biomagnetic Phenomena

Biomagnetic phenomena have been reported in the past on a variety of biological systems (163). However, these phenomena have often been received with skepticism by scientific communities. The magneto-orientation effect due to oriented diamagnetic anisotropy in the rod outer segments and in the *Chlorella* cells has been established experimentally and explained theoretically. This effect serves no physiological purpose, since the required strength of the magnetic field exceeds the terrestrial magnetic field ($\sim$0.5 G) by several orders of magnitude. The magneto-orientation effect is merely a manifestation of the long-range order in the structures of the rod outer segments and the *Chlorella* cells. However, as we have pointed out (24), this effect can be used to prepare oriented biological samples for other experimental purposes. Results which are difficult to obtain in a randomly oriented sample can be achieved (117, 158–162, 164, 165).

It has been repeatedly suggested for over a century that birds and other animals that migrate over long distances might use the terrestrial magnetic field as a clue for navigation. However, reproducible results emerged only in the last few years (166–169). Using Helmholtz coils to control the magnetic field, Wiltschko and Wiltschko (170) have obtained data suggesting that the "magnetic compass" of European robins does not use the polarity of the magnetic field for detecting the north direction. The birds obtain their sense of the north direction from the inclination of the axial direction of the magnetic field lines in space, i.e., the birds combine the cues of both the magnetic field and the gravitational field of the earth. This observation suggests that a sensory organ with parallel multiple membranous structures based on diamagnetic anisotropy may be a good candidate. Other magnetic sensory mechanisms such as ferromagnetism (171) or magnetic-field-induced membrane currents (172) can detect the polarity of the magnetic field in addition to the axial direction. A further test of Wiltschko and Wiltschko's interpretation under more physiological conditions would be to perform homing or orientation experiments in the southern hemisphere with birds trained in the northern hemisphere at a comparable latitude. So far these and other results on a "magnetic compass" in birds have not been received with confidence, hence any notion of special sensory organs remains speculative. Nevertheless, the climate in scientific communities has changed; no longer can the biomagnetic phenomena be brushed aside on theoretical grounds alone.

V. SUMMARY

Biological membranes are highly ordered structures which are ubiquitous in a living cell. The lipid bilayer configuration of membranes provides the first level of order by establishing a membrane phase and two interfaces. Oriented and asymmetric arrangement of integral proteins and other membrane-bound components constitute the second and higher level of order. Stacked coin-like arrays of multiple membranes such as occurring in rod outer segments and in chloroplasts make up the third and long-range order of membranous structures.

The first level of order makes possible vectorial photochemistry, which is underlying photoelectric effects in pigmented membranes. The asymmetry which is required to make these effects externally observable is provided either by a chemical gradient in the two aqueous phases, or by a voltage gradient across the membrane. In natural photosystems, it could also be provided by an intrinsic asymmetry due to oriented pigment within the membrane (second level of order). Photoelectric effects in pigmented membranes can be studied in terms of a dc component and an ac component. The dc photoelectric responses involve a net transfer of charge across the membrane but are electrically in parallel with membrane ionic currents. This photoinduced net proton transfer across photosynthetic membranes and purple membranes provides the proton gradient which is utilized in ATP synthesis, in accordance with Mitchell's hypothesis. The ac photoelectric response is a manifestation of charging and discharging of a novel chemical capacitance without a net transfer of charge across the membrane. We believe that the ERP in visual photoreceptors and other fast photoresponses in pigmented membranes belong to this category.

The magneto-orientation effect in the rod outer segments and in *Chlorella* is a manifestation of the second and the third level of order in membrane structures. It is due to summed diamagnetic anisotropy of oriented membrane components. This effect can be used as a tool to orient these biological samples for experiments, but serves no physiological purposes in the retina and in the chloroplasts. However, evidence has accumulated that certain birds might use the terrestrial magnetic field as an orientational cue in space. Summed diamagnetic anisotropy may be a candidate for the mechanism; however, no definite sensory receptors have been found.

ACKNOWLEDGMENTS

The author wishes to express his deep gratitude to his mentors Professor Chen-Yuan Lee of Taiwan University College of Medicine and Professor David Mauzerall of The Rockefeller University. The author thanks Drs. Nicholas Chalazonitis of Institut de Neurophysiologie et Psychophysiologie, Marseille, and Nicholas Geacintov of New York University for communicating their results of research prior to publication. The author is also obliged to Dr. Elias Greenbaum and Mr. Richard Piccioni of The Rockefeller University for discussion on photosynthesis and for critical reading of the manuscript, to Dr. John Cheng-Po Yu of Harvard Biological Laboratories for discussion on membrane structures and for valuable criticism of the manuscript, to Drs. Donald Griffin and Ronald Larkin of The Rockefeller University for discussion on bird navigation and orientation, and to Miss Virginia Genther and Mrs. Sharon Silverman for typing the first and the final drafts of manuscript, respectively. This work has been supported in part by Grant GM-20729 from the National Institutes of Health.

REFERENCES

1. SMITH, H. W., *Circulation* **26**, 987 (1962).
2. JACOBS, M. H., *Circulation* **26**, 1013 (1962).
3. FLEISCHER, S., AND PACKER, L. (Eds.), "Methods in Enzymology, Vols., XXXI and XXXII: Biomembranes." Academic Press, New York, 1974.
4. SINGER, S. J., *in* "Structure and Function of Biological Membranes" (L. I. Rothfield, Ed.), p. 145. Academic Press, New York, 1971.
5. SINGER, S. J., AND NICOLSON, G. L., *Science* **175**, 720 (1972).
6. ENGELMAN, D. M., *J. Mol. Biol.* **47**, 115 (1970).
7. WILKINS, M. H. F., BLAUROCK, A. E., AND ENGELMAN, D. M., *Nature New Biol.* **230**, 72 (1971).
8. STEIM, J. M., TOURTELLOTTE, M. E., REINERT, J. C., MCELHANEY, R. N., AND RADAR, R. L., *Proc. Nat. Acad. Sci. USA* **63**, 104 (1969).
9. HUBBELL, W. L., AND MCCONNELL, H. M., *Proc. Nat. Acad. Sci. U.S.A.* **61**, 12 (1968).
10. GLASER, M., SIMPKINS, H., SINGER, S. J., SHEETZ, M., AND CHAN, S. I., *Proc. Nat. Acad. Sci. U.S.A.* **65**, 721 (1970).
11. CHAN, S. I., FEIGENSON, G. W., AND SEITER, C. H. A., *Nature (London)* **231**, 110 (1971).
12. LEE, A. G., BIRDSALL, N. J. M., AND METCALFE, J. C., *Biochemistry* **12**, 1650 (1972).
13. STECK, T. L., *J. Cell Biol.* **62**, 1 (1974).
14. ZWAAL, R. F. A., ROELOFSEN, B., AND COOLEY, C. M., *Biochim Biophys. Acta* **300**, 159 (1973). (1974).
15. EDIDIN, M., *Annu. Rev. Biophys. Bioeng.* **3**, 179 (1974).
16. NICOLSON, G. L., *Biochim. Biophys. Acta* **457**, 57 (1976).
17. HÖBER, R., "Physikalische Chemie der Zelle under der Gewebe," 6th ed., ENGELMANN, Leipzig, 1926; also "Physical Chemistry of Cells and Tissues." Blakiston, Philadelphia, 1945.
18. TIEN, H. T. (Ed.), Photoelectric bilayer lipid membranes, *Photochem. Photobiol.* **24**, 95 (1976).
19. MUELLER, P., RUDIN, D. O., TIEN, H. T., AND WESCOTT, W. C., *Nature (London)* **194**, 979 (1962).
20. BROWN, K. T., *Vision Res.* **8**, 633 (1968).
21. WITT, H. T., *Quart. Rev. Biophys.* **4**, 365 (1971).
22. STOECKENIUS, W., *Sci. Amer.* **234**, 38 (1976).
23. CHALAZONITIS, N., CHAGNEUX, R., AND ARVANITAKI, A., *C. R. Acad. Sci. Ser. D.* **271**, 130 (1970).
24. HONG, F. T., MAUZERALL, D., AND MAURO, A., *Proc. Nat. Acad. Sci. U.S.A.* **68**, 1283 (1971).
25. CHAGNEUX, R., AND CHALAZONITIS, N., *C. R. Acad. Sci. Ser. D.* **274**, 317 (1972).
26. GEACINTOV, N. E., VAN NOSTRAND, F., POPE, M., AND TRINKEL, J. B., *Biochim. Biophys. Acta* **226**, 486 (1971).
27. GEACINTOV, N. E., VAN NOSTRAND, F., BECKER, J. F., AND TRINKEL, J. B., *Biochim. Biophys. Acta* **267**, 65 (1972).
28. TIEN, H. T., *Nature (London)* **219**, 272 (1968).
29. TIEN, H. T., *J. Phys. Chem.* **72**, 4512 (1968).
30. TIEN, H. T., AND VERMA, S. P., *Nature (London)* **227**, 1232 (1970).
31. TIEN, H. T., AND KOBAMOTO, N., *Nature (London)* **224**, 1107 (1969).
32. KOBAMOTO, N., AND TIEN, H. T., *Biochim. Biophys. Acta* **241**, 129 (1971).
33. HUEBNER, J. S., AND TIEN, H. T., *Biochim. Biophys. Acta* **256**, 300 (1972).
34. MILLER, T. E., AND TIEN, H. T., *Bioenergetics* **6**, 1 (1974).
35. SHIEH, P. K., AND TIEN, H. T., *Bioenergetics* **6**, 45 (1974).
36. ULLRICH, H.-M., AND KUHN, H., *Z. Naturforsch.* **24b**, 1342 (1969).
37. ULLRICH, H.-M., AND KUHN, H., *Biochim. Biophys. Acta* **266**, 584 (1972).
38. TRISSL, H.-W., AND LÄUGER, P., *Z. Naturforsch.* **25b**, 1059 (1970).
39. TRISSL, H.-W., AND LÄUGER, P., *Biochim. Biophys. Acta* **282**, 40 (1972).
40. LUTZ, H. W., TRISSL, H.-W., AND BENZ, R., *Biochim. Biophys. Acta* **345**, 257 (1974).

41. ILANI, A., AND BERNS, D. S., *J. Membr. Biol.* **8**, 333 (1972).

42. MANGEL, M., BERNS, D. S., AND ILANI, A., *J. Membr. Biol.* **20**, 171 (1975).

43. CHEN, C.-H., AND BERNS, D. S., *Proc. Nat. Acad. Sci. U.S.A.* **72**, 3407 (1975).

44. HONG, F. T., AND MAUZERALL, D., *Biochim. Biophys. Acta* **275**, 479 (1972).

45. HONG, F. T., AND MAUZERALL, D., *Nature New Biol.* **240**, 154 (1972).

46. HONG, F. T., AND MAUZERALL, D., *Proc. Nat. Acad. Sci. U.S.A.* **71**, 1564 (1974).

47. SCHADT, M., *Biochim. Biophys. Acta* **323**, 351 (1973).

48. FROEHLICH, O., AND DIEHN, B., *Nature (London)* **248**, 802 (1974).

49. DRACHEV, L. A., JASAITIS, A. A., KAULEN, A. D., KONDRASHIN, A. A., LIBERMAN, E. A., NEMECEK, I. B., OSTROUMOV, S. A., SEMENEV, A. YU., AND SKULACHEV, V. P., *Nature (London)* **249**, 321 (1974).

50. DRACHEV, L. A., KAULEN, A. D., OSTROUMOV, S. A., AND SKULACHEV, V. P. *FEBS Lett.* **39**, 43 (1974).

51. MONTAL, M., AND KORENBROT, J. I., *Nature (London)* **246**, 219 (1973).

52. MONTAL, M., *in* "Molecular Aspects of Membrane Phenomena" (H. R. Kaback, G. K. Radd, and H. Neurath, Eds.), p. 316. Springer-Verlag, Berlin, 1975.

53. HUEBNER, J. S., *Biochim. Biophys. Acta.* **406**, 178 (1975).

54. NASTUK, W. L. (Ed.), Electrophysiological methods, Vols. V and VI of "Physical Techniques in Biological Research." Academic Press, New York, 1963 and 1964.

55. HONG, F. T., Ph.D. Dissertation, The Rockefeller University, New York, 1973.

56. HONG, F. T., *Photochem. Photobiol.* **24**, 155 (1976).

57. HONG, F. T., AND MAUZERALL, D., *J. Electrochem. Soc.* **123**, 1317 (1976).

58. ILANI, A., AND BERNS, D. S., *Biophysik* **9**, 209 (1973).

59. TIEN, H. T., AND HUEBNER, J. S., *J. Membr. Biol.* **11**, 57 (1973).

60. TRISSL, H.-W., *Z. Naturforsch.* **30c**, 124 (1975).

61. TIEN, H. T., *Photochem. Photobiol.* **16**, 271 (1972).

62. TIEN, H. T., *in* "Biennial Review of Surface Chemistry and Colloids. MTP International Review of Science" (M. Kerker, Ed.), p. 25. Butterworths, London; University Park Press, Baltimore, 1972.

63. TIEN, H. T., AND CHEN, V. K., *in* "Progress in Surface and Membrane Science" (D. A. Cadenhead, J. F. Danielli, and M. D. Rosenberg, Eds.), Vol. 8, p. 119. Academic Press, New York, 1974.

64. TIEN, H. T., "Bilayer Lipid Membranes: Theory and Practice." Dekker, New York, 1974.

65. MAUZERALL, D., AND HONG, F. T., *in* "Porphyrins and Metalloporphyrins" (K. M. Smith, Ed.), 2nd ed., p. 701. Elsevier, Amsterdam, 1975.

66. TIEN, H. T., *Photochem. Photobiol.* **24**, 97 (1976).

67. BERNS, D. S., *Photochem. Photobiol.* **24**, 117 (1976).

68. STRAUSS, G., *Photochem. Photobiol.* **24**, 141 (1976).

69. FUHRHOP, J.-H., AND MAUZERALL, D., *J. Amer. Chem. Soc.* **90**, 3875 (1968).

70. FUHRHOP, J.-H., AND MAUZERALL, D., *J. Amer. Chem. Soc.* **91**, 4174 (1969).

71. CARAPELLUCCI, P.A., AND MAUZERALL, D., *Ann. N. Y. Acad. Sci.* **244**, 214 (1975).

72. HONG, F. T., *Fed. Proc.* **33**, 1268, Abstr. No. 249 (1974).

73. VERWEY, E. J. W., AND OVERBEEK, H. Th. G., "Theory of the Stability of Lyophobic Colloids." Elsevier, Amsterdam, 1948.

74. EVERITT, C. T., AND HAYDON, D. A., *J. Theor. Biol.* **18**, 371 (1968).

75. WORTHINGTON, C. R., *Annu. Rev. Biophys. Bioeng.* **3**, 53 (1974).

76. MENKE, W., *in* "Biochemistry of Chloroplasts" (T. W. Goodwin, Ed.), Vol. I, p. 3. Academic Press, New York, 1966.

77. TREBST, A., *Annu. Rev. Plant Physiol.* **25**, 423 (1974).

78. ANDERSON, J. M., *Biochim. Biophys. Acta* **416**, 191 (1975).

79. FALK, J. E., "Porphyrins and Metalloporphyrins." Elsevier, Amsterdam, 1964.

80. SMITH, K. M. (Ed.), "Porphyrins and Metalloporphyrins," 2nd ed., Elsevier, Amsterdam, 1975.

81. WITT, H. T., *in* "First Reactions and Primary Processes in Chemical Kinetics," (S. Claesson, Ed.), Nobel Symp. V., p. 261. Interscience, New York; Almqvist and Wiksell, Stockholm, 1967.

82. BULYCHEV, A. A., ANDRIANOV, V. K., KURELLA, G. A., AND LITVIN, F. F., *Nature (London)* **236**, 175 (1972).

83. VREDENBURG, W. J., AND TONK, W. J. M., *Biochim. Biophys. Acta* **387**, 580 (1975).

84. BULYCHEV, A. A., AND VREDENBERG, W. J., *Biochim. Biophys. Acta* **423**, 548 (1976).

85. BULYCHEV, A. A., ANDRIANOV, V. K., KURELLA, G. A., AND LITVIN, F. F., *Biochim. Biophys. Acta* **430**, 336 (1976).

86. VREDENBERG, W. J., AND BULYCHEV, A. A., *Plant Sci. Lett.* **7**, 101 (1976).

87. FOWLER, C. F., AND KOK, B., *Biochim. Biophys. Acta* **357**, 308 (1974).

88. BECKER, J. F., Ph.D. Dissertation, New York University, New York, 1975.

89. MITCHELL, P., *Nature (London)* **191**, 144 (1961).

90. MITCHELL, P., *Biol. Rev.* **41**, 445 (1966).

91. JAGENDORF, A. T., AND HIND, G., *in* "Photosynthetic Mechanisms of Green Plants," National Academy of Science—National Research Council Publication 1145, p. 599, 1963.

92. JAGENDORF, A. T., AND URIBE, E., *Proc. Nat. Acad. Sci. U.S.A.* **55**, 170 (1966).

93. RUMBERG, B., AND SCHRÖDER, H., *in* "Proceedings of the 1st European Biophysical Congress, Baden, Austria" (E. Broda, A. Locker, and H. Springer-Lederer, Eds.), Vol. IV, p. 57. Verlag der Wiener Med. Akad., 1971.

94. SCHRÖDER, H., MUHLE, H., AND RUMBERG, B., *in* "Proceedings of the 2nd International Congress on Photosynthesis Research, Stresa 1971" (G. Forti, M. Avron, and A. Melandri, Eds.), Vol. II, p. 919. Junk, The Hague, 1972.

95. RUMBERG, B., AND SIGGEL, U., *Z. Naturforsch.* **23b**, 239 (1968).

96. JUNGE, W., RUMBERG, B., AND SCHRÖDER, H., *Eur. J. Biochem.* **14**, 575 (1970).

97. JOLIOT, P., AND DELOSME, R., *Biochim. Biophys. Acta* **357**, 267 (1974).

98. WALD, G., *Science* **162**, 230 (1968).

99. COHEN, A. I., *in* "Physiology of Photoreceptor Organs: Handbook of Sensory Physiology (M. G. F. Fuortes, Ed.), Vol. VII/2" Springer–Verlag, Berlin, 1972.

100. SCHMIDT, W. J., *Kolloidzeitschrift* **85**, 137 (1938).

101. DENTON, E. J., *J. Physiol.* **124**, 16P (1954).

102. DENTON, E. J., *Proc. Roy. Soc. London Ser. B* **150**, 78 (1959).

103. LIEBMAN, P. A., *Biophys. J.* **2**, 161 (1962).

104. BROWN, P. K., *Nature New Biol.* **236**, 35 (1972).

105. CONE, R. A., *Nature New Biol.* **236**, 39 (1972).

106. POO, M.-M. AND CONE, R. A., *Exp. Eye Res.* **17**, 503 (1973).

107. POO, M.-M., AND CONE, R. A., *Nature (London)* **247**, 438 (1974).

108. GRAS, W. J., AND WORTHINGTON, C. R., *Proc. Nat. Acad. Sci. U.S.A.* **63**, 233 (1969).

109. DEWEY, M. M., DAVIS, P. K., JR., BLASIE, J. K., AND BARR, L., *J. Mol. Biol.* **39**, 395 (1969).

110. DRATZ, E. A., GAW, J. E., SCHWARTZ, S., AND CHING, W. M., *Nature (London)* **237**, 99 (1972).

111. DRATZ, E. A., AND SCHWARTZ, S., *Nature (London)* **242**, 212 (1973).

112. STEINEMANN, A., AND STRYER, L., *Biochemistry* **12**, 1499 (1973).

113. YEAGER, M., SCHOENBORN, B. D., ENGELMAN, D., MOORE, P., AND STRYER, L., *Fed. Proc.* **33**, 1575 (1974).

114. YEAGER, M., *Fed. Proc.* **34**, 583 (1975).

115. YEAGER, M., *Brookhaven Symp. Biol.* **27**, III-3 (1976).

116. SCHOENBORN, B. P., *Biochim. Biophys. Acta* **457**, 41 (1976).

117. CHABRE, M., *Biochim. Biophys. Acta* **382**, 322 (1975).

118. WU, C.-W., AND STRYER, L., *Proc. Nat. Acad. Sci. U.S.A.* **69**, 1104 (1972).

119. EBREY, T., AND HONIG, B., *Quart. Rev. Biophys.* **8**, 129 (1975).

120. BROWN, K. T., AND MURAKAMI, M., *Nature (London)* **201**, 626 (1964).

121. CONE, R. A., *Nature (London)* **204**, 736 (1964).

122. BROWN, K. T., AND MURAKAMI, M., *Nature (London)* **204**, 739 (1964).

123. PAK, W. L., *Cold Spring Harbor Symp. Quant. Biol.* **30**, 493 (1965).

124. BRINDLEY, G. S., AND GARDNER-MEDWIN, A. R., *J. Physiol.* **182**, 185 (1966).

125. ARDEN, G. B., BRIDGES, C. D. B., IKEDA, H., AND SIEGEL, I. M., *Vision Res.* **8**, 3 (1968).

126. CONE, R. A., AND BROWN, P. K., *Science* **156**, 536 (1967).

127. PENN, R. D., AND HAGINS, W. A., *Nature (London)* **223**, 201 (1969).

128. HAGINS, W. A., PENN, R. D., AND YOSHKIAMI, S., *Biophys. J.* **10**, 380 (1970).

129. PENN, R. D., AND HAGINS, W. A., *Biophys. J.* **12**, 1073 (1972).

130. TOMITA, T., *Quart. Rev. Biophys.* **3**, 179 (1970).

131. HAGINS, W. A., *Annu. Rev. Biophys. Bioeng.* **1**, 131 (1972).

132. CONE, R. A., *Science* **155**, 1128 (1967).

133. HAGINS, W. A., AND McGAUGHEY, R. E., *Science* **157**, 813 (1967).

134. FALK, G., AND FATT, P., *in* "Photochemistry of Vision" (H. J. A. Dartnall, Ed.), p. 200. Springer–Verlag, Berlin, 1972.

135. MATHEWS, R. G., HUBBARD, R., BROWN, P. K., AND WALD, G., *J. Gen. Physiol.* **47**, 215 (1963).

136. FALK, G., AND FATT, P., *J. Physiol.* **183**, 211 (1966).

137. WALD, G., BROWN, P. K., AND GIBBONS, I. R., *J. Opt. Soc. Amer.* **53**, 20 (1963).

138. CONE, R. A., *Cold Spring Harbor Symp. Quant. Biol.* **30**, 483 (1965).

139. PAK, W. L., *Photochem. Photobiol.* **8**, 495 (1968).

140. YOSHIKAMI, S., AND HAGINS, W. A., *Biophys. J.* **11**, 47 a (1971).

141. YOSHIKAMI, S., AND HAGINS, W. A., *in* "Biochemistry and Physiology of Visual Pigments" (H. Langer, Ed.), p. 245. Springer–Verlag, Berlin, 1973.

142. HAGINS, W., AND YOSHIKAMI, S., *Exp. Eye Res.* **18**, 299 (1974).

143. ARMSTRONG, C. M., AND BENZANILLA, F., *J. Gen. Physiol.* **63**, 533 (1974).

144. OESTERHELT, D., AND STOECKENIUS, W., *Nature New Biol.* **233**, 149 (1971).

145. BOGOMOLNI, R. A., AND STOECKENIUS, W., *J. Supramol. Struct.* **2**, 775 (1974).

146. BLAUROCK, A. E., AND STOECKENIUS, W., *Nature New Biol.* **233**, 152 (1971).

147. HENDERSON, R., AND UNWIN, P. N. T., *Nature (London)* **257**, 28 (1975).

148. OESTERHELT, D., AND STOECKENIUS, W., *Proc. Nat. Acad. Sci. U.S.A.* **70**, 2853 (1973).

149. DANON, A., AND STOECKENIUS, W., *Proc. Nat. Acad. Sci. U.S.A.* **71**, 1234 (1974).

150. STOECKENIUS, W., AND LOZIER, R. H., *J. Supramol. Struct.* **2**, 769 (1974).

151. LOZIER, R. H., BOGOMOLNI, R. A., AND STOECKENIUS, W., *Biophys. J.* **15**, 955 (1975).

152. BOGOMOLNI, R. A., BAKER, R. A., LOZIER, R. H., AND STOECKENIUS, W., *Biochim. Biophys. Acta* **440**, 68 (1976).

153. LOZIER, R. H., NIEDERBERGER, W., BOGOMOLNI, R. A., HWANG, S.-B., AND STOECKENIUS, W., *Biochim. Biophys. Acta* **440**, 545 (1976).

154. RACKER, E., AND STOECKENIUS, W., *J. Biol. Chem.* **249**, 662 (1974).

155. LEWIS, A., SPOONHOWER, J., BOGOMOLNI, R. A., LOZIER, R. H., AND STOECKENIUS, W., *Proc. Nat. Acad. Sci. U.S.A.* **71**, 4462 (1974).

156. NILSSON, S. E. G., *J. Ultrastruct. Res.* **12**, 207 (1965).

157. PARK, R. B., *in* "The Chlorophylls" (L. P. Vernon and G. R. Seely, Eds.), p. 283. Academic Press, New York, 1966.

158. BECKER, J. F., GEACINTOV, N. E., VAN NOSTRAND, F., AND VAN METTER, R., *Biochem. Biophys. Res. Commun.* **51**, 597 (1973).

159. BRETON, J., MICHEL-VILLAZ, M., AND PAILLOTIN, G., *Biochim. Biophys. Acta* **314**, 42 (1973).

160. BRETON, J., BECKER, J. F., AND GEACINTOV, N. E., *Biochem. Biophys. Res. Commun.* **54**, 1403 (1973).

161. GEACINTOV, N. E., VAN NOSTRAND, F., AND BECKER, J. F., *Biochem. Biophys. Acta* **347**, 443 (1974).

162. BECKER, J. F., BRETON, J., GEACINTOV, N. E., AND TRENTACOSTI, F., *Biochim. Biophys. Acta* **440**, 531 (1976).

163. BARNOTHY, M. F. (Ed.), "Biological Effects of Magnetic Fields," Vols. 1 and 2, Plenum, New York, 1964 and 1969.

164. CHABRE, M., *Brookhaven Symp. Biol.* **27**, III-77 (1976).

165. WORCESTER, D. L., *Brookhaven Symp. Biol.* **27**, III-37 (1976).

166. KEETON, W. T., *in* "Advances in the Study of Behavior" (D. Lehrman, R. Hinde, and E. Shaw, Eds.), Vol. 5, p. 47. Academic Press, New York, 1974.

167. WALCOTT, C., *Amer. Sci.* **62**, 542 (1974).

168. EMLEN, S. T., *in* "Avian Biology" (D. S. Farner and J. R. King, Eds.), Vol. 5, p. 129. Academic Press, New York, 1975.

169. GRIFFIN, D. R., *Harvey Lect.*, in press, 1977.

170. WILTSCHKO, W., AND WILTSCHKO, R., *Science* **176**, 62 (1972).

171. BLACKMORE, R., *Science* **190**, 377 (1975).

172. KALMIJN, A. J., *in* "Electro-Receptors and Other Specialized Receptors in Lower Vertebrates, Handbook of Sensory Physiology" (A. Fessard, Ed.), Vol. III/3, p. 147. Springer–Verlag, Berlin, 1974.

Surface Thermodynamics

Surface Thermodynamics

R. DEFAY, I. PRIGOGINE, AND A. SANFELD

Chimie-Physique II, Free University of Brussels, Campus Plaine, C.P.231, Bd. Triomphe, Brussels 1050, Belgium

Received May 24, 1976; accepted November 1, 1976

A general survey of the thermodynamics of surfaces is presented. Three domains are investigated: (a) the case where the entropy production is reduced to adsorption and chemical reaction, with the equilibrium as a special case; (b) the linear nonequilibrium thermodynamics initiated by Onsager; (c) the nonlinear region characterized by an excess entropy production. In some cases, far from equilibrium, the surface may become unstable after a threshold. Self-organization such as space and temporal dissipative structures may then occur in the surface.

INTRODUCTION

Surface thermodynamics provides interesting investigative possibilities because of the increase of parameters and variables as compared to bulk phenomena. Thus, additional possibilities of behavior develop in capillary systems.

As is well known, classical thermodynamics for continuous systems, as well as for surfaces, is concerned with equilibrium situations. However, there exists today a great interest in nonequilibrium processes arising in various disciplines such as elementary particle physics, physical chemistry, and biology. Peculiarly, the role played by the surfaces in far-from-equilibrium conditions, both in physical chemistry and in cell biology, has led to very new situations in recent years.

We shall treat both equilibrium and nonequilibrium situations in a unified modern formulation. For this reason, we shall start with the second law of thermodynamics as a balance equation of entropy. We may then consider three special important domains:

(a) The case where the entropy production is reduced to adsorption and chemical reaction (1), with equilibrium (2) as a special case.

(b) The linear nonequilibrium thermodynamics initiated by Onsager (3), and

(c) The nonlinear region as developed in recent years (4).

Because of the great interest currently in nonequilibrium situations, we cannot include an exhaustive historical survey. We start, therefore, essentially with the Gibbs–De Donder method (5, 6). In nonequilibrium states, the Gibbs method was extended by the Brussels school of thermodynamics (1).

The modern formulation (7) for describing nonequilibrium processes is

$$dS = d_eS + d_iS, \qquad [1]$$

where d_eS is the entropy flow and d_iS the entropy production. This last term is the central quantity of thermodynamics of irreversible processes. It is zero when the system undergoes reversible (or equilibrium) transformations only, but it is positive if the system is subject to irreversible processes.

In the frame of the assumption of local equilibrium to which we shall return, the explicit form of the entropy production (1) by unit time is the bilinear expression

$$P[S] = \frac{d_iS}{dt} = \int \sigma(S)dV$$

$$= \int \sum_k J_k X_k dV \geqslant 0, \quad [2]$$

where

$$\sigma(S) = \sum_k J_k X_k \quad [3]$$

is the local entropy production, J_k are the generalized fluxes of the irreversible process, and X_k the corresponding generalized thermodynamic forces, which may be either gradients or chemical affinities.

In capillary systems, for partial equilibrium (including thermal and mechanical equilibrium), the global entropy production (Eq. [2]) leads immediately to the idea of the affinity of adsorption (1):

$$\frac{d_iS}{dt} = T^{-1} \sum_{\gamma\alpha} A_\gamma{}^\alpha \frac{d\xi_\gamma{}^\alpha}{dt}$$

$$+ T^{-1} \sum_\rho A_\rho \frac{d\xi_\rho}{dt} \geqslant 0, \quad [4]$$

where $A_\gamma{}^\alpha$ is the affinity of adsorption of component γ from phase α to the surface, A_ρ the affinity of the chemical reaction ρ, $\xi_\gamma{}^\alpha$ and ξ_ρ are, respectively, the extent of adsorption and of reaction, and $d\xi_\gamma{}^\alpha/dt$, $d\xi_\rho/dt$ are, respectively, the velocity of adsorption and of chemical reactions.

Let us now comment on the assumption of local equilibrium. Usually, it means that the local entropy depends on the local macroscopic variables:

$$s = s(u, v, x_\gamma), \quad [5]$$

where u is the internal energy density, v the specific volume and, x_γ the mass fraction of component γ. This assumption implies that the collisional effects must be sufficiently dominant to exclude large deviations from statistical equilibrium. Chemical reactions are

included under the condition that they proceed not so quickly as to perturb in an exaggerated way the statistical equilibrium (8).

Unfortunately, for surface phases in many cases, it is not possible to assume that the surface energy is completely determined, as for equilibrium, by variables associated only with the surface. In a surface layer out of adsorption equilibrium, it is not possible to neglect the interactions between molecules in the surface and in the bulk phases. Consequently, the surface energy depends not only on the composition of the surface layer, but also on the composition of the bulk phases. It is this dependence which is the origin of the dynamic surface tension (1, 2). Bulk phases are said to be autonomous while surface phases are nonautonomous. However, it is easy to complete the local formulation by introducing, besides the ordinary chemical potentials, quantities called cross-chemical potentials, $\epsilon_\gamma{}^\alpha$, which are the partial derivatives of the surface free energy per unit area f^Ω with respect to the concentrations of component γ in the adjacent bulk phases $C_\gamma{}^\alpha$:

$$\epsilon_\gamma{}^\alpha = \partial f^\Omega/\partial C_\gamma{}^\alpha, \quad [6]$$

with

$$f^\Omega = f^\Omega(T, C_1^1 \cdots {}_\gamma^\alpha, \Gamma_{1\cdots\gamma}). \quad [7]$$

The extension of Eq. [5] for surfaces is then

$$s^\Omega = s^\Omega(u^\Omega, \Gamma_{1\cdots\gamma}, C_1^1 \cdots {}_\gamma^\alpha). \quad [8]$$

The Gibbs surface law for non-physico-chemical equilibrium can be written

$$d\sigma = -s^\Omega dT - \sum_\gamma \Gamma_\gamma d\mu_\gamma{}^\Omega + \sum_{\gamma\alpha} \epsilon_\gamma{}^\alpha dC_\gamma{}^\alpha, \quad [9]$$

where σ is the surface tension, s^Ω the surface entropy per unit area, Γ_γ the Gibbs adsorption of component γ, $C_\gamma{}^\alpha$ the mass concentration, and $\mu_\gamma{}^\Omega$ the surface chemical potential of component γ defined by

$$\mu_\gamma{}^\Omega = \partial f^\Omega/\partial \Gamma_\gamma. \quad [10]$$

When the system is in equilibrium, the assumption of local equilibrium in its usual form remains valid ($\epsilon_\gamma{}^\alpha = 0$) and Eq. [9] reduces to the well-known Gibbs equation

$$d\sigma = -s^\Omega dT - \sum_\gamma \Gamma_\gamma d\mu_\gamma, \qquad [11]$$

where μ_γ is the chemical potential in the bulk phase.

Adsorption phenomena in the Gibbs model are related to a geometrical dividing surface, parallel to the surface of tension defined by Young (9). The two bulk phases are supposed to remain homogeneous up to this Gibbs dividing surface. The values of the adsorptions and of other thermodynamic quantities which have to be associated with the surface to make the model equivalent to the physical system depend markedly, however, on the exact position of the surface. To overcome this problem, use is made of the relative adsorption $\Gamma_{\gamma 1}$ and other corresponding thermodynamic quantities, defined so that their values are invariant with respect to the choice of the position of the dividing surface.

The hypothesis of local equilibrium for surfaces holds however in various special cases: for example, when two immiscible pure fluids without surface mass density are separated by a possibly moving interface. Bedeaux *et al.* (3) used the following relation:

$$\frac{d^\Omega}{dt} s^\Omega = T_\Omega^{-1} \frac{d^\Omega}{dt} u^\Omega, \qquad [12]$$

where d^Ω/dt is the total time derivative at the surface, s^Ω and u^Ω are, respectively, the entropy and the internal energy per unit area on the surface. These authors derived the local surface entropy production connected on the one hand to nonuniformities in the surface velocity and in the surface temperature, and on the other hand to discontinuities in the velocities and temperature fields across the surface. They obtained the well-known bilinear form (Eq. [3]) for the momentum flow and the heat flux, and their corresponding forces in the surface and in its surroundings.

Another example of the applicability of the local equilibrium in the surface was worked out by Steinchen and Sanfeld (4). These authors

wrote the entropy production for moving insoluble monolayers between two immiscible fluids undergoing surface chemical reactions, momentum flow, and heat fluxes.

In numerous other examples, a particular equilibrium assumption for surfaces remains valid even when transfer of matter through the surface occurs. The rate of adsorption between the sublayer and the surface has to be, then, much more rapid than the diffusion from the bulk to the sublayer. This hypothesis is due to Ward and Tordai (10).

It is, however, not valid for bipolar molecules for which an adsorption barrier is often observed (11, 12). The hypothesis of local equilibrium for surfaces was recently discussed by Kerszberg (13).

Returning now to the local entropy production of Eq. [3] for continuous systems, three main situations may occur:

(a) The equilibrium where both the forces and flows vanish:

$$J_k{}^{\text{eq}} = 0 \quad \text{and} \quad X_k{}^{\text{eq}} = 0. \qquad [13]$$

(b) The "linear region" beyond equilibrium corresponding to the linear relations between flows and forces:

$$J_k = \sum_l L_{lk} X_l. \qquad [14]$$

In this region, two general theorems have been established, the Onsager reciprocity laws (14)

$$L_{lk} = L_{kl}, \qquad [15]$$

and the theorem of minimum entropy production (7). For stationary nonequilibrium states (with time-independent boundary conditions)

$$P[S] = \min. \qquad [16]$$

This theorem shows that the behavior of matter in the linear region corresponds to the natural "continuation" of equilibrium. At equilibrium, P vanishes; for stationary states near equilibrium, it takes its minimum value compatible with the given boundary conditions. Near equilibrium, $P[S]$ acts as a kind of universal thermodynamic potential. This region includes Fourier's, Fick's, and Ohm's laws, coupling effects like electrokinetic phenomena, thermal diffusion, Knudsen and thermomechanical effects, and coupling between linear chemical reactions.

(c) The "nonlinear range," far from equilibrium, where a universal evolution principle

$$d_X P = \int_V dV \sum_k J_k dX_k \leqslant 0 \qquad [17]$$

characterizes the change in time of the part of the entropy production due to the changes in the generalized forces (15).

It is in this region that completely new and unexpected types of behavior can occur. These effects are due to the appearance of an instability on the "thermodynamic branch" starting from equilibrium (15). This instability is related to the sign of the "excess entropy production"

$$\delta_X P = \sum_k \delta J_k \delta X_k, \qquad [18]$$

where δJ_k and δX_k are, respectively, the excess flows and forces due to the deviation of the state of the system from the reference regime.

Near equilibrium, this quantity is always positive. At sufficient distance from equilibrium, the excess entropy production may change sign (taking into account all constraints imposed on the system). An instability may then arise for hydrodynamic phenomena as well as for chemical reactions involving catalytic effects. The new feature is that there exists, then, critical points beyond which the system may present macroscopic phenomena of self-organization in space and time (15, 16). For surfaces, these three situations, equilibrium, linear, and nonlinear regions, may also be investigated in the frame of the same general thermodynamic theory. Moreover, let us mention that in far from equilibrium conditions, surfaces exhibit new types of instabilities.

EQUILIBRIUM AND PARTIAL EQUILIBRIUM

Let us deal now with some examples of the applications of both classical-thermodynamics and statistical-mechanics model calculations for surfaces. First, Gibbs' surface law will be used to derive some macroscopic properties. A microscopic formulation of the surface tension may be introduced with the aid of the pressure tensor. This formulation is based on the molecular distribution function (17–19). An alternative method of finding σ is through the partition function (20, 21). This last method is often used in cell model calculations (22, 23). For solutions, very crude statistical models have given a qualitative picture of the interfacial region, for mono- or multilayer. Let us go back to the equilibrium macroscopic properties of the surface obtained experimentally through the use of the Gibbs law [11]. For a dilute ideal solution of two components in the presence of air, one obtains the classical relation (1, 24, 25)

$$(\partial \sigma / \partial \ln C_2)_{T,p} = -RT\Gamma_{21}, \qquad [19]$$

where C_2 is the bulk concentration of component 2.

For nonideal solutions, more sophisticated equations were derived (26). When $\Gamma_{\gamma 1}$ is positive, component γ is "surface active." Traube (1, 27) showed that the surface activity $\partial \sigma / \partial C_\gamma$ of normal fatty acids in aqueous solutions is approximately tripled for each additional $-CH_2-$ group in the molecule (Traube's law).

In the same way as done by Gibbs for volume phases, Defay (2) investigated the variance of capillary systems and derived the surface phase rule. The variance is given by different equations depending upon whether the surfaces are curved or plane. For curved surfaces, c components, r' chemical reactions, φ bulk phases separated by s types of surfaces, each comprising one or several surface phases, the total of surface phases being ψ, the variance is

$$\omega = 1 + (c - r') - (\psi - s). \qquad [20]$$

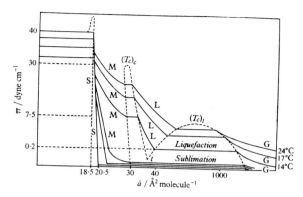

FIG. 1. Pressure–area phase diagram for a fatty acid film spread on a water surface.

For a droplet of a pure liquid in the presence of its vapor, $\omega = 2$ and the vapor pressure depends on the temperature and on the radius of the droplet (Kelvin's equation (1)).

Many authors have extensively investigated the phenomena of equilibrium surface phase transitions (28–30). These phase transformations in the surface give rise to two-dimensional equilibrium structures. As for volume phases, in the neighborhood of critical points of surface phase transitions (liquid–vapor, liquid–liquid condensed, liquid–solid with formation of crystal or floating lens), small thermal fluctuations are amplified, and attain a macroscopic level and drive the system to a new surface phase. The evolution to the new surface phase occurs as an abrupt change beyond an instability point of the reference state. In the critical region around the instability, the system exhibits a markedly coherent behavior which is frequently combined with an increase in spatial order.

Equilibrium structures are formed and are maintained through reversible transformations implying no appreciable deviation from equilibrium. The response of the system to its own spontaneous fluctuations is the central point of the stability problem at equilibrium and nonequilibrium. The stability conditions of a given process become, then, the conditions for the regression of fluctuations. For equilibrium situations far from phase transitions,

the system is stable with respect to its own fluctuations.

Glansdorff and Prigogine (15) worked out a new formulation of the equilibrium stability criterion for continuous volume phases. For small perturbations, the stability depends on the sign of the second derivative of the entropy $\delta^2 S$. Systems in electric and electromagnetic fields were also investigated (31, 32).

The stability criterion has been recently extended to Gibbs surfaces (4), using the curvature of surface entropy S^{Ω} in a characteristic quadratic form:

$$\delta^2 S^{\Omega} = -\int \frac{\Gamma}{T} \left[\frac{C_{\Omega}}{T} (\delta T)^2 + \frac{\Gamma}{E} (\delta\omega)^2_{x_\gamma} \right.$$

$$\left. + \sum_{\gamma\gamma'} \frac{\partial \mu_\gamma}{\partial x_{\gamma'}{}^{\Omega}} \delta x_{\gamma}{}^{\Omega} \delta x_{\gamma'}{}^{\Omega} \right] d\Omega \leqslant 0, \quad [21]$$

with Γ the total adsorption, C_{Ω} the surface heat capacity, $E = d\sigma/d \ln \Omega$ the Gibbs elasticity, ω the area per unit mass, and Ω the total area of the surface. The surface equilibrium stability conditions can be split into:

$C_{\Omega} > 0$, thermal surface stability;

$E > 0$, mechanical stability of the surface;

$\sum_{\gamma\gamma'} (\partial\mu_\gamma/\partial x_{\gamma'}{}^{\Omega})\delta x_{\gamma}{}^{\Omega}\delta x_{\gamma'}{}^{\Omega} > 0$, stability with respect to surface diffusion.

The form of Eq. [21] is similar to the one

obtained by Glansdorff and Prigogine for continuous volume phases. The Gibbs elasticity E plays here the same role as the compressibility coefficient in a volume phase. The necessary and sufficient condition for the stability of equilibrium with respect to small fluctuations requires that $\delta^2 S^\Omega$ be a negative definite quadratic form in the neighborhood of equilibrium.

The Gibbs equation [11] allows also for the discussion of the extreme values of the surface tension (1) and their role in the formation of micellar aggregates.

Other interesting problems treated by means of surface equilibrium thermodynamics, and based on the statements of Young, Laplace, Kelvin, and Gibbs, are the wetting and adhesion properties (25, 33–36), the effect of curvature on the surface tension (1, 2, 6, 23, 37, 38), on the vapor pressure (39), on capillary condensation (1, 25, 40–42), on retardation of boiling (1), on the heat of vaporization (1), and on the position of the triple heat point (1, 2).

The equilibrium conditions for the coexistence of three bulk phases, when two among them have very small sizes (very small drop and very small crystal), and the influence of the size of the small phases on the temperature of the triple point, are of importance in meteorological processes, as in the formation of clouds (43). The triple point problem is also applied to the freezing of liquids in porous solids (42, 44). When treating the surface tension of solids as a mechanical force, many difficulties are encountered (1, 45) due to the anisotropic character of the system (46).

Thermodynamics theories have been also used to describe the interactions between colloidal particles (47, 48), to explain the disjoining pressure in thin films (49), and to calculate the thickness of the surface layer (50), and to develop double layer properties (51). Recently, Everett (52) developed a unifying thermodynamic framework with which he treated the effects of adsorption on the interactions between solid particles.

For partial equilibrium (including me-

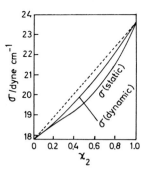

FIG. 2. Static and dynamic surface tension of mixtures of diethyl ether and acetone versus the mole fraction of acetone.

chanical and thermal equilibrium), the main purpose has been to calculate the surface tension, whatever be the matter distribution in the layer. Different approaches can be used; the continuous model provides the Bakker (54) relation between surface tension and the pressure tensor:

$$\sigma = \int_{-\infty}^{\infty} (p_N - p_T)dz, \qquad [22]$$

where p_N and p_T are, respectively, the normal and tangential component of the pressure tensor. The tangential pressure p_T can be related to the molecular field in the layer by

$$p_T = p^u + \tfrac{1}{2} \sum \rho_\gamma \psi_\gamma{}^e, \qquad [23]$$

where p^u is the pressure in a homogeneous phase at the same temperature and concentrations ρ_γ as in the real system, and $\psi_\gamma{}^e$ is the excess chemical potential of the molecular field related to the difference of molecular interactions between a homogeneous medium and a very strongly inhomogeneous region like the interface (54–57).

The functional character of $\psi_\gamma{}^e$ was already stressed by Van der Waals (58), Cahn–Hilliard (59), and Hart (60) in terms of series expansions with the concentration and its derivatives. For a continuous equidistant molecular planes model, the calculation of p_T in terms of the local concentrations and their

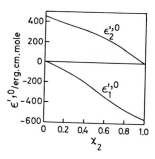

FIG. 3. Cross-chemical potentials of diethyl ether and acetone versus the mole fraction of acetone for a freshly formed surface.

derivatives leads to

$$\sigma = e^3 \int_{-\infty}^{\infty} \sum_{\gamma\gamma'} k_{\gamma\gamma'} \left(\frac{dC_\gamma}{dz}\right)\left(\frac{dC_{\gamma'}}{dz}\right) dz, \quad [24]$$

where e is the distance between two neighboring molecular planes, $k_{\gamma\gamma'}$ are positive coefficients related to interactions between molecules γ and γ' in adjacent planes. If the slopes of concentrations are of opposite sign, there is a possibility of vanishing surface tension. This situation could account for the transitory states of zero surface tension observed in spontaneous emulsification (61).

Partial equilibrium may not always be described by the local equilibrium condition

$$s^\Omega = s^\Omega(u^\Omega, \Gamma_\gamma). \quad [25]$$

As seen before, it has to be replaced by the Defay equation (1)

$$ds^\Omega = T^{-1}du^\Omega - T^{-1}\sum_\gamma \mu_\gamma{}^\Omega d\Gamma_\gamma$$

$$- T^{-1}\sum_{\gamma\alpha} \epsilon_\gamma{}^\alpha dC_\gamma{}^\alpha, \quad [26]$$

giving Eq. [9] for the surface tension.

The cross-chemical potentials $\epsilon_\gamma{}^\alpha$ (Eq. [6]) have been calculated by means of very simple molecular models. For a regular solution of two components, the pure dynamic surface tension (freshly formed surface) is (1)

$$\sigma_{dyn} = \sigma_1 x_1{}^l + \sigma_2 x_2{}^l - (\alpha m/a)x_1{}^l x_2{}^l, \quad [27]$$

where σ_1 and σ_2 are the surface tensions of both pure liquids, $x_1{}^l$ and $x_2{}^l$ the mole fractions in bulk, α is the coefficient of interaction, m the fraction of first neighbors in the sublayer for each molecule of the surface layer, and a is the molar surface. Figure 2 gives the static and pure dynamic surface tension of mixtures of diethyl ether and acetone versus the mole fraction of acetone. From the curve of pure dynamic surface tension of binary mixtures, it is easy to calculate the cross-chemical potentials of both components (Fig. 3). Other model calculations allowed calculation of the dynamic surface tensions for any aging of the surface (1) (Fig. 4). To account for the change of orientation of molecules at the interface, orientation variables were introduced in [11]. The local equilibrium assumption, Eq. [25], then has to be generalized by adding dipole moments as new variables (51). An illustration of this effect is the adsorption of bipolar molecules at fresh interfaces when there is a barrier of potential (11).

NONEQUILIBRIUM

We will now quote a few results obtained in the linear region of nonequilibrium thermodynamics. We will deal here only with systems where the basic assumption of the local equilibrium, Eq. [25], remains valid:

$$\delta s^\Omega = T^{-1}\delta u^\Omega - T^{-1}\sum_\gamma \mu_\gamma{}^\Omega \delta\Gamma_\gamma. \quad [29]$$

Different types of systems will be treated:

(1) Systems with no slip of temperature and velocity.

(2) Systems with slip conditions.

For systems without slip conditions, the surface entropy source $\sigma(S^\Omega)$ becomes (4)

$$\sigma(S^\Omega) = T_{\Omega,\alpha}{}^{-1}j^\alpha + v_\beta{}^\alpha \pi^{\alpha\beta}T_\Omega{}^{-1}$$

$$- \sum_\gamma (\mu_\gamma{}^\Omega T_\Omega{}^{-1})_\alpha \Gamma_\gamma \Delta_\gamma{}^\alpha + \sum F_\gamma{}^\alpha T_\Omega{}^{-1}\Gamma_\gamma \Delta_\gamma{}^\alpha$$

$$+ \sum_\rho A_\rho{}^\Omega V_\rho{}^\Omega, \quad [30]$$

where α and β are the coordinates in the surface, j^α is the heat flux per unit length in the surface, $\pi^{\alpha\beta}$ the viscous surface stress tensor, $F_\gamma{}^\alpha$ the α component of the external force on γ, v^α the α component of the surface velocity, $\Delta_\gamma{}^\alpha$ the diffusion velocity along α in the surface, $A_\rho{}^\Omega$ the affinity of the reaction in the surface, and $V_\rho{}^\Omega$ the reaction rate.

The nonequilibrium properties of two immiscible fluids separated by a moving interface were also investigated by Slattery (62), Waldmann (63), and Georgescu–Moldovan (64), and more recently by Bedeaux *et al.* (3).

Waldmann derived the surface entropy production, taking into account heat and momentum currents along the interface. However he did not consider the possibility of energy or entropy densities in the surface. In his model, the surface appears only as a geometrical surface without mass and energy content. In this way, no direct connection could be made with the classical surface equilibrium thermodynamics. On the other hand, Bedeaux *et al.* set up a more realistic frame for treating nonequilibrium surface thermodynamics, including surface currents and surface-energy and -entropy densities, but they restricted their analysis to systems without surface mass density. Waldmann, as well as Bedeaux, allowed for thermal and velocity slip conditions at the surface.

Bedeaux *et al.* obtained the following linear relations between fluxes and thermodynamic forces in the surface:

$$\pi^{\alpha\beta} = \epsilon[v_{,\beta}{}^\alpha + v_{,\alpha}{}^\beta - (v_{,\alpha}{}^\alpha + v_{,\beta}{}^\beta)] \quad [31]$$

and

$$\tfrac{1}{2}T_r\overline{\overline{\pi}} = K(v_{,\alpha}{}^\alpha + v_{,\beta}{}^\beta) + L_{12}\Delta_s\left[\frac{1}{T}\right]$$

$$+ L_{13}\left[\frac{1}{T_\Omega} - \left\{\frac{1}{T}\right\}\right], \quad [32]$$

where ϵ and K are, respectively, the surface shear and dilatational viscosity coefficients of Newtonian surface fluids (65), L_{12} and L_{13} are the Onsager cross coefficients, $\{1/T\}$ is mean value of $1/T$ across the interface

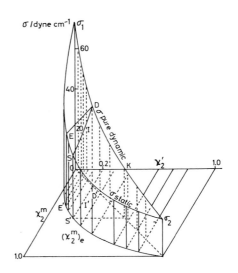

FIG. 4. Dynamic and static tensions versus mole fraction in bulk liquid and in surface phase. Curves calculated for an athermal mixture of molecules of different size.

$\{=\tfrac{1}{2}(1/T_1 + 1/T_2)\}$, and Δ_s is the difference of the bracket quantity across the interface.

For systems without thermal slip,

$$\Delta_s[1/T] = 0$$

and

$$1/T_\Omega = \{1/T\} = 2/(T_1 + T_2),$$

and Eqs. [31] and [32] are the classical hydrodynamical Newtonian constitutive relations for the surface as given by R. Aris (66, 67).

The constitutive equation obtained by Bedeaux *et al.* for the heat surface flux is

$$j^\alpha = -\lambda_\Omega T_{\Omega,\alpha} - \alpha_{12}\Delta_s[v^\alpha], \quad [33]$$

where λ_Ω is the surface heat conductivity and α_{12} is the Onsager cross-phenomenological coefficient. These various phenomenological laws then yield a Fourier's law in the surface, including terms of heat conduction along the surface as well as heat conduction orthogonal to the surface, and thermal effects due to the expansion of the surface and of the fluids on either side of the surface. The reciprocity laws in the surface were also discussed.

Fick's surface law

$$\frac{\partial \Gamma}{\partial t} = \sum_{\alpha} \mathfrak{D}_s \frac{\partial^2 \Gamma}{\partial \alpha^2} \qquad [34]$$

resulting from the linear relation between the surface gradient of chemical potential $(\mu^{\Omega}_{\gamma,\alpha})$ and the surface diffusion flow $(\Gamma_\gamma \Delta_\gamma{}^\alpha)$ has been investigated at solid (68) and liquid interfaces (69, 70).

Surface diffusion plays an important role in biological processes arising in cell membranes (capping) (71). In the Singer–Nicolson fluid mosaic model of biological membranes (72), surface diffusion is responsible for the mobility of lipidic and proteic entities that ensure the fluidity of the membrane.

Another interesting example of phenomenological laws in the surface is the transport equation of charges under the influence of an electric field $F_\gamma{}^\alpha$ along the surface (73). These surface diffusion–migration laws are now under investigation in relation to phenomena arising at the metal–solution interface in corrosion processes.

The diffusion–migration in the surface is of interest for the interpretation of biological phenomena such as the transmission of the nervous influx (74, 75), and the polarization of the amoeba proteus during the monopodial expansion (76).

Finally, cooperation between diffusion in a diffusion layer and surface chemical reactions in the linear region has been investigated by Prigogine (77), Fischbeck (78), and Damkohler (79).

STABILITY AND DISSIPATIVE STRUCTURES IN SURFACES

Equilibrium structures are well understood in terms of classical thermodynamics and equilibrium statistical mechanics. But in addition, there exists a new class of structures which may originate in conditions far from equilibrium. Such "dissipative structures" (15) correspond to an amplification of fluctuations which are stabilized by the flows of energy and matter from the surroundings.

This principle is called "order through fluctuations." The formation of a fluctuation of a given type is fundamentally a stochastic process. If after an instability threshold the fluctuations become macroscopic to drive the system to another macroscopic state, the system is said to be unstable.

The stability conditions for states far from equilibrium have been given by Glansdorff and Prigogine (15). These authors showed that the theorem of minimum entropy production at a steady state breaks down for states far from thermodynamic equilibrium. An extension of the minimum entropy production property to the nonlinear domain of irreversible processes is given by inequality [17]. However, this inequality does not imply the stability of the steady state because $d_x P$ is not the differential of a state function in the general case. Glansdorff and Prigogine established, then, a universal stability condition for non-equilibrium states, that is,

$$\sum_k \delta J_k \delta X_k \geqslant 0. \qquad [35]$$

Numerous examples of instabilities with "creation" of order (80) are now available in physical chemistry (81, 82), biochemistry (83, 84), ecology (85), as well as in fluid dynamics (86, 87).

Under appropriate boundary conditions, dissipative structures like spatio-temporal organization and dissipative phase transitions may occur in the system.

We will only give here a pure thermodynamic frame for the stability analysis of surfaces (4) as an extension of the Glansdorff–Prigogine theory.

For surfaces without convective effects, the surface stability conditions are

$$\sigma(\delta S^\Omega) = \delta T^{-1}{}_{\Omega,\alpha} \delta j\alpha$$
$$+ \sum_\gamma \delta(\mu_\gamma{}^\Omega T_\Omega{}^{-1})_{,\alpha} \delta(\Gamma_\gamma \Delta_\gamma{}^\alpha)$$
$$+ \sum \delta(F_\gamma{}^\alpha T_\Omega{}^{-1}) \delta(\Gamma_\gamma \Delta_\gamma{}^\alpha)$$
$$+ \sum_\rho \delta(A_\rho{}^\Omega T_\Omega{}^{-1}) \delta V_\rho{}^\Omega \geqslant 0. \qquad [36]$$

The only destabilizing contribution arises from the chemical reaction term $\delta A_p{}^\Omega T^{-1} \delta V_p{}^\Omega$ that can change the sign of the surface excess entropy production for auto- or cross-catalytic surface reactions.

It can be shown very simply by the following example of chemical kinetics. Consider the surface reaction

$$X + Y \xrightarrow{k} 2X, \qquad [37]$$

where the concentration of Y is assumed to remain constant. For a totally irreversible reaction, the perturbed reaction rate is

$$\delta V_p{}^\Omega = kY\delta X, \qquad [38]$$

and the associated affinity perturbation for constant temperature and pressure is

$$\delta A_p{}^\Omega = -RT(\delta X/X). \qquad [39]$$

The part of the excess entropy production related to that chemical process, then, is

$$\delta A_p{}^\Omega \delta V_p{}^\Omega = -kRT(Y/X)(\delta X)^2 < 0, \qquad [40]$$

showing a destabilizing effect of the auto-catalytic process [37]. For a linear step, it is easy to show in the same way that $\delta A^\Omega \delta V^\Omega$ is always positive.

An interesting experimental example of surface chemical instability giving rise to spatial structures was recently analyzed by A. T. Winfree (88) on the well-known Zhabotinsky–Belousov reaction (89). Its re-action scheme was exhaustively studied by Field, Noyes, and Körös (90). It involves many auto- and cross-catalytic steps. The surface structure shown by Winfree is a typical example of symmetry breaking insta-bility (91) in the chemical surface reaction. On metal surfaces, oxidation reactions and electro-chemical processes (in this case, $F_\gamma{}^\alpha$ are the ponderomotive electrical forces) are able, under particular conditions, to induce temporal oscillations of current and potential (92, 93).

An example of surface instability in corrosion–passivation phenomena is under investigation by Fabiani and Sanfeld. A new variety of instabilities arises from the difference of the dielectric constant and diffusion coeffi-cients between the reacting layer on the metal and the liquid.

For convective surfaces, new additional terms due to the onset of convection appear in the generalized surface excess entropy production. Some of them are responsible for well-known typical surface instabilities. Among them is the Marangoni effect (94–96), whose contribution to the surface entropy produc-tion is

$$-\{T_\Omega^{-1}\delta v_\alpha \Delta_s \delta P^{\alpha z} + T_\Omega^{-1}(\delta v_\alpha \delta \sigma^{\alpha\beta})_{,\beta}\}. \qquad [41]$$

These terms include the fluctuations of the gradient of surface tension along the surface as well as terms due to surface viscosity $(\delta\sigma^{\alpha\beta})_{,\beta}$ and terms due to the difference of viscosities of the adjacent bulk phases $\Delta_s \delta P^{\alpha z}$. The Marangoni effect plays an important role in the Bénard phenomena for free surfaces (97, 98); the hexagonal shape of the convection cells observed at free surfaces depends markedly on the surface tension. Another typical surface convective effect is the Rayleigh–Taylor instability of two superposed fluids of different densities. Its contribution to the generalized surface entropy production is

$$- \left\{ T_\Omega^{-1}\delta v_z \Delta_s \delta p^{zz} + T_\Omega^{-1}\delta v_z \delta(\Gamma F_z) \right.$$
$$- T_\Omega^{-1}\delta v_z \delta\left[\sigma\left(\frac{1}{R_1}+\frac{1}{R_2}\right)\right]$$
$$\left. + T_\Omega^{-1}F_z\Delta_s\rho\delta v_z\delta z^\Omega \right\}, \qquad [42]$$

where the competition between the gravi-tational effect $F_z = g$ on the media of different densities $\Delta_s\rho$ with the surface tension σ and the mean curvature $(1/R_1 + 1/R_2)$ as well as the viscosities of the adjacent bulk phases $\Delta_s\delta p^{zz}$ appears clearly, in agreement with the hydrodynamic stability analysis of Chandrasekhar. Dissipative structures induced by diffusion through interfaces and related to the Marangoni effect were already observed experimentally and were analyzed theoretically (96, 100–103). Coupling between surface reactions, diffusion, and convection has been recently investigated by several authors (103–108). Interesting results were obtained showing that an instability in a surface chemical reaction can induce a deformation of the interface with space and temporal structures and, conversely, that a hydrodynamic surface instability is able to induce an instability in surface chemical reactions. The biological phenomena of deformation of the cell membrane by chemical and electrical signals (for example, during the phagocytosis) could be due to such coupling (109).

Thermodynamic surface stability criteria provide a qualitative description of the destabilizing effects. For a quantitative treatment, we need a kinetic approach based on the macroscopic equations of evolution of the mechanical and physicochemical variables of the system (velocity, density, concentrations).

The thermodynamic stability theory or its kinetic counterpart deals with the response of the system to an externally imposed small (but macroscopic) perturbation. In the vicinity of an instability such perturbations are amplified, as the system evolves away from the reference state. It becomes, therefore, essential to complete the macroscopic analysis with a detailed study of the *fluctuations*, i.e., of the spontaneous deviation from the reference state induced by the system itself (16).

Such investigations have been conducted recently by our group (110, 111). One of the most interesting and unexpected results has been to reveal some analogies between nonequilibrium instabilities and the more familiar phenomena of phase transitions and nucleation at equilibrium (112). In particular, the evolution away from the unstable state toward an ordered configuration is associated with the formation of a local fluctuation whose range exceeds some critical threshold. Beyond this threshold the processes responsible for the growth of the fluctuation (e.g., autocatalysis) dominate over the processes responsible for the decay (e.g., diffusion) and result in a macroscopic change to the new regime.

Currently, a large number of concrete problems in such widely different areas as chemical kinetics, enzyme regulation, morphogenesis, immunology, or brain research can be analyzed within the framework described in the present paper. For instance, malignant growth can be viewed as a nucleation of a macroscopic fluctuation responsible for the transformation of a metastable state corresponding to "microcancer" toward a stable "cancerous" state (113). We find it fascinating that such complex problems can at least be formulated in terms of mathematical modeling.

REFERENCES

1. DEFAY, R., PRIGOGINE, I., BELLEMANS, A., AND EVERETT, D. H., "Surface Tension and Adsorption." Longmans, Green, London, 1966.
2. DEFAY, R., "Etude Thermodynamique de la Tension Superficielle." Villars, Paris, 1934.
3. BEDEAUX, D., ALBANO, A. M., AND MAZUR, P., *Physica* **82A**, 438 (1976).
4. STEINCHEN, A., AND SANFELD, A., Euchem Conference, Collioure, France, 1976.
5. DE DONDER, TH., "L'Affinité" (new edition, P. Van Rysselberghe). Hermann, Paris, 1936.
6. GIBBS, J. W., "Collected Works," 2 Vols., Longmans, Green, New York, 1928.
7. PRIGOGINE, I., "Introduction to Thermodynamics of Irreversible Processes." Thomas, New York, 1955.
8. NICOLIS, G., WALLENBORN, J., AND VELARDE, M. G., *Physica* **43**, 263 (1969).
9. YOUNG, T., *Phil. Trans. Roy. Soc.*, **95**, 65 (1805).
10. WARD, A. F. H., AND TORDAI, L., *J. Chem. Phys.* **14**, 453 (1946).
11. DEFAY, R., AND PETRÉ, G., *in* "Surface and Colloid Science" (E. Matijevic, Ed.), Vol. 3, Wiley, New York, 1971.
12. SANFELD, A., STEINCHEN A., AND DEFAY, R., *J. Phys. Chem.* **73**, 4047 (1969).
13. KERSZBERG, M., Dissertation, Free University of Brussels, 1975.
14. ONSAGER, L., *Phys. Rev.* **37**, 405 (1931).
15. GLANSDORFF, P., AND PRIGOGINE, I., "Thermodynamics of Structure, Stability and Fluctuations." Wiley, New York, 1971.
16. NICOLIS, G., AND PRIGOGINE, I., "Self-Organization in Non-Equilibrium Systems." Wiley, New York, 1977.
17. KIRKWOOD, J. G., AND BUFF, F. P., *J. Chem. Phys.* **17**, 338 (1949).
18. BELLEMANS, A., *Physica* **28**, 493, 617 (1962).

19. STILLINGER, F. H., *J. Chem. Phys.* **47**, 2513 (1967).
20. BUFF, F. P., *Z. Elektrochem.* **56**, 311 (1952).
21. HARASIMA, A., *Advan. Chem. Phys.* **1**, 203 (1958).
22. PRIGOGINE, I., AND SARAGA, L., *J. Chem. Phys.* **49**, 399 (1952).
23. ONO, S., AND KONDO, S., *in* "Handbuch der Physik" (S. Flügge, Ed.), Vol. X. Springer, Berlin, 1960.
24. ADAM, N. K., "The Physics and Chemistry of Surfaces," 3rd ed. Oxford Univ. Press, London, 1941.
25. HARKINS, W. D., "The Physical Chemistry of Surface Films." Interscience, New York, 1952.
26. HILDEBRAND, J. H., "Solubility." Rheinhold, New York, 1926.
27. FREUNDLICH, H., "Kapillarchemie." Akad. Verlags., Leipzig, 1923.
28. PRIGOGINE, I., AND SARAGA, L., "Changements de Phases." Soc. Chim. Phys., Paris, 1952.
29. DERVICHIAN, D., AND LACHAMPT, F., *Bull. Soc. Chim. Fr.* **13**, 486 (1946).
30. CRISP, D. J., *in* "Surface Chemistry," Proceedings of the Joint Meeting of the Faraday Society, London, 1949.
31. SANFELD, A., AND STEINCHEN, A., *Bull. Acad. Sci. Belg.* **LVII**, 684 (1971).
32. STEINCHEN, A., AND SANFELD, A., *Experientia* (Suppl.) **18**, 599 (1971).
33. GUASTALLA, J., AND COSTE, J. F., *J. Chem. Phys.* **72**, 1007 (1975).
34. TER-MINASSIAN-SARAGA, L., *R.G.C.D.* **47**, 1043 (1970).
35. JORDAN, D. O., AND LANE, J. E., *Amst. J. Chem.* **17**, 7 (1964).
36. DETTREE, R. H., AND JOHNSON, R. E., *J. Phys. Chem.* **69**, 1507 (1965).
37. BUFF, F. P., *J. Chem. Phys.* **19**, 1591 (1951).
38. HILL, T. L., *J. Phys. Chem.* **56**, 526 (1952).
39. LA MER, V. K., AND GRUEN, R., *Trans. Faraday Soc.* **48**, 410 (1952).
40. DE BOER, J. H., "The Dynamical Character of Adsorption." Oxford Univ. Press, London, 1953.
41. EVERETT, D. H., *Trans. Faraday Soc.* **51**, 1551 (1955).
42. VIGNES-ADLER, M., Dissertation, Paris, 1975.
43. DUFOUR, L., AND DEFAY, R., "Thermodynamics of Clouds." Academic Press, New York, 1963.
44. EVERETT, D. H., AND HAYNES, J. M., *RILEM Bull.*, n° 27 64 (1965).
45. GHEZ, R., Dissertation, Ecole Polytechnique, University of Lausanne, France, 1968.
46. WULFF, G., *Z. Kristallogr.* **34**, 449 (1901).
47. OVERBEEK, J. TH. G., *in* "Colloid Science" (H. R. Kruyt, Ed.), Vol. 1. Elsevier, Amsterdam, 1952.
48. DUKKIN, S. S., AND DERJAGUIN, B. V., *in* "Surface and Colloid Science" (E. Matijevic, Ed.), Vol 7. Wiley, New York, 1974.
49. SHELUDKO, A., *in* "Advances in Colloid and Interface Science," p. 392. Elsevier, Amsterdam, 1967.
50. RUSANOV, A. J., *in* "Progress in Surface and Membrane Science," Vol. 4, p. 57. Academic Press, New York, 1971.
51. SANFELD, A., "Introduction to Thermodynamics of Charged and Polarized Layers." Wiley, London, 1968.
52. EVERETT, D. H., Euchem Conference, Collioure, France, 1976.
53. BAKKER, G., *in* "Handbuch der Experimentalphysik," Vol. VI. Akad. Verlags., Leipzig, 1928.
54. STEINCHEN, A., Dissertation, Free University of Brussels, 1970.
55. STEINCHEN, A., DEFAY, R., AND SANFELD, A., *J. Chem. Phys.* **68**, 518, 835, 1241, 1323 (1971); **69**, 1374, 1380 (1972).
56. DEFAY, R., AND SANFELD, A., *J. Chim. Phys.* **70**, 895 (1973); *Ann. Quim.* **71**, 856 (1975).
57. SANFELD, A., STEINCHEN, A., AND DEFAY, R., *J. Phys. Chem.* **73**, 4047 (1969).
58. VAN DER WAALS, J. D., AND KOHNSTAMM, PH., "Lehrbuch der Thermostatik." Leipzig, 1927.
59. CAHN, J. W., AND HILLIARD, J. E., *J. Chem. Phys.* **31**, 6 (1959).
60. HART, E. W., *J. Chem. Phys.* **39**, 3075 (1963).
61. GERBACIA, W., AND ROSANO, H. L., *J. Colloid Interface Sci.* **44**, 242 (1973).
62. SLATTERY, J. C., *IEC Fundam.* **6**, 108 (1967).
63. WALDMANN, L., *Z. Naturforsch. A.* **22**, 1269 (1967).
64. GEORGESCU, L., AND MOLDOVAN, R., *Surface Sci.* **15**, 177 (1969).
65. GOODRICH, F. C., *Proc. Roy. Soc. London Ser. A.* **260**, 481 (1961).
66. ARIS, R., "Vector, Tensor and the Basic Equations of Fluid Mechanics," Chap. X. Prentice–Hall, Englewood Cliffs, N. J., 1969.
67. BARRÈRE, M., AND PRUD'HOMME, R., "Equations Fondamentales de l' "Aérothermochimie." Masson, Paris, 1973.
68. BLAKELY, J. M., *in* "Progress in Materials Sciences" (B. Chalmers, Ed.), Vol. 10, p. 6395. Macmillan, New York, 1963.
69. KAKIKARA, Y., HIMMELBLAU, D. M., AND SCHECHTER, R. S., *J. Colloid Interface Sci.* **30**, 200 (1969).
70. KUMMER, J., Dissertation, Free University of Brussels, 1972.
71. TAYLOR, R., DUFFUS, P., RAFF, M., AND DE PETRIS, S., *Nature (London)* **233**, 225 (1971).
72. NICOLSON, G. L., *Biochim. Biophys. Acta* **457**, 57 (1976).
73. MILLER, C. A., *J. Fluid Mech.* **55**, 641 (1972).

74. TASAKI, I., "Nerve Excitation." Thomas, Springfield, Ill., 1968.
75. LEFEVER, R., AND CHANGEUX, J. P., *C. R. Acad. Sci. Series D* 275, 591 (1972).
76. KOMNICK, H., STOCKEN, W., AND WOHLFARTH-BOTTERMAN, K. E., *Int. Rev. Cytol.* 34, 169 (1973).
77. PRIGOGINE, I., "Etude Thermodynamique des Phénomènes Irréversibles." Desoer, Liège, Belgium, 1947.
78. DAMKÖHLER, C., *in* "Der Chemie-Ingenieur" (Euchem-Jacob, Ed.), Vol. III. Leipzig, 1937.
79. FISHBECK, K., *in* "Der Chemie-Ingenieur" (Euchen-Jacob, Ed.), Vol. III. Leipzig, 1937.
80. EBELING, W., "Strukturbildung bei Irreversible Prozessen," Mathematisch-Naturwissenschaftliche Bibliotek Band 60, BSBBG, Teubner Verlags., 1976.
81. NICOLIS, G., AND LEFEVER, R., *Eds.* Membrane, dissipative structures and evolution, *Advan. Chem. Phys.* 29, 1 (1975).
82. WINFREE, A. T., *in* "AMS–SIAM Symposium," Vol. 8. Amer. Math. Soc., Providence, R. I., 1974.
83. BOITEUX, A., GOLDBETER, A., AND HESS, B., *Proc. Nat. Acad. Sci. U.S.A.* 72, 3829 (1975).
84. WILSON, H. R., AND COWAN, J. D., *Biophys. J.* 12, 1 (1972).
85. DENEUBOURG, J. L., *in* "Insectes Sociaux" (H. Montagnier, Ed.). Masson, Paris, 23, 329–342 (1976).
86. PRIGOGINE, I., AND RICE, S. A., Eds., Instability and dissipative structures in hydrodynamics, *Advan. Chem. Phys.* 32, (1975).
87. SCHECHTER, R. S., VELARDE, M. G., AND PLATTEN, J. K., *Advan. Chem. Phys.*, Vol. XXVI, 265 (1975).
88. WINFREE, A. T., *Science* 175, 634 (1972).
89. ZAIKIN, A., AND ZHABOTINSKY, A., *Nature (London)* 225, 535 (1970).
90. FIELD, R. J., NOYES, R. M., AND KÖRÖS, E., *J. Amer. Chem. Soc.* 94, 8649 (1972).
91. PRIGOGINE, I., *Nature (London)* 246, 67 (1973).
92. VETTER, K. J., "Electrochemical Kinetics." Academic Press, New York, 1967.
93. WOJTOWICZ, J., AND CONWAY, B. E., *J. Chem. Phys.* 52, 1407 (1970).
94. LEVICH, V., "Physico-Chemical Hydrodynamics." Prentice–Hall, Englewood Cliffs, N. J., 1967.
95. SCRIVEN, L. E., AND STERNLING, C. V., *J. Fluid Mech.* 19, 321 (1964).
96. SCHWARTZ, E., AND LINDE, L., *Phys. Fluids*, 4, 535 (1963).
97. LEMOINE, C., Memoire, Mons State University, Belgium, 1975.
98. PEARSON, J. R. A., *Fluid Mech.* 4, 489 (1958).
99. CHANDRASEKHAR, S., "Hydrodynamic and Hydromagnetic Stability," Chap. X. Oxford Univ. Press, London, 1961.
100. STERNLING, C. V., AND SCRIVEN, L. E., *AIChE J.* 5, 514 (1959).
101. THIESSEN, D., C. I. D., "Proceedings of the IVth, International Congress (Gordon and Breach, Ed.)." Brussels, 1964.
102. GOUDA, J., AND JOOS, P., *Chem. Eng. Sci.* 30, 521 (1974).
103. LINDE, H., AND SEHRT, B., *Monatsberichte* 7, 34 (1965).
104. SANFELD, A., AND STEINCHEN A., *Biophys. Chem.* 247, 156 (1974).
105. HENNENBERG, M., SØRENSEN, T. S., STEINCHEN, A., AND SANFELD, A., *J. Chim. Phys.* 72, 1202 (1975).
106. SØRENSEN, T. S., HENNENBERG, M., STEINCHEN, A., AND SANFELD, A., *J. Colloid Interface Sci.*, 56, 191 (1976).
107. JOOS, P., AND VAN BOCKSTAELE, M., *Ann. Quim.* 71, 889 (1975).
108. NACACHE, E., AND DUPEYRAT, M., *C. R. Acad.* 277, Series C, 599 (1973).
109. VAN OSS, C. J., AND GOLLMAN, C. F., *J. Reticulo Soc.* 12, 497 (1972).
110. NICOLIS, G., MALEK-MANSOUR, M., VAN NYPELSEER, A., AND KITAHARA, K., *J. Statist. Phys.* 14, 417 (1976).
111. PRIGOGINE, I., NICOLIS, G., HERMAN, R., AND LAM, T., *Collect. Phenom.* 2, 103 (1975).
112. ZETTLEMOYER, A. C. (Ed.)., "Nucleation." Dekker, New York, 1969.
113. LEFEVER, R., AND GARAY, R., *in* "Theoretical Immunology" (G. I. Bell, G. Pimbley, and A. Perelson, Eds.). Dekker, Basel, Switzerland, in press (1977).

Mobility of Physically Adsorbed Hydroxylic Molecules on Surfaces Made from Oxygen Atoms

J. J. FRIPIAT

*Centre de Recherche sur les Solides à Organisation Cristalline Imparfaite, Rue de la Férollerie,
45045 Orleans Cedex, France*

Received January 30, 1976; accepted October 19, 1976

Measurements of correlation times, and especially assignment of the correlation times to specified motions, though often ambiguous, give important information that helps in the understanding of thermodynamic data obtained for adsorption processes. This is illustrated by correlation-time measurements carried out by the NMR technique on methanol adsorbed by silica gels and an NaY zeolite and on water adsorbed by layer lattice silicates. Surface diffusion coefficients, rotation, and tumbling of the adsorbed species are studied with respect to degree of coverage and temperature. These results are compared with the nature of the equation of state applicable to the adsorbed species and also to the loss of entropy arising from the adsorption. It is shown that there is generally good agreement between these two types of information.

INTRODUCTION

The study of the motions experienced by physically adsorbed molecular species on various kinds of surfaces is of great interest, since to some extent they rule the adsorption entropy as well as the equation of state of the adsorbate. If π is the bidimensional pressure, namely, the free enthalpy per unit surface area, $\pi = -(\partial F/\partial A)$, σ the area occupied by the adsorbed molecule, and T the absolute temperature, the function φ,

$$\pi = \varphi(\sigma, T), \qquad [1]$$

depends on the mobility in the condensed phase, on the frequency of the various modes of vibration with respect to the surface, on the interactions between neighbors, and so on.

The correlation function $G(\tau)$ defined as

$$G(\tau) = \langle f(t)f^*(t + \tau) \rangle, \qquad [2]$$

describes the evolution of the system adequately if function f contains the information about the molecular motions. Random reorientation or translational jumps are generally represented by $G = \langle f(0)f^*(0) \rangle \exp{-(\tau/\tau_c)}$,

where τ_c is the correlation time; τ_c is either the time between jumps or the time needed to reorient the molecule by an angle of 2π rad (1a).

The correlation time or times, if several kinds of motions occur, may be measured in the range 10^{-4} to 10^{-10} sec by pulse nuclear magnetic resonance because the longitudinal spin–lattice relaxation rate T_1^{-1} is proportional to the Fourier transform of G,

$$T_1 = \int_0^{\infty} G(\tau) \cos{\omega\tau}d\tau, \qquad [3]$$

at the nuclear resonance $\omega = 2\pi\nu$. τ_c defines the time scale of the microscopic event which causes relaxation.

The aim of this paper is to review the correlation times obtained experimentally in that way for several systems in which the temperature, the degree of coverage, and the pore structure of the adsorbent have been varied. The systems are methanol adsorbed by silica gels (1b, 2, 3) or occluded in the NaY near-faujasite molecular sieves (4) and water adsorbed in the bidimensional interlamellar space of layer lattice silicates (5–7).

Details on the NMR techniques are given in the above references. The problems arising from the proton exchange process within the adsorbate or with the surface are reviewed elsewhere (8).

Here, particular emphasis will be placed on the physical significance of the correlation times and on the parameters derived therefrom and relevant to a dynamic description of the physical adsorption of methanol and water.

The quoted surfaces are mainly composed of oxygen atoms or hydroxyl groups and, in the case of the zeolite and layer silicates, the surface contains in addition cations balancing the negative charge of the framework.

The surface heterogeneity is often marked by a distribution of correlation times

$$P(\tau_c)d\tau_c = \beta^{-1}\pi^{-\frac{1}{2}} \exp[-(Z/\beta)^2 dZ], \quad [4]$$

$$Z = \ln \tau_c/\tau_m, \quad [5]$$

where β is the spreading coefficient of the distribution function, and τ_m the average correlation time

$$\tau_m = \tau_0 \exp(\bar{H}/RT), \quad [6]$$

and where $\bar{H}$ is the average activation enthalpy.

1. THE "ISOTROPIC" MOTIONS ON SILICAGELS SURFACES OR IN ZEOLITIC CAGES

Here, isotropic means that the adsorbent is made of a collection of surfaces randomly oriented and confined in an intricate network of pores. The adsorption of methanol has been studied for two gels. The first, called Xerogel, is characterized by pores with an average diameter of 17.5 Å and the other, called Aerogel, contains pores unavailable to N_2 but available to dipolar water or methanol molecules (9). The diameter is probably smaller than 10 Å.

In order to assign the correlation time to some motion, information must be obtained about the magnitude of the local magnetic field acting on the proton and arising from either other protons in the same or from other molecules or any other magnetic nuclei or paramagnetic centers. In the case where the number of paramagnetic centers is small (say, for instance, less than 50 ppm Fe^{3+} impurities) and there are no magnetic nuclei other than the protons, the measurements of the proton second moment (the average quadratic local magnetic field) permits one to decide what the possible motions are to which the measured correlation time(s) may be assigned because the motions modulate the local field and provoke the relaxation.

This is the case for the methanol–silica gel and methanol–zeolite systems from which the data shown in Figs. 1 and 2 are derived. In the Xerogel, independently of the degree of coverage (θ), the second moment at a temperature of the order of $-140°C$ corresponds to a molecule in which the CH_3 group is already reorienting rapidly around the C_3 symmetry axis. By contrast, at that temperature, there is no free rotation of the CH_3 group in the Aerogel. From the viewpoint of motions, in the NaY zeolite at $-140°C$, the methanol filling the large cavities is in a state quite comparable with that on the surface of Xerogel.

When the linear relationships shown in Fig. 1 and relative to these three situations are compared, it is clearly apparent that the average activation enthalpy (Eq. [6]) is of a comparable magnitude in the situations described by the Arrhenius plots 2, 3 and 5 whereas for plot 4, (Aerogel), it is much less. In solid methanol O'Reilly et al. (10) observed that the activation enthalpy for the rotation is 1.6 kcal mole^{-1}, whereas in the liquid state the activation enthalpy for diffusion is 3.2 kcal mole^{-1}.

This remark and also what has been said about the low-temperature values of the second moment suggest that correlation times 2, 3, and 5 in Fig. 1 are those of translational jumps, whereas correlation time 4 is that of the methyl group rotation.

In the larger pores of Xerogel and in the temperature range -140 to $+50°C$, the

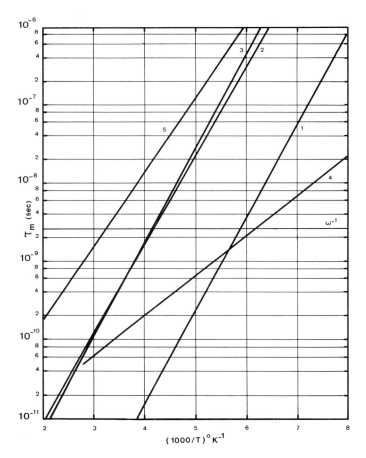

FIG. 1. Correlation times observed at the coverage $\theta = 1.3$ for various systems. (1) ^{2}H resonance in the CD$_3$OH–XOH system, $\beta = 3$ and $\bar{H} = 5.4$ kcal mole^{-1}; (3) ^{1}H resonance in the same system, $\beta = 4$ and $\bar{H} = 5.2$ kcal mole^{-1}; (2) ^{1}H resonance in the CH$_3$OD–XOD system, $\beta = 3.25$ and $\bar{H} = 5.5$ kcal mole^{-1}; (4) ^{1}H resonance in the CH$_3$OH–AOH system, $\beta = 0.8$ and $\bar{H} = 2.32$ kcal mole^{-1}; (5) ^{1}H resonance in the CH$_3$OD–ZNa system, $\beta = 2.5$ and $\bar{H} = 4.4$ kcal mole^{-1}, full zeolitic cage. X, Xerogel (average pore diameter: 17.5 Å); A, Aerogel (average pore diameter < 10 Å); ZNa: NaY (near faujasite) molecular sieve; ω, proton resonance frequency in the 14-kgauss field used in the NMR instrument.

methanol would thus diffuse while the methyl group is rotating freely. In the narrower pores of Aerogel and in the same temperature range, diffusion would not occur. The thermal activation results in a progressively freer rotation of the methyl group. In Aerogel at decreasing θ, the methyl group rotation itself becomes progressively hindered while in Xerogel, as shown by three examples in Fig. 2, the translational correlation time decreases with θ.

The activation enthalpy for diffusion obtained at different degrees of coverage is shown in the enclosure. It increases from about 4 to about 6 kcal mole^{-1} in passing from half to the complete monolayer content and then it decreases progressively toward the value

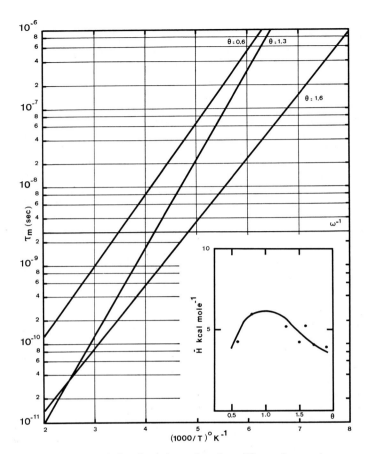

Fig. 2. Variation of the correlation times observed at three different degrees of coverage. ^1H resonance in the $CH_3OD–XOD$ systems. In the enclosure the values of $\bar{H}$ observed at seven degrees of coverage.

obtained for the free liquid at $\theta > 2$. This indicates that the effect of the surface on the diffusional motions is still felt by molecules separated by more than two "statistical" layers from the solid wall.

It is also interesting to point out that in agreement with de Boer's (11) deduction, the activation enthalpy is approximately half the isosteric heat of adsorption obtained from

$$q_{st} = -RT^2[(\delta \ln p)/\delta T]_\theta. \qquad [7]$$

Indeed (1), between $\theta = 0.7$ and $\theta = 1$, q_{st} increases from 10 to 14 kcal mole^{-1} and then it decreases for 14 to 12 kcal mole^{-1} in going from $\theta = 1$ to $\theta = 1.3$.

It has been shown (9) that the molecular area of methanol on the Xerogel and Aerogel surfaces is about 25.5 Å^2 at $\theta = 1$. If this value is considered as the quadratic diffusional jump distance $\langle l^2 \rangle$ and if the surface diffusion coefficient is approximated by

$$D = \langle l^2 \rangle / 6\tau_m, \qquad [8]$$

then the surface diffusion coefficients shown by the solid line in Fig. 3 are obtained for the Xerogel at $+25°C$. Two values obtained at the full and one-third-filled zerolitic cages are shown at $\theta = 1$ and $\theta = \frac{1}{3}$. As confirmed recently for water (12), filling the zeolitic cages diminishes the diffusion coefficient, while for methanol on Xerogel between the half-mono-layer and the monolayer content a rapid increase in that coefficient is observed. For Aerogel (2), the diffusion coefficient is probably smaller than 10^{-10} cm^2 sec^{-1} since the translational motion is outside the range of observation, e.g., $\tau_m > 10^{-6}$ sec.

It is interesting to compare the equations of state for mobile and immobile films (13) with the information about the motions obtained so far.

A mobile film (van der Waals equation) is expected for Xerogel; therefore,

$$\ln \frac{\theta}{1-\theta} + \frac{\theta}{1-\theta} - \ln p = \frac{2a}{bkT} - \ln K. \quad [9]$$

An immobile film (Fowler–Guggenheim equation) should be observed for Aerogel:

$$\ln \frac{\theta}{1-\theta} - \ln p = \frac{C\omega}{kT} - \ln K \quad [10]$$

In [9], a and b are the usual van der Waals coefficients, whereas in [10], $C\omega$ is the interaction energy.

In order to fit the adsorption data the technique proposed by Ross and Olivier (14) has been applied. A Gaussian distribution of adsorption energy U is assumed:

$$\theta = \int \psi(p, U)\phi(U - \bar{U})dU, \quad [11]$$

where

$$\phi(U - \bar{U}) \propto \exp[-\gamma(U - \bar{U})^2]. \quad [12]$$

γ is the spreading coefficient of the Gaussian distribution and is a measure of the surface heterogeneity.

$$K \propto \exp(-U/kT). \quad [13]$$

$\bar{U}$ may be approximated by $\overline{q_{st}}$: for Xerogel, q_{st} is 12.5 kcal for $\theta < 1$. As already shown by

Thompson and Resing (15), there must be a relationship between γ and β, the spreading coefficient of the translational (average) activation enthalpy $\bar{H}$. It may easily be shown that

$$\beta = 1/(2kT\gamma^{\frac{1}{2}}), \quad [14]$$

assuming $\bar{U} = 2\bar{H}$. In Figs. 4A and 4X, it is shown that a good fitting between the equations of state and the observed experimental points may be obtained by taking $\beta = 3$, $\gamma = 0.07$ for Xerogel, and $\gamma = 0.344$ and $C\omega/RT = -0.77$ for Aerogel. It seems indeed reasonable to observe an apparently less heterogeneous surface (higher γ and smaller β) for Aerogel than for Xerogel since, in the former, the translational contribution is negligible and therefore the adsorbed molecules do not experience the various situations that they meet in diffusing on the surface of the latter.

Let us come back to the correlation times and especially to line 1 in Fig. 1. It represents the correlation time obtained from the spin–lattice correlation time at the deuterium resonance for the CD_3OH–XOH systems at three degrees of coverage: $\theta = 0.8$, 1.3, and 1.7, respectively. There is in that case practically no influence by the degree of coverage. This is not surprising because the quadrupole–inner electrical field gradient interaction (the so-called quadrupole coupling constant, QCC) is the main contribution to the deuterium nuclear relaxation. In that case the correlation time has been assigned to molecules tumbling in a surface potential well. Indeed, this motion should imply an average activation enthalpy similar to that of diffusion, i.e., that of breaking hydrogen bonds, and it should be coverage independent since it does not include any cooperative effect in opposition to diffusion.

Finally it is interesting to point out the good agreement between correlation times 2 and 3 in Fig. 1. Correlation time 3 has been derived from the diffusional contribution to the proton spin–lattice relaxation time measured for the CD_3OH–XOH system, after the proton exchange contribution has been removed, whereas 2 has been obtained in a

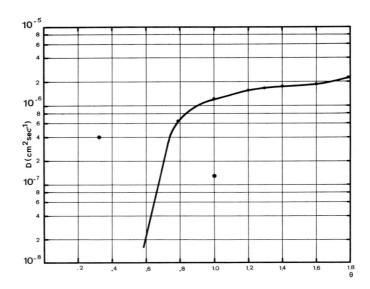

Fig. 3. Variation of the diffusion coefficients with respect to the degree of coverage, methanol in Xerogel, and with respect to the filling factor, (●) methanol in an NaY zeolite. In the liquid phase, the diffusion coefficient is about 2.6×10^5 cm² sec⁻¹.

straightforward manner from the CH_3OD–XOD system.

2. ANISOTROPIC MOTIONS OF WATER IN THE INTERLAMELLAR SPACE OF LAYER LATTICE SILICATES

Van Olphen (16) has shown that the water adsorption isotherm for an Na (Llano) vermiculite is characterized by two well-defined steps corresponding to d_{001} X-ray spacings of 11.8 and 14.8 Å for the one-layer and two-layer bidimensional hydrate, respectively.

Using the variation of the splitting of the doublet-shaped NMR signal with respect to the orientation of the C^* crystal axis it has been shown (6) that the two-layer hydrate has a particularly simple structure as shown schematically in Fig. 5. Each Na^+ is the center of a regular octahedron composed of six water molecules. These water molecules reorient rapidly around their C_2 symmetry axis at temperatures as low as $-70°C$.

Simultaneously the octahedron reorients around the C_3 axis, passing through the centers

of the upper and lower equilateral triangle formed by the two sets of three water molecules. The rotations around C_3 and C_2 occur at a frequency much higher than the doublet splitting, i.e., $\gg 2.3 \times 10^4$ Hz.

Two different motions bring about their contributions in two well-separated temperature domains. Their correlation times are shown in Fig. 6 by two linear functions 1 and 2. Neutron inelastic scattering data obtained by Olejnik et al. (17) have produced a proton (either belonging to the water molecule or as a "free" species) diffusion coefficient of the order of 1.7×10^{-6} cm² sec⁻¹ at 25°C. Assuming an average diffusional jump distance of 10 Å² corresponding to the molecular packing in the adsorbed phase, the correlation time calculated by using Eq. [8] is 10^{-10} cm² sec⁻¹. This is almost exactly the value obtained at room temperature for function 2 in Fig. 6. Therefore the corresponding motion may be assigned to diffusional jumps. The problem of assigning correlation time 1 to a specified motion is more complicated.

The rotation of the hydration shell around the C_3 axis, if fast enough, would account for the doublet splitting. Such a rotation could be at the origin of the diffusion of the hydrated cations. Assuming for instance, $\langle l^2 \rangle$ of the order of 50 Å², the diffusion coefficient deduced from correlation time at room temperature would be about 10^{-8} cm² sec⁻¹.

The Na^+ autodiffusion coefficients measured by Calvet (18) in homoionic montmorillonites at 20° by the radiotracer technique are between 10^{-8} and 10^{-7} cm² sec⁻¹ for water contents between the one and the two-layer hydrate, respectively. Because the main interaction contributing to the spin–lattice relaxation time is that between the proton and the paramagnetic centers (Fe^{3+}) randomly distributed in the lattice, the assignment of the correlation times is more indirect. This is

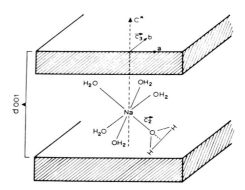

Fig. 5. Schematic structure of the two-layer hydrate of the Na vermiculite according to Ref. (5).

generally the case for molecules adsorbed on solids with an iron content of the order of or higher than 1000 ppm. The correlation time derived from T_1 observed for the Na one-layer and Ca two-layer hydrate of an iron-"rich" Camp–Berteau montmorillonite is represented by function 2. Recently Kadi-Hanifi (7) obtained for the one-layer hydrate of a Li-hectorite the correlation times shown by functions 3 and 4. This magnesium smectite has an iron content lower than 100 ppm and thus the main interaction contributing to T_1 is that between protons.

It is remarkable that functions 2, 4, and 5 have about the same value, e.g., 10^{-10} sec, at room temperature, in spite of differences in hydration, structure, etc. The model adopted for this one-layer Li-hectorite hydrate is easy to represent. In Fig. 5, replace Na^+ by Li^+ and remove one layer of water: the d_{001} spacing is 12.5 Å.

It is difficult to understand why the spacing of the interlamellar space where the bidimensional hydrate is squeezed would not affect the diffusion of molecular water. Therefore, we are inclined to believe that correlation times 2, 4, and 5, in Fig. 6 are those of free protons diffusing in a network of less mobile water molecules. Functions 1 and 3 could then represent the diffusion of water molecules independently of the cation. However, it is difficult to understand why the orientation of

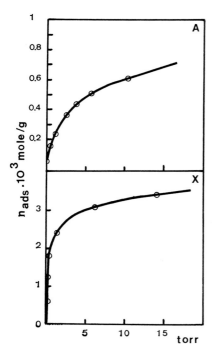

Fig. 4. Methanol adsorption isotherm for an immobile film (A = Aerogel) and for a mobile film (X = Xerogel). Solid line, calculated; open circles, experimental data at 293°K. In (A), $\gamma = 0.344$; in (X), $\gamma = 0.07$.

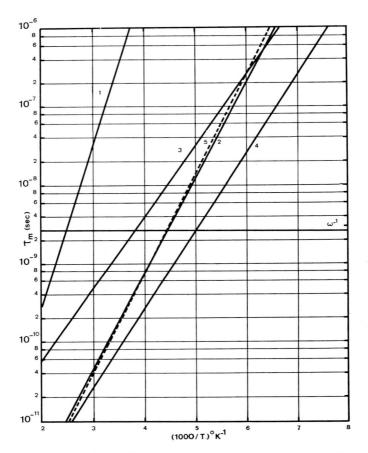

FIG. 6. Correlation times: (1) and (2), two-layer hydrate of a sodium vermiculite; (3) and (4), one-layer hydrate of a lithium hectorite. For (1), $\beta = 3$ and $\bar{H} = 8.5$ kcal mole^{-1}; for (2), $\beta = 1.5$ and $\bar{H} = 5.5$ kcal mole^{-1}, for (3) and (4), $\beta = 0$, $H = 4.4$ kcal mole^{-1}. The results shown by line (5) are the average values obtained for the one-layer hydrate and the two-layer hydrate of Na and Ca montmorillonites.

the C_2 axis would be maintained under these conditions.

Van Olphen (16) has measured the isosteric heat of adsorption and the entropy of adsorption of water for the Na-vermiculite.

The isosteric heats of adsorption range between 10 and 15 kcal and between 21 and 32 kcal per gram of cation for the one- and two-layer hydrates, respectively. Integral entropies of hydration are lower than the entropy of liquid water in both cases. Hence the water molecules in the adsorbed phase should have a higher degree of order than that existing in the bulk liquid.

For sodium and calcium montmorillonites, entropy measurements have shown (21) that the entropy loss ΔS_a could correspond with that expected for a bidimensional fluid in which the molecules have one degree of freedom for rotation for water contents above that required by the hydration of cations. The first step of solvating these cations, e.g., for

$\theta < 0.2$, is marked by a higher entropy loss, suggesting less mobile and more ordered water molecules, as shown in Fig. 7.

There is thus a good general agreement between these thermodynamic data and the models obtained from the study of the microdynamic behavior of adsorbed water.

DISCUSSION

Hydroxyl groups on the surface of silica gel and sodium cations on the surface of silicates have a strong influence on the motions of either adsorbed methanol or adsorbed water. Formation of hydrogen bonds with surface silanols or coordination to the cation are of course responsible for to the restriction in mobility but also for ordering on the

Comparison between the Observed and Calculated Diffusion Coefficients $D(\times 10^5 \text{ cm}^2 \text{ sec}^{-1})$ for Free Water and Water Filling ($\theta = 1$) the Zeolitic Cages of an Na 13Y Molecular Sieve.[a]

Temperature (°C)	D (free water)		D (water in zeolite)	
	Obsd	Calcd	Obsd	Calcd
35	2.7	2.9	0.10	0.13
12	1.5	1.5	0.05	0.062
−6	0.85	0.85	0.025	0.020
−23	0.32	0.32	0.005	0.0033
−38	0.25	0.25	0.002	0.002

[a] The calculated D are obtained through Eqs. [18] [17], respectively.

surface. The pore size and the nature of porosity also play an important role on the magnitude of the surface diffusion.

Since these factors rule the adsorption enthalpy and entropy, it may be expected that the measurement of correlation times and especially the assignment of correlation times to specified motions should provide interesting information helping the interpretation of thermodynamic data.

More experimental and theoretical research aiming to correlate these two kinds of data appears desirable. From a theoretical point of view, the model proposed by Angell and Rao (19) to explain glass transition could possibly be used. The system of molecules associated by hydrogen bonds could be represented as a pseudolattice locked into a ground-state configuration at low temperature. Each "bond lattice" element may exist in an excited state characterized by an excitation enthalpy and entropy, ΔH and ΔS, respectively. The fraction N_x of excited bonds is given by

$$N_x = \{1 + \exp(\Delta H - T\Delta S)/RT\}^{-1}. \quad [15]$$

The structural rearrangements in the system are ruled by local fluctuations in the fraction of excited bonds. The probability of a mass-transporting event, $W(T)$, is related to the

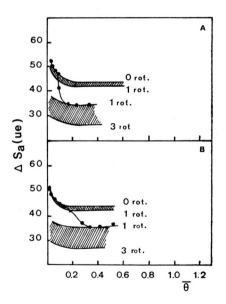

FIG. 7. Entropy loss ΔS_a (e.u.) of water molecules as a function of the averaged coverage degree $\bar{\theta}$: black dots, experimental values; upper shaded range, theoretical entropy loss for an immobile film; the upper and lower limits of this range correspond, respectively, to the loss of three and two rotational degrees; lower shaded range, theoretical entropy loss for a mobile film characterized by one (upper limit) and three (lower limit) rotational degrees; (A) calcium montmorillonite at 27°; (B) sodium montmorillonite at 27°.

fraction of excited bonds such as

$$W(T) = Z \exp(-B/N_x), \qquad [16]$$

where Z is a frequency factor and B is a parameter close to but smaller than unity. A derivation of Eq. [16] using simple statistical arguments was given recently (20) and it was extended to adsorbed phases (22). An interesting example of this application is obtained for water occluded in zeolite. The experimental diffusion coefficients for this example have been obtained from the correlation time reported by Resing (23). The diffusion coefficients for water and supercooled water are those published by Gillen *et al.* (24).

It may be shown that for adsorbed water

$$D = D_0 \exp(-1/N_x) \exp(-E/RT), \qquad [17]$$

whereas for bulk water

$$D = D_0 \exp(-1/N_x), \qquad [18]$$

where E is an Arrhenius activation energy arising from the confinement in the zeolitic cage.

Table I gives a comparison between the observed and experimental diffusion coefficients. For free water $\Delta H = 2$ kcal mole^{-1} and $\Delta S = 5.4$ cal mole^{-1} $^\circ$K^{-1}; for occluded water $\Delta H = 2.7$ kcal mole^{-1} and $\Delta S = 7.9$ cal mole^{-1} $^\circ$K^{-1}; $E = 2.07$ kcal mole^{-1}.

Obviously in the two situations ΔH and ΔS are close to each other. This suggests that in a full zeolitic cage the influence of the zeolite surface on the network of hydrogen bonds is not too important. Indeed, it is remarkable that it is possible to pass from the liquid to the adsorbed phase diffusion coefficients by applying a "translation factor" $\exp(-E/RT)$ to the liquid phase data.

REFERENCES

1a. PFEIFFER, H., "Advances in Nuclear Magnetic Resonance," Vol. 55, Springer, New York, 1973.
1b. CRUZ, M. I., STONE, W. E. E., AND FRIPIAT, J. J., *J. Phys. Chem.* **76**, 3078 (1972).
2. CRUZ, M. I., VAN CANGH, L., AND FRIPIAT, J. J., *Acad. R. Belg. Bull. Cl. Sci.*, **58**, 439 (1972).
3. SEYMOUR, S., CRUZ, M. I., AND FRIPIAT, J. J., *J. Phys. Chem.* **77**, 2847 (1973).
4. SALVADOR, P., AND FRIPIAT, J. J., *J. Phys. Chem.* **79**, 1842 (1975).
5. TOUILLAUX, R., SALVADOR, P., VANDERMEERSCHE, C., AND FRIPIAT, J. J., *Isr. J. Chem.* **6**, 337 (1968).
6. HOUGARDY, J., STONE, W. E. E., AND FRIPIAT, J. J., *J. Chem. Phys.* **64**, 3840 (1976).
7. KADI-HANIFI, unpublished results, this laboratory.
8. FRIPIAT, J. J., "Magnetic Resonance in Colloid and Interface Science," A.C.S. Symposium Series, No. 34, p. 261, Amer. Chem. Soc., Washington, D. C., 1976.
9. CRUZ, M. I., ANDRE, J., VERDINNE, K., AND FRIPIAT, J. J., *Quimica* **69**, 895 (1973).
10. O'REILLY, D. E., AND PETERSON, E. M., *J. Chem. Phys.* **55**, 215 (1971).
11. DE BOER, J. M., "The Dynamical Character of Adsorption." Oxford Univ. Press, London, 1953.
12. MURDAY, J. J., PATTERSON, R. L., RESING, H. A., THOMPSON, J. K., AND TURNER, N. H., *J. Phys. Chem.* **79**, 2676 (1975).
13. FRIPIAT, J. J., CHAUSSIDON, J., AND JELLI, A., "Chimie-Physique des Phénomènes de Surface." Masson, Paris 1971.
14. ROSS, S., AND OLIVIER, J. P., "On Physical Adsorption." Interscience, New York, 1964.
15. THOMPSON, J. K., AND RESING, H. A., *J. Chem. Phys.* **43**, 3853 (1965).
16. VAN OLPHEN, H., *J. Colloid. Sci.* **20**, 822 (1965); and *in* "Proceedings of the International Clay Conference, Tokyo," Vol. 1, p. 649. Israël Univ. Press, Jerusalem, 1969.
17. OLEJNIK, S., STIRLING, G. S., AND WHITE, J. H., *Spec. Discuss. Faraday Soc.* **1**, 188 (1970).
18. CALVET, R., Thèse, Faculté des Sciences de Paris, 1972.
19. ANGELL, C. A., AND RAO, K. J., *J. Chem. Phys.* **57**, 470 (1972).
20. VAN DAMME, H., AND FRIPIAT, J. J., *J. Chem. Phys.* **62**, 3365 (1975).
21. FRIPIAT, J. J., JELLI, A., PONCELET, G., AND ANDRE, J., *J. Phys. Chem.* **69**, 2185 (1965).
22. FRIPIAT, J. J., AND VAN DAMME, H., *Acad. R. Belg. Bull. Cl. Sci.* **60**, 568 (1974).
23. RESING, H. A., *Advan. Mol. Relaxation Processes* **3**, 199 (1972).
24. GILLEN, K. T., DOUGLAS, D. C., AND KOCH, M. J. R., *J. Chem. Phys.* **55**, 2155 (1971).

Statistical Mechanics of Chain Molecules at an Interface

A. BELLEMANS

Université Libre de Bruxelles, Brussels, Belgium

Received June 14, 1976; accepted July 19, 1976

A new technique, derived from Nagle's method for the bulk phase, is applied to a monomer–dimer mixture on a regular lattice, in order to obtain the surface tension and adsorption (only the athermal case is considered). The method is described in some detail; it involves a zeroth-order approximation plus a sum of corrections related to a certain class of subgraphs derived from the lattice. Numerical results are presented for the simple cubic lattice. Further applications of the same kind of technique to other problems are discussed, including hydrogen-bonded mixtures.

1. INTRODUCTION

Some time after I communicated the title of this paper to the organizers of the Meeting, I realized that it was far too general and that it would be necessary to narrow the subject rather drastically. So I finally decided to limit myself to the case of *short chain* molecules at an interface, very short molecules indeed as most of my time will be spent on *dimers*. Also this problem will be examined only within the frame of the conventional *lattice model*. The advantage of this admittedly somewhat simple-minded model, is that lattice statistics, on account of its wide application to many problems, can be tackled by a choice of powerful combinatorial and graph-counting techniques.

Hence, in this paper I will merely emphasize what I believe to be a new kind of approach for deriving boundary effects in lattice models, by considering a mixture of monomers and dimers.

2. GENERAL FORMULATION OF THE MONOMER–DIMER PROBLEM (ATHERMAL CASE)

The so-called monomer–dimer problem is a classic example of the theory of mixtures (1). Two kinds of molecules, "1" and "2," respectively occupy a single site and a pair of neighboring sites of a regular lattice; energies ϵ_{11}, ϵ_{12}, and ϵ_{22} are associated with the various types of nearest-neighbor contacts. We shall limit ourselves here to the *athermal case*, i.e., $\epsilon_{12} = (\epsilon_{11} + \epsilon_{22})/2$. Since the pioneering work of Fowler and Rushbrooke (2), extensive series expansions have been worked out for the bulk thermodynamic properties of this model (3, 4) and even exact solutions have been found for two-dimensional lattices completely covered with dimers (5, 6). The first attempt to study the surface properties of a monomer–dimer mixture seems to be due to Prigogine (7). In this work as well as in subsequent ones, sometimes considering longer molecules (8–13), the system was sliced into monolayers and the approximate Flory–Huggins combinatorial formula was used in each of these for describing the mixing entropy.

It is actually possible to extend some of the exact techniques developed in lattice statistics for coping with bulk properties, in order to cover surface properties as well (14). The results which follow are based on the method devised by Nagle for homogeneous systems (3). We shall start with the most general case of the monomer–dimer problem, by considering a connected (but otherwise arbitrary) graph G with V sites numbered $1, \ldots, i, \ldots, V$ and N_E edges, and by requiring each site to

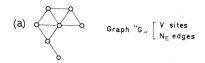

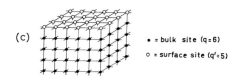

(a) Graph "G." $\left[\begin{array}{l} \text{V sites} \\ N_E \text{ edges} \end{array}\right.$

(b)

(c) • = bulk site (q = 6)
○ = surface site (q' = 5)

Fig. 1. (a) Example of graph G. (b) Particular configuration on G. (c) Simple cubic lattice with plane (1, 0, 0) as boundary layer ($q = 6$, $q' = 5$).

be either occupied by one monomer or by one extremity of a dimer disposed along any one of the edges incident with this site. (See Figs. 1a, 1b). Calling q_i the number of edges incident with site i (= degree of site i in G), we obviously have

$$N_E = \tfrac{1}{2} \sum_1^V q_i. \qquad [1]$$

Given the chemical activities z_1 and z_2 of the two components and using Nagle's machinery, it is then possible to express the grand partition function of the system in the following form (15):

$$\Xi_G(z_1, z_2)$$

$$= z_1{}^V \left(\frac{\exp(-\epsilon_{11}/kT)}{1 + x}\right)^{N_E} \prod_1^V (1 + q_i \alpha u^{q_i - 1})$$

$$\times [1 + \sum_{G' \subseteq G} W(G')], \qquad [2]$$

where α is an auxiliary variable related to the activities

$$\alpha = [(z_2/z_1{}^2)x \exp(\epsilon_{11}/kT)]^{\frac{1}{2}} \qquad [3]$$

and

$$u = \exp(\epsilon_{11} - \epsilon_{22})/2kT. \qquad [4]$$

The sum of Eq. [2] includes all subgraphs G' of the graph G. (We recall here that, generally speaking, a graph G' is a subgraph of G when its vertex set is contained in the vertex set of G and all edges of G' are edges of G.) These subgraphs G' need not to be singly connected but they may not contain isolated vertices. The function $W(G')$ is a weight which can be factorized as

$$W(G') = \prod_{i \subset G'} w_i(d), \qquad [5]$$

where $w_i(d)$ is associated to site i, depending on its degree d in G',

$$w_i(d) = x^{d/2} \left\{1 - d \frac{(1 + x)\alpha u^{q_i - 1}}{x[1 + q_i \alpha u^{q_i - 1}]}\right\}. \qquad [6]$$

The parameter x involved in [2] and [6] can be freely chosen at this stage and its value will be fixed later, in order to reduce the sum over the subgraphs of G as much as possible.

Let us now consider a lattice graph G suitable for treating surface effects, i.e., such that its V sites can be divided into two classes with regard to their degree in G:

  (a) V_B bulk sites with degree q,
  (b) V_S surface sites with degree q' ($< q$).

(See, e.g., Fig. 1c.) One has, of course, $V_B + V_S = V$, with $V_B \gg V_S$ and

$$N_E = \tfrac{1}{2}qV_B + \tfrac{1}{2}q'V_S$$
$$\equiv \tfrac{1}{2}qV - \tfrac{1}{2}(q - q')V_S. \qquad [7]$$

It will of course be necessary to distinguish surface sites and bulk sites in the subgraphs G' and to consider two kinds of weights $w_B(d)$ and $w_S(d)$ for individual sites when evaluating $W(G')$. At this point a very convenient choice of the parameter x is to impose $w_B(1) = 0$ (3). This cancels the contributions of all subgraphs involving bulk sites of degree 1 and therefore considerably reduces the number of significant terms of the sum involved in Eq. [2]. One

552

finds

$$\alpha u^{q-1} = x[1 - (q - 1)x]^{-1}, \qquad [8]$$

$$z_2 e^{(q-1)\,\epsilon_{22}/kT}/z_1^2 e^{q\,\epsilon_{11}/kT}$$

$$= x[1 - (q - 1)x]^{-2}, \qquad [9]$$

$$w_B(d) = x^{d/2}(1 - d), \qquad [10]$$

$$w_s(d)$$

$$= x^{d/2}\left(1 - d\,\frac{1 + x}{q'x + K[1 - (q - 1)x]}\right),$$

$$[11]$$

with

$$\ln K = (q - q')(\epsilon_{11} - \epsilon_{22})/2kT. \qquad [12]$$

Equation [9] clearly displays the relationship between x and the chemical activities of the two components; in particular, x equals 0 and $(q - 1)^{-1}$ for pure monomers and for pure dimers, respectively.

The final step for separating bulk and surface terms in Ξ_G proceeds as follows. We first notice that

$$1 + \sum_{G' \subseteq G} W(G') \equiv 1 + \sum_{g} (g; G)W(g), \qquad [13]$$

where g is any graph isomorphic with certain subgraphs of G and $(g; G)$ is the number of such subgraphs in G, known as the weak lattice constant of g in G (16). In the present case $(g; G)$ is a polynomial in both V and V_s, the total degree of which is equal to the number of disconnected parts of g. Following a theorem due to Domb (17) one has

$$1 + \sum_{g} (g; G)W(g)$$

$$= \exp \sum_{g} [(g; G)_{1,0}V + (g; V)_{0,1}V_s]W(g),$$

$$[14]$$

where $(g; G)_{1,0}$ and $(g; G)_{0,1}$, respectively, denote the coefficients of linear terms in V and V_s in $(g; G)$. Hence the final form of Ξ_G,

taking into account [8], [12], and [14], is

$$\Xi_G = \left(\frac{z_1 e^{-q\,\epsilon_{11}/2kT}}{(1 + x)^{(q/2)-1}[1 - (q - 1)x]}\right.$$

$$\times \exp \sum_{g} (g; G)_{1,0}W(g)\Big)^V$$

$$\times \left(e^{(q-q')\,\epsilon_{11}/2kT}\,\frac{1 + x(q'K^{-1} - q + 1)}{(1 + x)^{1+\frac{1}{2}(q'-q)}}\right.$$

$$\times \exp \sum_{g} (g; G)_{0,1}W(g)\Big)^{V_s}. \qquad [15]$$

On the other hand it follows from statistical thermodynamics that

$$\Xi_G = \exp(pvV - \gamma a V_s)/kT, \qquad [16]$$

where p, γ stand, respectively, for pressure and surface tension and v, a represent the volume per bulk site and the area per surface site. Hence the expression for γ is

$$\gamma a/kT = -(q - q')\epsilon_{11}/2kT$$
$$-\ln\,[1 + x(q'K^{-1} - q + 1)]$$
$$+ [1 + \tfrac{1}{2}(q' - q)]\ln\,(1 + x)$$
$$-\Phi_s, \qquad [17]$$

with

$$\Phi_s = \sum_{g} (g; G)_{0,1}W(g). \qquad [18]$$

For pure monomers, x equals 0 and all terms of [17] vanish except the first one, which is consequently directly related to the surface tension γ_1 of pure monomers.

In a similar way the relative adsorption of dimers follows from the straightforward application of Gibbs formula; one finds

$$\Gamma a = x\,\frac{1 - (q - 1)x}{1 + (q - 1)x}\left(\frac{q'K^{-1} - q + 1}{1 + x(q'K^{-1} - q + 1)}\right.$$

$$\left.-\frac{1 + \tfrac{1}{2}(q' - q)}{1 + x} + \frac{\partial \Phi_s}{\partial x}\right). \qquad [19]$$

One will notice that on account of the front factors, this expression correctly vanishes at both limits $x = 0$ and $x = (q - 1)^{-1}$.

TABLE I

Some of the Relevant Graphs g with the Corresponding Weights and Lattice Constants for the Simple Cubic Lattice, with a (1,0,0) Plane as Boundary Layer[a]

g	W(g)	$(g;G)_{1,0}$	$(g;G)_{0,1}$
	$w_B(2)^4$	3	-4
	$w_B(2)^6$	22	-33
	$w_S(1)w_B(2)^3 w_B(3)$	0	8
	$w_S(1)^2$	0	2
	$w_S(2)^2 w_B(2)^2$	0	2
	$w_S(1)^2 w_B(2)^2$	0	2
	$w_S(1)^2 w_B(2)^3$	0	6
	$w_S(1)^2 w_S(2)$	0	6
	$w_S(2)^4$	0	1
	$w_S(1)^2 w_S(2)^2$	0	18
	$w_S(1)^4$	0	-7

[a] Bulk and surface sites are represented by plain and open circles, respectively.

Finally, x itself may be related to the bulk concentration of dimers by applying standard expressions of statistical mechanics. We obtain

$$\varphi = q\,\frac{x}{1+x} + 2x\,\frac{1-(q-1)x}{1+(q-1)x}\,\frac{\partial \Phi_B}{\partial x}, \quad [20]$$

where φ is the volume fraction of dimers in the bulk and

$$\Phi_B = \sum_g (g;G)_{1,0} W(g). \quad [21]$$

This completes the formalism. Given a particular lattice, what remains to be done (apart from numerical work) is to determine the weak lattice constants $(g;G)_{1,0}$ and $(g;G)_{0,1}$; actually many of these have already been published or can be obtained from

existing data (17). As an illustration some of the relevant graphs for the simple cubic lattice are listed in Table I, together with their corresponding weight and lattice constants.

An interesting feature of the formalism is that all properties such as γ and Γ split into two parts, the first one given by a close formula and the second one by an infinite sum of graphs. It is certainly tempting to consider the first part as a kind of *zeroth* approximation, to be subsequently corrected by including more and more terms of the second part. This point of view will be discussed in the next section for a particular case.

3. SOME NUMERICAL RESULTS FOR THE SURFACE PROPERTIES OF MONOMER–DIMER SYSTEMS (ATHERMAL CASE)

There are clearly two factors influencing γ and Γ. One is an entropic effect resulting from the reduced orientational freedom of dimers located near the surface and which is related to the ratio $q'/q < 1$. The second one is an energetic effect (or mainly so) described by the parameter K which is the Boltzmann factor associated with the replacement of a monomer by half a dimer on a surface site. Both effects will of course be coupled in general. We shall quote here some numerical results for the simple cubic lattice with a plane (1, 0, 0) as boundary layer i.e., $q = 6$, $q' = 5$.

In the first place let us investigate the *pure entropic effect* by putting $K = 1$ (i.e., $\epsilon_{11} = \epsilon_{22}$), which obviously cancels the energetic effect completely. We then have, for the particular case considered (on account of (17), (19), (20) and (11)):

$$(\gamma - \gamma_1)a/kT$$
$$= \tfrac{1}{2}\ln(1+x) - \Phi_S, \quad [22]$$

$$\Gamma a = x\,\frac{1-5x}{1+5x}\left(-\frac{1}{2(1+x)} + \frac{\partial \Phi_S}{\partial x}\right), \quad [23]$$

$$\varphi = \frac{6x}{1+x} + 2x\,\frac{1-5x}{1+5x}\,\frac{\partial \Phi_B}{\partial x}, \quad [24]$$

$$w_S(d) = x^{d/2}[1 - d(1+x)]. \quad [25]$$

All weights $W(g)$ turn out to be polynomials in x, starting with x^{e+h} as lowest power, where e is the number of edges in g and h is the number of surface sites of degree 1. By a tedious (but otherwise straightforward) enumeration of graphs, we found up to x^8

$$\Phi_B(x) = 3x^4 + 22x^6 + 72x^7$$
$$- 28\tfrac{1}{2}x^8 + \cdots, \quad [26]$$

$$\Phi_S(x) = 2x^3 - 7x^4 + 24x^5 - 30x^6$$
$$+ 182x^7 - 206\tfrac{1}{2}x^8 + \cdots$$
$$(0 < x < \tfrac{1}{5}). \quad [27]$$

In spite of some irregularity the values of these series can be reasonably estimated by the method of Padé approximants and the numerical results for γ and Γ are shown in Figs. 2 and 3, respectively. As a first point we notice that the zeroth approximation obtained by neglecting all graph contributions is not unreasonable. A second point is that the entropy effect, causing a negative adsorption and a rise of γ, is rather small; for pure dimers we found

$$(\gamma_2 - \gamma_1)a/kT \simeq 0.078, \quad [28]$$

which for ordinary liquids at room temperature amounts to ~1 dyn/cm.

We now turn to the more general case with $K \neq 1$ so that the energetic effect comes into

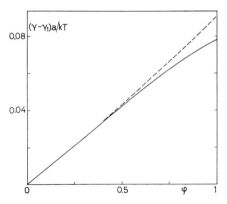

Fig. 2. Pure entropic effect (simple cubic lattice). Plot of γ vs φ; plain line: estimated from [26] and [27]; broken line: zeroth approximation.

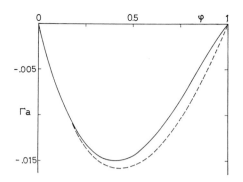

Fig. 3. Pure entropic effect (simple cubic lattice). Plot of Γ vs φ; plain line: estimated from [26] and [27]; broken line: zeroth approximation.

play as well. Specializing the formulas of the preceding section to the simple cubic lattice, we obtain

$$(\gamma - \gamma_1)a/kT$$
$$= -\ln\left[1 + 5x(K^{-1} - 1)\right]$$
$$+ \tfrac{1}{2}\ln(1 + x) - \Phi_S, \quad [29]$$

$$\Gamma a = x\,\frac{1 - 5x}{1 + 5x}\left(\frac{5(K^{-1} - 1)}{1 + 5x(K^{-1} - 1)}\right.$$
$$\left. - \frac{1}{2(1 + x)} + \frac{\partial\Phi_S}{\partial x}\right), \quad [30]$$

$$w_S(d) = x^{d/2}\left(1 - d\,\frac{1 + x}{1 + (1 - 5x)(K - 1)}\right).$$
$$[31]$$

No changes occur in [24] or in [26] and the enumeration of graphs remains the same. Their classification in Φ_S becomes, however, less obvious as $W(g)$ now depends on two parameters, x and $(K - 1)$. The simplest procedure is then to collect all graphs up to a given number of bonds and to compute the corresponding value of Φ_S. We have verified that, for reasonable values of K, i.e., $\tfrac{2}{3} < K < \tfrac{3}{2}$, the successive values obtained for Φ_S by increasing the number of bonds converge rather well. Let us also notice that for pure dimers, K disappears in $w_S(d)$ as in this

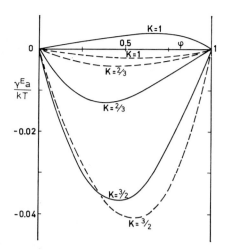

FIG. 4. Excess surface tension vs φ as defined by Eq. [33] (simple cubic lattice). Solid lines: estimated from the exact expression [29]; broken lines: zeroth approximation.

situation x equals $\frac{1}{5}$. It then follows from [29] that the surface tension of pure dimers is simply given by the value obtained above for $K = 1$, augmented by $\ln K$, i.e.,

$$(\gamma_2 - \gamma_1)a/kT \simeq 0.078 + \ln K. \quad [32]$$

Some numerical results are presented in Fig. 4 for what might perhaps be called an *excess* surface tension

$$\gamma^E = \gamma - (1 - \varphi)\gamma_1 - \varphi\gamma_2, \quad [33]$$

for two particular values of K, $\frac{3}{2}$ and $\frac{2}{3}$, corresponding roughly to $|\gamma_2 - \gamma_1| \simeq 5$–$10$ dyn/cm for ordinary liquids at room temperature; the case $K = 1$ is also shown for comparison. The difference between the zeroth approximation and the actual curves appears more impressive in this kind of presentation; it should, however, be clear that when translated into values corresponding to ordinary liquids, 0.01 on Fig. 4 scales as 0.1–0.2 dyn/cm. In this sense the zeroth approximation remains reasonably good.

4. CONCLUSIONS AND FURTHER EXTENSIONS OF THE METHOD TO OTHER PROBLEMS

The results of the preceding section are encouraging: the zeroth approximation looks rather reasonable and it can be corrected to any degree of accuracy by including the contributions of more and more graphs. Previous treatments of the same problem usually considered a finite number of parallel monolayers to the surface, with varying compositions, different from that of the bulk phase; apart from the difficulty of getting these compositions, these treatments involve the combinatorial factor of Flory or of Huggins–Miller–Guggenheim at some step, which is only a first approximation (8–13).

Let us now briefly discuss the possible extensions of the method described here to other problems. First of all one might consider the case of nonathermal solutions by removing the assumption $\epsilon_{12} = (\epsilon_{11} + \epsilon_{22})/2$. In this respect it is interesting to note that the model of regular solutions (or equivalently the Ising model) can be reformulated by means of Nagle's method (18). Hence if one is ready to pay the price by complicating somewhat the graphs and their corresponding weights, the method described above may be adapted to nonathermal monomer–dimer mixtures.

On the other hand the extension to longer chains does not offer much hope. It is true that the case of trimers was treated by Nagle's method for what concerns the bulk phase (19) and it seems therefore possible to adapt it for handling surface effects as well. But the general case of r-mers is certainly unpractical.

Perhaps the most interesting and straightforward extension of the method lies in the field of associated solutions, involving hydrogen-bonded complexes. It seems at present that there is no satisfactory treatment of the surface properties of such mixtures. For example, the well-known model of ideal association, treating each complex as a chemical species, is difficult to use in this respect because one does not know how to modify the equilibrium constants near the surface. Recently Abraham and Heilmann (20) developed a rather simple and attractive lattice model of hydrogen-bonded mixtures. In short, they consider a regular lattice with coordination number q, covered with two kinds of molecules A and B, each occupying a single site. Two neighboring molecules A *may* bond together, the formation of such a bond giving rise to an increment of free energy η; each molecule A can however be engaged into *two bonds at most*. This model allows the formation of a large variety of cyclic as well as linear complexes, controlled by η and q (or more correctly by the geometry of the lattice). We recently showed that Nagle's method for dimers can be adapted rather simply to this model and that for the bulk phase, at least, some kind of zeroth approximation, disregarding all subgraph contributions works reasonably well (21). (Surprisingly this turns out to be a much easier problem than an athermal mixture of monomers and r-mers with r fixed.) Preliminary investigations show that surface properties of this model can be studied by the same method, the essential effect being that near the surface the complexes are less easily formed due to the reduction of the lattice coordination number. Further work on this topic is in progress.

REFERENCES

1. GUGGENHEIM, E. A., "Mixtures," Chaps. 10 and 11, Oxford Univ., Press, London/New York, 1952.
2. FOWLER, R. H., AND RUSHBROOKE, G. S., *Trans. Faraday Soc.* 33, 1272 (1937).
3. NAGLE, J. F., *Phys. Rev.* 152, 190 (1966).
4. GAUNT, D. S., *Phys. Rev.* 179, 174 (1969).
5. FISHER, M. E., *Phys. Rev.* 124, 1664 (1961).
6. KASTELEYN, P. W., *Physica* 27, 1209 (1961); *J. Math. Phys.* 4, 287 (1963).
7. PRIGOGINE, I., *J. Chim. Phys.* 47, 3 (1950).
8. PRIGOGINE, I., AND SAROLÉA, L., *J. Chim. Phys.* 47, 807 (1950).
9. MARÉCHAL, J., *Trans. Faraday Soc.* 48, 601 (1952).
10. PRIGOGINE, I., AND MARÉCHAL, J., *J. Colloid Sci.* 7, 122 (1952).
11. BELLEMANS, A., *Acad. R. Belg. Bull. Cl. Sci.* 46, 157 (1960).
12. EVERETT, D. H., *Trans. Faraday Soc.* 61, 2478 (1965).
13. ASH, S. G., EVERETT D. H., AND FINDENEGG, G. H., *Trans. Faraday Soc.* 64, 2639 (1968); 64, 2645 (1968); 66, 708 (1970).
14. BELLEMANS, A., AND FUKS, S., *Physica* 50, 348 (1970); BELLEMANS, A., *Physica* 65, (1973).
15. BELLEMANS, A., submitted for publication in *Physica*.
16. ESSAM, J. A., AND FISHER, M. E., *Rev. Mod. Phys.* 42, 271 (1970).
17. DOMB, C., *Advan. Phys.* 9, 149 (1960); see p. 221 particularly.
18. NAGLE, J. F., *J. Math. Phys.* 9, 1007 (1968).
19. VAN CRAEN, J., AND BELLEMANS, A., *J. Chem. Phys.* 56, 2041 (1972).
20. ABRAHAM, D. B., AND HEILMANN, O. J., *J. Statist. Phys.* 4, 15 (1972); 13, 461 (1975).
21. RYCKAERT, J-P., Doctoral Dissertation, University of Brussels, 1976.

On the Relationship between the Electrostatic and the Molecular Component of the Adhesion of Elastic Particles to a Solid Surface

B. V. DERJAGUIN, YU. P. TOPOROV, V. M. MULLER,
AND I. N. ALEINIKOVA

The Institute of Physical Chemistry of the USSR Academy of Sciences, Moscow, USSR

Received October 12, 1976; accepted October 13, 1976

It is well known that the sticking of solids is conditioned by the joint action of forces of various kinds. If a homeopolar chemical bond does not form at the interface of the solids, the sticking force may be represented in the following form:

$$F = F_m + F_c + F_e, \qquad [1]$$

where F_m is the molecular component of the sticking force, F_c is the component dependent on the previous electrification of bodies that are brought in contact with one another, F_e is the attractive force which is conditioned by the double electric layer appearing in the zone of a contact on the formation of the latter and determined by the density of its charges.

Let us limit ourselves to examining the case where there is no previous electrification and, hence, $F_c = 0$.

In calculating the molecular component of the sticking force in the case of an elastic contact, the following circumstances will have to be taken into account. When a plastic contact is broken, a variation in the energy of interaction of two bodies may be connected only with a variation in the distance between them. However, in the case of breaking the elastic contact, a change in the spacing of interacting bodies occurs simultaneously with a change in the form between separate areas of the surfaces of those bodies, as well as in the size of the interaction zone.

Therefore, in the case when an elastic contact is broken the interaction should be characterized by a generalized force which can be determined as a derivative of the interaction energy with respect to deformation (1–3).

For the sake of simplicity, let us consider the case of contact between an elastic ball and an absolutely hard surface. The complete energy of the molecular interaction of the ball with a plane is equal to a sum (4):

$$W = W_H + W_s, \qquad [2]$$

where W_H is the volume energy of the elastic deformation of a sphere, and W_s is the surface energy.

W_s is the sum of the surface energies in contact and outside the contact

$$W_s = W_s' + W_s''. \qquad [3]$$

According to the principle of virtual displacements, the force which should be applied in the equilibrium process of the breaking of an elastic contact and of the recovery of the initial form of the sphere will be equal to

$$F = -\frac{dW}{d\alpha} = -\frac{dW_H}{d\alpha} - \frac{dW_s}{d\alpha}$$

$$= F_H + F_s. \qquad [4]$$

(α is the approach of the center of the ball to the plane, which is characteristic of the deformation of the ball in contact.)

559

The elastic reaction F_H may be calculated according to the Hertz theory. For the case of the contact of the ball of radius R with the plane (5)

$$F_H = 4R^{\frac{1}{2}}E\alpha^{\frac{3}{2}}/3(1 - \eta^2), \quad [5]$$

where E is the modulus of elasticity, and η is the Poisson coefficient of the ball material. The generalized surface force is equal to

$$F_s = -\frac{dW_s}{d\alpha} = -\frac{dW_s'}{d\alpha} - \frac{dW_s''}{d\alpha}$$

$$= F_s' + F_s'', \quad [6]$$

where F_s', F_s'' are the components of the force of molecular interaction in contact and outside the contact, respectively. Calculations (1–4), made on the assumption that the breaking off is effected in a thermodynamically equilibrium manner, have shown that component F_s' practically does not depend on deformation and is determined by an expression

$$F_s' = \pi R\varphi(\epsilon), \quad [7]$$

where $\varphi(\epsilon)$ is the interaction energy per unit area of a flat surface. Extrapolation of the theoretical dependence (6) of the interaction energy φ on distance for small gaps yields

$$\varphi(\epsilon) = -A/12\pi\epsilon^2. \quad [8]$$

(ϵ is the least possible distance, of the order of the diameter of a molecule, between the surfaces of the sphere and the plane, and A is the Hamaker constant.) Hence, the component F_s is equal to

$$F_s' = -AR/12\epsilon^2. \quad [9]$$

The force F_s'' was calculated on the basis of the relationship determining the shape of the surface of the deformed ball dependent on the value α. The component F_s'' increases monotonically in the process of tearing off. The character of its variation is determined by a relationship between values α and ϵ. At $\alpha \gg \epsilon$, the force (4)

$$F_s'' \simeq -\frac{\pi^2 AR}{54\epsilon^2(3)^{\frac{1}{3}}(6\pi)^{\frac{1}{3}}} \left\{ \left(\frac{\epsilon}{\alpha}\right)^{\frac{3}{2}} + \cdots \right\}, \quad [10]$$

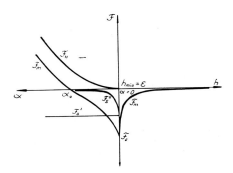

FIG. 1. Schematic drawing showing the curves of the molecular interaction of a sphere with a plane in the process of breaking an elastic contact.

and it may be neglected as compared with F_s'.

At $\alpha \ll \epsilon$

$$F_s'' \simeq -\frac{AR}{12\epsilon^2}\left[1 - \frac{3(2)^{\frac{1}{2}}}{4}\left(\frac{\alpha}{\epsilon}\right)^{\frac{1}{2}} + \cdots\right], \quad [11]$$

and at $\alpha \to 0$ (a point contact)

$$F_s'' = F_s' = -AR/12\epsilon^2. \quad [12]$$

Formulas [5]–[12] show that the generalized force of the molecular interaction at the interface between the elastic sphere and the plane has its maximum absolute value at the point contact (at zero deformation, when simultaneously $F_H = 0$) and is equal to

$$|F| = AR/6\epsilon^2. \quad [13]$$

The same value is also possessed by the molecular component of the adhesion force. What has been stated here may be illustrated by Fig. 1. (α designates the approach of the center of a ball to a plane, and h is the spacing between the surface of the sphere and the plane.)

When tearing off, it is obvious that the interaction of the charges of a double layer will also be overcome in the annular zone arising and continuously increasing around the contact. In a manner similar to that used in calculating the molecular component of the sticking (1, 2), F_e may be determined as a derivative with respect to value α, of the

560

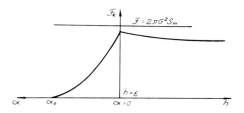

FIG. 2. Schematic diagram showing the curve of the electrostatic interaction of the sphere with the plane in the process of breaking the elastic contact.

energy of interaction between the charges on the deformed portion of the sphere surface and the charges on the plane, taking into account the Hertz theory. This energy is determined by a relationship

$$W_e = 8\pi\sigma^2 a_m^3 \int_0^1 \int_0^1 \frac{xy\,dx\,dy}{[(x + y)^2 + t^2]^{\frac{1}{2}}}$$

$$\times K\left(\frac{4xy}{(x + y)^2 + t^2}\right)^{\frac{1}{2}}, \quad [14]$$

where K is the complete elliptic integral of the first kind, $a_m = (\alpha_m R)^{\frac{1}{2}}$ is the maximum radius of contact, α_m is the maximum deformation,

$$t = \epsilon/a_m \quad \text{at} \quad 0 \leqslant x \leqslant a/a_m, \quad [15]$$

$$t = \frac{a_m}{\pi R}\left[\frac{a}{a_m}\left(x^2 - \frac{\alpha}{\alpha_m}\right)^{\frac{1}{2}}\right.$$

$$\left. - \left(2\frac{\alpha}{\alpha_m} - x^2\right)\text{arctg}\left(x^2\frac{\alpha_m}{\alpha} - 1\right)^{\frac{1}{2}}\right]$$

$$+ \frac{\epsilon}{a_m}, \quad \text{at } a/a_m < x \leqslant 1,$$

$$a = (\alpha R)^{\frac{1}{2}}, \quad 0 \leqslant \alpha \leqslant \alpha_m,$$

$$t = \frac{a_m}{R}\left(\frac{x^2}{2} + \frac{h}{a_m}\right) \quad \text{at } \epsilon \leqslant h \leqslant \infty.$$

h is the distance between the sphere and the plane. The calculation was carried out on the assumption that here no neutralization of the charges of the electric double layer occurs, when

the sphere is being torn off the plane. (Such an assumption is valid for the case where small particles of dielectrics are torn off a solid substrate (7, 8).)

In accordance with the calculation, a variation in F_e with α has a form which is represented in Fig. 2. When α decreases the contact area decreases, but the area of the annular zone increases, and it is here that the separation of the interacting charges of the double layer takes place. In the process of breaking the elastic contact as α decreases, the component F_e increases up to its maximum value at the moment of the point contact, i.e., $|F_e|_{max} \leqslant F' = 2\pi\sigma^2 S_m$, where S_m is the starting contact area. The value F' corresponds to the attractive force acting between the plates of a flat condenser.

Since $S_m = \pi\alpha_0 R$ (5), and the equilibrium value α_0 corresponds to the conditions that the molecular and elastic forces are equal:

$$\frac{AR}{12\epsilon^2} \simeq \frac{4(R)^{\frac{1}{2}}E}{3(1 - \eta^2)}\alpha^{\frac{3}{2}} \quad [16]$$

is equal to

$$\alpha_0 \simeq \frac{(2)^{\frac{1}{3}}A^{\frac{2}{3}}(1 - \eta^2)^{\frac{2}{3}}R^{\frac{1}{3}}}{8^{\frac{2}{3}}E^{\frac{2}{3}}}. \quad [17]$$

Then

$$F_e \simeq 2\pi\sigma^2 S_m \simeq \frac{(2)^{\frac{1}{3}}\pi^2\sigma^2 A^{\frac{2}{3}}(1 - \eta^2)^{\frac{2}{3}}R^{\frac{4}{3}}}{4\epsilon^{\frac{2}{3}}E^{\frac{2}{3}}}. \quad [18]$$

According to the formulas given, the ratio of components F_e and F_m is equal to

$$\frac{F_e}{F_m} = \frac{3\pi^2(2)^{\frac{1}{3}}}{2}(1 - \eta^2)^{\frac{2}{3}}\sigma^2\left(\frac{R\epsilon^2}{AE^2}\right)^{\frac{1}{3}}. \quad [19]$$

Hence, it is obvious that this relationship is determined mainly by the density of the charges of the double layer, to a smaller extent by the elastic properties of the material of the particles, and to a very slight extent by the radius of the particles and the constant of molecular forces.

Let us carry out a numerical comparison of values F_e and F_m for the case of the contact between spherical particles of polystyrene [$E = 3 \times 10^{10}$ dyn/cm^2, $\eta = 0.33$, $A = 6 \times 10^{-13}$ erg (4)] and the steel surface. Assume $R = 10$ μm, $\epsilon = 5$ Å, than for $\sigma = 10^3$ CGSE $F_e/F_m \sim 1$, but for $\sigma = 10^4$ CGSE $F_e/F_m \sim 10^2$. Since the order of 10^3–10^4 CGSE for the density of the charges of the double layer is completely realistic, which has been established experimentally (3), the ratios obtained prove a considerable, often prevailing role of electrostatic forces in the process of the elastic breaking of an adhesion contact.

It will have to be noted that the order of magnitude of F_e calculated as the attractive force acting between the plates of a flat condenser may actually prove to be underrated. Indeed, it has been shown in (8) that if the thickness of a double electric layer H ($H \geqslant \epsilon$) is smaller or of the same order of magnitude as the characteristic spacing l between the charges on the condenser plates, then a simplified model of a flat condenser with the charges that are uniformly smeared over the surface must be improved taking into consideration the discrete structure of the double layer. For densities $\sigma = 10^3$–10^4 CGSE the spacing $l = (e/\sigma)^{\frac{1}{2}}$, where e is the elementary charge, amounts to (5–15) ϵ. In this case, according to (8), taking account of the discreteness of the charges of the double layer may lead to an increase in the value of F_e by several times. Thus, the relative contribution F_e to the sticking force may actually prove to be prevalent.

Now, let us compare the work to be done in the process of tearing an elastic sphere off the solid surface, for overcoming electrostatic forces (A_e) with that for overcoming molecular forces (A_m).

The ratio of the values A_e and A_m is obviously equal to that of the interaction energies in the state of equilibrium (prior to the application of a tearing-off force), i.e., to W_e and W_m, respectively.

The value represents the energy of interaction of two uniformly charged (at the density of charges σ) dielectric disks having radius a, which are disposed at spacing ϵ

from one another, and is determined formulas [14] and [15] as follows:

$$W_e = -(16/3)\pi\sigma^2 a^3. \qquad [20]$$

The energy of the molecular interaction forces for the contact of an elastic sphere with the plane is equal to

$$-W_m = \int_{\infty}^{\epsilon} F_m' dh + \int_0^\alpha F_m'' d\alpha, \qquad [21]$$

where the first term is the work to be done in making the sphere and plane approach point contact, and the second term is the work to be done in the formation of an elastic contact. From what has been said above, it follows that

$$F_m' = -AR/6h^2, \qquad [22]$$

$$F_m'' = F_s' + F_s'' + F_H. \qquad [23]$$

Then, taking into account that $\alpha_0 \gg \epsilon$, that is, neglecting the force F_s'' (see Fig. 1), we find

$$W_m \simeq \int_0^\alpha (F_s' + F_H) d\alpha = F_s'(\alpha_0)\alpha_0 + \tfrac{2}{5} F_H(\alpha_0)\alpha_0. \qquad [24]$$

Since in the equilibrium $F_H(\alpha_0) = -F_s'(\alpha_0)$, we obtain

$$W_m \simeq \tfrac{3}{5} F_s'(\alpha_0)\alpha_0 = -(AR/20\epsilon^2)\alpha_0. \qquad [25]$$

Hence,

$$\frac{A_e}{A_m} = \frac{|W_e|}{|W_m|}$$

$$\simeq \frac{160\pi}{3(2)^{\frac{1}{3}}} (1 - \eta^2)^{\frac{1}{3}}\sigma^2 \left(\frac{\epsilon^4 R^2}{A^2 E}\right)^{\frac{1}{3}}. \qquad [26]$$

If this relationship is evaluated for the case of a contact between polystyrene particles and the steel surface, then for $\sigma = 10^3$ CGSE it is equal to about 10 and amounts to a value of the order of 10^3 for $\sigma = 10^4$ CGSE.

Similar relationships for the case of breaking the adhesion contact between a flat surface

TABLE I

Ratio of the Electrostatic and the Molecular Component of the Adhesion Force (F_e/F_m) and of the Tearing-off Work (A_e/A_m)

Contact	$\sigma \sim 10^3$ CGSE	$\sigma \sim 10^4$ CGSE
Sphere–plane (elastic)	$F_m \sim F_e$ $A_e/A_m \sim 10^1$	$F_m < F_e$ $A_e/A_m \sim 10^3$
Plane–plane (plastic)	$A_e/A_m \sim 10^3$ $F_m > F_e$	$A_e/A_m \sim 10^5$ $F_m \sim F_e$

essentially differ from those given above. Thus, for example, in this case the ratio of the contributions of the electrostatic and the molecular components of the sticking force, which are equal to $F_e = 2\pi\sigma^2 S$ and $F_m = AS/6\pi\epsilon^3$, respectively, will be

$$F_e/F_m = 24\pi^2\sigma^2\epsilon^3/A. \qquad [27]$$

Even for $\sigma = 5 \times 10^3$ CGSE this ratio is equal to about 0.1—that is, the contribution of F_e to the total adhesion force is small, and becomes comparable with F_m only at the maximum values of $\sigma \sim 10^4$ CGSE.

Let us consider the ratio of the amounts of work to be consumed for overcoming the electrostatic and molecular forces in breaking the adhesion contact of flat surfaces.

In the case of a contact between flat surfaces, the specific adhesion work, A_e, per unit area of the contact is defined as

$$A_e = F_e h_d = 2\pi\sigma^2 S h_d, \qquad [28]$$

where $h_d \sim 10^{-4}$ cm is the discharge gap in air (2).

In this case, the specific adhesion work A_m is equal to

$$A_m = W_m(\epsilon) - W_m(h_d), \qquad [29]$$

where $W_m(x)$ is the energy of the molecular interaction of planes at spacing x from one

another, i.e.,

$$A_m = \frac{A}{12\pi\epsilon^2} - \frac{A}{12\pi h_d^2} \simeq \frac{A}{12\pi\epsilon^2}, \qquad [30]$$

for $h_d \gg \epsilon$. Hence, it follows that

$$A_e/A_m \simeq 24\pi^2\sigma^2\epsilon^2 h_d/A. \qquad [31]$$

At $\sigma = 5 \times 10^3$ CGSE and $A = 10^{-12}$ erg the ratio $A_e/A_m \sim 10^3$.

What has been stated above is illustrated by Table I. The physical sense of the relations represented in the table is quite clear. In breaking the sphere–plane elastic contact, the molecular forces rapidly decrease with distance, and are overcome not at once, but consecutively, on different areas of the contact zone. This decreases their role as compared with the case of breaking the plastic contact of particles or with the case of breaking the plane–plane contact.

The far-range electrostatic forces are determined, both in the case of the sphere–plane contact (elastic or plastic) and in the case of the plane–plane contact by the initial contact area; these forces do not decrease if a noticeable neutralization of electric double layer charges does not occur during the tearing off.

Also, it is easy to explain the greater contribution of the electrostatic component to the adhesion work in the case of breaking the contact of two planes as compared with the case of breaking the sphere–plane contact. In the first case, the component F_e remains constant until a gas discharge starts.

In the second case, the interaction force of the double electric layer charges concentrated on small areas rapidly decreases in accordance with Coulomb's law.

Thus, the particular features of the breaking of the elastic contact between particles and a solid surface are attributable to the peculiarities of the elastic recovery of the shape of particles in combination with the far-range character of electrostatic forces. Those features lead to a considerable variation in the ratio of the electrostatic and the molecular components of the interaction force and the adhesion work as compared with the case of an extensive contact.

REFERENCES

1. DERJAGUIN, B. V. *Kolloid. Z.* **69**, 155 (1934).
2. DERJAGUIN, B. V., AND KROTOVA, N. A., "Adhesion" (in Russian), Moscow published by the USSR Academy of Sciences, 1949, Plenum, New York, 1976.
3. DERJAGUIN, B. V., KROTOVA, N. A., AND SMILGA, V. P., "Adhesion of Solids" (in Russian), Nauka, Moscow, 1973.
4. DERJAGUIN, B. V., MULLER, V. M., AND TOPOROV, YU. P., *J. Colloid and Interface Sci.* **53**, 314 (1975).
5. LANDAU, L. D., AND LIFSHITS, E. M., "Theory of Elasticity" (in Russian), Nauka, Moscow, 1970.
6. LIFSHITS, E. M., *Zhurn. Eksp. Teor. Fiz.* **29**, 94 (1955).
7. ALEINIKOVA, I. N., DERJAGUIN, B. V., AND TOPOROV, YU. P., *Kolloid. Z.* **30**, 128 (1968).
8. ZHIGULEVA, I. S., AND SMILGA, V. P., "Surface Forces in Thin Films and Disperse Systems" (in Russian), p. 220. Nauka, Moscow, 1974.

Liquid Crystals

Structures and Properties of the Liquid Crystalline State of Matter

GLENN H. BROWN

Liquid Crystal Institute, Kent State University, Kent, Ohio 44242

Received June 9, 1976; accepted August 18, 1976

Liquid crystals were discovered in 1888 but have been studied seriously during the past 20 years. The kinds of molecules that form liquid crystals on heating and by mixing two or more substances are explained. Properties of liquid crystals and related structural characteristics are summarized. The properties considered include optical characteristics, textures, and the role of water in lyotropic systems. The structures of nematic and smectic liquid crystals are explained and systemized. Determination of the tilt angle of the molecules in the smectic C compounds terephthal-bis-(4,n-butylaniline) (TBBA and its C_5 and C_6 homologs) is discussed. The molecular packings in the crystalline forms of three nematogenic compounds are explained.

I. INTRODUCTION

Liquid crystalline materials have been observed for over a century but were not recognized as such until the 1880's. Papers written in the 1850's describe systems that fit the properties we now ascribe to liquid crystals. We recognize the beginning of liquid crystal sciences with observations of Reinitzer (1) in 1888. He prepared cholesteryl benzoate and found two interesting properties when he heated the compound. He pointed out that the crystal lattice collapsed at 145°C to form a turbid liquid (liquid crystal). On further heating to 179°C the turbid liquid disappeared and the isotropic liquid was formed. On cooling the process was reversed. Reinitzer's second and most interesting observation was that the turbid liquid changed color as the temperature changed, going from red to blue with an increase in temperature and reversing the color pattern as the system was cooled.

The term "liquid crystals" is both intriguing and confusing; while it appears self-contradictory, the designation is really an attempt to describe a particular state of matter. The liquid crystalline state of matter has the optical properties of solids and the flow characteristics of liquids. This state of matter mixes the properties of both the liquid and solid states and shows some properties that are intermediate between the two; it can show properties unique to itself. Liquid crystals combine a kind of long-range order (in the sense of a solid) with the ability to form droplets and to pour (in the sense of water-like liquids).

Liquid crystals may be defined as condensed fluid states with spontaneous anisotropy. For details on liquid crystals the reader is directed to books and reviews (2–9).

Many interesting developments have taken place in liquid crystal science in the past 10 years. A list of scientific advances include (1) structure studies and the use of X-ray methods to classify smectic liquid crystals, (2) structure analysis of lyotropic liquid crystals, (3) use of liquid crystals as solvents to study chemical reactions, (4) application as stationary phases in chromatograph, (5) use as solvents to study the structure of solute molecules, and (6) proof that cholesteric liquid crystals are really a special kind of nematic liquid crystal (10).

Liquid crystals have found use (1) in numeric displays; (2) in the preparation of fibers;

TABLE I

SOME THERMOTROPIC LIQUID CRYSTALLINE COMPOUNDS

Formula	Name	Liquid crystalline range [°C]
Some classical nematic liquid crystals:		
H_3C-O—⟨⟩—$\overset{H}{C}=N$—⟨⟩—C_4H_9-n	4-methoxybenzylidene-4'-n-butylaniline (MBBA)	21-47°
H_3C-O—⟨⟩—$N=\overset{O}{\overset{\uparrow}{N}}$—⟨⟩—$C_4H_9$-$n$	4-methoxy-4'-n-butylazoxybenzene	19-76° (mixture of isomers)
H_3C-O—⟨⟩—$N=\overset{O}{\overset{\uparrow}{N}}$—⟨⟩—$O-CH_3$	p-azoxyanisole (PAA)	117-137°
n-$H_{13}C_6$—⟨⟩—⟨⟩—CN	4-n-hexyl-4'-cyanobiphenyl	14-28°
Spontaneously twisted nematic liquid crystals (cholesteric):		
$CH_3(CH_2)_7-\overset{O}{\overset{\parallel}{C}}-O$— (cholesterol structure)	cholesteryl nonanoate	145-179°
H_3C-O—⟨⟩—$\overset{H}{C}=N$—⟨⟩—$C=C-C-O-CH_2-\overset{CH_3}{\overset{\mid}{C}}-C_2H_5$	(-) 2-methyl-4-(p-n-ethoxybenzylideneamino) cinnanate	76-125°
Smectic A:		
⟨⟩—⟨⟩—$\overset{H}{C}=N$—⟨⟩—$COOC_2H_5$	p-phenylbenzal-p-aminoethylbenzoate	121-131°
Smectic B:		
H_5C_2O—⟨⟩—$\overset{H}{C}=N$—⟨⟩—$CH=CH-COOC_2H_5$	ethyl-4-ethoxybenzol-4'-aminocinnamate	77-116°
Smectic C:		
nH$_{17}$C$_8$O—⟨⟩—COOH	p-n octyloxybenzoic acid	108-147°
Smectic E:		
C_2H_5OOC—⟨⟩—⟨⟩—⟨⟩—$COOC_2H_5$	diethyl-p-terphenyl-4,4''-carboxylate	173-189°
Smectic F and G·		
nH$_{11}$C$_5$-O—⟨⟩—⟨N⟩—⟨⟩—$C_5H_{11}n$	2-(4-n-pentylphenyl)- 5-(4-n-pentyloxyphenyl)- pyrimidine	SF 103-114° SG 79-103°

and (3) for temperature sensing. Liquid crystals are found in life processes in certain diseases and in cell membrane structure.

Thousands of organic compounds on heating will form the liquid crystalline state. Certain hydrated metallic oxides (11) have been identified as liquid crystalline. Pitch is now known to exhibit liquid crystallinity. Liquid crystals are found in mammals, including man.

The idea of the hexagonal packing of molecules in thermotropic liquid crystalline structures was first presented by Gulrich and

Brown (12) in 1968. Since that first report a good number of other hexagonal packings have been found in different kinds of liquid crystalline phases.

Much of the contents of this paper will be addressed to the structure of liquid crystals. Most of the scientific results which I discuss were obtained in our laboratories. Therefore, I will make a general presentation of the results without going into great detail.

Before we get into the classification, structure, and properties of liquid crystals, a look at the kind of molecules that form liquid crystals seems in order. For convenience liquid crystals are divided into two categories, namely thermotropic and lyotropic. Thermotropic liquid crystals are subdivided into nematics and smectics and the lyotropic ones are lamellar, hexagonal, cubic, and tetragonal.

II

A. Thermotropic Liquid Crystals

General statements can be made about the geometry of molecules that form liquid crystals on heating. These guidelines should not be taken too literally but should be considered an aid in selecting liquid crystalline compounds. For a detailed discussion the reader is referred to Gray (9). The sketch of a hypothetical molecule about which a number of statements on molecular geometry will be made can be represented as

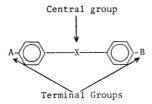

Some guidelines for selection of organic thermotropic liquid crystals may be summarized as follows. (1) The majority of liquid crystalline compounds have aromatic nuclei which are polarizable, planar, and rigid; (2) the central

group (X) in the molecule usually contains a multiple bond or a system of conjugated double bonds, or involves a dimerization of carboxyl groups; (3) the central group (X) connecting two benzene rings should constitute a lath or rodlike core of the molecule; (4) the length of the molecule should be greater than its diameter (assuming a cylindrical geometry); (5) a strong polar group near the center of the molecule and along the molecular axis generally enhances liquid crystallinity; and (6) weak polar groups (A and B) at the extremities of the molecule enhance liquid crystallinity.

Cholesteric–nematic liquid crystals require the existence of a center of chirality within the molecule. Cholesteric–nematic liquid crystals may be obtained by the addition of chiral solutes, which may or may not be liquid crystals, to nematic liquids. Resolution of the enantiomers of a nematogenic compound also results in a cholesteric–nematic liquid crystal.

A selected list of thermotropic liquid crystalline compounds is recorded in Table I.

Molecules with shapes other than cylindrical will form liquid crystals. Condensed ring systems which form disk-shaped molecules are known to be liquid crystalline.

B. Lyotropic Liquid Crystals

Amphiphiles are the most common type of molecule which form lyotropic liquid crystals. An amphiphilic compound has a polar head which dissolves in water (hydrophilic) and a water insoluble organic tail. The molecular geometries found in amphiphilic compounds are of two common types. Type 1 is commonly found in molecules such as sodium stearate. In this type of molecule the polar head is attached to a long hydrophobic tail. Type 2 has the polar head attached to two hydrophobic tails. The hydrophobic groups generally lie side by side and form a "clothespin" structure, or the groups can lie at an acute angle to each other, i.e., the molecules are peg-shaped. Examples of Type 2 molecules are Aerosol OT and phospholipids. The formula

TABLE II

Polymorphic Forms of Liquid Crystals

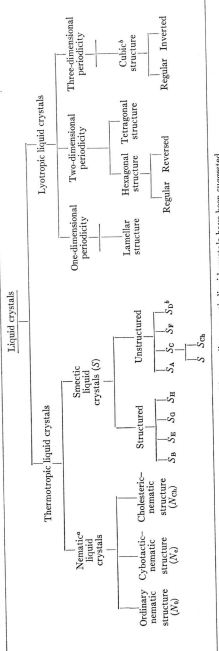

[a] Classes such as skewed-cybotactic nematic liquid crystals and intermediate nematic liquid crystals have been suggested.
[b] Isotropic.

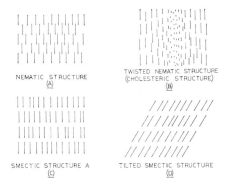

NEMATIC STRUCTURE
(A)

TWISTED NEMATIC STRUCTURE
(CHOLESTERIC STRUCTURE)
(B)

SMECTIC STRUCTURE A
(C)

TILTED SMECTIC STRUCTURE
(D)

FIG. 1. Schematic arrangements of molecular packing in several liquid crystalline structures.

for Aerosol OT is:

$$
\begin{array}{c}
C_2H_5 \\
| \\
CH_3-(CH_2)_3-CH-CH_2-OOC-CH_2 \\
| \\
CH_3-(CH_2)_3-CH-CH_2-OOC-CH(SO_3^-)Na^+ \\
| \\
C_2H_5
\end{array}
$$

The nature of the polar group is important in the preparation of lipophilic systems. Polar groups that are hydrophilic may be classified as (1) ionic (anionic and cationic), or (2) nonionic (zwitterionic, semipolar, and polyoxy).

Hydrated oxides of certain metals (Fe, Mo, V, Al) in an aqueous environment form liquid crystals.

III. CLASSIFICATION, STRUCTURE, AND PROPERTIES OF THERMOTROPIC LIQUID CRYSTALS

Thermotropic liquid crystals are classified as nematic and smectic. These classes can be subdivided to give many polymorphic forms (See Table II). A brief discussion of the most common classes follows.

A. Ordinary Nematic Structure

The arrangement of molecules in two dimensions in the ordinary nematic liquid crystal is represented schematically in Fig. 1A. Two

features of ordinary nematic liquid crystals are:

(1) There is a long-range orientation order, i.e., the long axes of the molecules are essentially parallel.
(2) The nematic structure is fluid; i.e., there is no long-range correlation of the molecular center of mass positions.

The direction of the principal axis $\hat{n}$ (the director) is arbitrary in space. The only structural restriction in the ordinary nematic liquid crystal is that the long axes of the molecules maintain a parallel or nearly parallel arrangement. The molecules are mobile in three directions and can rotate about one axis (Fig. 1A).

Bulk samples of nematic liquid crystals are turbid. In films greater than 0.1 mm thick, they show threadlike disclinations between crossed polarizers. In thinner films, a schlieren texture with point-like singularities can be obtained. These singularities are vertical threads and may be characterized by the number of dark brushes that appear when observed between crossed polarizers. Points with two or four brushes are commonly found. By simultaneous rotation of the polarizer and analyzer, positive and negative points can be distinguished, depending on whether the brushes rotate in the same (positive) or opposite sense (negative).

Molecules in a film of a nematic liquid can be oriented by surface action. If the surface is rubbed, the molecules tend to align with their long axes parallel to the direction of rubbing. Other treatments, such as certain surfactants, may orient the molecules so that their long axes stand perpendicular to the surface. If the orientation is complete, a pseudoisotropic texture results.

The nematic phase is the highest-temperature mesophase. It is transformed on heating to the isotropic liquid. This transition is first order; the enthalpy of the transition generally lies between 0.1 and 1.0 kcal/mole. Nematic liquids have an infinite-fold symmetry axis and are, therefore, uniaxial. The orientation of the molecules in a nematic liquid crystal is incomplete. The measure of the degree of orientation can be reasonably expressed by a single-order parameter (S)

$$S = \tfrac{1}{2}(3\cos^2\theta - 1),$$

where θ denotes the angle between the long molecular axis and the nematic symmetry axis. Experimental values of S (order parameter) range from near 0.4 at the nematic–isotropic point to near 0.8 in the lower-temperature range of the nematic liquid.

B. Cybotactic Nematic Liquid Crystal Structure

The ordinary nematic and the cybotactic nematic liquid crystals can be distinguished by their X-ray patterns. In a cybotactic liquid crystal the molecules are clustered in small domains which have molecular order with the long molecular axes parallel. For a randomly oriented sample, the most obvious difference lies in the relative intensities of the two diffusion rings. Generally, the inner ring intensity is less for the ordinary nematic liquid crystal. For the cybotactic nematic liquid crystal the intensity of the inner ring is greater than that of the outer ring.

C. Cholesteric–Nematic Structure

Cholesteric–nematic liquid crystals were first observed with cholesteryl esters. Non-steroidal molecules which exhibit optical activity may show the cholesteric–nematic structure. To distinguish these compounds from the cholesteric esters, we will call them chiral nematic liquid crystals.

When molecular layers in a cholesteric–nematic liquid crystal are uniformly aligned in such a way that the helical axis is normal to the substrate, the system exhibits unique optical properties. The most striking feature of such a structure is the selective light reflection which is the origin of the brilliant colors.

The molecular organization in a cholesteric liquid crystal is such that there is a unidirectional alignment of molecules within a given layer. The individual layers are stacked so that

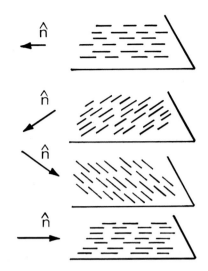

FIG. 2. Schematic sketch of cholesteric–nematic structure. The cholesteric director $\hat{n}$ follows the form of a helix.

the direction of the long axes of molecules in one layer is displaced slightly from the direction of the long axes in the adjacent layer tracing out a helical structure shown in Fig. 2. The helical structure may be right-handed or left-handed depending on the nature of the compound.

If a cholesteric–nematic liquid crystal is placed on the stage of a polarizing microscope and the analyzer rotated, a color change will occur. The rotation of the light changes sign at a wavelength λ_0 (Fig. 3). In a small wavelength range about λ_0 a light beam incident parallel to the helical axis is split into two circularly polarized components. One of these components is transmitted through the sample and the other is reflected. Maximum reflection occurs when

$$\lambda_0 = \bar{n}p,$$

where $\bar{n} = (n_{\shortparallel} + n_\perp)/2$, the mean refractive index within a normal to the helical axis and p is the pitch. The width of the reflection band can be defined as $\Delta\lambda = \Delta np$, where $\Delta n = n_{\shortparallel} - n_\perp$. Δn is the anisotropy of the refractive index.

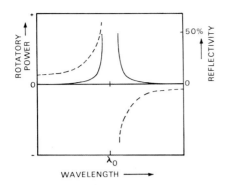

FIG. 3. Schematic representation of optical properties of a *dextro* cholesteric liquid crystal (incident beam linearly polarized).

Outside the reflection band, cholesteric–nematic liquid crystals exhibit a very strong optical rotatory power which can be as large as 20,000° or 50 complete rotations per millimeter. The sign of the rotatory power is different below and above the reflection band.

The pitch p and hence the reflection color is strongly temperature-dependent. Thin homogeneously aligned layers of these cholesteric–nematic compounds are good temperature sensors. Usually the pitch decreases as the temperature increases and the reflection band is shifted to shorter wavelengths. Reverse situations are known. The temperature sensitivity is extremely strong in the neighborhood of a cholesteric to smectic phase transition.

Observation of Figs. 1A and 1B shows that one can go from the ordinary nematic to the cholesteric–nematic liquid crystal by a mechanical twist of the ordinary nematic packing.

Mechanically twisted nematic liquid crystals can be generated by rubbing two glass surfaces in the same direction and then arranging the two glass plates so that the direction of rubbing of one plate is perpendicular to that of the other. This arrangement will rotate polarized light through 90°.

D. Smectic Structures

Eight smectic structures have been described in the literature. They are identified as smectic A through smectic H. The term "smectic" is not exactly specific, as we use it at the present time, but covers all thermotropic liquid crystals that are not nematic.

In most smectic structures, the molecules are arranged in strata; depending on the molecular order within the strata, we can differentiate between smectics with structured and unstructured strata. The structured smectic liquid crystals have long-range order in the arrangement of the molecules in layers and form a regular two-dimensional lattice.

E. Comments on the Structure of Three Smectic Liquid Crystals

(1) Smectic B structure. The most common of the structured smectic liquid crystals is the smectic B. The smectic B structure has the well-ordered layers of molecules and, in addition, orderly packing of the molecules in the layers. X-Ray patterns of a smectic B structure show two sharp rings, as one would expect from the molecular packing characteristics. The smectic B structure has two different symmetries, $D_{\infty h}$ and C_{2v}. The first of these has a hexagonal packing, with the molecular axis perpendicular to the layers, and is optically uniaxial. The second smectic B has its molecules tilted in the layers; because of its lower symmetry, it is biaxial. The texture of the structured smectic is a modification of the fan and schlieren textures and of the mosaic texture. The mosaic texture has optically uniform birefringent areas.

(2) Smectic A structure. The molecules in an S_A liquid crystal are packed in strata with the long axes of the molecule perpendicular to the plane of the strata and the molecules in a stratum are randomly arranged. A two-dimensional schematic arrangement is illustrated in Fig. 1C. The strata show a repeat distance between centers of gravity of molecules in adjacent strata, and an X-ray pattern of the structure shows a sharp ring characteristic of this packing pattern. The layer thickness is essentially identical to a full molecular length. A second ring on the X-ray pattern at about

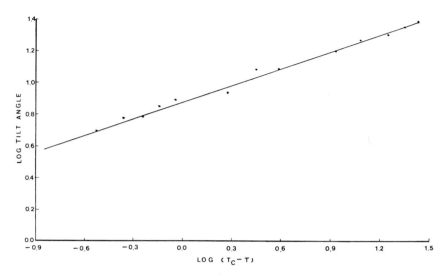

FIG. 4. Log tilt angle versus log T_c for terephthal-bis-(p,n-pentylaniline).

10° Bragg angle is diffuse in nature, thus showing that the molecules in a stratum are randomly packed.

At thermal equilibrium the S_A phase is optically uniaxial due to the infinite-fold rotational symmetry about an axis parallel to the layer normal. The typical texture of the $D_{\infty h}$ symmetry is focal conic; its modification is the fan texture.

(3) *Smectic C structure.* Another type of unstructured smectic liquid crystal has C_{2v} symmetry and is optically biaxial. The most common smectic liquid crystal of this type is smectic C, which has layers that are monomolecular. The molecules in the layer are tilted. The textures of C_{2v} symmetry are of the same kind as those of $D_{\infty h}$ but are often more complicated. The typical C_{2v} textures are broken focal conic and broken fan textures.

X-Ray patterns and microscopic studies support the idea that the smectic C structure (Fig. 1D) has a uniform tilting of the molecular axes with respect to the layer normal. The spacing between strata, as determined by X-ray studies, is considerably less than the molecular length, and the difference between these values indicates molecular tilt. The

smectic C structure is optically biaxial, which supports the idea of molecular tilt. The tilt angle is temperature dependent. Not all smectic C structures show a change in tilt angle with temperature.

After recognizing that the molecules in the smectic C structure have a tilt relative to the boundary plane, the question arises about the value of the tilt angle. We have made such a study, which I report in an abbreviated way. The compounds we choose are terephthal-bis-(4,n-butylaniline) (TBBA) and its C_5 and C_6 homologs. The formula for TBBA is

$$nC_4H_9-\!\!\bigcirc\!\!-N{=}\underset{H}{C}-\!\!\bigcirc\!\!-\underset{H}{C}{=}N-\!\!\bigcirc\!\!-nC_4H_9$$

This compound shows the following phase changes: Crystal ↔ smectic B ↔ smectic C ↔ smectic A ↔ nematic ↔ isotropic liquid. We are particularly interested in the phase change from smectic C to smectic A. De Gennes (13) predicted that liquid crystals which exhibit the smectic C–smectic A transition will show a tilt angle θ that varies as $(T_c - T)^{0.35}$ where T_c is the transition temperature and T is the measured temperature in degrees Kelvin. Doane *et al.* (14) studied the tilt angle of TBBA as a function of temperature using NMR techniques. Their data showed a dependence of $(T_c - T)^{0.40\pm0.04}$.

We used X-ray techniques to establish the tilt angle for the molecules in TBBA and its C_5 and C_6 homologs. A plot of $\log_{10}\theta$ versus $\log_{10}\Delta T(T_c - T)$ gives a measured dependence of $(T_c - T)^{0.35\pm0.01}$ for the C_5 derivative (see Fig. 4).

One of the real difficulties in measuring the critical constant for the S_C–S_A transition is the accurate determination of T_c. The range of the

TABLE III

Classification of Smectic Liquid Crystals

Symbol	Brief description of X-ray pattern	Molecular arrangement	Optical properties	Texture
		I. Structured Smectic liquid crystals		
S_B	Two rings, both well defined	Molecules in layers with repeat distances between layers; molecules in layers in orderly packing	Uniaxially or biaxially positive	Mosaic; stepped drops; pseudoisotropic; homogeneous; schlieren
S_E	Two strong outer rings	Evidence not conclusive but there may be two forms, one with the molecules perpendicular to the plane of the layers and another where the molecules are tilted to the boundary planes	Uniaxially positive	Mosaic; pseudoisotropic
S_G	Two strong outer rings	Similar to smectic E; smectic E and smectic G may differ in detail but not in kind	Uniaxially positive	Mosaic
S_H	Many sharp reflections originating from well-defined lattice planes		Biaxially positive	Mosaic
		II. Unstructured smectic liquid crystals		
S_A	Two rings: sharp inner ring; diffuse outer ring	Long axes of molecular perpendicular to the boundary planes	Uniaxially positive	Focal conic (fan-shaped or polygon) stepped drops; homogeneous; pseudoisotropic
S_C	Two rings: sharp inner	Long axes of the molecules are tilted to the boundary planes; the angle of tilt is dependent on the composition of the compound and the temperature	Biaxially positive	Broken focal conic; schlieren; homogeneous
S_D	Inner ring is split into six spots in an approximate hexagonal arrangement	Molecules are packed in spherical arrangement and the spheres then pack in a cubic order	Isotropic	Isotropic; mosaic
S_F	No definitive data yet	Presumably somewhat similar to smectic A and smectic C; layer packing	Uniaxially positive	Schlieren; broken focal conic with concentric axes

CLASS γ SMECTIC E   CLASS β SMECTIC B   CLASS α SMECTIC A

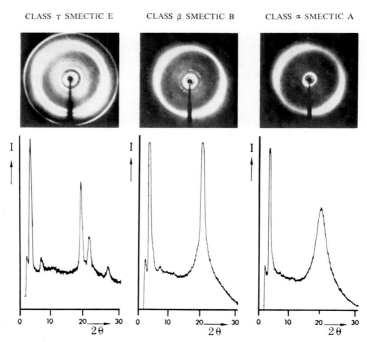

FIG. 5. Diffraction patterns and densitometer traces of S_E, S_B, and S_A (left to right). Photographs on same scale.

tilt-angle values are generally between 0 and 35°, with some larger angles occurring occasionally. The TBBA molecule and its C_5 and C_6 homologs gave maximum tilt angles of about 25°.

Information on these three smectic liquid crystals and others is collected in Table III.

Figure 5 from A. de Vries (15) shows the kind of X ray data one can get from structural analysis on liquid crystals and how one can distinguish one liquid crystal from another by their X ray patterns.

IV. CLASSIFICATION, STRUCTURE, AND PROPERTIES OF LYOTROPIC LIQUID CRYSTALS

A. Properties of Lyotropic Liquid Crystals

A typical lyotropic liquid crystal can be prepared by mixing water and sodium stearate.

A lyotropic liquid crystal which has properties similar to a nematic liquid crystal can be made by mixing water, n-decanol, decyl sulfate, and a salt such as potassium chloride.

We will consider only two-component systems composed of water and an amphiphile. Multicomponent systems are common not only in the laboratory but in the human anatomy.

Polymorphic mesophases may be formed by mixing an amphiphile and water. The common mesophases exhibit lamellar, cubic, and hexagonal molecular packings. Removing water will reverse the order of mesophase formation. Lyotropic liquid crystals are responsive to concentration changes and to changes in temperature. Phase changes can come about by change in either concentration or temperature, or by a change of both simultaneously.

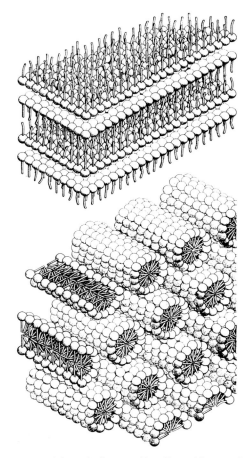

likewise, nonpolar units associate to form hydrocarbon layers in the lamellar packing. The forces directing these molecular packings are van der Waals forces and hydrophobic interactions in the hydrocarbon part of the molecule. Electrostatic forces dominate in the polar portion of the molecule.

In the case of lipids (a kind of amphiphile), those that have polar groups have melting points of 200–300°C. There are lipids that have no polar unit in the molecule; these melt around 70°C.

In lyotropic structures, the polar group seeks water and the hydrocarbon part of the amphiphilic molecule seeks others of its kind. Further, it should be noted that the molecules in the cubic structure are packed in a spherical design and the spheres are then packed in a cubic pattern.

B. Lyotropic Phases and Their Structure (Amphiphile Plus Water)

(1) The "neat" or "G" phase. It is generally agreed that this phase is smectic in character, and the amphiphilic molecules with water form a lamellar packing. A sketch of this packing is found in Fig. 6 and a description of the phase is given in Table IV. A recent study (16) shows that the molecules in the lamellar packing can exhibit a tilt angle which is evidently temperature sensitive.

(2) The "middle" or "M₁" phase. This phase is stable at higher water concentrations than the G phase in those cases in which both of these phases are formed from the same components. X-Ray diffraction studies show that the amphiphilic molecules are grouped into rod-like clusters (Fig. 6) of indefinite length, which, in turn, are arranged side by side in a hexagonal packing (middle phase). It has been proposed that in each rod the molecules are arranged radially around the rod axis with the polar groups on the outside. A schematic picture of the molecular packing of this phase is presented in Fig. 6 and a description of the phase is found in Table V.

FIG. 6. Schematic diagram of lamellar and hexagonal molecular packings of lyotropic liquid crystals. The upper view shows a lamellar packing of the amphiphilic molecules and water. The organic portions of the molecules dissolve in each other. The resulting structure is a bilayer unit and water is a part of the structure. The cylindrical packing in the lower view has water dispersed among the cylinders to give the structure stability.

Amphiphilic molecules associate in such a pattern that there is a minimum of free energy and the molecular aggregates in both the dry and wet forms are not fundamentally different. In the packing of amphiphilic molecules, like units of the molecules interact with each other, i.e., the polar units associate with each other;

TABLE IV

Lamellar Packing of Lyotropic Liquid Crystals

Designation	Optical properties	Basic structure	Description of proposed structure
Neat phase (soap Boiler's Neat Soap); lamellar	Anisotropic	Lamellar double layers	Double layers of amphiphiles with polar groups in the interfaces with intervening layers of water
Single layered neat phase	Anisotropic	Lamellar single layers	Single layers of amphiphile molecules oriented with polar groups toward opposite interfaces with intervening layers of water.
Mucous woven phase	Slightly anisotropic	Lamellar double layers	Same as above, neat phase

1. Structural arrangement displaying Bragg spacing ratio $1:\frac{1}{2}:\frac{1}{3}$
2. One-dimensional periodicity
3. Layer structure

(3) *The viscous "isotropic" or "V_1" phase.* This phase appears in some systems at concentrations of amphiphile intermediate between those within which the G and M_1 phases are stable. Ordinary optical observations can give no information on the structure of the phase beyond showing that it is isotropic. X ray diffraction studies indicate that the molecules pack in spheres and the spheres then pack in a face-centered cubic lattice. See Table VI for a description of this phase.

(4) *"Inverse" phases ("V_2" and "M_2").* In some systems at concentrations of the amphiphile greater than those at which the G phase is stable, another viscous isotropic phase, V_2, occurs. This phase is followed, with a further increase of concentration of amphiphile, by another middle phase, M_2. The M_2 phase has a structure like the M_1 phase, but with the polar groups directed inward and enclosing a water core; the medium between the rods is of hydrocarbon composition.

(5) *Isotropic "S_{1c}" phase.* This phase has a higher water concentration than the M_1 phase; the compound *n*-decyltrimethylammonium chloride, for example, forms the isotropic S_{1c} phase.

If all the mesophases described above occurred for a two-component system at a given temperature, the sequence of their appearance with an increase of amphiphile concentration is as follows:

$$S_{1c} \rightarrow M_1 \rightarrow V_1 \rightarrow G \rightarrow V_2 \rightarrow M_2.$$

No system has been found that exhibits all mesophases. The common ones are M_1, V_1, and G, V_2, and M_2. The phases that exist occur in the order cited above, i.e.,

$$M_1 \rightarrow V_1 \rightarrow G$$

Increasing amphiphile concentration

V. STRUCTURE OF NEMATOGENIC COMPOUNDS

A. General Comments

Relatively few crystal analyses have been done on nematogenic compounds. Most of these studies have been done by my colleagues, my graduate students, and me. Until a recent structure determination by Carpenter (17) in our Institute only imbracated molecular packing in nematogenic compounds was known.

Molecular packing in crystals in which the end of one molecule is located near the middle of its nearest neighbor is identified as imbracated. In this type of packing the long axes of the molecules are essentially parallel but with no defined organization of the ends of the molecules.

TABLE V

Hexagonal Packing of Lyotropic Liquid Crystals

Designation	Optical properties	Basic structure	Description of proposed structure
Middle phase, normal	Anisotropic	Two-dimensional hexagonal	Long, mutually parallel rods in hexagonal array; amphiphilic molecules in rods are essentially in radial pattern
Middle phase	Anisotropic	Two-dimensional hexagonal	
Hexagonal complex phase, normal	Anisotropic	Two-dimensional hexagonal	Indefinitely long, mutually parallel rods in hexagonal array

1. Structural arrangement displaying Bragg spacing ratio $1:\frac{1}{3}:\frac{1}{4}:\frac{1}{7}$
2. Two-dimensional periodicity
3. Molecules packed in rodlike pattern

B. Crystal Analysis of 2,2'-Dibromo-4,4'-bis(p-methoxybenzylideneamino) biphenyl (MBAB)

The crystal structure of 2,2'-dibromo-4,4'-bis (p-methoxybenzylideneamino) biphenyl (MBAB) has been established (18). Crystal data on the compound are: $a = 7.631(7)$, $b = 11.55(2)$, $c = 15.99(3)$ Å, $\alpha = 114.39(4)$, $\beta = 105.58(2)$, and $\gamma = 92.16(4)$. The crystal belongs to the triclinic class and is of the space group $P\bar{1}$. There are two molecules per unit cell. The molecular axis is almost parallel to the c axis and neighboring molecules are shifted with respect to each other, such that the middle of one molecule is near the end of its nearest neighbor. This is an imbracated packing.

The angle between the two benzene rings of the biphenyl segment is 80.1°, yielding a herringbone packing in the plane perpendicular to the molecular axis. The details of this structure analysis have been published elsewhere (18).

C. Crystal Analysis of di-n-Propyl-4'-terphenyl-4,4''-carboxylate (DPTC)

The compound di-n-propyl-4'-terphenyl-4,4''-carboxylate (DPTC) is a smectic E liquid crystal (19). The formula for DPTC is:

$$C_3H_7O \underset{O}{\overset{O}{C}} - \bigcirc - \bigcirc - \bigcirc - \underset{O}{\overset{O}{C}} OC_3H_7$$

The crystal data on DPTC are:

$a = 22.191 \pm 0.014$ Å,
$b = 5.960 \pm 0.001$ Å,
$c = 10.951 \pm 0.004$ Å,
$\alpha = 129.651 \pm 0.016°$,

Space group is $P1$ or $P\bar{1}$.

The crystal is of the triclinic class.

TABLE VI

Cubic Packing of Lyotropic Liquid Crystals

Designation	Optical properties	Basic structure	Description of proposed structure
Optical isotropic mesophase, normal viscous isotropic phase	Isotropic	Body-centered	Packing of spheres
Optical isotropic mesophase, reversed viscous	Isotropic	Body-centered	Packing of spheres

1. Phases characterized by three-dimensional periodicity
2. Structural arrangement displaying cubic symmetry

The benzene rings in DPTC are not coplanar. The terminal benzene rings make a small angle with the inner benzene ring. The molecules form layers and within the layers the molecules form a herringbone packing and are packed in an imbracated fashion. Details of the structure will be published elsewhere.

D. Crystal Analysis of di(p-Methoxyphenyl)-trans-cyclohexane-1,4-dicarboxylate (DMCD)

The structure of the nematogenic compound (17) di(p-methoxyphenyl)-trans-cyclohexane-1,4-dicarboxylate (DMCD) has modified our thinking about the molecular packing of this type of compound. Prior to the study of DMCD, all crystal structures of nematogenic compounds were known to exhibit imbracated molecular packing. The imbracated packing of nematogenic compounds was first proposed by Bernal in 1933. The DMCD crystal belongs to the monoclinic system with space groups $P2_1/c$. The unit cell dimensions are $a = 11.6212 \pm 0.0005$ Å, $b = 9.2439 \pm 0.0006$ Å, $c = 9.257 \pm 0.001$ Å, and $\alpha = 92.04° \pm 0.03°$. There are two molecules per unit cell.

The compound DMCD has a regular arrangement of the ends of the molecules with the ends of the molecules lying in a plane which, in turn, is parallel to the plane encompassing the atoms in the other end of the molecule. This is seen in Fig. 7 if we realize that there are 16 molecules illustrated with only eight visible because of the overlap in the projection. In Fig. 7 two different levels are illustrated and a slight offsetting of packing layers occurs from one level to the next. Figure 7 illustrates the packing of the molecules in the crystal structure of DMCD. One can emphasize that the rodlike (DMCD) molecules are packed in a two-dimensional hexagonal close-packed arrangement. The layer packing is characteristic of smectic liquid crystals. Details of the structure will be published elsewhere.

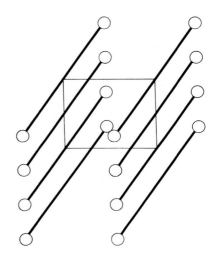

FIG. 7. Molecular packing for the nematogenic compound di(p-methoxyphenyl)-trans-cyclohexane-1,4-dicarboxylate (DMCD).

ACKNOWLEDGMENTS

Part of the research reported in this article was supported by the National Science Foundation under Grant Number DMR-74-13173.

The author acknowledges permission of the editor of Pramana ["X-Ray Studies of Liquid Crystals. V. Classification of Thermotropic Liquid Crystals and Discussions of Intermolecular Distances" by Adriaan de Vries, Suppl. 1, 93–113 (1975)] and of Dr. A. de Vries to reproduce Fig. 5. Permission to use Fig. 6 from Enciclopedia della Chimica, Volume IV, USES, Firenze, 1975, is gratefully acknowledged.

REFERENCES

1. REINITZER, F., Monatsh. Chem. 9, 421 (1888).
2. BROWN, G. H., AND SHAW, W. G., Chem. Rev. 57, 1049 (1957).
3. BROWN, G. H., DOANE, J. W., AND NEFF, V. D., "Review of the Structure and Properties of Liquid Crystals," Chemical Rubber Co., Cleveland, Ohio, 1971.
4. EKWALL, P., in "Advances in Liquid Crystals" (G. H. Brown, Ed.), Vol. 1, p. 1. Academic Press, New York, 1975.
5. CHISTYAKOV, I., in "Advances in Liquid Crystals" (G. H. Brown, Ed.), Vol. 1, p. 143. Academic Press, New York, 1975.

6. SKOULIOS, A., *in* "Advances in Liquid Crystals" (G. H. Brown, Ed.), Vol. 1, p. 169. Academic Press, New York, 1975.

7. SMITH, G. W., *in* "Advances in Liquid Crystals" (G. H. Brown, Ed.), Vol. 1, p. 189. Academic Press, New York, 1975.

8. KLEMAN, M., *in* "Advances in Liquid Crystals" (G. H. Brown, Ed.), Vol. 1, p. 267. Academic Press, New York, 1975.

9. GRAY, G. W., AND WINSOR, P. A., "Liquid Crystals and Plastic Crystals," Vols. 1 and 2, Halsted, London, 1974.

10. ROBINSON, C., *Trans. Faraday Soc.* **52**, 571 (1956).

11. ZOCHER, H., *in* "Liquid Crystals 2" (G. H. Brown, Ed.), Part I, p. 115. Gordon and Breach Science Publishers, 1969.

12. GULRICH, L., AND BROWN, G. H., *Mol. Cryst.* **3**, 493 (1968).

13. DE GENNES, P., *C. R. Acad. Sci. (Paris)* **274**, 758 (1952).

14. WISE, R. A., SMITH, D. H., AND DOANE, J. W., *Phys. Rev. A* **7**, 1366 (1973).

15. DE VRIES, A., *Pramana, Suppl. 1*, 93 (1975).

16. BIRRELL, G. B., AND GRIFFITH, O. H., *Arch. Biochem. Biophys.* **172**, 455 (1976).

17. CARPENTER, R. A., Doctoral Dissertation, Kent State University, 1975.

18. LESSER, D. P., DE VRIES, A., REED, J. W., AND BROWN, G. H., *Acta Crystallogr. B* **31**, 653 (1975).

19. CHUNG, D., Doctoral Dissertation, Kent State University, 1975.

Textures, Deformations, and Structural Order of Liquid Crystals

ALFRED SAUPE

Liquid Crystal Institute, Kent State University, Kent, Ohio 44242

Received June 28, 1976; accepted August 18, 1976

Characteristic differences are described between textures of uniaxial liquid crystals, comparing systems with a two-dimensional translational periodicity, e.g., middle soap, systems with a one-dimensional periodicity, e.g., neat soap and smectic A, and nematic liquid crystals which have only orientational order. It is pointed out how the textures are correlated to structural order and that characteristic differences in textures reflect characteristic differences in structure. Studies on myelin figures are reported and a theory is suggested to explain their formation and their tendency to coil. The explanation is based on the assumption that the lamellae have an intrinsic curvature caused by a concentration gradient. On the same basis the stability of cylindrical vesicles is discussed to explain an experimentally observed decay of cylindrical structures to strings of spheres.

I. INTRODUCTION

Liquid crystals may be optically uniaxial or biaxial like solid crystals but they have unique optical features and display characteristically different textures in thin films under the polarizing microscope. The reason is the existence of a variety of macroscopic structures corresponding in a way to nearly unlimited large deformations. Their features are related to the molecular structure of the liquid crystal. Such relations were first used successfully by Friedel to deduce the molecular structure of simple liquid crystals. Most of the general features of static textures are now well understood and many of them were explained by Lehman and Friedel at the beginning of this century. There remain, however, many details which need further study.

Microscopic methods are commonly used for a convenient classification of thermotropic liquid crystals. They are not widely used for liquid crystals formed in mixtures of amphiphilic compounds and water. The distinctive features that separate neat and middle soap were worked out only relatively late by Rosevear (1). His subsequent studies also revealed that amphiphilic systems may form a nematic phase (2). Such amphiphilic nematic phases are now of much interest as solvents for NMR spectroscopy and are commonly used for that purpose. Their existence as a phase different from the smectic lamellar phase and the hexagonal phase has so far not been established by X-ray studies.

In this paper we focus attention on three common types of liquid crystals of interest for amphiphilic systems: lamellar smectic phases, hexagonal phases, and nematic liquid crystals. Nematic textures have been extensively studied on thermotropic systems. The textures of amphiphilic nematic systems have the same characteristic features. Lamellar smectic phases also show the same characteristic features as the extensively studied smectic A textures of thermotropics. Detailed descriptions are conveniently accessible in recent books (3, 4). We describe here only a few basic features and emphasize the points of relevance for distinction between nematic and smectic structures. The correlations between textures and structures of hexagonal phases and other phases of a two-dimensional periodicity are not so well known but are particularly simple.

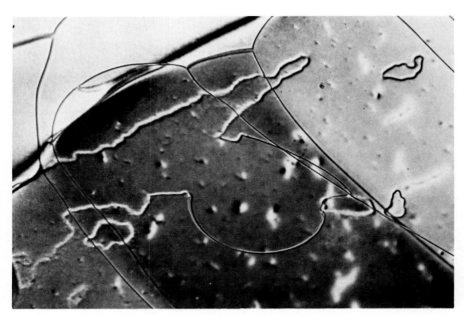

Fig. 1. Nematic Schlieren texture, methoxybenzalbutylaniline; 28× crossed polarizers.

The phase of most interest for biological systems is perhaps the lamellar phase. It was, in fact, on a system of biological interest forming such a phase that a liquid crystalline state was first seen. Tubelike structures called myelin figures were observed in 1854 by Virchow while mixing myelin and water. The details of these figures are very complicated. We discuss some of their features and show that their existence may be understood on the basis of an elastic theory. We also describe and analyze theoretically some related observations on cylindrical vesicles formed in a lecithin–water mixture.

II. NEMATIC LIQUID CRYSTALS

Nematic liquid crystals are best described as anisotropic liquids or anisotropic solutions in particular in the case of amphiphilic systems as anisotropic micellar solutions. They are macroscopically distinct from all smectic liquid crystals due to the fact that the field defined by the optical axis, the director field, assumes all kinds of deformations with about equal ease. This point deserves particular notice since it proves the absence of a translational periodicity in the equilibrium structure. Smectic liquid crystals are characterized as such by the existence of a periodicity in one or two dimensions.

A particular nematic texture commonly observed is the Schlieren texture. Two examples are shown in Figs. 1 and 2. The only constraints on the director field are those imposed by boundary conditions which determine the finally assumed equilibrium texture. Lyotropic nematics, as far as we studied them, have a perpendicular surface orientation on glass surfaces so that finally a pseudoisotropic texture results.

The smoothly curved threadlike structures in Fig. 1 represent line singularities (nematic threads) in the bulk of the nematic. There exist different kinds of line singularities. Thicker, usually somewhat diffuse lines cor-

FIG. 2. Nematic Schlieren texture, and pseudoisotropic texture (dark plots); parts of the sample are in the lamellar phase; 48× crossed polarizers; decylammoniumchloride and water.

respond to disclinations of strength 1, the sharply defined lines to disclinations (3–7) of strength $\frac{1}{2}$.

No disclination lines are recognizable in the lyotropic nematic of Fig. 2 although loops of half-numbered lines are present. They are difficult to observe in amphiphilic nematics because of the very low birefringence. Figure 2 demonstrates that different kinds of deformations occur in the director field and that the field is not subjected to severe constraints. The lyotropic nematic phase was, however, close to a transition to a lamellar smectic phase. This has some quantitative effects on the texture in that preferably splay deformations occur.

III. UNIAXIAL SMECTIC LIQUID CRYSTALS WITH A ONE–DIMENSIONAL PERIODICITY

The class of uniaxial smectics with one translational periodicity includes smectic A liquid crystals and the lamellar smectic phases of amphiphilic systems (neat soap). The general features of the texture are obtained by admitting only deformations which are compatible with a perfect layered structure. Since the layers are liquid-like and slip against each other they can be bent and curved in various ways. However, in states of low energy, the layers form a family of parallel surfaces or at least will not deviate significantly from this ideal picture. It is known that the field defined by the normals in this case consists of straight lines. The director field coincides with the field of the surface normals; therefore, no bend deformations are allowed in this field. As can be directly seen, a bend in the director field violates the equidistance and requires the formation of dislocations which are defects of high energy. In nematic systems, when the dimensions of the amphiphilic aggregates are finite, this difficulty does not exist.

The third kind of deformation, called "twist," is measured by $\mathbf{L} \cdot \mathrm{curl}\ \mathbf{L}$, where $\mathbf{L}$ denotes the unit vector of the director field. For a field

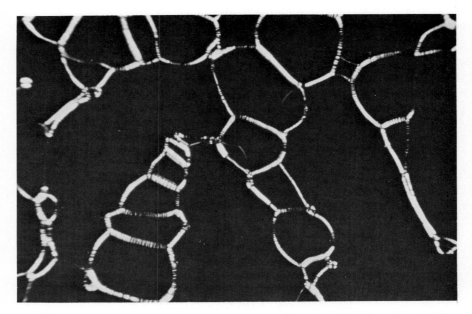

FIG. 3. Pseudoisotropic texture with oily streaks, lamellar phase; 160× crossed polarizers; lecithin (vegetable) and water.

defined by surface normals $\mathbf{n} \cdot \text{curl } \mathbf{n} = 0$. The constraint that the two fields fall together therefore eliminates twist deformations. A family of parallel surfaces of particular interest is the Dupin cyclides. They are very often formed by smectic layers. The director field in this case has a pair of line disclinations forming focal conic sections. (For a detailed discussion, see Ref. (8).) Lamellar amphiphilic phases usually form pseudoisotropic textures with oily streaks (Fig. 3). The streaks themselves are probably connected with focal conic domains but may consist of bundles of line dislocations (5).

The pseudoisotropic texture can readily be disturbed by temperature changes or by the evaporation of water. As the equilibrium distance decreases, at constant sample thickness and at a constant number of layers, a dilation is produced which leads to the crumpling of layers (3, 9). Finally, regular patterns may result, as illustrated in Fig. 4.

IV. UNIAXIAL PHASES WITH TWO-DIMENSIONAL PERIODICITY

The best-known representative for a system with a twofold periodicity is the hexagonal phase or middle soap and the inverse hexagonal phase. For the sake of simplicity we consider here the normal hexagonal phase, but the results are equally applicable to any liquid crystal with a twofold periodicity.

In the normal hexagonal phase the amphiphiles form cylindrical aggregates of infinite length which are packed hexagonally but separated by water. The optical axis and therefore the director field is parallel to the cylinder axis. Structures of low energies are those in which the hexagonal packing remains undisturbed and where the distance between cylinders remains constant. A plane that cuts one of the cylinders vertically therefore also cuts all other cylinders of the same domain vertically. In other words, the director field is vertical to a family of planes. Intersections of

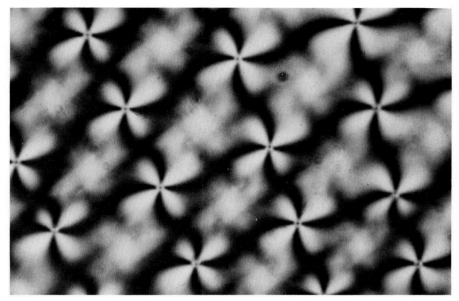

Fɪɢ. 4. Disturbed pseudoisotropic lamellar phase texture; 400× crossed polarizers; cesium decylsulfate 0.5 g, methylammoniumdecylsulfate 0.37 g, decanol 0.14 g and water.

planes correspond to disclinations. There exist no point disclinations. Disclination lines are straight and there cannot be more than one disclination line per domain. In such a domain the field lines (and the cylinder axes) form concentric circles or fractions of circles around the vertical disclination line.

Figure 5 shows an example of a middle soap texture which is formed by domains containing vertical disclination lines. In general, different domains are separated by sharp boundaries. Other textures may show less regular domains without disclination lines. The individual domains usually show a division by radial lines so that a fanlike pattern results. The primary reason for them is probably similar to that for the crumpling of layers. Because of changes of temperature and concentration, equilibrium distances and the lengths of cylinders change. The resulting elastic tension may then be reduced by an undulation of the cylinders that finally may lead to a subdivision, as indicated by the observed pattern (Fig. 5). In the beginning when a domain grows from a dis-

clination center it is completely smooth; once the subdivision has formed, it grows with the domain (Fig. 6).

V. MYELIN FIGURES

Myelin figures are tubelike structures formed by dissolving amphiphilic compounds in water. They occur in systems which do not form a normal hexagonal phase so that the lamellar phase can be in direct contact with water. As mentioned, myelin figures are the first liquid crystalline structures described in the literature. Figure 7 shows an example. The formation of the structures can be understood on the basis of deformation–energy considerations.

Optical studies suggest that the lamella form concentric cylinders around the axes of tubelike structures. After some time the cylinders tend to widen and gradually a transition to a planar texture occurs. This indicates that the structures are stable only as long as a concentration gradient exists within the lamellar phase. The water concentration in a cylinder

587

FIG. 5. Fan texture in middle soap; 160× crossed polarizers. 64°C, dec; ammoniumchloride and water.

FIG. 6. Growing fan texture of middle soap; 48× crossed polarizers, 67°C, decylammoniumchloride, and water.

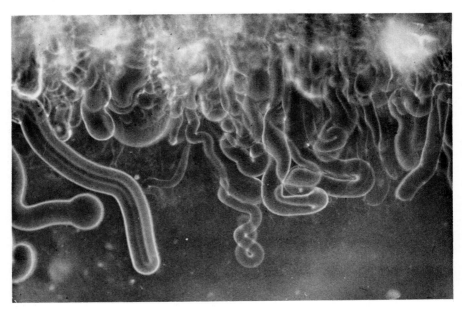

FIG. 7. Myelin figures; 48×, darkfield; lecithin (vegetable) and water.

will be highest close to the surface and decrease toward the center. Therefore the bilayers have two nonequivalent sides and accordingly possess an intrinsic curvature. The curvature is directed outside since the surface area per polar group is larger at higher water concentration (see, for instance, Refs. (10, 11)). The existence of an intrinsic curvature can stabilize the cylinders. A small calculation will be illuminating.

For simplicity we assume that there are no dilation effects. The energy is therefore just a function of the layer curvature. There are two invariants on which the free-energy density can depend, the mean curvature $\frac{1}{2}[(1/R_1) + (1/R_2)]$ and the Gaussian curvature $(1/R_1R_2)$. We can write in second order

$$g = k_1 \left(\frac{1}{R_1} + \frac{1}{R_2}\right) + \tfrac{1}{2}k_{11} \left(\frac{1}{R_1{}^2} + \frac{1}{R_2{}^2}\right)$$

$$+ k_{12} \frac{1}{R_1R_2}. \quad [1]$$

Here R_1 and R_2 denote the principal normal curvature radii. An equation of this type is also used by Helfrich (12) to describe elastic properties of bilayers and in a simplified form is used by Canham (13) for the red cell membrane.

The term proportional to the Gaussian curvature $1/R_1R_2$ may be generally disregarded. Its integral over a simply connected surface is 2π minus the line integral of the geodesic curvature over the border (Gauss–Bonnet). Therefore, it generally makes no significant contribution to the bulk energy. In the case of a straight cylinder the Gaussian curvature disappears anyway.

For a straight cylinder we set $R_1 = r$, the distance from the center, and $R_2 = 0$. The integration of Eq. [1] over the volume, assuming that k_1 and k_{11} are constants, gives for the energy per unit volume

$$E = 2\frac{k_1}{a} + \frac{k_{11}}{a^2} \log (a/a_0) + \frac{E_{\text{core}}}{a^2}. \quad [2]$$

Here a denotes the radius of the cylinder. We assume that Eq. [1] is good down to a distance a_0 from the center. The energy of the inner part is accounted for by E_{core}.

The coefficient k_1 is negative according to the argument given above. For sufficiently large a the energy will therefore become negative. (We have not included a surface-energy term because it may be best to consider it incorporated into the k_1 of the surface layer.) In general, the coefficients will depend on the distance from the center since the composition changes with it, but that does not change the essential result. It only means that the coefficients have to be replaced by some average value.

The energy of the planar texture, according to Eq. [1], is zero. The result, therefore, is that the cylinders are favored when their radius is sufficiently large. Setting $a_0 = 30$ Å and neglecting E_{core}, this is estimated to require $a > -nk_{11}/2k_1$, where n in general will be between 7 and 9.

The formation of the cylinders becomes understandable when this requirement is fulfilled. The study of the spontaneous bending and coiling remains. It will be explained when we can show that the straight cylinders are unstable to bend.

Cylinders can be bent without change of surface area and volume and without change of the circular cross section. We assume that the cylinder axis is curved with a curvature radius b. Since we are only interested in stability we can assume $b \gg a$. In this approximation and neglecting any change in the core energy, we find that the energy per unit volume changes, because of bending, by

$$\Delta E_b^{(2)} = k_{11}/4b^2. \qquad [3]$$

The coefficient k_1 makes no contribution. The energy change is positive and it stabilizes, therefore, the straight cylinder. The bending cannot be explained on the basis of Eq. [1]. However, since the curvature in the cylinder itself is rather high, we have to consider higher-order terms also.

The contribution of interest, the third-order term to the free-energy density, is

$$g^{(3)} = \tfrac{1}{2}k_{211}\left(\frac{1}{R_1 R_2{}^2} + \frac{1}{R_2 R_1{}^2}\right). \qquad [4]$$

Calculating this contribution as before and neglecting higher-order terms in $1/b$ gives

$$\Delta E_b^{(3)} = k_{211}/2b^2 a. \qquad [5]$$

We see that the cylinder will coil when $k_{211} < 0$ and when "a" is sufficiently small. The inclusion of a higher-order term allows, therefore, an explanation for the spontaneous bending of cylinders.

There are now two inequalities which have to be fulfilled simultaneously:

$$-\frac{2k_{211}}{a} > k_{11} \quad \text{and} \quad -\frac{k_1}{4a} > k_{11}.$$

The second relation is the condition for the formation of cylinders (with $n = 8$). It was obtained neglecting higher-order terms. It is not crucial there since the higher-order contributions come mainly from the inner part of the cylinder and may be included in the core energy. On the other hand, under bending these contributions to the core energy vary and must be considered explicitly. At present, no values for elastic coefficients are available and we cannot decide whether the inequalities impose reasonable requirements on the materials.

VI. VESICLES

Lamellar lyotropic phases, in particular the system composed of lecithin and water, can form vesicles. These are closed smectic membranes forming bodies of different shapes. The thickness of the membrane varies. It may be as thin as one double layer. On very thin membranes the thermal fluctuations are directly observable.

The vesicles formed in the lecithin–water mixture assume in final equilibrium a spherical shape. However, cylindrical structures may also be present (Fig. 8). A string of beads can

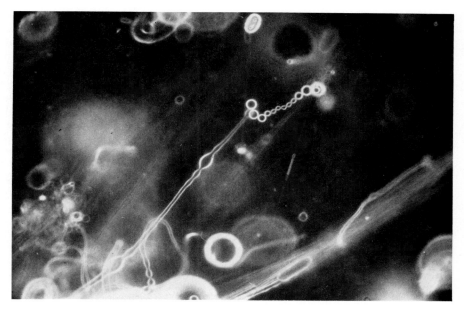

FIG. 8. Myelin figures and vesicles; $16 \times 10/3$ darkfield, lecithin (vegetable) and water.

be seen in the center of Fig. 8 which formed by the decay of a cylindrical structure. The manner in which this transformation occurs suggests that the structures are cylindrical vesicles. They seem to have a tendency to become unstable and to transform to spheres. The spheres do not separate completely but remain connected by a small tube as long as there is no outside disturbance.

Observations show that the instability begins with a periodic variation of the cylinder diameter. Some estimates of the ratio repeat-distance to cylinder diameter were made in ranges where the periodicity was quite regular. The ratio was found to be about 4:1. However, in general the variations are rather irregular and this value may not be meaningful.

In view of this observation it is of some interest to analyze the stability of cylindrical vesicles on the basis of an elastic theory. For a theoretic discussion of the shape of vesicles, the following constraints have to be taken into account. The area of the membrane and the included volume is constant. The energy as a function of shape is then dependent on curvature only. In second order it is again given by Eq. [1], where g now denotes the energy per unit area.

To study the instability we consider a disturbed cylinder given by

$$x = r \cos \phi, \quad y = r \sin \phi,$$
$$r = r_0[1 + \epsilon \sin (2\pi/p)z], \qquad [6]$$

where ϵ is a small quantity.

The undisturbed cylinder may have the radius a. The volume-to-area ratio is $a/2$. The same ratio must be valid for the deformed cylinder. This gives a relation between r_0, ϵ, p, and a. We set

$$p = \alpha \pi a \qquad [7]$$

and find in second order

$$a/r_0 = 1 + (\epsilon^2/2)[1 - (2/\alpha^2)]. \qquad [8]$$

The final result for the energy per unit area of

the smectic membrane is

$$E = \frac{1}{2a}\left(2k_1 + \frac{k_{11}}{a}\right) + \frac{\epsilon^2}{2a}\left[k_1\right.$$

$$\left. + \frac{k_{11}}{a}\left(\frac{16}{\alpha^4} - \frac{4}{\alpha^2} + 1\right)\right]. \quad [9]$$

The sign of the derivative

$$\frac{dE}{d(\epsilon^2)} = \frac{1}{2a}\left[k_1 + \frac{k_{11}}{a}\left(\frac{16}{\alpha^4} - \frac{4}{\alpha^2} + 1\right)\right] \quad [10]$$

determines the stability. For negative values the tube becomes unstable. Since $k_{11} > 0$ while $k_1 < 0$, the instability occurs first for $\alpha = 2(2)^{\frac{1}{2}}$ where $(16/\alpha^4) - (4/\alpha^2) + 1$ has a minimum. The corresponding period is $p = 2(2)^{\frac{1}{2}}\pi a$ and

$$\frac{dE}{d(\epsilon^2)} = \frac{1}{2a}\left(k_1 + \frac{3}{4}\frac{k_{11}}{a}\right). \quad [11]$$

An instability occurs, therefore, only when $k_1 < -\frac{3}{4}(k_{11}/a)$.

The calculated repeat distance $p = 8.9a$ agrees fairly well with the observations that suggest $p = 8a$. It is of interest to notice that for $p = 8.9a$ in addition to area and volume the length of the tube is not affected by the disturbance. For the bending instability of the tube the same equations apply as for the solid cylinder (Eqs. [3] and [4]). Accordingly, a tube of small diameter tends to become unstable to bending while a tube of larger diameter will be unstable to thickness variations.

It should be emphasized that these calculations are good only for tubes with thin membranes. Dilation terms will be of importance for thicker membranes (p depends on a). They can be expected to stabilize the tube with respect to thickness variations but they have no influence on the bending instability. Concluding, it should be emphasized that further studies are needed to establish the validity of this approach although the qualitative agreement at this stage is good.

REFERENCES

1. ROSEVEAR, F. B., *J. Amer. Oil Chem. Soc.* **31**, 628 (1954).
2. ROSEVEAR, F. B., *J. Soc. Cosmet. Chem.* **19**, 581 (1968).
3. DE GENNES, P. G., "The Physics of Liquid Crystals." Clarendon, Oxford, 1974.
4. GRAY, G. W., AND WINSOR, P. A. (Eds.), "Liquid Crystals and Plastic Crystals." Horwood, Chichester, England 1974.
5. KLEMAN, M., *in* "Advances in Liquid Crystals" (G. H. Brown, Ed.), Vol. 1, p. 267. Academic Press, New York, 1975.
6. SAUPE, A., *Mol. Cryst. Liq. Cryst.* **21**, 211 (1973).
7. KLEMAN, M., AND FRIEDEL, J., *J. Phys. (Paris)* **30**, C4 (1969).
8. BOULIGAND, Y., *J. Phys.* **33**, 525 (1972).
9. BROCHARD, F., AND DE GENNES, P. G., *Pramama, Suppl.* **1**, 1 (1975).
10. LUZZATTI, V., *in* "Biological Membranes" (D. C. Chapman, Ed.). Academic Press, New York.
11. EKWALL, P., *in* "Advances in Liquid Crystals" (G. H. Brown, Ed.), Vol. 1, p. 1. Academic Press, New York, 1975.
12. HELFRICH, W., *Z. Naturforsch.* C **28**, 693 (1973); CC **29**, 182 (1974); CC **29**, 510 (1974); C **30**, 841 (1975).
13. CANHAM, P. B., *J. Theor. Biol.* **26**, 61 (1970).

Electro-Optical Applications of Liquid Crystals

J. DAVID MARGERUM AND LEROY J. MILLER

Hughes Research Laboratories, Malibu, California 90265

Received June 4, 1976; accepted August 18, 1976

The unique properties of thermotropic liquid crystals have led to the development of many new electro-optical devices, particularly for display applications. Basic properties and surface alignments of liquid crystals are reviewed with regard to these applications. Electro-optical effects based on conductivity and field effect alignment are described for both nematic and cholesteric materials. Three applications of nematics are selected for more detailed discussion: a flat panel television display using dynamic scattering activated by a semiconductor matrix, a watch display using polarization modulation of twisted nematic cells on transparent segment electrodes, and a large screen projection system using tunable birefringence in a photoactivated light valve.

INTRODUCTION

The intermolecular forces responsible for the mesomorphic phases also lead to a cooperative behavior between large numbers of molecules in regard to their surface alignment, field alignment, and flow alignment phenomena. Because liquid crystals have highly anisotropic properties such as birefringence, conductivity anisotropy, and dielectric anisotropy, large effects in thin cells can result from subtle surface pretreatments combined with relatively low applied electrical fields and current levels. Liquid crystals are particularly attractive for display applications such as wrist watches, calculators, message boards, flat panel television, and large screen projection systems. Other devices include page composers, electronic reticles, real-time optical data processing systems, waveguide switches, and graphic arts duplicating devices. The applications are based on the resulting advantages of one or more factors, particularly: flat panel cells, low operational voltage, low power consumption, portability, viewability in high ambient light, good resolution capability, rapidity of response, image storage, and stability.

In this paper we summarize the liquid crystal and surface alignment properties most pertinent to display device applications and we review the general scope of electro-optic device effects. We also describe three distinctly different types of nematic liquid crystal display systems in more detail as illustrative of the broad range of applications being used. To avoid excessive length, we have limited this paper to description, without equations for device characteristics. Readers seeking more information should find adequate leads in the references cited, and also should consult some of the recent reviews and books on electrooptical effects and applications of liquid crystals (1–6).

BULK LIQUID CRYSTAL PROPERTIES

Type of liquid crystal. Of primary importance is the type of internal long-range order that the liquid crystal possesses, i.e., whether it is a nematic, cholesteric, or smectic liquid crystal. The uniaxial optical property of nematics and their relatively low viscosity are major reasons for their being used most widely in electrooptic devices. However, the higher viscosity of cholesterics and smectics is turned to advantage for storage devices, in which a lack of mobility is required. A property peculiar to the cholesterics is the helical pitch, which can be varied continuously from that of the pure compounds to infinity (at which point it becomes a nematic) by admixture of a nematic, or by combining cholesterics of opposite pitches in the desired proportions (7). When the pitch is twice the wavelength of the incident light, some light will be reflected from the recurring layers of equivalent refractive indices, but only circularly polarized light having the same handedness as the cholesteric helix. This reflected light is an important ingredient of some electro-optical displays.

Temperature range. Obviously the material must exist in the desired mesomorphic state over the operating temperature range. However, the nematic–isotropic transition temperature (i.e., the clearpoint, T_c) has additional significance and usually should be well above the operating temperature. The reason is its influence on the order parameter S, which is defined by $S = \frac{1}{2}(3\overline{\cos^2\theta} - 1)$ where θ is the angle between the long molecular axis and the director L (L is the preferred average orientation of the axis for the aggregate); $\overline{\cos^2\theta}$ is the averaged value over time and space. S is a measure of the extent to which the inherent anisotropy of the molecules is reproduced in the anisotropy of the liquid crystalline state, and therefore it affects birefringence, dielectric anisotropy, viscosity, etc. S varies as a function of the reduced temperature $[(T - T_c)/T_c]$ approximately as described by a theoretical curve for all nematics (8).

It is usually necessary to minimize the melting temperature of the liquid crystal. The customary approach is to formulate a suitable eutectic mixture from several compounds having the desired combination of other properties.

Dielectric anisotropy. The dielectric anisotropy is defined as $\Delta\epsilon = \epsilon_{\parallel} - \epsilon_{\perp}$, where $\epsilon_{\parallel}$ and $\epsilon_{\perp}$ denote the dielectric constants parallel and perpendicular to the director L. The dielectric constants include contributions due both to permanent dipoles and to molecular and atomic polarizabilities. In the absence of a permanent dipole moment, $\Delta\epsilon$ is slightly positive. Strong dipole moments (such as from p-cyano groups) directed along the molecular axis can readily increase $\Delta\epsilon$ to between $+10$ and $+20$. However, it is more difficult to introduce strong dipoles perpendicular to the axis and maintain liquid crystallinity; consequently, negative values of $\Delta\epsilon$ are usually limited to about -1 to -2.

The sign of $\Delta\epsilon$ is crucial for most electro-optic devices because it determines the orientation of the molecules when they are aligned by the field $\mathcal{E}$: if $\Delta\epsilon > 0$, $L \parallel \mathcal{E}$; if $\Delta\epsilon < 0$, $L \perp \mathcal{E}$. The magnitude of $\Delta\epsilon$ is also important, since the threshold voltage for most devices is inversely proportional to $|\Delta\epsilon|^{\frac{1}{2}}$.

Relaxation frequencies for $\epsilon_{\perp}$ are typically 10^7 to 10^8 Hz, while those for $\epsilon_{\parallel}$ can occur over a wide range between 10^3 and 10^7 Hz. The lowest relaxation frequencies for $\Delta\epsilon$ are characteristic of nematics with an exceptionally long major axis. Such materials can have $\Delta\epsilon > 0$ at low frequencies and $\Delta\epsilon < 0$ at readily available high frequencies. Depending upon frequency, they can be field-aligned with $L \parallel \mathcal{E}$ or with $L \perp \mathcal{E}$, and therefore they can be driven in each direction by an applied field. This can be used to speed up the decay time; however, the crossover frequency is quite sensitive to temperature.

Birefringence. Birefringence is given by $\Delta n = n_{\parallel} - n_{\perp} = n_e - n_o$, in which $n_{\parallel} = n_e$ = the extraordinary refractive index determined with polarized light having its electric vector parallel to L, and $n_{\perp} = n_o$ = the ordinary refractive index seen by light having its electric vector perpendicular to L. Unlike $\Delta\epsilon$, Δn is always positive. Except for those few devices involving absorption by a dye dopant, all of the useful optical effects of liquid crystal devices stem from their birefringence. The importance of the magnitude of Δn has not been recognized widely. Increasing the value of Δn makes it possible to produce the same optical effect with a thinner film of liquid crystal. This can greatly increase the speed of a device, since the response time is usually proportional to the square of the thickness.

Conductivity. Thermotropic liquid crystals generally have nonionic chemical structures and have low conductivities ($\sigma < 10^{-11}$ ohm^{-1} cm^{-1}) when they are adequately purified. Such low conductivity materials are usually advantageous in electro-optical devices using strictly ac field effect phenomena. However, other devices depend upon conduction effects and the resulting electrohydrodynamic flow of the liquid crystal. In order to utilize these effects, the liquid crystal must contain conductivity dopants (either as additives or impurities) which will provide the required level of ionic conduction at reasonable voltages. The typical range is about $\sigma = 10^{-8}$ to 10^{-10} ohm^{-1} cm^{-1}. The choice of dopants can affect the conductivity, conductivity anisotropy, electrochemical stability, and alignment properties of the liquid crystal.

Conductivity anisotropy. The conductivity anisotropy can be described by the ratio $\sigma_{11}/\sigma_{\perp}$, where σ_{11} and $\sigma_{\perp}$ are the conductivities parallel and perpendicular, respectively, to the liquid crystal director. In nematics $(\sigma_{11}/\sigma_{\perp}) > 1$ because ions can move more easily parallel to than across the rod-like molecules. This property is partially responsible (9, 10) for some of their unique electrohydrodynamic electro-optical effects, such as Williams domains (11) and dynamic scattering (12). Recently, we have found that the magnitude of the conductivity anisotropy is a property of both the liquid crystal and the dopant structures (Margerum *et al.*, submitted for publication). It is also temperature dependent and the ratio decreases sharply to unity when a liquid crystal is heated to its isotropic state. Ions in smectic liquid crystals can move more easily between molecular layers than through them, so that contrary to the usual nematics $(\sigma_{11}/\sigma_{\perp}) < 1$. Recent studies on cybotactic nematics (13–16), which have short-range smectic order, show unusual conductivity anisotropy effects for nematics because the $\sigma_{11}/\sigma_{\perp}$ ratio increases with temperature over certain ranges (17–19). We expect that this property will be found to have a large effect on the electrohydrodynamic properties (e.g., dynamic scattering) of such cybotactic nematics.

Viscosity coefficients and elastic constants. Due to anisotropic order, nematics are characterized by five independent viscosity coefficients (20–22) and three elastic constants (23). Combinations of these enter the equations for the response times of electro-optic devices. It would be desirable to acquire an understanding of the relationships between these property constants and molecular structure, but at the present time the usual approach in device development is to establish empirical relationships between response times and chemical composition.

Stability. Consideration must be given to the thermal, hydrolytic, oxidative, electrochemical, and photochemical stability of the liquid crystal. Thermal stability is generally not a problem, and using the liquid crystal in a sealed device will prevent hydrolysis and oxidation. Suitable dopants and electrode material may be necessary to prevent undue electrochemical damage. Careful selection of the liquid crystal compounds and the use of light filters may be essential to limit the rate of photolysis.

DOPANT EFFECTS

Conductivity dopants. Electro-optical effects based on the electrohydrodynamic behavior of liquid crystals require the presence of conductivity dopants. In many cases, impurities in the liquid crystal have inadvertently played the role of dopants and it had generally been assumed that the conductivity anisotropy was an inherent property of the liquid crystal. Our own work and other recent studies (24–26) show that substantial differences occur in the conductivity anisotropy $(\sigma_{11}/\sigma_{\perp})$ and in the dynamic scattering threshold voltage of a given liquid crystal when dopants of various structures are used.

The dc-activated conductivity effects and electro-optical effects in liquid crystals are complex because electrochemical reactions occur at the electrodes and these reactions can intro-

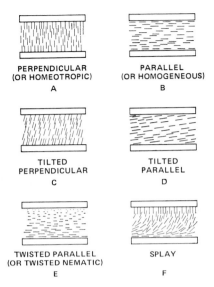

PERPENDICULAR
(OR HOMEOTROPIC)
A

PARALLEL
(OR HOMOGENEOUS)
B

TILTED
PERPENDICULAR
C

TILTED
PARALLEL
D

TWISTED PARALLEL
(OR TWISTED NEMATIC)
E

SPLAY
F

FIG. 1. Nematic liquid crystal surface alignment modes.

duce new charge carriers. Also, if the reactions are not electrochemically reversible, they can cause decomposition of the electrodes, liquid crystal, and dopants. When inert salt-type conductivity dopants are used in dc-activated cells, then the liquid crystal itself undergoes oxidation and reduction reactions which generally lead to rather rapid decomposition effects. Since some devices can only be operated in the dc-mode and others have better dc than ac response characteristics, it is desirable to have dc-activated liquid crystal cells that are stable. We have shown that liquid crystals containing a redox dopant pair not only have a high conductivity anisotropy but also have long lifetimes of dc operation (27).

Surface alignment dopants. Some aligning agents are soluble in liquid crystals and can be introduced as dopants rather than being used to pretreat the cell surfaces. For example, surface-perpendicular alignment can usually be obtained by adding lecithin or hexadecyltrimethylammonium bromide (28) to a liquid crystal and letting it equilibrate with the surface. The latter salt dopant illustrates a problem that can exist when ionic aligning agents

are used, since they will affect both the alignment and the conductivity characteristics.

Pleochroic dyes. Pleochroic dyes (whose absorption spectrum is a function of their molecular orientation with respect to the polarization of incident light) are used as dopants whose absorptivity is dependent upon the alignment direction of the surrounding liquid crystal molecules. Thus a field effect alignment change of a liquid crystal "host" also changes the absorption behavior of its "guest" dye (29–32). The maximum effect is obtained with dyes that have a long cylindrical-like shape which gives them a high order parameter for alignment in the liquid crystal, and that have a strong absorption band along their long molecular axis (31). Such dyes have conjugated double bonds in liquid crystal-like structures, such as azo-type dyes (30, 31) and merocyanic-type dyes (32). Of course, to be useful in guest/host devices, these dyes must also be soluble in the liquid crystal and have good electrochemical, thermal, and photochemical stabilities. Generally, nonionic dye structures are preferred to avoid increasing the conductivity of the liquid crystal with ionic dyes, which also may introduce solubility and electrochemical problems.

SURFACE EFFECTS

Types of alignment. The alignment of the liquid crystal in the off-state of the device is determined solely by the interaction with the confining surfaces. The orientation of the surface layers is propagated through the bulk to minimize the total energy. Alignment is extremely important because it determines the off-state optical properties and the manner in which the molecules will be reoriented by the applied electrical field.

Some of the alignment modes possible for a nematic held between two parallel electrode surfaces are indicated diagrammatically in Fig. 1. The perpendicular or homeotropic alignment, A, and the parallel or homogeneous alignment, B, are in a sense two extreme positions. In practice, it is usually preferable to

| PLANAR OR GRANDJEAN | FINGERPRINT FOCAL CONIC | DISORDERED FOCAL CONIC |

FIG. 2. Cholesteric liquid crystal alignment modes.

have alignments that are uniformly slightly tilted away from these extremes, as shown in diagrams C and D. The off-state tilt determines the direction in which the molecules will tilt further when the field is applied. The twisted nematic mode illustrated in E is a very important case in which the alignment is parallel to both surfaces but with the director L at one surface perpendicular to L at the other. In this mode L describes a quarter-turn helical twist in passing through the cell. Since both right- and left-handed twists are equally probable for a twist of 90°, a twist of slightly less than 90° is usually preferred in order to make one twist direction energetically more favorable than the other. The splay alignment in F demonstrates that alignment can be surface-parallel at one surface and surface-perpendicular at the other.

All of the alignment modes shown in Fig. 1, and other variations as well, can be considered to derive from the existence of two independent variables at each of the two surfaces. These variables are the polar angle θ between the director and the normal to the surface, and the azimuth angle ϕ, which is the angle between an arbitrarily fixed x axis on the surface and the projection of the director onto the surface.

Figure 2 describes the cholesteric alignment modes (or "textures") most important for electro-optic devices. The planar alignment or Grandjean texture is transparent and has the axis of the cholesteric helix oriented perpendicular to the electrode surfaces. The focal conic texture has been applied loosely to include the special case in which the axis of the helix is parallel to surfaces and the more general case in which the axes of numerous helical domains are oriented more or less at random throughout the bulk. The fingerprint focal conic has

periodic variations in refractive index as a result of continuous variations in the ϕ angle of the helix axis. The disordered focal conic texture is highly scattering and may be irridescent if the pitch is appropriate in length.

Only two alignment modes of smectics are of significance to existing devices. One is a transparent homeotropic alignment that has the molecular axes roughly perpendicular to the electrodes and the smectic layers parallel to the surfaces. The other is a highly scattering disordered focal conic texture.

Alignment Techniques. A nematic will spontaneously orient its molecular axis parallel to a high-energy surface, but the angle ϕ will vary across the surface in a manner that bears the imprint of the flow direction when the material was first applied. The uniform ϕ angle implied by Fig. 1B can be achieved most easily by rubbing (33) the surface with a paper tissue or cotton swab, or with an abrasive such as diamond dust (34–37). Grooves are created either in the surface or in an adventitious layer of organic material. Calculations demonstrated that the elastic strain energy in the bulk is minimized by having the director oriented approximately parallel to the grooves (34–36).

A more uniform parallel alignment can be induced by an oblique vapor-deposited film of a material such as silicon monoxide or gold (38). By varying the angle of deposition, one can achieve either (a) a truly parallel alignment at right angles to the deposition direction, or (b) a tilted alignment parallel to the deposition direction, with a tilt of about 20 to 45° out of the surface toward the vapor source (39–41). The tilt can be reduced to 2° by also rubbing along the deposition direction (42). Another method of obtaining uniform parallel

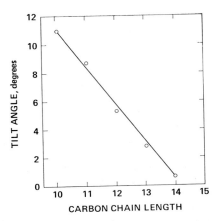

FIG. 3. Control of liquid crystal tilt angle θ (tilt off normal) by the length of alkyl groups bound to an SiO₂ surface.

alignment is to ion beam etch a silica surface at an oblique angle (43). The surface topography after SiO deposition (44) and after ion beam etching differ in some important details, as shown by differences in the angle of tilt, but each appears to produce surfaces that are grooved on a very small scale. A groove-free, chemical method involves the unidirectional deposition of a surfactant by rapidly pulling the substrate through the surface of the surfactant solution in a specified concentration range (45).

The angle θ is also in ill-defined function of the surface energy and the properties of the liquid crystal. All of the methods of obtaining a perpendicular alignment involve decreasing the surface energy of a high-energy surface by coating it with a polymer or a surfactant or by attaching organic groups in some other manner. For example, materials used include hexadecyltrimethylammonium bromide (45), lecithin (37), polyamide resin (46), chemisorbed silicones derived from trifunctional silanes with long alkyl substituents (47, 48), plasmapolymerized hexamethyldisiloxane (49), sputtered polytetrafluoroethylene (50), tetrafunctional carboxylatochromium complexes (51), and even the condensed vapors from the pyrolysis of paper (52).

It has been postulated (37, 53) that when the critical surface tension of the surface exceeds the surface tension of the liquid crystal, the alignment is parallel, and when the reverse is true, the alignment is perpendicular. Several investigators (50, 54) have found pronounced exceptions to this rule, however. An experimentally more difficult approach to predicting alignment mode involves calculations of the polar and dispersion contributions to the free energy of adhesion (55). Neither theory provides for tilted perpendicular alignment, which has been noted or implied by the reports of several groups (40, 43, 52, 54). We have found that when alkyl groups are chemically attached to an ion beam etched SiO₂ surface the angle θ increases as the chain length decreases (see Fig. 3). This provides a simple means for adjusting the tilt to the angle desired (43).

Alignment methods that induce parallel alignment in nematics tend to induce planar alignment in cholesterics. Methods that promote perpendicular alignment in nematics tend to align cholesterics in the fingerprint texture and produce homeotropic alignment in smectics.

Cell Thickness Effects. The cell thickness, e.g., the thickness of the liquid crystal layer between the parallel surfaces of conductive glass electrodes, has major effects on the electro-optical properties. The periodicity of conductive flow patterns and the phase retardation due to birefringence are each proportional to the cell thickness. In the latter case thickness variations within a cell will cause differences in the color transmission characteristics. The maximum spatial resolution of electro-optical effects in various devices is generally no better than the cell thickness. However, the most important influence is on the rise and decay times of electro-optical effects, which are approximately proportional to the square of the cell thickness. Thus a 2-μm cell will operate about 40 times as fast as the conventional 12.7-μm ($\frac{1}{2}$ mil) cell. While thin cells can be used to obtain faster response times, the uniformity of spacing becomes increasingly important in thin cells. For example, a thickness difference of only

500 nm (one wavelength of visible light) gives a 25% thickness difference in a 2-μm cell, which would cause about a 50% difference in its response times. In the 2-μm-thick light valves, 0.5-in.-thick polished optical flats are used to obtain thickness uniformity over a 1 in. × 1 in. aperture (43, 56). In large area flat panel displays the uniformity of even 12- to 25-μm thicknesses is difficult to achieve with just perimeter-type spacers. Special designs or techniques are often used in large area cells including the addition of tiny glass beads of the appropriate mesh size to the liquid crystal.

CATEGORIES OF ELECTRO-OPTICAL DEVICE EFFECTS

Liquid crystal electro-optic devices have been or are being developed from such a rich variety of physical structures, addressing modes, material compositions, and optical characteristics that any categorization is somewhat arbitrary. We choose to discuss them in terms of the type of liquid crystal employed (i.e., nematic, cholesteric, or smectic), the methods by which information is recorded in the liquid crystal, and the type of optical change involved.

Nematics: conductivity effects. When an electrical field is applied to a liquid crystal, the material is reoriented in space by both the field and the ionic conduction effects. If the conductivity of the liquid crystal is sufficient and the electrical frequency is below the cut-off frequency (f_c), conduction effects will predominate, causing characteristic patterns of motion due to electrohydrodynamic instabilities. These flow patterns are affected markedly by alignment mode and applied voltage. The general appearance is very similar for both dc and low-frequency ac, although there are differences in the mechanistic origin of the effects (57). Nematics with negative or only slightly positive dielectric anisotropies are employed in these displays, although optical effects also occur in more strongly positive dielectric anisotropy materials.

With nematics having a homogeneous or twisted nematic alignment, the first optical effect to appear as the voltage is increased is *diffraction* due to the generation of stripes by cylindrical rotation of the liquid crystal (11, 25, 58–62). The stripes, which are commonly known as Williams domains, derive from the periodic variation in refractive index due to vortical flow. To a first approximation the spacing is controlled by the thickness of the liquid crystal layer. Williams domains form a diffraction grating that may be utilized for nearly real-time recording of phase-type holograms (63).

At higher voltages in these cells, and almost immediately at low voltages in homeotropically aligned nematics, the material undergoes increasingly complex patterns of motion that cause *light scattering* to be the most striking optical effect. This phenomenon has been termed dynamic scattering (12, 64–66) and has been widely used in displays in which the image can either be viewed directly or projected in both the reflective or transmissive modes. The transition from diffraction to scattering is indistinct and gradual (61, 67–72), and the optical effects during this transition may be enhanced or modified by the use of a single polarizer (60). By placing the dynamic scattering layer between crossed polarizers, one may also exploit the *depolarization* of polarized light which is an additional consequence of the scattering state (73). Dynamic scattering ceases spontaneously in nematics and the material returns to its original alignment after the field is removed.

The behavior of nematics with $\Delta\epsilon < 0$ changes drastically above the cut-off frequency f_c for the dynamic scattering mode. Above f_c the space charge in the liquid crystal can no longer follow the oscillations of the fields (74). Theoretically, f_c is a complex function of the conductivities and the dielectric constants both parallel and perpendicular to the major axis. In the dielectric regime above f_c flow patterns occur and construct domains that may be related to conductivity effects. Typically these assume the appearance of chevrons at high voltages and have a spatial frequency several times that of the corresponding Williams domains (75–77). They will follow an ac field with response times of <5 msec (78). Surprisingly, if a high-purity, high-resistivity liquid crystal is used and if the layer thickness is 6 μm or less, the same type of domains can be established with low-frequency and even dc fields (79). Unlike the Williams domains, the periodicity is controllable by means of the voltage and the frequency (79–81). This provides the properties of an electrically controlled *variable diffraction* grating. This effect could find application in color displays, light deflectors, and controllable optical filters (79). It has been used in conjunction with a photoconductor layer as an image amplifier (63, 82).

Nematics: field effects. In the absence of significant conduction effects, the nematic usually is simply reoriented by the field. Positive dielectric anisotropy nematics align parallel and negative dielectric anisotropy materials align perpendicular to the field. Off-state alignment is exceedingly important in such devices, since it fixes a limit for the spatial reorientation by the field and determines the uniformity of response. Polarizers are usually required to render the field-induced realignment visible or to achieve maximum contrast. Nematic field effect displays often use low voltage and low current, often requiring one or more orders of magnitude less power than dynamic scattering devices. This provides a strong incentive for employing field effect devices in small battery-operated displays.

The effect that is commericially most significant at present is the *rotation of polarized light* by a twisted nematic alignment and the disappearance of this effect when a field is applied (83). A positive dielectric anisotropy is essential and very strongly positive values are preferred, since the threshold voltage is inversely proportional to the square of this property (83, 84). Both polarizer and analyzer are necessary. To achieve maximum contrast, the plane of polarization must be either parallel or perpendicular to the director at the point where the light enters the liquid crystal, and

the analyzer must be positioned to either transmit or reject the rotated plane at the exit. Both white-on-black and black-on-white displays are possible, and the information may be viewed or projected in either the reflective or the transmissive mode.

A change in the *phase retardatation of polarized light* may be induced by reorienting the nematic in the field to alter the effective birefringence of the liquid crystal layer (85–89). If the material has a positive dielectric anisotropy, the alignment should be essentially surface-parallel or tilted-parallel. With a parallel alignment the maximum change occurs when the polarizer and analyzer are crossed and the plane of polarization of the entering light beam forms an angle of 45° with the director. Colored light is transmitted in both the off and on state (except at extremely high field strengths), and the color changes in response to variations in voltage. If the nematic has a negative anisotropy, the alignment should be perpendicular or tilted-perpendicular. When viewed with crossed polarizers, the off-state black proceeds through shades of gray to white, yellow, orange, magenta, blue, green, pink, etc., as the voltage is increased above the threshold voltage.

The optical data processing light valve (56, 90) operates in a hybrid field effect mode that incorporates both the rotation of polarization and the phase retardation effects. The combination of effects is due to a 45° twisted parallel alignment.

A change in *refractive index* can be utilized to produce a display if it is accompanied by a change from a transmitting state to one of total internal reflection (91, 92). The device has been reported to be capable of very rapid response, with rise times on the order of tens of microseconds, in contrast to the milliseconds to seconds required for most of the other liquid crystal displays (92). Refractive index changes have also been studied for use in beam deflectors (93) and waveguide switches (94, 95).

The reorientation by the field may be used to bring about a *change in light absorption* by incorporating a pleochroic dye as a "guest" species in the nematic "host" (29–32). Both surface-parallel and surface-perpendicular alignments can be used, and polarizers are not necessarily required, particularly in cells containing chiral additives (31).

Erasure is always spontaneous in the nematic field effect devices. A noteworthy variation, however, employs a nematic with a frequency-dependent dielectric anisotropy and a two-frequency address system. One frequency records the information and the other accelerates erasure (96–98).

The optical effect of these devices can be altered by incorporating additional passive optical elements in the display system. For example, the twisted nematic device can be used with birefringent films to provide a two-color birefringent display (99, 100). Two or more such devices can be cascaded to produce additional colors. Similarly, a twisted nematic display can be used with a film containing a pleochroic dye to form a two-color display (99) or it can be combined with a quarter-wave plate and a cholesteric film to achieve the asethetically pleasing reflective colors of the cholesteric (101).

Cholesterics: conductivity effects. Conductivity effects in cholesterics can produce diffraction and dynamic scattering effects (102, 103) somewhat similar to those of nematics, but these effects are secondary in importance to the *stored light scattering* images that are formed (104–108). Typically the liquid crystal is a mixture of a cholesteric and a nematic and has a pitch length of 0.5 to 10 μm. The dielectric anisotropy must be negative. The molecules are aligned parallel to both electrode surfaces to provide a transparent planar texture in the off-state. When a low-frequency field ($< f_c$, or ~100 Hz) is applied with voltages just above threshold, conduction effects generate characteristic square array patterns of motion capable of diffracting light. At higher voltages the liquid crystal proceeds through more complex patterns to a turbulent scattering state (103). Removal of the field leaves the liquid crystal in the semistable, light-scattering focal conic texture. If reconversion

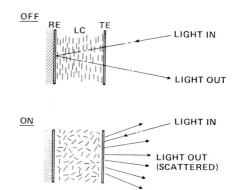

FIG. 4. Diagram of a dynamic scattering cell (RE = reflective electrode, TE = transparent electrode, LC = liquid crystal).

to the planar texture is allowed to take place spontaneously, the required time is exponentially dependent on the ratio of the thickness to the pitch (109), and in some cases it can be measured in years. The images can be erased by applying an audio frequency ac field that returns the molecules to the planar texture by means of a field effect. The cut-off frequency ($f_{c'}$) that marks the division between writing and erasing is a function of the conductivity of the material (106).

Cholesterics: field effects. Most important of the field effects of cholesterics is the *elimination of light scattering* that accompanies the transition from a cholesteric to a field-aligned nematic structure (110–117). A positive dielectric anisotropy is necessary for the effect (115). The cholesteric may originally be aligned either in the planar or the focal conic texture, since the field initially will convert a planar alignment to a metastable focal conic texture with the helical axes roughly parallel to the electrodes. Upon increasing the field strength, the molecules are pulled out of the helical twist to create the equivalent of a nematic in the transparent homeotropic orientation. After removing the field, the liquid crystal reforms the cholesteric helix in the light-scattering focal conic texture. Generally, a mixture of cholesterics or cholesterics and

DISPLAY ELEMENT CIRCUIT

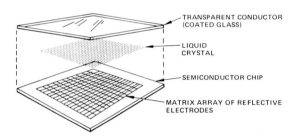

FIG. 5. Module construction of flat panel TV matrix display.

nematics is formulated to obtain a long pitch length (p) and a large positive dielectric anisotropy, since the critical field strength is inversely proportional to p and $\Delta\epsilon^{\frac{1}{2}}$ (115–116). Short pitch lengths may be desired, on the other hand, to decrease the decay time (115) or to impart irridescence to the scattering state (114). The dynamics of relaxation to the cholesteric phase are also dependent on electrode surface conditions and may be nucleated by surface inhomogeneities (118, 119). Rise times in the range of tens of microseconds can be realized at high field strengths (115), but decay times are typically tens of milliseconds or longer.

If the cholesteric has a pitch length suitable for the Bragg reflection of visible light (120), there will be a *change in the wavelength of reflectively scattered light* as the cholesteric helix is rotated and the pitch is dilated by the field. The phenomenon is not well understood, and both blue and red shifts have been reported. A

shift from blue to red with increasing field strengths was said to take place reversibly with response times of 1 to 2 msec (121), and a red to blue shift occurring at 20 cycles/sec forms the basis for a patented display (122).

Cholesterics and smectics: thermal address and field effect erase. Several rather sophisticated light valve projection displays (123–127) have been developed that use an intensity-modulated infrared laser beam to record and store information in a liquid crystal film and an electrical field to erase. Both cholesteric and smectic liquid crystals can be used to obtain the *stored light scattering* effect. The cholesteric has a negative dielectric anisotropy and is originally aligned planar, whereas the smectic has a positive dielectric anisotropy and is aligned homeotropically in the transparent film. When the light valve is addressed by the laser, absorption of the IR beam raises the temperature locally above the clearpoint of the liquid crystal. After the beam has passed on,

602

the isotropic liquid quickly cools to the meso-phase but assumes the highly scattering focal conic texture. Continuous gray scale is achieved in a smectic by controlling the voltage at the moment when the material cools to the meso-phase (127). The information can be erased in cholesterics by means of an audio frequency field (123), or in smectics by either the field (127) or a combination of a field and the thermal effect of the same scanning laser beam (124). The latter method makes selective erasure possible.

DEVICE APPLICATIONS

There are a great variety of liquid crystal devices, which have been derived from the various electro-optical effects combined with specific methods of addressing and reading out these effects. Several major categories of electrical addressing techniques can be distinguished. *Direct address of electrode segments* is the simplest, and it is widely used in digital displays. *Multiplexing methods* reduce the number of electrical leads to the segments in digital and bar-graph displays. Various techniques for the *direct matrix address* (128–136) of liquid crystals are employed for alpha-numeric, graphic, and pictorial displays. *Matrix addresses of storage elements* (137–144) is used for liquid crystal pictorial displays with the fast response required for a large array of elements. *Photoactivation of a photoconductor film* (43, 56, 63, 90, 145–152) also serves indirectly as a storage element address technique for high resolution, fast response liquid crystal pictorial displays. We have selected three of the interesting nematic display devices with which we are familiar for further discussion. Each of them utilizes distinctly different electro-optical effects and addressing techniques.

Dynamic scattering/TV matrix display. The dynamic scattering mode has been used commercially for segment displays, such as in digital watches and electronic calculators as well as in large signs and advertisement dis-

Fig. 6. Liquid crystal TV matrix display cell mounted on a glass plate.

ONE INCH

FIG. 7. Photograph of a flat panel TV matrix cell in operation.

plays. A schematic diagram of a reflective-type display cell is shown in Fig. 4. In the off-state the liquid crystal is uniformly aligned (shown here as surface-parallel) and is transparent. In the on-state the current causes turbulent motion of the liquid crystal which results in light scattering—primarily small angle scattering in the direction of light propagation rather than backscattering. In this type of cell the light is scattered by the liquid crystal, reflected off the rear electrode, and scattered more in passing back through the liquid crystal. With the proper light-trap arrangement for the off-

state, a direct view shows a dark off-state and a white on-state. The threshold voltage (V_{th}) is approximately independent of the cell thickness and is determined by the applied voltage rather than the applied field. Dynamic scattering is a conduction-induced effect in which V_{th} is highly dependent upon the conductivity anisotropy of the liquid crystal–dopant system as well as its resistivity and the frequency of the applied signal. Typically, nematics of negative dielectric anisotropy are used with dopants added to give resistivities in the range of 10^8 to 10^{10} ohm cm. Threshold voltages are in the range of about 1 to 10 V dc and 5 to 15 V ac (at low frequencies), while the maximum scattering levels require about 15 V for the lowest threshold and 40 V or more for the higher thresholds. The scattering versus voltage curves are not sharp, and it is difficult to obtain full contrast without crosstalk in a simple $X-Y$ addressed matrix, although large matrix displays (136) with relatively slow up-date times have been made with more complex multiplexing schemes. A superimposed higher frequency (kHz range) ac voltage increases V_{th} for the dc or low-frequency ac signal (129, 130) and this has also been used in a two-frequency addressing scheme to suppress crosstalk in a direct address $X-Y$ matrix display without gray scale. The signal on-time (delay time plus rise time) for dynamic scattering is typically in the range of 10 to 100 msec, al-though shorter on-times are possible because they are proportional to the square of thickness/voltage. However, using practical thickness and voltage levels, a large number of matrix display elements (or rows) cannot be addressed serially within the short time period of a TV frame rate.

A TV rate liquid crystal display system has been made (142–144) by using a circuit (137) on a semiconductor substrate to address in-dividually a great many (10,000/in.²) display elements every 1/30 sec. The single-element display circuit is shown at the top of Fig. 5. Associated with each picture element is a transistor switch (MOSFET) and a capacitor which can be charged to various levels in a microsecond by an incoming video signal pulse. The capacitor holds the dc signal on the liquid crystal for the milliseconds required to acti-vate dynamic scattering. This pulse-stretching circuit makes it possible to build a real-time display using the comparatively slow response of liquid crystals because the overall light scattering picture from many elements is de-veloped in nearly a parallel mode. The 1-in. square display modules have 10,000 individual reflective metal electrodes (100 rows by 100 columns), each of which covers its own circuit element on a common silicon chip, as indicated

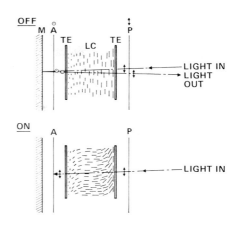

FIG. 8. Diagram of twisted nematic cell (P = polari-zer, A = analyzer, M = mirror, TE = transparent electrode, LC = liquid crystal).

in Fig. 5. The module is mounted on a glass plate and is used as the activating substrate for a reflective liquid crystal cell. A picture of such a mounted module is shown in Fig. 6 and a picture of an actual TV dynamic scattering display is shown in Fig. 7. The driving signal was taken from a TV camera operated through a signal converter. The drive circuitry de-velops a 20-V (maximum) video signal for each column and a 28-V sweep signal pulse for each row of the display. The power con-sumption of the display itself is only about 0.5 mW, although of course outside illumina-tion (from sunlight or a lamp) is required.

Off-the-air TV is readily viewable in direct sunlight with this small flat panel display, al-though there is a slight smear of the fastest image motions. At present it shows a contrast ratio of 15:1, with six to eight shades of gray, a scattering rise time of 100 msec and a decay time of 60 msec. Four of these modules have been combined to give a 2 in. $\times$ 2 in. display with 40,000 elements (144), and studies are in progress regarding the combination of larger numbers of such modules to give displays with greater than 500×500 elements. These dis-plays are particularly of interest for the presentation of pictorial, graphic and symbolic information in airplane cockpits subjected to

FIG. 9. Picture of a lab-model liquid crystal watch.

high ambient light conditions. The liquid crystal materials for this display must be stable over a wide temperature and need to be optimized for dc operation, particularly for stable resistivity levels over long lifetimes. Recent studies (26) indicate that long operational lifetimes for dc dynamic scattering (>18,000 hr) are feasible with redox-doped phenyl benzoate liquid crystals. The practicality of these displays is also dependent upon development of economical techniques for manufacturing defect-free semiconductor substrate modules in reasonable yield.

Twisted nematic/watch display. The largest market for liquid crystal electro-optical devices has been as segmented digital displays for wrist watches. The low voltage and low power requirements of a twisted nematic field effect (83) display permit continuous operation for several years from a small battery. The basic operation of a twisted nematic cell with crossed polarizers is illustrated in Fig. 8. In the off-state the twisted surface-parallel alignment of the liquid crystal cell causes it to act as a polarizing element that gives a 90° twist to the incident plane-polarized light, thus causing light to be transmitted by the analyzer behind the cell. The light is then reflected off the back "mirror," is twisted back 90° by the liquid crystal, and passes out to the viewer as a bright background. In the on-state the field applied to the positive dielectric liquid crystal causes it to align perpendicular to the surfaces, thus removing the polarization rotation effect. In the on-state the incident plane-polarized light passes through the liquid crystal without modification and is absorbed by the analyzer so that the viewer sees a dark background.

606

The displays use biphenyl, phenyl benzoate, or Schiff base liquid crystals with *para*-cyano end groups, giving highly positive dielectric anisotropies (153–155). Their twisted nematic cells have threshold voltages as low as the 1-V range, and nearly complete realignment (untwisting) of the liquid crystal occurs with 2 to 3 V. Since it is a field effect, ac signals (typically in the 30 to 1 kHz range) and liquid crystals with high resistivities ($>10^{10}$ ohm cm) are used to give long operational lifetimes. In a typical 13 μm thick cell operated at 3 V rms the turn-on times are about 100 to 200 msec and the decay times are 200 to 400 msec. The optimum contrast ratio is high ($>50:1$), but the contrast is much lower at wide viewing angles, particularly when viewing from the direction of the surface tilt angle.

A picture of a twisted nematic watch is shown in Fig. 9. It continuously displays the hours and minutes, and the seconds are indicated with a flashing colon. It also has push-button controls to read the date, read the seconds, or illuminate the hour/minute display at night.

Tunable birefringence/projection light valves. The large anisotropy of refractive index in

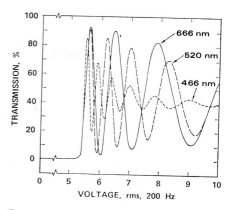

Fig. 11. Transmission characteristics of a tunable birefringence cell between crossed polarizers (Schiff base mixture, surface-perpendicular alignment).

nematic liquid crystals can be applied directly to various field effect display devices. The diagram in Fig. 10 illustrates a tunable birefringence display cell in which a nematic of negative dielectric anisotropy is aligned with its director essentially perpendicular to the electrode surfaces. In the off-state plane-polarized incident light travels down the optic axis of the liquid crystal, is reflected back and is absorbed by a crossed polarizer (analyzer). When a field above a critical threshold voltage (typically 3 to 10 V) is applied, the liquid crystal begins to turn away from the direction of the applied field, and then it no longer appears isotropic to the normal incident light. The liquid crystal film shows a birefringence effect that increases with the applied voltage. This introduces a large phase retardation so that a given wavelength of light passes through the analyzer in a series of maxima and minima as the voltage is increased. This is shown in Fig. 11 for the transmission of three different wavelengths in a test cell containing a Schiff base nematic mixture of N-(p-methoxybenzylidene)-p-butylaniline (MBBA) and N-(p-ethoxybenzylidene)-p-butylaniline (EBBA). The different colors are not separated in their first transmission peak above threshold, but become better separated at higher voltages. As a result, with incident white light the viewer

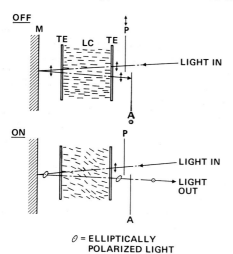

Fig. 10. Diagram of a tunable birefringence field effect cell (P = polarizer, A = analyzer, TE = transparent electrode, M = mirror, LC = liquid crystal).

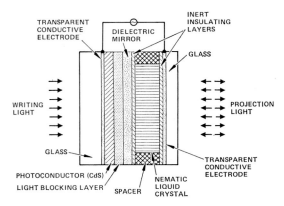

FIG. 12. Diagram of ac photoactivated liquid crystal light valve.

sees no light below V_{th}, then gray, white, and a sequential series of Newton colors appear as the cell voltage increases. The optimum color and transmission effects are obtained when the liquid crystal director is realigned uniformly by the field with its ϕ angle at 45° with respect to the incident polarized light. This can be accomplished by introducing a slight pretilt of the liquid crystal in that direction in the off-state.

Several different displays have utilized this effect, and we are working on its applications with photoactivated ac light valves (43, 151). This type of liquid crystal light valve was first described (150) as an improvement over dc light valves (63, 145, 146) for use with dynamic scattering liquid crystals. Its general construction is shown in Fig. 12. The writing light activates a CdS photoconductor, which in turn switches an ac signal across the liquid crystal. A dielectric mirror and a light blocking layer of CdTe protect the photoconductor from an intense projection light readout, which passes twice through the liquid crystal. The projection system is diagrammed in Fig. 13. The ac light valve is filled with a negative $\Delta\epsilon$ liquid crystal with a tilted-perpendicular alignment. It can be used with a 4- to 12-μm thickness of liquid crystal as a color symbology projection display (151) or with about a 2-μm thickness as a black and white projection television display (43). In both cases the writing light source is a small cathode ray tube (CRT) coated with a P-1 phosphor. It has a peak output at 525 nm, which matches the maximum sensitivity of the CdS photocon-

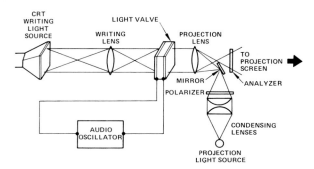

FIG. 13. Diagram of optical system for color symbology or TV liquid crystal light valves.

ductor, and the phosphor has a relatively slow decay time (24 msec to 10%), which provides a signal period of suitable length for activation of the CdS and the liquid crystal. In the color symbology cell the CdS photoconductivity response has about a 20 msec rise time and an 80 msec decay time. Thus the signal switched onto the liquid crystal by the CdS is greatly expanded in time compared to the rapidly scanned electron beam in the CRT—allowing an essentially parallel development of the liquid crystal response over the entire cell. Typically, a 4-μm thickness of the Schiff base mixture MBBA/EBBA is operated with about a 10-V ac drive frequency (e.g., 140 kHz) across the light valve. The colors that are obtained in the projection display depend upon the intensity of the CRT input light. This is indicated in Table I, which also shows some typical response times. There is an intensity-dependent gray scale range (about six shades of gray) for the black and white mode, which shows a resolution of >40 lines/mm in the light valve (corresponding to more than 500 ×500 separate resolution elements projected from a one inch square device). The mirror and light blocking layer adequately protect the photoconductor so that an intense projection light can be used. Images containing more than 350 lm have been projected from the light valves to give 40 ft-L of brightness on a large screen.

Figure 14 shows a photograph taken from the light valve projection onto a 5 × 5 ft screen. In this case a variable intensity input optical signal was taken from a lamp passing through a slide negative combined with pieces of neutral density filters. The lowest intensity shows a black and white region with gray scale, while higher intensities give yellow, magenta, and blue colors. This illustrates the capabilities of these light valves, which are designed to convert signals from a small (1″ or 2″ diameter) computer-driven CRT into large screen color symbology display with real-time update capabilities. Problem areas being studied include projection techniques for defect-free substrates, higher birefringent liquid crystal

TABLE I

Light Valve Performance (151)

Light valve[a] output	Input intensity (μW/cm²)	Turn-on time[b] (msec)	Turn-off time[c] (msec)
Threshold	3	—	—
White[d]	12	400	400
Magenta	36	400	400
Blue	64	300	400
Green	90	250	350

[a] Four-micrometer thickness of MBBA/EBBA with 1° pretilt.
[b] Activation time from black to stable color.
[c] Decay time from color to black.
[d] White on black contrast ratio of 80:1.

layers for thin cells with improved response times and better resolution in the color region, and long operational lifetimes with intense projection light.

SUMMARY AND OUTLOOK

A great variety of interesting electro-optical devices result from the proper combinations of surface treatments, liquid crystal materials, addressing signals, and optical components. Liquid crystals are finding wide acceptance in digital watches, in which the display is on continuously and can be read in bright sunlight. New digital color displays may be made feasible by utilizing improved guest/host dyes (31). Advances in the control of surface effects, the development of liquid crystal materials tailored to specific device requirements, and improved addressing techniques are expected to result in the increased use of electro-optical devices. Multiplexing techniques are being applied to bar-graph (156) and alphanumeric (131) displays for use in instrument panels, portable message boards, etc. Similarly, multiplex address of liquid crystal matrix systems is being developed for large area flat panel displays (136) and for real-time page-composer devices (157). Semiconductor matrix addressing circuits can be used to provide small flat panel, TV-rate, pictorial displays (143, 144). Such portable panels will have widespread military and commerical applica-

Fig. 14. B/W photo of a 5 X 5 ft screen filled with the projected image from a color symbology valve.

tions when economical manufacturing techniques are perfected. Photoactivated light valves, in which the liquid crystal acts as a high resolution medium for temporary image-storage, are expected to have widespread applications for large screen TV projection displays (43, 151) and real-time optical data processing systems (56, 90).

ACKNOWLEDGMENTS

The authors are indebted to the Directorate of Chemical Sciences, Air Force Office of Scientific Research, Contract F44620-72-C-0075 for partial financial support of our research studies. We are also indebted to our many colleagues in the Hughes Aircraft Co. who have carried out advanced work on liquid crystal applications in watches, matrix displays, reticles, and light valves.

REFERENCES

1. MEIER, G., SACKMANN, E., AND GRABMAIER, J. G., "Applications of Liquid Crystals," Springer–Verlag, Berlin, 1975.
2. BLINOV, L. M., Sov. Phys. Usp. 17, 658 (1975).
3. KALLARD, T., "Liquid Crystals and Their Applications," Optosonic Press, New York, 1970.
4. KALLARD, T., "Liquid Crystal Devices," Optosonic Press, New York, 1973.
5. TOBIAS, M., "International Handbook of Liquid Crystal Displays 1975–1976," Ovum, London, 1975.
6. SCHEFFER, T. J., AND GRULER, H., in "Molecular Electro-Optics" (C. T. O'Konski, Ed.), Dekker, New York, in press.
7. ADAMS, J., AND HAAS, W., Mol. Cryst. Liq. Cryst. 30, 1 (1975).
8. SAUPE, A., AND MAIER, W., Z. Naturforsch. 16a, 816 (1961).
9. CARR, E. F., Mol. Cryst. Liq. Cryst. 7, 253 (1969).
10. HELFRICH, W., J. Chem. Phys. 51, 4092 (1969).
11. WILLIAMS, R., J. Chem. Phys. 39, 384 (1963).
12. HEILMEIER, G. H., ZANONI, L. A., AND BARTON, L. A., Appl. Phys. Lett. 13, 46 (1968); Proc. IEEE 56, 1162 (1968).
13. DE VRIES, A., Mol. Cryst. Liq. Cryst. 10, 219 (1970).
14. CHISTYAKOV, I. G., AND CHAIKOWSKY, W. M., Mol. Cryst. Liq. Cryst. 7, 269 (1969); and in "Liquid Crystals" (G. H. Brown, Ed.), Part II, p. 803. Gordon and Breach, London, 1969.
15. CHISTYAKOV, I. G., AND CHAIKOWSKY, W. M., Sov. Phys. Crystallogr. 18, 181 (1973).
16. DE VRIES, A., J. Phys. (Paris) Colloq. C1 36, 1 (1975).
17. RONDELEZ, F., Solid State Commun. 11, 1675 (1972).
18. MIRCEA-ROUSSEL, A., LÉGER, L., RONDELEZ, F., AND DEJEU, W. H., J. Phys. (Paris) Colloq. C1 36, 93 (1975).
19. HEPPKE, G., AND SCHNEIDER, F., Z. Naturforsch. 309, 316 (1975).
20. ERICKSEN, J. L., Arch. Ration. Mech. Anal. 23, 266 (1966).
21. LESLIE, F. M., Quart. J. Mech. Appl. Math. 19, 357 (1966).
22. PARODI, O., J. Phys. (Paris) 31, 581 (1970).
23. FRANK, F. C., Discuss. Faraday Soc. 25, 19 (1958).
24. CHANG, R., in "Liquid Crystals and Ordered Fluids" (J. F. Johnson and R. S. Porter, Eds.), Vol. 2, p. 367. Plenum, New York, 1974.
25. BARNIK, M. I., BLINOV, L. M., GREBENKIN, M. F., PIKIN, S. A., AND CHIGRINOV, V. G., Phys. Lett. A. 51, 175 (1975).
26. LIM, H. S., AND MARGERUM, J. D., Appl. Phys. Lett. 28, 478 (1976).
27. LIM, H. S., AND MARGERUM, J. D., J. Electrochem. Soc., 123, 837 (1976).
28. HALLER, I., AND HUGGIN, H. A., U.S. Pat. 3,656,834, April 18, 1972.
29. HEILMEIER, G. H., AND ZANONI, L. A., Appl. Phys. Lett. 13, 91 (1968).
30. HEILMEIER, G. H., COSTELLANO, J. A., AND ZANONI, L. A., Mol. Cryst. Liq. Cryst. 8, 293 (1969).
31. WHITE, D. N., AND TAYLOR, G. N., J. Appl. Phys. 45, 4718 (1974).
32. MORITA, M., IMAMURA, S., AND YATABE, K., Japan. J. Appl. Phys. 14, 315 (1975).
33. CHATELAIN, P., Bull. Soc. Fr. Mineral. Cristallogr. 66, 105 (1943).
34. BERREMAN, D. W., Phys. Rev. Lett. 28, 1683 (1972).
35. WOLFF, U., GREUBEL, W., AND KRÜGER, H., Mol. Cryst. Liq. Cryst. 23, 187 (1973).
36. BERREMAN, D. W., Mol. Cryst. Liq. Cryst. 23, 215 (1973).
37. CREAGH, L. T., AND KMETZ, A. R., Mol. Cryst. Liq. Cryst. 24, 59 (1973).
38. JANNING, J. L., Appl. Phys. Lett. 21, 173 (1972).
39. GUYON, E., PIERANSKI, P., AND BOIX, M., Lett. Appl. Eng. Sci. 1, 19 (1973).
40. URBACH, W., BOIX, M., AND GUYON, E., Appl. Phys. Lett. 25, 479 (1974).
41. MEYERHOFER, D., Phys. Lett. A 51, 407 (1975).
42. RAYNES, E. P., Electron. Lett. 10, 141 (1974).
43. JACOBSON, A. D., GRINBERG, J., BLEHA, W. P., MILLER, L. J., FRAAS, L. M., AND BOSWELL, D. D., SID Int. Symp. Dig. 6, 26 (1975).

W. E., Nelson, K. F., Stephany, J. F., and Tuihasi, S., SID International Symposium, Beverly Hills, California, May 1976.

153. Gray, G. W., Harrison, K. J., Nash, J. A., Constant, J., Hulme, D. S., Kirton, J., and Raynes, E. P., *in* "Liquid Crystals and Ordered Fluids" (J. F. Johnson and R. S. Porter, Eds.), Vol. 2, p. 617. Plenum, New York, 1974.

154. Boller, A., Scherrer, H., Schadt, M., and Wild, P., *Proc. IEEE* **60**, 1002 (1972).

155. Alder, C. J., and Raynes, E. P., *J. Phys. D: Appl. Phys.* **6**, 46 (1973).

156. Carl, W. L., and Stein, C. R., *SID Int. Symp. Dig.*, Beverly Hills, California, 40, (May 1976).

157. Labrunie, G., Robert, J., and Borel, J., *Rev. Phys. Appl.* **10**, 143 (1975).

Liquid Crystals in Living and Dying Systems[1]

DONALD M. SMALL

Boston University Medical School, Boston, Massachusetts 02118

Received September 20, 1976; accepted October 14, 1976

I. INTRODUCTION

Liquid crystals may be defined as states of matter having characteristics of both liquids and crystalline solids, in that they have some degree of order but also some degree of fluidity. Interest in liquid crystals has expanded greatly in industry, particularly in the field of display systems and temperature sensors. However, little attention has been paid to the liquid crystalline nature of certain arrays of molecules in biological systems and even less to the accumulation of liquid crystalline deposits in human disease states. In healthy living systems, a liquid crystalline order of molecules occurs in membraneous structures such as myelin, the rods and cones of the eye, chloroplasts, certain other cellular organelles, and serum lipoproteins. While this paper will be directed particularly toward the liquid crystal systems primarily involving lipids, I should point out that there are a large number of protein systems which might be classified as liquid crystalline, for instance the major components of muscle are arranged in ordered liquid crystalline-like arrays which allow movement to occur when the appropriate energy is supplied.

I will first review the physical properties of lipids of biological importance, the types of liquid crystals which they form, and the types of interactions between various classes of lipids. I will then progress to a discussion of liquid crystals in living systems concentrating specifically on lipoproteins and upon the plasma membrane as a prototype of liquid crystals in

[1] Supported by HL 18623 and GM 00176 from the National Institutes of Health.

all living systems. Finally, I will show that liquid crystals often accumulate in specific human diseases.

In such diseases, the delicate metabolic balance of a certain lipid species is disturbed, often by the lack of a specific catabolic enzyme, leading to the accumulation of a particular lipid species and disruption of cellular function by the accumulation. Certain lipid species while occurring in trace amounts in normal tissue, become quite toxic when accumulated in larger amounts. However, other more innocuous lipid species cause problems only because of mass effects which compromise the cell and the organ in which the lipids accumulate.

II. GENERAL CONSIDERATIONS—LIQUID CRYSTALLINE STATES

A. THERMOTROPIC MESOMORPHISM

When a given molecule is heated it may not melt directly into an isotropic liquid, but instead passes through intermediate states, called mesophases or liquid crystals, which are characterized by less order than the crystal, but more order than the liquid. Such molecules undergo thermotropic mesomorphism. The chemical characteristics necessary to produce liquid crystalline states have been reviewed by Gray and Winsor, and by Brown and others (1–4). The physical properties of liquid crystals have been reviewed by de Gennes (5) and McMillan (6). In general, the molecules are somewhat longer than they are wide and have a polar or aromatic part somewhere along the length of the molecule. The shape and polar–polar or aromatic interaction permits the molecules to

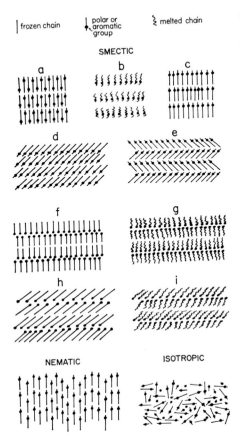

| frozen chain | polar or aromatic group | melted chain |

SMECTIC

a b c

d e

f g

h i

NEMATIC **ISOTROPIC**

FIG. 1. Schematic representation of a few of the many thermotropic liquid crystalline phases. Straight line indicates an immobile, stiff hydrocarbon aliphatic chain and a wiggly line indicates a melted chain. Polar or aromatic groups may be at the end of the molecule (f–i) or more centrally located (a–e). Above, smectic or layered phases with chains normal to layer plane (a, b, c, f, g) or tilted (d, e, h, i). Note that the letters a–i are not related to the common smectic A–H nomenclature (2, 4). Structures f–i are bilayered structures. Structure g would be equivalent to L_α and h to L_β' of Luzatti (7).

as well as those with one-dimensional order (7–11). These structures are characteristic of molecules with a polar group at one end of the molecule. Liquid crystals with long-range order in the direction of the long axis of the molecule are called smectic, layered, or lamellar liquid crystals. These states may be formed by molecules above the melting point of the crystals (stable states) or on undercooling (metastable states). If the molecule contains a long aliphatic chain, the transition from crystal to the smectic state is often related to the partial or total melting of the aliphatic chain (7–11). A number of different kinds of layered states are possible (Figs. 1a–i, 2a), including those composed of single or double layers of molecules, those with molecules normal or tilted or to the plane of the layer, those with frozen or melted aliphatic chains, and those with twisted structure (Fig. 2).

In the nematic state the molecules are aligned side by side but not in specific layers (Fig. 1). This state has particular optical properties but because the molecules have no long-range order, X-ray analysis does not give much information about the intermolecular structures. A special type of nematic phase is the cholesteric phase where each molecule is slightly displaced in relation to the next giving rise to a helical arrangement of the molecules (Figs. 2b, c). Since this state was first described in cholesterol esters, it has been called the cholesteric liquid crystalline phase. It might be more simply called the twisted nematic phase since the addition of a molecule of the right-handed pitch to one of the left-handed pitch will produce an untwisted nematic phase.

B. LYOTROPIC LIQUID CRYSTALS

Lyotropic liquid crystals are defined as liquid crystals forming in the presence of a solvent. A solvent penetrates a crystal liquifying a part of the molecule, thus, producing a liquid crystal state. Although both polar and nonpolar solvents may produce liquid crystals, I will confine my discussion to water as the solvent.

align in partially ordered arrays. Several structural types of thermotropic liquid crystals have been described by X-ray diffraction: three-dimensional cubic, two-dimensional rectangular, tetragonal, or hexagonal symmetry

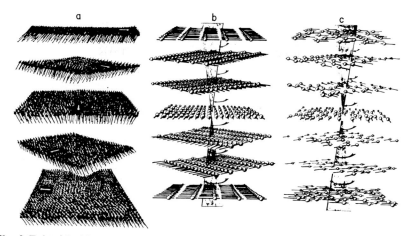

FIG. 2. Twisted liquid crystalline structures. (a) twisted smectic, (b) and (c) twisted nematic (cholesteric) structure.

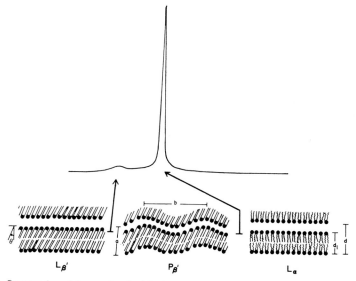

FIG. 3. Structural models associated with thermal transitions in synthetic lecithins. Dimyristoyl lecithin. Above DSC trace showing two transitions (see arrows), the low-enthalpy transition at 11°C and a large one at 23°C. d is the interlamellar repeat, d_l the bilayer thickness, and θ the angle of tilt. a and b are the cell parameters for the two-dimensional lattice P_{β}'. Below 11°C the structure is L_{β}', between the two transitions (11–23°C) the structure is P_{β}', and above 23°C the structure is L_{α}. (Taken from Ref. (13).)

Molecules that form liquid crystals with water must have a polar part which can attract water. Lyotropic liquid crystalline phases include the "gel phase" and the "lamellar liquid crystalline phase" which have been termed L_{β} and L_{α}, respectively, by Luzatti (7, 8). The β term[1] refers to the aliphatic chains being packed in a stiff, nearly crystalline hexagonal array, and

[1] β' is used if the chains are tilted with respect to the plane of the bilayer.

617

TABLE I

Classification of Biologically Active Lipids

Class	Surface properties[a]	Bulk properties[b]	Examples
Nonpolar	Will not spread to form monolayer	Insoluble	Long-chain, saturated or unsaturated, branched or unbranched, aliphatic hydrocarbons with or without aromatic groups, e.g., dodecane, octadecane, hexadecane, paraffin oil, phytane, pristane, carotene, lycopene, gadusene, squalene.
			Large aromatic hydrocarbons, e.g., cholestane, benzpyrenes, coprostane, benzphenantrocenes.
			Esters and ethers in which both components are large hydrophobic lipids, e.g., sterol esters of long-chain fatty acids, waxes of long-chain fatty acids and long chained normal monoalcohols, ethers of long chained alcohols sterol ethers, long chained triethers of glycerol.
Polar			
Class I insoluble nonswelling amphiphilic lipids	Spread to form stable monolayer	Insoluble or solubility very low	Triglycerides, diglycerides, long chained protonated fatty acids, long chained normal alcohols, long chained normal amines, long chained aldehydes, phytols, retinols, vitamin A, vitamin K, vitamin E, cholesterol, desmosterol, sitosterol, vitamin D, unionized phosphatidic acid, sterol esters of very short chain acids, waxes in which either acid or alcohol moiety is less than 4 carbon atoms long (e.g., methyl oleate).
Class II insoluble swelling amphiphilic lipids	Spread to form stable monolayer	Insoluble but swells in water to form lyotropic liquid crystals	Phosphatidyl choline, phosphatidyl ethanolamine, phosphatidyl inositol, sphingomyelin, cardiolipin, plasmalogens, ionized phosphatidic acid, cerebrosides, phosphatidyl serine, monoglycerides, "acid-soaps," alpha hydroxy fatty acids, monoethers of glycerol, mixtures of phospholipids and glycolipids extracted from cell membranes or cellular organelles (glycolipids and plant sulfolipids).
Class IIIA soluble amphiphiles with lyotropic mesomorphism	Spreads but forms unstable monolayer due to solubility in aqueous substrate	Soluble; form micelles above a CMC. At low water concentrations forms liquid crystals.	Sodium and potassium salts of long chained fatty acids, many of the ordinary anionic, cationic, and nonionic detergents, lysolecithin, palmotyl coenzyme A and other long chained thioesters of coenzyme A, gangliosides, sulfo cerebrosides.
Class IIIB soluble amphiphiles, no lyotropic mesomorphism	Spreads but forms unstable monolayer due to solubility in aqueous substrate	Forms micelles but not liquid crystals	Conjugated and free bile salts, sulfated bile alcohols, sodium salt of fusidic acid, rosin soaps, saponins, sodium salt of phenanthrene sulfonic acid, penicillins, phenothiazines.

[a] At air–water interface.
[b] In aqueous systems.

the α refers to the chains being in a melted or more liquid-like state. In hydrated synthetic lecithins, Janiak et al. (13) has shown that the L_β' phase changes to a lamellar phase with a periodic ripple (P_β') and at higher temperatures to the L_α phase (Fig. 3). The transition between the L_β' and the P_β' is a low-energy first-order transition while the transition from P_β' to L_α is a high-energy first-order transition. Thus, the major energy change involves chain melting.

The transition temperature from β to α chains (frozen to liquid-like) is influenced by the amount of water present (the more water the lower the transition), by the number of carbons in the fatty acid chains of the lipid (the more the higher the transition), and by the number of double bonds and other chain substitutents, most of which decrease the transition temperature. Thus, lyotropic systems also undergo thermotropic transitions. Finally, there are a

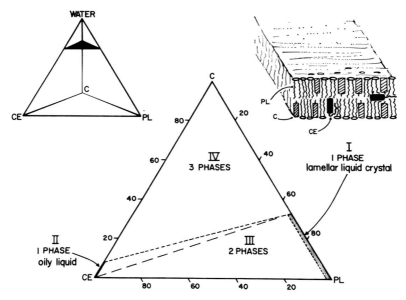

FIG. 4. The three-component system phospholipid (PL)–cholesterol ester (CE)–cholesterol (C) at constant water content. The tetrahedron at the upper left shows the position of the section containing the four-component system with 70% water by weight. This section is shown enlarged below, and is dealt with as a three-component system PL–CE–C (see text). Zone I, a lamellar liquid-crystalline phase containing varying amounts of C and CE (shown diagrammatically in the upper right); Zone II, a phase of CE containing from 0% when cholesteryl ester is crystalline (lower dashed line) to almost 8% when the ester is in a liquid state (upper dashed line); Zone III, a two-phase zone consisting of the lamellar liquid-crystalline phase and the oily liquid phase of CE; Zone IV, an invariant zone of three phases, the lamellar liquid-crystalline phase saturated with C and CE, the oily CE phase saturated with C, and C crystals. (Taken in part from Ref. (19).)

number of other liquid crystalline structures which occur with specific lipid classes, such as the cubic phase of certain phospholipids, the orthorhombic of phospholipids and soaps, and the hexagonal phases of phospholipids, soaps, and detergents. Many of these phases have been carefully described in the laboratory of Luzatti (7, 8) and presumably even other phases are possible as suggested by Mabis (14).

III. CLASSIFICATION OF BIOLOGICALLY
ACTIVE LIPIDS

Biologically active lipids have been classified according to their surface and bulk properties ((15, 16); Table I). Nonpolar molecules are defined as molecules which do not spread at an air–water interface and are insoluble in the

bulk. From the point of view of mammalian pathobiology the most important molecules of the nonpolar class are the cholesterol esters such as cholesteryl oleate and cholesteryl linoleate (17). *Polar* lipids fall into three general classes; (I) those which form stable monolayers but are virtually insoluble in the bulk (nonswelling amphiphiles); (II) those which undergo lyotropic liquid crystal formation, but are quite insoluble in the bulk (swelling amphiphiles); and (III) those which form unstable monolayers, have a high solubility in the aqueous phase, and form micelles (soluble amphiphiles). Important members of these classes include cholesterol and triglyceride (Class I), phospholipids and cerebrosides (Class II), and typical detergents, soaps, sulfocerebrosides,

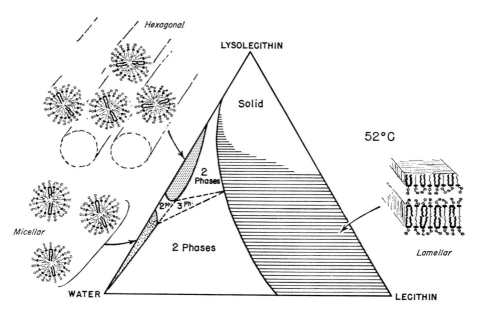

FIG. 5. Lecithin–lysolecithin–water ternary-phase diagram at 52°C. The structures of the lamellar, hexagonal, and micellar phases are indicated by the insets. The lecithin molecules have been drawn in darker characters so that they will stand out. The zones of hexagonal phase and micellar phase formed by lysolecithin are small, showing that these phases become saturated with small amounts of lecithin. Lecithin in excess of about 1 molecule to 10 of lysolecithin separates as a lamellar liquid crystal. (Reproduced from Ref. (15, Fig. 11), with permission of the Journal of the American Oil Chemical Society.)

and gangliosides (Class III). This classification is not rigorous, thus, the classification of a given lipid may depend on the outside influences such as temperature, pH, ionic strength, etc. For instance, certain ionic lipids may fall in Class III when fully ionized, but revert to Class II or I when they become partly ionized or unionized.

IV. THE INTERACTIONS BETWEEN LIPID SPECIES

The general interactions between lipids of different class in aqueous systems were first outlined by Dervichian (18) and later described in a number of phase diagrams (15). The interaction between phospholipids, cholesterol, and cholesterol esters in 70% water (19) are depicted in Fig. 4. The Zone 1 on the right-hand side consists of a single phase made up of varying amounts of phospholipid and cholesterol and small amounts (less than 2%) of choles-

terol ester. The dimensions of this phase will vary depending upon its composition as well as on the specific phospholipid chosen. On the left-hand side there is a single phase (Zone II) consisting of cholesterol ester into which is incorporated a few percent of free cholesterol. This phase can exist, depending upon the temperature, as a crystalline phase which excludes all cholesterol (lower dashed line), or as a smectic liquid crystalline phase, or as a cholesteric liquid crystalline phase or finally, as an isotropic oil. The absolute amount of cholesterol solubilized in these phases varies from zero when the ester is crystalline up to almost 8% in a liquid phase at body temperature. Mixtures falling in Zone III, that is, between Zones I and II, will separate into two phases along tie lines connecting the composition with appropriate points on the phase boundaries. Mixtures falling in Zone IV contain excess cholesterol which, at equilibrium, precipitate as cholesterol monohydrate.

The relationship between Class II and Class III molecules in water is illustrated in Fig. 5 showing the interrelation of lecithin, a swelling amphiphile, with lysolecithin, a biological detergent with a low critical micellar concentration (15). At low lecithin concentrations, the detergent molecules are able to incorporate a small amount of lecithin into a hexagonal liquid crystalline phase and with excess water and transorm this hexagonal liquid crystalline phase into a micellar phase. However, it takes quite a large amount of lysolecithin, perhaps 20 molecules to one of lecithin to form a mixed micellar solution. As the amount of lecithin is increased in the phase diagram, the lysolecithin is incorporated into the lamellar liquid crystalline phase. Since the polar groups are not charged, the lamellar phase does not swell excessively. However, with a more charged detergent the liquid crystalline phase might swell up to as high as 95% water with very large spacings between the sheets of lamellar liquid crystal. Thus, the interaction of detergent-like molecules and other amphiphiles depends upon: (1) the weight ratio between the two molecules, (2) the presence of charge, etc. Large ratios of detergent completely solubilize insoluble amphiphilic lipids but smaller ratios are incorporated into lyotropic liquid crystalline phases.

V. LIQUID CRYSTALS IN LIVING SYSTEMS

A. MEMBRANES

A single plasma membrane or an intracellular organelle membrane such as mitochondrial membrane or nuclear membrane cannot strictly be considered liquid crystalline. They do, however, have the characteristics of certain liquid crystalline systems in that at least some of the lipids are in a fluid-like state and at least part of the membrane has a bilayer-like configuration. This has been established in a large number of biological membranes ranging from plasma membranes to mitochondrial membranes. Figure 6 shows plots of the compositions of plasma membranes from a large number of species and of intracellular organelle membranes. The composition indicates that there is very little cholesterol ester (and for that matter triglyceride) and that the predominant lipids making up virtually all of the membrane systems are free cholesterol and phospholipids. Only the golgi apparatus, the sites of packaging cholesterol esters into lipoproteins, has an appreciable quantity of cholesterol ester.

Membranes also contain varying amounts of proteins (20). Figure 7 presents Singer and Nicholsons' (21) concept of the fluid mosaic membrane. It illustrates proteins, the large potato-like masses, floating in a sea of bilayer. Some of the proteins penetrate through the membrane and some merely penetrate part way. This interesting concept gives no indication of the differences in lipid species which are actually present in the membrane. While phospholipids and free cholesterol are the major membrane lipids, there are a large number of other species of lipids which are present in small amounts in most membranes and certain other components present in relatively large amounts in specific membranes. These other membrane lipids include, for instance, all of the glycosphingolipids from gangliosides to ceramides, other sterols, aliphatic and aromatic alcohols, minor quantities of fatty acids, mono, di, and triglycerides, and cholesterol ester. Thus, while the lipid composition is made up of about 90% phospholipid and free cholesterol, other species, nonpolar and polar lipid, are present in small amounts within membranes. This point is illustrated in a schematic view of the membrane drawn to illustrate the heterogeneity of lipid molecules and potential states of molecules existing in a hypothetical plasma membrane, in Fig. 8. It is probably realistic to suggest that most lipids are in a more or less fluid state; therefore, the continuous phase of the membrane, if one were looking down on the surface, would be "fluid." Those areas rich in cholesterol are somewhat less fluid than those areas relatively free of cholesterol. However, it is also quite possible that parts of the membrane are relatively solid with the chains existing in the β

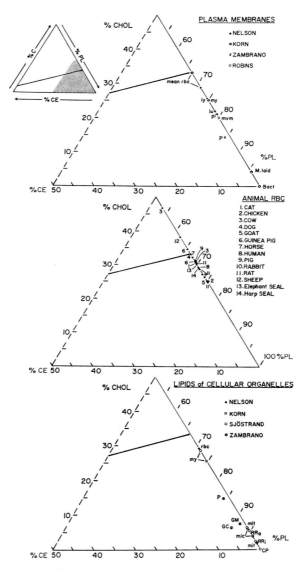

Fig. 6. Composition of cellular membranes plotted on triangular coordinates as in Fig. 4. Plasma membranes from different cells contain varying amounts of cholesterol. However, in mammals most plasma membranes e.g., red blood cells (rbc), white blood cells (lymphocytes, ly; leucocytes; lu), myelin (my), intestinal microvillous membranes (mvm), and liver cell membranes (P) contain appreciable cholesterol. Lower forms such as mycoplasma (M.laid) and most bacteria contain little or no cholesterol. On the other hand, intracellular organelle membranes (lower triangle) contain less cholesterol than plasma membranes. Only the golgi complex (GC) contains appreciable cholesteryl ester. GM, golgimembranes; mit, mitochondria; CP, chloroplasts; RR$_o$ and RR$_i$, retinal receptor cells outer and inner segments; mic, microsomes. Data taken from: (1) Nelson, G. J. (1972), Lipid composition and

form. These may form ridges extending across the bilayer which may be barriers to lateral diffusion of molecules within the more fluid parts of the membrane. They may occur around protein molecules as boundary lipid, and in highly ordered areas of the membrane, they may be the predominant species. Furthermore, there may be islands on one side or the other of the membrane which have a β chain configuration due to the lateral phase separations or to the interaction of the polar groups of certain lipids with proteins or cations such as calcium and magnesium (22). Thus, while lipids and some proteins may undergo diffusion within a membrane, it is quite possible that one-half of the membrane might be able to move independent of the other half, that is, the liquid monolayer of the membrane flowing past the solid part and vice versa. Furthermore, if a change in the surface pressure occurs at one part of the membrane, for instance, by removal of a specific lipid by biochemical reaction, a pressure gradient exists in the membrane and channels of flow may exist within that pressure gradient. Therefore, lipids are shown flowing, not just diffusing. The important feature of this concept is that the membrane is made up of a large number of lipids, mostly swelling amphiphile phospholipids which carry a host of minor compounds. These include certain other lipids which may be in more solid form with the chains in the β form, some nonpolar and polar nonswelling lipids which are solubilized and carried along in the membrane and finally, some soluble lipids such as the gangliosides, carried in the membrane in trace amounts, probably for purposes of recognition. As will be shown later, it is this heterogeneity of lipid composition which can give rise to abnormal liquid crystalline deposits occurring in certain metabolic disorders.

There are specific well-known membrane-like structures which are ordered within in concentric lamellae or stacked one on top of another, which can be specifically described as liquid crystalline. These include the rods and cones of the eye, the myelin sheaths of nerves and white matter in the brain, perhaps the lamellar bodies of dipalmitoyl lecithin of the lung.

B. Serum Lipoproteins

A second kind of liquid crystal which can occur normally in humans occurs within a small particle called low-density lipoprotein. This particle has been shown by Deckelbaum *et al.* to undergo a transition from a smectic to a less ordered state at about body temperature (23–25). The postulated structure is given in Fig. 9. Below a specific temperature (the transition temperature), the lipoprotein exists as a shell of phospholipid-free cholesterol and protein surrounding a core of cholesterol esters and a small amount of triglyceride. The core is arranged in a smectic layered structure. While the exact position of the molecules is not known, a rough estimate of the layered structure is shown in Fig. 9 (24). If the lipoprotein is heated, it undergoes a transition with an enthalpy ΔH, equal to that for the extracted esters of the lipoprotein going from smectic to the cholesteric phase ($\Delta H = 0.7$ calories per gram of ester), and the sharp fringe seen by X ray at 36 Å due to the ester is seen only below the transition. These and other facts strongly indicate the presence of a smectic-like liquid crystalline phase in low-density lipoprotein. Since body temperature varies from approximately 38°C in the core of the body to as low as 26°C in the skin, it is quite probable that the low-density lipoproteins in humans are periodically undergoing transitions as they pass from the center of the body to the

metabolism of erythrocytes, Chap. 7 in "Blood Lipids and Lipoproteins: Quantitation, Composition, and Metabolism" (Gary J. Nelson, Ed.), Interscience; (2) Korn, E. D. (1966) *Science* **153**, 1491–1498; (3) Zambrano, E., *et al.* (1975), *Biochim. Biophys. Acta* **380**, 357–369; (4) Robins, S. J. (1971), *Biochim. Biophys. Acta* **233**, 550–561; (5) Sjöstrand, F. S. (1969), Morphological aspects of lipoprotein structures, Chap. 3 in "Structure and Functional Aspects of Lipoproteins in Living Systems" (E. Tria and A. M. Scanu, Eds.), Academic Press, New York.

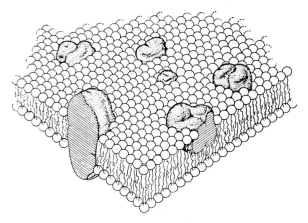

Fig. 7. The lipid–globular protein mosaic model with a lipid matrix (the fluid mosaic model); schematic three-dimensional and cross-sectional views. The solid bodies with stippled surfaces represent the globular integral proteins, which at long range are randomly distributed in the plane of the membrane. At short range, some may form specific aggregates, as shown (By permission from Singer and Nicholson, Ref. (21).)

periphery. The pathological significance of this transition has not yet been established.

VI. LIQUID CRYSTALS IN PATHOLOGICAL CONDITIONS

Many, but not all, of the diseases which accumulate liquid crystals involve a specific block in the catabolism of a particular lipid species. Under normal circumstances, when all of the appropriate metabolic apparatus is present in the system, all lipids are turned over within the cell and remain at relatively fixed concentrations. The synthesis or intake of lipids from lipoproteins guarantees an adequate content of the appropriate lipids within the cell and the breakdown, principally by enzymes

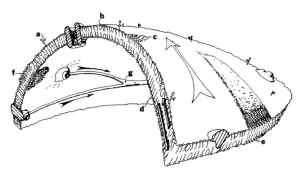

Fig. 8. Lipid heterogeneity in membranes. Schematic diagram of a membrane to show differences in lipids within the membrane. Proteins are similar to those shown in Singer's model (Fig. 7). Arrows indicate possible directions of movement of lipids in the plane of the membrane. Unidirectional lipid flow may occur down pressure gradients on either or both sides of the membrane. Movement of one monolayer in opposition to the other is also possible. (a) Gangliosides (b) cholesterol (c) a patch of lipid with stiff chains on outer face of bilayer (d) a patch of stiff-chained lipids on the inner half of the bilayer (e) a line of stiff-chained lipids creating a boundary within the membrane (f) a small patch of stiff-chained lipids (g) the cyto-skeletal system.

A   B

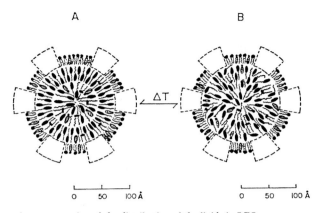

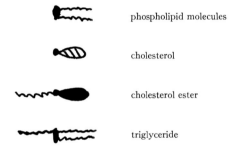

Fig. 9. Schematic representation of the distribution of the lipids in LDL:

phospholipid molecules

cholesterol

cholesterol ester

triglyceride

A. Below the transition in intact LDL, at 10°, the cholesterol esters are arranged in 4 concentric layers with a periodicity of about 36 Å in a smectic-like state. B. Above the transition, at 45°, this layered arrangement is lost, but the organization of cholesterol ester molecules is not totally random, and some degree of motional restriction persists. In both A and B, cholesterol esters are the dominant lipid in the particle core. All LDL triglycerides and about 15% of LDL free cholesterol are dissolved in this core. Around the neutral lipid core are the polar lipids of LDL, phospholipids, and most of the free cholesterol. Since these two lipid classes cannot cover all the surface of the neutral lipid core, we suggest that some aproprotein is adjacent to neutral lipid. Thus, the outer shell is made up of apoprotein and polar lipids mutually interacting with each other at the surface as well as with the neutral lipids in the LDL particle core. The water of hydration of the LDL particle also forms part of this outer shell. (Taken from Ref. (24).)

within the lysosomes (the digestive organ of the cell) and by specific removal mechanisms carried out by certain lipoproteins, control the level of lipid within the cell. When a specific catabolic enzyme is lacking, the lipid which would be broken down by that enzyme accumulates within the cell. It first accumulates in the sites of the cell in which it is prevalent or being synthesized. Then, as the cell recognized that excess of this lipid is present, intracellular lysosomes engulf the excess lipid and attempt to break it down. However, since the specific enzyme is deficient the lipid accumulates. Thus, the cell can be affected in at least two ways, one by a specific malfunction of the organelles of the cell by an excess of the specific lipid, or second by massive engorgement within lysosomes of the uncatabolized lipid species. It is obvious that any of the different kinds of lipid species depicted in the model membrane in Fig. 8 may accumulate if the enzyme necessary to catabolize that particular lipid is lacking in a

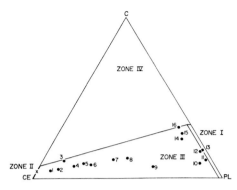

FIG. 10. Lipid composition of tissues in cholesterol ester storage disease, Tangier disease, and control. Points 1 and 2 are liver tissues from patients with cholesterol ester storage disease (39). Points 3–9 are tissues from patients with Tangier disease: 3, lymph tissue (43); 4, tonsil tissue (43); 5, spleen tissue (46, 47); 6 and 7, tonsil tissue (43); 8, spleen tissue (43); and 9, liver tissue (47). Points 10–16 are tissues from control patients: 10, liver (39); 11, lymph (39, 43); 12, spleen (43); 13, tonsil (43); 14, spleen (43); 15, splenic artery (47); and 16, spleen (47). The point (x) (47) in Zone II is lipids isolated from droplets found in the spleen tissue from a patient with Tangier disease.

given cell population. Furthermore, the *characteristics of the lipid deposits will depend upon the physical properties of the lipid which is accumulating, and its interaction with other adjacent lipids.* Some lipids merely accumulate as relatively harmless droplets within the cell while others affect specific functions such as nervous conduction enzymatic action, membrane permeability, etc.

A. DISEASES ACCUMULATING NONPOLAR LIPIDS

Cholesterol ester storage disease. Cellular metabolism of cholesterol is complex, but the tissue-culture work of Bailey (26) and Rothblat (27) and recently that of Brown and Goldstein on fibroblasts (28–30) have done much to elucidate the metabolism of cholesterol in peripheral tissue. The work of Dietschy and others (36, 37) has helped us to understand cholesterol metabolism in the liver. In general, many peripheral tissue cells have a receptor for low-

density lipoprotein and when cellular stores of cholesterol begin to fall, the receptor takes up the lipoprotein. The lipoprotein is taken into the lysosome, and the cholesterol esters are hydrolyzed to free cholesterol which then moves out of the lysosome and resupplies the cell with cholesterol. This free cholesterol then decreases the number of receptor sites and shuts off the cholesterol synthesis within the cell (which was trying to compensate for the previous low level) to achieve a new steady state. Any excessive cholesterol ingested may be removed by a particular tissue fluid carrier, probably the serum high-density lipoprotein, which circulates through the extracellular tissues. Thus, a beautiful balance is maintained between uptake, synthesis, utilization, and removal. The liver probably also takes up low-density lipoprotein but this organ is also responsible for removing the large amounts of cholesterol arriving via the chylomicron remnants (39), that is, the lipoproteins formed in the intestine which contain ingested cholesterol. Normally, the liver seems to have an obligate step taking the cholesterol-rich remnants from the blood. In the liver cells lysosomes exist to break down the ingested cholesterol, which in turn regulates cholesterol synthesis within the liver cell, and perhaps controls the excretion of cholesterol into the bile either as free cholesterol or as bile acid, the major metabolic product of cholesterol. In a very rare disease called *cholesterol ester storage disease* (39, 41) it was discovered that the lysosomal enzyme for the breakdown of cholesterol ester is deficient throughout the tissue (41). These patients therefore take up cholesterol to some extent in the peripheral tissue but much more in the liver and cannot break down the ester that is retained within the lysosome of the liver cell; thus it accumulates in large amounts. For illustration, the composition of liver cells is shown in Fig. 10 on the ternary phase diagram. Normal liver (point 10) has very little cholesterol ester although there is more than the pure membrane fraction due to the presence of small amounts of cholesterol ester-rich lipoprotein in the liver. However, the liver from patients with choles-

a   b   c

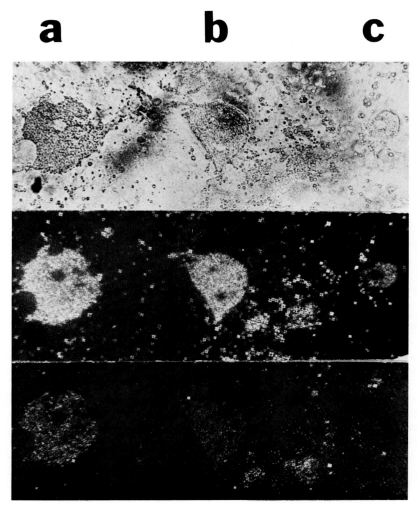

FIG. 11. Individual spleen foam cells from a patient with Tangier disease. Spaces in cells are nuclei. Same field of 400×: top, ordinary light; middle, crossed polars 20°C; bottom, crossed polars 48°C. At 48°C, droplets in cell on left (a) are still birefringent, those in the middle cell (b) are nearly melted and those in the cell on right (c), are completely melted. Birefringent droplets are smectic liquid crystal phases of cholesteryl esters. Different melting temperatures of droplets in different cells indicate a different cholesteryl ester mixture in each cell (from Ref. (47)).

terol ester storage disease contain nearly 90% of their total lipid as cholesterol ester. Such patients may carry up to a kilogram of cholesterol ester in their liver. Since the ester is not a particularly toxic substance as it exists within the liver, it seems to create damage within the liver cells only by its mass effect. In 1967 the author examined a sample of unfixed liver from patient TH of Dr. Schiff (41). The cells were full of smectic liquid crystals

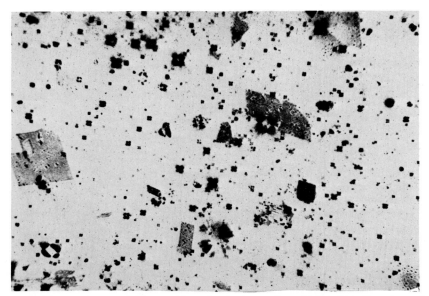

Fig. 12. Smear of center of a human atherosclerotic plaque. Crossed polars 90×. Crystals are cholesteryl monohydrate and birefringent droplets are cholesteryl ester phase in the smectic liquid crystalline phase.

which melted sharply at 35°C. The birefringent droplets were isolated and shown to be 98% cholesterol ester. Thus the deposits in this patient were almost pure cholesterol ester capable of forming smectic liquid crystals at temperatures very close to body temperature.

Tangier disease. Tangier disease is a rare, inherited disorder discovered in the population of Tangier Island, and discussed in great detail by D. S. Frederickson and associates at the National Institutes of Health (42, 45). These patients lack the ability to synthesize adequate quantities of high-density apoprotein; therefore they have about a 600-fold decrease in the major apoprotein of high-density lipoprotein and consequently extremely low levels in the serum. Thus, they lack one of the major acceptor systems for cholesterol in the tissue. Patients with Tangier disease were recognized as having very large tonsils full of cholesterol esters (42). Furthermore, a number of other organs containing reticuloendothelial cells, those cells which scavenge debris within the body, were found to contain large quantities of

cholesterol ester. Figure 10 shows that the composition of the lipids (42–47), taken from a number of organs in the patient with Tangier disease, is extremely high in cholesterol ester. However, many tissues did not appear to have massive amounts of esters and only the reticuloendothelial system appears to accumulate the lipids. Thus it has been suggested that high-density lipoproteins are particularly necessary in removing cholesterol from those specific cells, the macrophages of the reticuloendothelial system, which scavenge the cholesterol containing remnants of lipoproteins and cell membranes which have lost their use in the body. These patients, while sometimes having enormous spleens and tonsils full of cholesterol ester, rarely develop serious symptoms and apparently live a relatively normal life. Figure 11 shows live cells taken from the spleen of a patient with Tangier disease. Over 80% of all of the cells are birefringent at 37° and have a uniaxial positive sign of birefringence within each droplet (46, 47). The cells as well as the droplets isolated from tissue give a smectic A

X-ray diffraction pattern at 37°C with the major narrow angle fringe at 36 Å (46, 47). Thus most of the lipids are in the smectic liquid crystalline phase at body temperature. The chemical composition of the esters from such cells is very rich in oleic acid and their mean transition temperature from smectic to isotropic is at about 41°C which is similar to the temperature for the smectic–cholesteric transition of cholesterol oleate (17). The composition of the droplets isolated from the cell is plotted on Fig. 10 (47) and shows that they are almost pure cholesterol ester.

Atherosclerosis. The etiology of atherosclerosis is not so simple. This widespread disease may have several etiologic factors, several relating to abnormalities in the metabolism of the serum lipoproteins (48–50). The lesions are often small but occur in strategic places such as the coronary vessels or arteries going to the brain and therefore they cause problems by restricting the blood flow to vital organs. Figure 12 shows a smear from an atherosclerotic plaque of a patient who had died of atherosclerosis. This plaque is typical of advanced lesion and is loaded with crystals of cholesterol monohydrate and also with optically positive birefringent droplets. These isolated droplets contain 94% cholesterol ester and 6% free cholesterol, that is, they are the cholesterol ester phase saturated with cholesterol. Plaques also contain many droplets which are isotropic. The lack of birefringence is probably due to high concentrations of cholesterol linoleate and other more unsaturated esters as well as the presence of small amounts of triglyceride in the cholesterol ester phase. All these low-melting constituents (17) lower the transition temperature of the higher melting esters.

Atherosclerosis apparently starts at a very early age with excessive cholesterol beginning to accumulate in the membranes of the arterial intima (19). Thus as the membrane becomes saturated with cholesterol it begins to change its characteristics and to accumulate cholesterol esters. These esters appear to be accumulated within cells both in lysosomes and also in droplets not bounded by lysosomal

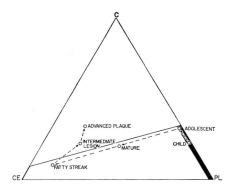

FIG. 13. Possible progression of the arterial intima from normal child to advanced plaque. The chemical compositions of the child and adolescent are taken from Refs. (52, 53), as presented by Small and Shipley (19). The chemical data of the mature intima, the fatty streak, the intermediate lesion, and the advanced plaque are taken from (51). The lipid composition of the intima in children appears to consist of a single lamellar liquid crystalline phase, that is like cellular membranes. By adolescence, the membrane system appears to be saturated with cholesterol. The mature intima (i.e., from people 45–70) appear normal but microscopically contain a second phase of cholesterol ester. The fatty streak is an abnormal lipid deposit consisting of a large quantity of cholesteryl ester in the smectic liquid crystalline phase. The intermediate lesion (51) looks like a fatty streak but has become supersaturated with cholesterol. Finally, the advanced lesion is a necrotic lesion with many dead cells containing many cholesterol monohydrate crystals, the smectic and isotropic phase of cholesteryl esters as well as a lamellar membrane phase of phospholipid.

membrane. When these lesions become visible to the maked eye they are called fatty streaks. Furthermore, the esters tend to be richer in oleate than in linoleate suggesting that some of the esters are synthesized within the cell or that polyunsaturated esters are selectively hydrolyzed. Thus the early stage of the fatty-streak lesion of atherosclerosis is characterized by the deposition of cholesteryl oleate-rich smectic liquid crystals primarily within the cells. Advanced lesions of atherosclerosis called plaques, are lipid-rich lesions, full of cholesterol crystals (see Fig. 12), which protrude into the lumen of large arteries. They also contain many

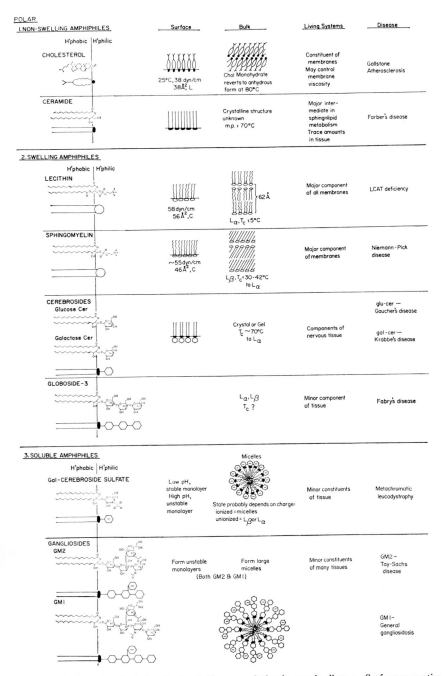

FIG. 14. Physical characteristics of polar lipids accumulating in certain diseases. Surface properties at ambient temperature include: collapse pressure of monolayer (dynes/cm), limiting surface area (Å²/molecules) and film viscosity, L = liquid, C = condensed. Ceremides and cerebrosides form solid film.

TABLE II

Class	Major lipid accumulating	Disease	Tissue affected	Composition of deposit	Physical state of deposit
Nonpolar	CE[a]	Atherosclerosis	Arterial intima	40% Chol., 40% CE, 20% PL	Cholesterol monohydrate, smectic LC of CE, liquid oil, L_α of PL
		CE storage disease	Liver cells	>90% CE	Smectic CE
		Tangier	RE cells of liver, spleen nodes, etc.	>90% CE	Smectic CE
Polar	1—Nonswelling Cholesterol	Cholesterol gallstones	Bile	Chol.	Crystalline cholesterol monohydrate
	Ceramide	Farber's	Brain, kidney, etc.	?	? Crystalline ceramide
	2—Swelling Lecithin	LCAT deficiency	Blood and other tissues	70% lecithin, 30% cholesterol	
	Sphingomyelin	Neimann–Pick	RE cells of liver, spleen, lymph nodes, kidney, lung, CNS	45 mole% sphingomyelin, 20 mole% other PL, 35 mole% cholesterol	? L_α 70 Å
	Glucosyl cerebroside	Gauchers (glucosyl ceramidosis)	RE cells of liver, spleen, bone marrow, lung, lymph nodes (CNS)	59% glucosyl cerebroside 11% lecithin 30% cholesterol	? L_α 60 Å
	Galactosyl cerebroside	Krabbe's (galactosyl ceramidosis)	CNS, kidney	Galactosyl cerebroside	? L_α
	Globoside 3 (gal-gal-glu-cer)	Fabrys	RE system, endothelium, kidney, nervous tissue, heart, lung	Lysosomal bodies high content of globoside 3.	? L_α
	3—Soluble Cerbroside sulfate	Metachromatic leucodystrophy	Myelin, white matter, kidney	Myelin–54% PL, 18% chol., 21% cer sulfate, 7% other lipids	Swollen L_α or L_β ? mixed micelles
	GM₂	Tay Sachs	Nervous tissue, liver, lung, spleen	30–40% GM₂, rest of lipid is cholesterol PL and other glyco-sphingolipids	Swollen L_α or L_β, ? mixed micelles with other lipids
	GM₁	General gangliosidosis	Brain, liver, spleen, kidney, bone	High amount of GM₁, varying amounts of other lipids	Swollen L_α or L_β, ? mixed micelles with other lipids

[a] Abbreviations: CE = cholesterol esters; chol. = cholesterol; PL = phospholipids; Cer = ceramide; GM = gangliosides; RE = reticuloendothelial system; L_α = lamellar liquid crystal with "fluid" chains; L_β = lamellar liquid crystal with "frozen" chains.

droplets and a few smectic liquid crystals at body temperature (19, 51).

The mean lipid compositions of the intima are shown in Fig. 13. The young child's arterial intima is mainly phospholipid and contains only a single phase unsaturated with cholesterol or cholesterol esters (19, 52, 53). The early lesion, the fatty streak, is extremely rich in cholesterol esters (51, 54) and many of these are in the smectic state (46, 51, 55, 56). The advanced lesion moves into the zone when three phases are present, and indeed these phases have been physically identified (19, 51) and partially isolated (56). It is this lesion which compromises arterial blood flow and causes myocardial infarctions and strokes. Katz *et al.* have found an intermediate lesion which looks like a fatty streak but is supersaturated with cholesterol and has properties intermediate between plaques and normal fatty streaks (51). It is possible that certain enzymatic processes involved in the breakdown of stored cholesterol ester initiates supersaturation of free cholesterol within the cell and precipitation of cholesterol monohydrate. This solid intracellular phase initiates destruction and death of the cell which in turn gives rise to the development of the advanced necrotic plaque.

B. Diseases Accumulating Polar Lipids

A number of specific enzyme deficiencies lead to the accumulation of various polar lipids in the body (See Ref. 57, Chapters 27, 29–35 for an excellent review). Figure 14 gives a list of the various polar lipids which accumulate in certain diseases and their physical properties as far as is known. Table II lists the various diseases in which these lipids accumulate, the major organs affected, the best estimate of the composition of the accumulation, and speculation concerning the physical state of the accumulation. Each disease will be discussed by lipid class.

1. Nonswelling Amphiphilic Lipid Accumulation

Farbers disease. Farbers disease is a very rare disorder in which the central molecule in sphingolipid metabolism accumulates because an enzyme required for its breakdown is deficient (58). Ceramide contains two rigid chains and a small polar group. It has a melting point above 79°C and does not swell in water at 25°C (D. M. Small, unpublished observations). The physical state and precise composition of the lesions have not been determined but one would predict that if ceramide separates as a phase that it would form a crystalline deposit.

Cholesterol gallstone disease. Cholesterol gallstone disease is caused by the secretion into bile of excess cholesterol such that it precipitates in the gallbladder in the form of cholesterol monohydrate (59–65). Bile lipids form a number of lyotropic liquid crystals (59–63) which may be implicated in the gallstone formation (63). These crystals grow to form stones of almost pure cholesterol monohydrate which then can block the biliary tract and cause serious disease (65). Presumably mechanisms for cholesterol gallstone disease involve either inadequate secretion of bile salts to solubilize the cholesterol or excessive cholesterol secretion (65–66). The exact metabolic reasons for this have not yet been discovered athough obesity is a factor in the secretion of excess cholesterol (66). An enzymatic defect in the conversion of cholesterol to bile acid has been suggested as a possibility for the decreased bile salt secretion in certain gallstone patients (67).

Cholesterolosis of the gallbladder. Certain patients with gallstones occasionally develop the deposition of lipids, possibly cholesterol esters, in the wall of the gallbladder. These lipids appear to be present in the form of smectic liquid crystals (68). Dogs fed high doses of cholesterol develop similar lesions containing liquid crystals (68).

2. Swelling Amphiphilic Lipid Accumulation

Lecithin cholesterol acyl transferase deficiency. While there has been no lysosomal enzyme discovered leading to an accumulation of lecithin within the cells, there is a specific disease involving a serum enzyme necessary for lecithin catabolism. This serum enzyme, lecithin cholesterol acyl transferase (LCAT), catalizes the transfer of a fatty acid from lecithin to free cholesterol to produce cholesterol ester and lysolecithin (69). The site of the reaction is the surface of certain serum lipoproteins. The lysolecithin is removed from the particle by albumin and the cholesterol ester is deposited within the core of the lipoprotein. In lecithin cholesterol acyl transferase deficiency normal lipoproteins cannot be formed because the cholesterol esters are not formed. Therefore, the serum contains lipoproteins extremely rich in lecithin and free cholesterol (70–73) in a ratio of 1:1 which forms stacks of small fragments of lecithin–cholesterol lamellar liquid crystals. Furthermore, such deposits have been found in tissues within cells as lamellar bodies.

Neimann–Pick disease. Sphingomyelin is a ubiquitous sphingophospholipid which has properties rather similar to egg lecithin (74) except that its chain melting transition is higher (75). The transition from L_β to L_α is from about 32 to 42°C (75). In Neimann–Pick disease the enzyme necessary for the degradation of sphingomyelin is absent and this molecule accumulates in large amounts in many tissues (76–78). Several forms of the disease are present, among them a severe juvenile form in which the patient dies of massive accumulation of sphingomyelin in the brain and a later form in which the patient lives much longer and the sphingomyelin accumulates mainly in nonneural tissues (78). The aggregates formed as shown by electron microscopy consist of lamellar aggregates containing cholesterol, sphingomyelin and lecithin (79).

Cerebrosides (Gaucher's disease and Krabbe's disease). Each of two major cerebrosides, that is, glucosyl ceramide and galactosyl ceramide, is involved in a specific disease. Specific enzymes necessary for the breakdown of one or the other of these molecules to the parent compound, ceramide, are lacking. In Gaucher's disease (57), glucosyl ceramide accumulates in massive amounts in the liver and spleen and other tissues of the body. Glucosyl ceramide has rather rigid hydrocarbon chains, has a high transition temperature of about 70°C and forms a solid monolayer at 37°C (80). The deposits contain a large quantity of the glucocerebroside and smaller amounts of free cholesterol and other phospholipids. The tissues have been studied by X-ray diffraction by Lee *et al.* (81), and have been shown to be basically an L_α liquid crystal with a possible twist of the plane of the lamellae. The treatment of the tissue however, prior to the X-ray diffraction studies may have distorted the true structure.

Krabbe's disease is characterized by the deposition of the galactoceramide and the morphology of the lesions is very similar to that of Gaucher's disease; however, there is no physicochemical data concerning the structure of the deposits. It is interesting to note the studies on mongalactosyl diglyceride by Shipley *et al.* (82). This molecule is similar to galactoceramide, but it forms an inverted hexagonal phase when hydrated with water.

Ceramide trihexosidosis (Fabry's disease). Ceramide trihexoside (Globoside III) that is, the globoside with three sugars accumulates in Fabry's disease (57, 83). In this disease, the enzyme necessary to remove the last sugar is absent (84). Liquid crystalline deposits are seen in many tissues in this disease including the kidney, liver, spleen, bone marrow, etc. Ceramide trihexoside also accumulates in blood (85) cells and renal deposits appear to compromise renal function. While the exact nature of the deposit is not known, their optical properties suggest that they are lamellar liquid crystals of either L_α or L_β form.

3. Soluble Amphiphilic Lipid Accumulation

Metachromatic leucodystrophy. Cerebroside sulphate, a detergent-like molecule with a charged sulfate group on the sugar molecule, has been shown to form unstable monolayers and to form micelles in high concentrations of water (80). At lower water concentrations or at low pH it probably forms L_β and L_α forms at the appropriate temperatures. It is a minor constituent of most living tissues but accumulates in a disease called metachromatic leucodystrophy, in which the enzyme necessary for the removal of the sulfate is absent (57, 86). The disease affects myelin and other nervous tissues often causing severe symptoms in infants and early death in those afflicted with this disease. The liquid crystalline deposits appear lamellar by EM but their specific structural characteristics and composition are not yet known.

Gangliosidoses (Tay Sachs GM-2 disease and generalized G-1 gangliosidosis). Gangliosides are sphingolipids with 4 or more sugar-like polar groups and at least 1 scialic acid residue. In Tay Sachs disease ganglioside GM-2 with four sugars cannot be broken down and it accumulates in the brain (57, 87–92). Gangliosides have been shown to act as detergents in water systems (93–95), and also to solubilize phospholipids such as lecithin in model systems (96). In the brain the liquid crystalline deposits appear by EM to be highly swollen lamellar aggregates (97). It is perhaps because of the charge on the ganglioside and the large, bulky polar head group that the lamellae are separated by large amounts of fluid. This particular factor along with the disruption of other cellular functions due to excess gangliosides throughout the cell, causes the cells in the brain to accumulate not only large amounts of lipid, but also excessive amounts of water. The brains of children dying of this disease contain excess contents of water (57). General GM-1 gangliosidosis is characterized by the deposition of more complex gangliosides containing 5 or more sugar residues which cannot be broken down (57). The characteristics of general gangliosidosis as well as the lesions are similar to those of Tay Sachs disease.

I have gathered together a group of diseases in which liquid crystals have been implicated in their pathogenesis, or which accumulate specific lipid deposits having liquid crystalline characteristics. These deposits may behave as either thermotropic or lyotropic liquid crystals or both. It is evident in many cases that these deposits arise from the inability of certain cells to catabolize a certain lipid species and therefore the species becomes a predominant lipid and separates from the membranes, either as the pure lipid or in combination with other membrane lipids. The physical properties of the accumulation determines the nature of cell dysfunction and the ultimate fate of the individual afflicted with the disease.

ACKNOWLEDGMENTS

The author wishes to thank D. Armitage, W. Curatolo, M. Juniak, V. Liepkalms and others of the Biophysics Institute for their helpful criticism of this manuscript.

REFERENCES

1. GRAY, G. W., "Molecular Structure and the Properties of Liquid Crystals." Academic Press, New York, 1962.
2. GRAY, G. W., AND WINSOR, P. A. (Eds.), "Liquid Crystals and Plastic Crystals." Horwood, Chichester, England, 1974.
3. BROWN, G. H., AND SHAW, W. G., Chem. Rev. 57, 1049 (1957).
4. BROWN, G. H., DOANE, J. W., AND NEFF, V. D., "A Review of the Structure and Physical Properties of Liquid Crystals." CRC Press, 1971.
5. DEGENNES, P. G., "The Physics of Liquid Crystals." Oxford Univ. Press, New York, 1974.
6. McMILLAN, J., Phys. Rev. A4, 1238 (1971); A6, 936 (1972); A9, 1419 (1974).
7. TARDIEU, A., LUZZATI, V., AND REMAN, F. C., J. Mol. Biol., 75, 711 (1973).
8. RANCK, J. L., MATEU, L., SADLER, D. M., TARDIEU, A., GULIK-KRYZWICKI, T., AND LUZATTI, V., J. Mol. Biol. 85, 249 (1974).
9. DE VRIES, A., Mol. Cryst. Liq. Cryst. 10, 31 (1970).
10. CHAPMAN, D. (Ed), "The Structure of Lipids." Methuen, London, 1965.
11. CHAPMAN, D., AND WALLACH, D. F. H. (Eds.), "Biological Membranes," Vol. 2. Academic Press, New York, 1973.
12. SHIPLEY, G. G., in "Biological Membranes," D. Chapman and D. F. H. Wallach (Eds.), Vol. 2, Chapt. 1. Academic Press, New York, 1973.
13. JANIAK, M. J., SMALL, D. M., AND SHIPLEY, G. G., Biochemistry 15, 4575 (1976).
14. MABIS, A. J., Acta Crystallog. 15, 1152 (1962).
15. SMALL, D. M., J. Amer. Oil Chem. Soc. 45, 108 (1968).
16. SMALL, D. M., Fed. Proc. 29, 1320 (1970).
17. SMALL, D. M. in "Surface Chemistry of Biological Systems" (M. Blank, Ed.), p. 55. Plenum Press, New York, 1970.
18. DERVICHIAN, D. G., Trans. Faraday Soc. 42B, 180 (1946).
19. SMALL, D. M., AND SHIPLEY, G. G., Science 185, 222 (1974).
20. GUIDOTTI, G., Annu. Rev. Biochem. 41, 731 (1972).
21. SINGER, S. J., AND NICOLSON, G. L., Science 175, 7188 (1972).
22. PAPAHADJOPOULAS, D., MOSCARELLO, M., EYLAR, E. H., AND ISAC, T., Biochim. Biophys. Acta 401, 317 (1975).
23. DECKELBAUM, R. J., SHIPLEY, G. G., SMALL, D. M., LEES, R. S., AND GEORGE, P. K., Science 190, 392 (1975).
24. DECKELBAUM, R. J., SHIPLEY, G. G., AND SMALL, D. M., J. Biol. Chem. 252, 744 (1977).
25. SEARS, B., DECKELBAUM, R. J., JANIAK, M. J.,

SHIPLEY, G. G., AND SMALL, D. M., *Biochemistry* **15**, 4151 (1976).

26. BAILEY, J. M., *Ciba Found. Symp.* **12**, 63 (1973).

27. ROTHBLAT, G. H., ARBOGAST, L., KRITCHEVSKY, D., AND NOFTULIN, M., *Lipids* **11**, 97 (1976).

28. GOLDSTEIN, J. L., AND BROWN, M. S., *Proc. Natl. Acad. Sci. USA* **70**, 2804 (1973).

29. BROWN, M. S., AND GOLDSTEIN, J. L., *Science* **191**, 150 (1976).

30. GOLDSTEIN, J. L., BASU, S. K., BRUNSCHEDE, G. Y., *et al.*, *Cell* **7**, 85 (1976).

31. FOGELMAN, A., *et al.*, *J. Biol. Chem.* **250**, 2045 (1975).

32. KANDUTSCH, A. A., AND CHEN, H. W., *J. Biol. Chem.* **249**, 6057 (1974).

33. SIMONS, L. A., REICHL, D., MYANT, N. B., *et al.*, *Atherosclerosis* **21**, 283 (1975).

34. STEIN, O., WEINSTEIN, D. B., STEIN, Y., *et al.*, *Proc. Natl. Acad. Sci. USA* **73**, 14 (1976).

35. WERB, Z., AND COHAN, Z., *J. Exp. Med.* **134**, 15 (1971).

36. DIETSCHY, J. M., AND WILSON, J. D., *N. Engl. J. Med.* **282**, 1128, 1179, 1241 (1970).

37. SIPERSTEIN, M. D., *Curr. Top. Cell Regul.* **2**, 65 (1970).

38. REDGRAVE, T., *J. Clin. Invest.* **49**, 465 (1970).

39. SLOAN, H. R., AND FREDERICKSON, D. S., *in* "Metabolic Basis of Inherited Disease" (J. B. Stanbury, J. B. Wyngarden, and D. S. Frederickson, Eds.), Chapt. 36. McGraw–Hill, New York, 1972.

40. INFANTE, R., POLONOVSKI, J., AND CAROLI, J., *Presse Med.* **75**, 2829 (1967).

41. SCHIFF, L., SCHUBERT, W. K., McADAMS, A. J., SPIEGEL, E. L., AND O'DONNELL, J. F., *Amer. J. Med.* **44**, 538 (1968).

42. FREDERICKSON, D. S., AND ALTROCCHI, P. H., *in* "Cerebral Sphingolipidoses" (S. M. Aronson and B. W. Volk, Eds.), p. 343. Academic Press, New York, 1962.

43. FREDERICKSON, D. S., *in* "The Metabolic Basis of Inerited Disease" (J. B. Stanbury, J. B. Wyngaarden, and D. S. Frederickson, Eds.), 2nd ed., p. 486. McGraw–Hill, New York, 1966.

44. FREDERICKSON, D. S., GOTTO, A. M., AND LEVY, R. I., *in* "The Metabolic Basis of Inherited Disease" (J. B. Stanbury, J. B. Wyngaarden, and D. S. Frederickson, Eds.), 3rd ed., p. 493. McGraw–Hill, New York, 1972.

45. FERRANS, V. J., AND FREDERICKSON, D. S., *Am. J. Pathol.* **78**, 101 (1975).

46. KATZ, S. S., SMALL, D. M., BROOKS, J. G., AND LEES, R. S., *Circulation*, in press.

47. KATZ, S. S., SMALL, D. M., BROOKS, J. G., AND LEES, R. S., *J. Clin. Invest.* (1977), in press.

48. FREDERICKSON, D. S., AND LEVY, R. I., *in* "The Metabolic Basis of Inherited Disease" (J. B.

Stanbury, J. B. Wyngaarden, and D. S. Frederickson, Eds.), 3rd ed., p. 545. McGraw–Hill, New York, 1972.

49. GOLDSTEIN, J. L., HAZZARD, W. R., SCHROTT, H. G., *et al.*, *J. Clin. Invest.* **52**, 1533 (1973).

50. HAVEL, R. J., *in* "Textbook of Medicine" (P. B. Beeson and W. McDermott, Eds.), 14th ed., p. 1629. Saunders, Philadelphia, 1975.

51. KATZ, S. S., SHIPLEY, AND SMALL, D. M., *J. Clin. Invest.* **58**, 200 (1976).

52. DAY, A. J., AND WAHLQUIST, M. L., *Exp. Mol. Pathol.* **13**, 199 (1970).

53. SMITH, E. B., EVANS, P. H., AND DOWNHAM, M. D., *J. Atheroscler. Res.* **7**, 171 (1967).

54. LANG, P. D., AND INSULL, W., *J. Clin. Invest.* **49**, 1479 (1970).

55. HATA, Y., HOWER, J., AND INSULL, W. JR., *Amer. J. Pathol.* **75**, 423 (1974).

56. KATZ, S. S., SMALL, D. M., SHIPLEY, G. G., AND ROGERS, E. L., *Circulation* **50**, 6a (1974).

57. STANBURY, J. B., WYNGAARDEN, J. B., AND FREDERICKSON, D. S. (Eds.), "The Metabolic Basis of Inherited Disease" 3rd ed. McGraw–Hill, New York, 1972.

58. SUGITA, M., DULANEY, J. T., AND MOSER, H. W., *Science* **176**, 1100 (1972).

59. SMALL, D. M., BOURGES, M., AND DERVICHIAN, D. G., *Nature* **211**, 816 (1966).

60. SMALL, D. M., AND BOURGES, M., *Mol. Cryst. Liq. Cryst.* **1**, 541 (1966).

61. SMALL, D. M., *in* "The Bile Acids—Chemistry, Physiology and Metabolism" (P. P. Nair and D. Kritchevsky, Eds.), Vol. 1, Chapt. 8, p. 247. Plenum, New York, 1971.

62. SMALL, D. M., *in* "The Metabolism of Bile Acids" (L. Schiff, Ed.), Chapt. 19, p. 223. Thomas, Springfield, Ill., 1969.

63. SMALL, D. M., *Advan. Intern. Med.* **16**, 243 (1970).

64. ADMIRAND, W. H., AND SMALL, D. M., *J. Clin. Invest.* **47**, 1045 (1968).

65. SMALL, D. M., *in* "Controversy in Internal Medicine II" (F. J. Ingelfinger, R. V. Ebert, M. Finland, and A. 'S. Relman, Eds.), Chapt. 21, p. 545. Saunders, Philadelphia, 1974.

66. BENNION, L., AND GRUNDY, S., *J. Clin. Invest.* **56**, 996 (1975).

67. NICOLAU, G., *et al.*, *Gastroenterologia* **64**, 887 (1973).

68. HOLZBACH, R. T., MARSH, M., TANG, P., *in* "Advances in Bile Acid Research" (S. Matern, J. Hackenschmidt, P. Back, and W. Gerok, Eds.), p. 339. F. K. Schattauer Verlag (1974).

69. GLOMSET, S., *J. Lipid Res.* **9**, 155 (1968).

70. GLOMSET, J. A., NORUM, K. R., NICHOLS, A. V., *et al.*, *Scand. J. Clin. Lab. Invest.* **35**, 3 (1975).

71. NORUM, K. R., GLOMSET, J. A., NICHOLS, A. V., *et al.*, *Scand. J. Clin. Lab. Invest.* **35**, 31 (1975).

72. GLOMSET, J. A., AND NORUM, K. R., *Advan. Lipid Res.* **11**, 1 (1973).

73. NORUM, K. R., GLOMSET, J. A., NICHOLS, A. V., AND FORTE, T., *J. Clin. Invest.* **50**, 1131 (1971).

74. SMALL, D. M., *J. Lipid Res.* **8**, 551 (1967).

75. SHIPLEY, G. G., AVECILLA, L., AND SMALL, D. M., *J. Lipid Res.* **15**, 124 (1974).

76. KLENK, E., *Z. Physiol. Chem.* **229**, 151 (1934).

77. BRADY, R. O., KANFER, J. N., MOCK, M. B., AND FREDERICKSON, D. S., *Proc. Natl. Acad. Sci. USA* **53**, 366 (1966).

78. FREDERICKSON, D. S., AND SLOAN, H. R., *in* "The Metabolic Basis of Inherited Disease" (J. B. Stanbury, J. B. Wyngaarden, and D. S. Frederickson, Eds.), 3rd ed., Chapt. 35. McGraw–Hill, New York, 1972.

79. CROCKER, A. C., *in* "Lipid Storage Diseases" (J. Bernsohn and H. J. Grossman, Eds.), Chapt. 1, p. 27. Academic Press, New York, 1971.

80. ABRAHAMSSON, S., PASCHER, I., LARSSON, K., AND KARLSSON, K-A., *Chem. Phys. Lipids* **8**, 152 (1972).

81. LEE, R. E., WORTHINGTON, C. R., AND GLEW, R. H., *Arch. Biochem. Biophys.* **159**, 259 (1973).

82. SHIPLEY, G. G., GREEN, J. P., AND NICHOLS, B. W., *Biochim. Biophys. Acta* **311**, 531 (1973).

83. SWEELEY, C. C., SNYDER, P. D. JR., AND GRIFFIN, C. E., *Chem. Phys. Lipids* **4**, 393 (1970).

84. BRADY, R. O., GAL, A. E., BRADLEY, R. M., *et al.*, *N. Engl. J. Med.* **276**, 1163 (1967).

85. CLARKE, J. T. R., STOLTZ, J. M., AND MULCAHEY, M. R., *Biochim. Biophys. Acta* **431**, 317 (1976).

86. JATZKEWITZ, H., AND MEHL, E., *J. Neurochem.* **16**, 19 (1969).

87. SVENNERHOLM, L., *Biochem. Biophys. Res. Commun.* **9**, 436 (1962).

88. KLENK, E., LIEDTKE, U., AND GIELEN, W., *Z. Physiol. Chem.* **334**, 186 (1963).

89. OKADA, S., AND O'BRIEN, J., *Science* **165**, 698 (1969).

90. HUTBERG, B., *Lancet* **2**, 1195 (1969).

91. KOLODNY, E. H., BRADY, R. O., AND VOLK, B. W., *Biophys. Biochem. Res. Commun.* **37**, 526 (1969).

92. SANDHOFF, K., *FEBS Lett.* **4**, 351 (1969).

93. GAMMACK, D. B., *Biochem. J.* **88**, 373 (1963).

94. HOWARD, R. E., AND BURTON, R. M., *Biochim. Biophys. Acta* **84**, 435 (1964).

95. CURATOLO, W., SHIPLEY, G. G., AND SMALL, D. M., *Biochim. Biophys. Acta* (1977), in press.

96. HILL, M. W., AND LESTER, R., *Biochim. Biophys. Acta* **282**, 18 (1972).

97. TERRY, R. D., *in* "Lipid Storage Diseases" (J. Bernsohn and H. J. Grossman, Eds.), Chapt. 1, p. 3. Academic Press, New York, 1971.